Proceedings of the 10th International Conference

DGA 2007

Differential Geometry and its Applications

Proceedings of the 10th International Conference

DGA 2007

Olomouc, Czech Republic

27 – 31 August 2007

Differential Geometry and its Applications

Edited by

Oldřich Kowalski
Charles University, Czech Republic

Demeter Krupka
Palacky University, Czech Republic

Olga Krupková
Palacky University, Czech Republic

Jan Slovák
Masaryk University, Czech Republic

World Scientific

NEW JERSEY • LONDON • SINGAPORE • BEIJING • SHANGHAI • HONG KONG • TAIPEI • CHENNAI

Published by

World Scientific Publishing Co. Pte. Ltd.
5 Toh Tuck Link, Singapore 596224
USA office: 27 Warren Street, Suite 401-402, Hackensack, NJ 07601
UK office: 57 Shelton Street, Covent Garden, London WC2H 9HE

British Library Cataloguing-in-Publication Data
A catalogue record for this book is available from the British Library.

DIFFERENTIAL GEOMETRY AND ITS APPLICATIONS
Proceedings of the 10th International Conference on DGA2007

ISBN-13 978-981-279-060-6
ISBN-10 981-279-060-8

Printed in Singapore by World Scientific Printers

PREFACE

The *International Conference on Differential Geometry and its Applications* is a series of scientific meetings which have been held in the Czech Republic every three years since 1980. The most recent of these, DGA 2007, took place in Olomouc on August 27-31, and was dedicated to the 300th anniversary of the birth of Leonhard Euler.

This book begins with an address by Prof. R.J. Wilson *Leonhard Euler – 300 Years On*, surveying the life, labours and legacy of the great mathematician. The remainder of the book contains selected conference contributions covering contemporary differential geometry, global analysis, and geometric methods in physics, presented in three sessions:

A Riemannian geometry and submanifolds (chair O. Kowalski),
B Geometric structures (chair J. Slovák),
C Global analysis and Geometric methods in physics (chair D. Krupka)

or in the poster session. We have also included a list of participants and a list of previous conference proceedings.

The main organizer of the Conference was Palacký University in Olomouc. The Scientific Committee comprised Demeter Krupka (chairman); Olga Krupková and Josef Mikeš (Palacký University in Olomouc); Josef Janyška, Ivan Kolář, Jana Musilová and Jan Slovák (Masaryk University in Brno); Oldřich Kowalski (Charles University in Prague); and JiříVanžura (Mathematical Institute of the Czech Academy of Sciences, branch Brno).

The editors would like to thank all the authors for their contributions, and also the referees for their hard work under strict time limits. They also appreciate the indispensable help of Petr Volný in organizing the refereeing process and preparing the electronic version of the manuscript.

Olomouc April 10, 2008 The Editors

ACKNOWLEDGMENTS

A number of people and institutions participated in the organization of the conference, and I would like to thank all of them on behalf of the Scientific Committee of the DGA 2007.

The conference took place under the auspices of Prof. RNDr. Lubomír Dvořák, CSc., Rector of the Palacký University in Olomouc, RNDr. Ivan Kosatík, President of the Olomouc Region, and Martin Novotný, Mayor of the City of Olomouc. Our thanks should be extended to all of them for their support and interest in this international scientific event. We especially acknowledge the awards presented by the Rector to five conference participants, S. Gindikin, I. Kolář, O. Kowalski, D. Krupka, A.M. Vinogradov, appreciating their research in differential geometry, initiatives connected with the establishing of the journal Differential Geometry and its Applications, as well as long lasting activities related to previous conference organization.

Special thanks should go to Prof. R.J. Wilson for his address *Leonhard Euler – 300 Years On.*

In particular, I appreciate very much the work of the session chairmen Oldřich Kowalski and Jan Slovák.

I thank all the members of the Local Organizing Committee, but especially young researchers from the Research Group of Global Analysis and its Applications, Jan Brajerčík, Marie Chodorová, Zdeněk Dušek, Dana Smetanová, Zbyněk Urban and Petr Volný for their effort and work and many useful ideas making the conference successful.

The help of the Economical Department of the Faculty of Science, namely by Ing. Dagmar Kopecká, is also highly appreciated.

The organizers gratefully acknowledge a financial support of the conference activities from the following institutions:

- Olomoucký kraj (the Olomouc district)
- Statutární město Olomouc (the City of Olomouc)
- Laboratory Imaging s.r.o., Prague

- Inženýring dopravních staveb a.s., Prague
- Stavoprojekt Olomouc a.s.
- Alfaprojekt a.s., Olomouc
- Ecological consulting a.s.

Last but not least, we thank to the Czech Science Foundation and the Czech Ministry of Education, Youth and Sports for their grant support of many research articles published in this book. We also thank the Publisher, World Scientific Publishing Company, for an outstanding collaboration during the preparation of this volume to print.

Olomouc April 10, 2008

Demeter Krupka
Chairman of the Scientific Committee and
the Organizing Committee

CONTENTS

Part 3 Global analysis 383

Differential Geometry and its Applications
Proc. Conf., in Honour of Leonhard Euler, Olomouc, August 2007

Leonhard Euler – 300 Years On

Robin Wilson

The Open University,
United Kingdom
E-mail: R.J.Wilson@open.ac.uk

2007 marked the 300th anniversary of the birth of Leonhard Euler. This article presents a selection of his achievements.

Euler was the most prolific mathematician of all time. He wrote more than 500 books and papers during his lifetime, with 400 further publications appearing posthumously. His collected works and correspondence, still not completely published, fill over seventy large volumes comprising tens of thousands of pages. He worked in an astonishing variety of areas, ranging from pure mathematics (the theory of numbers, the geometry of a circle and musical harmony), via infinite series, logarithms, the calculus and mechanics, to practical topics (optics, astronomy, the motion of the Moon and the sailing of ships). He originated so many ideas that his successors have been kept busy trying to follow them up ever since. Indeed, his influence was such that several concepts were later named after him: Euler's constant, Euler's polyhedron formula, the Euler line of a triangle, Euler's equations of motion, Eulerian graphs, Euler's pentagonal formula for partitions, to name but a few.

His life can be divided into four periods. He was born in Basel, Switzerland, on 15 April 1707, where he grew up and went to university. At the age of 20 he went to Russia, to the St Petersburg Academy, where he became head of the mathematics division. In 1741 he went to Berlin, where he stayed for twenty-five years. In 1766 he returned to St Petersburg where he spent the rest of his life, dying on 7 September 1783.

Basel, Switzerland

Leonhard Euler's father was a Calvinist pastor who wished his son to follow him into the ministry. On entering the University of Basel at the age of 14, the young Euler duly studied theology and Hebrew, law and philosophy. While there, he encountered Johann Bernoulli, possibly the finest mathematician of his day, who was impressed with his mathematical abilities and gave him private teaching every Saturday, quickly realising that his pupil was highly talented. Euler also became close friends with Johann's sons, Daniel and Nicholas.

Euler took his Master's degree in 1724, at the age of 17, and entered divinity school to train for the ministry, but made little progress since mathematics was proving to be such a distraction. Eventually, Bernoulli persuaded Euler's reluctant father that his son was destined to become a great mathematician, and Euler abandoned his theological training.

Euler's first significant mathematical achievement occurred when he was just 20. The Paris Academy had proposed a prize problem involving the placing of masts on a sailing ship in order to combine speed with stability. Euler's memoir, while not gaining the prize, received an honourable mention; later, he won the Paris prize on twelve occasions.

Euler next applied for the Chair of Mathematics at the University of Basel. Because of his young age he was unsuccessful, but meanwhile Daniel Bernoulli had taken up a position at the St Petersburg Academy in Russia, and invited Euler to join him there. The only available position was in medicine and physiology, but jobs were scarce so Euler learned these subjects, and his study of the ear led him to investigate the mathematics of sound and the propagation of waves.

St Petersburg, Russia

Unfortunately, on the very day that Euler arrived in Russia, Empress Catherine I, who had set up the Academy, died. Her heir was still a boy, and the faction that ruled on his behalf regarded the Academy as a luxury. Euler quietly got on with his work, while working closely with Daniel Bernoulli.

In 1733, Daniel Bernoulli had had enough of the problems of the

Academy and returned to an academic position in Switzerland. Euler, still aged only 26, replaced him in the Chair of Mathematics, and determined to make the best of a difficult situation. The 1730s were indeed very productive years for him, with substantial advances in number theory, the summation of series, and mechanics. At the same time he was acting as a scientific consultant to the government – preparing maps, advising the Russian navy, testing designs for fire engines, and writing textbooks for the Russian schools.

An area to which Euler contributed throughout his life was the theory of numbers. In December 1729, he received a letter from his St Petersburg colleague Christian Goldbach, who is best remembered for the still-unproved *Goldbach conjecture* that every even number can be written as the sum of two prime numbers. Goldbach's letter was concerned with the *Fermat numbers*:

$$2^1 + 1 = 3,\ 2^2 + 1 = 5,\ 2^4 + 1 = 17,\ 2^8 + 1 = 257,\ 2^{16} + 1 = 65,537, \ldots .$$

Are *all* such numbers prime? Fermat had conjectured that they are, but Euler found an ingenious argument to prove that the next one $(2^{32} + 1)$, a ten-digit number, is divisible by 641. Since then, *no* other 'Fermat number' has been shown to be prime, so Fermat's conjecture was unfortunate.

Euler's prodigious calculating abilities were legendary. One day, two students were trying to sum a complicated progression and disagreed over the 50th decimal place: Euler simply calculated the correct value in his head to settle the argument. Another challenge given to Euler was to find four different numbers, the sum of any two of which is a perfect square: he produced the quartet: 18530, 38114, 45986 and 65570.

A different preoccupation in the 1730s was his work on infinite series. He first became interested in the 'harmonic series'

$$1 + \tfrac{1}{2} + \tfrac{1}{3} + \tfrac{1}{4} + \tfrac{1}{5} + \ldots,$$

which does not converge, and noticed that the sum of the first n terms is very close to $\log n$. In particular, he proved that as n becomes large, the difference between them tends to a fixed number $0.5772\ldots$, now known as *Euler's constant*.

Another problem on infinite series, known as the *Basel problem*, exercised many minds at the time. It was to find the sum of the reciprocals of the perfect squares:

$$1 + \tfrac{1}{4} + \tfrac{1}{9} + \tfrac{1}{16} + \tfrac{1}{25} + \ldots .$$

One of Euler's earliest achievements was to show that this sum is $\pi^2/6$, and this brought him international fame. He also extended his calculations to find the sum of the reciprocals of the 4th powers, the 6th powers, and so on, up to the 26th powers. This led him to investigate what is now known as the *Riemann zeta function.*

We next turn to a recreational puzzle that Euler solved in 1735: the problem of the *bridges of Königsberg.* The medieval city of Königsberg consisted of four areas of land linked by seven bridges, and the problem was to find a route crossing each bridge just once and returning to the starting point. Euler discovered a counting argument involving the number of bridges emerging from each land area, and proved that no such route exists. Furthermore, he obtained for any arrangement of land areas and bridges a corresponding rule for deciding when such a route is possible – namely,

- if there are no areas with an odd number of bridges, then a route exists starting anywhere and ending in the same place;
- if there are two areas with an odd number of bridges, then a route exists, starting in one area and ending in the other;
- if there are more than two such areas (as in Königsberg), then there is no such route.

Euler's solution of the Königsberg bridges problem is considered as the earliest contribution to graph theory, and is now solved by looking at a network with four points representing the land areas and seven lines representing the bridges. But Euler never did this – the network that represents this puzzle was not drawn for 150 years.

Around the same time, Euler published *Mechanica*, his first treatise on the dynamics of a particle. However, his most important work in this area came in 1750 with his work on the motion of rigid bodies – free, or rotating about a point. By choosing the point as the origin of coordinates, and with axes aligned along the principal axes of inertia of the body, he obtained what are now called *Euler's equations of motion*; the concept of *moment of inertia* was also due to him. Even later, in 1776, he proved that any rotation of a rigid body about a point is equivalent to a rotation about a line through that point. Much of his work in this area used differential equations, an area to which he had himself contributed a great deal.

In the late 1730s, Euler went blind in his right eye. Although he attributed this to overwork, it was probably due to an eye infection. However, this did not diminish his productivity: he continued to write on acoustics, musical harmony, ship-building, prime numbers, and much more besides.

Berlin, Germany

In 1741, with his fame preceding him, Euler received an invitation from Prussia's Frederick the Great to join the newly vitalised Berlin Academy. With the political situation in Russia still uncertain, he accepted it and remained in Berlin for 25 years.

At first, he got on well with Frederick, but later, especially after the seven years war between Germany and Russia, things began to cool as Frederick started to take more and more interest in the workings of the Academy. Frederick considered himself cultured and witty, and found Euler unsophisticated; in return, Euler found Frederick pretentious, petty and rude.

Even so, Euler managed to work on a dazzling range of topics, writing works in the 1740s and 1750s on the theory of tides, the calculus of variations, the motion of the moon, hydrodynamics, and the wave motion of vibrating strings.

His most important work from this period was the *Introductio in Analysin Infinitorum* ('Introduction to the analysis of the infinite'), published in 1748. It was here that he presented some of his earlier work on the number $\mathrm{e} = 2.718281\ldots$, defined as

$$1 - \tfrac{1}{1!} + \tfrac{1}{2!} + \tfrac{1}{3!} + \ldots \text{ or as } \lim_{n\to\infty} \left(1 + \tfrac{1}{n}\right)^n,$$

and the related exponential function e^x.

In the *Introductio*, Euler expressed certain well-known functions as infinite series, and then introduced his great masterstroke. He knew how the functions e^x, $\sin x$ and $\cos x$ could be expanded in powers of x, but at first they seem to have nothing in common. However, on introducing the complex number i, and manipulating the power series, he deduced the fundamental formula linking them,

$$\mathrm{e}^{\mathrm{i}x} = \cos x + \mathrm{i} \sin x.$$

Although he did not explicitly write down the simple consequence $\mathrm{e}^{\mathrm{i}\pi} = -1$, often described as his most famous result, he would surely have known it.

There were many other interesting things in the *Introductio*. Over the 100 years since Descartes there had been a gradual movement from geometry towards algebra, and this reached its climax when Euler actually defined the conic sections (ellipse, parabola and hyperbola) by their algebraic equations, rather than geometrically as sections of a cone. In particular, starting with the equation $y^2 = \alpha + \beta x + \gamma x^2$, he showed that we get an ellipse if $\gamma < 0$, a parabola if $\gamma = 0$, and a hyperbola if $\gamma > 0$. He then extended his algebraic arguments to three dimensions, to the seven types of *quadrics*, and discovered the *hyperbolic paraboloid* in the process.

Yet another interesting topic in the *Introductio* is *partitions*, or 'divulsions of integers', as Leibniz had called them in a letter to Bernoulli. In how many ways can we split up a positive integer n into smaller ones?

Let $p(n)$ be this number – for example, $p(4) = 5$, corresponding to the five partitions $4,\ 3+1,\ 2+2,\ 2+1+1$ and $1+1+1+1$ (the order doesn't matter). So we can draw up a table of values – but how can we show that $p(200) = 3{,}972{,}999{,}029{,}388$?

To investigate partitions, Euler introduced the generating function

$$p(x) = 1 + p(1)x + p(2)x^2 + p(3)x^3 + \dots,$$

and used it to derive what is now called *Euler's pentagonal number formula* involving the 'generalised pentagonal numbers' $\frac{1}{2}k(3k \pm 1)$:

$$p(n) = p(n-1) + p(n-2) - p(n-5) - p(n-7) + p(n-12) + \dots .$$

This yields $p(n)$ by iteration, and is still the most efficient way of finding $p(n)$.

A particularly nice result on partitions, which appears in the *Introductio*, concerns odd and distinct partitions. In an *odd partition* all the parts are odd, and in a *distinct partition* all of them are different: for example, 9 has eight odd partitions (9, 7+1+1, 5+3+1, etc.) and eight distinct partitions (9, 8 + 1, 7 + 2, etc.). Using generating functions, Euler proved that for *any* number, the number of odd partitions is the same as the number of distinct partitions.

Another preoccupation was mentioned in a letter to Goldbach, in 1750. Euler had been looking at *polyhedra*, and observed that the numbers of vertices, edges and faces are always related by the formula:

$$(\textit{no. of faces}) + (\textit{no. of vertices}) = (\textit{no. of edges}) + 2.$$

This formula, now known as *Euler's polyhedron formula*, has been incorrectly credited to Descartes, who did not have the terminology or motivation to derive it: indeed, it was Euler who introduced the concept of an *edge*. However, Euler's proof was deficient – a complete proof was not given until 40 years later, by the algebraist and number-theorist Legendre.

Euler's most popular and best-selling book was his *Letters to a German Princess*. Euler was always an extremely clear writer, and this was a multi-volume masterpiece of exposition that he produced when he was asked to give elementary science lessons to the Princess of Anhalt-Dessau. The resulting collection had over 200 'letters' that Euler wrote on a range of scientific topics, including gravity, astronomy, light, sound, magnetism, logic, and much else besides. He wrote about why the sky is blue, why the moon looks larger when it rises, and why the tops of mountains are cold (even in the tropics). It was one of the best books ever written on popular science.

The last of Euler's Berlin books was his 1755 massive tome on the differential calculus. This contained all the latest results, many due to him, and presented the calculus in terms of the basic idea of a function – indeed, it was Euler who introduced the notation f for a function. Other notations he introduced at various times were $\sum$ (for summation), i (the square root of -1) and e (the exponential number). He also popularised the notation for π, although that had actually been introduced by William Jones in 1706. He followed his book on differentiation in 1768–1770 with a three-volume treatise on the integral calculus.

St Petersburg, revisited

After his difficulties with Frederick the Great, Euler must have felt very relieved when in 1766, at the age of 59, he received an invitation from Catherine the Great of Russia to return to St Petersburg. Things there had improved greatly, thanks to the enlightened Empress, and he was received royally.

He continued to work enthusiastically, soon producing a delightful result in pure geometry. In any triangle three particular points of interest are the *orthocentre* (the meeting point of the perpendiculars from the vertices to the opposite sides), the *centroid* (the meeting point of the three lines joining a vertex to the midpoint of the opposite side), and the *circumcentre* (the centre of the circle passing through the vertices of the triangle). By calculating their coordinates, Euler proved the attractive result that these three points always lie in a straight line – now called the *Euler line* of the triangle – and that the centroid always lies exactly one-third of the distance between the other two.

Euler's life-long interest in number theory continued into his later years, when he extended some results associated with Fermat – in particular, *Fermat's last theorem* that, for any $n > 2$, there are no positive numbers a, b, c satisfying $a^n + b^n = c^n$. In his number theory book of 1770, Euler proved for the first time that the sum of two cubes cannot equal another cube ($n = 3$).

Another connection with Fermat was *Fermat's little theorem*, which states that, if a is any number that is not divisible by a given prime number

p, then p divides $a^{p-1} - 1$; for example, on taking $p = 29$ and $a = 48$, we deduce that $48^{28} - 1$ is divisible by 29. In 1760 Euler extended this result to numbers other than primes, introducing the *Euler φ-function* and proving that, for any numbers a and n, $a^{\varphi(n)} - 1$ is always divisible by n.

Yet another result in number theory concerned *perfect numbers* – numbers whose proper divisors add up to the number itself; for example, 28 is a perfect number because its proper divisors are 1, 2, 4, 7 and 14, which sum to 28. In his *Elements*, Euclid had proved that every number of the form $2^{n-1} \times (2^n - 1)$ is perfect, when $2^n - 1$ is prime. Euler proved that every *even* perfect number has this form – but it is still not known whether any *odd* perfect numbers exist.

The last few years of Euler's life, though more peaceful than his earlier ones, saw many personal tragedies. In 1771 his house burned down, with the loss of his library and almost his life, but fortunately his manuscripts were saved. Shortly after this, his beloved wife died and he remarried. And finally he lost most of the sight in his other eye – but again, his productivity remained undiminished as he wrote on slates with his sons and friends as amenuenses; indeed, shortly after he went blind, he produced a 700-page volume on the motion of the moon.

In a book on recreational mathematics in 1725, Jacques Ozanam had shown how to lay out the sixteen court cards so that each row and column contains each suit and each value (J, Q, K, A). It is also possible to make a similar arrangement with 25 cards (five suits and five values), but what about 36 cards? In the year before he died, in a paper mainly on magic squares (another interest of his), Euler posed this as the 36 officers problem:

Arrange 36 officers, one each of 6 ranks from 6 regiments, in a square array so that each row or column has one officer of each rank and one officer of each regiment.

Euler believed that this cannot be done, and this was eventually confirmed around 1900 by Gaston Tarry, essentially by enumerating all the possibilities. Euler also claimed that the corresponding problem has a solution for any number of ranks and regiments, except when this number is of the form $4k + 2$: that is, 6, 10, 14, 18, ... He was right about 6, but wrong about *all* the others, although his error was not demonstrated until almost 300 years later.

Conclusion

Euler worked until the very end. In a Eulogy by the Marquis de Condorcet, we read about his final afternoon:
On the 7th of September 1783, after amusing himself with calculating on a slate the laws of the ascending motion of air balloons, the recent discovery of which was then making a noise all over Europe, he dined with Mr Lexell and his family, talked of Herschel's planet (Uranus), and of the calculations which determine its orbit. A little after, he called his grandchild, and fell a playing with him as he drank tea, when suddenly the pipe, which he held in his hand, dropped from it, and he ceased to calculate and to breathe. The great Euler was no more.

Acknowledgments

This article is adapted from the author's article 'Read Euler, read Euler, he is the master of us all', *Plus* (Online mathematics magazine), Mathematics Millennium Project (University of Cambridge, England), 42, March 2007.

Further reading

A good introductory book on Euler's life and works is:
William Dunham, *Euler: the master of us all*, Mathematical Association of America, 1999.

To celebrate the 300th anniversary of his death, the Mathematical Association of America has also published a series of five books in 2007 on the life and works of Leonhard Euler.

Further information can also be found in standard reference works, such as:
The Dictionary of Scientific Biography, Scribner, New York, 1970–1990.

PART 1

Riemannian geometry, submanifolds

Differential Geometry and its Applications
Proc. Conf., in Honour of Leonhard Euler, Olomouc, August 2007

Curvature logarithmic derivatives of curves and isometric immersions

Toshiaki Adachi

Department of Mathematics, Nagoya Institute of Technology,
Nagoya 466-8555, Japan
E-mail: adachi@nitech.ac.jp

In this note we study isometric immersions by extrinsic shapes of some curves and extend Nomizu-Yano's characterization of extrinsic spheres. We give a characterization of non-totally geodesic isotropic immersions in the sense of O'Neill by the amount of curvature logarithmic derivatives of curves preserved by these isometries. Also we give a characterization of totally umbilic submanifolds with parallel normalized mean curvature vector in terms of curvature logarithmic derivatives of curves of order 2 preserved by immersions.

Keywords: Isotropic immersion, curvature logarithmic derivative, totally umbilic, parallel normalized mean curvature vector, extrinsic shapes of curves, derived maps.

MS classification: 53C40.

1. Introduction

Let $f : M \to \widetilde{M}$ be an isometric immersion. For a smooth curve $\gamma : I \to M$ parameterized by its arc-length, we call the curve $f \circ \gamma$ the *extrinsic shape* of γ. It is an interesting problem to characterize isometric immersions by some properties of extrinsic shapes of smooth curves. In this area Nomizu-Yano's characterization on extrinsic spheres by extrinsic shapes of circles is one of the most fundamental results. A Riemannian submanifold M is said to be an extrinsic sphere in $\widetilde{M}$ if it is totally umbilic and has parallel mean curvature vector. We call a smooth curve γ parameterized by its arc-length *circle* if it satisfies $\nabla_{\dot\gamma}\dot\gamma = \kappa Y_\gamma$, $\nabla_{\dot\gamma} Y_\gamma = -\kappa\dot\gamma$ with some positive constant κ and a unit vector field Y_γ along γ. In Ref. 5, Nomizu and Yano showed that M is an extrinsic sphere in $\widetilde{M}$ if and only if the extrinsic shape of each circle is also a circle. In this note we give characterizations of isotropic immersions and totally umbilic immersions by extending their idea from a bit different point of view. This also refines Sugiyama and the author's results (Refs. 1,7,8).

2. Derived maps of curves

For a Riemannian manifold M we consider a family $\mathcal{C}(M) = \{\gamma : I_\gamma \to M\}$ of smooth curves on M which are parameterized by their arc-length and whose domains contain the origin. An isometry $f : M \to \widetilde{M}$ causes a correspondence $\mathcal{F} : \mathcal{C}(M) \in \gamma \mapsto f \circ \gamma \in \mathcal{C}(\widetilde{M})$ between the families of curves. In previous papers which characterized some isometric immersions (Refs. 3,5,6, for example), properties of $\mathcal{F}$ are taken into account: An isometry is characterized by the extrinsic shape $\mathcal{F}(\mathcal{S})$ of a "nice" family $\mathcal{S}$.

In this note by changing our viewpoint we shall consider maps derived from the extrinsic shape correspondence $\mathcal{F}$. For a smooth curve $\gamma \in \mathcal{C}(M)$, we define a function κ_γ along γ by $\kappa_\gamma = \|\nabla_{\dot\gamma}\dot\gamma\|$ and call it the first *geodesic curvature* of γ. When it does not have inflection points, that is $\kappa_\gamma > 0$, we set $\ell_\gamma = \kappa_\gamma'/\kappa_\gamma$ and call it the *curvature logarithmic derivative* of γ. For each circle, this function is identically zero. As a first level we consider the product $UM \times [0,\infty)$ of the unit tangent bundle of M and a half line. We then have a canonical projection ϖ_1 given as $\mathcal{C}(M) \ni \gamma \mapsto \big(\dot\gamma(0), \kappa_\gamma(0)\big) \in UM \times [0,\infty)$ which shows initial vector and initial geodesic curvature. For an isometric immersion $f : M \to \widetilde{M}$ we see by use of Gauss formula that

$$\kappa^2_{\mathcal{F}(\gamma)} = \kappa_\gamma^2 + \sigma_f(\dot\gamma,\dot\gamma)^2 \tag{1}$$

with the second fundamental form σ_f of f. We hence find that f induces a derived map $f_1 : UM \times [0,\infty) \to U\widetilde{M} \times [0,\infty)$ of the first level.

As for a second level we consider the product $\mathcal{O}_2M \times [0,\infty) \times \mathbb{R}$ of the set $\mathcal{O}_2M$ of orthonormal pairs of tangent vectors and a half plane. If we restrict ourselves on the subfamily $\mathcal{C}_2(M)$ of smooth curves whose initials are not inflection points, we have a canonical projection ϖ_2 which is given as

$$\mathcal{C}_2(M) \ni \gamma \mapsto \Big(\big(\dot\gamma(0), Y_\gamma(0)\big),\ \kappa_\gamma(0), \ell_\gamma(0)\Big) \in \mathcal{O}_2M \times [0,\infty) \times \mathbb{R},$$

where Y_γ is a unit vector field along γ defined by $Y_\gamma = (1/\kappa_\gamma)\nabla_{\dot\gamma}\dot\gamma$. By differentiating both sides of the equality (1) we have the following:

Lemma 2.1 (Ref. 1). *Curvature logarithmic derivatives of a curve γ and its extrinsic shape $\mathcal{F}(\gamma)$ are related as follows*:

$$\begin{aligned}\kappa_\gamma^2\big(\ell_{\mathcal{F}(\gamma)} - \ell_\gamma\big) + \ell_{\mathcal{F}(\gamma)}\|\sigma(\dot\gamma,\dot\gamma)\|^2 &= \kappa^2_{\mathcal{F}(\gamma)}\big(\ell_{\mathcal{F}(\gamma)} - \ell_\gamma\big) + \ell_\gamma\|\sigma(\dot\gamma,\dot\gamma)\|^2 \\ &= \big\langle\big(\overline{\nabla}_{\dot\gamma}\sigma\big)(\dot\gamma,\dot\gamma), \sigma(\dot\gamma,\dot\gamma)\big\rangle + 2\big\langle\sigma\big(\nabla_{\dot\gamma}\dot\gamma,\dot\gamma\big),\sigma(\dot\gamma,\dot\gamma)\big\rangle.\end{aligned} \tag{2}$$

Here the covariant differentiation $\overline{\nabla}$ *of* σ_f *is defined by*

$$(\overline{\nabla}_X\sigma_f)(Y,Z) = \nabla^{\perp}_X\big(\sigma_f(Y,Z)\big) - \sigma_f(\nabla_X Y, Z) - \sigma_f(Y,\nabla_X Z)$$

for vector fields X, Y, Z *on* M.

This lemma guarantees that an isometry f induces a derived map $f_2 : \mathcal{O}_2M \times [0,\infty) \times \mathbb{R} \to \mathcal{O}_2\widetilde{M} \times [0,\infty) \times \mathbb{R}$ of the second level. Instead of properties of extrinsic shape correspondence, we make use of properties of these derived maps. Since the derived map of the first level is too simple to take its properties, we shall study the derived map of the second level. We pay attention to the last component of the second level, which corresponds to curvature logarithmic derivatives of curves, and characterize some isometries by properties of the derived map of the second level.

3. Isotropic immersions

An isometric immersion $f : M \to \widetilde{M}$ is said to be *isotropic* at $x \in M$ in the sense of O'Neill if the norm $\|\sigma_f(u,u)\|$ of the second fundamental form does not depend on the choice of a unit tangent vector $u \in U_xM$. It is clear that if f is umbilic at x then it is isotropic at this point. We say an isometric immersion f is *geodesic* at x if $\sigma_f(u,u) = 0$ for every $u \in U_xM$. Trivially, if f is geodesic at x then it is isotropic at this point. In this section we characterize isotropic immersions.

For an orthonormal pair $(u,v) \in \mathcal{O}_2M_x$ of tangent vectors at $x \in M$, we denote by $\mathcal{C}(u,v)$ $\big(\subset \mathcal{C}(M)\big)$ the subfamily of curves whose initial velocity vector is u and initial principal normal is v. We define the sets of curvature logarithmic derivatives preserved by f as

$$\mathcal{A}_f(u,v) = \big\{\ell_\gamma(0) \bigm| \gamma \in \mathcal{C}(u,v),\ \ell_{\mathcal{F}(\gamma)}(0) = \ell_\gamma(0)\big\},$$
$$\mathcal{A}_f^+(u) = \bigcup_{v\,(\perp u)} \mathcal{A}_f(u,v), \qquad \mathcal{A}_f^-(u) = \bigcap_{v\,(\perp u)} \mathcal{A}_f(u,v).$$

If we need to specify the geodesic curvature at initial, we consider the family $\mathcal{C}(u,v;\kappa) = \big\{\gamma \in \mathcal{C}(u,v) \bigm| \kappa_\gamma(0) = \kappa\big\}$ for a positive constant κ, and define the sets $\mathcal{A}_f(u,v;\kappa)$, $\mathcal{A}_f^+(u;\kappa)$, $\mathcal{A}_f^-(u;\kappa)$ in the same manner for this family. If we interpret the set $\mathcal{A}_f(u,v;\kappa)$ in terms of the derived map of the second level, it is the set of fixed values of the function $f_2((u,v),\kappa,*) : \mathbb{R} \to \mathbb{R}$. It should be noted that $\mathcal{A}_f(u,v;\kappa) \neq \emptyset$ for each $(u,v) \in \mathcal{O}_2M$ and each positive κ for an arbitrary isometric immersion f. Hence the condition of existence of curvature logarithmic derivatives preserved by an isometry is too weak. We hence consider the quantity of such curvature logarithmic derivatives and characterize isotropic immersions in the following manner:

Theorem 3.1. *For an isometric immersion $f : M \to \widetilde{M}$ the following conditions at a point $x \in M$ are mutually equivalent to each other:*

1) *At this point f is isotropic in the sense of O'Neill and is not geodesic;*
2) *The set $\mathcal{A}_f(u,v)$ consists of a single value for every $(u,v) \in \mathcal{O}_2M_x$;*
3) *the set $\mathcal{A}_f^+(u)$ consists of a single value for every $u \in U_xM$;*
4) *For every $u \in U_xM$ there is a positive κ_u satisfying that the set $\mathcal{A}_f^+(u;\kappa_u)$ consists of a single value;*
5) *The set $\mathcal{A}_f^-(u)$ consists of a single value for every $u \in U_xM$;*
6) *For every $u \in U_xM$ there is a positive κ_u satisfying that the set $\mathcal{A}_f^-(u;\kappa_u)$ consists of a single value.*

Theorem 3.2. *For an isometric immersion $f : M \to \widetilde{M}$ the following conditions at a point $x \in M$ are mutually equivalent to each other:*

1) *f is geodesic at this point;*
2) *There exist an orthonormal basis $\{u_1,\dots,u_n\}$ of the tangent space T_xM and $v_j \in U_xM$ $(j = 1,\dots,n)$ satisfying $v_j \perp u_j$ and $\mathcal{A}_f(u_j,v_j) = \mathbb{R}$ for all j;*
3) *There exist a basis $\{u_1,\dots,u_n\}$ of the tangent space T_xM by unit vectors and pairs (v_j,κ_j) $(j = 1,\dots,n)$ of a unit tangent vector and a positive constant satisfying $v_j \perp u_j$ and $\mathcal{A}_f(u_j,v_j;\kappa_j) = \mathbb{R}$ for all j.*

Proof of Theorems 3.1 and 3.2. By Lemma 2.1, we see $\gamma \in \mathcal{C}(u,v)$ satisfies $\ell_{\mathcal{F}(\gamma)}(0) = \ell_\gamma(0)$ if and only if

$$\ell_\gamma(0) = \|\sigma_f(u,u)\|^{-2}\Big(\langle(\widetilde{\nabla}_u\sigma_f)(u,u),\sigma_f(u,u)\rangle + 2\kappa_\gamma(0)\langle\sigma_f(u,u),\sigma_f(u,v)\rangle\Big)$$

when $\sigma_f(u,u) \neq 0$. Hence the feature of the set $\mathcal{A}_f(u,v)$ is classified into the following 3 cases:

1) $\mathcal{A}_f(u,v) = \mathbb{R}$ if and only if $\sigma_f(u,u) = 0$;
2) $\mathcal{A}_f(u,v)$ consists of a single value if and only if $\sigma_f(u,u) \neq 0$ and $\langle\sigma_f(u,u),\sigma_f(u,v)\rangle = 0$;
3) $\mathcal{A}_f(u,v)$ is an open half line if and only if $\langle\sigma_f(u,u),\sigma_f(u,v)\rangle \neq 0$.

It is clear that in the second case the value $\mathcal{A}_f(u,v)$ does not depend on the choice of v. Since f is isotropic at x if and only if $\langle\sigma_f(u,u),\sigma_f(u,v)\rangle = 0$ for every $(u,v) \in \mathcal{O}_2M_x$, we get our conclusions. $\square$

Remark 3.1. Our proof of Theorem 3.1 also shows that f is isotropic at x if and only if for every $(u,v) \in \mathcal{O}_2M_x$ there is a positive $\kappa_{u,v}$ with $\mathcal{A}_f(u,v:\kappa_{u,v}) \cap \mathcal{A}_f(u,-v;\kappa_{u,v}) \neq \emptyset$ (see Ref. 1).

We are now interested in the single value of preserved curvature logarithmic derivatives. When f is isotropic everywhere, we just call it isotropic and define a function of isotropy $\lambda_f : M \to \mathbb{R}$ as the norm of the second fundamental form at each point.

Remark 3.2. If $f : M \to \widetilde{M}$ is isotropic, then we have $\mathcal{A}_f^+(u) = \mathcal{A}_f^-(u) = \{u(\lambda_f)/\lambda_f(x)\}$ for every $u \in U_xM$ at an arbitrary point $x \in M$.

As a consequence of Theorem 3.1 we can characterize Veroneze embeddings. For a positive integer k, we denote by $f_k : \mathbb{C}P^n(c/k) \to \mathbb{C}P^{N(k)}(c)$ the k-th Veronese embedding which is given by

$$\big[z_i\big]_{0\leq i\leq n} \to \Big[\sqrt{k!/(k_0!\cdots k_n!)}\; z_0^{k_0}\cdots z_n^{k_n}\Big]_{k_0+\cdots+k_n=k}$$

with homogeneous coordinate $[*]$, where $N(k) = (n+k)!/(n!k!) - 1$. This embedding is isotropic whose function of isotropy is a constant function $\lambda_{f_k} \equiv c(k-1)/(2k)$.

Theorem 3.3 (c.f. Ref. 1). *Let $f : M \to \mathbb{C}M^N(c)$ be a non-totally geodesic Kähler isometric full immersion of a Kähler manifold M of complex dimension $n \geq 2$ into a complex space form. Then the following conditions are mutually equivalent:*

1) *There is a positive integer k satisfying that $N = N(k)$, the ambient space is $\mathbb{C}P^N(c)$, the manifold M is locally congruent to $\mathbb{C}P^n(c/k)$ and f is locally equivalent to f_k;*
2) *The set $\mathcal{A}_f^+(u)$ (resp. $\mathcal{A}_f^-(u)$) consists of a single value for every unit tangent vector $u \in TM$.*
3) *The set $\mathcal{A}_f(u,v)$ consists of a single value for every $(u,v) \in \mathcal{O}_2M$;*
4) *For every $u \in U_xM$ there is a positive κ_u satisfying that the set $\mathcal{A}_f^+(u;\kappa_u)$ (resp. $\mathcal{A}_f^-(u;\kappa_u)$) consists of a single value.*

We should note that such characterization does not hold for the case $\dim_{\mathbb{C}} M = 1$ because all holomorphic curves are isotropic.

4. Totally umbilic immersions

In this section we study isometric immersions with umbilic points. When $\gamma \in \mathcal{C}(M)$ does not have inflection points, the differential of Y_γ is of the form

$\nabla_{\dot\gamma}Y_\gamma = -\kappa_\gamma\dot\gamma + Z_\gamma$ with some vector field Z_γ along γ which is orthogonal to both $\dot\gamma$ and Y_γ. We say γ is of (proper) order 2 at initial if $\kappa_\gamma \neq 0$ and $Z_\gamma(0) = 0$. Clearly every circle is of order 2 at each point. For an orthonormal pair $(u,v) \in \mathcal{O}_2M_x$ we define a subset $\mathcal{B}_f(u,v)$ of $\mathcal{A}_f(u,v)$ as

$$\mathcal{B}_f(u,v) = \left\{ \ell_\gamma(0) \;\middle|\; \begin{array}{l} \gamma \in \mathcal{C}(u,v),\ \ell_{\tilde\gamma}(0) = \ell_\gamma(0), \\ \mathcal{F}(\gamma) \text{ is of order 2 at } f(x) \end{array} \right\}.$$

We put $\mathcal{B}_f(u) = \bigcup_{v\,(\perp u)} \mathcal{B}_f(u,v)$. The set $\mathcal{B}_f(u,v)$ can be interpreted in the following manner. We consider a subfamily

$$\mathcal{S} = \big\{\gamma \in \mathcal{C}_2(M) \;\big|\; \mathcal{F}(\gamma) \text{ is of order 2 at } \mathcal{F}(\gamma)(0)\big\}$$

and study the restriction of the derived map f_2 of the second level onto the set $\varpi_2(\mathcal{S})$. The set $\mathcal{B}_f(u,v)$ consists of fixed values of this map with respect to the last component. We can also explain this set in another way. We denote by $\mathcal{O}_3M$ the set of triplets of orthonormal tangent vectors and set $\widetilde{\mathcal{O}_3M} = \mathcal{O}_3M \oplus \big(\mathcal{O}_2M\times\{0\}\big)$. As for a third level we consider the product $\widetilde{\mathcal{O}_3M} \times [0,\infty) \times \mathbb{R} \times [0,\infty)$ and define a projection ϖ_3 of $\mathcal{C}_2(M)$ onto this level by

$$\gamma \mapsto \Big(\big(\dot\gamma(0), Y_\gamma(0), Z_\gamma(0)/\|Z_\gamma(0)\|\big),\ \kappa_\gamma(0), \ell_\gamma(0), \|Z_\gamma(0)\|\Big),$$

where in case $\|Z_\gamma(0)\| = 0$ we regard $Z_\gamma(0)/\|Z_\gamma(0)\|$ as a null tangent vector. By the following Lemma 4.1 we get a derived map

$$f_3 : \widetilde{\mathcal{O}_3M} \times [0,\infty) \times \mathbb{R} \times [0,\infty) \to \widetilde{\mathcal{O}_3\widetilde{M}} \times [0,\infty) \times \mathbb{R} \times [0,\infty)$$

of the third level from an isometry $f : M \to \widetilde{M}$. Since $\gamma \in \mathcal{C}_2(M)$ belongs to $\mathcal{S}$ if and only if the last component of $f_3\big(\varpi_3(\gamma)\big)$ is 0, we can interpret $\mathcal{B}_f(u,v)$ in terms of this derived map f_3.

Lemma 4.1. *For $\gamma \in \mathcal{C}_2(M)$ the vector field $Z_{\mathcal{F}(\gamma)}$ along $\mathcal{F}(\gamma)$ is given as*

$$\begin{aligned}\kappa_{\mathcal{F}(\gamma)}Z_{\mathcal{F}(\gamma)} = \|\sigma_f(\dot\gamma,\dot\gamma)\|\,\dot\gamma - A_{\sigma_f(\dot\gamma,\dot\gamma)}\dot\gamma + \kappa_\gamma\big(\ell_\gamma - \ell_{\mathcal{F}(\gamma)}\big)Y_\gamma + \kappa_\gamma Z_\gamma \\ - \ell_{\mathcal{F}(\gamma)}\sigma_f(\dot\gamma,\dot\gamma) + 3\kappa_\gamma\sigma_f(\dot\gamma,Y_\gamma) + \big(\overline{\nabla}_{\dot\gamma}\sigma_f\big)(\dot\gamma,\dot\gamma).\end{aligned} \tag{3}$$

We now study isometric immersions with umbilic points. The quantity of $\mathcal{B}_f(u,v)$ shows the umbilic property of isometries.

Theorem 4.1. *Let $f : M \to \widetilde{M}$ be an isometric immersion.*

(1) *If $\mathcal{B}_f(u,v) \neq \emptyset$ for every $(u,v) \in \mathcal{O}_2M_x$, then f is umbilic at $x \in M$.*
(2) *If we suppose moreover there is a basis $\{u_1,\dots,u_n\}$ of T_xM by unit vectors which satisfies the property that $\mathcal{B}_f(u_j)$ contains at least two values for all j, then f is geodesic at this point x.*

Proof. If $\mathcal{F}(\gamma)$ for $\gamma \in \mathcal{C}(u,v)$ is of order 2 at the origin, by paying attention on the tangential and normal components of the equality (3), we find the following hold:

$$\begin{cases} \kappa_\gamma(0)\big(\ell_\gamma(0) - \ell_{\mathcal{F}(\gamma)}(0)\big) = \langle \sigma_f(u,u), \sigma_f(u,v)\rangle, \\ \ell_{\mathcal{F}(\gamma)}(0)\sigma_f(u,u) = 3\kappa_\gamma(0)\sigma_f(u,v) + \big(\overline{\nabla}_u\sigma_f\big)(u,u). \end{cases} \tag{4}$$

Thus if $\mathcal{B}_f(u,v) \neq \emptyset$ then $\langle \sigma_f(u,u), \sigma_f(u,v)\rangle = 0$. Since we also have $\mathcal{B}_f(u,-v) \neq \emptyset$ there is a smooth curve $\rho \in \mathcal{C}(u,-v)$ whose extrinsic shape $\mathcal{F}(\rho)$ is of order 2 at $f(x)$. We therefore have

$$\ell_{\mathcal{F}(\rho)}(0)\sigma_f(u,u) = -3\kappa_\rho(0)\sigma_f(u,v) + \big(\overline{\nabla}_u\sigma_f\big)(u,u).$$

Thus we obtain

$$\big(\ell_{\mathcal{F}(\gamma)}(0) - \ell_{\mathcal{F}(\rho)}(0)\big)\sigma_f(u,u) = 3\big(\kappa_\gamma(0) + \kappa_\rho(0)\big)\sigma_f(u,v),$$

which shows $\sigma_f(u,v) = 0$ holds. As $\sigma_f(u,v) = 0$ holds for an arbitrary $(u,v) \in \mathcal{O}_2M_x$, we can conclude x is an umbilic point. □

Since the condition in Theorem 4.1 is not a necessary and sufficient condition, we here study more about it. We shall say a submanifold has parallel normalized mean curvature vector if its mean curvature vector $\mathfrak{h}$ is of the form $\mathfrak{h} = H\xi$ with its mean curvature $H = \|\mathfrak{h}\|$ and some unit vector field ξ which is parallel with respect to $\nabla^\perp$.

Theorem 4.2. *An isometric immersion $f : M \to \widetilde{M}$ is totally umbilic and has parallel normalized mean curvature vector on the outside of the set of geodesic points if and only if $\mathcal{B}_f(u,v) \neq \emptyset$ for every orthonormal pair $(u,v) \in \mathcal{O}_2M$.*

Proof. By Equality (3) we can conclude the following: When f is totally umbilic, for a smooth curve γ we have

$$\kappa_{\mathcal{F}(\gamma)} Z_{\mathcal{F}(\gamma)} = \kappa_\gamma\big(\ell_{\mathcal{F}(\gamma)} - \ell_\gamma\big)Y_\gamma + \kappa_\gamma Z_\gamma + \ell_{\mathcal{F}(\gamma)}\mathfrak{h}_\gamma - \nabla^\perp_{\dot\gamma}\mathfrak{h}. \tag{5}$$

If $\mathcal{B}_f(u,v) \neq \emptyset$ for every $(u,v) \in \mathcal{O}_2M$, we see by Theorem 4.1 that f is totally umbilic. On the outside of the set of geodesic points, we consider a unit vector field $\xi = (1/H)\mathfrak{h}$. If $\mathcal{F}(\gamma)$ of $\gamma \in \mathcal{C}(u,v)$, $(u,v) \in \mathcal{O}_2M_x$ is of order 2 at $f(x)$. we find by Equation (5) that $\ell_{\mathcal{F}(\gamma)}(0)\mathfrak{h}_x = \nabla^\perp_u\mathfrak{h}$, which is equivalent to $\ell_{\mathcal{F}(\gamma)}(0)H_x\xi_x = u(H)\xi_x + H_x\nabla^\perp_u\xi$. As $\langle \nabla^\perp_u\xi, \xi_x\rangle = 0$, we see $\nabla^\perp_u\xi = 0$. Thus we obtain M has parallel normalized mean curvature vector on the outside of the set of geodesic points.

On the other hand, we suppose f is totally umbilic and has parallel normalized mean curvature vector on the outside of the set of geodesic points. For $(u,v) \in \mathcal{O}_2M_x$ at non-geodesic point $x \in M$ we take a smooth curve $\gamma \in \mathcal{C}(u,v)$ which is of order 2 at x and satisfies $\ell_\gamma(0) = u(H)/H_x$. We then see by Lemma 2.1 that

$$\kappa_\gamma(0)\big(\ell_{\mathcal{F}(\gamma)}(0) - \ell_\gamma(0)\big) + \ell_\gamma(0)H_x^2 = \langle \nabla_u \mathfrak{h}, \mathfrak{h}\rangle = u(H)H_x,$$

hence $\ell_{\mathcal{F}(\gamma)}(0) = \ell_\gamma(0)$. Therefore we find by Equality (5) that $\mathcal{F}(\gamma)$ is of order 2 at $f(x)$. Thus we see $u(H)/H_x \in \mathcal{B}_f(u,v)$. □

We here study isometric immersions of Kähler manifolds into real space forms. We denote by $\mathbb{R}M^n(c)$ a real space form of constant sectional curvature c, which is either a standard sphere $S^n(c)$, a Euclidean space $\mathbb{R}^n$, or a real hyperbolic space $\mathbb{R}H^n(c)$.

Theorem 4.3. *Let $f : (M,J) \to \mathbb{R}M^N(\tilde{c})$ be an isometric immersion of a connected Kähler manifold of complex dimension n with complex structure J into a real space form.*

(1) *If $\mathcal{B}_f(u,Ju) \neq \emptyset$ and $\mathcal{B}_f(u,-Ju) \neq \emptyset$ hold for every $u \in UM$, then f has parallel second fundamental form.*
(2) *If we suppose moreover that $\mathcal{B}_f(u,Ju) \cup \mathcal{B}_f(u,-Ju)$ contains a non-zero value for every $u \in UM$, then f is totally geodesic. That is, it is locally equivalent to an embedding $\mathbb{C}^n \to \mathbb{R}^N$.*

Proof. By the proof of Theorem 4.1 we see $\sigma_f(u,Ju) = 0$ for every $u \in UM$. Thus for arbitrary $u, w \in TM$, considering $(u+w)/\|u+w\|$ we get $\sigma_f(u,Jw) = -\sigma_f(Ju,w)$. Since M is Kähler, this leads us to $(\overline{\nabla}_X\sigma_f)(Y,JZ) = -(\overline{\nabla}_X\sigma_f)(JY,Z)$ for arbitrary vector fields X, Y, Z on M. By use of Codazzi equation we can conclude $(\overline{\nabla}_X\sigma_f)(Y,JZ) = 0$ (see Ref. 7). When $\mathcal{B}_f(u,Ju) \cup \mathcal{B}_f(u,-Ju)$ contains a non-zero value, the second equality in (4) shows that $\sigma_f(u,u) = 0$. Thus we get the conclusion. □

At the last stage, we shall mention that there are some studies on the geometry of frame bundles (Refs. 2,4). The author suspects that there might be some relationship between the geometry of $\mathcal{O}_2M$ and that of a base manifold M through the geometry of curves of order 2.

Acknowledgments

The author is partially supported by Grant-in-Aid for Scientific Research (C) (No. 17540072) JSPS.

References

1. T. Adachi and T. Sugiyama, A characterization of isotropic immersions by extrinsic shapes of smooth curves, to appear in *Diff. Geom. its Appl.*
2. O. Kowalski and M. Sekizawa, On the geometry of orthonormal frame bundles, to appear in *Math. Nachr.*
3. S. Maeda, A characterization of constant isotropic immersions by circles, *Arch. Math. (Basel)* **81** (2003) 90–95.
4. K.P. Mok, On the differential geometry of frame bundles of Riemannian manifolds, *J. reine angew Math.* **302** (1978) 16–31.
5. K. Nomizu and K. Yano, On circles and spheres in Riemannian geometry, *Math. Ann.* **210** (1974) 163–170.
6. K. Sakamoto, Planer geodesic immersions, *Tôhoku Math. J.* **29** (1977) 25–56.
7. T. Sugiyama, Totally geodesic Kähler immersions in veiw of curves of order two, to appear in *Geom. Dedicata.*
8. T. Sugiyama and T. Adachi, Totally umbilic isometric immersions and curves of order 2, *Monatsh. Math.* **150** (2007) 73–81.

Differential Geometry and its Applications
Proc. Conf., in Honour of Leonhard Euler, Olomouc, August 2007

Global behaviour of maximal surfaces in Lorentzian product spaces

Alma L. Albujer
Departamento de Matemáticas, Universidad de Murcia,
E-30100 Espinardo, Murcia, Spain
E-mail: albujer@um.es

In this paper we report on some recent results about maximal surfaces in a Lorentzian product space of the form $M^2 \times \mathbb{R}_1$, where M^2 is a connected Riemannian surface and $M^2 \times \mathbb{R}_1$ is endowed with the Lorentzian metric $\langle , \rangle = \langle , \rangle_M - dt^2$. In particular, if the Gaussian curvature of M is non-negative, we establish new Calabi-Bernstein results for complete maximal surfaces immersed into $M^2 \times \mathbb{R}_1$ and for entire maximal graphs over a complete surface M. We also construct counterexamples which show that our Calabi-Bernstein results are no longer true without the hypothesis $K_M \geq 0$. Finally, we introduce two local approaches to our global results. We do not provide here with detailed proofs of our results. For further details, we refer the reader to the original papers Ref. 1–3.

Keywords: Maximal surfaces, Lorentzian product spaces, Parabolicity.

MS classification: 53C42, 53C50.

1. Introduction

The study of maximal surfaces, that is, spacelike surfaces with zero mean curvature, has been of increasing interest in recent years from both physical and mathematical points of view. Here by spacelike we mean that the induced metric from the ambient Lorentzian metric is a Riemannian metric on the surface and the term *maximal* comes from the fact that these surfaces locally maximize area among all nearby surfaces having the same boundary.

One of the most important global results in Lorentzian geometry is the Calabi-Bernstein theorem for maximal surfaces in the Lorentz-Minkowski space $\mathbb{R}^3_1$. The Calabi-Bernstein theorem states, in a parametric version that the only complete maximal surfaces in $\mathbb{R}^3_1$ are the spacelike planes This result can also be seen in a non-parametric form, and it establishes that the only entire maximal graphs in $\mathbb{R}^3_1$ are the spacelike planes; that is, the only entire solutions to the maximal surface equation

$$\mathrm{Div}\left(\frac{Du}{\sqrt{1-|Du|^2}}\right)=0, \quad |Du|^2<1$$

on the Euclidean plane $\mathbb{R}^2$ are affine functions. The Calabi-Bernstein theorem was first proved by Calabi,[7] and extended later to the general n-dimensional Lorentz-Minkowski space by Cheng and Yau.[8]

The aim of this paper is to report on some of our recent results, obtained jointly with Alías, about maximal surfaces immersed into a Lorentzian product space of the form $M^2 \times \mathbb{R}$, where M^2 is a connected Riemannian surface and $M^2 \times \mathbb{R}$ is endowed with the Lorentzian metric

$$\langle , \rangle = \pi_M^*(\langle , \rangle_M) - \pi_{\mathbb{R}}^*(dt^2).$$

Here π_M and $\pi_{\mathbb{R}}$ denote the projections from $M \times \mathbb{R}$ onto each factor, and $\langle , \rangle_M$ is the Riemannian metric on M. For simplicity, we will write

$$\langle , \rangle = \langle , \rangle_M - dt^2,$$

and we will denote by $M^2 \times \mathbb{R}_1$ the 3-dimensional product manifold $M^2 \times \mathbb{R}$ endowed with that Lorentzian metric.

In the first section, we introduce some Calabi-Bernstein type results for maximal surfaces in $M^2 \times \mathbb{R}_1$. In that sense, our first main result, Theorem 2.1, establishes that any complete maximal surface Σ immersed into a Lorentzian product space $M^2 \times \mathbb{R}_1$, where M is a (necessarily complete) Riemannian surface with non-negative Gaussian curvature, is totally geodesic. Moreover, if M is non-flat then Σ must be a *slice* $M \times \{t_0\}$, $t_0 \in \mathbb{R}$. Here by *complete* it is meant, as usual, that the induced Riemannian metric on Σ from the ambient Lorentzian metric is complete. We also present a non-parametric version of this result. In concrete, Theorem 2.2 states that any entire maximal graph in $M^2 \times \mathbb{R}_1$ must be totally geodesic and, as a consequence, the only entire solutions to the maximal surface equation on any complete non-flat Riemannian surface M with non-negative Gaussian curvature are the constant functions. Observe that in these results the assumption on the Gaussian curvature is necessary as shown by the fact that when $M = \mathbb{H}^2$ is the hyperbolic plane there exist examples of complete maximal surfaces in $\mathbb{H}^2 \times \mathbb{R}_1$ which are non-totally geodesic, as well as examples of non-trivial entire maximal graphs over $\mathbb{H}^2$ (see examples in Section 3). These examples are obtained in two different ways. Firstly, we can obtain some examples in an implicit way via a simple but nice duality result between solutions to the minimal surface equation in a Riemannian product $M^2 \times \mathbb{R}$ and solutions to the maximal surface equation in a Lorentzian product $M^2 \times \mathbb{R}_1$, Theorem 3.1. On the other hand, we can also obtain new examples of non-trivial complete and non-complete entire maximal graphs over $\mathbb{H}^2$ by solving directly the maximal surface equation for particular type of functions.

In the last two sections, we introduce local approaches to these Calabi-Bernstein type results. In concrete in Section 4 we make an incursion on the study of parabolicity properties of Riemannian surfaces with non-empty boundary, giving in Theorem 4.1 an interesting parabolicity criterium for these surfaces in $M^2 \times \mathbb{R}_1$, being M^2 a complete Riemannian surface with non-negative Gaussian curvature. Finally, in Section 5 we establish a local integral inequality for the squared norm of the second fundamental form of a maximal surface in $M^2 \times \mathbb{R}_1$, Theorem 5.1. As a consequence of these local results, when we consider complete entire maximal graphs we provide alternative proofs of Theorem 2.2. We do not provide here with detailed proofs of our results. For further details, we refer the reader to the original papers.[1–3]

The author would like to heartily thank Luis J. Alías for some useful comments during the preparation of this work.

2. A Calabi-Bernstein result for maximal surfaces

Firstly, it is worth pointing out the following remarkable topological property of $M^2 \times \mathbb{R}_1$. If $M^2 \times \mathbb{R}_1$ admits a complete spacelike surface $f : \Sigma^2 \to M^2 \times \mathbb{R}_1$, then M has to be necessarily complete and the projection $\Pi = \pi_M \circ f : \Sigma \to M$ is a covering map. This follows from the fact that Π is a local diffeomorphism which increases the distance between the Riemannian surfaces Σ and M, that is $\Pi^*(\langle , \rangle_M) \geq \langle , \rangle$. Then the assertion follows recalling that if a map, from a connected complete Riemannian manifold M_1 into another connected Riemannian manifold M_2 of the same dimension, increases the distance, then it is a covering map and M_2 is complete. In particular, if $M^2 \times \mathbb{R}_1$ admits a compact spacelike surface, then M is necessarily compact (see Ref. 6, Proposition 3.2 (i)).

Under completeness assumption, we establish a parametric version of a Calabi-Bernstein result for maximal surfaces in $M^2 \times \mathbb{R}_1$.

Theorem 2.1 (Ref. 2, Theorem 3.3). *Let M^2 be a (necessarily complete) Riemannian surface with non-negative Gaussian curvature, $K_M \geq 0$. Then any complete maximal surface Σ^2 in $M^2 \times \mathbb{R}_1$ is totally geodesic.*

In addition, if $K_M > 0$ at some point on M, then Σ is a slice $M \times \{t_0\}$, $t_0 \in \mathbb{R}$.

As a direct consequence of Theorem 2.1 we have the following.

Corollary 2.1 (Ref. 2, Corollary 3.4). *Let M^2 be a complete non-flat Riemannian surface with non-negative Gaussian curvature, $K_M \geq 0$. Then the only complete maximal surfaces in $M^2 \times \mathbb{R}_1$ are the slices $M \times \{t_0\}$, $t_0 \in \mathbb{R}$.*

Take into account that if $M^2 = \mathbb{R}^2$ is the flat Euclidean plane, then $M^2 \times \mathbb{R}_1 = \mathbb{R}^3_1$ is nothing but the 3-dimensional Lorentz-Minkowski space, and any spacelike affine plane in $\mathbb{R}^3_1$ which is not horizontal determines a complete totally geodesic surface which is not a slice. On the other hand, the assumption $K_M \geq 0$ is necessary as shown by the fact that there exist examples of non-totally geodesic complete maximal surfaces in $\mathbb{H}^2 \times \mathbb{R}_1$, where $\mathbb{H}^2$ is the hyperbolic plane (see examples in Section 3).

Although we are not going to give here a proof of Theorem 2.1, we want to remark that the main idea of the proof is to show that Σ is a *non-hyperbolic* surface in the sense that any non-positive subharmonic function on the surface must be constant. Then, considering a suitable function we derive the conclusions of the theorem, (for the details, see the proof of Theorem 3.3 in Ref. 2).

Let $\Omega \subseteq M^2$ be a connected domain. Every smooth function $u \in \mathcal{C}^\infty(\Omega)$ determines a graph over Ω given by $\Sigma(u) = \{(x, u(x)) : x \in \Omega\} \subset M^2 \times \mathbb{R}_1$. The metric induced on Ω from the Lorentzian metric on the ambient space via $\Sigma(u)$ is given by

$$\langle , \rangle = \langle , \rangle_M - du^2. \tag{1}$$

Therefore, $\Sigma(u)$ is a spacelike surface in $M^2 \times \mathbb{R}_1$ if and only if $|Du|^2 < 1$ everywhere on Ω, where Du stands for the gradient of u in Ω and $|Du|$ denotes its norm, both with respect to the original metric $\langle , \rangle_M$ on Ω. If $\Sigma(u)$ is a spacelike graph over a domain Ω, then it can be seen that the mean curvature $H(u)$ of $\Sigma(u)$ is given by

$$2H(u) = \mathrm{Div}\left(\frac{Du}{\sqrt{1 - |Du|^2}}\right),$$

where Div stands for the divergence operator on Ω with respect to the metric $\langle , \rangle_M$. In particular, $\Sigma(u)$ is a maximal graph if and only if the function u satisfies the following partial differential equation on the domain

Ω,

$$\text{Maximal}[u] = \text{Div}\left(\frac{Du}{\sqrt{1-|Du|^2}}\right) = 0, \quad |Du|^2 < 1. \tag{2}$$

A graph is said to be entire if $\Omega = M$. It is known that when M is a complete Riemannian surface which is simply connected, then every complete spacelike surface in $M^2 \times \mathbb{R}_1$ is an entire graph. In fact, since M is simply connected then the projection Π is a diffeomorphism between Σ and M, and hence Σ can be written as the graph over M of the function $u = h \circ \Pi^{-1} \in \mathcal{C}^\infty(M)$. Here $h \in \mathcal{C}^\infty(\Sigma)$ is defined as the height function of the surface Σ, that is, $h = \pi_M \circ f : \Sigma^2 \to M^2 \times \mathbb{R}_1$. However, in contrast to the case of graphs into a Riemannian product space, an entire spacelike graph in $M^2 \times \mathbb{R}_1$ is not necessarily complete. To see it, in Section 3 we construct examples of non-complete entire maximal graphs in $\mathbb{H}^2 \times \mathbb{R}_1$.

For that reason, the Calabi-Bernstein result given at Theorem 2.1 does not imply in principle that, under the same hypothesis on M, any entire maximal graph in $M^2 \times \mathbb{R}_1$ must be totally geodesic. However, we can prove the following non-parametric version of our Calabi-Bernstein theorem.

Theorem 2.2 (Ref. 2, Theorem 4.3). *Let M^2 be a complete Riemannian surface with non-negative Gaussian curvature, $K_M \geq 0$. Then any entire maximal graph $\Sigma(u)$ in $M^2 \times \mathbb{R}_1$ is totally geodesic. In addition, if $K_M > 0$ at some point on M, then u is constant.*

And as a direct consequence we get,

Corollary 2.2 (Ref. 2, Corollary 4.4). *Let M^2 be a complete non-flat Riemannian surface with non-negative Gaussian curvature, $K_M \geq 0$. Then the only entire solutions to the maximal surface equation (2) are the constant functions.*

The proof of Theorem 2.2 is obtained as a consequence of the following result, (we refer the reader to Ref. 2, Section 4 for a detailed proof).

Theorem 2.3 (Ref. 2, Theorem 4.1). *Let M^2 be a (non necessarily complete) Riemannian surface with non-negative Gaussian curvature, $K_M \geq 0$. Then any maximal surface Σ^2 in $M^2 \times \mathbb{R}_1$ which is complete with respect to the metric induced from the Riemannian product $M^2 \times \mathbb{R}$ is totally geodesic. In addition, if $K_M > 0$ at some point on Σ, then M is necessarily complete and Σ is a slice $M \times \{t_0\}$.*

3. Non-trivial examples of entire maximal graphs in $\mathbb{H}^2 \times \mathbb{R}_1$

In this section we give examples of complete and non-complete entire maximal graphs, different to slices, in $\mathbb{H}^2 \times \mathbb{R}_1$. These graphs provide counterexamples which show that our Calabi-Bernstein results are no longer true without the assumption $K_M \geq 0$. On the other hand, is well known that every complete spacelike surface with constant mean curvature which is a closed subset in the Lorentz-Minkowski space $\mathbb{R}^3_1$ is necessarily complete. Therefore, the existence of non-complete entire maximal graphs in $\mathbb{H}^2 \times \mathbb{R}_1$ points out a curious difference between the topology of entire maximal graphs in $\mathbb{R}^3_1$ and in $\mathbb{H}^2 \times \mathbb{R}_1$.

Calabi[7] first gave a simple but nice duality result between solutions to the minimal surface equation in the Euclidean space $\mathbb{R}^3$ and solutions to the maximal surface equation in the Lorentz-Minkowski space $\mathbb{R}^3_1$. By regarding $\mathbb{R}^3$ as the Riemannian product space $\mathbb{R}^2 \times \mathbb{R}$ and $\mathbb{R}^3_1$ as the Lorentzian product space $\mathbb{R}^2 \times \mathbb{R}_1$, we observe that the same duality holds in general between solutions to the minimal surface equation in a Riemannian product space $M \times \mathbb{R}$ and solutions to the maximal surface equation in a Lorentzian product space $M \times \mathbb{R}_1$. First of all, recall that a smooth function u on a connected domain $\Omega \subseteq M^2$ defines a minimal graph $\Sigma(u)$ in $M \times \mathbb{R}$ if and only if u satisfies the following partial differential equation on Ω,

$$\text{Minimal}[u] = \text{Div}\left(\frac{Du}{\sqrt{1+|Du|^2}}\right) = 0, \tag{3}$$

where, as in (2), Div and Du stand for the divergence operator and the gradient of u in Ω with respect to the metric $\langle , \rangle_M$, respectively. Following the ideas in Ref. 4 we can prove the following general result.

Theorem 3.1 (Ref. 2, Theorem 5.1). *Let $\Omega \subseteq M^2$ be a simply connected domain of an oriented Riemannian surface M^2. There exists a non-trivial solution u to the minimal surface equation on Ω,*

$$\text{Minimal}[u] = 0,$$

if and only if there exists a non-trivial solution w to the maximal surface equation on Ω,

$$\text{Maximal}[w] = 0, \quad |Dw|^2 < 1.$$

Here *non-trivial solutions* correspond to non-totally geodesic graphs.

The interest of Theorem 3.1 relays on the fact that it allows us to construct new solutions to the maximal surface equation from known solutions to the minimal surface equation, and vice versa. Let us consider the half-plane model of the hyperbolic plane $\mathbb{H}^2$; that is,

$$\mathbb{H}^2 = \{x = (x_1, x_2) \in \mathbb{R}^2 : x_2 > 0\}$$

endowed with the complete metric

$$\langle , \rangle_{\mathbb{H}^2} = \frac{1}{x_2^2}(dx_1^2 + dx_2^2),$$

conformal to the flat Euclidean metric. This allows us to express the hyperbolic differential operators in terms of the Euclidean ones, so the minimal surface equation of a graph over the hyperbolic plane yields

$$\text{Minimal}[u] = \frac{x_2^2 \Delta_o u}{\sqrt{1+x_2^2|D_o u|_o^2}} - \frac{x_2^2}{(1+x_2^2|D_o u|_o^2)^{\frac{3}{2}}} \left(x_2 u_{x_2}|D_o u|_o^2 + x_2^2 Q(u)\right), \quad (4)$$

where D_o and Δ_o stand for the Euclidean gradient and Laplacian, respectively, $|\cdot|_o$ denotes the Euclidean norm and

$$Q(u) = u_{x_1}^2 u_{x_1 x_1} + 2u_{x_1} u_{x_2} u_{x_1 x_2} + u_{x_2}^2 u_{x_2 x_2}.$$

Example 3.1 (Ref. 2, Example 5.2). From (4) is a straightforward computation to show that the function

$$u(x_1, x_2) = \log(x_1^2 + x_2^2) \quad (5)$$

defines a non-trivial entire minimal graph in $\mathbb{H}^2 \times \mathbb{R}$. Therefore, from our Theorem 3.1 and the entire minimal graph defined by the function (5) we know that there exists a smooth function $w \in C^\infty(\mathbb{H}^2)$ which determines a non-trivial entire maximal graph in $\mathbb{H}^2 \times \mathbb{R}_1$. This shows that the assumption $K_M \geq 0$ in Theorem 2.2 and Corollary 2.2 is necessary. Moreover, the entire maximal graph determined by w is also complete. Actually, if we denote by $\langle , \rangle$ the induced metric on $\mathbb{H}^2$ via the graph, which is given by

$$\langle , \rangle = \langle , \rangle_{\mathbb{H}^2} - dw^2,$$

then it holds that

$$\langle , \rangle \geq \frac{1}{5} \langle , \rangle_{\mathbb{H}^2}.$$

As a consequence, the metric $\langle , \rangle$ is complete on $\mathbb{H}^2$. This shows that the assumption $K_M \geq 0$ in Theorem 2.1 and Corollary 2.1 is also necessary.

Example 3.2 (Ref. 2, Example 5.3). It is also an immediate fact that the function

$$u(x_1, x_2) = \frac{x_1}{x_1^2 + x_2^2} \quad (6)$$

satisfies (4). Therefore, it defines another non-trivial entire minimal graph in $\mathbb{H}^2 \times \mathbb{R}$ which gives rise via Theorem 3.1 to another non-trivial entire maximal graph in the Lorentzian product $\mathbb{H}^2 \times \mathbb{R}_1$. In contrast to Example 3.1, this example is non-complete. To see it, let $w \in \mathcal{C}^\infty(\mathbb{H}^2)$ stands for the smooth function defining this entire maximal graph, which we denote by $\Sigma(w)$. Then, the curve $\alpha : (0,1) \to \Sigma(w)$ given by

$$\alpha(s) = (0, s, w(0, s))$$

defines a divergent curve in $\Sigma(w)$ with finite length. As a consequence, $\Sigma(w)$ is non-complete.

The above examples are given in an implicit way, since we do not obtain any explicit expression for the functions which determine both graphs. However, in Ref. 1 we provide new examples of complete and non-complete entire maximal graphs in $\mathbb{H}^2 \times \mathbb{R}_1$ by solving directly the maximal surface equation in $\mathbb{H}^2$ for particular types of functions,

4. A parabolicity criterium for maximal surfaces.

A Riemannian surface (Σ, g) with non-empty boundary, $\partial\Sigma \neq \emptyset$, is said to be *parabolic* if every bounded harmonic function on Σ is determined by its boundary values, which is equivalent to the existence of a proper non-negative superharmonic function on Σ. Fernández and López[9] have recently proved that properly immersed maximal surfaces with non-empty boundary in the Lorentz-Minkowski spacetime $\mathbb{R}^3_1$ are parabolic if the Lorentzian norm on the maximal surface in $\mathbb{R}^3_1$ is eventually positive and proper. Motivated by their work, we study similar parabolicity criteria for maximal surfaces immersed into $M^2 \times \mathbb{R}_1$.

Given any Riemannian surface M^2, consider the function $\hat{r} : M \to \mathbb{R}$ defined by $\hat{r}(x) = \mathrm{dist}_M(x, x_0)$ where $x_0 \in M$ is a fixed point. Then, a natural generalization of the Lorentzian norm on a surface in $\mathbb{R}^3_1$ to immersed surfaces into $M^2 \times \mathbb{R}_1$ is the function $\phi = r^2 - h^2 \in \mathcal{C}^\infty(\Sigma)$ where the function $r : \Sigma \to \mathbb{R}$ is given by $r = \hat{r} \circ \pi_M \circ f$. In that context, we prove the following result (see Ref. 3 for details of the proofs).

Theorem 4.1 (Ref. 3, Theorem 2). *Let M^2 be a complete Riemannian surface with non-negative Gaussian curvature, $K_M \geq 0$. Consider Σ a maximal surface in $M^2 \times \mathbb{R}_1$ with non-empty boundary, $\partial\Sigma \neq \emptyset$, and assume that the function $\phi : \Sigma \to \mathbb{R}$ defined by*

$$\phi(p) = r^2(p) - h^2(p)$$

is eventually positive and proper. Then Σ is parabolic.

As is usual, by eventually we mean here a property that is satisfied outside a compact set.

It is interesting to look for some natural conditions under which the assumptions of Theorem 4.1 are satisfied. In that sense we can state the following proposition.

Proposition 4.1 (Ref. 3, Corollary 4). *Let M^2 be a complete Riemannian surface with non-negative Gaussian curvature. Then every proper maximal immersion $f : \Sigma^2 \to M^2 \times \mathbb{R}_1$ with non-empty boundary which eventually lies in*

$$\mathcal{W}_a = \{(x,t) \in M^2 \times \mathbb{R}_1 : |t| \leq a\hat{r}(x)\},$$

for some $0 < a < 1$, is parabolic.

It is worth pointing out that a Riemannian surface (Σ, g) without boundary, $\partial\Sigma = \emptyset$, is non-hyperbolic if and only if for every non-empty open set $O \subset \Sigma$ with smooth boundary $\Sigma \setminus O$ is parabolic (as a surface with boundary). This follows from the observation that a Riemannian manifold without boundary is non-hyperbolic precisely when almost all Brownian paths are dense in the manifold. Therefore, as a direct consequence of Theorem 4.1 we derive

Corollary 4.1 (Ref. 3, Corollary 5). *Let M^2 be a complete Riemannian surface with non-negative Gaussian curvature, $K_M \geq 0$, and let Σ be a maximal surface in $M^2 \times \mathbb{R}_1$ without boundary, $\partial\Sigma = \emptyset$. If the function $\phi(p) = r^2(p) - h^2(p)$ is eventually positive and proper on Σ, then Σ is non-hyperbolic.*

Another interesting situation is when we consider entire maximal graphs $\Sigma(u)$. Observe that we can always assume, up to a vertical translation which is an isometry in $M^2 \times \mathbb{R}_1$, that for a fixed point $x_0 \in M$, $u(x_0) = 0$. In that case, we can also conclude that $\phi : M \to \mathbb{R}$ is a positive function for every $x \neq x_0$ and proper. Therefore,

Corollary 4.2 (Ref. 3). *Every entire maximal graph defined over a complete Riemannian surface M^2 with non-negative Gaussian curvature is non-hyperbolic.*

As a consequence of Corollary 4.2 we can derive an alternative proof of Theorem 2.2, (see Ref. 3 for the details).

5. A local estimate for maximal surfaces

Alías and Palmer[5] introduced a new approach to the Calabi-Bernstein theorem in the Lorentz-Minkowski space $\mathbb{R}^3_1$, based on a local integral inequality for the Gaussian curvature of a maximal surface in $\mathbb{R}^3_1$ which involved the local geometry of the surface and the image of its Gauss map. In this section, we generalize that local approach to the case of maximal surfaces in a Lorentzian product space $M^2 \times \mathbb{R}_1$.

Theorem 5.1 (Ref. 2, Theorem 6.1). *Let M^2 be an analytic Riemannian surface with non-negative Gaussian curvature, $K_M \geq 0$, and let $f : \Sigma^2 \rightarrow M^2 \times \mathbb{R}_1$ be a maximal surface in $M^2 \times \mathbb{R}_1$. Let p be a point of Σ and $R > 0$ be a positive real number such that the geodesic disc of radius R about p satisfies $D(p, R) \subset\subset \Sigma$. Then for all $0 < r < R$ it holds that*

$$0 \leq \int_{D(p,r)} \|A\|^2 \mathrm{d}\Sigma \leq c_r \frac{L(r)}{r \log (R/r)}, \tag{7}$$

where $L(r)$ denotes the length of the geodesic circle of radius r about p, and

$$c_r = \frac{\pi^2(1+\alpha_r^2)^2}{4\alpha_r \arctan \alpha_r} > 0.$$

Here

$$\alpha_r = \sup_{D(p,r)} \cosh \theta \geq 1,$$

where θ denotes the hyperbolic angle between N and ∂_t along Σ.

In particular, when Σ is complete then it is not difficult to prove that the local integral inequality (7) implies Theorem 2.1, (see Ref. 2, Section 6).

Acknowledgments

A.L. Albujer was partially supported by FPU Grant AP2004-4087 from Secretaría de Estado de Universidades e Investigación, MEC Spain and MEC/FEDER Grant MTM2004-04934-C04-02.

References

1. A.L. Albujer, New examples of entire maximal graphs in $\mathbb{H}^2 \times \mathbb{R}_1$, to appear in *Differential Geom. Appl.*
2. A.L. Albujer and L.J. Alías, Calabi-Bernstein results for maximal surfaces in Lorentzian product spaces, preprint 2006; available at http://arxiv.org/abs/0709.4363.

3. A.L. Albujer and L.J. Alías, Parabolicity of maximal surfaces in Lorentzian product spaces, preprint 2007.
4. L.J. Alías and B. Palmer, A duality result between the minimal surface equation and the maximal surface equation, *An. Acad. Bras. Ciênc.* **73** (2001) 161–164.
5. L.J. Alías and B. Palmer, On the Gaussian curvature of maximal surfaces and the Calabi-Bernstein theorem, *Bull. London Math. Soc.* **33** (2001) 454–458.
6. L.J. Alías, A. Romero and M. Sánchez, Uniqueness of complete spacelike hypersurfaces of constant mean curvature in generalized Robertson-Walker spacetimes, *Gen. Relativity Gravitation* **27** (1995) 71–84.
7. E. Calabi, Examples of Bernstein problems for some nonlinear equations 1970 *In: Global Analysis* (Proc. Sympos. Pure Math., Vol. XV, Berkeley, Calif., 1968, Amer. Math. Soc., Providence, R.I.) 223–230.
8. S.Y. Cheng and S.T. Yau, Maximal space-like hypersurfaces in the Lorentz-Minkowski spaces, *Ann. of Math. (2)* **104** (1976) 407–419.
9. I. Fernández and F. López, Relative parabolicity of zero mean curvature in $\mathbb{R}^3$ and $\mathbb{R}^3_1$, preprint.
10. A. Grigor'yan, Analytic and geometric background of recurrence and non-explosion of the Brownian motion on Riemannian manifolds, *Bull. Amer. Math. Soc.* **36** (1999) 135–249.
11. J.E. Marsden and F.J. Tipler, Maximal hypersurfaces and foliations of constant mean curvature in general relativity, *Phys. Rep.* **66** (1980) 109–139.
12. W. Meeks and J.Pérez, Conformal properties on classical minimal surface theory, *In: Surveys in Differential Geometry* (Volume IX: Eigenvalues of Laplacians and other geometric operators, (A. Grigor'yan and S.T. Yau, Eds.) International Press).
13. J. Pérez, Parabolicity and minimal surfaces, *Proceedings of the 2002 Summer School "The global theory of minimal surfaces"* (Clay Mathematical Institute & University of California at Berkeley, MSRI), to appear.

Differential Geometry and its Applications
Proc. Conf., in Honour of Leonhard Euler, Olomouc, August 2007

Invariant Einstein metrics on certain Stiefel manifolds

A. Arvanitoyeorgos

Department of Mathematics, University of Patras,
Rion, GR-26500, Greece
E-mail: arvanito@math.upatras.gr
www.math.upatras.gr/~ arvanito

V. V. Dzhepko and Yu.G. Nikonorov

Rubtsovsk Industrial Institute,
ul. Traktornaya, 2/6, Rubtsovsk, 658207, Russia
E-mail: J_Valera_V@mail.ru, nik@inst.rubtsovsk.ru

A Riemannian manifold (M, ρ) is called Einstein if the metric ρ satisfies the condition $Ric(\rho) = c \cdot \rho$ for some constant c. This paper is devoted to the investigation of G-invariant Einstein metrics with additional symmetries, on some homogeneous spaces G/H of classical groups. As a consequence, we obtain new invariant Einstein metrics on the Stiefel manifolds $SO(2k + l)/SO(l)$.

Keywords: Riemannian manifolds, Homogeneous spaces, Einstein metrics, Stiefel manifolds.

MS classification 53C25, 53C30.

1. Introduction

A Riemannian manifold (M, ρ) is called Einstein if the metric ρ satisfies the condition $Ric(\rho) = c \cdot \rho$ for some real constant c. A detailed exposition on Einstein manifolds can be found in the book of A. Besse,[1] and more recent results on homogeneous Einstein manifolds can be found in the survey of M. Wang.[2] General existence results are hard to obtain. Among the first important attempts are the works of G. Jensen[3] and M. Wang, W. Ziller.[4] Recently, a new existence approach was introduced by C. Böhm, M. Wang, and W. Ziller.[5,6]

In the present work we prove existence of invariant Einstein metrics on certain Stiefel manifolds $SO(n)/SO(n-k)$. The simplest case $S^{n-1} = SO(n)/SO(n-1)$ is an irreducible symmetric space, therefore it admits up to scale a unique invariant Einstein metric. It was S. Kobayashi[7] who proved first the existence of an invariant Einstein metric on the unit sphere bundle $T_1S^n = SO(n)/SO(n-2)$. In Ref. 8 A. Sagle proved that the Stiefel manifolds $SO(n)/SO(n-k)$ admit at least one homogeneous invariant Einstein metric. For $k \geq 3$ G. Jensen[9] found a second metric. Einstein metrics on $SO(n)/SO(n-2)$ are completely classified. If $n = 3$ the group $SO(3)$ has a unique Einstein metric. If $n \geq 5$ it was shown by A. Back and W.Y. Hsiang[10] that $SO(n)/SO(n-2)$ admits exactly one homogeneous invariant Einstein metric. The same result was obtained by M. Kerr.[11] The Stiefel manifold $SO(4)/SO(2)$ admits exactly two invariant Einstein metrics which follows from the classification of 5-dimensional homogeneous Einstein manifolds due to D.V. Alekseevsky, I. Dotti, and C. Ferraris.[12] We also refer to Ref. 5, p.727-728 for further discussion. For $k \geq 3$ there is no obstruction for existence of more than two homogeneous invariant Einstein metrics on Stiefel manifolds $SO(n)/SO(n-k)$.

In this paper we investigate G-invariant metrics on G/H with additional symmetries. Let G be a compact Lie group and H a closed subgroup so that G acts almost effectively on G/H. Let K be a closed subgroup of G with $H \subset K \subset G$, and suppose that $K = L' \times H'$, where $\{e_{L'}\} \times H' = H$. It is clear that $K \subset N_G(H)$, the normalizer of H in G. If we denote $L = L' \times \{e_{H'}\}$, then the group $\widetilde{G} = G \times L$ acts on G/H by $(a,b) \cdot gH = agb^{-1}H$, and the isotropy subgroup at eH is $\widetilde{H} = \{(a,b) : ab^{-1} \in H\}$.

It will be shown that the set $\mathcal{M}^{\widetilde{G}}$ of $\widetilde{G}$-invariant metrics on $\widetilde{G}/\widetilde{H}$ is a subset of $\mathcal{M}^G$, the set of G-invariant metrics on G/H. Therefore, it would be simpler to search for invariant Einstein metrics on $\mathcal{M}^{\widetilde{G}}$. In this way we obtain existence results for Einstein metrics for certain quotients.

We apply the above method for the homogeneous space $SO(k_1 + k_2 + k_3)/SO(k_3)$, and we show the following:

Theorem 1.1. *If $l > k \geq 3$ then the Stiefel manifold $SO(2k+l)/SO(l)$ admits at least four $SO(2k+l) \times SO(k) \times SO(k)$-invariant Einstein metrics, two of which are Jensen's metrics.*

Our method can be extended to other homogeneous spaces of classical Lie groups (cf. Ref. 13).

2. The main construction

Let G be a compact Lie group and H a closed subgroup so that G acts almost effectively on G/H. Let $\mathfrak{g}$, $\mathfrak{h}$ be the Lie algebras of G and H, and let $\mathfrak{g} = \mathfrak{h} \oplus \mathfrak{p}$ be a reductive decomposition of $\mathfrak{g}$ with respect to some $Ad(G)$-invariant inner product of $\mathfrak{g}$. The orthogonal complement $\mathfrak{p}$

can be identified with the tangent space $T_{eH}G/H$. Any G-invariant metric ρ of G/H corresponds to an $Ad(H)$-invariant inner product $(\cdot,\cdot)$ on $\mathfrak{p}$ and vice-versa. For G semisimple, the negative of the Killing form B of $\mathfrak{g}$ is an $Ad(G)$-invariant inner product on $\mathfrak{g}$, therefore we can choose the above decomposition with respect to this form. We will use such a decomposition later on. Moreover, the restriction $\langle\cdot,\cdot\rangle = -B|_{\mathfrak{p}}$ is an $Ad(G)$-invariant inner product on $\mathfrak{p}$, which generates a G-invariant metric on G/H called *standard*.

The normalizer $N_G(H)$ of H in G acts on G/H by $(a, gH) \mapsto ga^{-1}H$. For a fixed a this action induces a G-equivariant diffeomorphism $\varphi_a : G/H \to G/H$. Note that if $a \in H$ this diffeomorphism is trivial, so the action of the gauge group $N_G(H)/H$ is well defined. However, it is simpler from technical point of view to use the action of $N_G(H)$. Let ρ be a G-invariant metric of G/H with corresponding inner product $(\cdot,\cdot)$. Then the diffeomorphism φ_a is an isometry of $(G/H,\rho)$ if and only if the operator $Ad(a)|_{\mathfrak{p}}$ is orthogonal with respect to $(\cdot,\cdot)$.

Let K be a closed subgroup of G with $H \subset K \subset G$ such that $K = L'\times H'$, where $\{e_{L'}\}\times H' = H$, and consider $L = L'\times\{e_{H'}\}$. It is clear that $K \subset N_G(H)$. The group $\widetilde{G} = G\times L$ acts on G/H by $(a,b)\cdot gH = agb^{-1}H$, and the isotropy at eH is given as follows:

Lemma 2.1. *The isotropy subgroup $\widetilde{H}$ is isomorphic to K.*

Proof. It is clear that $\widetilde{H} = \{(a,b)\in G\times L : ab^{-1}\in H\}$. Let $i : K \hookrightarrow G$ be the inclusion of K in G. Then $i(\{e_{L'}\}\times H') = H$ and $i(L'\times\{e_{H'}\}) = L$. Let $(a,b)\in G\times L$ be such that $ab^{-1} = h \in H$. Then $a = hb$, so

$$(a,b)=(hb,b)=(i(b',e_{H'})i(e_{L'},h'),i(b',e_{H'}))=((b',h'),b')\in K\times L'=L'\times H'\times L'.$$

Thus $\widetilde{H}$ is identified with a subgroup of $L'\times H'\times L'$, and it is then obvious that $\widetilde{H}$ is isomorphic to $L'\times H' = K$. □

The set $\mathcal{M}^G$ of G-invariant metrics on G/H is finite dimensional. We consider the subset $\mathcal{M}^{G,K}$ of $\mathcal{M}^G$ corresponding to $Ad(K)$-invariant inner products on $\mathfrak{p}$ (and not only $Ad(H)$-invariant).

Let $\rho\in\mathcal{M}^{G,K}$ and $a\in K$. The above diffeomorphism φ_a is an isometry of $(G/H,\rho)$. The action $\widetilde{G}$ on $(G/H,\rho)$ is isometric, so any metric form $\mathcal{M}^{G,K}$ can be identified a metric in $\mathcal{M}^{\widetilde{G}}$ and vice-versa. Therefore, we may think of $\mathcal{M}^{\widetilde{G}}$ as $\mathcal{M}^{G,K}$, which is a subset of $\mathcal{M}^G$.

Since metrics in $\mathcal{M}^{G,K}$ correspond to $Ad(K)$-invariant inner products on $\mathfrak{p}$, we call these metrics *$Ad(K)$-invariant metrics on G/H.*

We will apply the above construction for $G/H = SO(k_1 + k_2 + k_3)/SO(k_3)$, and prove existence of Einstein metrics in the set $\mathcal{M}^{G,K}$, where $K = L' \times H' = (SO(k_1) \times SO(k_2)) \times SO(k_3)$.

3. $Ad(K)$-invariant metrics on $SO(k_1 + k_2 + k_3)/SO(k_3)$

Let $n = k_1 + k_2 + k_3$, and $\mathfrak{p}_i$ be the subalgebra $so(k_i)$ in $\mathfrak{g} = so(n)$, $i = 1, 2$. Note that the submodules $\mathfrak{p}_i$ of $\mathfrak{p}$ are $Ad(K)$-invariant and $Ad(K)$-irreducible submodules. For $1 \leq i < j \leq 3$ we denote by $\mathfrak{p}_{(i,j)}$ the $Ad(K)$-invariant and $Ad(K)$-irreducible submodule of $\mathfrak{p}$ which is determined by the equality $so(k_i + k_j) = so(k_i) \oplus so(k_j) \oplus \mathfrak{p}_{(i,j)}$, where $\mathfrak{p}_{(i,j)}$ is orthogonal to $so(k_i) \oplus so(k_j)$ with respect to the Killing form B.

Denote by d_i and $d_{(i,j)}$ the dimensions of the modules $\mathfrak{p}_i$ and $\mathfrak{p}_{(i,j)}$ respectively. It is easy to obtain that $d_i = \frac{k_i(k_i-1)}{2}$, $d_{(i,j)} = k_i k_j$.

We have a decomposition of $\mathfrak{p}$ into a sum of $Ad(K)$-invariant and $Ad(K)$-irreducible submodules:

$$\mathfrak{p} = \mathfrak{p}_1 \oplus \mathfrak{p}_2 \oplus \mathfrak{p}_{(1,2)} \oplus \mathfrak{p}_{(1,3)} \oplus \mathfrak{p}_{(2,3)}. \tag{1}$$

Lemma 3.1 (Ref. 13). *Assume that $k_1, k_2, k_3 \geq 2$, and at most one of k_1, k_2 is equal to 2. Then there are no pairwise $Ad(K)$-isomorphic submodules among $\mathfrak{p}_1, \mathfrak{p}_2$ and $\mathfrak{p}_{(i,j)}$ $(1 \leq i < j \leq 3)$.*

If the assumptions of Lemma 3.1 are satisfied, then we have a complete description of all $Ad(K)$-invariant metrics on G/H.

Let ρ be any $Ad(K)$-invariant metric on G/H with corresponding $Ad(K)$-invariant inner product $(\cdot,\cdot)$ on $\mathfrak{p}$.

Lemma 3.2. *If there are no pairwise $Ad(K)$-isomorphic submodules among p_i and $p_{(i,j)}$, then*

$$\begin{aligned}(\cdot,\cdot) = x_1 \cdot \langle\cdot,\cdot\rangle|_{\mathfrak{p}_1} + x_2 \cdot \langle\cdot,\cdot\rangle|_{\mathfrak{p}_2} + x_{(1,2)} \cdot \langle\cdot,\cdot\rangle|_{\mathfrak{p}_{(1,2)}} \\ + x_{(1,3)} \cdot \langle\cdot,\cdot\rangle|_{\mathfrak{p}_{(1,3)}} + x_{(2,3)} \cdot \langle\cdot,\cdot\rangle|_{\mathfrak{p}_{(2,3)}}\end{aligned} \tag{2}$$

for positive constants $x_1, x_2 > 0$ and $x_{(i,j)} > 0$, where $\langle\cdot,\cdot\rangle = -B|_{\mathfrak{p}}$. Therefore, the set of $Ad(K)$-invariant metrics on $SO(k_1 + k_2 + k_3)/SO(k_3)$ depends on 5 parameters.

4. The scalar curvature and the Einstein condition

Let $\{e_\alpha^j\}$ be an orthonormal basis of $\mathfrak{p}_\alpha$ with respect to $\langle\cdot,\cdot\rangle$, where $1 \leq j \leq d_\alpha$ (here α means any of the symbols of type i or (k,i)). We define the numbers (cf. Ref. 4) $[\alpha\beta\gamma]$ by the equation

$$[\alpha\beta\gamma] = \sum_{i,j,k} \langle \left[e^i_\alpha, e^j_\beta\right], e^k_\gamma \rangle^2,$$

where i, j, k vary from 1 to $d_\alpha, d_\beta, d_\gamma$ respectively. The symbols $[\alpha\beta\gamma]$ are symmetric with respect to all three indices, as follows from the $Ad(G)$-invariance of $\langle\cdot,\cdot\rangle$.

According to Ref. 4, the scalar curvature S of $(\cdot,\cdot)$ is given by

$$S((\cdot,\cdot)) = \frac{1}{2}\sum_\alpha \frac{d_\alpha}{x_\alpha} - \frac{1}{4}\sum_{\alpha,\beta,\gamma} [\alpha\beta\gamma]\frac{x_\gamma}{x_\alpha x_\beta},$$

where α, β and γ are arbitrary symbols of the type i ($i = 1, 2$) or of the type (i,j) ($1 \le i < j \le 3$).

By performing explicit computations of $[\alpha\beta\gamma]$ (cf. Ref. 13), the scalar curvature for the metric (2) takes the following form.

Proposition 4.1. *The scalar curvature S of an $Ad(K)$-invariant metric (2) is given by*

$$\begin{aligned} S = &\frac{k_1(k_1-1)(k_1-2)}{8(n-2)}\cdot\frac{1}{x_1} + \frac{k_2(k_2-1)(k_2-2)}{8(n-2)}\cdot\frac{1}{x_2} \\ &+\frac{1}{2}\left(\frac{k_1k_2}{x_{(1,2)}} + \frac{k_1k_3}{x_{(1,3)}} + \frac{k_2k_3}{x_{(2,3)}}\right) - \frac{1}{8(n-2)}\left(k_1k_2(k_1-1)\frac{x_1}{x^2_{(1,2)}}\right. \\ &\left.+k_1k_3(k_1-1)\frac{x_1}{x^2_{(1,3)}} + k_2k_3(k_2-1)\frac{x_2}{x^2_{(2,3)}} + k_1k_2(k_2-1)\frac{x_2}{x^2_{(1,2)}}\right) \\ &-\frac{1}{4(n-2)}k_1k_2k_3\left(\frac{x_{(1,2)}}{x_{(1,3)}x_{(2,3)}} + \frac{x_{(1,3)}}{x_{(1,2)}x_{(2,3)}} + \frac{x_{(2,3)}}{x_{(1,2)}x_{(1,3)}}\right). \end{aligned} \tag{3}$$

Denote by $\mathcal{M}^G_1$ the set of all G-invariant metrics with a fixed volume element on the space G/H. The following variational principle for invariant Einstein metrics is well known.

Proposition 4.2 (Ref. 1). *Let G/H be a homogeneous space, where G and H are compact. Then the G-invariant Einstein metrics on the homogeneous space G/H are precisely the critical points of the scalar curvature functional S restricted to $\mathcal{M}^G_1$.*

For the general construction as described in Section 2, the above variational principle implies the following:

Proposition 4.3. *Let $\mathcal{M}_1^{G,K}$ be the subset of $\mathcal{M}^{G,K}$ with fixed volume element. Then a metric in $\mathcal{M}_1^{G,K}$ is Einstein if and only if it is a critical point of the scalar curvature functional S restricted to $\mathcal{M}_1^{G,K}$.*

Proof. The set $\mathcal{M}_1^{G,K}$ is precisely the set of $\widetilde{G}$-invariant metrics with fixed volume element on $\widetilde{G}/\widetilde{H}$. □

The volume condition for the metric (2) takes the form

$$x_1^{d_1} x_2^{d_2} x_{(1,2)}^{d_{(1,2)}} x_{(1,3)}^{d_{(1,3)}} x_{(2,3)}^{d_{(2,3)}} = constant. \tag{4}$$

By using Proposition 4.3 the problem of searching for $Ad(K)$-invariant Einstein metrics on G/H reduces to a Lagrange-type problem for the scalar curvature functional S under the restriction (4).

5. Invariant Einstein metrics on Stiefel manifolds

In this section we will investigate $SO(k_1+k_2+k_3) \times SO(k_1) \times SO(k_2)$-invariant Einstein metrics on the space $SO(k_1+k_2+k_3)/SO(k_3)$. Here $L' = SO(k_1) \times SO(k_2)$. By Lemma 3.2 these metrics depend on 5 parameters.

We apply Proposition 4.1, and by the Lagrange method we obtain the following system of equations

$$\begin{aligned}
&x_2 x_{23}^2\Big((k_1-2)x_{12}^2x_{13}^2 + k_2x_1^2x_{13}^2 + k_3x_1^2x_{12}^2\Big) = x_1x_{13}^2\Big((k_2-2)x_{12}^2x_{23}^2 \\
&\quad + k_3x_2^2x_{12}^2 + k_1x_2^2x_{23}^2\Big), \\
&x_{13}\Big((k_2-2)x_{12}^2x_{23}^2 + k_3x_2^2x_{12}^2 + k_1x_2^2x_{23}^2\Big) = x_2x_{23}\Big(2(k_1+k_2+k_3 \\
&\quad -2)x_{12}x_{13}x_{23} - (k_1-1)x_1x_{13}x_{23} - (k_2-1)x_2x_{13}x_{23} + k_3x_{12}^3 \\
&\quad - k_3x_{12}x_{13}^2 - k_3x_{12}x_{23}^2\Big), \\
&x_{13}\Big(2(k_1+k_2+k_3-2)x_{12}x_{13}x_{23} - (k_1-1)x_1x_{13}x_{23} - (k_2-1) \\
&\quad \cdot x_2x_{13}x_{23} + k_3x_{12}^3 - k_3x_{12}x_{13}^2 - k_3x_{12}x_{23}^2\Big) = x_{12}\Big(2(k_1+k_2+k_3 \\
&\quad -2)x_{12}x_{13}x_{23} - (k_1-1)x_1x_{12}x_{23} + k_2x_{13}^3 - k_2x_{12}^2x_{13} - k_2x_{13}x_{23}^2\Big), \\
&x_{23}\Big(2(k_1+k_2+k_3-2)x_{12}x_{13}x_{23} - (k_1-1)x_1x_{12}x_{23} + k_2x_{13}^3 \\
&\quad - k_2x_{12}^2x_{13} - k_2x_{13}x_{23}^2\Big) = x_{13}\Big(2(k_1+k_2+k_3-2)x_{12}x_{13}x_{23} \\
&\quad - (k_2-1)x_2x_{12}x_{13} + k_1x_{23}^3 - k_1x_{12}^2x_{23} - k_1x_{13}^2x_{23}\Big).
\end{aligned} \tag{5}$$

If $x_{13} = x_{23} = z$ then system (5) reduces to the following:

$$\begin{aligned}
&x_{12}z^2\Big((k_1-k_2)x_{12}+(k_2-1)x_2-(k_1-1)x_1\Big)=0,\\
&z^2\Big(\Big(((k_1-2)x_2-(k_2-2)x_1)x_{12}^2+(k_2x_1-k_1x_2)x_1x_2\Big)z^2\\
&\quad+k_3(x_1-x_2)x_1x_2x_{12}^2\Big)=0,\\
&z\Big(\Big((k_2-2)x_{12}^2+(k_1+k_2-1)x_2^2-2(k_1+k_2-2)x_2x_{12}\\
&\quad+(k_1-1)x_1x_2\Big)z^2+k_3(x_2-x_{12})x_2x_{12}^2\Big)=0,\\
&z\Big(\Big(2(k_1+k_2-2)x_{12}-(k_2-1)x_2-(k_1-1)x_1\Big)z^2\\
&\quad+\Big((k_2+k_3)x_{12}-2(k_3+k_2+k_1-2)z+(k_1-1)x_1\Big)x_{12}^2\Big)=0.
\end{aligned}\tag{6}$$

From the first equation of system (6) we obtain that $x_1 = \frac{(k_1-k_2)x_{12}+(k_2-1)x_2}{k_1-1}$, and substituting to the two other equations we obtain

$$\begin{aligned}
&z(x_2-x_{12})\Big(((2k_2+k_1-2)x_2-(k_2-2)x_{12})z^2+k_3x_2x_{12}^2\Big)=0,\\
&z\Big(\Big((k_1-3k_2-4)x_{12}-2(k_2-1)x_2\Big)z^2+\Big((k_3+k_1)x_{12}\\
&\quad-2(k_1+k_2+k_3-2)z+(k_2-1)x_2\Big)x_{12}^2\Big)=0,\\
&(k_1-k_2)z^2(x_{12}-x_2)\Big(\Big((1-k_1)(k_2-2)x_{12}^2+(k_1(k_1-2)\\
&\quad+k_2(k_1-k_2)+1)x_{12}x_2+(k_1(k_2-1)+k_2(k_2-2)+1)x_2^2\Big)z^2\\
&\quad+k_3\Big((k_1-k_2)x_{12}+(k_2-1)x_2\Big)x_2x_{12}^2\Big)=0.
\end{aligned}\tag{7}$$

If $x_2 = x_{12}$ then system (7) reduces to the equation

$$\Big((k_1+3k_2-4)x_2-2(k_2-1)x_2\Big)z^2-2(n-2)x_2^2z+(n-1)x_2^3=0.$$

which has two solutions, that were found by G. Jensen.[9] So assume that $x_2 \neq x_{12}$. Then from the first equation of system (7) we get $x_2 =$

$\frac{(k_2-2)x_{12}z^2}{k_3x_{12}^2+(2k_2+k_1-2)z^2}$. Substituting to (7) we obtain the following system:

$$\begin{aligned}
&z^4x_{12}^3(k_2-1)(k_2-2)(k_1-k_2)\Big((k_1+k_2-1)z^2+k_3x_{12}^2\Big)\\
&\quad\cdot\Big((k_1+k_2)z^2+k_3x_{12}^2\Big)^2=0,\\
&(k_3^2+k_1k_3)x_{12}^4+(4k_3-2k_2k_3-2k_1k_3-2k_3^2)zx_{12}^3+(k_1^2-6k_3\\
&\quad+5k_2k_3+2k_1k_3-2k_1+k_2^2+2k_1k_2-3k_2+2)z^2x_{12}^2+(8k_1\\
&\quad-6k_1k_2-4k_2k_3+12k_2-2k_1^2+4k_3-4k_2^2-2k_1k_3-8)z^3x_{12}\\
&\quad+(5k_1k_2-8k_2+k_1^2+4k_2^2-6k_1+4)z^4=0.
\end{aligned}\tag{8}$$

From the first equation of (8) we see that a solution exists only when $k_1=k_2$.

Let $k_1=k_2=k$ $(k\geq 3)$, $k_3=l$, $x_1=x_2=x$, $x_{12}=1$, $x_{13}=x_{23}=z$. Then the original system (5) reduces to the following:

$$\begin{aligned}
&z(x-1)\Big(((3k-2)x-k+2)z^2+lx\Big)=0,\\
&z\Big((2kx-2x-4k+4)z^2+(4k+2l-4)z-(k-1)x-k-l\Big)=0.
\end{aligned}\tag{9}$$

If $x=1$ we obtain Jensen's solutions, so assume that $x\neq 1$. Then from the first equation of (9) we find that

$$x=\frac{(k-2)z^2}{(3k-2)z^2+l},$$

which implies in particular, that $0<x<1$ for any positive z. Substituting this expression to the second equation of (9), the Einstein equation reduces to $F(z)=0$, where

$$\begin{aligned}
F(z)=&\,2(5k^2-7k+2)z^4-2(6k^2+3kl-10k-2l+4)z^3\\
&+(4k^2+7kl-5k-6l+2)z^2-22k+l-2)z+l(k+l).
\end{aligned}\tag{10}$$

If the equation $F(z)=0$ has a positive solution, then we obtain a new $SO(2k+l)\times SO(k)\times SO(k)$-invariant Einstein metric on $SO(2k+l)/SO(l)$. The numbers of new Einstein metrics for some values of k,l are shown in Table 1.

However, we can show that there exists an infinite series of new homogeneous Einstein manifolds, as the next proposition shows.

Proposition 5.1. *If $l>k\geq 3$ then the Stiefel manifold $SO(2k+l)/SO(l)$ admits at least four $SO(2k+l)\times SO(k)\times SO(k)$-invariant Einstein metrics.*

Table 1. The number of positive solutions of the equation $F_{SO}(z) = 0$ for various (k, l). (New $SO(2k + l) \times SO(k) \times SO(k)$-invariant Einstein metrics on $SO(2k + l)/SO(l)$).

k / l	3	4	5	6	7	8	9	10	11	12	13	14	15	16	17	18	19	20
1	0	0	0	0	0	0	0	0	0	0	0	0	0	0	0	0	0	0
2	0	0	0	0	0	0	0	0	0	0	0	0	0	0	0	0	0	0
3	2	0	0	0	0	0	0	0	0	0	0	0	0	0	0	0	0	0
4	2	2	2	0	0	0	0	0	0	0	0	0	0	0	0	0	0	0
5	2	2	2	2	2	0	0	0	0	0	0	0	0	0	0	0	0	0
6	2	2	2	2	2	2	2	0	0	0	0	0	0	0	0	0	0	0
7	2	2	2	2	2	2	2	2	2	2	0	0	0	0	0	0	0	0
8	2	2	2	2	2	2	2	2	2	2	2	2	0	0	0	0	0	0
9	2	2	2	2	2	2	2	2	2	2	2	2	2	2	0	0	0	0
10	2	2	2	2	2	2	2	2	2	2	2	2	2	2	2	2	2	0
11	2	2	2	2	2	2	2	2	2	2	2	2	2	2	2	2	2	2
12	2	2	2	2	2	2	2	2	2	2	2	2	2	2	2	2	2	2
13	2	2	2	2	2	2	2	2	2	2	2	2	2	2	2	2	2	2
14	2	2	2	2	2	2	2	2	2	2	2	2	2	2	2	2	2	2
15	2	2	2	2	2	2	2	2	2	2	2	2	2	2	2	2	2	2
16	2	2	2	2	2	2	2	2	2	2	2	2	2	2	2	2	2	2
17	2	2	2	2	2	2	2	2	2	2	2	2	2	2	2	2	2	2
18	2	2	2	2	2	2	2	2	2	2	2	2	2	2	2	2	2	2
19	2	2	2	2	2	2	2	2	2	2	2	2	2	2	2	2	2	2
20	2	2	2	2	2	2	2	2	2	2	2	2	2	2	2	2	2	2

Proof. We consider the polynomial (10). Then $F(0) = l(k+l) > 0$, $F(z) \to \infty$ as $z \to \infty$, and $F(1) = 2k^2 - 2kl + k - 2 - l^2 + 2l < 0$ for $l > k$, so $F(z) = 0$ has two positive solutions. From the above discussion these solutions are Einstein metrics, which are different from Jensen's Einstein metrics. Thus the result follows. □

Acknowledgments

The first author was partially supported by the C. Carathéodory grant # C.161 (2007-10), University of Patras. The second and the third author are supported in part by RFBR (grant N 05-01-00611-a) and by the Council on grants of the President of Russian Federation for supporting of leading scientific schools of Russian Federation (grants NSH-8526.2006.1) The third author is supported by the Council on grants of the President of Russian Federation for supporting of young Russian scientists (grant MD-5179.2006.1)

References

1. A. Besse, *Einstein Manifolds* (Springer, Berlin Heidelberg New York, 1987).
2. M. Wang, Einstein metrics from symmetry and bundle constructions, *In: Surveys in differential geometry: essays on Einstein manifolds* (Surv. Differ. Geom., VI, Int. Press, Boston, MA 1999) 287–325.
3. G. Jensen, The scalar curvature of left invariant Riemannian metrics, *Indiana J. Math.* **20** (1971) 1125–1144.
4. M. Wang and W. Ziller, Existence and Non-existence of Homogeneous Einstein Metrics, *Invent. Math.* **84** (1986) 177–194.
5. C. Böhm, M. Wang and W. Ziller, A variational approach for compact homogeneous Einstein manifolds, *Geom. Func. Anal.* **14** (2004) 681–733.
6. C. Böhm, Homogeneous Einstein metrics and simplicial complexes, *J. Diff. Geom.* **67** (2004) 79–165.
7. S. Kobayashi, Topology of positive pinched Kähler manifolds, *Tôhoku Math. J.* **15** (1963) 121–139.
8. A. Sagle, Some homogeneous Einstein manifolds, *Nagoya Math. J.* **39** (1970) 81–106.
9. G. Jensen, Einstein metrics on principal fiber bundles, *J. Diff. Geom.* **8** (1973) 599–614.
10. A. Back and W.Y. Hsiang, Equivariant geometry and Kervaire spheres, *Trans. Amer. Math. Soc.* **304** (1987) 207–227.
11. M. Kerr, New examples of homogeneous Einstein metrics, *Michigan J. Math.* **45** (1998) 115–134.
12. D.V. Alekseevsky, I. Dotti and C. Ferraris, Homogeneous Ricci positive 5-manifolds, *Pacific J. Math.* **175** (1996) 1–12.
13. A. Arvanitoyeorgos, V.V. Dzhepko and Yu.G. Nikonorov, Invariant Einstein metrics on some homogeneous spaces of classical Lie groups, to appear in *Canadian J. Math.*; arXiv:math.DG/0612504v2.

Differential Geometry and its Applications
Proc. Conf., in Honour of Leonhard Euler, Olomouc, August 2007

Constant mean curvature submanifolds in $(\alpha,\beta)-$Finsler spaces

V. Balan

Department of Mathematics I, Faculty of Applied Sciences,
University Politehnica of Bucharest, Bucharest, RO-060042, Romania
E-mail: vbalan@mathem.pub.ro

The explicit equations of CMC submanifolds immersed in certain (α,β) Finsler spaces - which extend the known results from[3] and include the Randers, Kropina and Euclidean cases, are determined. In particular, the minimal surfaces of revolution and the graphs immersed in Kropina Finsler spaces, are characterized.

Keywords: CMC hypersurface; Minimal surface; Mean curvature, Finsler structure, Randers metric, Kropina metric.

MS classification: 53C60, 53B40, 53C42.

1. Introduction

Recent advances in the theory of Finsler surfaces have been provided after 2002 by Z.Shen - who has intensively studied 2-dimensional Finsler metrics and their flag curvature. Recently Z. Shen,[1] and further M. Souza and K. Tenenblat[2,3] have investigated minimal surfaces immersed in Finsler spaces from differential geometric point of view. Still, earlier rigorous attempts using functional analysis exist in the works of G. Bellettini and M. Paolini after 1995. In 1998, based on the notion of Hausdorff measure, Z. Shen[1] has introduced the notion of mean curvature on submanifolds of Finsler spaces as follows.

Theorem 1.1 (Shen, 1998). *Let $(\tilde{M},\tilde{F})$ be a Finsler structure, and let $\varphi:(M,F)\to(\tilde{M},\tilde{F})$ be an isometric immersion, with F induced by $\tilde{F}$. Then the mean curvature of M is given by*

$$H_\varphi(X)=\frac{1}{G}\left(G_{;x^i}-G_{;z^i_a z^j_b}\varphi^j_{;u^a u^b}-G_{;x^j z^i_a}\varphi^j_{;u^a}\right)X^i,$$

where lower indices stand for corresponding partial derivatives and:

- $(u^a,v^b)_{a,b\in\overline{1,n}}$ *are local coordinates in TM $(\dim M=n)$;*

- $(x^i, y^j)_{i,j\in\overline{1,m}}$ *are local coordinates in* $T\tilde{M}$ $(\dim \tilde{M} = m)$*;*
- z^i_a *are the entries of the Jacobian matrix* $[J(\varphi)] = (\partial\varphi^i/\partial u^a)_{a=\overline{1,n},i=\overline{1,m}}$*;*
- $\varphi_t : M \to \tilde{M}, t \in (-\varepsilon, \varepsilon), \varphi_0 = \varphi$*, is the variation of the surface;*
- X *is the vector field* $X_x = \frac{\partial \varphi_t}{\partial t}|_{t=0}(x)$ *induced along* φ *attached to the variation;*
- G *is the Finsler induced volume form*

$$G_{\tilde{e}}(z) = \frac{vol[B^n]}{vol\{(v^a) \in \mathbb{R}^n \mid \tilde{F}(v^a z^i_a \tilde{e}_i) \leq 1\}}, \tag{1}$$

where $z = (z^i_a)_{a=\overline{1,n},i=\overline{1,m}} \in GL_{m\times n}(\mathbb{R})$, $\tilde{e} = \{\tilde{e}_i\}_{i=\overline{1,m}}$ *is an arbitrary basis in* $\mathbb{R}^m$ *and* $B^n \subset \mathbb{R}^n$ *is the standard Euclidean ball.*

It was proved that the variation of the volume of M reaches a minimum for $H_\varphi = 0$.[1] Recent advances in constructing minimal surfaces ($n = 2$) based on (1) were provided in Ref. 2,3, by characterizing the minimal surfaces of revolution in Finsler $(\alpha, \beta)-$spaces $(\tilde{M} = \mathbb{R}^3, \tilde{F})$ with the Randers fundamental function

$$\tilde{F}(x,y) = \alpha(x,y) + \beta(x,y), \quad \alpha(x,y) = \sqrt{a_{ij}(x)^i y^j}, \beta(x,y) = b_i(x)y^i,$$

for the particular case when $a_{ij} = \delta_{ij}$ (the Euclidean metric) and $\beta = b\cdot dx^3$, with $b \in [0,1)$.

In the following, we determine the Busemann-Hausdorff mean curvature for hypersurfaces of (α, β) Finsler spaces which allow algebraic indicatrix closure of second order. In particular, we shall explicitly derive the equations of CMC and minimal graphs and surfaces of revolution for Finsler Kropina spaces.

We further consider the case when $\dim \tilde{M} = m = n + 1$. Let $\tilde{F} : T\tilde{M} \to \mathbb{R}$ be a positive 1-homogeneous locally Minkowski Finsler fundamental function.[4,5] Continuing the research from Refs. 6,7 we shall study the Busemann-Hausdorff mean curvature for certain (α, β)-metrics, which include the subcases:

a) the Randers space $(\tilde{M}, \tilde{F})$ with the fundamental function

$$\tilde{F}(x,y) = \sqrt{\delta_{ij} y^i y^j} + b_s y^s, \quad b_s \in \mathbb{R}, s = \overline{1,n},$$

which is reducible to the case studied by M. Souza and K. Tenenblat,[2,3]

$$\tilde{F}(x,y) = \sqrt{\sum_{i=1}^{n+1} (y^i)^2} + by^{n+1}, \; b \in [0,1). \tag{2}$$

b) the Kropina space $(\tilde{M}, F)$ with the fundamental function

$$\tilde{F}(x,y) = \frac{\delta_{ij}y^i y^j}{b_s y^s}, \quad b_s \in \mathbb{R}, s = \overline{1,n}, \quad \sum_{i=1}^{n+1} b_i^2 > 0,$$

which is reducible to

$$\tilde{F}(x,y) = \sum_{i=1}^{n+1} (y^i)^2/(by^{n+1}), \quad b \in (0,1). \tag{3}$$

2. The volume form of a Finsler hypersurface

Let $H = Im\ \varphi$, $\varphi : D \subset \mathbb{R}^n \to \tilde{M} = \mathbb{R}^{n+1}$ be a simple hypersurface. We denote $z^i_\alpha = \dfrac{\partial \varphi^i}{\partial u^\alpha}$, $u = (u^1, \ldots, u^n) \in D$. We shall further determine the volume of the body $Q \subset T_{\varphi(u)}H$ bounded by the induced on $T_{\varphi(u)}H$ indicatrix from $\tilde{M}$

$$\partial Q = T_{\varphi(u)}H \cap \{y \in T_{\varphi(u)}\mathbb{R}^{n+1} | \tilde{F}(y) = 1\}.$$

If $v = v^\alpha \dfrac{\partial}{\partial u^\alpha} \in T_u D$, then $\varphi_{*,u}(v) = z^i_\alpha v^\alpha \left.\dfrac{\partial}{\partial y^i}\right|_{\varphi(u)} \in T_{\varphi(u)}H$ and hence at some fixed point $u \in D$, Q is given by

$$Q = \{v \in T_u D \mid \tilde{F}(\varphi(u), \varphi_{*,u}(v)) \leq 1\}.$$

In particular, Q is explicitly described as follows:

a) In the Randers case (2) we have

$$Q_R : \begin{cases} (y^1)^2 + \cdots + (y^{n+1})^2 \leq (1 - by^{n+1})^2 \\ y^i = z^i_\alpha v^\alpha, \ i = \overline{1, n+1} \end{cases} \Leftrightarrow$$

$$\Leftrightarrow \sum_{i=1}^{n} (z^i_\alpha v^\alpha)^2 + (1 - b^2)(z^{n+1}_\alpha v^\alpha)^2 + 2bz^{n+1}_\alpha v^\alpha - 1 \leq 0.$$

b) In the Kropina case (3) we obtain

$$Q_K : \begin{cases} (y^1)^2 + \ldots (y^{n+1})^2 - by^{n+1} \leq 0 \\ y^i = z^i_\alpha v^\alpha, \ i = \overline{1, n+1} \end{cases} \Leftrightarrow \sum_{i=1}^{n+1} (z^i_\alpha v^\alpha)^2 - bz^{n+1}_\alpha v^\alpha \leq 0.$$

We note that in both cases body Q has the generic form

$$Q : \sum_{i=1}^{n} (z^i_\alpha v^\alpha)^2 + \mu(z^{n+1}_\alpha v^\alpha)^2 + 2\nu z^{n+1}_\alpha v^\alpha + \rho \leq 0, \tag{4}$$

where $\mu, \nu, \rho \in \mathbb{R}$ and that one may specify Q to the Randers body Q_R (for $\mu = 1 - b^2, \nu = b, \rho = -1$) and to the Kropina body Q_K (for $\mu = 1, \nu = -b/2, \rho = 0$).

In order to find the volume form of $H \subset \tilde{M}$, we need to determine first $G_{\tilde{e}}(z) = \dfrac{vol(B^n)}{vol[Q]}$. To this aim, we use

Lemma 2.1. *Consider the matrix* $\tilde{H} = (\tilde{h}_{ij})_{i,j\in\overline{1,n}}$, $\tilde{h}_{ij} = h_{ij} + l_i l_j$, *with* $H = (h_{ij})_{i,j\in\overline{1,n}}$ *non-degenerate. Then:*

a) The inverse of $\tilde{H}$ *has the coefficients* $\tilde{h}^{ij} = h^{ij} - (1 + l_s l^s)^{-1} l^i l^j$, *where* $l^i = h^{is} l_s$.

b) We have $\det(\tilde{H}) = \det(H) \cdot (1 + l_s l^s)$.

We note that (4) briefly rewrites

$$Q : v^t \tilde{H} v + 2\nu b v + \rho \leq 0 \quad \Leftrightarrow \quad (v^t, 1)\hat{H}\begin{pmatrix} v \\ 1 \end{pmatrix} \leq 0 \tag{5}$$

where $v = (v^1, \ldots, v^n)^t$, $\tilde{H} = C^t C$, $C = \begin{pmatrix} (z^i_\alpha)_{i,\alpha=\overline{1,n}} \\ \sqrt{\mu} b \end{pmatrix}$, $b = (z_1^{n+1}, \ldots, z_n^{n+1})$ and

$$\hat{H} = \begin{pmatrix} \tilde{H} & \nu b^t \\ \nu b & \rho \end{pmatrix} = \begin{pmatrix} (\tilde{h}_{ab})_{a,b=\overline{1,n}} & \nu(z_\alpha^{n+1})_\alpha \\ \nu(z_\alpha^{n+1})_\alpha^t & \rho \end{pmatrix}. \tag{6}$$

After performing an orthogonal transformation, the bounding quadric ∂Q has a canonic analytic equation; hence

$$Q : \sum_{a=1}^n \lambda_a (v^a)^2 + \frac{\Delta}{\delta_*} \leq 0 \Leftrightarrow \sum_{a=1}^n \frac{(v^a)^2}{(\sqrt{-\Delta/(\delta_* \lambda_a)})^2} \leq 1,\ (v^1, \ldots, v^n) \in \mathbb{R}^n,$$

where $\lambda_1, \ldots, \lambda_n$ are the eigenvalues of $\tilde{H}$, $\delta_* = \lambda_1 \cdot \cdots \cdot \lambda_n = \det \tilde{H}$ and $\Delta = \det \hat{H}$. Hence

$$Vol(Q) = Vol(B_n) \cdot \sqrt{\frac{(-\Delta)^n}{\delta_*^n \cdot \lambda_1 \cdot \cdots \cdot \lambda_n}} = \frac{Vol(B_n) \cdot (-\Delta)^{n/2}}{\delta_*^{(n+1)/2}} \tag{7}$$

or, denoting $\eta_* = \rho - \nu^2 z_\alpha^{n+1} z_\beta^{n+1} \tilde{h}^{\alpha\beta}$,

$$Vol(Q) = \frac{Vol(B_n) \cdot (-\eta_*)^{n/2}}{\sqrt{\delta_*}}. \tag{8}$$

We note that the coefficients of the matrix $\tilde{H}$ which provide δ_* and contribute to η_* are of the form

$$\tilde{h}_{ab} = \sum_{i=1}^{n} z_a^i z_b^i + \mu z_a^{n+1} z_b^{n+1} = h_{ab} + l_a l_b,$$

where $h_{ab} = \sum_{i=1}^{n+1} z_a^i z_t^i$ and $l_a = \sqrt{\mu - 1} z_a^{n+1}$. Using the Lemma, and denoting

$$\tau = l_a l^a, \ \delta = \det(h_{ab})_{a,b=\overline{1,n}}, \tag{9}$$

we get

$$\begin{cases} \tau = (\mu - 1) z_a^{n+1} z_b^{n+1} h^{ab} \\ \tilde{h}^{ab} = h^{ab} - (1+\tau)^{-1} l^a l^b \\ \delta_* = \det \tilde{H} = \delta(1+\tau), \end{cases}$$

and hence a more refined form of the volume of the body bounded by Q given by (4) is

$$Vol(Q) = \frac{Vol(B_n) \cdot (\nu^2 z_\alpha^{n+1} z_\beta^{n+1} \tilde{h}^{\alpha\beta} - \rho)^{n/2}}{\sqrt{\delta(1+\tau)}}. \tag{10}$$

Two distinct cases occur:

a) For $\mu \neq 1$ (and hence for $\tau \neq 0$) we infer

$$\begin{aligned} \eta_* &= \rho - \nu^2 z_a^{n-1} z_b^{n+1} \left(h^{ab} - (1+\tau)^{-1} l^a l^b\right) \\ &= \rho - \frac{\nu^2}{\mu - 1} \cdot \left(l_a l_b h^{ab} - \frac{\tau^2}{1+\tau}\right) = \rho - \frac{\nu^2 \tau}{(\mu-1)(1+\tau)}. \end{aligned} \tag{11}$$

b) For $\mu = 1$ we have $\tau = 0$ and

$$\eta_* = \rho - \nu^2 z_a^{n+1} z_b^{n+1} h^{ab}. \tag{12}$$

Hence, making use of (11) and (12) in (8), we finally obtain:

Theorem 2.1. *The volume of the body Q in (4) is given by*

$$Vol(Q) = \begin{cases} \dfrac{Vol(B_n)}{\sqrt{\delta} \cdot (1+\tau)^{(n+1)/2}} \cdot \left(\dfrac{\nu^2 \tau}{\mu - 1} - \rho(1+\tau)\right)^{n/2}, & \text{for } \mu \neq 1 \\ \dfrac{Vol(B_n) \cdot (-\rho + \nu^2 z_a^{n+1} z_b^{n+1} h^{ab})^{n/2}}{\sqrt{\delta}}, & \text{for } \mu = 1, \end{cases} \tag{13}$$

where τ and δ are given by (9).

In particular, we obtain the following result:

Corollary 2.1. *a) In the Randers case (2) we obtain the known result (Ref. 2, p.627; Ref. 3),*

$$Vol(Q_R) = \frac{Vol(B^n)}{\sqrt{\delta}(1 - b^2 z_\alpha^{n+1} z_\beta^{n+1} h^{\alpha\beta})^{(n+1)/2}}.$$

b) In the Kropina case (3) we have

$$Vol(Q_K) = \frac{Vol(B^n)}{\sqrt{\delta}} \left(\frac{b^2}{4} z_\alpha^{n+1} z_\beta^{n+1} h^{\alpha\beta} \right)^{n/2}.$$

Proof. We use (13) with $\mu = 1-b^2, \nu = b, \rho = -1, \tau = -b^2 z_\alpha^{n+1} z_\beta^{n+1} h^{\alpha\beta}$ in the Randers case, and with $\mu = 1, \nu = -b/2, \rho = \tau = 0, \delta^* = \delta, \tilde{h}^{\alpha\beta} = h^{\alpha\beta}$ in the Kropina case. □

Remark 2.1. In the Kropina case, the function G from (1) has the expression

$$G_K = \frac{Vol(B_n)}{Vol(Q_K)} = \frac{\sqrt{\delta}}{(z_a^{n+1} z_b^{n+1} h^{ab} \cdot b^2/4)^{n/2}} = C \cdot \left(\frac{2}{\sqrt{B}} \right)^n,$$

where we have used the notations

$$B = b^2 z_a^{n+1} z_b^{n+1} h^{ab}, \quad C = \sqrt{\delta}, \quad \delta = \det(h_{\alpha\beta}), \quad h_{\alpha\beta} = \sum_{i=1}^{n+1} z_\alpha^i z_\beta^j. \tag{14}$$

Then the mean curvature field has the components

$$\bar{H}_i = \frac{1}{G} \left(\frac{\partial^2 G}{\partial z_\varepsilon^i z_\eta^j} \cdot \frac{\partial^2 \varphi^j}{\partial u^\varepsilon \partial u^\eta} \right), i = \overline{1, n+1}, \tag{15}$$

and the volume form of a hypersurface H is $dV_F = G_K du^1 \wedge \cdots \wedge du^n$.

3. CMC and minimal surfaces in Kropina spaces

Having in view the previous remark, we are able to state the following main result

Theorem 3.1. *The mean curvature field of a hypersurface H in the Kropina space $\tilde{M} = \mathbb{R}^{n+1}$ with the fundamental function (3) has the following expression in terms of B and C given by (14)*

$$
\begin{aligned}
H_i = -\frac{1}{B^2 C}\Bigg[& \frac{\partial^2 C}{\partial z^i_\varepsilon \partial z^j_\eta} B^2 + \frac{n(n+2)}{4} C \frac{\partial B}{\partial z^i_\varepsilon}\frac{\partial B}{\partial z^j_\eta} \\
& -\frac{nB}{2}\left(\frac{\partial C}{\partial z^i_\varepsilon}\frac{\partial B}{\partial z^j_\eta} + \frac{\partial C}{\partial z^j_\eta}\frac{\partial B}{\partial z^i_\varepsilon} + C\frac{\partial^2 B}{\partial z^i_\varepsilon \partial z^j_\eta}\right)\Bigg] \frac{\partial^2 \varphi^j}{\partial u^\varepsilon \partial u^\eta}, i = \overline{1, n+1}.
\end{aligned} \tag{16}
$$

Proof. We subsequently obtain

$$
\begin{aligned}
K^{\varepsilon\eta}_{ij} \equiv \frac{\partial^2 G}{\partial z^i_\varepsilon \partial z^j_\eta} &= 2^n \cdot \frac{\partial}{\partial z^j_\eta}\left(\frac{\partial C}{\partial z^i_\varepsilon} B^{-n/2} - \frac{n}{2} C \frac{\partial B}{\partial z^i_\varepsilon} B^{-(n+2)/2}\right) \\
&= 2^n B^{-(n+4)/2}\Bigg[\frac{\partial^2 C}{\partial z^i_\varepsilon \partial z^j_\eta} B^2 + \frac{n(n+2)}{4} C \frac{\partial B}{\partial z^i_\varepsilon}\frac{\partial B}{\partial z^j_\eta} - \\
&-\frac{nB}{2}\left(\frac{\partial C}{\partial z^i_\varepsilon}\frac{\partial B}{\partial z^j_\eta} + \frac{\partial C}{\partial z^j_\eta}\frac{\partial B}{\partial z^i_\varepsilon} + C\frac{\partial^2 B}{\partial z^i_\varepsilon \partial z^j_\eta}\right)\Bigg],
\end{aligned}
$$

whence, using $G = 2^n C B^{-n/2}$, the claim follows. □

Corollary 3.1. *The mean curvature field of a* surface H *in the Kropina space* $\tilde{M} = \mathbb{R}^3$ *with the fundamental function (3) has the following expression*

$$
\begin{aligned}
H_i = \frac{-1}{E^2 C^2}\Bigg[& 6E^2 \frac{\partial C}{\partial z^i_\varepsilon}\frac{\partial C}{\partial z^j_\eta} + 2C^2 \frac{\partial E}{\partial z^i_\varepsilon}\frac{\partial E}{\partial z^j_\eta} - C^2 \frac{\partial^2 E}{\partial z^i_\varepsilon \partial z^j_\eta} \\
& -3CE\left(\frac{\partial E}{\partial z^i_\varepsilon}\frac{\partial C}{\partial z^j_\eta} + \frac{\partial E}{\partial z^j_\eta}\frac{\partial C}{\partial z^i_\varepsilon}\right) + 3E^2 C \frac{\partial^2 C}{\partial z^i_\varepsilon \partial z^j_\eta}\Bigg] \frac{\partial^2 \varphi^j}{\partial u^\varepsilon \partial u^\eta}, i = \overline{1,3},
\end{aligned} \tag{17}
$$

where

$$
E = b^2 \sum_{k=1}^{3} \sum_{\alpha,\beta=1}^{2} (-1)^{\alpha+\beta} z^k_{\tilde{\alpha}} z^k_{\tilde{\beta}} z^3_\alpha z^3_\beta, \quad \tilde{\alpha} = 3 - \alpha. \tag{18}
$$

Proof. For $n = 2$, we infer $B = EC^{-2}$. Then tedious computation leads to the stated result. □

Remark 3.1. Using $\frac{\partial^2 C}{\partial z^i_\varepsilon \partial z^j_\eta} = (2C)^{-2}\left(\frac{\partial^2 C^2}{\partial z^i_\varepsilon \partial z^j_\eta} - 2\frac{\partial C}{\partial z^i_\varepsilon}\frac{\partial C}{\partial z^j_\eta}\right)$, we infer that the mean curvature field of a surface H in the Kropina space $\tilde{M} = \mathbb{R}^3$

with the fundamental function (3) is $H_* = H_i X^i$, with

$$H_i = \frac{-1}{E^2C^2}\left[3E^2\frac{\partial C}{\partial z^i_\varepsilon}\frac{\partial C}{\partial z^j_\eta} + 2C^2\frac{\partial E}{\partial z^i_\varepsilon}\frac{\partial E}{\partial z^j_\eta} - C^2\frac{\partial^2 E}{\partial z^i_\varepsilon \partial z^j_\eta}\right.$$
$$\left. -3CE\left(\frac{\partial E}{\partial z^i_\varepsilon}\frac{\partial C}{\partial z^j_\eta} + \frac{\partial E}{\partial z^j_\eta}\frac{\partial C}{\partial z^i_\varepsilon}\right) + \frac{3}{2}E^2\frac{\partial^2 C^2}{\partial z^i_\varepsilon \partial z^j_\eta}\right]\frac{\partial^2 \varphi^j}{\partial u^\varepsilon \partial u^\eta}, i = \overline{1,3}, \tag{19}$$

where C and E are provided respectively by (14) and (18),

$$X = ||N||^{-1}_{\tilde{F},X} \cdot N, \quad N^i = \varepsilon^{ijk}(G_* Z^1)_j (G_* Z^2)_k,$$

where $Z^1 = (z^1_1, z^2_1, z^3_1)$, $Z^2 = (z^1_2, z^2_2, z^3_2)$, and $G_* v$ is defined by the equality

$$(G_{*,X}v)(v') = \langle v, v' \rangle_{F,X} = \frac{1}{2}\frac{\partial F^2}{\partial y^i \partial y^j} v^i v'^j,$$

"$*$" is the Euclidean Hodge operator and ε^{ijk} is the skew-symmetrization symbol. We note that $N \equiv *((G_* Z^1) \wedge (G_* Z^2))$ and

$$X \in Ker(G_* Z^1) \cap Ker(G_* Z^2) \cap \{y \in T_{\varphi(u)}\tilde{M} \mid \tilde{F}(y) = 1\}. \tag{20}$$

Remark 3.2. The relation (19), leads to the general form of the mean curvature:

$$H_i X^i = \frac{-1}{E^2C^2}\left(3E^2\frac{\partial C}{\partial z^i_\varepsilon}X^i \cdot \frac{\partial C}{\partial z^j_\eta}\frac{\partial^2 \varphi^j}{\partial u^\varepsilon \partial u^\eta} + \frac{3}{2}E^2\frac{\partial^2 C^2}{\partial z^i_\varepsilon \partial z^j_\eta}\frac{\partial^2 \varphi^j}{\partial u^\varepsilon \partial u^\eta}X^i\right.$$
$$-3CE\left(\frac{\partial E}{\partial z^i_\varepsilon}X^i \cdot \frac{\partial C}{\partial z^j_\eta}\frac{\partial^2 \varphi^j}{\partial u^\varepsilon \partial u^\eta} + \frac{\partial C}{\partial z^i_\varepsilon}X^i \cdot \frac{\partial E}{\partial z^j_\eta}\frac{\partial^2 \varphi^j}{\partial u^\varepsilon \partial u^\eta}\right) \tag{21}$$
$$\left. +2C^2\frac{\partial E}{\partial z^i_\varepsilon}X^i\frac{\partial E}{\partial z^j_\eta}\frac{\partial^2 \varphi^j}{\partial u^\varepsilon \partial u^\eta} - C^2\frac{\partial^2 E}{\partial z^i_\varepsilon \partial z^j_\eta}\frac{\partial^2 \varphi^j}{\partial u^\varepsilon \partial u^\eta}X^i\right).$$

3.1. *CMC surfaces of revolution*

Corollary 3.2. *Let $M = \Sigma = Im\,\varphi$ be a surface of revolution described by*

$$\varphi(t,\theta) = (f(t)\cos\theta, f(t)\sin\theta, t), \quad (t,\theta) \in D = \mathbb{R} \times [0, 2\pi).$$

Then M is of constant mean curvature iff the function f satisfies the ODE

$$\frac{-1}{f^3b^2}\left[\frac{3f^3b^2(1+2g^2)}{1+g^2}h + 2g^2(b^2f^2 - 2) + 3b^2f^2 - 2\right] = k, \quad k \in \mathbb{R}. \tag{22}$$

Proof. Let v be the transversal vector to $\Sigma = Im\, r$, $X = (X^1, X^2, X^3) = (-\sin\theta, -\cos\theta, f'(t))$, such that Z_1, Z_2 and X are linear independent. Then the minimality of M is equivalent with the vanishing of the value of $\bar{H} = H_i dy^i|_{\varphi(u)}$ on X. Assuming $f(t) > 0, \forall t \in \mathbb{R}$, for the surface of revolution Σ we have

$$A = (h_{\alpha\beta})_{\alpha,\beta=\overline{1,2}} = \begin{pmatrix} 1+g^2 & 0 \\ 0 & f^2 \end{pmatrix}, \quad C = \det A = f\sqrt{1+g^2}, \quad E = b^2 f^2,$$

where we have denoted for brevity $f = f(t), g = f'(t), h = f''(t)$. Then, according to Ref. 2, Lemma 5, p.632, we have:

$$\begin{cases} \dfrac{\partial E}{\partial z^i_\varepsilon} X^i = 2b^2 f^2 g \delta_{\varepsilon 1}, \quad \dfrac{\partial E}{\partial z^j_\eta} \dfrac{\partial^2 \varphi^j}{\partial u^\varepsilon \partial u^\eta} = 2b^2 f g \delta_{\varepsilon 1}, \\ \dfrac{\partial^2 E}{\partial z^i_\varepsilon \partial z^j_\eta} \dfrac{\partial^2 \varphi^j}{\partial u^\varepsilon \partial u^\eta} X^i = 2b^2 f(1+2g^2), \quad \dfrac{\partial C}{\partial z^i_\varepsilon} X^i = 0 \\ \dfrac{\partial^2 C^2}{\partial z^i_\varepsilon \partial z^j_\eta} \dfrac{\partial^2 \varphi^j}{\partial u^\varepsilon \partial u^\eta} X^i = 2f(-fh + 1 + g^2), \\ \dfrac{\partial C}{\partial z^j_\eta} \dfrac{\partial^2 \varphi^j}{\partial u^\varepsilon \partial u^\eta} = \dfrac{g}{\sqrt{1+g^2}} (fh + 1 + g^2) \delta_{\varepsilon 1} \end{cases}$$

and hence from (21) we infer

$$H_i X^i = \frac{-1}{f^3 b^2} \left[\frac{3f^3 b^2 (1+2g^2)}{1+g^2} h + 2g^2(b^2 f^2 - 2) + 3b^2 f^2 - 2 \right],$$

whence the claim follows. □

Remark 3.3. As consequence of (22), the ODE which characterizes the generator curve of minimal surfaces of revolution in the 3-dimensional Kropina space with metric (3), has the explicit form:

$$\frac{3f^3 b^2(1+2g^2)}{1+g^2} h + 2g^2(b^2 f^2 - 2) + 3b^2 f^2 - 2 = 0. \tag{23}$$

3.2. *CMC graphs*

In the case of a graph $f : D \subset \mathbb{R}^2 \to \mathbb{R}$ immersed in the Kropina space $\tilde{M} = \mathbb{R}^3$ with metric (3), the relation which defines the mean curvature considerably simplifies. Considering the associated parametrization $\varphi : D \to \mathbb{R}^3$,

$$\varphi(u^1, u^2) = (u^1, u^2, f(u^1, u^2)) = (u^1 \delta_{i1} + u^2 \delta_{i2} + f \delta_{i3})_{i=\overline{1,3}}, \tag{24}$$

we yield $\varphi^i_\varepsilon = z^i_\varepsilon = \delta_{1\varepsilon}\delta_{i1} + \delta_{i3} f_\varepsilon$ and $\varphi^j_{\varepsilon\eta} = \delta_{j3} f_{\varepsilon\eta}$, where the lower indices denote partial derivatives w.r.t. the corresponding domain-variables. As well, we shall choose

$$X = \varphi_1 \times \varphi_2 = (-\delta_{i1} f_1 - \delta_{i2} f_2 + \delta_{i3})_{i=\overline{1,3}}. \tag{25}$$

we obtain the following computational result:

Corollary 3.3. *Let $M = \Sigma = Im\ \varphi$ be a graph described by (24) immersed in the 3-dimensional Kropina space with metric (3). Then M is of constant mean curvature iff the function f satisfies the ODE*

$$\begin{aligned} &\frac{-1}{E^2C^2}\{[3b^4(C^2-1)^2 - 2b^2C^2][(1+f_1^2)f_{22} - 2f_1f_2f_{12} + (1+f_2)^2 f_{11}] \\ &+ [8b^2C^2 - 6b^4(C^2-1)](f_1^2 f_{11} + 2f_1f_2f_{12} + f_2^2 f_{22})\} = k,\ k \in \mathbb{R}, \end{aligned} \tag{26}$$

where $C = \sqrt{\det A} = \sqrt{1 + f_1^2 + f_2^2}$, $E = z^3_\alpha z^3_\beta h^{\alpha\beta} = BC^2$ *and where* $B = b^2(f_1^2 + f_2^2)C^{-2}$.

Proof. Computational, using the shape of the terms within the CMC equation infered by (21), for the case of graphs,[8] namely:

$$\frac{\partial E}{\partial z^i_\varepsilon} X^i = 2b^2(\delta_{\varepsilon 1} f_1 + \delta_{\varepsilon 2} f_2), \quad \frac{\partial E}{\partial z^j_\eta}\frac{\partial^2 \varphi^j}{\partial u^\varepsilon \partial u^\eta} = 2b^2(f_1 f_{1\varepsilon} + f_2 f_{2\varepsilon}),$$

$$\frac{\partial C}{\partial z^j_\eta}\frac{\partial^2 \varphi^j}{\partial u^\varepsilon \partial u^\eta} = (f_1 f_{1\varepsilon} + f_2 f_{2\varepsilon})C^{-1}, \quad \frac{\partial C}{\partial z^i_\varepsilon} X^i = 0,$$

$$\frac{\partial^2 C^2}{\partial z^i_\varepsilon \partial z^j_\eta}\frac{\partial^2 \varphi^j}{\partial u^\varepsilon \partial u^\eta} X^i = 2[(1+f_1^2)f_{22} - 2f_1f_2f_{12} + (1+f_2^2)f_{11}]$$

$$\frac{\partial^2 E}{\partial z^i_\varepsilon \partial z^j_\eta}\frac{\partial^2 \varphi^j}{\partial u^\varepsilon \partial u^\eta} X^i = 2b^2[(1+f_1^2)f_{22} - 2f_1f_2f_{12} + (1+f_2^2)f_{11}]. \qquad \square$$

Remark 3.4. Equation (26) yields as consequence of (22), the PDE which characterizes the function of a minimal graph in the 3-dimensional Kropina space with metric (3):

$$\begin{aligned} &\{[3b^4(C^2-1)^2 - 2b^2C^2][(1+f_1^2)f_{22} - 2f_1f_2f_{12} + (1+f_2)^2 f_{11}] \\ &+ [8b^2C^2 - 6b^4(C^2-1)](f_1^2 f_{11} + 2f_1f_2f_{12} + f_2^2 f_{22})\} = 0. \end{aligned} \tag{27}$$

Acknowledgments

The author is grateful to Yi-Bing Shen and to V.S. Sabău for useful discussions and support in developing the present topic. Warm thanks are addressed to the organizers of the Conference DGA-2007 for their concern and hospitality.

References

1. Z. Shen, On Finsler geometry of submanifolds, *Mathematische Annalen* **311** (1998) 549–576.
2. M. Souza, Superficies Minimas en Espacos de Finsler com Uma Metrica de Randers, Ph.D. Thesis, University of Brasil, 2001.
3. M. Souza and K. Tenenblat, Minimal surface of rotation in Finsler space with a Randers metric, *Math. Ann.* **325** (2003) 625.
4. A. Bejancu, Special immersions of Finsler spaces, *Stud.Cerc.Mat.* **39** (1987) 463.
5. R. Miron and M. Anastasiei, *The Geometry of Vector Bundles. Theory and Applications* (Kluwer Acad. Publishers, FTPH, no.59, 1994).
6. V. Balan, DPW method for CMC surfaces in Randers spaces, *In: Modern Trends in Geometry and Topology* (Cluj Univ. Press, Romania, 2006).
7. V. Balan, Moving frames of Riemannian surfaces in GL^3 spaces, *Algebras, Groups and Geometries, Hadronic Press* **17** (2000) 273.
8. M. Souza, J. Spruck and K. Tenenblat, A Bernstein type theorem on a Randers space, *Math. Annalen* 329 (2004) 291–305.

Differential Geometry and its Applications
Proc. Conf., in Honour of Leonhard Euler, Olomouc, August 2007

Generalized Einstein manifolds

Cornelia-Livia Bejan

Seminar matematic, Universitatea "Al.I. Cuza",
Iasi, 700506, România
E-mail: bejan@math.tuiasi.ro

Tran Quoc Binh

Department of Mathematics, University of Debrecen,
H-4010, Debrecen, P.O. Box 12, Hungary
E-mail: binh@math.klte.hu

We give here necessary and sufficient conditions for a quasi–Einstein manifold to be η–Einstein. Then we deal with the locally decomposable Riemannian manifolds and we generalize a result obtained by K. Yano and M. Kon in Ref. 15. The special case of quasi–constant sectional curvature is considered.

Keywords: Quasi–Einstein manifolds, quasi–constant curvature, locally decomposable Riemannian manifolds.

MS classification: 53C25, 53C15.

1. Some classes of generalized Einstein manifolds

The field equations for the interaction of gravitation with other fields, proposed by A. Einstein in Lorentzian framework, take on a Riemannian manifold (M, g) the form:

$$Q - \alpha I = T, \tag{1}$$

where Q, I and T are respectively the Ricci operator, the identity and the energy–momentum tensor. The self–adjoint operator T is particularized on certain manifolds as follows:

I). The "Ricci constant condition" ($Q = \alpha I$) yields $T = 0$, which defines Einstein metrics. When $dimM > 2$, the function α is necessarily constant.

II). Einstein manifolds can be generalized by

Dedicated to Prof. Lajcs Tamassy for the 85th anniversary of his birthday.

Definition 1.1. A Riemannian manifold (M, g) is called quasi–Einstein, provided it carries a globally defined unit vector field ξ (which we refer to as the generator) and its dual form η w.r.t. g

$$\eta = g(\xi, \cdot) \tag{2}$$

such that

$$T = \beta\eta \otimes \xi, \tag{3}$$

where α, β are functions on M.[6]

In particular, if α, β are constant, it is natural to say that M is η–Einstein, by extending this notion from the context of contact geometry.[13]

From (3), or equivalently:

$$g(TX, Y) = \beta\eta(X)\eta(Y), \forall X, Y \in \Gamma(TM),$$

one can see that T is self-adjoint and the Ricci (0,2)–tensor field Ric satisfies

$$Ric(X, Y) = g(QX, Y) = \alpha g(X, Y) + \beta\eta(X)\eta(Y), \forall X, Y \in \Gamma(TM),$$

which gives the scalar curvature r:

$$r = n\alpha + \beta. \tag{4}$$

As any η–Einstein manifold is quasi–Einstein, we study now when the converse is true.

We recall that on a Riemannian manifold (M, g), a 1–form η is called harmonic if η is both closed ($d\eta = 0$) and coclosed ($\delta\eta = trace(\nabla.\eta)\cdot = 0$).

Theorem 1.1. *Let (M^n, g, ξ), $n \geq 2$, be a quasi–Einstein manifold and assume that the dual form η of ξ is harmonic. Then M is η–Einstein if and only if its scalar curvature is constant.*

Proof. As M is quasi–Einstein, from (1) and (3) we have

$$Q = \alpha I + \beta\eta \otimes \xi. \tag{5}$$

We apply the Levi–Civita connection ∇ to (5) and we obtain:

$$(\nabla_Z Q)(W) = (\nabla_Z \alpha)(W) + (\nabla_Z \beta)\eta(W)\xi + \beta[(\nabla_Z \eta)(W)\xi + \eta(W)\nabla_Z \xi], \forall Z, W \in \Gamma(TM).$$

We may chose an orthonormal frame $\{e_i, i = \overline{1, n}\}$ such that $e_1 = \xi$. Then we sum over repeated indices, as follows:

$$g((\nabla_{e_i} Q)(W), e_i) = g((\nabla_{e_i} \alpha)(W), e_i) + (\nabla_{e_i} \beta)\eta(W)\eta(e_i) + \beta[(\nabla_{e_i} \eta)(W)\eta(e_i) + \eta(W)g(\nabla_{e_i} \xi, e_i)], \forall W \in \Gamma(TM).$$

Equivalently, we have:

$$g((\nabla_{e_i}Q)(W), e_i) = \nabla_W\alpha + (\nabla_\xi\beta)\eta(W) + \beta[(\nabla_\xi\eta)(W) - \eta(W)\eta(\nabla_{e_i}e_i)], \forall W \in \Gamma(TM). \tag{6}$$

The second Bianchi identity gives:

$$Wr = 2g((\nabla_{e_i}Q)(W), e_i). \tag{7}$$

Now we use that η is coclosed:

$$\begin{aligned} 0 = \delta\eta &= trace(\nabla.\eta)\cdot = (\nabla e_i\eta)e_i \\ &= e_i(\eta(e_i)) - \eta(\nabla_{e_i}e_i) = -\eta(\nabla_{e_i}e_i) \end{aligned} \tag{8}$$

and that η is closed:

$$\begin{aligned} (\nabla_\xi\eta)(W) &= \xi\eta(W) - \eta(\nabla_\xi W) \\ &= \xi\eta(W) - \eta(\nabla_W\xi + [\xi, W]) = \xi\eta(W) - \eta([\xi, W]) \\ &= d\eta(\xi, W), \forall W \in \Gamma(TM), \end{aligned} \tag{9}$$

since from (2) we have on one side $\eta(\xi) = 1$ and therefore $W(\eta(\xi)) = 0$ and on the other side $\eta(\nabla_W\xi) = g(\nabla_W\xi, \xi) = 0$. By replacing (7), (8) and (9) in (6) we obtain:

$$Wr = 2W\alpha + 2(\xi\beta)\eta(W), \forall W \in \Gamma(TM). \tag{10}$$

If we suppose that M is η–Einstein, it follows from (10) that r is constant.

Conversely, if we suppose that r is constant, then from (4) and (10) we obtain the system:

$$\begin{cases} nW\alpha + W\beta = 0 & \text{and} \\ W\alpha + (\xi\beta)\eta(W) = 0, \forall W \in \Gamma(TM) \end{cases} \tag{11}$$

If we take $W = \xi$, then

$$\begin{cases} n\xi\alpha + \xi\beta = 0 \quad \text{and} \\ \xi\alpha + (\xi\beta) = 0, \end{cases}$$

which imply $\xi\alpha = \xi\beta = 0$. If in (11) we take W orthogonal to ξ, then we get $W\alpha = W\beta = 0, \forall W \in \xi^\perp$.

From the last two relations, we obtain $W\alpha = W\beta = 0, \forall W \in \Gamma(TM)$, that is α and β are constant and therefore M is η–Einstein. □

Remark 1.1. If (M, g, ξ) is an η–Einstein manifold, then from (5) it follows that any two of the following conditions imply the third: (i) η is coclosed; (ii) Q is divergence–free (i.e. $\delta Q = trace(\nabla.Q)\cdot = 0$) and (iii) ξ is autoparallel (i.e. $\nabla_\xi\xi = 0$).

Example 1.1. In our terminology, the main theorem in Ref. 5 states that the unit tangent bundle of a 4–dimensional Einstein manifold, equipped with the canonical contact metric structure, is η–Einstein if and only if the base manifold is a space of constant sectional curvature 1 or 2. Therefore, the unit tangent bundle $T_1(S^4)$ of the unit sphere S^4, endowed with the canonical contact metric structure stands for an example of Theorem 1.1.

III). A Kähler manifold (M, J, g) is called Kähler–Einstein if its metric g is Einstein. This notion has a natural generalization, as follows:

IV). A Kähler manifold (M, J, g) endowed with a generator ξ and its dual form η is called Kähler η–Einstein, provided

$$T = \beta[\eta \otimes \xi - (\eta \circ J) \otimes J\xi],$$

where α, β are functions on M.[1]

The first author and V. Oproiu gave in Ref. 3 the conditions when some natural metrics obtained on TM by lifting procedure, provide a particular class of Kähler η–Einstein metrics, where η is the dual form of the Liouville vector field. For the study of TM and classification theorems, see Ref. 10.

V). From now on, let $M = M_1 \times M_2$ be a locally decomposable Riemannian manifold, that is a class of Riemannian almost product manifolds (M, P, g).[12] Precisely, $P^2 = I$, $P \neq \pm I$, P is self adjoint w.r.t. g and parallel w.r.t. the Levi–Civita connection of g.

For this type of manifolds, we introduce now a notion which corresponds to the above cases I, II, III, IV, i.e. we consider another particular case for the energy–momentum tensor T.

2. Product Einstein manifolds

Definition 2.1. Let (M, P, g) be a locally decomposable Riemannian manifold. We say that M is:

(a) product–Einstein, if in (1.1) we have

$$T = \beta P, \tag{12}$$

with α, β constant;

(b) product quasi–Einstein, provided M carries a generator ξ and its dual form η such that

$$T = \beta P + \gamma[\eta \otimes \xi + (\eta \circ P) \otimes P\xi] + \nu[\eta \otimes P\xi + (\eta \circ P) \otimes \xi], \tag{13}$$

where α, β, γ, ν are functions on M.

If moreover these functions are constant, then we call M a product η–Einstein manifold.

Examples

(1) Any Riemannian product $M = M(c_1) \times M(c_2)$ of two manifolds of constant sectional curvature is a product–Einstein manifold.
(2) If M_1 is an Einstein manifold and M_2 is a canal hypersurface in $\mathbb{R}^n$, i.e. the envelope of a 1–parameter family of hyperspheres,[8] then the Riemannian product $M = M_1 \times M_2$ is a product quasi–Einstein manifold which is not product–Einstein.

Theorem 2.1. *Let $M = M_1 \times M_2$ be a locally decomposable Riemannian manifold with a generator ξ. Then M is product quasi–Einstein (resp. product η–Einstein) if and only if:*

(i) both components are quasi–Einstein (resp. η–Einstein) provided ξ is nowhere tangent to only one of them;
(ii) M_1 is quasi–Einstein (resp. η–Einstein) and M_2 is Einstein, provided ξ is tangent to M_1 only, while $dim M_2 > 2$.

Proof. Some geometric objects on M are decomposed on M_1, M_2 as follows:

$$Q = Q_1 \times Q_2, \xi = \xi' + \xi'', \eta = \eta' + \eta'', \tag{14}$$

where $\xi' = (\xi + P\xi)/2$, $\xi'' = (\xi - P\xi)/2$ and η', η'' are their dual, respectively.

(i) Since in the first case ξ' and ξ'' are nowhere vanishing, we may take $\xi_1 = \xi'/\|\xi'\|$, $\xi_2 = \xi''/\|\xi''\|$ and let η_i be the dual of ξ_i, $i = 1, 2$.
If we suppose that M is a product quasi–Einstein (resp. product η–Einstein) manifold, then

$$Q = \alpha I + \beta P + \gamma[\eta \otimes \xi + (\eta \circ P) \otimes P\xi] + \nu[\eta \otimes P\xi + (\eta \circ P) \otimes \xi], \tag{15}$$

where α, β, γ, ν are functions (resp. constant).
From (14) and (15) one gets:

$$Q_i = \alpha_i I + \beta_i \eta_i \otimes \xi, i = 1, 2, \tag{16}$$

where $\alpha_1 = \alpha + \beta$, $\alpha_2 = \alpha - \beta$, $\beta_1 = 2(\gamma + \nu)$, $\beta_2 = 2(\gamma - \nu)$ are functions (resp. constant).
Conversely, let suppose M_1, M_2 are quasi–Einstein (resp. η–Einstein). From (16) one gets (15), where $\alpha = (\alpha_1 + \alpha_2)/2$, $\beta = (\alpha_1 - \alpha_2)/2$, $\gamma = (\beta_1 + \beta_2)/4$, $\nu = (\beta_1 - \beta_2)/4$ are functions (resp. constant) and

therefore M is a product quasi–Einstein (resp. product η–Einstein) manifold.

(ii) In this case $\xi_1 = \xi' = \xi$, $\eta_1 = \eta$, $\xi'' = 0$, $\eta'' = 0$. We obtain the equivalence between (15) and (16), where the coefficients are related as in the case (i), but here α_2 is constant (since $dimM_2 > 2$), $\gamma = \nu$, hence $\beta_2 = 0$, which complete the proof. □

Theorem 2.1 generalizes Theorem 2.4 obtained by K. Yano, M. Kon in Ref. 15 (pp. 421), which in view if Definition 2.1 can be restated as follows:

Corollary 2.1. *A locally decomposable Riemannian manifold $M = M_1^p \times M_2^q$ $(p, q > 2)$ is a product–Einstein manifold if and only if M_1 and M_2 are both Einstein.*

3. Quasi–constant sectional curvature

The following notion was defined independently by different authors:

Definition 3.1. A Riemannian manifold (M, g) endowed with a generator ξ and its dual form η is called of quasi–constant sectional curvature if it satisfies one of the following conditions:

(a) The curvature of any plane σ depends only on the point and on the angle between σ and ξ;

(b) The curvature (0,4)–tensor field R takes the form

$$R = \alpha U + \beta S, \tag{17}$$

where α, β are functions on M and U, S are (0,4)–tensor fields defined by

$$U(X, Y, Z, V) = g(X, V)g(Y, Z) - g(Y, V)g(X, Z), \tag{18}$$

$$\begin{aligned} S(X, Y, Z, V) &= g(X, V)\eta(Y)\eta(Z) - g(X, Z)\eta(Y)\eta(V) \\ &\quad + g(Y, Z)\eta(X)\eta(V) - g(Y, V)\eta(X)\eta(Z), \\ &\qquad \forall X, Y, Z, V \in \Gamma(TM). \end{aligned}$$

(c) M is of almost constant curvature.

Comment. The notion of quasi–constant sectional curvature was introduced geometrically in Boju and Popescu's paper[4] by the statement (a). Independently, Chen and Yano defined the same notion in Ref. 7, by the algebraic condition (b). On the other hand, Vranceanu defined in Ref. 14 the almost constant curvature. It turns out that (a), (b) and (c) are equivalent.

The equivalence (a) $\Leftrightarrow$ (b) (resp. (b) $\Leftrightarrow$ (c)) was proved in Ref. 9 (resp. in Ref. 11).

Some types of manifolds of quasi–constant sectional curvature are discussed in Ref. 2.

Theorem 3.1. *Let $M = M_1 \times M_2$ be a locally decomposable Riemannian manifold with M_i endowed by the unit form η_i for $i = 1, 2$. Then M has both components of quasi–constant sectional curvature if and only if its (0,4)–tensor field R is of the form:*

$$\begin{aligned} R(X,Y,Z,V) &= \alpha[U(X,Y,Z,V) + U(PX,PY,Z,V)] \\ &+ \beta[U(PX,PY,PZ,V) + U(PX,PY,Z,PV)] \\ &+ \gamma[S(X,Y,Z,V) + S(PX,PY,PZ,PV) \\ &+ S(PX,PY,Z,V) + S(X,Y,PZ,PV)] \\ &- \nu[S(X,Y,Z,PV) + S(X,Y,PZ,V) \\ &- S(PX,PY,Z,PV) + S(PX,PY,PZ,V)], \\ &\qquad \forall X,Y,Z,V \in \Gamma(TM), \end{aligned} \tag{19}$$

where α, β, γ, ν are functions on M and $\eta = (\eta_1 + \eta_2)/2$.

Proof. If we suppose that M_i is of quasi–constant sectional curvature, then from (17), R_i takes the form

$$R_i = \alpha_i U_i + \beta_i S_i, i = 1, 2, \tag{20}$$

where U_i and S_i satisfy (18) with g_i and η_i, $i = 1, 2$.

Then $R = R_1 + R_2$ will satisfy (19) with $\alpha = (\alpha_1 + \alpha_2)/4$, $\beta = (\alpha_1 - \alpha_2)/4$, $\gamma = (\beta_1 + \beta_2)/8$ and $\nu = (\beta_1 - \beta_2)/8$.

Conversely, if we suppose that R is given by (19), then $R = R_1 + R_2$, where R_i satisfies (17) on $M_i, i = 1, 2$, with $\alpha_1 = 2(\alpha + \beta)$, $\alpha_2 = 2(\alpha - \beta)$, $\beta_1 = 4(\gamma + \nu)$ and $\beta_2 = 4(\gamma - \nu)$, which complete the proof. □

Acknowledgments

The authors are indebted to professor O. Kowalski for valuable suggestions.

References

1. C.L. Bejan, Kähler manifolds of quasi–constant holomorphic sectional curvature, to appear in *J. Geom.*

2. C.L. Bejan, Types of manifolds of quasi–constant curvature, *Sci. Annals UASVM Iasi*, Tom XLIX, V. **2** (2006) 155–158.
3. C.L. Bejan and V. Oproiu, Tangent bundles of quasi–constant holomorphic sectional curvatures, *J. Geom. Appl.* **11** (1) (2006) 11–22.
4. V. Boju and M. Popescu, Espaces à courbure quasi–constante, *J. Diff. Geom.* **13** (1978) 373–383.
5. Y.D. Chai, S.H. Chun, J.H Park and K. Sekigawa, Remarks on η–Einstein unit tangent bundles, 10 Aug 2007; arXiv:0708.1400v1[math.DG].
6. M. Chaki and R. Maity, On quasi Einstein manifolds, *Publ. Math. Debrecen* **57** (2000) 297–306.
7. B.Y. Chen and K. Yano, Hypersurfaces of a conformally flat space, *Tensor N.S.* **26** (1972) 318–322.
8. B.Y. Chen and K. Yano, Special conformally flat spaces and canal hypersurfaces, *Tohoku Math. J.* **25** (1973) 177–184.
9. G. Ganchev and V. Mihova, Riemannian manifolds of quasi–constant sectional curvatures, *J. Reine Angew. Math.* **522** (2000) 119–141.
10. O. Kowalski and M. Sekizawa, Natural transformations of Riemannian metrics on manifolds to metrics on tangent bundles – a classification –, *Bull. Tokyo Gakugei Univ.* **40** (4) (1988) 1–29.
11. A.L. Mocanu, *Les variètès à courbure quasi–constant de type Vranceanu* (Lucr. Conf. Nat. Geom., Targoviste, 1987).
12. A.M. Naveira, A classification of Riemannian almost product manifolds, *Rend. Mat. Appl. Math.*, VII, ser. 3 (1983) 577–592.
13. M. Okumura, On infinitesimal conformal and projective transformations of normal contact spaces, *Tohoku Math. J.* II, ser. 14 (1962) 394–412.
14. Gh. Vranceanu, *Leçons des Geometrie Differential* (vol. 4, Ed. Acad., Bucharest, 1968).
15. K. Yano and M. Kon, *Structures on manifolds* (World Scientific, Singapore, 1984).

Differential Geometry and its Applications
Proc. Conf., in Honour of Leonhard Euler, Olomouc, August 2007

Canonical almost geodesic mappings of type $\tilde{\pi}_1$ onto pseudo-Riemannian manifolds

V.E. Berezovski

Department of Mathematics, University of Uman,
Institutskaya 1, Uman, Ukraine
E-mail: berez.volod@rambler.ru

J. Mikeš and A. Vanžurová

Department of Algebra and Geometry, Faculty of Science, Palacký University,
Tomkova 40, 779 00 Olomouc, Czech Republic
E-mail: josef.mikes@upol.cz, alena.vanzurova@upol.cz

Our aim is to examine almost geodesic mappings of affine manifolds and give necessary and sufficient conditions for existence of the so-called $\tilde{\pi}_1$ mappings (canonical almost geodesic mappings of type π according to Sinyukov) of a manifold endowed with a linear connection onto pseudo-Riemannian manifolds. The conditions take the form of a closed system of PDE's of first order of Cauchy type. Our result is a generalization of some previous theorems of N.S. Sinyukov.

Keywords: Linear connection, affine manifold, pseudo-Riemannian space, geodesic curve, almost geodesic curve, geodesic mapping, almost geodesic mapping, deformation tensor.

MS classification: 53B20, 53B30, 53B35.

1. Introduction

Unless otherwise specified, all objects under consideration are supposed to be differentiable of a sufficiently high class (mostly, differentiability of the class C^3 is sufficient).

Let $A_n = (M, \nabla)$ be an n-dimensional (C^k, C^∞ or C^ω) manifold endowed with a linear connection ∇ (an "affine manifold"). Let $c : I \to M$, $t \mapsto c(t)$ defined on an open interval $I \subset \mathbb{R}$ be a (C^k, or smooth) curve on M satisfying the regularity condition

$$c'(t) = dc(t)/dt \neq 0 \quad \text{for all} \quad t \in I.$$

Denote by ξ the corresponding (C^{k-1}, or smooth) tangent vector field along c ("velocity field"), $\xi(t) = (c(t), c'(t))$, $t \in I$, and let

$$\xi_1 = \nabla(\xi;\xi) = \nabla_\xi \xi, \qquad \xi_2 = \nabla^2(\xi;\xi,\xi) = \nabla_\xi \xi_1. \tag{1}$$

Geodesics $c(s)$, parametrized by canonical affine parameter (given up to affine transformations $s \mapsto as + b$), are characterized by $\nabla_\xi \xi = 0$ while unparametrized geodesic curves (i.e. arbitrarily parametrized, called also *pregeodesics* in the literature) can be characterized by the formula $\nabla_\xi \xi = \lambda \xi$ where $\lambda(t) : I \to \mathbb{R}$ is a real function.

Let $D = \mathrm{span}\,(X_1, X_2)$ (i.e. vector fields X_1, X_2 along c form a basis of D). Recall that D is parallel (along c) if and only if covariant derivatives along c of basis vector fields belong to the distribution (the property is independent on reparametrization of the curve).[1–3]

As a generalization of (an unparametrized) geodesic, let us introduce an *almost geodesic curve* as a curve c satisfying: there exists a two-dimensional (differentiable) distribution D parallel along c (relative to ∇) such that for any tangent vector of c, its parallel translation along c (to any other point) belongs to the distribution D.

Equivalently, c is almost geodesic if and only if there exist vector fields X_1, X_2 parallel along c (i.e. satisfying $\nabla_\xi X_i = a^j X_j$ for some differentiable functions $a_i^j(t) : I \to \mathbb{R}$) and differentiable real functions $b^i(t)$, $t \in I$ along c, such that $\xi = b^1 X_1 + b^2 X_2$ holds. For almost geodesic curves, the vector fields ξ_1 and ξ_2 belong to the corresponding distribution D. If the vector fields ξ and ξ_1 are independent at any point (and hence the (local) curve c is not a geodesic one), we can write $D = \mathrm{span}\,(\xi, \xi_1)$. So we get another equivalent characterization: a curve is almost geodesic if and only if $\xi_2 \in \mathrm{span}\,(\xi, \xi_1)$.

2. Almost geodesic mappings

Geodesic mappings of manifolds with linear connection (in short, affine manifolds) are (C^k)-diffeomorphisms characterized by the property that all geodesics are send onto (unparametrized in general) geodesic curves. The classification of geodesic mappings is more or less known.

Recall that even for Riemannian spaces, there is a lack of a nice simple criterion for decision when a given Riemannian space admits non-trivial geodesic mappings.

Let $A_n = (M, \nabla)$, $\bar{A}_n = (\bar{M}, \bar{\nabla})$ be n-dimensional affine manifolds, $n > 2$, endowed with torsion-free linear connections.

We may ask which (C^k-)diffeomorphisms of affine manifolds send almost geodesic curves onto almost geodesic ones again. The answer is: such mappings reduce to geodesic ones, since there are "too many" almost geodesic curves. It appears that the following definition is more acceptable.

We say that a (C^κ-)diffeomorphism $f\colon M \to \bar{M}$ is *almost geodesic* if any geodesic curve of (M, ∇) is mapped under f onto an almost geodesic curve in $(\bar{M}, \bar{\nabla})$.

This concept of an almost geodesic mapping was introduced by N.S. Sinyukov,[1] and before by V.M. Chernyshenko,[4] from a rather different point of view. The theory of almost geodesic mappings was treated in Ref. 1–3.

Due to the fact that f is a diffeomorphism we can accept the useful convention that both linear connections ∇ and $\bar{\nabla}$ are in fact defined on the same underlying manifold M, so that we can consider their difference $P = \bar{\nabla} - \nabla$. That is, P is a $(1,2)$-tensor, called sometimes a *deformation tensor* of the given connections under f,[2] given by $\bar{\nabla}(X,Y) = \nabla(X,Y) + P(X,Y)$ for $X, Y \in \mathcal{X}(M)$. Since the connections are symmetric, P is also symmetric in X, Y. Of course, we identify objects on M with their corresponding objects on $\bar{M}$: a curve c on M identifies with its image $\bar{c} = f \circ c$, its tangent vector field $\xi(t)$ with the corresponding vector field $\bar{\xi}(t) = Tf(\xi(t))$ etc.

Besides the deformation tensor, we will use type $(1,3)$ tensor field, denoted by the same symbol P, introduced by

$$P(X,Y,Z) = \sum_{CS(X,Y,Z)} \nabla_Z P(X,Y) + P(P(X,Y),Z), \quad X, Y, Z \in \mathcal{X}(M),$$

where $\sum_{CS(,,)}$ means the cyclic sum on arguments in brackets (i.e. symmetrization without coefficients).

Almost geodesic diffeomorphisms $f\colon (M,\nabla) \to (M,\bar{\nabla})$ are characterized by the following condition on the type $(1,3)$ tensor P:

$$P(X_1,X_2,X_3) \wedge P(X_4,X_5) \wedge X_6 = 0, \quad X_i \in \mathcal{X}(M), \quad i = 1,\dots,6;$$

$X \wedge Y$ means the exterior product of X and Y, the decomposable bivector.

N.S. Sinyukov[1–3] distinguished three kinds of almost geodesic mappings, namely π_1, π_2, and π_3, characterized, respectively, by the conditions for the deformation tensor:

$$\pi_1\colon \nabla_X P(X,X) + P(P(X,X),X) = a(X,X)\cdot X + b(X)\cdot P(X,X), X \in \mathcal{X}(M),$$

where $a \in S^2(M)$ is a symmetric type $(0,2)$ tensor field and b is a 1-form;

$$\pi_2\colon P(X,X) = \psi(X)\cdot X + \varphi(X)\cdot F(X), \quad X \in \mathcal{X}(M),$$

where ψ and φ are 1-forms, and F is a type $(1,1)$ tensor field satisfying

$$\nabla_X F(X) + \varphi(X)\cdot F(F(X)) = \mu(X)\cdot X + \varrho(X)\cdot F(X), \quad X \in \mathcal{X}(M)$$

for some 1-forms μ, ϱ;

$$\pi_3\colon P(X,X) = \psi(X)\cdot X + a(X,X)\cdot Z, \quad X \in \mathcal{X}(M)$$

where ψ is a 1-form, $a \in S^2(M)$ is a symmetric bilinear form and $Z \in \mathcal{X}(M)$ is a vector field satisfying

$$\nabla_X Z = h\cdot X + \theta(X)\cdot Z$$

for some scalar function $h\colon M \to \mathbb{R}$ and some 1-form θ. Remark that the above classes are not disjoint.

3. Canonical almost geodesic mappings $\tilde{\pi}_1$

We are interested here in a particular subclass of π_1-mappings, the so-called $\tilde{\pi}_1$-*mappings*, or *canonical* almost geodesic mappings, distinguished by the condition $b = 0$. That is, $\tilde{\pi}_1$-mappings are just morphisms satisfying

$$\nabla_X P(X,X) + P(P(X,X),X) = a(X,X)\cdot X, \quad a \in S^2(M),\ X \in \mathcal{X}(M).$$

In local coordinates, the condition reads

$$P^h_{(ij,k)} = a_{(ij}\delta^h_{k)} - P^h_{\alpha(i}P^\alpha_{jk)}. \tag{2}$$

Here and after the comma "," denotes covariant derivative with respect to ∇, δ^h_i is the Kronecker delta, the round bracket denote the cyclic sum on indices.

Any geodesic mapping is a π_1-mapping (the characterizing condition can be checked), and any π_1-mapping can be written as a composition of a geodesic mapping followed by a $\tilde{\pi}_1$-mapping. So we can consider geodesic mappings as trivial almost geodesic mappings, and we will omit them in further considerations; they have been analysed in Ref. 5. It was proven by Sinyukov[2] that the basic partial differential equations (PDE's) of $\tilde{\pi}_1$-mappings of an affine manifold (M,∇) onto Ricci-symmetric ($\bar{\nabla}\bar{\mathrm{Ric}} = 0$, the Ricci tensor is parallel) pseudo-Riemannian spaces $(\bar{M},\bar{g})$ (of arbitrary signature) can be transformed into (an equivalent) closed system of PDE's of first order of Cauchy type. Hence the solution (if it exists) depends on a finite set of parameters. Consequently, for an affine manifold with a symmetric connection admitting $\tilde{\pi}_1$-mappings onto Ricci-symmetric spaces, the set of all Ricci-symmetric spaces $(\bar{M},\bar{g})$ which can serve as images of the given affine manifold (M,∇) under $\tilde{\pi}_1$-mappings is finite. The cardinality r of such a set is bounded by the number of free parameters.

On the other hand, geodesic mappings form a subclass among $\tilde{\pi}_1$-mappings (they obey the definition). Basic equations describing geodesic mappings of affine manifolds do not form a closed system of Cauchy type (the general solution depends on n arbitrary functions; if the given manifold admits geodesic mappings, the cardinality of the set of possible images is big). It follows that the conditions (2) describing $\tilde{\pi}_1$-mappings of affine manifolds, in general, cannot be transformed into a closed system of Cauchy type. But if we choose a suitable subclass of images and restrict ourselves (for the given manifold) only onto mappings with co-domain in the appropriate subclass we might succeed to get an equivalent closed system of Cauchy type. If this is the case then the given manifold admits either non (if the system is non-integrable) or a finite number of $\tilde{\pi}_1$-images in the given class.

Our aim is to analyse $\tilde{\pi}_1$-mappings of affine manifolds onto affine manifolds in general, and to use the reached results for examining $\tilde{\pi}_1$-mappings of affine manifolds onto (pseudo-)Riemannian spaces (in general, without any restrictive conditions onto Ricci tensor), which will generalize the above result by Sinyukov. In the rest, we will omit "pseudo".

All $\tilde{\pi}_1$-mappings $f\colon M \to M$ can be described by the following system of (differential) equations:[2,3]

$$3(\nabla_Z P(X,Y) + P(Z, P(X,Y))) =$$

$$\sum_{CS(X,Y)} (R(Y,Z)X - \bar{R}(Y,Z)X) + \sum_{CS(X,Y,Z)} a(X,Y)Z.$$

In the rest, we prefer to express our equalities in local coordinates (with respect to a map (U,φ) on M) since the invariant formulas are rather complicated. The above formula has the local expression

$$3(P^h_{ij,k} + P^h_{k\alpha}P^\alpha_{ij}) = R^h_{(ij)k} - \bar{R}^h_{(ij)k} + a_{(ij}\delta^h_{k)}, \tag{3}$$

where P^h_{ij}, a_{ij}, R^h_{ijk}, $\bar{R}^h_{ijk}$ are local components of tensors P, a, R, and $\bar{R}$.

Assuming (3) as a system of PDE's for functions P^h_{ij} on M, the corresponding integrability conditions read

$$\begin{aligned}\bar{R}^h_{(ij)[k,\ell]} &= R^h_{(ij)[k,\ell]} + \delta^h_{(i}a_{jk),\ell} - \delta^h_{(i}a_{j\ell),k} + 3(P^\alpha_{ij}\bar{R}^h_{\alpha k\ell} - P^h_{\alpha(j}R^\alpha_{i)k\ell}) \\ &\quad - P^h_{\alpha k}(R^\alpha_{(ij)\ell} - \bar{R}^\alpha_{(ij)\ell}\delta^\alpha_{(i}a_{j\ell)}) + P^h_{\alpha\ell}(R^\alpha_{(ij)k} - \bar{R}^\alpha_{(ij)k}\delta^\alpha_{(i}a_{jk)})\,.\end{aligned}$$

Passing from $\nabla\bar{R}$ to $\bar{\nabla}\bar{R}$ on the left hand side we get integrability conditions of the system (3) in the form

$$\bar{R}^h_{(ij)[k;\ell]} = \delta^h_{(i}a_{jk),\ell} - \delta^h_{(i}a_{j\ell),k} + \Theta^h_{ijk\ell}\,, \tag{4}$$

where we denoted

$$\Theta^h_{ijk\ell} = R^h_{(ij)[k,\ell]} + 3(P^\alpha_{ij}\bar{R}^h_{\alpha k\ell} - P^h_{\alpha(j}R^\alpha_{i)k\ell}) - P^h_{\alpha k}(R^\alpha_{(ij)\ell} + \delta^\alpha_{(i}a_{j\ell)})$$
$$+P^h_{\alpha\ell}(R^\alpha_{(ij)k} + \delta^\alpha_{(i}a_{jk)}) - P^\alpha_{\ell(i}\bar{R}^h_{|\alpha|j)k} - P^\alpha_{\ell(i}\bar{R}^h_{j)\alpha k} + P^\alpha_{k(i}\bar{R}^h_{|\alpha|j)\ell} + P^\alpha_{k(i}\bar{R}^h_{j)\alpha\ell},$$

where ";" denotes covariant derivative with respect to $\bar{\nabla}$.

If we apply covariant differentiation with respect to $\bar{\nabla}$ to the integrability conditions (4) of the system (3), and then pass from covariant derivation $\bar{\nabla}$ to ∇, we get

$$\bar{R}^h_{(ij)k;\ell m} - \bar{R}^h_{(ij)\ell;mk} = \delta^h_{(i}a_{jk),\ell m} - \delta^h_{(i}a_{j\ell),km} + T^h_{ijk\ell m}, \tag{5}$$

where we denoted

$$T^h_{ijk\ell m} = \bar{R}^h_{\alpha mk}\bar{R}^\alpha_{(ij)\ell} - \bar{R}^\alpha_{\ell mk}\bar{R}^h_{(ij)\alpha} - \bar{R}^\alpha_{jmk}\bar{R}^h_{(i\alpha)\ell} - \bar{R}^\alpha_{imk}\bar{R}^h_{(j\alpha)\ell}$$
$$- P^h_{m\alpha}\delta^\alpha_{(i}a_{jk),\ell} - P^\alpha_{mj}\delta^h_{(i}a_{\alpha k),\ell} - P^\alpha_{mi}\delta^h_{(\alpha}a_{jk),\ell} - P^\alpha_{mk}\delta^h_{(\alpha}a_{ij),\ell}$$
$$- P^\alpha_{ml}\delta^h_{(i}a_{jk),\alpha} - P^h_{m\alpha}\delta^\alpha_{(i}a_{j\ell),k} + P^\alpha_{mi}\delta^h_{(\alpha}a_{j\ell),k} + P^\alpha_{mj}\delta^h_{(i}a_{\alpha\ell),k}$$
$$+ P^\alpha_{mk}\delta^h_{(i}a_{j\ell),\alpha} - P^\alpha_{ml}\delta^h_{(i}a_{j\alpha),k} - \Theta^h_{ijk\ell,m} + P^h_{\alpha m}\Theta^\alpha_{ijk\ell} - P^\alpha_{mi}\Theta^h_{\alpha jk\ell}$$
$$- P^\alpha_{mj}\Theta^h_{i\alpha k\ell} - P^\alpha_{mk}\Theta^h_{ij\alpha\ell} - P^\alpha_{m\ell}\Theta^h_{ijk\alpha}.$$

Alternating (5) in ℓ, m yields

$$\begin{aligned}\bar{R}^h_{(ij)m;\ell k} - \bar{R}^h_{(ij)\ell;mk} &= \delta^h_{(i}a_{jm),k\ell} - \delta^h_{(i}a_{j\ell),km} + T^h_{ijk[lm]}\\ &+ \bar{R}^h_{(i|\alpha k|}\bar{R}^\alpha_{j)m\ell} + \bar{R}^h_{(ij)\alpha}\bar{R}^\alpha_{km\ell} - \bar{R}^\alpha_{(ij)k}\bar{R}^h_{\alpha m\ell} + \bar{R}^h_{\alpha(i|k|}\bar{R}^\alpha_{j)m\ell}\\ &+ \delta^h_{(\alpha}a_{jk)}R^\alpha_{i\ell m} + \delta^h_{(\alpha}a_{ik)}R^\alpha_{j\ell m} + \delta^h_{(i}a_{j\alpha)}R^\alpha_{k\ell m} - \delta^h_{(i}a_{jk)}R^\alpha_{\alpha\ell m}.\end{aligned} \tag{6}$$

Using properties of the Riemannian tensor, we rewrite (6) as

$$\bar{R}^h_{im\ell;jk} + \bar{R}^h_{jm\ell;ik} = \delta^h_{(i}a_{j\ell),km} - \delta^h_{(i}a_{jm),k\ell} - N^h_{ijk\ell m}, \tag{7}$$

where the last term is

$$N^h_{ijk\ell m} = T^h_{ijk[\ell m]} + \bar{R}^\alpha_{im\ell}\bar{R}^h_{(\alpha j)k} + \bar{R}^\alpha_{jm\ell}\bar{R}^h_{(\alpha i)k} + \bar{R}^\alpha_{km\ell}\bar{R}^h_{(ij)\alpha}$$
$$- \bar{R}^h_{\alpha m\ell}\bar{R}^\alpha_{(ij)k} + \delta^h_{(\alpha}a_{jk)}R^\alpha_{i\ell m} + \delta^h_{(\alpha}a_{ik)}R^\alpha_{j\ell m} + \delta^h_{(\alpha}a_{ij)}R^\alpha_{k\ell m} - a_{(ij}R^h_{k)\ell m}.$$

Alternating (7) in j, k we get

$$\begin{aligned}\bar{R}^h_{jm\ell;ik} - \bar{R}^h_{km\ell;ij} &= \delta^h_{(i}a_{j\ell),km} - \delta^h_{(i}a_{jm),k\ell} - \delta^h_{(i}a_{k\ell),jm} + \delta^h_{(i}a_{km),j\ell}\\ &- N^h_{i[jk]\ell m} + \bar{R}^h_{\alpha m\ell}\bar{R}^\alpha_{ikj} + \bar{R}^h_{i\alpha\ell}\bar{R}^\alpha_{mkj} + \bar{R}^h_{im\alpha}\bar{R}^\alpha_{\ell kj} - \bar{R}^\alpha_{im\ell}\bar{R}^h_{\alpha kj}.\end{aligned} \tag{8}$$

Let us change mutually i and k in (7), and then use (8). We evaluate

$$\begin{aligned}2\bar{R}^h_{jm\ell;ik} &= \delta^h_{(i}a_{j\ell),km} - \delta^h_{(i}a_{jm),k\ell} - \delta^h_{(k}a_{jm),i\ell}\\ &+ \delta^h_{(i}a_{km),j\ell} - \delta^h_{(i}a_{k\ell),jm} + \delta^h_{(j\ell}a_{k),im} + \Omega^h_{ijk\ell m},\end{aligned} \tag{9}$$

where we used the notation

$$\Omega^h_{ijk\ell m} = -N^h_{ijk\ell m} - N^h_{k[ij]k\ell m} - \bar{R}^h_{\alpha m\ell}\bar{R}^\alpha_{(kj)i} + \bar{R}^h_{j\alpha\ell}\bar{R}^\alpha_{mik} + \bar{R}^h_{jm\alpha}\bar{R}^\alpha_{\ell ik}$$
$$-\bar{R}^h_{\alpha i(j}\bar{R}^\alpha_{k)m\ell} + \bar{R}^h_{j\alpha\ell}\bar{R}^\alpha_{mik} + \bar{R}^h_{jm\alpha}\bar{R}^\alpha_{\ell ik} - \bar{R}^h_{\alpha m\ell}\bar{R}^\alpha_{ikj} - \bar{R}^h_{i\alpha\ell}\bar{R}^\alpha_{mkj} + \bar{R}^\alpha_{im[\ell}\bar{R}^h_{\alpha]kj}.$$

On the left side of (9), let us pass from covariant derivation $\bar{\nabla}$ to ∇:

$$\begin{aligned} 2\bar{R}^h_{jm\ell,ik} &= \delta^h_{(i}a_{j\ell),km} - \delta^h_{(i}a_{jm),k\ell} - \delta^h_{(k}a_{jm),i\ell} \\ &\quad + \delta^h_{(i}a_{km),j\ell} - \delta^h_{(i}a_{k\ell),jm} - \delta^h_{(k}a_{j\ell),im} + S^h_{ijk\ell m}, \end{aligned} \tag{10}$$

where

$$\begin{aligned} S^h_{ijk\ell m} &= \Omega^h_{ijk\ell m} - 2\,[\bar{R}^\alpha_{jm\ell,i}P^h_{\ell k} - \bar{R}^h_{\alpha m\ell,i}P^\alpha_{jk} - \bar{R}^h_{j\alpha\ell,i}P^\alpha_{mk} \\ &\quad - \bar{R}^h_{jm\alpha,i}P^\alpha_{\ell k} - \bar{R}^h_{jm\ell,\alpha}P^\alpha_{ik} \\ &\quad + (\bar{R}^\alpha_{jm\ell}P^\beta_{\alpha i} - \bar{R}^h_{\alpha m\ell}P^\alpha_{ij} - \bar{R}^h_{j\alpha\ell}P^\alpha_{im} - \bar{R}^h_{jm\alpha}P^\alpha_{i\ell})P^h_{\beta k} \\ &\quad - (\bar{R}^\alpha_{jm\ell}P^h_{\alpha\beta} - \bar{R}^h_{\alpha m\ell}P^\alpha_{\beta j} - \bar{R}^h_{j\alpha\ell}P^\alpha_{\beta m} - \bar{R}^h_{jm\alpha}P^\alpha_{\beta\ell})P^\beta_{ik} \\ &\quad - (\bar{R}^\alpha_{\beta m\ell}P^h_{\alpha i} - \bar{R}^h_{\alpha m\ell}P^\alpha_{\beta i} - \bar{R}^h_{\beta\alpha\ell}P^\alpha_{im} - \bar{R}^h_{\beta m\alpha}P^\alpha_{i\ell})P^\beta_{jk} \\ &\quad - (\bar{R}^\alpha_{j\beta\ell}P^h_{\alpha i} - \bar{R}^h_{\alpha\beta\ell}P^\alpha_{ji} - \bar{R}^h_{j\alpha\ell}P^\alpha_{\beta i} - \bar{R}^h_{j\beta\alpha}P^\alpha_{i\ell})P^\beta_{km} \\ &\quad - (\bar{R}^\alpha_{jm\beta}P^h_{\alpha i} - \bar{R}^h_{\alpha m\beta}P^\alpha_{ji} - \bar{R}^h_{j\alpha\beta}P^\alpha_{mi} - \bar{R}^h_{jm\alpha}P^\alpha_{\beta i})P^\beta_{k\ell}]\,. \end{aligned} \tag{11}$$

Let there exist a $\tilde{\pi}_1$-mapping of an affine manifold $A_n = (M, \nabla)$ onto a Riemannian manifold $\bar{V}_n = (M, \bar{g})$ where $\bar{g} \in T^0_2 M$ is a metric tensor with components $\bar{g}_{ij}$. Recall that the Riemannian tensor $\bar{R}_{hijk} = \bar{R}^\alpha_{ijk}\bar{g}_{\alpha h}$ of type $(0,4)$ satisfies

$$\bar{R}_{hijk} + \bar{R}_{ihjk} = 0. \tag{12}$$

In (9), let us apply the metric tensor $\bar{g}_{h\beta}$ and then use symmetrization with respect to h and j. According to (12) we get

$$\begin{aligned} &\bar{g}_{ih}(a_{m[k,j]l} + a_{l[j,k]m}) + \bar{g}_{ij}(a_{m[k,h]l} + a_{l[h,k]m}) + \bar{g}_{kh}(a_{m[i,j]l} \\ &+ a_{l[j,i]}) + \bar{g}_{kj}(a_{m[i,h]l} + a_{l[h,i]m}) + \bar{g}_{mh}(a_{k[i,j]l} - a_{ij,kl}) \\ &+ \bar{g}_{mj}(a_{k[i,h]l} - a_{ih,kl}) + \bar{g}_{lj}(a_{kh,il} - a_{i(h,k)m}) \\ &+ 2\bar{g}_{jh}(a_{k(l,i)m} - a_{m(i,k)l}) + \bar{g}_{lh}(a_{k[j,i]m} - a_{ij,km}) = -\Omega^\alpha_{i(j|klm}\bar{g}_{\alpha|h)}. \end{aligned} \tag{13}$$

Contraction of the last formula with the dual tensor $\bar{g}^{jh}$ ($\|\bar{g}^{ij}\| = \|\bar{g}_{ij}\|^{-1}$) gives

$$a_{kl,im} - a_{im,kl} - a_{km,il} + a_{il,km} = -\frac{2}{n+1}\Omega^\alpha_{i\alpha klm}\,. \tag{14}$$

Let us symmetrize the above formula in k and l. From (14) we get

$$\begin{aligned} 2a_{kl,im} - 2a_{im,kl} &= 2a_{\alpha m}R^\alpha_{lik} + a_{\alpha i}R^\alpha_{mlk} + a_{\alpha k}R^\alpha_{mil} + a_{\alpha l}R^\alpha_{mik} \\ &+ \frac{2}{n+1}(\Omega^\alpha_{i\alpha kim} - \Omega^\alpha_{i\alpha(kl)m}). \end{aligned} \tag{15}$$

Using (14) and (15) the equation (13) reads

$$
\begin{aligned}
&2\bar{g}_{ih}(a_{km,jl}-a_{jm,kl})+2\bar{g}_{ij}(a_{km,hl}-a_{hm,kl})+2\bar{g}_{kh}(a_{im,jl}-a_{jm,il})\\
&\quad+2\bar{g}_{kj}(a_{im,hl}-a_{hm,il})+\bar{g}_{mk}(a_{ki,jl}-a_{kj,il}-a_{ij,kl})\\
&\quad+\bar{g}_{mj}(a_{ki,hl}-a_{kh,il}-a_{ih,kl})+\bar{g}_{lj}(a_{kh,im}-a_{i(h,k)m})\\
&\quad+\bar{g}_{lh}(a_{kj,im}-a_{i(k,j)m})=C_{ijkhl},
\end{aligned}
\tag{16}
$$

where

$$
\begin{aligned}
C_{ijkhl}&=-\Omega^{\alpha}_{i(j|klm}\bar{g}_{\alpha|h)}+\frac{2}{n+1}\Omega^{\alpha}_{i\alpha klm}\bar{g}_{jh}-\bar{g}_{kh}a_{\alpha l}R^{\alpha}_{mij}\\
&+\bar{g}_{ih}(\frac{2}{n+1}\Omega^{\alpha}_{m\alpha ljk}-a_{\alpha k}R^{\alpha}_{(ml)j}-a_{\alpha j}R^{\alpha}_{(l|k|m)}-a_{\alpha m}R^{\alpha}_{lkj}-a_{\alpha l}R^{\alpha}_{mkj})\\
&+\bar{g}_{ij}(\frac{2}{n+1}\Omega^{\alpha}_{m\alpha lhk}-a_{\alpha k}R^{\alpha}_{(ml)h}-a_{\alpha h}R^{\alpha}_{(l|k|m)}-a_{\alpha m}R^{\alpha}_{lkh}-a_{\alpha l}R^{\alpha}_{mkh})\\
&+\bar{g}_{kh}(\frac{2}{n+1}\Omega^{\alpha}_{m\alpha lji}-a_{\alpha i}R^{\alpha}_{(ml)j}-a_{\alpha j}R^{\alpha}_{(l|i|m)}-a_{\alpha m}R^{\alpha}_{lij}+a_{\alpha l}R^{\alpha}_{mij})\\
&+\bar{g}_{kj}(\frac{2}{n+1}\Omega^{\alpha}_{m\alpha lhi}-a_{\alpha i}R^{\alpha}_{(ml)h}-a_{\alpha h}R^{\alpha}_{(l|i|m)}-a_{\alpha m}R^{\alpha}_{lih}+a_{\alpha l}R^{\alpha}_{mih}).
\end{aligned}
$$

If we contract (16) with the dual $\bar{g}^{ij}$ of the metric tensor, use (15) and the Ricci identity we get

$$
a_{km,hl}-a_{kl,hm}=\tfrac{1}{2(n+3)}(\bar{g}_{hm}\mu_{kl}-\bar{g}_{hl}\mu_{km})+B_{kmhl},
\tag{17}
$$

where $\mu_{km}=a_{\alpha\beta,km}\bar{g}^{\alpha\beta}$, and

$$
\begin{aligned}
B_{kmhl}&=C_{\alpha\beta kmhl}\bar{g}^{\alpha\beta}+3a_{m\alpha}R^{\alpha}_{lhk}+\frac{3}{2}(a_{h\alpha}R^{\alpha}_{mkl}+a_{k\alpha}R^{\alpha}_{mhl}+a_{l\alpha}R^{\alpha}_{mhk})\\
&+\frac{3}{n+1}(\Omega^{\alpha}_{l\alpha khm}-\Omega^{\alpha}_{h\alpha(kl)m})-\frac{1}{2}(a_{m\alpha}R^{\alpha}_{lkm}+a_{k\alpha}R^{\alpha}_{mhl}+a_{h\alpha}R^{\alpha}_{mkl}+a_{l\alpha}R^{\alpha}_{mkh})\\
&-\frac{1}{n+1}(\Omega^{\alpha}_{l\alpha hkm}-\Omega^{\alpha}_{k\alpha(hl)m})-a_{\alpha(h}R^{\alpha}_{k)lm}+\frac{1}{2}(a_{k\alpha}R^{\alpha}_{lmh}+a_{h\alpha}R^{\alpha}_{lkm}+a_{m\alpha}R^{\alpha}_{lkh}).
\end{aligned}
$$

Now contract (16) with $\bar{g}^{ih}$. According to (17) we get

$$
\bar{g}_{kl}\mu_{jm}-\bar{g}_{jl}\mu_{km}+\bar{g}_{km}\mu_{jl}-\bar{g}_{jkm}\mu_{kl}=\frac{n+3}{n+1}C_{kljm},
\tag{18}
$$

where

$$
\begin{aligned}
C_{kljm}&=C_{\alpha jkl(m|\beta|l)}\bar{g}^{\alpha\beta}-2(n+1)(B_{k(ml)j}-a_{\alpha(l}R^{\alpha}_{m)jk}\\
&\quad+a_{j\alpha}R^{\alpha}_{(m|k|l)}+a_{k\alpha}R^{\alpha}_{(lm)j}).
\end{aligned}
$$

Contracting (18) with $\bar{g}^{k\ell}$ and using the notation $K=\mu_{\alpha\beta}\bar{g}^{\alpha\beta}$ we obtain components of the tensor μ:

$$
\mu_{jm}=\frac{1}{n}K\bar{g}_{jm}+\frac{n+3}{n(n+1)}C_{\alpha\beta jm}\bar{g}^{\alpha\beta}.
\tag{19}
$$

Using (19) we can rewrite (17) in the form

$$a_{km,hl} - a_{hm,kl} = \frac{K}{2n(n+3)}(\bar{g}_{mh}\bar{g}_{kl} - \bar{g}_{lh}\bar{g}_{km}) + A_{kmhl}, \tag{20}$$

where

$$A_{kmhl} = B_{kmhl} + \frac{1}{2n(n+1)}(\bar{g}_{hm}C_{\alpha\beta kl}\bar{g}^{\alpha\beta} - \bar{g}_{hl}C_{\alpha\beta km}\bar{g}^{\alpha\beta}).$$

Combining (16) and (20) we get

$$\begin{aligned}\bar{g}_{jl}a_{ih,km} + \bar{g}_{hl}a_{ij,km} - \bar{g}_{jm}a_{ih,kl} - \bar{g}_{hm}a_{ij,kl} = \\ -\frac{K}{n(n+3)}(\bar{g}_{ih}\bar{g}_{kl}\bar{g}_{jm} - \bar{g}_{ih}\bar{g}_{km}\bar{g}_{jl} + \bar{g}_{ij}\bar{g}_{kl}\bar{g}_{hm} - \bar{g}_{ij}\bar{g}_{km}\bar{g}_{hl} \\ + 3\bar{g}_{kh}\bar{g}_{il}\bar{g}_{jm} - 3\bar{g}_{kh}\bar{g}_{jl}\bar{g}_{im} + 3\bar{g}_{kj}\bar{g}_{il}\bar{g}_{hm} - 3\bar{g}_{lh}\bar{g}_{jk}\bar{g}_{im}) + A_{ijkmhl},\end{aligned} \tag{21}$$

where we have denoted

$$A_{ijkmhl} = C_{ijkmhl} - 2(\bar{g}_{i(h}A_{|km|j)l} + \bar{g}_{k(h}A_{|im|jl)} - \bar{g}_{m(h}A_{|ki|j)l} - \bar{g}_{l(h}A_{|k|j)im})).$$

Finally, symmetrization of (21) in indices i, j, followed by contraction with $\bar{g}^{\ell h}$, enables us to express second covariant derivatives of the tensor a,

$$a_{ij,km} = \frac{K}{n(n+3)}(\bar{g}_{ij}\bar{g}_{km} + 3\bar{g}_{k(j}\bar{g}_{i)m}) + A_{(ij)km\alpha\beta}\bar{g}^{\alpha\beta}. \tag{22}$$

Now we can consider (22) as the system of PDE's (of first order) of Cauchy type relative to the tensor ∇a (i.e. in $a_{ij,k}$), find the integrability conditions and contract them with $\bar{g}^{ij}$ and $\bar{g}^{km}$, respectively. We calculate ∇K,

$$K_{,\beta} = \frac{n(n+3)}{n^2+5n-6}A_\beta, \tag{23}$$

where we denoted

$$\begin{aligned}A_\varrho = \Big[a_{\alpha(j,|k}R^\alpha_{i)m\varrho} + a_{ij,\alpha}R^\alpha_{km\varrho} - \frac{K}{n(n+3)}(\bar{g}_{ij,[\varrho}\bar{g}_{m]k} + \bar{g}_{ij}\bar{g}_{k[m,\varrho} \\ + 3\bar{g}_{kj,[\varrho}\bar{g}_{m]i} + 3\bar{g}_{kj}\bar{g}_{i[m,\varrho]} + 3\bar{g}_{ki,[\varrho}\bar{g}_{m]j} + 3\bar{g}_{ki}\bar{g}_{j[m,\varrho]}) \\ + A_{(ij)k[m|\alpha\beta|,\varrho]}\bar{g}^{\alpha\beta} + A_{(ij)k[m|\alpha\beta|}\bar{g}^{\alpha\beta}{}_{,\varrho]}\Big]\bar{g}^{ij}\bar{g}^{km}.\end{aligned}$$

We use $\bar{\Gamma}^h_{ij} = \Gamma^h_{ij} + P^h_{ij}$ and get

$$\bar{g}_{ij,k} = P^\alpha_{ik}\bar{g}_{\alpha j} + P^\alpha_{jk}\bar{g}_{\alpha i}. \tag{24}$$

Assume the tensors ∇a and $\nabla\bar{R}$, and denote their components by $a_{ijk} := a_{ij,k}$ and $\bar{R}^h_{ijk\ell} := \bar{R}^h_{ijk,\ell}$, respectively. Then (10) and (22) take the form

$$\begin{aligned}2R^h_{jmli,k} = \delta^h_{(i}a_{jl)k,m} - \delta^h_{(i}a_{jm)k,l} + \delta^h_{(k}a_{jl)i,m} - \delta^h_{(k}a_{jm)i,l} \\ + \delta^h_{(i}a_{km)j,l} - \delta^h_{(i}a_{kl)j,m} + S^h_{ijklm},\end{aligned} \tag{25}$$

$$a_{ijk,m} = \frac{K}{n(n+3)}(\bar{g}_{ij}\bar{g}_{km} + 3\bar{g}_{k(j}\bar{g}_{i)m}) + A_{(ij)km\alpha\beta}\bar{g}^{\alpha\beta}, \tag{26}$$

where covariant derivatives of the tensor a_{ijk} in (25) are supposed to be expressed according to (26), the tensor S was introduced componentwise in (11).

The formulas (3), (23)–(26) represent a closed system of Cauchy type for unknown functions

$$\bar{g}_{ij}(x),\ P^h_{ij}(x),\ a_{ij}(x),\ a_{ijk}(x),\ K(x),\ \bar{R}^h_{ijk}(x),\ R^h_{ijkl}(x), \tag{27}$$

which, moreover, must satisfy a finite set of algebraic conditions

$$\bar{g}_{[ij]} = P^h_{[ij]} = a_{[ij]} = a_{[ij]k} = \bar{R}^h_{i(jk)} = R^h_{i(jk)l} = 0,\ det\|\bar{g}_{ij}(x)\| \neq 0. \tag{28}$$

So we have proven:

Theorem 3.1. *The given affine manifold $A_n = (M, \nabla)$ admits $\tilde{\pi}_1$-mappings (i.e. canonical almost geodesic mappings of type π_1) onto Riemannian spaces $\bar{V}_n = (M, \bar{g})$ if and only if there exists solution of the mixed system of Cauchy type* (3), *(23)-(26), (28) for functions (27).*

As a consequence of the additional algebraic conditions, we get an upper boundary for the number r of possible solutions:

Corollary 3.1. *The family of all Riemannian manifolds $\bar{V}_n$ which can serve as images of the given affine manifold $A_n = (M, \nabla)$, depends on at most*

$$\frac{1}{2}n^2(n^2-1) + n(n+1)^2 + 1$$

parameters.

The above Theorem generalizes the result of Sinyukov[3] already mentioned as well as his results on geodesic mappings of Riemannian spaces.

Acknowledgments

Supported by grant No. 201/05/2707 of The Grant Agency of Czech Republic and by the Council of Czech Government MSM 6198959214.

References

1. N.S. Sinyukov, Almost geodesic mappings of affine-connected and Riemannian spaces. *DAN SSSR* **151** (4) (1963) 781–782 (in Russian).

2. N.S. Sinyukov, *Geodesic mappings of Riemannian spaces* (Nauka, Moscow, 1979).
3. N.S. Sinyukov, Almost geodesic mappings of affine-connected and Riemannian spaces. (Russian) *Itogi Nauki Tekh.*, Ser. Probl. Geom. **13** (1982) 3–26; Transl. in *J. Sov. Math.* **25** (1984) 1235–1249.
4. V.M. Chernyshenko, Affine-connected spaces with a correspondent complex of geodesics. *Collection of Works of Mech.-Math. Chair of Dnepropetrovsk Univ.* **6** (1961) 105–118.
5. V. Berezovsky and J. Mikeš, On a classification of almost geodesic mappings of affine connection spaces. *Acta Univ. Palacki. Olomuc., Fac Rerum Nat., Math.* **35** (1996) 21–24.

Differential Geometry and its Applications
Proc. Conf., in Honour of Leonhard Euler, Olomouc, August 2007

On minimal hypersurfaces of hyperbolic space $\mathbb{H}^4$ with zero Gauss-Kronecker curvature

U. Dursun
Istanbul Technical University,
Faculty of Science and Letters, Department of Mathematics,
34469 Maslak, Istanbul, Turkey
E-mail: udursun@itu.edu.tr

We determine minimal hypersurfaces of the hyperbolic space $\mathbb{H}^4(-1)$ with identically zero Gauss-Kronecker curvature. Such a hypersurface is the image of a subbundle spanned by a timelike vector field of the normal bundle of a totally geodesic surface of the de Sitter space $\mathbb{S}^4_1(1)$ under the normal exponential map. We also give some examples.

Keywords: Hyperbolic space, Minimal hypersurfaces, Gauss-Kronecker Curvature.

MS classification: 53C42.

1. Introduction

Minimal hypersurfaces in 4-dimensional space forms with vanishing Gauss-Kronecker curvature curvature has been studied in Refs. 1–5.

Hasanis, Savas-Halilaj and Vlachos[5] studied the classification of complete minimal hypersurfaces in the 4-dimensional hyperbolic space $\mathbb{H}^4(-1)$ with identically zero Gauss-Kronecker curvature. The Gauss-Kronecker curvature of a hypersurface is the product of the principal curvatures. Such hypersurfaces are closely related to stationary (maximal) spacelike surfaces in the de Sitter space $\mathbb{S}^4_1$, which is the Lorentzian unit sphere in the flat Lorentzian space $\mathbb{R}^5_1$. More precisely, if $f : M \to \mathbb{S}^4_1$ is spacelike stationary immersion, where M is a 2-dimensional manifold, and $\mathcal{N}^1$ is the timelike unit normal bundle of f, then the "polar map" $\psi_f(p,w) = w$, $(p,w) \in \mathcal{N}^1$, defines a minimal hypersurface in $\mathbb{H}^4(-1)$ with identically zero Gauss-Kronecker curvature, see Ref. 5. Also they provided a way to produce all spacelike stationary surfaces in $\mathbb{S}^4_1$ with normal curvature identically zero and without totally geodesic points.

In Ref. 6, Kimura determined minimal hypersurfaces M foliated by geodesics of a 4-dimensional space forms $\widetilde{M}^4$ that given by $M = \{\exp_p(t\xi)|p \in \Sigma,\, t \in \mathbb{R}\}$, where Σ is a minimal surface of a 4-dimensional space form $\widetilde{M}^4$ and ξ is a local unit normal vector field on Σ. In this work, motivated by Kimura's work we use totally geodesic spacelike surfaces of $\mathbb{S}^4_1$ to determine minimal hypersurfaces of the hyperbolic space $\mathbb{H}^4(-1)$ with identically zero Gauss-Kronecker curvature. We firstly determine minimal hypersurfaces of $\mathbb{H}^{m+2}(-1)$, that is, if $f : M \to \mathbb{S}^{m+2}_1$ is a totally geodesic isometric immersion from a connected m-dimensional Riemannian M into $\mathbb{S}^{m+2}_1$, and ξ is a non-parallel, timelike unit local normal vector field on M in $\mathbb{S}^{m+2}_1$, then by using normal exponential mapping of M in $\mathbb{S}^{m+2}_1$ in the direction ξ we define a map $F : M \times (\mathbb{R} - \{0\}) \to \mathbb{H}^{m+2}(-1)$ by $F(x,t) = \exp(x, t\xi)$, $(x,t) \in M \times (\mathbb{R} - \{0\})$, which is also called the suspension of f in $\mathbb{H}^{m+2}(-1)$,[5] The image $F(M \times (\mathbb{R} - \{0\}))$ is a spacelike hypersurface of $\mathbb{H}^{m+2}(-1)$ foliated by the geodesics of $\mathbb{H}^{m+2}(-1)$. We prove that the immersion F is minimal under some conditions on the normal connection form of f. Then we show that if $m = 2$, then the minimal immersion $F : M^2 \times (\mathbb{R} - \{0\}) \to \mathbb{H}^4(-1)$ has vanishing Gauss-Kronecker curvature. We also give some examples.

2. Preliminaries

Let $\widetilde{M}^m_q$ be an m-dimensional pseudo-Riemannian manifold with pseudo-Riemannian metric tensor $\tilde{g}$ of index q. Denoting by $\langle\, ,\, \rangle$ the associated nondegenerate inner product on $\widetilde{M}^m_q$, a tangent vector X to $\widetilde{M}^m_q$ is said to be *spacelike* if $\langle X, X\rangle > 0$ (or $X = 0$), *timelike* if $\langle X, X\rangle < 0$ or *lightlike* (*null*) if $\langle X, X\rangle = 0$ and $X \neq 0$.

Let M^m be a submanifold of a pseudo-Riemannian manifold $\widetilde{M}^{m+n}_q$. If the pseudo-Riemannian metric tensor $\tilde{g}$ of $\widetilde{M}^{m+n}_q$ induces a pseudo-Riemannian metric g on M^m, then M^m is called a pseudo-Riemannian submanifold of $\widetilde{M}^m_q$. If the index of g is zero, then M is called a spacelike submanifold.

Let X and Y be tangent vector fields on M^m and let ξ be a normal vector field on M^m in $\widetilde{M}^{m+n}_q$. Then the *Gauss* formula and the *Weingarten* formula are, respectively, given as

$$\widetilde{\nabla}_X Y = \nabla_X Y + h(X,Y) \quad \text{and} \quad \widetilde{\nabla}_X \xi = -A_\xi(X) + \nabla^\perp_X \xi,$$

where $\widetilde{\nabla}$ is the Riemannian connection of $\widetilde{M}^{m+n}_q$, ∇ and $\nabla^\perp$ are the induced Riemannian connection of M and the normal connection of M^m in $\widetilde{M}^{m+n}_q$,

h is the second fundamental form of M in $\widetilde{M}_q^{m+n}$ and A_ξ is the shape operator of M with respect to the normal vector ξ. Also the Gauss and Weingarten formulas yield

$$\langle A_\xi(X), Y\rangle = \langle h(X,Y), \xi\rangle. \tag{1}$$

Let M^m be a submanifold of a pseudo-Riemannian manifold $\widetilde{M}_q^{m+n}$. Let $\xi_1, \ldots, \xi_n$ be an orthonormal local basis for $T^\perp M$. Then the mean curvature vector is given by

$$H = \frac{1}{m}\sum_{i=1}^{n} \varepsilon_i(\text{trace} A_{\xi_i})\xi_i,$$

where $\varepsilon_i = \langle \xi_i, \xi_i\rangle = \pm 1$. A submanifold M of $\widetilde{M}_q^{m+n}$ is called stationary if $H = 0$ on M A stationary spacelike submanifold is called maximal.

Let $\mathbb{R}_1^m$ be an m-dimensional Lorentz-Minkowski space with metric tensor given by

$$\tilde{g} = -(dx_1)^2 + \sum_{i=2}^{m}(dx_i)^2,$$

where $(x_1, \ldots, x_m)$ is a rectangular coordinate system of $\mathbb{R}_1^m$. The de Sitter m-space $\mathbb{S}_1^m(1)$ is a pseudo-Riemannian m-manifold of constant sectional curvature 1 that can be realized as the hyperquadratic in $\mathbb{R}_1^{m+1}$:

$$\mathbb{S}_1^m(1) = \{x \in \mathbb{R}_1^{m+1} |\ \langle x, x\rangle = 1\}.$$

The hyperquadratic

$$\mathbb{H}^m(-1) = \{x \in \mathbb{R}_1^{m+1} |\ \langle x, x\rangle = -1 \text{ and } x_1 > 0\},$$

is the simply connected hyperbolic m-space of constant sectional curvature -1, where x_1 is the first component in $\mathbb{R}_1^{m+1}$.

Let $f : M^m \to \mathbb{S}_1^{m+2}(1)$ be a smooth isometric immersion from an m-dimensional connected Riemannian manifold M^m into an $(m+2)$-dimensional de Sitter space $\mathbb{S}_1^{m+2}(1)$. Let $\xi,\ \eta$ be a local orthonormal normal basis of M^m in $\mathbb{S}_1^{m+2}(1)$ with signatures $\varepsilon_1 = \langle \xi, \xi\rangle$ and $\varepsilon_2 = \langle \eta, \eta\rangle$. Let $X_1, \ldots, X_m$ be a local orthonormal tangent basis on M and s be the normal connection form for $\nabla^\perp$ defined by $s(X_i) = \langle \nabla^\perp_{X_i}\xi, \eta\rangle$. Since $\langle \xi, \eta\rangle = 0$, then we see that $\nabla^\perp_{X_i}\xi = \varepsilon_2 s(X_i)\eta$ and $\nabla^\perp_{X_i}\eta = -\varepsilon_1 s(X_i)\xi$. Here it is seen that if either ξ or η is parallel in the normal space, then the normal connection form for $\nabla^\perp$ is zero. We therefore suppose that ξ and η are nonparallel.

Denoting by s_i the components of the connection form s, the covariant derivative of the 1-form s is defined by

$$s_{ij} = (\nabla_{X_j} s)(X_i) = X_j(s_i) - s(\nabla_{X_j} X_i).$$

Then it is easily seen that

$$s_{ij} = \langle \nabla^{\perp}_{X_j} \nabla^{\perp}_{X_i} \xi - \nabla^{\perp}_{\nabla_{X_j} X_i} \xi, \eta \rangle.$$

As the ambient space is a space form, then the Ricci equation can be written as

$$\langle R^{\perp}(X,Y)\xi, \eta\rangle = \langle [A_\xi, A_\eta]X, Y\rangle, \quad \text{(Ref. 7 [p. 125])},$$

where $R^{\perp}$ denotes the normal curvature tensor of the normal connection $\nabla^{\perp}$ and $[A_\xi, A_\eta] = A_\xi A_\eta - A_\eta A_\xi$. So we can express the Ricci equation as

$$s_{ji} - s_{ij} = \langle R^{\perp}(X_i, X_j)\xi, \eta\rangle = \langle [A_\xi, A_\eta]X_i, X_j\rangle. \tag{2}$$

Let ξ be a local unit timelike normal vector field on M^m in $\mathbb{S}_1^{m+2}(1)$. Then η is spacelike, $\varepsilon_1 = -1$ and $\varepsilon_2 = 1$. The normal exponential mapping of M^m in $\mathbb{S}_1^{m+2}(1)$ in direction ξ is given by

$$\exp(x, t\xi) = \sinh t\, f(x) + \cosh t\, \xi(x),$$

where $x \in M$ and $t \in \mathbb{R}$.

3. Minimal Hypersurfaces with Zero Gauss-Kronecker Curvature

In this section we firstly determine minimal hypersurfaces of $\mathbb{H}^{m+2}(-1)$ based on a totally geodesic spacelike submanifold M^m of an $(m+2)$-dimensional de Sitter space $\mathbb{S}_1^{m+2}(1)$. Then we show that for $m = 2$ the minimal hypersurface of $\mathbb{H}^4(-1)$ has zero Gauss-Kronecker curvature.

Let $f : M^m \to \mathbb{S}_1^{m+2}(1)$ be a smooth isometric immersion from an m-dimensional connected Riemannian manifold M^m into an $(m+2)$-dimensional de Sitter space $\mathbb{S}_1^{m+2}(1)$. Let ξ, η be a local orthonormal normal basis of M^m in $\mathbb{S}_1^{m+2}(1)$ such that ξ is timelike and non-parallel. Then, by using the normal exponential mapping of M^m in $\mathbb{S}_1^{m+2}(1)$ in direction ξ we define a map $F : M \times (\mathbb{R} - \{0\}) \to \mathbb{H}^{m+2}(-1)$ by

$$F(x,t) = \exp(x, t\xi), \tag{3}$$

where $x \in M$ and $t \in \mathbb{R} - \{0\}$. The hypersurface $F(M \times (\mathbb{R} - \{0\}))$ of $\mathbb{H}^{m+2}(-1)$ is the part of the image of the subbundle, spanned by a non-parallel, timelike unit normal vector field ξ, of the normal bundle of a totally geodesic spacelike submanifold M of the de Sitter space $\mathbb{S}_1^{m+2}$ under the normal exponential mapping of M in $\mathbb{S}_1^{m+2}$.

Henceforth, for the sake of the simplicity of the calculations we take a local isothermal coordinate system $(x_1, \ldots, x_m)$ of M such that $\partial_i = \frac{\partial}{\partial x_i} =$

φX_i, $i = 1, \ldots, m$, where $X_1, \ldots, X_m$ form an orthonormal tangent basis on M and φ is a positive function on some open set in M. Thus the components of the first fundamental form g on M are $\langle f_i, f_j \rangle = \varphi^2 \delta_{ij}$, $i, j = 1, \ldots m$. In terms of the chosen tangent basis it is easily seen that

$$\nabla_{X_j} X_i = \sum_{k=1}^{m} \gamma_{ij}^k X_k, \quad \gamma_{ij}^k = -\frac{1}{\varphi}(X_j(\varphi)\delta_i^k - \Gamma_{ij}^k), \tag{4}$$

where Γ_{ij}^k are the Christoffel symbols of M, and hence $s(\nabla_{X_j} X_i) = \sum_{k=1}^{m} \gamma_{ij}^k s_k$. So we have

$$X_j(s_i) = s_{ij} + \sum_{k=1}^{m} \gamma_{ij}^k s_k. \tag{5}$$

The tangent vectors to the hypersurface at $(x_1, \ldots, x_m, t)$ are expressed as

$$F_i = \frac{\partial F}{\partial x_i} = \sinh t\, f_i + \cosh t\, \xi_i, \quad i = 1, \ldots, m,$$

and

$$F_t = \frac{\partial F}{\partial t} = \cosh t\, f + \sinh t\, \xi,$$

where F_i, F_t, f_i, ξ_i, ... denote the derivatives of F, f, and ξ with respect to x_i and t. Suppose that f is totally geodesic, that is, $A_\xi \equiv 0$ and $A_\eta \equiv 0$. Then,

$$\begin{aligned} F_i &= \varphi(\sinh t\, X_i + \cosh t D_{X_i}\xi) \\ &= \varphi(\sinh t\, X_i + \cosh t\, \nabla^{\perp}_{X_i}\xi) \\ &= \varphi(\sinh t\, X_i + s_i \cosh t\, \eta), \end{aligned}$$

where $i = 1, \ldots, m$, D is the covariant differentiation in $\mathbb{R}_1^{m+3}$. Hence

$$\langle F_i, F_j \rangle = \varphi^2(\sinh^2 t\, \delta_{ij} + s_i s_j \cosh^2 t), \quad \langle F_i, F_t \rangle = 0, \quad \langle F_t, F_t \rangle = 1,$$

where $i, j = 1, \ldots, m$. Therefore we have the metric G on M^* induced by F as

$$G = \left(\begin{array}{c|c} \varphi^2(\sinh^2 t\, \delta_{ij} + s_i s_j \cosh^2 t) & 0 \\ \hline 0 & 1 \end{array} \right).$$

We need the following Lemma[8] to show that the map F is an immersion.

Lemma 3.1. *Let $E = I + v^T v$ be an $m \times m$ matrix, where I is the $m \times m$ identity matrix and $v = (v_1, \ldots, v_m) \in \mathbb{R}^m$. Then E has two distinct eigenvalues 1 and $1 + \|v\|^2$ with multiplicities $m - 1$ and 1, respectively, and further $\det E = 1 + \|v\|^2$ and the matrix $I - \frac{1}{\det E} v^T v$ is the inverse of E.*

Proposition 3.1. *Let $f : M^m \to \mathbb{S}_1^{m+2}(1)$ be a smooth totally geodesic isometric immersion from an m-dimensional connected Riemannian manifold M^m into an $(m+2)-$dimensional de Sitter space $\mathbb{S}_1^{m+2}(1)$. Then the map $F : M \times (\mathbb{R} - \{0\}) \to \mathbb{H}^{m+2}(-1)$ defined by (3) is an immersion.*

Proof. As f is totally geodesic, using the Lemma 3.1 the determinant of G is calculated as

$$\begin{aligned}\det G =& \det(\varphi^2(\sinh^2 t\,\delta_{ij} + \cosh^2 t\; s_i s_j))\\ =&(\varphi^2 \sinh^2 t)^m \det(\delta_{ij} + \coth^2 t s_i s_j)\\ =&(\varphi^2 \sinh^2 t)^m\{1 + \coth^2 t\,(s_1^2 + \cdots + s_m^2)\}\\ =&\varphi^{2m}(\sinh t)^{2(m-1)}(\sinh^2 t + \hat{s}^2 \cosh^2 t),\end{aligned}$$

where $\hat{s}^2 = s_1^2 + \cdots + s_m^2$. Since φ is a positive function on M, then $\det G \neq 0$ if and only if $\sinh t \neq 0$, that is, $t \neq 0$. Therefore F is an immersion on $M \times (\mathbb{R} - \{0\})$. □

For the immersion F, by the Lemma 3.1 the inverse of G is obtained as

$$G^{-1} = \left(\begin{array}{c|c} \frac{1}{\alpha^2\varphi^2\sinh^2 t}(\alpha^2\delta_{ij} - \cosh^2 t\; s_i s_j) & 0 \\ \hline 0 & 1\end{array}\right), \tag{6}$$

where $\alpha^2 = \sinh^2 t + \hat{s}^2 \cosh^2 t$.

By considering (4) and (5) the second derivatives of F are calculated as

$$\begin{aligned}F_{ij} =& \frac{\partial^2 F}{\partial x_i \partial x_j} = \frac{\partial \varphi}{\partial x_j}(\sinh t X_i + s_i \cosh t\,\eta) + \varphi^2(\sinh t D_{X_j} X_i\\ &+ \cosh t X_j(s_i)\eta + s_i \cosh t\,\nabla^{\perp}_{X_j}\eta)\\ =& X_j(\varphi)F_i + \varphi^2\{\sum_{k=1}^{m}\gamma_{ij}^k(\sinh t X_k + s_k \cosh t\eta)\\ &+ \cosh t(s_i s_j \xi + s_{ij}\eta) - \sinh t\delta_{ij} f\}\\ =& \sum_{k=1}^{m}(X_j(\varphi)\delta_{ik} + \varphi\gamma_{ij}^k)F_k + \varphi^2 \cosh t\,(s_i s_j \xi - s_{ij}\eta) - \varphi^2 \sinh t\,\delta_{ij} f\end{aligned}$$

$$= \sum_{k=1}^{m}\Gamma_{ij}^k F_k + \varphi^2 \cosh t\,(s_i s_j \xi + s_{ij}\eta) - \varphi^2 \sinh t\,\delta_{ij} f,\; i,j = 1,\ldots,m, \tag{7}$$

$$F_{it} = \varphi(\cosh t\, X_i + s_i \sinh t\,\eta), \quad i = 1,\ldots,m$$

and

$$F_{tt} = \sinh t\, f + \cosh t\, \xi = F.$$

The unit normal vector N to F in $\mathbb{H}^{m+2}(-1)$ is given by

$$N = \frac{\coth t}{\varphi\alpha} \sum_{k=1}^{m} s_k F_k - \frac{\alpha}{\sinh t}\eta. \tag{8}$$

Let $\bar{h}^N$ denote the second fundamental form of F relative to N. For the coordinate vector fields $\partial_1, \ldots, \partial_m, \partial_t$, if we use the Gauss formula for F, then we have

$$\bar{h}^N(\partial_i, \partial_j) = \langle F_{ij}, N\rangle, \quad \bar{h}^N(\partial_i, \partial_t) = \langle F_{it}, N\rangle, \quad \bar{h}^N(\partial_t, \partial_t) = \langle F_{tt}, N\rangle$$

for the components of the second fundamental form $\bar{h}^N$.

We prove

Theorem 3.1. *Let $f : M^m \to \mathbb{S}_1^{m+2}(1)$ be a smooth totally geodesic isometric immersion from an m-dimensional connected Riemannian manifold M^m into an $(m+2)-$dimensional de Sitter space $\mathbb{S}_1^{m+2}(1)$. Then the immersion $F : M \times (\mathbb{R} - \{0\}) \to \mathbb{H}^{m+2}(-1)$ defined by (3) is minimal if and only if the components, s_i, of the normal connection form s of f satisfy the equations*

$$\sum_{i=1}^{m} s_{ii} = 0 \quad \text{and} \quad \sum_{i,j=1}^{m} s_i s_j s_{ji} = 0. \tag{9}$$

Proof. Since f is totally geodesic, then we can have (6), (7) and (8), and also from the Ricci equation (2) we get $s_{ij} = s_{ji}$. Thus, if we calculate the second fundamental form $\bar{h}^N$ by using (7) and (8) we obtain

$$\begin{aligned}
&\bar{h}^N(\partial_i, \partial_j) = \langle F_{ij}, N\rangle = -\frac{\varphi^2 \sinh 2t}{2\alpha} s_{ij}, \ i,j = 1, \ldots, m,\\
&\bar{h}^N(\partial_i, \partial_t) = \bar{h}^N(\partial_t, \partial_i) = \langle F_{it}, N\rangle = \frac{\varphi}{\alpha} s_i, \ i = 1, \ldots, m,\\
&\bar{h}^N(\partial_t, \partial_t) = \langle F_{tt}, N\rangle = 0.
\end{aligned} \tag{10}$$

From (1), (6) and (10), the shape operator $\bar{A}_N$ of F in the direction N according to the basis $\{\partial_1, \ldots, \partial_m, \partial_t\}$ is

$$\bar{A}_N = \frac{1}{\varphi\alpha^3}\left(\begin{array}{c|c} -\varphi \coth t(\alpha^2 s_{ij} - \cosh^2 t \sum_{k=1}^{m} s_{ik} s_k s_j) & \alpha^2 \varphi^2 s_i \\ \hline s_j & 0 \end{array}\right).$$

Therefore the mean curvature vector $\bar{H}$ of F in $\mathbb{H}^{m+2}(-1)$ is given by

$$\begin{aligned}\bar{H} &= \frac{1}{m+1}(\mathrm{trace}\bar{A}_N)N \\ &= -\frac{\coth t}{(m+1)\alpha^3}\left(\alpha^2\sum_{i=1}^{m} s_{ii} - \cosh^2 t \sum_{i,k=1}^{m} s_{ik}s_k s_i\right)N.\end{aligned}$$

As a result, F is minimal in $\mathbb{H}^{m+2}(-1)$, that is, $\mathrm{trace}\bar{A}_N = 0$ if and only if

$$\sinh^2 t \sum_{i=1}^{m} s_{ii} + \cosh^2 t\,(\hat{s}^2\sum_{i=1}^{m} s_{ii} - \sum_{i,k=1}^{m} s_{ki}s_i s_k) = 0,$$

from which we have the equations in (9) because $\sinh t$ and $\cosh t$ are linearly independent. □

Note that the hypersurface $F(M\times(\mathbb{R}-\{0\}))$ which is the part of the image of the subbundle, spanned by a timelike non-parallel unit normal vector field ξ, of the normal bundle of a totally geodesic spacelike submanifold M of a de Sitter space $\mathbb{S}_1^{m+2}$ under the normal exponential mapping of M in $\mathbb{H}^{m+2}(-1)$ is equivalent the following two conditions: (1) $F(M\times(\mathbb{R}-\{0\}))$ is foliated by the geodesic of $\mathbb{H}^{m+2}(-1)$, (2) m-dimensional distribution on $F(M\times(\mathbb{R}-\{0\}))$ orthogonal to the geodesics in (1) is locally integrable.

As the integral curves of $\frac{\partial}{\partial t}$ are the geodesics of $\mathbb{H}^{m+2}(-1)$ which are not complete on $M\times(\mathbb{R}-\{0\})$, then the immersion F is not complete.

Theorem 3.2. *Let $f : M^2 \to \mathbb{S}_1^4(1)$ be a smooth totally geodesic isometric immersion from a connected surface M^2 into the de Sitter space $\mathbb{S}_1^4(1)$. If the immersion $F : M\times(\mathbb{R}-\{0\}) \to \mathbb{H}^4(-1)$ defined by (3) is minimal in $\mathbb{H}^4(-1)$, then it has identically zero Gauss-Kronecker curvature and nowhere vanishing second fundamental form.*

Proof. As F is minimal, then, by Theorem 3.1, the components, s_i, of the normal connection form s of f satisfy the equations in (9). Since f is totally geodesic, then from the Ricci equation (2) we have $s_{ij} = s_{ji}$. By using (1) for the immersion F we have the Gauss-Kronecker curvature $K = \det\bar{A}_N = \det\bar{h}^N/\det G$. When we evaluate $\det\bar{h}^N$ by considering (10) for $m=2$ we obtain

$$\det\bar{h}^N = -\frac{\varphi^4\sinh 2t}{2\alpha^3}\left(2s_1s_2s_{12} - (s_1^2s_{22} + s_2^2s_{11})\right).$$

From the second equation in (9) we have $2s_1s_2s_{12} = -s_1^2s_{11} - s_2^2s_{22}$. Thus,

$$\det\bar{h}^N = -\frac{\hat{s}^2\varphi^4\sinh 2t}{2\alpha^3}(s_{11}+s_{22}),$$

which is zero because of the first equation in (9), that is, the Gauss-Kronecker curvature of the immersion F vanishes identically. Also, as the normal vector ξ is not parallel, then at least one the functions s_1, s_2 is not zero. Thus it follows form (10) that the second fundamental form of F never vanishes. □

4. Construction of Example

Here we construct some examples of minimal hypersurface, defined as in the previous section, of hyperbolic space $\mathbb{H}^{m+2}(-1)$ which has zero Gauss-Kronecker curvature when $m = 2$. We consider a totally geodesic isometric immersion $f : \mathbb{S}^m(1) \to \mathbb{S}_1^{m+2}(1)$ from an m-dimensional sphere $\mathbb{S}^m(1)$ into an $(m+2)-$dimensional de Sitter space $\mathbb{S}_1^{m+2}(1)$ defined by

$$f(x_1, \ldots, x_m) = \frac{1}{r^2}(0, (r^2 - 2), 2x_1, \ldots, 2x_m, 0), \tag{11}$$

where $x_1, \ldots, x_m \in \mathbb{R}$ and $r^2 = 1 + x_1^2 + \cdots + x_m^2$.

By a direct calculation the components of the induced first fundamental form on $\mathbb{S}^m(1)$ are obtained as $\langle f_i, f_j \rangle = \frac{4}{r^4}\delta_{ij}$, $i, j = 1, \ldots, m$, which means that the chosen coordinate system on $\mathbb{S}^m(1)$ is isothermal and $\varphi = \frac{2}{r^2}$. Thus, $X_i = \frac{r^2}{2}\frac{\partial}{\partial x_i}$, $i = 1, \ldots, m$, is a local orthonormal tangent basis on $\mathbb{S}^m(1)$. In terms of this metric the Christoffel symbols are obtained as

$$\Gamma_{ij}^k = -\frac{2}{r^2}(x_i\delta_{kj} + x_j\delta_{ik} - x_k\delta_{ij}). \tag{12}$$

For the normal bundle of $\mathbb{S}^m(1)$ in $\mathbb{S}_1^{m+2}(1)$, an orthonormal local basis can, generally, be chosen as

$$\xi = (\cosh\theta, 0, \cdots, 0, \sinh\theta), \quad \eta = (\sinh\theta, 0, \cdots, 0, \cosh\theta), \tag{13}$$

where $\theta = \theta(x_1, \ldots, x_m)$ is a smooth function on some open subset of $\mathbb{S}^m(1)$.

We will find θ which determines the unit, nonparallel, timelike normal vector ξ on $\mathbb{S}^m(1)$ such that the immersion F as defined in the previous section is minimal in $\mathbb{H}^{m+2}(-1)$. Now we will calculate the components s_i of the normal connection s of f and their covariant derivatives s_{ij}. From the definition of s_i we have,

$$s_i = \langle \nabla_{X_i}^{\perp}\xi, \eta \rangle = \langle D_{X_i}\xi, \eta \rangle = \frac{r^2}{2}\langle \frac{\partial \xi}{\partial x_i}, \eta \rangle = \frac{r^2}{2}\frac{\partial \theta}{\partial x_i}, \tag{14}$$

that is, $s_i = -\frac{r^2}{2}\theta_i$, $i = 1, \ldots, m$, and hence

$$X_j(s_i) = \frac{r^2}{2}\frac{\partial}{\partial x_j}(\frac{r^2}{2}\theta_i) = -\frac{r^2}{2}(x_j\theta_i + \frac{r^2}{2}\theta_{ij}).$$

Using the equations (4) and (12) we have $\gamma_{ij}^k = -(x_i\delta_{kj} - x_k\delta_{ij})$. Therefore, by considering (5) we obtain

$$s_{ij} = -\frac{r^2}{2}[x_j\theta_i + \frac{r^2}{2}\theta_{ij} + \sum_{k=1}^{m}(x_i\delta_{kj} - x_k\delta_{ij})\theta_k] = \frac{r^4}{4}(-\theta_{ij} + \sum_{k=1}^{m}\Gamma_{ij}^k\theta_k). \quad (15)$$

Here it is clear that $s_{ij} = s_{ji}$ if and only if $\theta_{ij} = \theta_{ji}$. Thus, by using (14) and (15) the first equation of (9) turns out to be

$$2(2-m)\tilde{\theta} + r^2\sum_{i=1}^{m}\theta_{ii} = 0, \quad (16)$$

and the second equation of (9) becomes

$$2\tilde{\theta}\hat{\theta}^2 + r^2\sum_{i,j=1}^{m}\theta_i\theta_j\theta_{ji} = 0, \quad (17)$$

where $\tilde{\theta} = \sum_{i=1}^{m} x_i\theta_i$ and $\hat{\theta}^2 = \sum_{i=1}^{m}\theta_i^2$.

In Ref. 6, Kimura obtained the general solutions of the partial differential equations (16) and (17) for $m = 2$. For instance,

$$\theta(x_1, x_2) = \arctan\frac{x_2}{x_1} \quad \text{and} \quad \theta(x_1, x_2) = \frac{1}{\sqrt{2}}\arctan\left(\frac{2\sqrt{2}x_1}{x_1^2 + x_2^2 + 2x_2 - 1}\right)$$

are two solutions of differential equations (16) and (17). For the first function if we change the coordinate system to $x_1 = u\cos v$, $x_2 = u\sin v$, where $(u, v) \in D_0 = \{(u, v) \in \mathbb{R}^2 |\ u > 0,\ |v| < \pi/2\}$, then the immersion F becomes

$$F(u, v, t) = \Big(\cosh t\cosh v, \frac{(u^2-1)\sinh t}{u^2+1}, \frac{2u\cos v\sinh t}{u^2+1}, \frac{2u\sin v\sinh t}{u^2+1}, \cosh t\sinh v\Big),$$

which is minimal with identically zero Gauss-Kronecker curvature on open connected set D_0.

For $m > 2$, some special solutions of the equations (16) and (17) were studied in Ref. 8. Let ℓ be a positive integer such that $\ell \leq m/2$, $m \geq 2$. The function

$$\theta(x_1, \ldots, x_m) = \sum_{i=1}^{\ell} C_i \arctan\frac{x_{2i}}{x_{2i-1}}, \quad (18)$$

is a solution of the differential equations on some open set of $\mathbb{R}^m$.

Also, from Ref. 8 we have another solution of the differential equations (16) and (17) as

$$\theta(x_1, \ldots, x_n, x_{n+1}, \ldots, x_m) = \arctan\left(\frac{C_1x_1 + \cdots + C_nx_n}{C_{n+1}x_{n+1} + \cdots + C_mx_m}\right), \quad (19)$$

in the open set $D = \{(x_1, \ldots, x_n, x_{n+1}, \ldots, x_m) \in \mathbb{R}^m \mid \sum_{i=n+1}^m C_ix_i \neq 0\}$ if $\sum_{i=1}^n C_i^2 = \sum_{i=n+1}^m C_i^2$, where $C_1, \ldots C_m \in \mathbb{R}$.

Using (3), (11) and (13) we can express the minimal immersion F for the solutions (18) and (19).

References

1. S.C. de Almeida and F.G.B. Brito, Minimal hypersurfaces of S^4 with constant Gauss-Kronecker curvature, *Math. Z.* **195** (1987) 99–107.
2. J. Ramanathan, Minimal hypersurfaces of S^4 with vanishing Gauss-Kronecker curvature, *Math. Z.* **205** (1990) 645–658.
3. T. Hasanis, A. Savas-Halilaj and T. Vlachos, Minimal hypersurfaces with zero Gauss-Kronecker curvature, *Illinois J. Math.* **49** (2005) 523–529.
4. T. Hasanis, A. Savas-Halilaj and T. Vlachos, Complete minimal hypersurfaces of $\mathbb{S}^4$ with zero Gauss-Kronecker curvature, *Math. Proc. Camb. Phil. Soc.* **142** (2007) 125–132.
5. T. Hasanis, A. Savas-Halilaj and T. Vlachos, Complete minimal hypersurfaces in the hyperbolic space $\mathbb{H}^4$ with vanishing Gauss-Kronecker curvature, *Trans. Amer. Math. Soc.* **359** (2007) 2799–2818.
6. M. Kimura, Minimal hypersurfaces foliated by geodesics of 4-dimensional space forms, *Tokyo J. Math.* **16** (1993) 241–260.
7. B. O'Neill, *Semi-Riemannian Geometry* (Academic Press, New York 1983).
8. U. Dursun, On minimal and Chen immersions in space forms, *J. Geom.* **66** (1999) 104–111.

Differential Geometry and its Applications
Proc. Conf., in Honour of Leonhard Euler, Olomouc, August 2007

Structure of geodesics in the flag manifold $SO(7)/U(3)$

Zdeněk Dušek

Department of Algebra and Geometry, Palacky University,
Tomkova 40, 77900 Olomouc, Czech Republic
E-mail: dusek@prfnw.upol.cz

The Riemannian flag manifold $SO(7)/U(3)$ is explicitly described and geodesic graph is constructed It is shown that the degree of this g.o. manifold is equal to 4.

Keywords: Riemannian homogeneous space, g.o. space, g.o. manifold, geodesic graph, degree of a g.o. manifold.

MS classification: 14M15, 14M17, 53C22, 53C30.

1. Introduction

Homogeneous geodesics in homogeneous Riemannian manifolds and *Riemannian g.o. manifolds* (*g.o. spaces*) were studied in Refs. 3,5,7-11. One of the methods for studying g.o. manifolds is based on the construction of *geodesic graphs*. It is well known, that a g.o. manifold is *naturally reductive*, if there exists a linear geodesic graph. The *degree* of a geodesic graph, the degree of a g.o. space and the degree of a g.o. manifold is the measure of nonlinearity of the geodesic graph. For linear geodesic graph, the degree is equal to zero. Hence, g.o. manifolds on which only nonlinear geodesic graphs exist are of special interest.

In dimension $n \leq 5$, every g.o. manifold is naturally reductive. Geodesic graphs on the examples of 6 and 7-dimensional g.o. manifolds which are not naturally reductive were described in Refs. 7,9. The degree of these g.o. manifolds is equal to 2. The example of a g.o. space of higher degree was given in Ref. 5. It is a 13-dimensional nilpotent Lie group which admits two presentations as a homogeneous space. For one of these spaces, the degree is equal to 6 and for the other, it is 3. The degree of the manifold itself is equal to 3.

In Ref. 1, the authors classify Riemannian flag manifolds which are g.o. (and not naturally reductive). There are two infinite series, namely

$SO(2n+1)/U(n)$ and $Sp(n)/Sp(n-1)U(1)$, where $n \geq 2$. For $n = 2$, these manifolds coincide, this example was described in Ref. 11 and the geodesic graph was constructed in Ref. 9. In the present paper, we investigate the next example in the first series, namely $SO(7)/U(3)$. We show that the degree of this g.o. manifold is equal to 4.

2. G.o. manifolds

Let M be a pseudo-Riemannian manifold. If there is a connected Lie group $G \subset I_0(M)$ which acts transitively on M as a group of isometries, then M is called a *homogeneous pseudo-Riemannian manifold.* Let $p \in M$ be a fixed point. If we denote by H the isotropy group at p, then M can be identified with the *homogeneous space* G/H. In general, there may exist more than one such group. For any fixed choice $M = G/H$, G acts effectively on G/H from the left. The pseudo-Riemannian metric g on M can be considered as a G-invariant metric on G/H. The pair $(G/H, g)$ is then called a *pseudo-Riemannian homogeneous space.*

If the metric g is positive definite, then $(G/H, g)$ is always a *reductive* homogeneous space in the following sense: we denote by $\mathfrak{g}$ and $\mathfrak{h}$ the Lie algebras of G and H respectively and consider the adjoint representation $\mathrm{Ad}\colon H \times \mathfrak{g} \to \mathfrak{g}$ of H on $\mathfrak{g}$. There exists a direct sum decomposition (*reductive decomposition*) of the form $\mathfrak{g} = \mathfrak{m} + \mathfrak{h}$, where $\mathfrak{m} \subset \mathfrak{g}$ is a vector subspace such that $\mathrm{Ad}(H)(\mathfrak{m}) \subset \mathfrak{m}$. If the metric g is indefinite, the reductive decomposition may not exist. For a fixed reductive decomposition $\mathfrak{g} = \mathfrak{m} + \mathfrak{h}$, there is a natural identification of $\mathfrak{m} \subset \mathfrak{g} = T_eG$ with the tangent space T_pM via the projection $\pi\colon G \to G/H = M$. Using this natural identification and the scalar product g_p on T_pM, we obtain the scalar product $\langle\, , \rangle$ on $\mathfrak{m}$. This scalar product is obviously $\mathrm{Ad}(H)$-invariant.

Now we give the general definition of homogeneous geodesics given in Ref. 4 and valid in the general pseudo-Riemannian situation.

Definition 2.1. Let $M = G/H$ be a reductive homogeneous pseudo-Riemannian space, $\mathfrak{g} = \mathfrak{m} + \mathfrak{h}$ a reductive decomposition and p the basic point of G/H. The geodesic $\gamma(s)$ through the point p defined in an open interval J (where s is an affine parameter) is homogeneous if there exists
1) a diffeomorphism $s = \varphi(t)$ between the real line and the open interval J;
2) a vector $X \in \mathfrak{g}$ such that $\gamma(\varphi(t)) = \exp(tX)(p)$ for all $t \in (-\infty, +\infty)$.
The vector X is then called a geodesic vector.

The basic formula characterizing geodesic vectors in the Riemannian situation was given in Ref. 11. The necessary generalization for the pseudo-

Riemannian case was derived in Ref. 4:

Lemma 2.1. *Let $M = G/H$ be a reductive homogeneous pseudo-Riemannian space, $\mathfrak{g} = \mathfrak{m} + \mathfrak{h}$ a reductive decomposition and p the basic point of G/H. Let $X \in \mathfrak{g}$. Then the curve $\gamma(t) = \exp(tX)(p)$ (the orbit of a one-parameter group of isometries) is a geodesic curve with respect to some parameter s if and only if*

$$\langle [X, Z]_{\mathfrak{m}}, X_{\mathfrak{m}} \rangle = k \langle X_{\mathfrak{m}}, Z \rangle \tag{1}$$

for all $Z \in \mathfrak{m}$, where $k \in \mathbb{R}$ is some constant.
Further, if $k = 0$, then t is an affine parameter for this geodesic. If $k \neq 0$, then $s = e^{-kt}$ is an affine parameter for the geodesic. The second case can occur only if the curve $\gamma(t)$ is a null curve in a (properly) pseudo-Riemannian space.

Remark 2.1. For the Riemannian homogeneous space, the diffeomorphism in Definition 2.1 is the identity map and the right-hand side of the formula (1) in Lemma 2.1 is zero. For more information about homogeneous geodesics in pseudo-Riemannian homogeneous spaces and further references we refer the reader to Refs. 2,4 or Ref. 6. From now on, we will concentrate only on the Riemannian situation.

Definition 2.2. A Riemannian homogeneous space $(G/H, g)$ is called a g.o. space if every geodesic of $(G/H, g)$ is homogeneous. A homogeneous Riemannian manifold (M, g) is g.o. manifold if $M = G/H$ is a g.o. space for $G = I_0(M)$. Here "g.o." means "geodesics are orbits".

Remark 2.2. For a homogeneous manifold M, it may happen that, for $G' \subsetneq G = I_0(M)$, G'/H' is not a g.o. space and G/H is a g.o. space (see Ref. 7). To investigate the properties of the *manifold* M, it is necessary to consider the full isometry group $G = I_0(M)$.

It is well known that all *naturally reductive* homogeneous spaces are g.o. spaces. Some decades ago, it was generally believed that also every g.o. space (and every g.o. manifold) is naturally reductive. The first counter-example comes from the work Ref. 8 by A. Kaplan. This is a six-dimensional Riemannian nilmanifold with a two-dimensional center, one of the so-called "generalized Heisenberg groups". The extensive study of Riemannian g.o. spaces started just with the Kaplan's paper. One of the techniques used for the characterization of g.o. spaces is based on the concept of "geodesic graph". The original idea (not using any explicit name) comes from the work Ref. 12 by J. Szenthe:

Definition 2.3. Let $(G/H, g)$ be a Riemannian g.o. space and let $\mathfrak{g} = \mathfrak{m} + \mathfrak{h}$ be an $\mathrm{Ad}(H)$-invariant decomposition of the Lie algebra $\mathfrak{g}$. A geodesic graph is an $\mathrm{Ad}(H)$-equivariant map $\eta\colon \mathfrak{m} \to \mathfrak{h}$ which is rational on an open dense subset U of $\mathfrak{m}$ and such that $X + \eta(X)$ is a geodesic vector for each $X \in \mathfrak{m}$.

According to Lemma 10 in Ref. 12, for a Riemannian g.o. space $(G/H, g)$, there exists at least one geodesic graph. The construction of canonical and general geodesic graphs is described in details in Ref. 9 or Ref. 5. In the present paper, we use only the canonical geodesic graph (which is usually denoted by ξ) and we explain the construction in Section 4.

On the open dense subset U of $\mathfrak{m}$ and with respect to the basis $E_1, \dots E_n$ of $\mathfrak{m}$ and the basis $F_1, \dots F_h$ of $\mathfrak{h}$, the components of a geodesic graph η are rational functions of the coordinates on $\mathfrak{m}$. They are of the form $\eta_k = P_k/P$, where P_k and P are homogeneous polynomials and $\deg(P_k) = \deg(P) + 1$.

Definition 2.4. Let $(G/H, g)$ be a Riemannian g.o. space and let $\mathfrak{g} = \mathfrak{m} + \mathfrak{h}$ be an $\mathrm{Ad}(H)$-invariant decomposition of the Lie algebra $\mathfrak{g}$. Let $E_1, \dots E_n$ and $F_1, \dots F_h$ be the bases of $\mathfrak{m}$ and $\mathfrak{h}$, respectively. Let $\eta\colon \mathfrak{m} \to \mathfrak{h}$ be a geodesic graph with the components $\eta_k = P_k/P$. The degree of the geodesic graph η is $\deg(\eta) = \deg(P)$. The degree of the g.o. space G/H is

$$\deg(G/H) = \min\{\deg(\eta) \colon \eta \text{ is a geodesic graph on } G/H\}.$$

Let (M, g) be a homogeneous Riemannian manifold. The degree of M is $\deg(M) = \min\{\deg(G/H) \colon M = G/H\}$.

Remark 2.3. According to Proposition 2.10 in Ref. 11, the manifold M is naturally reductive if there exists a linear geodesic graph (and, according to Definition 2.4 above, $\deg(M) = 0$).

For the examples of g.o. manifolds of degree 2, see Refs. 3,7,9. Up to now, there is just one example of a g.o. manifold of degree higher than 2. It is the 13-dimensional generalized Heisenberg group which admits 2 transitive groups of isometries, $G = I_0(M)$ and $G' \subsetneq G$. For this g.o. manifold, it holds $\deg(G'/H') = 6$, $\deg(G/H) = 3$ and $\deg(M) = 3$ (see Ref. 5).

In this paper, we are going to investigate the flag manifold $M = \mathrm{SO}(7)/\mathrm{U}(3)$ with the two-parameter family of Riemannian metrics and to show that $\deg(M) = 4$, with the exception of the metrics which are multiples of the standard one.

3. Flag manifolds

Definition 3.1 (see Ref. 1). Let G be a compact semisimple Lie group.

A homogeneous manifold $M = G/H$ is a flag manifold if the isotropy subgroup H is the centralizer of a torus in G.

For the description of flag manifolds using the painted Dynkin diagrams and also for further references, we refer the reader to Ref. 1. The flag manifolds of classical Lie groups are the following:

$$\begin{aligned}
A(\bar{n}) &= \mathrm{SU}(n)/\mathrm{S}(\mathrm{U}(n_1)\dots\mathrm{U}(n_s))\\
\bar{n} &= (n_1,\dots,n_s), n = n_1+\dots+n_s, n_1 \geq n_2 \geq \dots \geq n_s \geq 1\\
B(\bar{l}) &= \mathrm{SO}(2l+1)/(\mathrm{U}(l_1)\dots\mathrm{U}(l_k).\mathrm{SO}(2m+1))\\
C(\bar{l}) &= \mathrm{Sp}(l)/(\mathrm{U}(l_1)\dots\mathrm{U}(l_k).\mathrm{Sp}(m))\\
D(\bar{l}) &= \mathrm{SO}(2l)/(\mathrm{U}(l_1)\dots\mathrm{U}(l_k).\mathrm{SO}(2m))\\
\bar{l} &= (l_1,\dots,l_k,m), l=l_1+\dots+l_k+m, l_1 \geq l_2 \geq \dots \geq l_k \geq 1,\ k,m \geq 0.
\end{aligned}$$

Definition 3.2 (see Ref. 1). A flag manifold $M = G/H$ equipped with a G-invariant metric g is called a Riemannian flag manifold.

The main result in Ref. 1 claims that the only flag manifolds $M = G/H$ of a simple Lie group which admit non standard metrics with homogeneous geodesics are the manifolds $B(l,0) = \mathrm{SO}(2l+1)/\mathrm{U}(l)$ and $C(1,l-1) = \mathrm{Sp}(l)/\mathrm{U}(1).\mathrm{Sp}(l-1)$.

On these manifolds there is, up to a homothety, a 1-parameter family of invariant metrics with homogeneous geodesics. For one particular value of the parameter, the corresponding metric is the standard one, the full isometry group $\tilde{G}$ is $\mathrm{SO}(2l-2)$ (or, $\mathrm{SU}(2l)$, respectively) and the manifold with this metric is the symmetric space $\mathrm{SO}(2l+2)/\mathrm{U}(l+1)$ (or, $\mathrm{SU}(2l)/\mathrm{U}(2l-1)$, respectively). For other values of the parameter, the full isometry group of the corresponding metric is $\mathrm{SO}(2l+1)$ (or, $\mathrm{Sp}(l)$, respectively) and the manifold with this metric is not naturally reductive.

4. Riemannian flag manifold $SO(7)/U(3)$

Let us consider the algebra $\mathfrak{g} = \mathfrak{so}(7)$ and let us choose a basis $\{A,B,C,F,G,H,J,K,L,Z_1,\dots,Z_6,E_1,\dots,E_6\}$, such that every element $X \in \mathfrak{so}(7)$ whose coordinates with respect to the above basis are

$(a, b, c, f, g, h, j, k, l, z_j, x_i)$ is identified with the matrix

$$\begin{bmatrix} 0 & -a & -f-z_1 & -g-z_2 & -h-z_3 & -j-z_4 & x_1 \\ a & 0 & g-z_2 & -f+z_1 & j-z_4 & -h+z_3 & x_2 \\ f+z_1 & -g+z_2 & 0 & -b & -k-z_5 & -l-z_6 & x_3 \\ g+z_2 & f-z_1 & b & 0 & l-z_6 & -k+z_5 & x_4 \\ h+z_3 & -j+z_4 & k+z_5 & -l+z_6 & 0 & -c & x_5 \\ j+z_4 & h-z_3 & l+z_6 & k-z_5 & c & 0 & x_6 \\ -x_1 & -x_2 & -x_3 & -x_4 & -x_5 & -x_6 & 0 \end{bmatrix}. \tag{2}$$

We put $\mathfrak{h} = \text{span}\{A, B, C, F, G, H, J, K, L\}$. It can be easily verified that $\mathfrak{h} \simeq \mathfrak{u}(3)$. We obtain the reductive decomposition $\mathfrak{g} = \mathfrak{h} + \mathfrak{m}$. Now let us denote by $\mathfrak{z}, \mathfrak{v}$ the subspaces $\mathfrak{z} = \text{span}(Z_j), \mathfrak{v} = \text{span}(E_i)$ of $\mathfrak{m}$ and let us denote by A_{kl}, B_{kl} the operators on $\mathfrak{v}, \mathfrak{z}$ defined by the relations

$$A_{kl}E_i = \delta_{ik}E_l - \delta_{il}E_k, \qquad B_{kl}Z_j = \delta_{jk}Z_l - \delta_{jl}Z_k. \tag{3}$$

The adjoint action of the elements of $\mathfrak{h}$ on $\mathfrak{m} = \mathfrak{z} + \mathfrak{v}$ can be expressed via the operators (3) as

$$\begin{aligned} \text{ad}(A) &= B_{12} + B_{34} + A_{12}, \\ \text{ad}(B) &= B_{12} + B_{56} + A_{34}, \\ \text{ad}(C) &= B_{34} + B_{56} + A_{56}, \\ \text{ad}(F) &= B_{35} + B_{46} + A_{13} + A_{24}, \\ \text{ad}(G) &= B_{36} - B_{45} + A_{14} - A_{23}, \\ \text{ad}(H) &= -B_{15} - B_{26} + A_{15} + A_{26}, \\ \text{ad}(J) &= -B_{16} + B_{25} + A_{16} - A_{25}, \\ \text{ad}(K) &= B_{13} + B_{24} + A_{35} + A_{46}, \\ \text{ad}(L) &= B_{14} - B_{23} + A_{36} - A_{45}. \end{aligned} \tag{4}$$

For the Lie bracket on $\mathfrak{m}$ it holds

$$\begin{aligned} &[Z_1, Z_2] = 2(A+B), \\ &[Z_1, Z_3] = K, && [Z_2, Z_3] = -L, \\ &[Z_1, Z_4] = L, && [Z_2, Z_4] = K, && [Z_3, Z_4] = 2(A+C), \\ &[Z_1, Z_5] = -H, && [Z_2, Z_5] = J, && [Z_3, Z_5] = F, \\ &[Z_1, Z_6] = -J, && [Z_2, Z_6] = -H, && [Z_3, Z_6] = G, \\ &[Z_4, Z_5] = -G, && [Z_4, Z_6] = F, && [Z_5, Z_6] = 2(B+C), \end{aligned} \tag{5}$$

$$
\begin{aligned}
&[E_1,E_2]=A,\\
&[E_1,E_3]=1/2(F+Z_1),\ [E_2,E_3]=1/2(-G+Z_2),\\
&[E_1,E_4]=1/2(G+Z_2),\ [E_2,E_4]=1/2(F-Z_1),\ [E_3,E_4]=B,\\
&[E_1,E_5]=1/2(H+Z_3),\ [E_2,E_5]=1/2(-J+Z_4),[E_3,E_5]=1/2(K+Z_5),\\
&[E_1,E_6]=1/2(J+Z_4),\ [E_2,E_6]=1/2(H-Z_3),\ [E_3,E_6]=1/2(L+Z_6),\\
&[E_4,E_5]=1/2(-L+Z_6),[E_4,E_6]=1/2(K-Z_5),\ [E_5,E_6]=C.
\end{aligned}
\tag{6}
$$

The adjoint action of $\mathfrak{z}$ on $\mathfrak{v}$ can be described again via the operators as follows:

$$
\begin{aligned}
\mathrm{ad}(Z_1)|_{\mathfrak{v}} &= A_{13} - A_{24}, & \mathrm{ad}(Z_2)|_{\mathfrak{v}} &= A_{14} + A_{23},\\
\mathrm{ad}(Z_3)|_{\mathfrak{v}} &= A_{15} - A_{26}, & \mathrm{ad}(Z_4)|_{\mathfrak{v}} &= A_{16} + A_{25},\\
\mathrm{ad}(Z_5)|_{\mathfrak{v}} &= A_{35} - A_{46}, & \mathrm{ad}(Z_6)|_{\mathfrak{v}} &= A_{36} + A_{45}.
\end{aligned}
\tag{7}
$$

We introduce an $\mathrm{ad}(\mathfrak{h})$-invariant scalar product on $\mathfrak{m}$ by the orthogonal basis $\{Z_j, E_i\}$, where $\langle Z_j, Z_j\rangle = \beta$, $\langle E_i, E_i\rangle = \alpha$ and $\alpha, \beta > 0$. This scalar product induces a 2-parameter family of Riemannian metrics on the homogeneous space $G/H = \mathrm{SO}(7)/\mathrm{U}(3)$.

Now we are going to construct the canonical geodesic graph (which is the unique geodesic graph in this example). We write each vector $X \in \mathfrak{m}$ in the form

$$X = \sum_{i=1}^{6} x_i E_i + \sum_{j=1}^{6} z_j Z_j$$

and each vector $\xi(X) \in \mathfrak{h}$ in the form

$$\xi(X) = \xi_1 A + \xi_2 B + \xi_3 C + \cdots + \xi_9 L.$$

We consider the equation (1) in the form

$$\langle [X + \xi(X), Y]_{\mathfrak{m}}, X\rangle = 0, \tag{8}$$

where Y runs over all $\mathfrak{m}$. We have to determine the corresponding $\xi(X)$ to the given X. For $Y \in \mathfrak{m}$, we substitute, step by step, all 12 elements E_i, Z_j of the given basis into the formula (8). We obtain a system of 12 linear equations for the parameters $\xi_1, \ldots, \xi_9$, whose matrix $\mathbf{A}$ and the vector $\mathbf{b}$

of the right-hand sides are given by

$$\mathbf{A}=\begin{bmatrix} x_2 & 0 & 0 & x_3 & x_4 & x_5 & x_6 & 0 & 0 \\ -x_1 & 0 & 0 & x_4 & -x_3 & x_6 & -x_5 & 0 & 0 \\ 0 & x_4 & 0 & -x_1 & x_2 & 0 & 0 & x_5 & x_6 \\ 0 & -x_3 & 0 & -x_2 & -x_1 & 0 & 0 & x_6 & -x_5 \\ 0 & 0 & x_6 & 0 & 0 & -x_1 & x_2 & -x_3 & x_4 \\ 0 & 0 & -x_5 & 0 & 0 & -x_2 & -x_1 & -x_4 & -x_3 \\ z_2 & z_2 & 0 & 0 & 0 & -z_5 & -z_6 & z_3 & z_4 \\ -z_1 & -z_1 & 0 & 0 & 0 & -z_6 & z_5 & z_4 & -z_3 \\ z_4 & 0 & z_4 & z_5 & z_6 & 0 & 0 & -z_1 & z_2 \\ -z_3 & 0 & -z_3 & z_6 & -z_5 & 0 & 0 & -z_2 & -z_1 \\ 0 & z_6 & z_6 & -z_3 & z_4 & z_1 & -z_2 & 0 & 0 \\ 0 & -z_5 & -z_5 & -z_4 & -z_3 & z_2 & z_1 & 0 & 0 \end{bmatrix}, \mathbf{b}=\begin{bmatrix} c(z_1x_3+z_2x_4+z_3x_5+z_4x_6) \\ c(-z_1x_4+z_2x_3-z_3x_6+z_4x_5) \\ c(-z_1x_1-z_2x_2+z_5x_5+z_6x_6) \\ c(z_1x_2-z_2x_1-z_5x_6+z_6x_5) \\ c(-z_3x_1-z_4x_2-z_5x_3-z_6x_4) \\ c(z_3x_2-z_4x_1+z_5x_4-z_6x_3) \\ 0 \\ 0 \\ 0 \\ 0 \\ 0 \\ 0 \end{bmatrix},$$

where $c = \frac{\beta}{2\alpha} - 1$. The rank of this system is equal to 9. We select in a convenient way the 9 independent equations, for example, we omit the rows no. 1,2 and 7. Now we use the Cramer's rule and the computer (for this computation, the software Maple V, ©Waterloo Maple Inc., was used). We obtain the components of the solution of the above system in the form

$$\xi_k = \frac{\widetilde{P}_k}{\widetilde{P}}, \qquad k = 1\ldots 9, \tag{9}$$

where $\widetilde{P}_k$ and $\widetilde{P}$ are homogeneous polynomials in variables x_i and z_j. It holds $\deg(\widetilde{P}_k) - 1 = \deg(\widetilde{P}) = 9$. These polynomials have the common factor Q of degree 5. This common factor, as well as the polynomials $\widetilde{P}_k$ and $\widetilde{P}$, depends on our choice of the linearly independent rows from the above system. For our choice (for the omitted rows no. 1, 2 and 7) we have

$$\begin{aligned} Q = z_1 \big(& (-x_1 x_6 + x_2 x_5)(z_1 z_5 + z_2 z_6) \\ & + (x_1 x_5 + x_2 x_6)(z_1 z_6 - z_2 z_5) \\ & + (x_1 x_4 - x_2 x_3)(z_3 z_5 + z_4 z_6) \\ & + (x_1 x_3 + x_2 x_4)(-z_3 z_6 + z_4 z_5)\big). \end{aligned} \tag{10}$$

After cancelling out this common factor, we obtain the components of the canonical geodesic graph in the form

$$\xi_k = \left(\frac{\beta}{2\alpha} - 1\right)\frac{P_k}{P}, \qquad k = 1\ldots 9. \tag{11}$$

On the open dense subset $U \subset \mathfrak{m}$, where $P \neq 0$, the formula (11) describe the unique geodesic graph on G/H. For homogeneous polynomials P_k and P, it holds $\deg(P_k) - 1 = \deg(P) = 4$. The polynomial P in the denominators is an algebraic invariant with respect to the action of H on $\mathfrak{m}$. The Hilbert basis of the invariants is $\{I_1, I_2, I_3\}$, where

$$\begin{aligned} I_1 &= x_1^2 + x_2^2 + x_3^2 + x_4^2 + x_5^2 + x_6^2, \\ I_2 &= z_1^2 + z_2^2 + z_3^2 + z_4^2 + z_5^2 + z_6^2, \\ I_3 &= u_1^2 + u_2^2 \end{aligned} \tag{12}$$

and

$$\begin{aligned} u_1 &= x_1 z_6 + x_2 z_5 - x_3 z_4 - x_4 z_3 + x_5 z_2 + x_6 z_1, \\ u_2 &= x_1 z_5 - x_2 z_6 - x_3 z_3 + x_4 z_4 + x_5 z_1 - x_6 z_2 \end{aligned} \tag{13}$$

The denominator P can be then expressed as

$$P = I_1 I_2 - I_3. \tag{14}$$

We recall the conjecture stated in Ref. 5, that, for the components of a geodesic graph written in the form $\eta_k = P_k/P$, where P_k and P have no nontrivial common factor, the denominator P is *always* an algebraic invariant. The polynomials P_k are not invariants and their expressions are long, hence we write down only the terms which arise in the special case $x_4 = x_5 = x_6 = z_4 = z_5 = z_6 = 0$. We obtain

$$\begin{aligned} P_1 &= 2\,x_3\,x_2\,z_1^3 - 2\,x_3\,x_1\,z_2\,z_1^2 + \left(2\,x_3\,x_2\,z_2^2 + 2\,x_3\,x_2\,z_3^2\right) z_1 \\ &\quad - 2\,x_3\,x_1\,z_2^3 - 2\,x_3\,x_1\,z_3^2 z_2, \\ P_2 &= -2\,x_3\,x_2\,z_1^3 + 2\,x_3\,x_1\,z_2\,z_1^2 - 2\,x_3\,x_2\,z_2^2\,z_1 + 2\,x_3\,x_1\,z_2^3, \\ P_3 &= -2\,x_3\,x_2\,z_3^2\,z_1 + 2\,x_3\,x_1\,z_3^2 z_2, \\ P_4 &= \left(x_1^2 - x_2^2 + x_3^2\right) z_1^3 + 2\,x_2\,x_1\,z_2\,z_1^2 \\ &\quad + \left(\left(x_1^2 - x_2^2 + x_3^2\right) z_2^2 + \left(x_1^2 - x_2^2\right) z_3^2\right) z_1 \\ &\quad + 2\,x_2\,x_1\,z_2^3 + 2\,x_2\,x_1\,z_3^2 z_2, \\ P_5 &= -2\,x_2\,x_1\,z_1^3 + \left(x_1^2 - x_2^2 - x_3^2\right) z_2\,z_1^2 - 2\,x_2\,x_1\,\left(z_2^2 + z_3^2\right) z_1 \\ &\quad + \left(x_1^2 - x_2^2 - x_3^2\right) z_2^3 + \left(x_1^2 - x_2^2\right) z_3^2 z_2, \\ P_6 &= \left(x_1^2 - x_2^2 + x_3^2\right) z_3\,z_1^2 + \left(x_1^2 - x_2^2 - x_3^2\right) z_3\,z_2^2 \\ &\quad + \left(x_1^2 - x_2^2\right) z_3^3, \\ P_7 &= -2\,x_2\,x_1\,z_3\,z_1^2 - 2\,x_3^2 z_3\,z_2\,z_1 - 2\,x_2\,x_1\,\left(z_3\,z_2^2 - z_3^3\right), \\ P_8 &= -2\,x_3\,x_2\,z_3\,z_2\,z_1 + 2\,x_3\,x_1\,z_3\,z_2^2, \\ P_9 &= -2\,x_3\,x_2\,z_3\,z_1^2 + 2\,x_3\,x_1\,z_3\,z_2\,z_1 \end{aligned} \tag{15}$$

and

$$P = \left({z_1}^2 + {z_2}^2 + {z_3}^2\right)\left({x_1}^2 + {x_2}^2 + {x_3}^2\right) - {z_3}^2{x_3}^2. \tag{16}$$

It is clear that P_k and P have no longer a common factor.

The case $2\alpha = \beta$ corresponds to a multiple of the standard metric on the given manifold and the geodesic graph is the zero map in this case. Clearly, the manifold is naturally reductive. Moreover, according to the results in Ref. 1 mentioned earlier, it is the symmetric space. In the case $2\alpha \neq \beta$, the geodesic graph of degree 4 given by the formula (11) is the unique geodesic graph and we obtain $\deg(M) = 4$.

Acknowledgments

The author was supported by the grant GAČR 201/05/2707 and by the research project MSM 6198959214 financed by MŠMT.

References

1. D. Alekseevsky and A. Arvanitoyeorgos, Riemannian flag manifolds with homogeneous geodesics, *Trans. Amer. Math. Soc.* **359** (2007) 3769–3789.
2. Z. Dušek, Almost g.o. spaces in dimensions 6 and 7, to appear in *Adv. Geom.*
3. Z. Dušek, Explicit geodesic graphs on some H-type groups *Rend. Circ. Mat. Palermo*, Serie II, Suppl. **69** (2002) 77–88.
4. Z. Dušek and O. Kowalski, Light-like homogeneous geodesics and the Geodesic Lemma for any signature, *Publ. Math. Debrecen* **71** (2007) 245–252.
5. Z. Dušek and O. Kowalski, Geodesic graphs on the 13-dimensional group of Heisenberg type, *Math. Nachr.* **254-255** (2003) 87–96.
6. Z. Dušek and O. Kowalski, On 6-dimensional pseudo-Riemannian almost g.o. spaces, *J. Geom. Phys.* **57** (2007) 2014–2023.
7. Z. Dušek, O. Kowalski and S. Nikčević, New examples of Riemannian g.o. manifolds in dimension 7, *Differential Geom. Appl.* **21** (2004) 65–78.
8. A. Kaplan, On the geometry of groups of Heisenberg type, *Bull. London Math. Soc.* **15** (1983) 35–42.
9. O. Kowalski and S. Nikčević, On geodesic graphs of Riemannian g.o. spaces, *Archiv der Math.* **73** (1999) 223–234; Appendix: *Archiv der Math.* **79** (2002) 158–160.
10. O. Kowalski and J. Szenthe, On the existence of homogeneous geodesics in homogeneous Riemannian manifolds, *Geom. Dedicata* **81** (2000) 209–214, Erratum: *Geom. Dedicata* **84** (2001) 331–332.
11. O. Kowalski and L. Vanhecke, Riemannian manifolds with homogeneous geodesics, *Boll. Un. Math. Ital. B(7)* **5** (1991) 189–246.
12. J. Szenthe, Sur la connection naturelle à torsion nulle, *Acta Sci. Math. (Szeged)* **38** (1976) 383–398.

Differential Geometry and its Applications
Proc. Conf., in Honour of Leonhard Euler, Olomouc, August 2007

Global properties of the Ricci flow on open manifolds

Jürgen Eichhorn

Institut für Mathematik und Informatik, Universität Greifswald,
D-17487 Greifswald, Germany
E-mail: eichhorn@uni-greifswald.de

We study global properties of open Riemannian manifolds, like the existence of a spectral gap or the existence of L_p–characteristic numbers under the Ricci flow on open complete manifolds and show that these properties are in fact preserved.

Keywords: Ricci flow, spectral evolution, L_2-cohomology.

MS classification: 53C21, 53C20, 58J35.

1. Introduction

Since the spectacular work of Hamilton, Perelman and others the Ricci flow is in the focus of many differential geometers, global analysts and topologists. Given $(M^n, g = g(0) = g_0)$, the Ricci flow $g_t = g(t)$ is the solution of the evolution equation

$$\frac{\partial g(t)}{\partial t} = -2\,\mathrm{Ric}(g(t)), \quad g(0) = g_0, \tag{1}$$

in local coordinates

$$\frac{\partial g_{ij}(t)}{\partial t} = -2\,\mathrm{Ric}(g_{ij}(t)), \quad g_{ij}(0) = g_{0,ij}, \tag{2}$$

if it exists and is unique. As discussed already by Hamilton, the Ricci flow can be considered as the "natural" evolution of a Riemannian manifold. It is the solution of a good "forward" evolution equation, and this evolution is mostly transversal to the action of the diffeomorphism group.

For $n = 3$ and M^3 closed, the study of the Ricci flow leads to a proof of Thurston's geometrization conjecture, in particular to a proof of the famous Poincare conjecture.

For $n \geq 3$ arbitrary and M^n open, the analysis of (1) is much more complex and complicated. Already short time existence and uniqueness represent themselves as big challenges and to draw geometric and topological consequences similar to $n = 3$ and the closed case is even much more complicated.

Nevertheless, the Ricci flow remains the natural evolution for the metric and it is interesting for applications in global analysis, which global properties of (M^n, g_0) remain invariant under the Ricci flow. For this, we must fix those global properties, we have in mind. This concerns the character of the spectral value zero which is strongly connected with L_2–cohomology, and the L_2–property of the curvature tensor. Moreover the evolution of the infimum of the spectrum is of great interest.

The paper is organized as follows. In section 2, we present a collection of essential existence and uniqueness theorems which are needed in the sequel. Section 3 is devoted to the evolution of some important spectral properties of the character of the spectral value zero and the infimum of the spectrum. Finally in section 4, we prove that the L_2–property of the curvature is an invariant of the Ricci flow. This, in particular assures the existence and invariance of L_2–characteristic numbers under the Ricci flow.

2. Review of existence and uniqueness theorems

Not to speak in the forthcoming sections about the empty set, we need a well established existence and uniqueness for the Ricci flow on open manifolds. This has been done in particular by Shi,[9–11] Chen/Zhu[1] and others.

For clarity and common understanding we give here such a review. The first deep and substantial theorems have been established by Shi in 1989. For the proof we refer to Refs. 9–11. We denote by Rm the Riemannian curvature tensor, by Ric the Ricci tensor, by R scalar curvature and by K sectional curvature.

Theorem 2.1. *Let $(M^n, g = g_0)$ be open, complete $|Rm| \leq C$. Then there exists $T(n, c) > 0$ such that*

$$\begin{aligned} \frac{\partial g_{ij}(x,t)}{\partial t} &= -2R_{ij}(x,t) \\ g_{ij}(x,0) &= g_{0,ij}(x), \end{aligned} \tag{3}$$

has a smooth solution $g_{ij}(x,t)$ for $0 \leq t \leq T$ satisfying

$$\sup_{x \in M} |\nabla^k Rm(x,t)| \leq \frac{C_k}{t^k}, \quad k \geq 0, \quad 0 < t \leq T. \tag{4}$$

For the proof, Shi considers an exhaustion of M^n by compact domains D_k, with smooth boundary ∂D_k, $D_1 \subset D_2 \subset \cdots, \bigcup\limits_k D_k = M$, and considers the modified equation (Dirichlet problem)

$$\begin{aligned}\frac{\partial}{\partial t}\hat{g}_{ij}(k,x,t) &= -2\hat{R}_{ij}(k,x,t) + \nabla_i V_j \nabla_j V_i, \quad x \in D_k \\ \hat{g}_{ij}(k,x,0) &= g_{0,ij}(x) \text{ for } x \in D_k \\ \hat{g}_{ij}(x,t) &= g_{0,ij}(x,t) \text{ for } x \in \partial D_k \\ V_i &= \hat{g}_{ik} g^{\beta\gamma}(\hat{\Gamma}^k_{\beta\gamma} - \Gamma^k_{0,\beta\gamma}).\end{aligned} \tag{5}$$

(4) is a strictly parabolic system having a unique smooth solution $\hat{g}_{ij}(x,t)$ for $0 \leq t \leq t(n, C_0)$.

Defining $\varphi_t(x)$ by $\varphi_t(x) := y(x,t)$, $\frac{\partial y^\alpha}{\partial t} = \frac{\partial y^\alpha}{\partial}\hat{g}^{\beta\gamma}(\hat{\Gamma}^k_{\beta\gamma} - \Gamma^k_{0,\beta\gamma})$ and setting $g_t = (\varphi_t^*)^{-1}\hat{g}_t$, one obtains that $g_{ij}(x,t) = g_{ij}(k,x,t)$ satisfies the equation

$$\begin{aligned}\frac{\partial}{\partial t} g_{ij}(k,x,t) &= -2R_{ij}(k,x,t), \\ g_{ij}(k,x,0) &= g_{0,ij}(x).\end{aligned}$$

Now Shi establishes first and second order derivative estimates and obtains by an Arzela–Ascoli argument the desired g. We refer to Ref. 9 for details and the proof of (4).

Corollary 2.1. *If (M^n, g) is open, complete and has bounded sectional curvature, hen the Ricci flow produces for $t > 0$ metrics whose curvature is bounded up to an arbitrarily high order.*

Remark 2.1. If $T(n, C_0) > 1$ then for $t \longrightarrow T$, the bounds for $|\nabla^k Rm|$ become smaller and smaller with increasing k, i.e. the Ricci flow "smoothes out" the curvature tensor.

It is well known that the Riemannian curvature tensor Rm splits for $n \geq 3$ under the action of $O(n)$ as

$$\begin{aligned}Rm &= W + V + U \\ R_{ijkl} &= W_{ijkl} + V_{ijkl} + U_{ijkl},\end{aligned} \tag{6}$$

where

$$\begin{aligned}U_{ijkl} &= \frac{1}{n(n-1)} R(g_{ik}g_{jl} - g_{il}g_{jk}), \\ V_{ijkl} &= \frac{1}{n-2}(\mathring{R}_{ik}g_{jl} - \mathring{R}_{il}g_{jk} - \mathring{R}_{ij}g_{il} + \mathring{R}_{jl}g_{ik}), \\ \mathring{R}_{ij} &= R_{ij} - \frac{1}{n} R g_{ij}\end{aligned}$$

the traceless Ricci part, and

$$W_{ijkl} = R_{ijkl} - V_{ijkl} - U_{ijkl}$$

the Weyl conformal curvature tensor.

Set $\overset{\circ}{R}m = \{\overset{\circ}{R}_{ijkl}\} = \{R_{ijkl} - U_{ijkl}\}$.

Theorem 2.2. *Let $(M^n, g = g_0)$, $n \geq 3$, be open, complete. Suppose that for any $c_1, c_2, \delta > 0$ there exists $\varepsilon = \varepsilon(n, c_1, c_2, \delta) > 0$ such that*

a) $\mathrm{vol}(B_\gamma(x)) \geq c_1\gamma^n$ *for all* $x \in M$ *and*

b) $|\overset{\circ}{R}m|^2 \leq \varepsilon R^2$, $0 < R < c_2/\mathrm{dist}(x, x_0)^{2+\delta}$ *for all* $x \in M$.

Then

$$\frac{\partial g_{ij}}{\partial t} = -2R_{ij}(t),$$
$$g_{ij}(0) = g_{0,ij},$$

has a solution for all time $0 \leq t < \infty$ and $g_{ij}(t) \underset{t\to\infty}{\longrightarrow} g_{ij}(\infty)$, $g_{ij}(\infty)$ a smooth metric satisfying $R_{ijkl}(\infty) \equiv 0$ on M. Here convergence means C^∞-convergence.

We refer to Ref. 10 for the proof.

In the case $n = 3$ the Ricci flow even permits to indicate the topological type.

Theorem 2.3. *Let (M^3, g_0) be open, complete with $0 \leq \mathrm{Ric}(g_0) \leq k_0$. Then M^3 is diffeomorphic to a quotient space of one of the spaces $\mathbb{R}^3$,or $S^2 \times \mathbb{R}^1$ by a group of fixed point free isometries of the standard metrics.*

We refer to Ref. 11 for the proof.

Concerning complete open Kähler manifolds, we mention a short time and a long time existence result, established by Shi in Ref. 12.

Theorem 2.4. *Let (M^n, g_0) be open, complete with $|Rm(g_{0,\alpha\overline{\beta}})| \leq k_0$. Then the equation*

$$\frac{\partial g_{\alpha\overline{\beta}}}{\partial t}(x,t) = -R_{\alpha\overline{\beta}}(x,t), \quad x \in M, \quad t > 0$$
$$g_{\alpha\overline{\beta}}(x,0) = g_{0,\alpha\overline{\beta}}x), \quad x \in M,$$

has a maximal solution $g_{\alpha\overline{\beta}}(\cdot,t)$ on $[0, t_{\max}[$ with $t_{\max} > 0$ which remains Kähler. If additionally the initial metric has positive holomorphic bisectional curvature, the evolving metric has still positive holomorphic bisectional curvature.

Theorem 2.5. *Suppose $(M, g_{\alpha,\overline{\beta}})$ is a complete noncompact Kähler manifold with bounded and positive bisectional curvature. And suppose there exist positive constants C_2 and $0 < \theta < 2$ such that*

$$\frac{1}{\mathrm{vol}B_r(x_0, g(0))} \int\limits_{B_r(x_0,g(0))} R(x,0)dx \leq \frac{C_2}{(1+r)^\theta}$$

for all $x_0 \in M$, $0 \leq r < \infty$. Then the Ricci flow has a solution for all $t \in [0, \infty[$.

Concerning the uniqueness, Chen and Zhu proved in Ref. 1 the following theorem

Theorem 2.6. *Let (M^n, g_0) be open, complete with bounded sectional curvature. If $g(t)$ and $\overline{g}(t)$ are two solutions of the Ricci flow on $M^n \times [0,T]$ with $g(0) = g_0 = \overline{g}(0)$ and with bounded sectional curvature, then $g(t) = \overline{g}(t)$ on $M \times [0,T]$*

3. Evolution of some spectral properties

Before studying evolution of spectral properties, we need some facts concerning the evolution of curvature sign.

Hamilton proved already in Ref. 3, that on closed 3–manifolds the Ricci flow preserves the nonnegativity of the Ricci curvature and the sectional curvature. But in Ref. 8 Ni gives examples of complete manifolds (M^n, g_0), $n \geq 4$, such that the Ricci flow does not preserve the nonnegativity of the sectional curvature. In the very general case (admitting arbitrary solutions $g(t)$), a certain control of sectional curvature $K_{g(t)}$ is given by the following theorem of Kapovitch.[5]

Theorem 3.1. *Let (M^n, g_0) be open, complete with |sectional curvature $K_{g_0}| \leq 1$. Then for the Ricci flow $g(t)$ on $[0,T]$ there holds*

$$\inf K_{g_0} - C(n,T)t \leq K_{g_t} \leq \sup K_{g_0} + C(n,T)t.$$

It is absolutely clear and can be supported by simple examples that the spectrum $\sigma(\Delta_q)$ of the Laplace operators changes under the Ricci flow. On open manifolds, it is quite another question, whether the appearance of certain components of the spectrum like point spectrum σ_p, its purely discrete part σ_{pd}, continuous spectrum σ_c, absolutely continuous spectrum σ_{ac}, essential spectrum σ_e, is an invariant property under the Ricci flow. For application in PDE theory, the property of the spectral value 0 is of particular meaning.[2] For the sake of clarity, we recall a definition. Denote by

$$\sigma_{c,R}(\Delta_q) := \sigma(\Delta_q) \setminus \sigma_p(\Delta_q)$$

the resolvent continuous spectrum. It can be different from

$$\sigma_c(\Delta_q) = \sigma(\Delta_q|_{(L_2)_c}),$$

the spectrum of the Laplace operator restricted to the continuous subspace of L_2 which is defined as those $\omega \in L_2$ for which $d\langle E_\lambda \omega, \omega\rangle$ is a purely continuous measure.

There holds

$$\sigma_c \setminus (\sigma_c \cap \overline{\sigma}_p) \subset \sigma_{c,R}.$$

Theorem 3.2. *Let $(M^n, g_0 = g(0))$ be open, complete, $|Rm(g_0)| \le k_0$, and let $g(t)$ be the Ricci flow according to 2.1 and 2.5, $0 \le t \le T(n, k_0)$. Then there holds*

a) $0 \in \sigma_p(\Delta_q(g_0))$ *if and only if* $0 \in \sigma_p(\Delta_q(g(t))$,

b) $\Delta_q(g(o))$ *has a spectral gap above zero if and only if this holds for* $\Delta_q(g(t))$,

c) $0 \in \sigma_{c,R}\Delta_q(g(0))$ *if and only if this holds for* $\Delta_q(g(t))$.

Proof. According to theorem 2.1,

$$\sup_{x\in M} |\nabla^m Rm(g(t))|_x \le C_m/t^m, \quad 0 < t \le T(n, k_0).$$

In particular,

$$|Rm(g(t))|^2 \le C_0, \quad 0 \le t \le T(n, k_0).$$

Moreover

$$\begin{aligned}
&\left|\frac{\partial}{\partial t} g_{ij}\right|^2 = 4|R_{ij}|^2 \le 4n^2 |Rm(g(t))|^2 \le 4n^2 C_0, \\
&\left|\frac{\partial}{\partial t} g_{ij}\right| \le 2n\sqrt{C_0}, \quad -2n\sqrt{C_0} g_{ij} \le \frac{\partial}{\partial t} g_{ij} \le 2n\sqrt{C_0} g_{ij}, \quad 0 \le t \le T, \\
&e^{-2n\sqrt{C_0}t} g_{ij}(x,0) \le g_{ij}(x,t) \le e^{2n\sqrt{C_0}t} g_{ij}(x,0), \quad 0 \le t \le T,
\end{aligned} \tag{7}$$

in the sense of quadratic forms, i.e. $g(0)$ and $g(t)$ are quasi isometric. This implies for non–reduced and reduced L_2–cohomology

$$\begin{aligned}
H^{*,2}(M, g(0)) &= H^{*,2}(M, g(t)), \quad 0 \le t \le T, \\
\overline{H}^{*,2}(M, g(0)) &= \overline{H}^{*,2}(M, g(t)), \quad 0 \le t \le T,
\end{aligned}$$

and

$$\begin{aligned}&\overline{(\text{im }(d_q : D_{d_q} \longrightarrow L_2(\Omega^{q+1}))}/\text{im } d_q)(g(0))\\&= \overline{(\text{im }(d_q : D_{d_q} \longrightarrow L_2(\Omega^{q+1}))}/\text{im } d_q)(g(t)), \;\; 0 \le t \le T.\end{aligned}$$

But $0 \in \sigma_p(\Delta_q)$ if and only if $\overline{H}^{q,2} \neq 0$. This proves a). Δ_q has a spectral gap above zero if and only if im Δ_q = im $\Delta_q|_{(\ker\Delta_q)^\perp}$ is closed. But this is the case if and only if im d_{q-1}, im d_q are closed, hich is equivalent to $H^{q,2} = \overline{H}^{q,2}$, $H^{q+1,2} = \overline{H}^{q+1,2}$. This proves b). $\lambda \in \sigma_{c,R}(\Delta_q)$ if and only if $\overline{\text{im }\Delta_q - \lambda} = L_2$ and im $(\Delta_q - \lambda)$ is properly contained in L_2. But for $\lambda = 0$ this is equivalent to $\ker\Delta_q = 0$, im Δ_q not closed and $\overline{\text{im }\Delta_q} = L_2$. The latter is equivalent to $\overline{H}^{q,2} = 0$ and at least one of $\overline{\text{im } d_{q-1}}/\text{im } d_{q-1}$, $\overline{\text{im } d_q}/\text{im } d_q$ is $\neq \{0\}$. These 3 conditions are independent of t, according to (7), c) is done. □

As we already mentioned, for $n \ge 4$ the Ricci flow does not preserve positive sectional curvature (for open M^n). For applications in this section, we must consider this question more carefully and add some assumptions.

Following Ref. 10, we formulate

Assumption A. Suppose

$$\begin{aligned}&\frac{\partial g_{ij}}{\partial t}(x,t) = -2R_{ij}(x,t) \text{ on } M \times [0,T],\\&0 < R_{ijij}(x,0) \le k_0, \;\; x \in M, \;\; 0 \le t \le T,\\&|R_{ijkl}(x,t)|^2 \le c_1, \;\; x \in M, \;\; 0 \le t \le T.\\&\int_0^T |\nabla_p R_{ijkl}(x,t)|dt \le c_2, \;\; x \in M\end{aligned}$$

Proposition 3.1. *Under the assumption A.*

$$0 < R(x,t) \le n^2\sqrt{c_1} \text{ on } M \times [0,T],$$

i.e. the scalar curvature remains positive and bounded.

Main idea of proof. Apply the maximum principle 4.6 of Ref. 10, p. 318 for open manifolds to

$$\frac{\partial R}{\partial r} = \Delta R + 2|\operatorname{Ric}|^2.$$

We refer to Ref. 10, p. 327 for details.

Remark 3.1. Unfortunately this does not hold for the sectional curvature.

We recall the definition of the curvature operator $R^{op}m$,

$$R^{op}m(e_l^* \wedge e_k^*) = \sum_{1\le i,j\le n} (Rm(e_l, e_k)e_j, e_i)_g e_i^* \wedge e_j^*,$$

or, what is equivalent,

$$(R^{op}m\varphi, \psi)_g = R_{ijkl}\varphi_{ij}\psi_{kl}.$$

denote $R^{op}m \ge 0$ by $R^{op}_{ijkl} \ge 0$ or simply $R_{ijkl} \ge 0$.

Remark 3.2. $R^{op}m \ge c$ implies $K \ge 0$. The converse is wrong: $P^n(\mathbb{C})$, $P^n(\mathbb{H})$, $P^2(Ca)$ have $1 \ge K \ge \frac{1}{4}$ but the smallest eigenvalue of $R^{op}m$ is 0.

For later use, we recall some inequalities from Ref. 4 and Ref. 10.

Proposition 3.2. *Assume assumption A and $R^{op}_{ijkl}(g(0)) \ge 0$ on $(M^n, g(0))$. Then $R^{op}_{ijkl}(g(x,t)) \ge 0$ on $M \times [0,T]$. If even $R^{op}_{ijkl}(0) \ge 0$ then $R^{op}_{ijkl}(t) \ge 0$ on $M \times [0,T]$.*

We refer to Ref. 10 for the proof.

Remind $\mathring{R}m = Rm - U$.

Lemma 3.1. *If $|\mathring{R}m|^2 \le \delta_n(1-\varepsilon)^2 \frac{2}{n(n-1)} R^2$, where $\varepsilon > 0$, $\delta_3 > 0$, $\delta_4 = \frac{1}{5}$, $\delta_5 = \frac{1}{10}$ and $\delta_n = \frac{2}{(n-2)(n+1)}$ for $n \ge 6$, then*

$$R^{op}m \ge 2\varepsilon \frac{R}{n(n-1)}.$$

We refer to Ref. 4 for the proof.

Corollary 3.1. *Let $\beta_n \le \frac{\delta_n}{2n(n-1)}$ and suppose $|\mathring{R}m|^2 \le \beta_n R^2$. Then*

$$R^{op}m \ge \frac{R}{n(n-1)}.$$

Lemma 3.2. *$|\mathring{R}m|^2 \le \beta_n R^2$ implies*

$$|Rm|^2 \le [\beta_n + \frac{2}{n(n-1)}]R^2$$

on M.

Proof.

$$|Rm|^2 = |\mathring{R}m|^2 + \frac{2}{n(n-1)} R^2.$$

□

Theorem 3.3. *Let (M^n, g_0) be open, complete,*

$$|\mathring{R}m|^2 \le \beta_n R^2, \quad 0 < R \le c_0.$$

Then the Ricci flow exists for $0 < t < \infty$.

Outline of proof. First one has

$$|Rm(x,0)|^2 \le [\beta_n + \frac{2}{n(n-1)}]c_0^2$$

on M. Corollary 3.1 implies $R^{op}m(x,0) > 0$, altogether

$$0 < R_{ijij}(x,0) \le [\beta_n + \frac{2}{n(n-1)}]^{\frac{1}{2}} n^2 c_0.$$

From theorem 2.1 we infer the existence of a short time solution on $0 \le t \le T_0$, $\sup_{x \in M} |\nabla^m R_{ijkl}(x,t)|^2 \le c_{m+1}(n, c_0)/t^m$.

Lemma 3.3. *This solution satisfies assumption A.*

Thereafter one performs iteration, the procedure does not stop.

Corollary 3.2. *The solution according to lemma 3.3 still satisfies*

$$R^{op}_{ijkl}(x,t) > 0,$$
$$0 < R(x,t) \le C,$$
$$|\mathring{R}m|^2 \le \beta_n R^2,$$
$$\sup_{x \in M} |\nabla^m R_{ijkl}(x,t)|^2 \le C_m.$$

Outline of the proof of theorem 2.2. One shows: There exists $c_3(n, \varepsilon, c_1, c_2)$ such that

$$R(x,t) \le c_3 \left(\frac{1}{1+t}\right)^{1+\varepsilon_1}, \quad 0 \le t < \infty,$$

where $\varepsilon_1 = \frac{n\delta}{8(2+\delta)} > 0$ and $|\mathring{R}m| \le \varepsilon R^2$ by assumption. Hence $R(x,t) \underset{r \to \infty}{0}$, $Rm(x,t) \underset{r \to \infty}{0}$.

We come now to some questions concerning the spectral evolution.

Denote $\lambda_0 \equiv \lambda_0(\Delta(g(t))) := \inf \sigma(\Delta(t))) =$ smallest spectral value of the Laplace operator acting on functions. We ask: How evolves $\lambda_0(t)$? This is an extraordinary complicated question and we present here only some results for small n.

We proved in theorem 3.1.

Proposition 3.3. $\lambda_0(\Delta_q(0)) = 0$ *if and only if* $\lambda_0(\Delta_q(t))0$.

Corollary 3.3. *If* $\lambda_0(t^*) > 0$ *or some* t^* *then* $\lambda_0(t) > 0$ *for all* t, $0 \le t \le T$.

By the Raleigh–Ritz variational principle,

$$\lambda_0 = \inf_{\substack{\varphi \in C_c^\infty(M) \\ |\varphi|_{L_2}=1}} \frac{\langle \Delta\varphi, \varphi\rangle_{L_2}}{|\varphi|^2_{L_2}} = \inf_{\substack{\varphi \in C_c^\infty(M) \\ |\varphi|_{L_2}=1}} \frac{|\nabla\varphi|^2_{L_2}}{|\varphi|^2_{L_2}}.$$

Theorem 3.4. *Suppose* (M^2, g_0) *open, complete,* $K(M^2, g_0) \le -c$, $c > 0$, $\{g(t)\}_{t \ge 0}$ *the bounded Ricci flow and suppose* $\lambda_0(0) > 0$. *Then there exist* $\varepsilon > 0$ *such that* $\lambda_0(t)$ *is decreasing for* $t \in [0, \varepsilon]$.

Proof. Let $\Omega \subset M^2$ be a bounded domain with smooth boundary, $\lambda_0(0) = \lambda_0(\Omega, g_0)$ the first non–zero eigenvalue with Dirichlet boundary conditions, $\Delta\varphi_0 = \lambda_0\varphi_0$, $|\varphi_0|_{L_2} = 1$. Consider the equation $\Delta(t)\varphi_0(t) = \lambda_0(t)\varphi_0(t)$. An easy calculation (cf. Ref. 7) yields for

$$\lambda_0' = -2\int_\Omega R_{ij}\varphi_{0ij}\varphi_0 \,\mathrm{dvol}(g(t)).$$

In the case $n = 2$, $R_{ij} = \frac{1}{2}Rg_{ij}$, $R = 2K$, hence

$$\lambda_0' = -\int_\Omega Rg_{ij}\varphi_{0ij}\varphi_0 \,\mathrm{dvol} = \int_\Omega 2K\Delta\varphi_0 \cdot \varphi_0 \,\mathrm{dvol} = 2\lambda_0 \int_\Omega K|\varphi_0|^2 \,\mathrm{dvol},$$

which is ≥ 0 if $K \ge 0$, ≤ 0 if $K \le 0$. But, according to proposition 3.1,

$$\inf K(g_0) - C(n,T)t \le K(g(t)) \le \sup K(g_0) + C(n,T).$$

In our case, we have $K(g(t)) \le 0$ if $-c + C(n,T)t \le 0$, i.e. $t \le \frac{c}{C(n,T)} = \varepsilon$. Under this condition, $\lambda_0(\Omega, g(t))$ is decreasing for all Ω. But

$$\lambda_0(M^2, g(t)) = \inf_\Omega \lambda_0(\Omega, (t)). \qquad \square$$

Next we consider the case $n = 3$.

Theorem 3.5. *Let* (M^3, g_0) *be open, complete,* $|\mathring{R}m(g_0)| \le \frac{1}{12}R^2(g_0)$, $0 < R \le c_0$, *and let* $\{g(t)\}_t$ *be the bounded Ricci flow. Then* $R_{ij}(t) - \frac{R}{3}g_{ij} \ge 0$. *If in some time intervall* $[t_1, t_2]$

$$R_{ij}(g(t)) - \frac{R}{3}g_{ij}(t) \ge \frac{1}{6}Rg_{ij}(t)$$

then $\lambda_0(t)$ *is increasing,* $t_1 \le t \le t_2$.

Proof. We obtain from the preceding statements that $R^{op}m \geq \frac{1}{6}R$, $0 \leq t < \infty$, $0 < R < C$. Let again $\Omega \subset M^3$ be a compact domain with smooth boundary, $\lambda_0(t) = \lambda_C(\Omega, g(t))$. According to Ref. 6,

$$\frac{\lambda_0'}{2} = \frac{\lambda_0}{2} \int\limits_\Omega R\varphi_0^2 \,\mathrm{dvol} - \int\limits_\Omega (R_{ij} - \frac{R}{2} g_{ij})\varphi_{0i}\varphi_{0j} \,\mathrm{dvol},$$

$$R_{ij} - \frac{R}{2} g_{ij} = (R_{ij} - \frac{R}{3} g_{ij}) - \frac{1}{6} R g_{ij}.$$

The assumption yields $\lambda_0(\Omega, g(t))$ is increasing, the some holds for $\lambda_0(M^3, g(t))$. □

Remark 3.3. For n, q arbitrary, $\Omega \subset M^n$ compact with smooth boundary and standard boundary conditions, the following is clear.

$$\Delta_q \omega_0 = \Delta \omega_0 = \lambda_0 \omega_0$$

immediately implies

$$\lambda_0' = \int\limits_\Omega (\Delta' \omega_0, \omega_0)_{g(t)} \,\mathrm{dvol}.$$

The point is to find appropriate estimates for

$$(\Delta' \omega, \omega)_{g(t)}.$$

In the case $q = 1$,we have in local coordinates

$$\begin{aligned} \Delta' &= (-g^{kl}\nabla_k\nabla_l - R_i^i)' = -g'^{kl}\nabla_k\nabla_l - g^{kl}\nabla_k'\nabla_l - g^{kl}\nabla_k\nabla_l' + (R_j^i)' \\ &= -2R^{kl}\nabla_k\nabla_l - g^{kl}\nabla_k'\nabla_l' - g^{kl}\nabla_k\nabla_l' + 2R^{ik}R_{jk} - 2g^{ik}R_{jk}', \end{aligned}$$

where $\nabla_k' = (\Gamma_{ik}^j)' = g^{j\gamma}(\nabla_\gamma R_{ik} - \nabla_i R_{j\gamma} - \nabla_k R_{j\gamma}$. These expressions already indicate, how difficult it would be to establish meaningful results concerning monotonicity of $\lambda_0(\Delta_q)$, $q \geq 1$.

We conclude this section with some considerations concerning the evolution of L_2–cohomology classes and their norm.

Consider (M^n, g) open, complete, and the reduced and unreduced L_p–cohomology

$$\overline{H}^{q,p}(g) = Z^{q,p}(g)/\overline{B^{q,p}}(g), \quad H^{q,p}(g) = Z^{q,p}(g)/B^{q,p}(g).$$

Suppose that (M^n, g_0) has bounded curvature. Then there exists the bounded Ricci flow $\{g(t)\}_{0 \leq t \leq T}$, and there holds

$$\overline{H}^{q,p}(M^n, g(0)) = H^{q,p}(M^n, g(t)).$$

For $[\varphi] \in \overline{H}^{q,p}(g(t))$,

$$||[\varphi]||_{p,g(t)} = \inf_{\psi \in [\varphi]} |\psi|_{p,g(t)}$$

is a norm in the Banach space $\overline{H}^{q,p}(g(t))$. In the case $p = 2$

$$||[\varphi]||_{2,g(t)} = |\varphi_h|_{2,g(t)},$$

where $\varphi_h \in [\varphi]$ is the unique L_2–harmonic representative of $[\varphi]$, $\overline{H}^{q,p}(g(t)) \cong \mathcal{H}^{q,2}(g(t))$ canonically.

We remark that the single L_2–cohomology classes of $\overline{H}^{q,p}$ remain invariant under the bounded Ricci flow since $\overline{B^{q,2}L_2(g_0)} = \overline{B^{q,2}L_2(g(t))}$, i. e. the classes do not evolve. But the minimizer $\varphi_h \in [\varphi] \in \overline{H}^{q,p}$ can change. If $\varphi_h = \varphi_h(g(t_1))$ is harmonic, i.e. $d\varphi_h(g(t_1)) = 0 = *_{g(t_1)} d *_{g(t_1)} \varphi_h(g(t_1)) = 0$, then there is no argument visible that $*_{g(t_2)} d *_{g(t_2)} \varphi_h(g(t_2)) = 0$. Moreover the norm $||[\varphi]||_{2,g(t)}$ can evolve.

We construct an evolution for each single $\varphi \in [\varphi]$ and control its L_2–norm. This yields a control of the L_2–norm of $[\varphi]$.

Lemma 3.4. *Let $\varphi_0 = \varphi(0)$, $[\varphi_0] \in H^{q,2}(g(0))$, $(M^n, g(0))$ with bounded curvature and $\{g(t)\}_{0 \leq t \leq T}$ the bounded Ricci flow. Then the equation*

$$\frac{\partial \varphi}{\partial t} = -\Delta(g(t))\varphi, \quad \varphi(0) = \varphi_0 \tag{8}$$

has a unique solution in L_2 and

$$[\varphi(t)] = [\varphi(0)].$$

Proof. Consider the equation for the $(q-1)$–form f,

$$\frac{\partial f}{\partial t} = -\Delta(g(t))f - \delta(g(t))\varphi_0, \quad f(0) = 0.$$

According to proof of theorem 4.1, this equation has a unique solution f which is $\in L_2$. Set $\varphi = df + \varphi_0$. Then

$$\begin{aligned} \frac{\partial f}{\partial t} &= \frac{\partial}{\partial t}(\varphi_0 - df) = \frac{\partial}{\partial t} df = d\left(\frac{\partial f}{\partial t}\right) = d((-\Delta)f - \delta\varphi_0) \\ &= (-\Delta)(df + \varphi_0) = (-\Delta)\varphi \end{aligned}$$

and $\varphi(0) = 0$, $[\varphi] = [\varphi_0 + df] = [\varphi_0]$. □

Corollary 3.4. *The map $[\varphi_0] \ni \psi \longrightarrow \psi(t) \in [\varphi_0]$ is an affine isomorphism.*

Lemma 3.5. *Let $\{g(t)\}_{0\le t\le T}$ be the bounded Ricci flow as above, $\varphi_0 \in L_2(\Lambda^q, g(0))$ and $\varphi = \varphi(t)$ satisfying the differential equation*

$$\frac{\partial}{\partial t}\varphi = (-\Delta)\varphi, \quad \varphi(0) = \varphi_0.$$

If $u(t) = |\varphi(t)|^2_{g(t)}$ and $q = 1$, then $u(t)$ satisfies the differential inequality

$$\frac{\partial u}{\partial t} \le \left(-\frac{1}{2}\Delta\right) u.$$

If $q > 1$ then $u(t) = |\varphi(t)|^2_{g(t)}$ satisfies this differential inequality if $R^{op}m(g(t)) > 0$.

Proof. We recall the Weitzenboeck formulas,

$$q = 1, \quad (\Delta\varphi)_i = -\nabla^r\nabla_r\varphi_i + R\varphi.$$
$$q > 1, \quad (\Delta\varphi)_{i_1\ldots i_q} = -\nabla^r\nabla_r\varphi_{i_1\ldots i_q} + \sum_h R^r_{i_h}\varphi_{i_1\ldots r\ldots i_q} + \sum_{h<l} R^{r\ s}_{i_h\ i_l}\varphi_{i_1\ldots r\ldots s\ldots i_q}.$$

Hence we can write in the case $q = 1$

$$(\Delta\varphi, \varphi)_{g(t)} = (\nabla^*\nabla\varphi, \varphi)_{g(t)} + F_1(\varphi),$$

in the case $q > 1$,

$$(\Delta\varphi, \varphi)_{g(t)} = (\nabla^*\nabla\varphi, \varphi)_{g(t)} + F_{q,1}(\varphi) + F_{q,2}(\varphi),$$

where

$$F_1(\varphi) = R_{ij}\varphi^i\varphi^j = R^{ij}\varphi_i\varphi_j,$$
$$F_{q,1}(\varphi) = \frac{1}{(q-1)!}R^{rs}\varphi_r^{i_2\ldots i_q}\varphi_{si_2\ldots i_q}$$
$$\text{and } F_{q,2}(\varphi) = \frac{q-1}{2}R^{rsjk}\varphi_r^{rsi_3\ldots i_q}\varphi^{jh}_{i_3\ldots i_q}.$$

We set for $q > 1$ $F_q = F_{q,1} + F_{q,2}$. For any q–form φ there holds

$$\frac{1}{2}\Delta(|\varphi|^2) = (\Delta\varphi, \varphi) - |\nabla\varphi|^2 - F_q(\varphi).$$

Now we obtain for $q > 1$

$$\frac{1}{2}\frac{\partial}{\partial t} = \frac{1}{2}\frac{\partial}{\partial t}(\varphi,\varphi)_{g(t)} = \frac{1}{2}\frac{\partial}{\partial t}(g^{i_1 j_1}\cdots g^{i_q j_q}\varphi_{i_1\ldots i_q}\varphi_{j_1\ldots j_q})$$
$$= R^{i_1 j_1} g^{i_2 j_2}\cdots g^{i_q j_q}\varphi_{i_1\ldots i_q}\varphi_{j_1\ldots j_q} + \cdots + R^{i_q j_q} g^{i_1 j_1}\cdot g^{i_{q-1}, j_{q-1}}\varphi_{i_1\ldots i_q}\varphi_{j_1\ldots j_q}$$
$$+ \left(\frac{\partial}{\partial t}\varphi,\varphi\right)_{g(t)} = F_{q,1}(\varphi) + F_{q,2}(\varphi) - F_{q,2}(\varphi) - (\Delta\varphi,\varphi)_{g(t)}$$
$$= -[(\Delta\varphi,\varphi) - F_q(\varphi)] - F_{q,2}(\varphi) = -\frac{1}{2}\Delta|\varphi|^2 - |\nabla\varphi|^2 - F_{q,2}(\varphi)$$
$$\leq -\frac{1}{2}\Delta|\varphi|^2 = -\frac{1}{2}\Delta u,$$

if $R^{op}m(g(t)) \geq 0$ (since then $F_{q,2}(\varphi) \geq 0$). In the case $q = 1$, we obtain the same inequality by setting $F_{q,2} = 0$. □

Proposition 3.4. *Assume $|Rm(g(0))|$ bounded, $\{g(t)\}_{0\leq t\leq T}$ the bounded Ricci flow, $R(g(t)) \geq 0$. If $u(\cdot,t)$ is a non–negative solution of the scalar differential inequality*

$$\frac{\partial u}{\partial t} \leq -\Delta u$$

and $u(\cdot,0)$ is $\in L_1$, then $u(\cdot,t)$ remains $\in L_1$ and

$$u(t) \leq u(s)$$

for $t > s$.

We refer to Ref. 6 for the proof.

Remark 3.4. $R(g(t)) \geq 0$ is satisfied if we assume $R(g(0)) \geq 0$ and assumption A.

Theorem 3.6. *Suppose $(M^n, g(0))$ open, complete, $|Rm(g(0))|$ bounded, $R(g(0)) \geq 0$. Let $\{g(t)\}_{0\leq t\leq T}$ be the bounded Ricci flow and assume additionally $R(g(t)) \geq 0$. Let $0 \neq [\varphi] \in \bar{H}^{q,2}(M^n(g(0)))$ be an L_2–cohomology class. $[\varphi]$ remains invariant under the Ricci flow.*

a) *If $q = 1$, then*

$$|[\varphi]|_{L_2(g(t))} \leq |[\varphi]|_{L_2(g(s))}$$

for $t > s$.

b) *If $q > 1$ and $R^{op}m(g(t)) \geq 0$, then*

$$|[\varphi]|_{L_2(g(t))} \leq |[\varphi]|_{L_2(g(s))}$$

for $t > s$.

c) *If* $0 \neq [\varphi] \in \overline{H}^{q\,2}(g(0))$, $\varphi_h \in [\varphi]$ *is the unique harmonic representative and* φ_h *evolves according to*

$$\frac{\partial}{\partial t}\varphi_h(t) = -\Delta(g(t))\varphi_h(t), \quad \varphi_h(0) = \varphi_h,$$

then

$$|\varphi_h(t)|_{L_2(g(t))} \leq |\varphi_h(s)]|_{L_2(g(s))} \leq |\varphi_h|_{L_2(g(0))}$$

for $t > s$.

Proof. According to corollary 3.4, the map $[\varphi] \ni \psi(0) \longrightarrow \psi(t) \in [\varphi]$ is an affine isomorphism, which according to lemma 3.5, implies the differential inequality $\frac{\partial u}{\partial t} \leq -\frac{1}{2}\Delta u$ for $u(t) = |\psi(t)|^2_{g(t)}$. Proposition 3.4 then yields the norm inequality

$$|\psi(t)|_{L_2(g(t))} \leq |\psi(s)|_{L_2(g(s))}$$

for $t > s$. Taking the infimum over all elements of $[\varphi]$ finally yields the assertions. □

Corollary 3.5. *The assertions of theorem 3.6 are true in the following cases*

a) $|Rm(g(0))|^2 \leq \beta_n R^2(g(0))$, $0 < R^2(g(0)) \leq c$. *Then theorem 3.6 holds for* $0 \leq t < \infty$.

b) $n = 2$, $0 < K \leq c$.

4. Evolution of the L_2–property of the curvature

If M^n is closed, oriented then we have well–defined characteristic numbers. By the Chern–Weil construction, these can be expressed as an integral over an n–form with curvature coefficients.

In the case $n = 2m$ even, the Euler number $\chi(M^n)$ can be expressed e.g. as

$$\chi(M^n) = \chi(M^n, g) = \frac{2}{\mathrm{vol}(S^n)} \int e(g),$$

where

$$e(g) = \frac{1}{n!} \sum \varepsilon^{i_1 \dots i_n} \Omega^{i_1}_{i_2} \wedge \Omega^{i_3}_{i_4} \wedge \cdots \wedge \Omega^{i_{n-1}}_{i_n},$$

$\varepsilon^{i_1 \dots i_n}$ = sign of the permutation, Ω^i_j the curvature form, $R(\cdot,\cdot)E_i = \Omega^i_j E_j$. We use that $f \in L_p$ and bounded implies $f \in L_q$ for $q \geq p$. It is well known that in the closed case characteristic number (M^n, g) = characteristic number $(M^n, g_{)}$. In the open case, this is wrong. At first the integral version does not exist in general and moreover it can depend on g. Cheeger/Gromov gave an example of metrics g_0, g_1 on $\mathbb{R}^{2m}$, $m \geq 2$ such that

$$\chi(\mathbb{R}^{2m}, g_0) = 1 \neq 0 = \chi(\mathbb{R}^{2m}, g_1).$$

There arise the following natural questions.

1) If for (M^n, g_0) characteristic numbers are defined and $\{g(t)\}_{0 \leq t \leq T}$ is the bounded Ricci flow, are they defined also for $(M^n, g(t))$ and

2) do they coincide?

For $n \geq 4$ these questions are essentially equivalent to the following question.

Does $Rm(g(0)) \in L_2$ imply $Rm(g(t)) \in L_2$, with other words, does $\int\limits_M |Rm(g(0))|^2 \,\mathrm{dvol}(g(0))$ imply $\int\limits_M |Rm(g(t))|^2 \,\mathrm{dvol}(g(t)) < \infty$?

Theorem 4.1. *Let $(M^n, g(0))$ be open, complete, $Rm(g(0))$ bounded and $\in L_2$, $\{g(t)\}_{0 \leq t \leq T}$ the bounded Ricci flow. Then $Rm(g(t)) \in L_2$.*

Proof. The evolution equations for $|Rm|$ or $|Rm|^2$, resp., are

$$\frac{\partial}{\partial t}|Rm|^2 = [-\Delta|Rm|^2 - 2|\nabla Rm|^2 + 8R_{empq}R^{ejqh}R^{mp}_{jh} + 2R_{eqmp}R^{eqih}R^{mp}_{ih}] \quad (9)$$

and

$$\frac{\partial}{\partial t}|Rm| = \frac{1}{2}\frac{1}{|Rm|}[\ ]. \quad (10)$$

We introduce as usual the following abbreviation. If A and B are tensors, then we denote by $A*B$ a linear combination of terms formed by contraction on $A_{i\ldots j}B_{k\ldots l}$ using the g^{ik}.

Then one can write more general

$$\frac{\partial}{\partial t}\nabla^k Rm = -\Delta(\nabla^k Rm) + \sum_{i+j=k} \nabla^i Rm * \nabla^j Rm, \quad (11)$$

$$\frac{\partial}{\partial t}|\nabla^k Rm|^2 = -\Delta|\nabla^k Rm|^2 - 2|\nabla^{k+1} Rm|^2 + \sum_{i+j=k} \nabla^i Rm * \nabla^j Rm * \nabla^k Rm. \quad (12)$$

All these equations (9) – (12) are non–linear evolution equations of the type

$$\frac{\partial}{\partial t}u = -\Delta(t)u + \text{non–linear terms}.$$

If $Rm(g(0))$ is bounded, then there exists a solution of (9) – (12), according to theorem 2.1. Hence the point for us is not the existence of a solution but its properties. The whole problem consists in he fact that $\Delta = \Delta(g(t))$, i.e. in local coordinates the coefficients of the elliptic operator on the right hand side still (strongly) depend on t, i.e. the standard non–linear parabolic theory is not applicable.

To prove theorem 4.1, we must very briefly establish an appropriate framework. Let H be a Hilbert space and $V \hookrightarrow H \hookrightarrow V'$ be a Gelfand triple. Set

$$W_2^1(0,T) = \{f \in L_2(]0,T[,V) | \frac{df}{dt} \in L_2(]0,T[,V')\}$$

and

$$\langle f,g\rangle_W := \int_0^T \langle f(t),g(t)\rangle_V dt + \int_0^T \langle \tfrac{df}{dt}, \tfrac{dg}{dt}\rangle_{V'} dt.$$

Suppose for $t \in]0,T[$ to be given a sesquilinear form $a(t,\varphi,\psi)$ with
a) $a(t,\varphi,\psi)$ is for fixed $\varphi,\psi \in V$ measurable on $[0,T]$,
b) there exists $c > 0$ (independent of t) such that

$$|a(t,\varphi,\psi)| \le c \cdot |\varphi|_V \cdot |\varphi|_V \text{ for all } t,\varphi,\psi,$$

c) there exist $\alpha > 0$, $k_0 > 0$ (independent of t) such that

$$Re a(t,\varphi,\psi) + k_0|\varphi|_H^2 \ge a|\varphi|_V^2.$$

We infer from b) that there exist a representation $L(t)$ such that

$$a(t,\varphi,\psi) = \langle L(t)\varphi,\psi\rangle_H.$$

□

Lemma 4.1. *The assumptions a), b) imply that*

$$L(t) : L_2(]0,T[,V) \longrightarrow L_2(]0,T[,V')$$

is linear and continuous.

Let be $f \in L_2(]0,T[,V')$, $y_0 \in H$.

Proposition 4.1. *The problem*

$$L(t)y + \frac{dy}{dt} = f, \;\; y(0) = y_0, \tag{13}$$

has for $T < \infty$ a unique solution $y \in W_2^1(0,T)$ which continuously depends on f and y_0.

We refer to Ref. 13 for the proof which relies on a Picard–Lindelöf technique or the Banach fixed point theorem.

Now we set $y(t) = |Rm(g(x,t))|$, $f(t) = 0$, $L(t)y(t) = -$ r.h.s. of (10), $H = L_2(M, g(0)), H(t) = L_2(M, g(t)), 0 \leq t \leq T$. The quasiisometry of $g(0)$ and $g(t)$ imples that there are constants c_1, c_2 independent of t, such that

$$c_1|\ |_H \leq |\ |_{H(t)} \leq c_2|\ |_H, \quad 0 \leq t \leq T, \tag{14}$$

Set

$$\begin{aligned} V(t) &= \mathring{W}_2^1(M, \Delta(t)) = W_2^1(M, \Delta(t)) = W_2^1(M, \nabla^{g(t)}) \equiv W_2^1(t), \\ V'(t) &= W_2^{-1}(M, \Delta(t)) = W_2^{-1}(M, \nabla^{g(t)}). \end{aligned}$$

We need that $W_2^1(M, \nabla^{g(t)})$ is independent of t, i.e. there exist d_1, d_2 independent of t such that

$$d_1|\ |_{W_2^1(t_1)} \leq |\ |_{W_2^1(t_2)} \leq d_2|\ |_{W_2^1(t_1}, \quad 0 \leq t_1, t_2 \leq T. \tag{15}$$

Having in mind (14) and the quasiisometry of the $g(t)$s, (15) would follow if we could prove

$$|\nabla^{g(t)} - \nabla^{g(0)}| \equiv |\nabla(t) - \nabla(0)| \text{ bounded}, \quad 0 \leq t \leq T. \tag{16}$$

It remains to prove (16). $\nabla(t) - \nabla(0) = \Gamma_{ij}^k(t) - \Gamma_{ij}^k(c)$. The $\Gamma_{ij}^k(t)$ satisfy the system of ordinary differential equations

$$\frac{\partial}{\partial t}(\Gamma_{ij}^k(t) - \Gamma_{ij}^k(0)) = \frac{\partial}{\partial t}\Gamma_{ij}^k = -g^{kl}(\nabla_i R_{jl} + \nabla_j R_{il} - \nabla_l R_{ij}). \tag{17}$$

The right hand side of (17) is bounded on $0 \leq t \leq T$ and satisfies a Lipschitz condition w.r.t. t. The latter follows from

$$\frac{\partial}{\partial t} R_{ij} = -\Delta R_{ij} + 2R_{piqj}R^{pq} - 2g^{pq}R_{pi}R_{qj}.$$

Hence the system (4.9) for $\Gamma_{ij}^k(t) - \Gamma_{ij}^k(0)$ has a unique bounded solution $0 \leq t \leq T$. Set

$$a(t, \varphi, \psi) = \int_M \varphi\psi d\mu(t) + \int_M (\nabla\varphi, \nabla\psi)_{g(t)} d\mu(t).$$

There exist constants c_3, c_4 such that

$$c_3 d\mu(0) \leq d\mu(t) \leq c_4 d\mu(0). \tag{18}$$

Then we obtain from (14), (15), (18) with $H = L_2(M, g(0))$, $V = W_2^1(M, g(0))$

$$|a(t, \varphi, \psi)| \leq |\varphi|_{V(t)} \cdot |\psi|_{V(t)} \leq c_5|\varphi|_{V(0)} \cdot |\psi|_{V(0)} = c_5|\varphi|_V \cdot |\psi|_V$$

and for any $k_0 > 0$

$$a(t, \varphi, \psi) + k_0 \langle \varphi, \varphi \rangle_{L_2(g(t))} \geq |\varphi|^2_{V(t)} \geq c_6 |\varphi|^2_{V(0)} = c_6 |\varphi|^2_V.$$

Hence the problem (13) has a unique solution $\in W_2^1(0, T)$ and the solution of (10) is $\in L_2$.

Corollary 4.1. *Suppose* $(M^n, g(0))$ *open, complete,* $Rm(g(0))$ *bounded and* $\in L_2$ *and let* $\{g(t)\}_{0 \leq t \leq T}$ *be the bounded Ricci flow. Then* L_2*–characteristic numbers are defined and independent of* t, $0 \leq t \leq T$.

Proof. The existence immediately follows from theorem 4.1. The invariance follows from several other papers of the author. □

Applying exactly the same considerations to equation (12) we obtain

Theorem 4.2. *Suppose the hypotheses of theorem 4.1 and additionally* $\nabla^i(0) Rm(g(0)) \in L_2$, $1 \leq i \leq k$. *Then* $\nabla^i(g(t)) Rm(g(t)) \in L_2$, $1 \leq i \leq k$, $0 \leq t \leq T$.

Remark 4.1. In the same manner we get the unique solution of equation (2 in section 3) and $\varphi \in L_2$.

References

1. B.-L. Chen, X.-P. Zhu, Uniqueness of the Ricci flow on complete noncompact anifolds; arXiv:math.DG/0505447v3.
2. J. Eichhorn, The zero–in–the–spectrum conjecture and its modifications, preprint Greifswald University, to appear.
3. R.S. Hamilton, Three–manifolds with positive Ricci curvature, *J Diff. Geom.* **17** (1982) 255–306.
4. G. Huisken, Ricci deformation of the metric on an Riemannian manifold, *J. Diff. Geom.* **21** (1985) 47–62.
5. V. Kapovitch, Curvature bounds via Ricci smoothing; arXiv:math.DG/0405569v1.
6. L. Ma, Y. Yang, L_2–forms and Ricci flow with bounded curvature on complete non–compact manifolds; arXiv:math.DG/0509237v1.
7. L. Ma, Y. Yang, Eigenvalue monotonicity for the Ricci Hamilton flow, *Ann. Glob. Anal. Geometr.* **29** (2006) 287–297.
8. L. Ni, Ricci flow and nonnegativity of curvature; arXiv:math.DG/0305246v1.
9. W.-X. Shi, Deforming the metric on complete Riemannian manifolds, *J. Diff. Geom.* **30** (1989) 223–301.
10. W.-X. Shi, Ricci deformation of the metric on complete noncompact Riemannian manifolds, *J. Diff. Geom.* **30** (1989) 303–394.
11. W.-X. Shi, Complete noncompact three–manifolds with nonnegative curvature, *J. Diff. Geom.* **29** (1989) 353–360.

12. W.-X. Shi, Ricci flow and the uniformization on complete noncompact Kähler manifolds, *J. Diff. Geom.* **45** (1997) 94–220.
13. Wloka, *Partielle Differentialgleichungen* (Teubner Publishers, Stuttgart 1982).

Differential Geometry and its Applications
Proc. Conf., in Honour of Leonhard Euler, Olomouc, August 2007

Gauss maps and symmetric spaces

J.-H. Eschenburg

Institut für Mathematik, Universität Augsburg,
D-86135 Augsburg, Germany
E-mail: eschenburg@math.uni-augsburg.de
www.math.uni-augsburg.de

How much of the geometry of a full submanifold $M^m \subset \mathbb{R}^n$ is encoded by the induced metric and the Gauss map? We discuss this question in a number of cases. In particular we give a new proof for a theorem of Nikolaevskii claiming that $M^m \subset \mathbb{R}^n$ is rigid if $\mathbb{R}^n$ contains also a full, irreducible extrinsic symmetric submanifold M_o of the same dimension $m < n-1$ and the Gauss image of M is a subset of the Gauss image of M_o: then M (up to translation and scaling) is an open subset of M_o.

Keywords: Submanifolds, Fundamental Forms, Symmetric R-Spaces, Group Representations.

MS classification: 53A07, 53A10, 53C35, 53C24.

1. Introduction

In the last three decades, the theory of submanifolds in euclidean n-space gained momentum by two sources: the relationship between holonomy and symmetry (cf. Ref. 1) and the loop group methods in connection with the theory of integrable systems (cf. Ref. 16). In the present article we deal mainly with the first one which is connected to a classical problem: Give geometric characterizations of distinguished objects.

One of the classical tools for the theory of m-dimensional submanifolds M in euclidean n-space $\mathsf{E} = \mathbb{R}^n$ is the *Gauss map* N. It assigns to each point $x \in M \subset \mathsf{E}$ its normal space $N_x = (T_x)^\perp$ and takes values in the Grassmannian $\mathsf{Gr} = \mathsf{Gr}_k(\mathbb{R}^n)$ of all (possibly oriented) k-dimensional linear subspaces in $\mathbb{R}^n$ where $k = n - m$ is the codimension of M. We briefly discuss the question if the submanifold can be recovered from its Gauss map, and we ask for classes of submanifolds where this happens. Then we restrict attention to a class of submanifolds where much more is valid: these submanifolds are recovered just from the image of their Gauss maps.

In the hypersurface case $k = 1$, the range of the Gauss map is the m-sphere which has the same dimension as M. In higher codimension we always have $\dim M < \dim \mathsf{Gr}$, but sometimes the values of N actually lie in an m-dimensional totally geodesic subspace of Gr; e.g. this happens for complex hypersurfaces of $\mathbb{C}^p$ where N takes values in $\mathbb{CP}^{p-1} \subset \mathsf{Gr}_2(\mathbb{R}^{2p})$. Further, if $M \subset \mathsf{E}$ is an *extrinsic symmetric space*, i.e. it is invariant under the reflection at each of its normal spaces, then the Gauss image of M is also totally geodesic. According to Nikolaevskii,[15] there are no other totally geodesic Gauss images in Gr, and the last mentioned case is completely rigid:

Theorem 1.1. *Let $M_o \subset \mathsf{E}$ be an m-dimensional (full) irreducible extrinsic symmetric submanifold other than the sphere S^m, and let $N_o : M_o \to \mathsf{Gr}$ be its Gauss map. Let $M \subset \mathsf{E}$ be any submanifold of the same dimension m with non-constant Gauss map $N : M \to \mathsf{Gr}$. Assume $N(M) \subset N_o(M_o)$. Then up to translations and scaling, M is an open subset of M_o.*

In the present paper we give a conceptual proof of this part of Nikolaevskii's theorem (the original proof involves extended matrix computations with many different cases). This purely local theorem is very remarkable: It determines a (small piece of a) submanifold only from its Gauss image. A similar statement for real or complex hypersurfaces would be obviously false; the Gauss image is always contained in the sphere or complex projective space which does not restrict the shape of the submanifold.

The paper is organized as follows. In sections 2–5 we introduce the "main players": fundamental forms, Gauss maps and extrinsic symmetric spaces. The theorem is proved in Sections 6 and 7. One of the main tools for the proof is Lemma 2 which was also used in Ref. 2 and goes back to Naitoh (cf. Ref. 14). Note that all our considerations are local, therefore we do not have to distinguish between embeddings and immersions.

It is a pleasure to thank V. Matveev for hints and discussion.

2. Fundamental forms and Gauss map

Differential geometry began with the study of submanifolds, curves and surfaces in euclidean 3-space, and even today these are the main objects used to explain general geometric ideas. The theory of submanifolds M in euclidean n-space $\mathsf{E} = \mathsf{E}^n$ (cf. Ref. 1 for a recent approach) has two different aspects: inner and outer geometry. *Inner geometry* is based on the interior distance between points in M where the distance is measured by arc length of curves within M while *outer geometry* describes how tangent or normal spaces move from point to point inside the ambient space. They correspond to the two main invariants, the first and second fundamental forms g and α:

$$g(v,w) = \langle v,w\rangle, \quad \alpha^\xi(v,w) = \langle \partial_v w, \xi\rangle = -\langle w, \partial_v \xi\rangle = \langle A_\xi v, w\rangle$$

for any normal field ξ and any two tangent vector fields v, w on M. For the simplest objects of differential geometry, the planar curves, the two invariants are just numbers, arc length and curvature. C.F. Gauss derived a relation between the two fundamental forms which is now called *Gauss equation*:

$$R = \alpha \wedge \alpha \tag{1}$$

where R is the *Riemannian curvature tensor* (a second order expression in g). It was created by Gauss' student B. Riemann who isolated the first aspect and created an inner geometry on manifolds which no longer need to be embedded. The *existence and uniqueness theorem for submanifolds* says that one only needs to prescribe the two fundamental forms on M in order to encode the full submanifold geometry; in fact, the relations between the fundamental forms discovered by Gauss, Codazzi and Ricci are necessary and sufficient to warrant an embedding which is unique up to rigid motions.

Yet in some sense there is a more fundamental invariant for submanifolds in euclidean space, $M \subset \mathsf{E}$, expressing directly the point-dependence of the normal spaces. This is the *Gauss map* which has been studied before by Euler and others. It assigns to each point $x \in M$ its normal space $N_x := N_x M = (T_x M)^\perp \subset \mathsf{E}$. For surfaces in E^3 and more generally for m-dimensional submanifolds of E^{m+1} (hypersurfaces) the normal space is just a line, i.e. an element of real projective m-space $\mathbb{RP}^m$, and if the normal lines are orientated, the Gauss map takes values in the m-sphere S^m, the 2-fold cover of $\mathbb{RP}^m$. Gauss had a practical reason to study this map: Being the astronomer Royal of the kingdom of Hannover (now the German state Niedersachsen), he was responsible for the surveying of this country 1821–1825, and the Gauss map of the earth surface (the zenith direction) is important for surveying since it determines the geographical coordinates. The triangulation used by Gauss was shown on the last German 10 Mark bill.

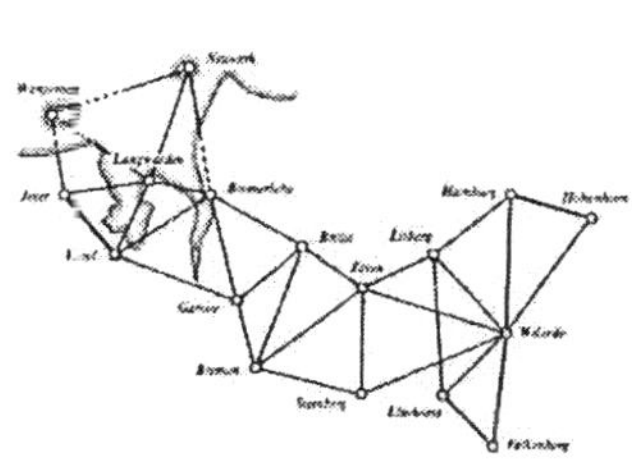

In general, the Gauss map of a submanifold $M^m \subset \mathsf{E}^n$ is a map into the Grassmannian $\mathsf{Gr} = \mathsf{Gr}_k(\mathsf{E}^n)$ of k-dimensional linear subspaces of E^n where $k = n - m$ is the codimension; if M is oriented, one may put Gr the *oriented* Grassmannian.[a] Since the derivative of this map $N : M \to \mathsf{Gr}$ measures how the normal space is moving inside the ambient space, it should be essentially the second fundamental form. In fact, the tangent vectors of Gr at $N_x \in \mathsf{Gr}$ are the linear maps $f : N_x \to N_x^\perp = T_x$,[b] and in particular, the tangent vector $\partial_v N \in T_{N_x}\mathsf{Gr}$ for any $v \in T_x$ is the linear map

$$\partial_v N : N_x \ni \xi \mapsto (\partial_v \xi)^T = -A_\xi v \in T_x. \tag{2}$$

3. "Inverting" the Gauss map?

We want to consider the following problem: To which extend does the Gauss map of a submanifold $M \subset \mathsf{E}$ determine its shape? More precisely, given the Gauss map $N : M \to \mathsf{Gr}$ together with the induced metric on M (first fundamental form), can we recover the embedding $M \hookrightarrow \mathsf{E}$? On the first sight, the answer seems easy: Since the second fundamental form is the derivative of N, both fundamental forms are given and the submanifold is determined by the existence and uniqueness theorem. But this is false! Counterexamples are obtained from *minimal surfaces*, i.e. surfaces in E^3 with $H := \frac{1}{2}\,\text{trace}\,\alpha = 0$. They allow an isometric deformation preserving the Gauss map; the best known example is the deformation of the catenoid into the helicoid.[c] Hence the same metric and Gauss map on the parameter manifold may allow several non-congruent isometric embeddings.[d]

What was wrong with the argument? The given data consist of an abstract Riemannian manifold M and a smooth map $N : M \to \mathsf{Gr}$. In order

---

[a]The *oriented Grassmannian* consists of the k-dimensional oriented subspaces of E (each subspace appears doubly, with the two possible orientations). It is a 2-fold cover of the ordinary Grassmannian; for $k = 1$ this is the covering $\mathsf{S}^m \to \mathbb{R}P^m$.

[b]Subspaces near N_x are graphs of linear maps $f : N_x \to N_x^\perp = T_x$.

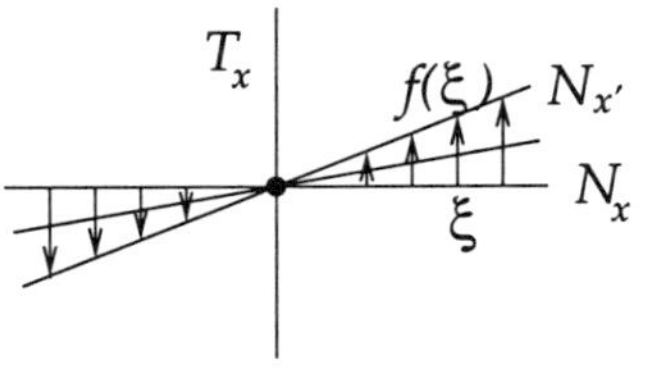

[c]http://en.wikipedia.org/wiki/Catenoid

[d]However, this cannot happen for non-minimal surfaces in E^3: The Weingarten map A (being a 2×2-matrix) satisfies $0 = A^2 - (\text{trace}\ A)A + (\det A)I = A^2 - 2HA + KI$. From g and N we obtain K and $A^2 = dN^t dN$ and $H^2 = \frac{1}{4}\,\text{trace}\,(A^2) + \frac{1}{2}K$, and if $H \neq 0$, we recover $A = \frac{1}{2H}(A^2 + KI)$. Thus the embedding is uniquely determined by g and N.

to obtain the second fundamental form from (2), we need to identify the abstract tangent bundle TM with the subbundle $N^\perp \subset M \times \mathsf{E}$. The corresponding orthogonal isomorphism $F : TM \to N^\perp$ needs to be parallel with respect to the Levi-Civita connection on TM and the projection connection on $N^\perp$, and moreover, the expression $(\partial_v(Fw))^\perp = \alpha(v,w)$ must be symmetric for any two tangent vector fields v, w on M.[e] For generic data it is impossible to construct such a map. The precise obstructions for a map $N : M \to \mathsf{Gr}$ to be the Gauss map of an embedding of M seem to be unknown.

We can find positive answers for certain classes of submanifolds. One such class is formed by the surfaces in E^3 with fixed nonzero constant mean curvature H, say $H = \frac{1}{2}$. Their Gauss map is harmonic,[18] and for each S^2-valued harmonic map N on a simply connected two-dimensional Riemann surface M there is (up to translations) exactly one immersion $M \to \mathsf{E}^3$ with $H = \frac{1}{2}$ and Gauss map N. Ironically it is the same isometric deformation which provides the counterexamples in the minimal surface case ($H = 0$) and which in the case $H = \frac{1}{2}$ is used to reconstruct the surface from its Gauss map.[f] In higher dimensions and codimensions we only know one class of examples which we are going to describe next.

If $M \subset \mathsf{E}$ is a hypersurface ($k = 1$), then M and $\mathsf{Gr} = \mathsf{S}^m$ have the same dimension m, a very important geometric property: recall that the Gauss-Kronnecker curvature of M is the determinant of dN which has many implications. This is no longer true for codimension $k \geq 2$ since $\dim \mathsf{Gr} = \dim \mathrm{Hom}(N, T) = km > m = \dim M$. But in some cases we are able to find a totally geodesic submanifold $Q \subset \mathsf{Gr}$ with $\dim Q = m$ and $N(M) \subset Q$. E.g. if $\mathsf{E} = \mathbb{R}^{2p} = \mathbb{C}^p$ and $M \subset \mathbb{C}^p$ is a complex submanifold with dimension $m = 2p - 2$ (*complex hypersurface*), then N_x is always a complex line and thus N takes values in complex projective space $\mathbb{CP}^{p-1}$ which is a totally geodesic m-dimensional submanifold of $\mathsf{Gr}_2(\mathbb{R}^{2p})$.

In fact, Nikolaevskii[15] has classified the situation where such Q exists. Besides real and complex hypersurfaces there is a third class which is related

---

[e]It is easy to see that these two requirements are also sufficient since the $\mathbb{R}^n$-valued 1-form F is closed and can be integrated. If (M, g) allows two different isometric embeddings with the same Gauss map N, the two parallel isomorphisms $F, \tilde{F} : TM \to N^\perp$ yield a parallel and orthogonal automorphism $F^{-1} \circ \tilde{F}$ of TM. Generically, i.e. if the holonomy group is $SO(m)$ with $m > 2$, such an automorphism must be trivial, thus an isometric embedding with prescribed Gauss map is unique. In the case of minimal surfaces described above, the automorphism is rotation by a constant angle. A similar phenomenon occurs in higher dimensions for pluri-minimal submanifolds, cf. Ref. 7.

[f]This is the famous *Sym-Bobenko Formula*; see Ref. 6 for details and generalization.

to *extrinsic symmetric spaces* (see next section). Using extended matrix computations, Nikolaevskii could show that this last case is completely rigid: the Gauss map image determines the submanifold uniquely. We wish to give a conceptual proof for this theorem (as stated in the introduction).

4. Extrinsic symmetric spaces

A *symmetric space* is a Riemannian manifold M with an isometric point reflection (*symmetry*) at every point $x \in M$, i.e. there is an isometry s_x of M with $s_x(x) = x$ and $(ds_x)_x = -I$. We are interested in a submanifold version of this notion which is defined as follows.

For every submanifold $M \subset \mathsf{E}$ and any point $x \in M$ we let τ_x be the *normal reflection* at x, i.e. the affine isometry of E fixing x whose linear part (differential) τ_{x*} has eigenvalues 1 on N_x and -1 on T_x. A submanifold $M \subset \mathsf{E}$ is called *extrinsic symmetric* if it is preserved by all these normal reflections τ_x, $x \in M$ (cf. Ref. 1, Section 3.7). Clearly, an extrinsic symmetric space with its induced Riemannian metric is a symmetric space where the symmetry at x is $\tau_x|_M$. Examples are euclidean spheres $\mathsf{S}^m \subset \mathsf{E}^{m+1}$, the orthogonal group $O(n) \subset \mathbb{R}^{n\times n}$ or the Grassmannians where a k-dimensional linear subspace is replaced with the orthogonal reflection at this subspace and the receiving space E consists of all symmetric $n \times n$-matrices.[g] In fact, most symmetric spaces (so called *symmetric R-spaces*[12]) allow such an embedding.

The *hermitian* symmetric spaces are of particular interest. These are symmetric spaces M with a Kähler structure J such that all symmetries are holomorphic. Then at any $x \in M$, the complex structure J_x on T_xM is a skew symmetric derivation of the curvature tensor at x and hence an element of the isotropy Lie algebra $\mathfrak{g}_x \subset \mathfrak{g}$ (see footnote n); the embedding $x \mapsto J_x : M \to \mathfrak{g}$ (*standard embedding*) is extrinsic symmetric (cf. Ref. 8, Ref. 6).

Extrinsic symmetric spaces are the most beautiful submanifolds, playing a similar rôle for submanifold geometry as symmetric spaces do for Riemannian geometry. While the latter spaces have parallel curvature tensor, the former ones have parallel second fundamental form which follows from the invariance of $\nabla\alpha$ under τ_{x*}.[h] The Gauss equations (1) show that α is even more fundamental than R. As symmetric spaces are stable in the sense of "intrinsic pinching" (a Riemannian manifold with $\nabla R \approx 0$ is already

[g] Cf. Ref. 3 for details and further examples.
[h] $\nabla\alpha(u, v, w) = \tau_{x*}\nabla\alpha(u, v, w) = \nabla\alpha(\tau_{x*}u, \tau_{x*}v, \tau_{x*}w) = -\nabla\alpha(u, v, w)$.

diffeomorphic to a locally symmetric space[10,13]), extrinsic symmetric submanifolds are stable in the sense of "extrinsic pinching":[17] any submanifold with $\nabla\alpha \approx 0$ is already close to an extrinsic symmetric one.

Now let us consider the Gauss map of an extrinsic symmetric space $M \subset \mathsf{E}$. We may assume that M is contained in the unit sphere[i] $\mathsf{S} \subset \mathsf{E}$. Then we have $x \in N_x$ for all $x \in M$ and thus τ_x is linear, $\tau_x = \tau_{x*}$. Since M is invariant under τ_x, the same is true for the set of all normal spaces of M. But Gr is a symmetric space whose symmetry at $N_x \in \mathsf{Gr}$ is precisely the reflection τ_x, therefore the Gauss image $Q = N(M)$ is a symmetric subspace[j] of Gr. Since the kernel of $dN = -A$ would be a parallel distribution on M whose leaves are affine subspaces of E, the kernel must be zero, thus $N : M \to Q$ is a covering map (in many cases a diffeomorphism). Moreover it is equivariant with respect to the group

$$K = \{A \in O(n);\ A(M) = M\}$$

acting transitively on M, and therefore N is an isometry (up to scaling) on each isotropy irreducible component[k] of M. In the case of the sphere $M = \mathsf{S}^m$, the Gauss map N identifies S^m with the Grassmannian of oriented one-dimensional subspaces in E^{m+1}.

5. Algebra of extrinsic symmetric spaces

According to work of Ferus,[4,8] each irreducible extrinsic symmetric space (up to congruence and rescaling) is a certain orbit of the isotropy representation of another symmetric space as follows.[l] Let G be a compact Lie group with two commuting involutions σ and τ. Then we have a

---

[i] Since α is parallel, the same holds for the mean curvature vector $\eta = \text{trace}\,\alpha$ and for the corresponding Weingarten map A_η. Hence the eigenspaces of A_η are also parallel and moreover invariant under any A_ξ, due to the parallelity of η and the Ricci equation which implies $[A_\xi, A_\eta] = 0$. The leaves of each eigenspace distribution are submanifolds in a subspace of E and M decomposes into such submanifolds which are the intersections of M with an orthogonal decomposition of E (*extrinsic splitting*). Each factor can be considered separately. If it is not an affine subspace, we have $A_\eta = \lambda I$ for some nonzero λ, and thus M lies in some sphere of radius $1/|\lambda|$. Normalizing we may assume that M is contained in the unit sphere S^{n-1}.

[j] If P is a symmetric space with symmetries σ_p, $p \in P$, and if $Q \subset M$ is a connected subset which is invariant under each σ_q, $q \in Q$, then Q is a symmetric subspace, i.e. a complete totally geodesic submanifold of P, see Ref. 9 or Ref. 3.

[k] In fact, if M does not split extrinsically as $M_1 \times M_2 \subset \mathsf{E}^{n_1} \times \mathsf{E}^{n_2} = \mathsf{E}$ with $M_i \subset \mathsf{E}^{n_i}$ (*extrinsic irreducibility*), it is also intrinsically isotropy irreducible, i.e. the isotropy representation is irreducible (Ref. 4, Theorem 4).

[l] Recently, J.R. Kim gave a similar characterization for extrinsic symmetric spaces with *indefinite* inner product, cf. Ref. 11, Ref. 5.

common decomposition of the Lie algebra

$$\mathfrak{g} = \mathfrak{k}_- + \mathfrak{k}_+ + \mathfrak{p}_- + \mathfrak{p}_+ \tag{3}$$

where $\mathfrak{k} = \mathfrak{k}_+ + \mathfrak{k}_-$ and $\mathfrak{g}_+ = \mathfrak{k}_+ + \mathfrak{p}_+$ are the Lie subalgebras fixed by σ_* and τ_*, respectively. Further we require that τ is inner and of a very simple type: $\tau_* = e^{\pi \mathrm{ad}(\eta)}$ for some $\eta \in \mathfrak{p}_+$ with $\mathrm{ad}(\eta)^3 = -\mathrm{ad}(\eta)$, i.e. the eigenvalues of $\mathrm{ad}(\eta)$ are 0 and $\pm i$. More precisely, since $\tau_* = e^{\pi \mathrm{ad}(\eta)} = \begin{cases} I & \text{on } \mathfrak{g}_+ \\ -I & \text{on } \mathfrak{g}_- \end{cases}$,

$$\mathrm{ad}(\eta) = \begin{cases} 0 \text{ on } \mathfrak{g}_+ \\ J \text{ on } \mathfrak{g}_- \end{cases} \tag{4}$$

where J is a complex structure on $\mathfrak{g}_+$ (eigenvalues $\pm i$). Let $K \subset G$ be the fixed group of σ. Then $M = \mathrm{Ad}(K)\eta \subset \mathfrak{p} =: \mathsf{E}$ is an extrinsic symmetric space whose normal reflection at the point η is τ_*. The tangent and normal spaces of M at η are $\mathfrak{p}_-$ and $\mathfrak{p}_+$. The ambient space $\mathsf{E} = \mathfrak{p}$ is a *Lie triple*, i.e. $[\mathfrak{p}, [\mathfrak{p}, \mathfrak{p}]] \subset \mathfrak{p}$. The corresponding trilinear map on $\mathfrak{p}$,

$$R(u, v)w = [w, [u, v]] \tag{5}$$

is called *Lie triple product*; it is the curvature tensor of the symmetric space G/K whose tangent space at eK can be identified with $\mathfrak{p}$. Both subspaces $\mathfrak{p}_+, \mathfrak{p}_-$ (being the fixed sets of the involutions τ_* and $\sigma_* \tau_*$ on $\mathfrak{p}$) are Lie subtriples, i.e. they are preserved by R. The Lie algebra $\mathfrak{k}$ consists of the derivations[m] of R, and $\mathfrak{k}_+$ (resp. $\mathfrak{k}_-$) contains those elements of $\mathfrak{k}$ which preserve (resp. reverse) the splitting $\mathfrak{p} = \mathfrak{p}_- + \mathfrak{p}_+$.

The same $\eta \in \mathfrak{p}_+ \subset \mathfrak{g}_+$ (cf. (4)) generates yet another extrinsic symmetric space: the full adjoint orbit $\hat{M} = Ad(G)\eta \subset \mathfrak{g}$ with $T_\eta \hat{M} = \mathfrak{g}_-$ and $N_\eta \hat{M} = \mathfrak{g}_+$. The complex structure J on $\mathfrak{g}_-$ defined in (4) makes $\hat{M}$ *hermitian* symmetric, and $\hat{M} \subset \mathfrak{g}$ is the standard embedding mentioned above. Thus we see that any extrinsic symmetric space M lies in a standard embedded hermitian symmetric space $\hat{M}$; in fact, M is a real form of $\hat{M}$ since it is the fixed set of $\sigma_*|_{\hat{M}}$ which acts as a complex conjugation on $\hat{M}$.

We can also view the hermitian case as a special case of the general construction of extrinsic symmetric spaces as follows. To avoid confusion with the notation in the general case we change symbols renaming $\mathfrak{g} \to \mathfrak{h}$, $J \to j$, $\eta \to \zeta$. The compact Lie group H (previous G) is considered as a symmetric space $H = G/K$ with $G = H \times H$ and $K = \{(h, h);\ h \in H\}$.

---

[m] *Derivations* of R are skew adjoint linear maps $A : \mathfrak{p} \to \mathfrak{p}$ with $A\, R(x, y)z = R(Ax, y)z + R(x, Ay)z + R(x, y)Az$. They form the Lie algebra of the automorphism group K of $\mathfrak{p}$.

Therefore we put $\mathfrak{g} = \mathfrak{h} \oplus \mathfrak{h}$ and

$$\mathfrak{p} = \{(X, -X);\ X \in \mathfrak{h}\}, \quad \mathfrak{k} = \{(X, X);\ X \in \mathfrak{h}\}. \tag{6}$$

The Cartan decomposition $\mathfrak{h} = \mathfrak{h}_+ + \mathfrak{h}_-$ extends to $\mathfrak{g} = \mathfrak{g}_+ + \mathfrak{g}_-$, and we have $\eta = (\zeta, -\zeta) \in \mathfrak{p}_+$ and $J = (j, -j)$.

6. The second fundamental form

Now let $M_o \subset \mathsf{E} = \mathfrak{p}$ be an irreducible extrinsic symmetric space and $M \subset \mathsf{E}$ any submanifold of the same dimension with Gauss image $N(M) \subset Q := N(M_o)$. We fix some point $x \in M$ and let $\eta \in M_o$ such that $N_x M = N_\eta M_o = \mathfrak{p}_+$ and $T_x M = T_\eta M_o = \mathfrak{p}_-$. For tangent and normal vector fields v, w and ξ on M we consider the Levi-Civita derivatives

$$\nabla_v w = (\partial_v w)^T, \quad \nabla_v \xi = (\partial_v \xi)^N \tag{7}$$

Together they form a covariant derivative on the trivial bundle $M \times \mathsf{E} = TM \oplus NM$. We show first that the constant Lie triple product R is parallel also with respect to this derivative:

Lemma 6.1. $\nabla R = 0$.

Proof. Let a, b, c, d be ∇-parallel vector fields along any curve $x(t)$ in M; we may assume that each of them is either tangent or normal. Put $R(a, b, c, d) := \langle R(a, b)c, d\rangle$. Denoting $' = \frac{d}{dt}$ we have to show $R(a, b, c, d)' = 0$. But

$$R(a, b, c, d)' = R(a', b, c, d) + R(a, b', c, d) + R(a, b, c', c) + R(a, b, c, d').$$

Since both the tangent and normal space, $\mathfrak{p}_-$ and $\mathfrak{p}_+$, are Lie subtriples, we have $R(a, b, c, d) = 0$ unless R has an even number of entries of each type (tangent vs. normal); e.g. $R(\mathfrak{p}_-, \mathfrak{p}_-, \mathfrak{p}_-, \mathfrak{p}_+) = 0$ since $R(\mathfrak{p}_-, \mathfrak{p}_-)\mathfrak{p}_- \subset \mathfrak{p}_- \perp \mathfrak{p}_+$, and by the curvature identities this remains true with permuted entries. But the derivative of a ∇-parallel tangent (resp. normal) field is normal (resp. tangent), hence each term on the right hand side contains an odd number of entries of each sort, thus it has to be zero. □

The difference tensor between the two derivatives ∇ and ∂ on $M \times E$,

$$L := \partial - \nabla \tag{8}$$

is essentially the second fundamental form; we have

$$L_v w = \alpha(v, w), \quad L_v \xi = -A_\xi v. \tag{9}$$

Since $\partial_v R = 0$ and $\nabla_v R = 0$ by the previous lemma, we get $L_v R = 0$, i.e. L_v is a derivation of R and therefore[n] $L_v \in \mathfrak{k}$, acting as $\mathrm{ad}(L_v)$. More precisely, L_v interchanges $\mathfrak{p}_-$ and $\mathfrak{p}_+$, hence $L_v \in \mathfrak{k}_-$. Thus we have defined a linear map

$$L : \mathfrak{p}_- \to \mathfrak{k}_-, \quad v \mapsto L_v = Lv. \tag{10}$$

Let $L^t : \mathfrak{k}_- \to \mathfrak{p}_-$ be the transposed of L with respect to an $\mathrm{Ad}(G)$-invariant inner product on $\mathfrak{g}$ extending the given $\mathrm{Ad}(K)$-invariant inner product on $\mathfrak{p} = \mathsf{E}$.

Lemma 6.2. *For all $\xi \in \mathfrak{p}_+$ we have*

$$L^t \circ \mathrm{ad}(\xi) = -\mathrm{ad}(\xi) \circ L, \tag{11}$$

$$L \circ \mathrm{ad}(\xi) = -\mathrm{ad}(\xi) \circ L^t. \tag{12}$$

Proof. The first equation follows from the symmetry of α which reads

$$[Lv, w] = [Lw, v] \tag{13}$$

for all $v, w \in \mathfrak{p}_-$. In fact, for any $\xi \in \mathfrak{p}_+$ we have

$$\begin{aligned} \langle [Lv, w], \xi \rangle &= \langle Lv, [w, \xi] \rangle = -\langle v, L^t \mathrm{ad}(\xi) w \rangle, \\ \langle [Lw, v], \xi \rangle &= \langle v, [\xi, Lw] \rangle = \langle v, \mathrm{ad}(\xi) Lw \rangle \end{aligned}$$

which proves (11). The second equation (12) is obtained by composing (11) from both sides with $\mathrm{ad}(\eta)$. From (4) we see that $\mathrm{ad}(\eta)$ and $\mathrm{ad}(\xi)$ commute (since $[\mathrm{ad}(\eta), \mathrm{ad}(\xi)] = \mathrm{ad}([\eta, \xi]) = 0$), and further $\mathrm{ad}(\eta)^2 = J^2 = -I$ on $\mathfrak{g}_-$. Using (11) for η in place of ξ,

$$\begin{aligned} \mathrm{ad}(\eta) L^t \mathrm{ad}(\xi) \mathrm{ad}(\eta) &= \mathrm{ad}(\eta) L^t \mathrm{ad}(\eta) \mathrm{ad}(\xi) = -\mathrm{ad}(\eta)^2 L\, \mathrm{ad}(\xi) = L\, \mathrm{ad}(\xi), \\ -\mathrm{ad}(\eta) \mathrm{ad}(\xi) L\, \mathrm{ad}(\eta) &= -\mathrm{ad}(\xi) \mathrm{ad}(\eta) L\, \mathrm{ad}(\eta) = \mathrm{ad}(\xi) L^t\, \mathrm{ad}(\eta)^2 = -\mathrm{ad}(\xi) L^t. \end{aligned}$$

The left hand sides are equal by (11), hence we have proved (12). □

The second fundamental form L defines a skew symmetric linear map $\hat{L} : \mathfrak{g}_- \to \mathfrak{g}_-$ interchanging $\mathfrak{p}_-$ and $\mathfrak{k}_-$:

$$\hat{L} = \begin{cases} L & \text{on } \mathfrak{p}_- \\ -L^t & \text{on } \mathfrak{k}_- \end{cases} \tag{14}$$

[n] Any Lie triple derivation $L : \mathfrak{p} \to \mathfrak{p}$ extends uniquely to a Lie algebra derivation $L : \mathfrak{g} \to \mathfrak{g}$ where for every $A \in \mathfrak{k}$ we let $L(A) \in \mathfrak{k}$ with $[L(A), x] = L[A, x] - [A, Lx]$ for any $x \in \mathfrak{p}$. Since $\mathfrak{g}$ is semisimple, each derivation is inner, $L = \mathrm{ad}(l)$ for some $l \in \mathfrak{g}$. In fact $l \in \mathfrak{k}$ since $L = \mathrm{ad}(l)$ preserves $\mathfrak{p}$ and $\mathfrak{k}$.

Lemma 6.3. *$\hat{L}$ commutes with the action of $\tilde{\mathfrak{g}}_+ = \tilde{\mathfrak{k}}_+ + \mathfrak{p}_+$ on $\mathfrak{g}_-$, where $\tilde{\mathfrak{k}}_+ = [\mathfrak{p}_+, \mathfrak{p}_+] \subset \mathfrak{k}_+$.*

Proof. From Equations (11) and (12) of the previous Lemma we obtain

$$\hat{L} \circ \mathrm{ad}(\xi) = \mathrm{ad}(\xi) \circ \hat{L} \tag{15}$$

for all $\xi \in \mathfrak{p}_+$. Hence $\hat{L}$ commutes with the action of $\mathfrak{p}_+$ on $\mathfrak{g}_-$, and consequently with the action of $[\mathfrak{p}_+, \mathfrak{p}_+] = \tilde{\mathfrak{k}}_+$. □

7. The group action

Lemma 7.1. *Let $M_o \subset \mathfrak{p}$ be extrinsic symmetric and irreducible, but $M_o \neq \mathsf{S}^m$. Then $[\mathfrak{p}_+, \mathfrak{p}_+] = \mathfrak{k}_+$ and hence $\tilde{\mathfrak{g}}_+ = \mathfrak{g}_+$.*

Proof. Note that any $A \in \mathfrak{k}_+$ with $A \perp [\mathfrak{p}_+, \mathfrak{p}_+]$ acts trivially on $\mathfrak{p}_+$. Hence $(\tilde{\mathfrak{g}}_+, \tilde{\mathfrak{k}}_+)$ is the effective pair[o] corresponding to the Lie triple $\mathfrak{p}_+$. We show by inspection of the tables for M_o (cf. Ref. 14, p. 241 f) that

$$\mathfrak{k}_+ = \tilde{\mathfrak{k}}_+ = [\mathfrak{p}_+, \mathfrak{p}_+] \tag{16}$$

in all cases but the one corresponding to $M_o = \mathsf{S}^m$ (Case No. 13 for $i = 1$). In fact, the symmetric pairs $(\mathfrak{k}, \mathfrak{k}_+)$ corresponding to $\mathfrak{k}_-$ are listed on p. 242, first column while the last column contains the pairs $(\tilde{\mathfrak{g}}_+, \tilde{\mathfrak{k}}_+)$ corresponding to $\mathfrak{p}_+$.[p] We have $\tilde{\mathfrak{k}}_+ = \mathfrak{k}_+$ in all cases except No. 13 for $i = 1$ where $\mathfrak{p}_+ = \mathbb{R}$ and the pair $(\mathbb{R} + \mathfrak{so}(n-2), \mathfrak{so}(n-2))$ is ineffective; the effective one would be $(\tilde{\mathfrak{g}}_+, \tilde{\mathfrak{k}}_+) = (\mathbb{R}, 0)$. □

Lemma 7.2.

$$\hat{L} = \lambda\, \mathrm{ad}(\eta) \tag{17}$$

for some real λ.

Proof. The vector space $\mathfrak{g}_-$ has a complex structure $J = \mathrm{ad}(\eta)|_{\mathfrak{g}_-}$ which commutes with the action of $\mathfrak{g}_+$ (by (4)) and of $\hat{L}$ (by (15)), i.e. these actions are complex linear.

---

[o]We call the pair of Lie algebras $(\mathfrak{g}, \mathfrak{k})$ with $\mathfrak{k} \subset \mathfrak{g}$ *ineffective* if there is a nonzero subalgebra $\mathfrak{l} \subset \mathfrak{k}$ with $[\mathfrak{l}, \mathfrak{g}] \subset \mathfrak{k}$; in other words, the subgroup $L \subset K$ corresponding to $\mathfrak{l}$ acts trivially on G/K.

[p]The table in Ref. 2, p. 735 f is very similar: It displays the pairs $(\mathfrak{k}, \mathfrak{k}_+)$ on p. 735, third column and the noncompact duals of $(\tilde{\mathfrak{g}}_+, \tilde{\mathfrak{k}}_+)$ on p. 736, second column

If M_o belongs to the cases 7–18 in Ref. 14 or Ref. 2, the Lie algebra $\mathfrak{g}$ is simple and hence the symmetric pair $(\mathfrak{g}, \mathfrak{g}_+)$ is irreducible.[q] Thus $\mathfrak{g}_+$ acts irreducibly on $\mathfrak{g}_-$, and $\hat{L}$ commutes with this action. By Schur's lemma, $\hat{L} = \alpha I$ for some $\alpha \in \mathbb{C}$. But $\hat{L}$ is antisymmetric as a real endomorphism, hence α is purely imaginary, $\alpha = \lambda i$ and $\hat{L} = \lambda J = \lambda \operatorname{ad}(\eta)|_{\mathfrak{g}_-}$.

The cases 1–6 are those where M_o is hermitian and $\mathfrak{p}$ itself a Lie algebra. More precisely, there is a simple compact Lie algebra $\mathfrak{h}$ with Cartan decomposition $\mathfrak{h} = \mathfrak{h}_+ + \mathfrak{h}_-$ such that $\mathfrak{g} = \mathfrak{h} \oplus \mathfrak{h}$ and

$$\mathfrak{p} = \{(X, -X);\ X \in \mathfrak{h}\}, \quad \mathfrak{k} = \{(X, X);\ X \in \mathfrak{h}\}.$$

Now $\mathfrak{g}_- = \mathfrak{h}_- \oplus \mathfrak{h}_-$ has two irreducible factors for the action of $\mathfrak{g}_+ = \mathfrak{h}_+ \oplus \mathfrak{h}_+$, and the complex structure $J = (j, -j)$ is given by the Lie bracket with $\eta = (\zeta, -\zeta) \in \mathfrak{p}_+$. By Schur's lemma, $\hat{L}$ has (imaginary) eigenvalues λi, μi on the two irreducible factors, and in particular

$$\hat{L}(X, X) = (\lambda j X, -\mu j X)$$

for all $X \in \mathfrak{h}_-$. But on the other hand $\hat{L}(\mathfrak{k}_-) \subset \mathfrak{p}_-$, hence

$$\hat{L}(X, X) = (Y, -Y)$$

for some $Y \in \mathfrak{h}_-$. Comparing these equations shows $\lambda = \mu$. □

Lemma 7.3.

$$\lambda = const \tag{18}$$

Proof. Locally we may view λ and η as smooth functions on M. Since η is the position vector of $M_o \subset \mathsf{S}$, its derivatives $\partial_v \eta$ lie in $T_\eta M_o = T_x M$, hence η is a parallel normal vector field on M. By Lemma 1 we have $\nabla R = 0$, and hence $\operatorname{ad}(\eta) = R(., \eta) : \mathfrak{p}_- \to \mathfrak{k}_- \subset \operatorname{Hom}(\mathfrak{p}_+, \mathfrak{p}_-)$ is ∇-parallel.[r] Codazzi equations show

$$(\nabla_v L)w = (\nabla_w L)v \tag{19}$$

for any two tangent vectors $v, w \in \mathfrak{p}_-$. Using (17) and (4) we obtain

$$(\partial_v \lambda) J w = (\partial_w \lambda) J v. \tag{20}$$

If v, w are linearly independent we have $\partial_v \lambda = 0 = \partial_w \lambda$. Thus $\lambda = const$.□

Proof of the Theorem. If $\lambda = 0$, then $L = 0$ and our submanifold M is affine. Otherwise we may assume $\lambda = \pm 1$ (up to scaling), and replacing η

[q]see Ref. 14, p. 241, third column or the dual in Ref. 2, p. 736, first column
[r]Remind that $\mathfrak{p}_-, \mathfrak{p}_+$ now depend on $x \in M$; they are subbundles of $M \times \mathsf{E}$.

by $-\eta$ if necessary, we have $\lambda = 1$. Since $\mathrm{ad}(\eta) : \mathfrak{p}_- \to \mathfrak{k}_- \subset \mathrm{Hom}(\mathfrak{p}_+, \mathfrak{p}_-)$ is a linear isometry up to scaling, being the differential of the (equivariant) Gauss map of M_o, the same holds for L. Thus both Gauss maps $N : M \to Q$ and $N_o : M_o \to Q$ are local isometries, and using these maps to identify the abstract Riemannian manifold M with an open subset of M_o, the second fundamental forms are the same ($\hat{L} = \mathrm{ad}(\eta)$). Now the theorem follows from the uniqueness part of the existence and uniqueness theorem for submanifolds. □

References

1. J. Berndt, S. Console and C. Olmos, *Submanifolds and Holonomy* (CRC Press 2003).
2. J. Berndt, J.-H. Eschenburg, H. Naitoh and K. Tsukada, Symmetric submanifolds associated with irreducible symmetric R-spaces, *Math. Ann.* **332** (2005) 721–737.
3. J.-H. Eschenburg, Lecture Notes on Symmetric Spaces, Preprint Augsburg 2000.
4. J.-H. Eschenburg and E. Heintze, Extrinsic symmetric spaces and orbits of s-representations, *manuscripta math.* **88** (1995) 517–524; Erratum *manuscripta math.* **92** (1997) 408.
5. J.-H. Eschenburg and J.R. Kim, Indefinite Extrinsic Symmetric Spaces, in preparation.
6. J.-H. Eschenburg and P. Quast, Pluriharmonic maps into Kähler symmetric spaces, Preprint Augsburg 2006.
7. J.-H. Eschenburg and R. Tribuzy, Associated families of pluriharmonic maps and isotropy, *manuscripta math.* **95** (1998) 295–310.
8. D. Ferus, Symmetric submanifolds of Euclidean space, *Math. Ann.* **246** (1980) 81–93.
9. S. Helgason, *Differential Geometry, Lie Groups and Symmetric Spaces* (Academic Press, 1978).
10. A. Katsuda, A pinching problem for locally homogeneous spaces. *J. Math. Soc. Japan* **41** (1989) 57–74.
11. J.R. Kim, Indefinite Extrinsic Symmetric Spaces, Thesis Augsburg 2005.
12. S. Kobayashi and T. Nagano, On filtered Lie algebras and geometric structures, I, *J. Math. Mech.* **13** (1964) 875–907.
13. M. Min-Oo and E.A. Ruh, Comparison theorems for compact symmetric spaces. *Ann. sc. Éc. Norm. Sup.(4)* **12** (1979) 335–353.
14. H. Naitoh, Symmetric submanifolds of compact symmetric spaces, *Tsukuba J. Math.* **10** (1986) 215–224.
15. Y.A. Nikolaevskii, Classification of Multidimensional Submanifolds in Euclidian Space with Totally Geodesic Gauss Image, *Russian Academy of Sciences Sbornik Mathematics* **76** (1993) 225–246.
16. C.-L. Terng (Ed.), *Integrable Systems, Geometry, and Topology* (AMS/IP Studies in Advanced Mathematics 2006).

17. P. Quast, A pinching theorem for extrinsically symmetric submanifolds of Euclidean space. *Manuscr. Math.* **115** (2004) 427–436.
18. E. Ruh and J. Vilms, The tension field of the Gauss map, *Trans. Amer. Math. Soc.* **149** (1970) 569–573.

Differential Geometry and its Applications
Proc. Conf., in Honour of Leonhard Euler, Olomouc, August 2007

Ricci flow on almost flat manifolds

Galina Guzhvina

Mathematisches Institut, Westfälische Wilhelms-Universität (WWU), 62 Einsteinstrasse, 48149 Münster, Germany
E-mail: guzhvina@math.uni-muenster.de

The present work studies how the Ricci flow acts on almost flat manifolds. We show that the Ricci flow exists on any ε-flat Riemannian manifold (M, g) with ε small enough for any $t \in \mathbb{R}_{\geq 0}$, that $\lim_{t\to\infty} |K|_{(M,g_t)} \cdot diam^2(M\ g_t) = 0$ along the Ricci flow and in the case when the fundamental group of (M, g_t) is (almost) Abelian we obtain the C^0-convergence of the metric to a flat limit metric. The cases of $\pi_1(M, g_t)$ Abelian and non-Abelian are handled in two different ways.

Keywords: Ricci flow, curvature pinching, nilmanifold.

MS classification: 51K10, 58J90.

1. Introduction

A compact Riemannian manifold M^n is called ε-flat if its curvature is bounded in terms of the diameter as follows:

$$|K| \leq \varepsilon \cdot d(M)^{-2},$$

where K denotes the sectional curvature and $d(M)$ the diameter of M. If one scales an $\varepsilon-$flat metric it remains $\varepsilon-$flat.

By almost flat we mean that the manifold carries ε-flat metrics for arbitrary $\varepsilon > 0$.

The (unnormalized) Ricci flow is the geometric evolution equation in which one starts with a smooth Riemannian manifold (M^n, g_0) and evolves its metric by the equation:

$$\frac{\partial}{\partial t} g = -2ric_g, \tag{1}$$

where ric_g denotes the Ricci tensor of the metric g.

The present paper studies how the Ricci flow acts on almost flat manifolds. We show that on a sufficiently flat Riemannian manifold (M, g_0) the

Ricci flow exists for all $t \in [0, \infty)$, $\lim_{t\to\infty} |K|_{g(t)} \cdot d(M, g(t))^2 = 0$ as $g(t)$ evolves along (1), moreover, if $\pi_1(M, g_0)$ is Abelian, $g(t)$ converges along the Ricci flow to a flat metric. More precisely, we establish the following result:

Theorem 1.1 (Main Th. (Ricci Flow on Almost Flat Manifolds)). *In any dimension n there exists an $\varepsilon(n) > 0$ such that for any $\varepsilon \leq \varepsilon(n)$ an ε-flat Riemannian manifold (M^n, g) has the following properties:*
(i) the solution $g(t)$ to the Ricci flow (1)

$$\frac{\partial g}{\partial t} = -2ric_g, \qquad g(0) = g,$$

exists for all $t \in [0, \infty)$,
(ii) along the flow (1) one has

$$\lim_{t\to\infty} |K|_{g_t} \cdot d^2(M, g_t) = 0$$

(iii) $g(t)$ converges to a flat metric along the flow (1), if and only if the fundamental group of M is (almost) Abelian (= Abelian up to a subgroup of finite index).

Note that in (iii) convergence is of class C^0. Actually, a little more can be said: the limit manifold (M, g_∞), where g_∞ is the limit of $g(t)$ along the Ricci flow (1), is isometric to a flat manifold.

Fundamental results concerning the algebraic structure of almost flat manifolds were obtained by Gromov at the end of 70's.

In fact, Gromov[1] showed that each nilmanifold (= compact quotient of a nilpotent Lie group) is almost flat. It means that almost flat manifolds which do not carry flat metrics exist and occur rather naturally. Moreover, the next theorem asserts that nilmanifolds are, up to finite quotients, the only almost flat manifolds.

Theorem 1.2 (Gromov). *Let M^n be an $\varepsilon(n)-$ flat manifold, where $\varepsilon(n) = \exp(-\exp(\exp n^2))$. Then M is finitely covered by a nilmanifold (compact quotient of a nilpotent Lie group). More precisely:*
(i) The fundamental group $\pi_1(M)$ contains a torsion-free nilpotent normal subgroup Γ of rank n,
(ii) The quotient $G = \pi_1(M)/\Gamma$ has finite order and is isomorphic to a subgroup of $O(n)$,
(iii) the finite covering of M with fundamental group Γ and deckgroup G is diffeomorphic to a nilmanifold N/Γ,

(iv) The simply connected nilpotent group N is uniquely determined by $\pi_1(M)$.

From this theorem it is not clear whether M is diffeomorphic to the quotient of N by a uniform discrete group of isometries for a suitable left invariant metric on N.

Ruh[7] proved a stronger version of this theorem. He showed that, under strict curvature assumptions, M itself, and not only the finite cover, possesses a locally homogeneous structure.

Theorem 1.3 (Ruh). *For a compact Riemannian manifold M^n and a suitably small number $\varepsilon = \varepsilon(n)$ the fact that M^n is ε−flat implies that M is diffeomorphic to the compact quotient N/Γ, where N is a simply connected nilpotent Lie group and Γ is an extension of a lattice $\bar{L}$ in N by a finite group H.*

2. Almost Flat Manifolds with Non-Abelian Fundamental Group

Gromov's Theorem establishes a close connection between the ε−flat manifolds with ε sufficiently small and nilmanifolds. The following Theorem shows that sequences of universal covers of ε_k−flat manifolds with $\varepsilon_k \to 0$ subconverge to nilmanifolds thus giving a kind of motivation to consider first the Ricci flow on nilmanifolds.

Theorem 2.1. *For any sequence of numbers $\varepsilon_k \to 0$ take a sequence of ε_k−flat Riemannian n−manifolds (M_k^n, g_k) such that $max_{M_k} \|R\|_k = 1$ and for any $i \in \mathbb{N} \cup \{0\}$ there exists a constant $c(i,n)$ such that $\|\nabla^i R\|_k \leq c(i,n)$ for all k. Consider the sequence $(\tilde{M}_k^n, \tilde{g}_k)$ of the universal coverings of (M_k^n, g_k) with the covering metrics.*

Then $(\tilde{M}_k^n, \tilde{g}_k)$ subconverges w. r. to Gromov-Hausdorff topology to a nilpotent Lie group with a left-invariant metric on it.

Interesting observations concerning the behaviour of the Ricci flow on almost flat manifolds were made by J.Lauret (cf. Ref. 4). From Ref. 4 follows (implicitly) the next property of the Ricci flow:

Theorem 2.2. *On a nilpotent Lie group Ricci flow (1) is a gradient flow of the functional $F = trRic^2$.*

To obtain the necessary estimates it often makes sense to consider instead of (1) the normalised Ricci flow (2):

$$\frac{\partial g}{\partial t} = -2ric_g - 2\|ric_g\|_g^2 g, \tag{2}$$

where $\|ric_g\|_g^2 = trRic_g^2$ and we normalise the scalar curvature $sc(g_0) = -1$.

The normalised Ricci flow (2) differs from the unnormalised one (1) only by parametrisation and scaling; moreover, the scalar curvature is preserved along (2) (cf. Ref. 4). It is not difficult to see that on a nilpotent Lie group the flow (2) preserves the scalar curvature. There is an analogue of Theorem 2.2 for (2):

Theorem 2.3. *On a nilpotent Lie group N with the Lie algebra η the normalised Ricci flow (2) is a gradient flow of the functional $F = trRic^2$ restricted to the sphere of $\Lambda^2\eta^\star \otimes \eta$, corresponding to the normalisation of the scalar curvature: $sc = -1$.*

$$S = \{\mu \in \Lambda^2\eta^\star \otimes \eta : \|\mu\|^2 = 4\}. \tag{3}$$

Lauret[4] obtained also the following characterisation of the critical points of flow (2):

Theorem 2.4 (Ref. 4, 4.2). *For a nilpotent Lie group structure μ on $S \cap \mathfrak{N} \in \Lambda^2\eta^\star \otimes \eta$, where $\mathfrak{N}$ can be regarded as the set of nilpotent algebra brackets on η, the following properties are equivalent:*
(i) $(N_\mu, < \cdot, \cdot >)$ is a Ricci soliton,
(ii) μ is a critical point of F restricted to $S = \mu \in \Lambda\eta^\star \otimes \eta : \|\mu\| = 1$,
(iii) μ is a critical point of F restricted to $S \cap Gl(\eta)\cdot\mu$, where $Gl(\eta)\cdot\mu$ can be identified with the set of all left-invariant metrics on N_μ.

Definition 2.1. A *steady* Ricci soliton is a special solution of the Ricci flow on a Riemannian manifold M which moves along the equation by diffeomorphisms, that is, where the metric $g(t)$ is the pull-back of the initial metric $g(0)$ by a one-parameter group of diffeomorphisms.

In general,a Ricci soliton is a special solution of the Ricci flow which moves along the equation by a diffeomorphism and also shrinks or expands by a factor at the same time: if φ_t is a one-parameter group of diffeomorphisms generated by some vector field $X \in \chi(M)$, $g(t) = e^{ct}\varphi_t^\star g$, $g(0) = g_0$, for some $c \in \mathbb{R}$.

The Ricci soliton (M, g) is called *expanding* if $c > 0$ and *shrinking* if $c < 0$.

It is not difficult to see that in case when the manifold N possesses the structure of a nilpotent Lie group we can consider the one-parameter group of diffeomorphisms φ_t in the definition of the homothetic Ricci soliton as a one-parameter group of automorphisms of N. If $\varphi_t = \exp(-\frac{t}{2}D)$, $D \in Der(\eta)$ then $\frac{\partial}{\partial t}|_0 \varphi_t^\star g = g(-D\cdot, \cdot)$.

Operators of the type D on nilmanifolds were studied in great detail by J. Heber.[3] In particular, he established the following property of their eigenvalues:

Theorem 2.5 (Ref. 3, 4.14). *For some positive multiple the operator $D_0 = \xi D$, has as eigenvalues positive integers with no common divisor.*

Let $\lambda_1 < ... < \lambda_m$ be the eigenvalues of D_0, d_i, $i = 1, ..., m$ be the corresponding multiplicities. Then we call the tuple

$$(\lambda_1 < ... < \lambda_m; d_1, ..., d_m)$$

the *eigenvalue type* of D.

Corollary 2.1 (Ref. 3, 4.11). *In every dimension, only finitely many eigenvalue types can occur.*

We are now ready to establish the key property of the nilsoliton metrics which serves as a starting point for the proof of the Main Theorem.

Theorem 2.6. *Every nilsoliton strongly contracts the metric. In other words, there exists a constant $\lambda > 0$, such that, if (N, g) is a Ricci nilsoliton, then along the flow (2), for any $t \geq 0$, $h > 0$, holds $g(h+t) < e^{-\lambda h} g(t)$, where g is considered as a symmetric operator on η.*

Now let U be a neighbourhood of all solitons in $\mathfrak{N}' := S \cap \mathfrak{N}$ such that on any manifold in U the corresponding left-invariant metric contracts along the flow (2). More precisely, there exists a $\lambda > 0$ such that for any $(N_\mu, g) \in U$, as long as $g(t) \in U$, $\forall t > 0, \forall h > 0$, holds: $g(t+h) < e^{-\frac{\lambda}{2}h} g(t)$ along the flow (2). Such a neighbourhood exists, as follows from the theory of the continuous dependence of solutions of ODE's on the initial data.

Thus theorem 2.4 permits us to obtain the following important result:

Proposition 2.1. *Choose a neighbourhood of the critical set as above. There exists a constant C such that for any nilmanifold $(N_\mu, g) \in \mathfrak{N}'$, along the normalised Ricci flow flow (2) the measure of the set $I := \{t : (N, g(t)) \notin U\}$ is less than or equal to C.*

Recall that up to now we considered nilmanifolds with scalar curvature normalised to -1, or, equivalently, with the norm of the algebraic structure constants normalised to 2. The sectional curvature function K can be expressed as a quadratic function of the structure constants μ_{ijk} (cf, for example, Ref. 5):

$$K(e_1, e_2) = \sum_k (\frac{1}{2}\mu_{12k}(-\mu_{12k} + \mu_{2k1} + \mu_{k12}) - \frac{1}{4}(\mu_{12k} - \mu_{2k1} + \mu_{k12})(\mu_{12k} + \mu_{2k1} - \mu_{k12}) - \mu_{k11}\mu_{k22}) \tag{4}$$

This means that, since along the flow (2) $sc_g \equiv -1$, there exists a constant $L(n) > 1$ depending only on the dimension such that

$$\frac{1}{\sqrt{L}} \leq \|R\| \leq \sqrt{L} \tag{5}$$

for the curvature tensor R. Thus along (2) for any $h > 0$, $t \geq 0$ holds:

$$\|R\|_{(t+h)} \leq L\|R\|_t. \tag{6}$$

On a compact manifold the flow (2) is equivalent to the flow (1) up to reparametrisation and scaling. More precisely,

$$g_{(1)}(\tilde{t}) = \psi(t) g_{(2)}(t), \tag{7}$$

$$\|R_{(1)}\|_{\tilde{t}} = \frac{1}{\psi(t)} \|R_{(2)}\|_t, \tag{8}$$

where $g_{(1)}, g_{(2)}$ and $R_{(1)}, R_{(2)}$ are metric and curvature tensors corresponding, accordingly, to the flows (2) and (1), and

$$\psi(t) = e^{2\int_0^t \|ric_g\|^2 ds}, \tag{9}$$

$$\tilde{t} = \int \psi(t) dt. \tag{10}$$

Remark 2.1. Note, that while the norm of the curvature $\|R\| \in [\frac{1}{\sqrt{L}}, \sqrt{L}]$ with L defined as above along the normalised Ricci flow (2), 8 and 9 show that $\|R\|_t \to 0$ along the Ricci flow (1).

For any $\tilde{t} \in \mathbb{R}_{\geq 0}$

$$\|R_{(1)}\|_{\tilde{t}} \leq \frac{\|R_{(2)}\|_t}{e^{\frac{2t}{n^2 L}}} \leq \frac{L}{e^{\frac{2t}{n^2 L}}} \|R_{(1)}\|_0, \tag{11}$$

hence for $\tilde{t} \geq \int_0^{n^2 L \cdot lnL} \psi(t) dt$, $\|R_{(1)}\|_{\tilde{t}} \leq \frac{1}{L} \|R_{(1)}\|_0$. Thus on a nilmanifold the curvature always shrinks along the Ricci flow.

Theorem 2.7. *In any dimension n there exist constants $c_1(n), c_2(n) \geq 1$ such that for any n-dimensional nilmanifold (N, g) along the Ricci flow*

$$\frac{\partial g}{\partial t} = -2ric_g$$

with $g(0) = g$ holds: if $\|R\|_0 \in [\frac{1}{10c_2(n)}, 10]$, then
(i)

$$\|R\|_t g(t) < \frac{1}{2c_2(n)} \|R\|_0 g(0) \tag{12}$$

for any $t > c_1(n)$,
(ii)

$$\|R\|_t g(t) < c_2(n) \|R\|_0 g(0) \tag{13}$$

for any $t > 0$.

Note also that the constants obtained in 2.7, are universal for all left-invariant metrics and a given algebraic structure, since $Gl(n)$ acts transitively on $Sym(\eta)$—the set of all symmetric bilinear forms on η.

Now we prove the main result on a segment. In the proof we make use of the ideas of Gromov-Hausdorff convergence. The structure of the limit space provides us with information about the behaviour of the spaces "close" to it. Note that under the norm of the tensor R at t (which is denoted as $\|R\|_t$) we understand the *sup*-norm of R on the manifold $(M, g(t))$.

Theorem 2.8. *In any dimension n and any $T > 0$ there exists an $\varepsilon(n)$ such that for any $\varepsilon \leq \varepsilon(n)$ and any ε - flat Riemannian manifold (M^n, g)*
(i) the solution of the Ricci flow (1)

$$\frac{\partial g}{\partial t} = -2ric_g$$

exists on M for all $t \in [0, T)$,
(ii) parametrise the curvature at $t = 0$ as $\|R\|_0 = 1$. Then, for c_1, c_2 defined as in 2.7 along the flow (1)

$$\|R\|_t g(t) < \frac{1}{2} \|R\|_0 g(0) \tag{14}$$

at $t = 2c_1$,

$$\|R\|_t g(t) < 2c_2 \|R\|_0 g(0) \tag{15}$$

for any $t \in [0, 2c_1]$, and

$$\|R\|_{2c_1} \leq \|R\|_0. \tag{16}$$

Consider an $\varepsilon-$flat manifold (M^n, g_0) with ε small enough for M to satisfy the hypotheses of 2.8 and with the curvature parametrised as $\|R\|_0 = 1$. Evolve the initial metric g_0 on M under the Ricci flow (1). From 2.8, at $t = 2c_1$ the manifold (M, g_t) is $\frac{\varepsilon}{2}-$flat. Moreover, from (2.1),

$$\|R\|_{2c_1} \leq \|R\|_0. \tag{17}$$

From (16) we can put $\|R\|_{2c_1} = \nu$, $0 < \nu \leq 1$. Applying theorem 2.8 to the rescaled manifold $(M, \frac{1}{\nu}g)$ shows that the Ricci flow continues to exist on the interval $[2c_1, 2c_1 + \frac{2c_1}{\nu}]$. At $t_2 = 2c_1 + \frac{2c_1}{\nu}$ the manifold (M, g_{t_2}) is $\frac{\varepsilon}{4}-$flat and

$$\|R\|_{t_2} \leq \|R\|_{t_1} \leq 1. \tag{18}$$

So, by induction, we obtain a sequence $t_i \to \infty$ of times such that

$$d(M, g_{t_i})^2 \cdot \|R\|_{t_i} < \frac{\varepsilon}{2i}, \tag{19}$$

Furthermore, for any $t \in [t_i, t_{i+1}]$

$$d(M, g_t)^2 \cdot \|R\|_t < \frac{c_2\varepsilon}{2i}. \tag{20}$$

So, the Ricci flow exists on the whole of $\mathbb{R}_{\geq 0}$ and $\lim_{t\to\infty} |K|_{g_t} \cdot d^2(M, g_t) = 0$. Points i) and ii) of the Main Theorem are proved.

3. Almost Flat Manifolds with (Almost) Abelian Fundamental Group

The Ricci flow can be regarded as a quasilinear parabolic PDE (see, for example, Chow-Knopf[2]) or a nonlinear heat equation for the metric. Moreover, the intrinsically defined curvatures of a Riemannian metric evolving by the Ricci flow all obey parabolic equations with quadratic non-linearities. The classical fact from PDE's is that parabolic equations possess certain smoothing properties. Therefore, it is reasonable to expect that appropriate bounds on the geometry of the given manifold (M^n, g_0) would induce a priori bounds on the geometry of the unique solution $g(t)$ of the Ricci flow such that $g(0) = g_0$. Moreover, we would expect the geometry to improve, at least for the short time.

Theorem 3.1 (Shi). *Let $(M^n, g_{ij}(x))$ be a compact manifold with its curvature tensor R_{ijkl} satisfying $|R_{ijkl}|^2 \leq k_0$ on M, $0 < k_0 < \infty$. Then there exists a constant $T(n, k_0) > 0$, s. t. the evolution equation (1)*

$$\frac{\partial(g_{ij}(x,t))}{\partial t} = -2ric_{ij} \tag{21}$$

on M,

$$g_{ij}(x,o) = g_{ij}(x) \tag{22}$$

for any x *on* M *has a smooth solution* $g_{ij}(x) > 0$ *on* M *for a short time* $0 \le t \le T(n,k_0)$ *and satisfies the following estimates: for any integer* $m \ge 0$ *there exist constants* $c(m,n)$, *depending only on* m *and* n *such that*

$$\|\nabla^m R_{ijkl}(x,t)\|^2 \le \frac{c(m,n)\cdot k_0}{t^m} \tag{23}$$

for any $t \in [0,T]$

Note that the estimates in theorem 3.1 follow the natural parabolic scaling in which time behaves as distance squared. Note too that the estimates are stated in a form that deteriorates as $t \to 0$. This is the best one can do without making further assumptions on the initial metric. Recall also, that the lifetime of a maximal solution is bounded below by $\frac{C(n)}{\sqrt{k_0}}$, where $C(n)$ is a universal constant depending only on the dimension.

Let now (M,g) be an $\varepsilon-$flat Riemannian manifold with Abelian fundamental group. Put $max_M|R|_0 = 1$, $\varepsilon_0 := \varepsilon$, $\delta_0 = \frac{\varepsilon^{\frac{1}{8}}}{\|R_0\|}$. For ε_0 sufficiently small, the Ricci flow a priori exists for $t \le \delta_0$ and, from 3.1, for $t \le \delta_0$

$$max_M|R|_t \le c(0,n). \tag{24}$$

Hence we have the following estimation for $\|Ric\|$ on the same segment:

$$\|Ric\|_t \le n^2\|R\|_t \le c(0,n)n^2\|R\|_0 \tag{25}$$

for any $t \le \delta_0$.

Lemma 3.1. *Define* δ_0 *as above and let* $g(t)$ *is the metric on* M *evolving along the Ricci flow* (1). *Then for any* $t \in [0,\delta_0]$,

$$e^{-2\int_0^t \|ric_s\|_{g(s)}ds} g(0) \le g(t) \le e^{2\int_0^t \|ric_s\|_{g(s)}ds} g(0). \tag{26}$$

this lemma is proved in Ref. 8.

We have the following estimate on $\int_0^{\delta_0} \|ric_t\|$ along the segment $[0,\delta_0]$:

$$\int_0^{\delta_0} \|ric_t\|_{g(t)} \le n^2 \cdot c(0,n)\cdot \delta_0\|R_0\| \le n^2 \cdot c(0,n)\cdot \varepsilon_0^{\frac{1}{8}}. \tag{27}$$

Which means that, for ε_0 sufficiently small and for any integers $i,j \le n$ and any $t \in [0,\delta_0]$,

$$\frac{1}{2}g_{ij}(0) \le g_{ij}(t) \le 2g_{ij}(0) : \tag{28}$$

The proof of the point iii) of the Main Theorem uses the following three inequalities:

$$\|R\|_{\delta_0}^{3/2} \leq c_1(n)\|\nabla R\|_{\delta_0}, \tag{29}$$

$$\|\nabla R\|_t \leq c_2(n) \cdot d(M, g_t)(\|\nabla^2 R\|_t + \|R\|_t^2), \tag{30}$$

$$\|\nabla^2 R\|_{\delta_0} \leq \frac{c(2,n)}{\delta_0}\|R\|_0, \tag{31}$$

where the last one follows from Theorem 3.1.

From these three inequalities we have finally

$$\|R\|_{\delta_0}^{3/2} \leq 2c_1(n)c_2(n) \cdot d(M, g_0)\|R\|_0 \left(c(0,n)\|R\|_0 + \frac{c(2,n)}{\varepsilon_0^{1/8}}\|R\|_0 \right),$$

thus there exists a constant $c(n)$ such that

$$\begin{aligned} &\|R\|_{\delta_0}^{3/2} \leq \frac{c(n)^{3/2} \cdot d(M, g_0) \cdot \|R_0\|^2}{\varepsilon_0^{1/8}} \leq c(n)^{3/2} \cdot \varepsilon_0^{3/8}\|R\|_0^{3/2}, \\ &\|R\|_{\delta_0} \leq c(n) \cdot \varepsilon_0^{\frac{1}{4}}\|R\|_0. \end{aligned} \tag{32}$$

So, after the time δ_0, the initial curvature diminishes by absolute value by $\varepsilon_0^{1/4}$. We have also the following estimation for the pinching constant ε_1 at $t = \delta_0$:

$$\varepsilon_1 = \|R\|_{\delta_0} \cdot d^2(M, d_{\delta_0}) \leq 4c(n) \cdot \varepsilon_0^{\frac{5}{4}}. \tag{33}$$

Define the sequence of points t_i on $\mathbb{R}_{\geq 0}$ such that $t_0 = 0$, $t_{i+1} = t_i + \delta_i$, where $\delta_i := \frac{\varepsilon_i^{\frac{1}{8}}}{\|R_{t_i}\|}$, $\varepsilon_i = d(M, g_{t_i})^2 \cdot \|R_{t_i}\|$.

Note that

$$\begin{aligned} \delta_i &= \frac{(d(M, g_{t_i})^2 \cdot \|R\|_{t_i})^{\frac{1}{8}}}{\|R\|_{t_i}} \geq \frac{d(M, g_{t_{i-1}})^{\frac{1}{4}}}{2^{\frac{1}{4}}c(n)^{\frac{7}{8}}\varepsilon_{i-1}^{\frac{7}{32}} \cdot \|R\|_{t_{i-1}}^{\frac{7}{8}}} \\ &\geq \frac{\varepsilon_{i-1}^{\frac{1}{8}}}{2^{\frac{1}{4}}c(n)^{\frac{7}{8}}\varepsilon_{i-1}^{\frac{7}{32}} \cdot \|R\|_{t_{i-1}}} = \frac{1}{2^{\frac{1}{4}}c(n)^{\frac{7}{8}}\varepsilon_{i-1}^{\frac{3}{32}} \cdot \|R\|_{t_{i-1}}}, \end{aligned}$$

It means that

$$\delta_i \geq \frac{\delta_{i-1}}{2^{\frac{1}{4}}c(n)^{\frac{7}{8}}\varepsilon_{i-1}^{\frac{7}{32}} \cdot \|R\|_{t_{i-1}}}, \tag{34}$$

hence the segments $[t_i, t_{i+1}]$ cover the whole of $\mathbb{R}_{\geq 0}$.

The last inequality and (25) permit us to estimate $\int_0^\infty \|ric\|_{g_s} ds$:

$$\begin{aligned}\int_0^\infty \|ric\|_{g_s} ds &\leq \Sigma_{i=0}^\infty \int_{t_i}^{t_{i+1}} \|ric\|_{g_s} ds \leq \Sigma_{i=0}^\infty \delta_i \cdot \|Ric\|_{t_i} \\ &\leq n^2 \Sigma_{i=0}^\infty \delta_i \cdot c(0,n)\|R\|_{t_i} \leq n^2 \Sigma_{i=0}^\infty \varepsilon_i^{\frac{1}{8}} \cdot c(0,n).\end{aligned} \tag{35}$$

From the same considerations as in (33) we get that

$$\varepsilon_i \leq 4c(n) \cdot \varepsilon_{i-1}^{\frac{5}{4}}, \tag{36}$$

so, the series $\Sigma_{i=0}^\infty \varepsilon_i^{\frac{1}{8}}$ is a geometric progression, therefore, for ε_0 small enough, it converges.

Hence the integral converges on the real line and curvature along the Ricci flow tends to zero.

Note, that the convergence of the metrics along the Ricci flow (1) is of class C^0, since from (26) we have

$$\|g_t - g_\infty\|_{g_t} \leq -1 + e^{2\int_t^\infty \|ric_{g_\tau}\| d\tau} \to 0$$

as $t \to \infty$.

The limit manifold (M, g_∞) is a Gromov-Hausdorff limit of the family (M, g_t) of Riemannian manifolds, where g_t evolves along (1). As $t \to \infty$ we have the convergence of the corresponding sectional curvatures to zero: $|K_t|_{t\to\infty} \to 0$. We can also show that the volumes of (M, g_t) remain bounded from below ($vol_t \geq v > 0$) for any t. Indeed, from (26), for any $t \in \mathbb{R}_{\geq 0}$ we have

$$g_t \geq e^{-2\int_0^t \|ric_{g_\tau}\| d\tau} g_0 \tag{37}$$

Therefore, for ε_0 small enough, $vol(M, g_t) \geq e^{-n\int_0^t \|ric_{g_\tau}\| d\tau} vol(M, g_0)$ for any t.

Now we can use the argument of Cheeger (cf., for example, Ref. 6):

Theorem 3.2 (Cheeger). *Given $n \geq 2$ and $v, k \in (0,\infty)$ and a compact n-manifold (M, g) with $|K| \leq k$, $volB(p,1) \geq v$ for all $p \in M$, then $injM \geq i_0$, where i_0 depends only on n, k and v.*

So, we can conclude that the injectivity radius of (M, g_t) is uniformly bounded from below and the Convergence Theorem of Riemannian Geometry[6] can be applied to this family of manifolds. We get that any sequence in this family subconverges in the Gromov-Hausdorff topology to a flat manifold.

Now, since the Gromov-Hausdorff limit is unique up to isometries, we can conclude that (M, g_∞) is isometric to a flat manifold.

So, we have shown that on any ε−flat Riemannian manifold with the Abelian fundamental group the Ricci flow (1) converges to a flat metric for ε small enough. Of course, this condition on the fundamental group is also necessary.

Now a few words about the proofs of the inequalities (29) and (30).

Concerning (29) we have the result

Theorem 3.3. *In any dimension n there exist an $\varepsilon(n)$ and a $c(n)$ such that for any $\varepsilon \le \varepsilon(n)$ and for any ε-flat manifold M we have*

$$\|R\|_M^{3/2} \le c(n)\|\nabla R_M\|. \tag{38}$$

Note that for 3.3 we do not need any conditions on the fundamental group. In the proof we make use of a standard compactness argument and theorem 2.1: if we argue by contradiction, the algebraic structure of the limit manifold will prove to be incompatible with the structure of a nilpotent Lie group.

More specific is the inequality (30).

The following statement holds:

Theorem 3.4. *In any dimension n there exists an $\varepsilon(n)$ such that for any $\varepsilon \le \varepsilon(n)$ and for any ε-flat manifold (M^n, g) with (almost) Abelian fundamental group, there exists a constant $c = c(n)$, depending only n such that*

$$\|\nabla R\| \le c(n) \cdot d(M)(\|\nabla^2 R\| + \|R\|^2), \tag{39}$$

where R is the curvature tensor of (M^n, g).

To prove it we consider the Taylor expansion of ∇R along a geodesic loop on M in the direction v s.t. $|\nabla_v R| := \|\nabla R\|$.

Recall that to each geodesic loop α at p we associate its holonomy motion

$$\begin{aligned} m(\alpha) &: T_pM \to T_pM, \\ m(\alpha)(x) &= r(\alpha)x + t(\alpha), \end{aligned} \tag{40}$$

where $r(\alpha)$ is a parallel transport around α (rotational part of $m(\alpha)$) and

$$t(\alpha) = \dot{\alpha}(0) \tag{41}$$

- the translational part of α.

This gives an embedding of the fundamental group $\pi_1(M)$ into $Iso(T_pM) = Iso(\mathbb{R}^n)$, whereby the equivalence class of each loop $[\gamma]$ is mapped into $m(\gamma)$. One of the important intermediary results of Gromov's

Theorem says that, if we put $\pi_\rho := \{\gamma \in \pi_1(M) : |\gamma| \leq \rho\}$, the map $i : \pi_\rho \to Iso(\mathbb{R}^n)$ is injective for ρ and ε sufficiently small.

Proposition 3.1 (Commutator estimates[1] 2.4.1). *Let (M, g) be a complete Riemannian manifold, $|K_M| < \Lambda^2$. Consider α, β at $p \in M$ such that $|\alpha| + |\beta| \leq \frac{\pi}{3\Lambda}$. Then*

$$d([r(\alpha), r(\beta)], r[\beta, \alpha]) \leq \frac{5}{3}\Lambda^2 \cdot |t(\alpha)| \cdot |t(\beta)| + \frac{5}{6}\Lambda^2 |t[\alpha, \beta]| (|t(\alpha)| + |t(\beta)|), \tag{42}$$

$$\begin{aligned} |t[m(\alpha), m(\beta)] - t[\beta, \alpha]| \leq \frac{10}{3}\Lambda^2 |t(\alpha)||t(\beta)|(|t(\alpha)| \\ + |t(\beta)|) + \frac{10}{6}|t([\beta, \alpha])|\Lambda^2 |t(\alpha)||t(\beta)|(|t(\alpha)| + |t(\beta)|). \end{aligned} \tag{43}$$

Proposition 3.1 shows that the Gromov product and commutator of geodesic loops are almost compatible with the easily computable product and commutator of the holonomy motions of the loops. The error is curvature controlled.

In case when the fundamental group is Abelian, more precise estimations can be obtained:

Theorem 3.5. *In any dimension n there exists an $\varepsilon(n) > 0$ such that for any $\varepsilon \leq \varepsilon(n)$ and for any ε-flat n-dimensional manifold (M, g) with (almost) Abelian fundamental group for a ρ s.t.*

$$\rho \leq c_2(n) d(M),$$

for $c_2(n)$ depending only on the dimension of M, for any geodesic loop α s.t. $|\alpha| < \rho$ we have that

$$\|r(\alpha)\| < 10^n c_2(n)^2 \varepsilon. \tag{44}$$

In other words, theorem 3.5 states that in its setting the holonomy angles of the geodesic loops are of the order ε, where ε is the pinching constant of M. Thus, the parallel transport of any tensor along such a loop is periodic "modulo" ε. This (informal) consideration concludes the proof of (30)

References

1. P. Buser and H. Karcher, Gromov's almost flat manifolds. *In: Périodique mensuel de la Société Mathématique de France* (1981, Astérisque 81).
2. B. Chow and D. Knopf, *The Ricci Flow: An Introduction* (Mathematical Surveys and Monographs, vol. 110, AMS, 2004).
3. J. Heber, Noncompact homogeneous Einstein spaces, *Invent. Math.* **133** (2) (1998) 279–352.

4. J. Lauret, Ricci soliton homogeneous manifolds. *Math. Ann.* **319** (4) (2001) 715–733.
5. J. Milnor, Curvatures of left-invariant metrics on Lie groups, *Advances in Math.* **21** (3) (1976) 293–329.
6. P. Petersen, *Riemannian Geometry* (Springer 171, 1962).
7. E. Ruh, Almost flat manifolds, *J. Differential Geom.* **17** (1) (1982) 1–14.
8. W.-X. Shi, Deforming the metric on complete Riemannian manifolds, *J. Differential Geom.* **30** (1) (1989) 223–301.
9. E.N. Wilson, Isometry groups on homogeneous manifolds, *Geom. Dedicata* **12** (3) (1982) 337–346.

Differential Geometry and its Applications
Proc. Conf., in Honour of Leonhard Euler, Olomouc, August 2007

Compact minimal CR submanifolds of a complex projective space with positive Ricci curvature

Mayuko Kon

*Department of Mathematics, Hokkaido University,
Kita 10 Nishi 8, Kita-ku, Sapporo,
Hokkaido 060-0810, Japan
E-mail: mayuko_k13@math.sci.hokudai.ac.jp*

n-dimensional minimal proper CR submanifold M immersed in a complex projective space CP^m with the complex structure J under the assumption that the Ricci curvature of M is equal or greater than $n-1$. Moreover, we classify compact n-dimensional minimal CR submanifolds whose Ricci tensor S satisfies $S(X,X) \geq (n-1)g(X,X)+kg(PX,PX)$, $k=0,1,2$, for any vector field X tangent to M, where PX is the tangential part of JX.

Keywords: Ricci curvature, CR submanifold, complex projective space.

MS classification: 53C40, 53C55.

1. Introduction

The purpose of the present paper is to study the pinching problem in terms of Ricci curvatures of minimal CR submanifolds immersed in a complex projective space.

Let CP^m denote the complex projective space of real dimension $2m$ (complex dimension m) with constant holomorphic sectional curvature 4 and Kähler structure (J,g). Let M be a real n-dimensional Riemannian manifold isometrically immersed in CP^m with induced metric g. If there exist a differentiable holomorphic distribution $H : x \mapsto H_x \subset T_x(M)$ and the complementary orthogonal anti-invariant distribution $H^\perp$, then M is called a CR submanifold. Especially, when M satisfies $JT_x(M)^\perp \subset T_x(M)$ for any point x of M, M is called a generic submanifold. Any real hypersurface is obviously generic.

Kon[6] proved that if the Ricci tensor S of a compact n-dimensional minimal CR submanifold M of CP^m satisfies $S(X,X) \geq (n-1)g(X,X) + 2g(PX,PX)$, then M is a real projective space RP^n, or a complex projective space CP^m, or a pseudo-Einstein real hypersurface $\pi(S^k(1/\sqrt{2}) \times S^k(1/\sqrt{2}))$ $(k = (n+1)/2)$ of some $CP^{(n+1)/2}$ in CP^m, where $S^k(r)$ is a k-dimensional sphere of radius r, π is the Hopf fibration and PX is the tangential part of JX (see also Kon[5]).

For a compact minimal real hypersurface M of CP^m, Maeda[7] studied the pinching problem in terms of Ricci curvatures of M of CP^m $(m \geq 3)$. He proved that if the Ricci tensor S of a minimal real hypersurface satisfies $(2m-2)g(X,X) \leq S(X,X) \leq 2mg(X,X)$, then it is congruent to $\pi(S^m(1/\sqrt{2}) \times S^m(1/\sqrt{2}))$.

On the other hand, Yamagata-Kon[12] proved that if the Ricci tensor S of a compact n-dimensional minimal generic submanifold M of CP^m, which is not totally real, satisfies $S(X,X) \geq (n-1)g(X,X)$, then M is a real hypersurface of CP^m, that is, $2m - n = 1$.

In this paper, we prove a reduction theorem of the codimension of a compact n-dimensional minimal proper CR submanifold M in CP^m. We prove that if the Ricci curvature of M is equal or greater than $n-1$, then M is a real hypersurface of some $CP^{(n+1)/2}$ in CP^m. Using this result, we classify compact n-dimensional minimal CR submanifold M immersed in CP^m whose Ricci tensor S satisfies $S(X,X) \geq (n-1)g(X,X) + kg(PX,PX)$, $k = 0,1,2$, for any vector field X tangent to M.

2. Preliminaries

Let CP^m denote the complex projective space of complex dimension m with constant holomorphic sectional curvature 4. We denote by J the complex structure, and by g the metric of CP^m.

Let M be a real n-dimensional Riemannian manifold isometrically immersed in CP^m. We denote by the same g the Riemannian metric on M induced from g, and by p the codimension of M, that is, $p = 2m - n$.

We denote by $T_x(M)$ and $T_x(M)^\perp$ the tangent space and the normal space of M at x, respectively.

Definition 2.1. A submanifold M of a Kähler manifold $\tilde{M}$ with complex strucature J is called a CR submanifold of $\tilde{M}$ if there exists a differentiable distribution $H : x \mapsto H_x \subset T_x(M)$ on M satisfying the following conditions:

(i) H is holomorphic, i.e., $JH_x = H_x$ for each $x \in M$, and

(ii) the complementary orthogonal distribution $H^\perp : x \mapsto H_x^\perp \subset T_x(M)$ is anti-invariant, i.e. $JH_x^\perp \subset T_x(M)^\perp$ for each $x \in M$.

In the following, we put $\dim H_x = h$ and $\dim H_x^\perp = q$. If $q = 0$ (resp. $h =$

0) for any $x \in M$, then the CR submanifold M is a complex submanifold (resp. totally real submanifold) of $\tilde{M}$. If $h > 0$ and $q > 0$, then a CR submanifold M is said to be *proper*.

We denote by $\tilde{\nabla}$ the operator of covariant differentiation in CP^m, and by ∇ that in M determined by the induced metric. Then the *Gauss and Weingarten formulas* are given respectively by

$$\tilde{\nabla}_X Y = \nabla_X Y + B(X,Y), \qquad \tilde{\nabla}_X V = -A_V X + D_X V$$

for any vector fields X and Y tangent to M and any vector field V normal to M, where D denotes the normal connection. We call both A and B the *second fundamental form* of M and are related by $g(B(X,Y),V) = g(A_V X, Y)$. The second fundamental forms A and B are symmetric with respect to X and Y.

The mean curvature vector field of M is defined to be the trace of the second fundamental form B, that is, $\mathrm{tr}B = \sum_i B(e_i, e_i)$, $\{e_i\}$ being an orthonormal basis of $T_x(M)$. If the mean curvature vector field vanishes identically, then M is said to be *minimal*.

The covariant derivative $(\nabla_X A)_V Y$ of A is defined by

$$(\nabla_X A)_V Y = \nabla_X(A_V Y) - A_{D_X V} Y - A_V \nabla_X Y.$$

If $(\nabla_X A)_V Y = 0$ for any vector fields X and Y tangent to M, then the second fundamental form of M is said to be *parallel in the direction of the normal vector* V. If the second fundamental form is parallel in any direction, it is said to be *parallel*. A vector field V normal to M is said to be *parallel* if $D_X V = 0$ for any vector field X tangent to M.

For $x \in M$, the *first normal space* $N_1(x)$ is the orthogonal complement in $T_x(M)^\perp$ of the set $N_0(x) = \{V \in T_x(M)^\perp : A_V = 0\}$. If $D_X V \in N_1(x)$ for any vector field V with $V_x \in N_1(x)$ and any vector field X of M at x, then the first normal space $N_1(x)$ is said to be parallel with respect to the normal connection.

In the sequel, we assume that M is a CR submanifold of CP^m. The tangent space $T_x(M)$ of M is decomposed as $T_x(M) = H_x + H_x^\perp$ at each point x of M. Similarly, we see that $T_x(M)^\perp = JH_x^\perp + N_x$, where N_x is the orthogonal complement of $JH_x^\perp$ in $T_x(M)^\perp$.

For any vector field X tangent to M, we put $JX = PX + FX$, where PX is the tangential part of JX and FX the normal part of JX. For any vector field V normal to M, we put $JV = tV + fV$, where tV is the tangential part of JV and fV the normal part of JV. Then we see that $FP = 0$, $fF = 0$, $tf = 0$ and $Pt = 0$.

We define the covariant derivatives of P, F, t and f by $(\nabla_X P)Y = \nabla_X(PY) - P\nabla_X Y$, $(\nabla_X F)Y = D_X(FY) - F\nabla_X Y$, $(\nabla_X t)V = \nabla_X(tV) - tD_X V$ and $(\nabla_X f)V = D_X(fV) - fD_X V$, respectively.

The Riemannian curvature tensor $\tilde{R}$ of a complex projective space CP^m is given by

$$\begin{aligned}\tilde{R}(X,Y)Z &= g(Y,Z)X - g(X,Z)Y + g(JY,Z)JX \\ &\quad - g(JX,Z)JY + 2g(X,JY)JZ\end{aligned}$$

for any vector fields X, Y and Z of CP^m. Then the *equation of Gauss* and the *equation of Codazzi* are given respectively by

$$\begin{aligned}R(X,Y)Z &= g(Y,Z)X - g(X,Z)Y + g(PY,Z)PX - g(PX,Z)PY \\ &\quad - 2g(PX,Y)PZ + A_{B(Y,Z)}X - A_{B(X,Z)}Y\end{aligned}$$

and

$$\begin{aligned}&g((\nabla_X A)_V Y, Z) - g((\nabla_Y A)_V X, Z) \\ &\quad = g(Y,PZ)g(X,tV) - g(X,PZ)g(Y,tV) - 2g(X,PY)g(Z,tV),\end{aligned}$$

where R is the Riemannian curvature tensor field of M.

We denote by S the Ricci tensor field of M. Then

$$\begin{aligned}S(X,Y) &= (n-1)g(X,Y) + 3g(PX,PY) \\ &\quad + \sum_a \mathrm{tr}A_a g(A_a X, Y) - \sum_a g(A_a^2 X, Y),\end{aligned} \tag{1}$$

where A_a is the second fundamental form in the direction of v_a, $\{v_1, \cdots, v_p\}$ being an orthonormal basis of $T_x(M)^\perp$, and tr denotes the trace of an operator. If the Ricci tensor S satisfies $S(X,Y) = \alpha g(X,Y)$ for some constant α, then M is called *Einstein*. When M is a real hypersurface of CP^m with a unit normal vector field V, if the Ricci tensor S satisfies $S(X,Y) = \alpha g(X,Y) + \beta g(X,tV)g(Y,tV)$ for some constants α and β, then M is said to be *pseudo-Einstein*.

We need the following examples of CR submanifolds in CP^m.

Example 2.1. An n-dimensional complete totally geodesic submanifold M of CP^m is either a complex projective space $CP^{n/2}$ or a real projective space RP^n of constant curvature 1. A real projective space RP^n is a totally real submanifold of CP^m (see Abe[1]).

Example 2.2. Let $z^0, z^1, \cdots, z^m$ be a homogeneous coordinates of CP^m. The *complex quadric* Q^{m-1} is a complex hypersurface of CP^m defined by the equation

$$(z^0)^2 + (z^1)^2 + \cdots + (z^m)^2 = 0.$$

Then Q^{m-1} is a Kähler manifold. Moreover, Q^{m-1} is an Einstein manifold with Ricci curvature $2(m-1)$ (see Smith[11]).

Example 2.3. For an integer k and for $0 < r < \pi/2$, we define $M(k,r)$ in S^{2m+1} by

$$\sum_{j=0}^{k} |z_j|^2 = \cos^2 r, \qquad \sum_{j=k+1}^{m} |z_j|^2 = \sin^2 r.$$

$M(k,r)$ is a standard product $S^{2k+1}(\cos r) \times S^{2l+1}(\sin r)$, $l = m-k-1$. We consider the Hopf fibration $\pi : S^{2m+1} \longrightarrow CP^m$, where S^{2m+1} denotes the unit sphere. Then $M^c(k,r) = \pi(M(k,r))$ is a real hypersurface in CP^m. For an integer $1 \leq k \leq m-2$, we see that $M^c(k,r)$ is the tube of radius r over CP^k (see Cecil and Ryan[3]).

When r satisfies $\cos r = \sqrt{(2k+1)/(2m)}$ and $\sin r = \sqrt{(2l+1)/(2m)}$, $M^c(k,r)$ is a minimal real hypersurface of CP^m. Moreover, we see that $M^c(k,r)$ is a pseudo-Einstein real minimal hypersurface of CP^m if and only if $k = l = (m-1)/2$ and $r = \pi/4$. Then the Ricci tensor S satisfies $S(X,Y) = (2m-2)(X,Y) + 2g(PX,PY)$.

3. Reduction of the codimension

In this section we prove the following reduction theorem of a codimension.

Theorem 3.1. *Let M be a compact n-dimensional minimal proper CR submanifold of a complex projective space CP^m. If the Ricci tensor S of M satisfies $S(X,X) \geq (n-1)g(X,X)$ for any vector X tangent to M, then M is a real hypersurface of some $CP^{(n+1)/2}$ in CP^m.*

First of all, using (1) and the formula for the Ricci tensor of the submanifold M (see Yano-Kon;[13] p.44), we have

Lemma 3.1. *Let M be a compact n-dimensional minimal CR submanifold of CP^m which is not a complex submanifold of CP^m. If the Ricci tensor S of M satisfies $S(X,X) \geq (n-1)g(X,X)$, then M is a real projective space RP^n or $q = 1$, that is, $\dim H_x^{\perp} = 1$.*

In the following, we shall prove that the first normal space of M is just $FH^{\perp}$ and is of dimension 1 when M is proper and satisfies the condition of Lemma 3.1. To prove this, we prepare some lemmas.

Lemma 3.2. *Let M be a compact n-dimensional minimal proper CR submanifold of CP^m. If the Ricci tensor S of M satisfies $S(X,X) \geq (n-1)g(X,X)$, then the following hold:*

(a) $\nabla f = 0$.

(b) *For any X tangent to M and any $V \in FH^{\perp}$, we have $D_X V \in FH^{\perp}$.*

(c) *For any X tangent to M and any $U \in N$, we have $D_X U \in N$.*

Lemma 3.3. *Let M be a compact n-dimensional minimal proper CR submanifold of CP^m. If the Ricci tensor S of M satisfies $S(X,X) \geq (n-1)g(X,X)$, then the second fundamental form A satisfies the following:*

(a) $A_v P A_v = P$, *where v is a unit vector field in $FH^{\perp}$.*

(b) $\|[P, A_v]\|^2 = 2\mathrm{tr}A_v^2 - 2(n-1)$, *where v is a unit vector field in $FH^{\perp}$.*

(c) $A_V A_U = A_U A_V$ *for any $V \in FH^{\perp}$ and $U \in N$.*

(d) $PA_U = A_{fU}$ *and* $PA_U + A_U P = 0$ *for any $U \in N$.*

We use the following theorem directly given by the results in Simons.[10]

Theorem 3.2. *Let M be an n-dimensional minimal submanifold of a complex projective space CP^m. Then we have*

$$\begin{aligned} &g(\nabla^2 A, A) \\ &\quad = (n-3)\sum_a \mathrm{tr}A_a^2 + 3\sum_a \mathrm{tr}A_{fa}^2 \\ &\qquad + 4\sum_{a,b}(g(A_a t v_b, A_b t v_a) - g(A_a t v_a, A_b t v_b)) - 8\sum_a \mathrm{tr}A_a A_{fa} P \\ &\qquad + 3\sum_a \|[P, A_a]\|^2 - \sum_{a,b}\|[A_a, A_b]\|^2 - \sum_{a,b}(\mathrm{tr}A_a A_b)^2. \end{aligned}$$

Using Theorem 3.2 and Lemma 3.3, we next compute the Laplacian for the square of the length of the second fundamental form of the minimal submanifold in CP^m whose Ricci tensor satisfies $S(X,X) \geq (n-1)g(X,X)$ for any tangent vector field X.

Lemma 3.4. *Let M be a compact n-dimensional minimal proper CR submanifold of CP^m. If the Ricci tensor S of M satisfies $S(X,X) \geq (n-1)g(X,X)$, then*

$$\begin{aligned} g(\nabla^2 A, A) &= (n+3)\mathrm{tr}A_v^2 + (n+4)\sum_a \mathrm{tr}A_{fa}^2 - 6(n-1) \\ &\quad - \sum_{a,b}\|[A_a, A_b]\|^2 - \sum_{a,b}(\mathrm{tr}A_a A_b)^2. \end{aligned}$$

From this, we obtain the following

Lemma 3.5. *Let M be a compact n-dimensional minimal proper CR submanifold of CP^m If the Ricci tensor S of M satisfies $S(X,X) \geq (n-1)g(X,X)$, then*

$$\sum_j g((\nabla^2 A)_v e_j, A_v e_j) = (n+3)\mathrm{tr}A_v^2 - 6(n-1) - (\mathrm{tr}A_v^2)^2,$$

$$\sum_{a\geq 2,j} g((\nabla^2 A)_a e_j, A_a e_j) = \sum_a \mathrm{tr}A_{fa}^2 - \sum_{a,b} \|[A_a, A_b]\|^2 - \sum_{a,b\geq 2} (\mathrm{tr}A_a A_b)^2,$$

where $\{e_i\}$ is an orthonormal basis of $T_x(M)$.

Next we give inequalities for $\sum_{a,b} \|[A_a, A_b]\|^2$ and $\sum_{a,b\geq 2}(\mathrm{tr}A_a A_b)^2$ in the equation in Lemma 3.5.

Lemma 3.6. *Let M be a compact n-dimensional minimal proper CR submanifold of CP^m. If the Ricci tensor S of M satisfies $S(X,X) \geq (n-1)g(X,X)$, then*

$$\sum_{a,b} \|[A_a, A_b]\|^2 \leq 4 \sum_a \mathrm{tr}A_{fa}^2,$$

$$\sum_{a,b\geq 2} (\mathrm{tr}A_a A_b)^2 \leq \frac{1}{2}(\sum_a \mathrm{tr}A_{fa}^2)^2.$$

Using Lemma 3.1-Lemma 3.6, we prove the following lemma.

Lemma 3.7. *Let M be a compact n-dimensional minimal proper CR submanifold of CP^m. If the Ricci tensor S of M satisfies $S(X,X) \geq (n-1)g(X,X)$, then $A_{fa} = 0$ for all a.*

From Lemma 3.2 and Lemma 3.7, the first normal space of M is of dimension 1 and parallel. By the similar method in the proof of Theorem 3.6 in Yano-Kon;[13] p.227, we see that M is a real hypersurface of some totally geodesic complex projective space $CP^{(n+1)/2}$ in CP^m.

4. Pinching theorems of the Ricci curvature

In the following, we take the unit normal vector field v of a real hypersurface M in CP^m, and we put $\xi = -Jv$. Then ξ is the unit tangent vector field of M and $P^2X = -X + g(X,\xi)\xi$, $P\xi = 0$. We also put $A_v = A$ to simplify the notation. Then $\nabla_X \xi = PAX$ for any vector field X tangent to M.

To prove our theorems, we need a well-known result given by T. E. Cecil and P. J. Ryan:

Proposition 4.1 (Ref. 3). *Let M be a real hypersurface (with unit normal vector v) of a complex projective space CP^m on which ξ is a principal curvature vector with principal curvature $\alpha = 2\cot 2r$ and the focal map ϕ_r has constant rank on M. Then the following hold:*

(a) M lies on a tube (in the direction $\eta = \gamma'(r)$, where $\gamma(r) = \exp_x(rv)$ and x is a base point of the normal vector v) of radius r over a certain Kähler submanifold N in CP^m.

(b) Let $\cot\theta$, $0 < \theta < \pi$, be a principal curvature of the second fundamental form A_η at $y = \gamma(r)$ of the Kähler submanifold N. Then the real hypersurface M has a principal curvature $\cot(r - \theta)$ at $x = \gamma(0)$.

And we use the following proposition given by S. Maeda:

Proposition 4.2 (Ref. 8). *Let M be a real hypersurface of a complex projective space CP^m. If $A\xi = 0$, except for the null set on which the focal map ϕ_r degenerates, M is locally congruent to one of the following:*

(a) a homogeneous real hypersurface which lies on a tube of radius $\pi/4$ over a totally geodesic CP^k $(1 \leq k \leq m-1)$,

(b) a nonhomogeneous real hypersurface which lies on a tube of radius $\pi/4$ over a Kähler submanifold N with nonzero principal curvatures $\neq \pm 1$.

From these results and Theorem 3.1, we have the following

Theorem 4.1. *Let M be a compact n-dimensional minimal CR submanifold of a complex projective space CP^m which is not a complex submanifold of CP^m. If the Ricci tensor S of M satisfies $S(X,X) \geq (n-1)g(X,X)$ for any vector X tangent to M, then M is congruent to one of the following:*

(a) a totally geodesic real projective space RP^n of CP^m,

(b) a pseudo-Einstein real hypersurface $M^c((n-1)/4, \pi/4)$ of some $CP^{(n+1)/2}$ in CP^m,

(c) a real hypersurface of some $CP^{(n+1)/2}$ in CP^m which lies on a tube of radius $\pi/4$ over certain Kähler submanifold N with principal curvatures $\cot\theta$, $0 < \theta \leq \pi/12$.

Remark 4.1. The author does not know examples of certain Kähler submanifold N having the properties required in Case (c) in Theorem 4.1.

Corollary 4.1. *Let M be a compact n-dimensional minimal proper CR submanifold of a complex projective space CP^m. If the Ricci tensor S of*

M satisfies $S(X,X) \geq (n-1)g(X,X)$, then M is congruent to one of the following:

(a) *a pseudo-Einstein real hypersurface $M^c((n-1)/4, \pi/4)$ of some $CP^{(n+1)/2}$ in CP^m,*

(b) *a real hypersurface of some $CP^{(n+1)/2}$ in CP^m which lies on a tube of radius $\pi/4$ over certain Kähler submanifold N with principal curvatures* $\cot\theta$, $0 < \theta \leq \pi/12$.

Using the theorem given by Maeda,[7] we have

Corollary 4.2. *Let M be a compact n-dimensional minimal proper CR submanifold of a complex projective space CP^m. If the Ricci tensor S satisfies $(n-1)g(X,X) \leq S(X,X) \leq (n+1)g(X,X)$, then M is congruent to a pseudo-Einstein real hypersurface $M^c((n-1)/4, \pi/4)$ of some $CP^{(n+1)/2}$ in CP^m.*

Moreover, we obtain the following theorem.

Theorem 4.2. *Let M be a compact n-dimensional minimal CR submanifold of a complex projective space CP^m. If the Ricci tensor S of M satisfies $S(X,X) \geq (n-1)g(X,X) + g(PX,PX)$ for any vector X tangent to M, then M is congruent to one of the following:*

(a) *a totally geodesic real projective space RP^n of CP^m,*

(b) *a totally geodesic complex projective space $CP^{n/2}$ of CP^m,*

(c) *a complex $(n/2)$ dimensional complex quadric $Q^{(n/2)}$ of some $CP^{n/2+1}$ of CP^m,*

(d) *a pseudo-Einstein real hypersurface $M^c((n-1)/4, \pi/4)$ of some $CP^{(n+1)/2}$ in CP^m,*

(e) *a real hypersurface of some $CP^{(n+1)/2}$ in CP^m which lies on a tube of radius $\pi/4$ over certain Kähler submanifold N with principal curvatures* $\cot\theta$, *where θ satisfies* $0 < \sin 2\theta \leq 1/3$.

Remark 4.2. About the proof above, in 1974, Chen and Ogiue[2] proved that if the Ricci curvature of n-dimensional Kähler submanifold of CP^m is everywhere equal to $n/2$, then M is locally Q^n in some CP^{n+1} in CP^m (see also Ogiue[9]).

We suppose that M is a compact n-dimensional minimal CR submanifold of a complex projective space CP^m. When the Ricci tensor S of M satisfies $S(X,X) \geq (n-1)g(X,X) + 2g(PX,PX)$ for any vector X tangent to M, the cases (c) and (e) in Theorem 4.2 do not occur. Thus we obtain a result by Masahiro Kon:

Theorem 4.3 (Ref. 6). *Let M be a compact n-dimensional minimal CR submanifold of a complex projective space CP^m. If the Ricci tensor S of M satisfies $S(X,X) \geq (n-1)g(X,X)+2g(PX,PX)$ for any vector X tangent to M, then M is equivalent to one of the following:*

(*a*) *a totally geodesic real projective space RP^n of CP^m,*

(*b*) *a totally geodesic complex projective space $CP^{n/2}$ of CP^m,*

(*c*) *a pseudo-Einstein real hypersurface $M^c((n-1)/4, \pi/4)$ of some $CP^{(n+1)/2}$ in CP^m.*

References

1. K. Abe, Applications of Riccati type differential equation to Riemannian manifolds with totally geodesic distribution, *Tôhoku Math. J.* **25** (1973) 425–444.
2. B-Y. Chen and K. Ogiue, A characterization of the complex sphere. *Michigan Math. J.* **21** (1974) 231–232.
3. T.E. Cecil and P.J. Ryan, Focal sets and real hypersurfaces in complex projective space, *Trans. Amer. Math. Soc.* **269** (1982) 481–499.
4. Masahiro Kon, On some complex submanifolds in Kaehler manifolds, *Canad. J. Math.* **26** (1974) 1442–1449.
5. Masahiro Kon, Generic minimal submanifolds of a complex projective space, *Bull. London Math. Soc.* **12** (1980) 355–360.
6. Masahiro Kon, Minimal CR submanifolds immersed in a complex projective space, *Geom. Dedicata* **31** (1989) 357–368.
7. S. Maeda, Real hypersurfaces of a complex projective space II, *Bull. Austral. Math. Soc.* **29** (1984) 123–127.
8. S. Maeda, Ricci tensors of real hypersurfaces in a complex projective space, *Proc. Amer. Math. Soc.* **122** (1994) 1229–1235.
9. K. Ogiue, Differential geometry of Kaehler submanifolds, *Advances in Math.* **13** (1974) 73–114.
10. J. Simons, Minimal varieties in riemannian manifolds, *Ann. of Math.* **88** (1968) 62–105.
11. B. Smith, Differential geometry of complex hypersurfaces, *Ann. of Math.* **85** (1967) 246–266.
12. M. Yamagata and Masahiro Kon, Reduction of the codimension of a generic minimal submanifold immersed in a complex projective space, *Coll. Math.* **74** (1997) 185–190.
13. K. Yano and Masahiro Kon, *Structures on manifolds* (World Scientific Publishing, Singapore, 1984).

Differential Geometry and its Applications
Proc. Conf., in Honour of Leonhard Euler, Olomouc, August 2007

Unique determination of domains

A.P. Kopylov

Sobolev Institute of Mathematics,
Akademik Koptyug Pr. 4, Novosibirsk, 630090, Russia;
Novosibirsk State University,
Pirogova street 2, Novosibirsk, 630090, Russia
E-mail: kopylov@math.nsc.ru

This paper is an extension of the author's survey lecture with the same title given at the 10th International Conference on Differential Geometry and its Applications dedicated to celebrations of the 300th anniversary of birth of Leonhard Euler, which was held in Olomouc (Czech Republic) in August 2007. The paper is devoted to two new trends in differential geometry connected with investigation of the classical problems concerning the uniqueness of determination of closed convex surfaces by their intrinsic metrics.

Keywords: Unique determination, intrinsic and relative metrics of boundary, relative conformal modulus, conformal-type unique determination.

MS classification: 53C24, 53A30.

1. Introduction

A classical theorem says (see, e.g., Ref. 1) that: *If two bounded closed convex surfaces in Euclidean 3-space are isometric in their intrinsic metrics, then they are equal, i.e., can be superposed by a motion.*

The problem of unique determination of closed convex surfaces by means of their intrinsic metrics dates back to Cauchy (1813) who proved *the unique determination of convex polyhedrons (in Euclidean 3-space $\mathbb{R}^3$) by their unfoldings*. Since then this problem became a subject of investigation by a number of mathematicians for about 150 years, e.g., it was studied by Minkowski, Hilbert, Weyl, Blashke, Cohn-Vossen, Aleksandrov, Pogorelov and other prominent mathematicians (see, for instance, a historical survey in Ref. 1, Chapter 3); finally, its complete solution, which is just the theorem we have cited at the beginning, was obtained by A. V. Pogorelov. Regarding generalizations of Pogorelov's result to higher dimensions see Ref. 2.

In Ref. 3 a new approach to the problem of unique determination of surfaces was suggested which provided an essentially larger framework for that problem. The following model situation fairly well illustrates the essence of that approach.

Let U_1 and U_2 be domains (i.e., open connected sets) in the real Euclidean n-space $\mathbb{R}^n$ whose closures cl U_j, where $j = 1, 2$, are Lipschitz manifolds (such that $\text{fr}\,(\text{cl}\,U_j) = \text{fr}\,U_j \neq \varnothing$ where fr E is the boundary of $E(\subset \mathbb{R}^n)$ in $\mathbb{R}^n$). This means that for each point $x \in \text{cl}\,U_j$, there are a neighborhood $V_{j,x}$ and a bi-Lipschitz mapping $h_{j,x} : V_{j,x} \to B(0,1)$ of this neighborhood onto the standard unit ball $B(0,1) = \left\{x = (x_1, x_2, \ldots, x_n) \in \mathbb{R}^n \;\middle|\; |x| = \left(\sum\limits_{s=1}^{n} (x_s)^2\right)^{1/2} < 1\right\}$ such that either $h_{j,x}(U_j \cap V_{j,x}) = B(0,1)$ or $h_{j,x}(U_j \cap V_{j,x}) = \{x \in B(0,1) \mid x_1 \le 0\}$. Assume that the boundaries fr U_1 and fr U_2 of these domains (which coincide with borders of manifolds cl U_1 and cl U_2) are isometric with respect to their relative metrics $\rho_{\text{fr}\,U_j, U_j}$ $(j = 1, 2)$. These metrics are the restrictions to the boundaries fr U_j of the extensions (by continuity) to cl U_j of the inner metrics of the domains U_j. The natural question arises: *Under what additional conditions are the domains U_1 and U_2 isometric*? We must stress that the problem of unique determination of closed convex surfaces mentioned at the beginning of the paper, is the most important particular case of the stated question. To see this, assume that S_1 and S_2 are closed convex surfaces in $\mathbb{R}^3$, i.e., are boundaries of two bounded convex domains $G_j \subset \mathbb{R}^3$ $(j = 1, 2)$. Let $U_j = \mathbb{R}^3 \setminus \text{cl}\,G_j$. Then, the intrinsic metrics on the surfaces $S_j = \text{fr}\,U_j$ coincide with the relative metrics $\rho_{\text{fr}\,U_j, U_j}$ on the boundaries of the domains U_j. This clearly shows that the problem of the unique determination of closed convex surfaces via their intrinsic metrics is indeed a particular case of the problem of the unique determination of domains via the relative metrics on their boundaries.

The generalization of the problem of the unique determination of surfaces ensuing from a new approach suggested in Ref. 3 manifests itself by the fact that the unique determination of domains by the relative metrics on their boundaries takes place not only when their complements are bounded and convex but also in the following cases. *Domain U_1 is bounded and convex, while domain U_2 is arbitrary* (A.P. Kopylov (see Theorems 2.1, 4.1 below)). *Domain U_1 is strictly convex, U_2 is arbitrary* (A.D. Aleksandrov (Theorem 3.1)). *Domains U_1, U_2 are bounded and have smooth boundaries* (V.A. Aleksandrov (Theorem 2.2)). *Domains U_1, U_2 have nonempty bounded complements, while their boundaries are $(n-1)$-dimensional con-*

nected C^1-manifolds without boundaries, $n \geq 3$ (V.A. Aleksandrov (Theorem 3.2)); etc.

The present paper can be divided into two parts.

The first one mainly contains a survey of results obtained since 1984 which pertain to the unique determination of domains by the relative metrics on their boundaries (see Sections 2-6 of the paper).

The second (very important) part is represented in Section 7. It contains a one more new approach in studying a unique determination of domains in Euclidean spaces. This approach is closely connected with the approach discussed in the first part and deals with the examination of a conformal-type unique determination problem. The main result in the second part is Theorem 7.1 which treats the unique determination of bounded convex polyhedral domains in terms of relative conformal moduli of their boundary condensers.

2. Theorems of unique determination of bounded domains

The first result concerning the problem of the unique determination of domains by the relative metrics of their boundaries was as follows (see Ref. 3).

Theorem 2.1. *Let $n \geq 2$ and $U_1 \subset \mathbb{R}^n$ be a bounded convex domain, and let $U_2 \subset \mathbb{R}^n$ be a bounded domain whose boundary is an $(n-1)$-dimensional C^0-manifold without boundary. Suppose also that U_2 has the property that the inner metric of U_2 may be extended by continuity to* cl U_2. *Furthermore, assume that the boundaries* fr U_1 *and* fr U_2 *of the domains U_1 and U_2 are isometric in their relative metrics, i.e., there exists a surjective and isometric in the relative metrics of boundaries, mapping f :* fr $U_1 \to$ fr U_2 $(\rho_{\mathrm{fr}\ U_2,U_2}(f(x'), f(x'')) = \rho_{\mathrm{fr}\ U_1,U_1}(x', x'')$ *if* $x', x'' \in$ fr $U_1)$. *Then U_1 and U_2 are isometric in Euclidean metrics.*

Remark 2.1. By a C^k-manifold $(k = 0, 1, 2, \ldots, \infty)$, real analytic manifold and Lipschitz manifold we mean corresponding submanifolds of $\mathbb{R}^n$ (regarding definition of Lipschitz n-dimensional manifold see Section 1; Lipschitz manifold of arbitrary dimension is defined analogously). The relative metric $\rho_{\mathrm{fr}\ U_j,U_j}$ $(j = 1, 2)$ of the boundary fr U_j of U_j is the metric being the restriction to fr U_j of the extension (by continuity) to cl U_j of the inner metric of U_j.

Remark 2.2. Although Theorem 2.1 may seem more general than Theorem 3 of Ref. 3, they are, in fact, equivalent. Moreover, Theorem 2.1 could be proved repeating verbatim the argumentations in Theorem 3 of Ref. 3.

We should stress that by virtue of Theorem 2.1, unlike classical Cauchy point of view on problems of unique determination of surfaces, a new approach to these problems (based on notions from Ref. 3; see also Definition 2.1 below) provides a unified and comprehensive examination of them even for $n = 2$.

Next Theorem by V.A. Aleksandrov[4] shows that a new approach permits to eliminate such assumptions as convexity of domains and connectivity of their boundaries what turns out to be redundant.

Theorem 2.2. *Let U_1 and U_2 be bounded domains in $\mathbb{R}^n$, $n \geq 2$, such that each* $\mathrm{cl}\, U_j$, $j = 1, 2$, *is an n-dimensional C^1-manifold with boundary* $\mathrm{fr}\, U_j$. *Suppose that* $\mathrm{fr}\, U_1$ *and* $\mathrm{fr}\, U_2$ *are isometric in their relative metrics. Then the domains U_1 and U_2 are isometric too.*

Note that Theorem 2.2 could be generalized to domains with piecewise smooth boundaries (see Ref. 4).

We will now state the propositions of that Section in a somewhat different form which will serve subsequently as a model pattern. To this end, denote by $\mathcal{A} = \mathcal{A}(n)$ the class of all bounded domains U in $\mathbb{R}^n$, $n \geq 2$, with the following properties. The closure $\mathrm{cl}\, U$ of U is an n-dimensional C^0-manifold with boundary $\mathrm{fr}\, U$, and the inner metric of U can be extended by continuity to its closure. Next we introduce the following notion which is one of the most important in our paper.

Definition 2.1. Let $\mathcal{A}_0 \subset \mathcal{A}$ and $U \in \mathcal{A}_0$. We will say that *a domain U is uniquely determined in the class $\mathcal{A}_0$ by the relative metric on its boundary* if the following condition is satisfied. Assume that a domain V belongs to the class $\mathcal{A}_0$ and there is a surjection $f : \mathrm{fr}\, V \to \mathrm{fr}\, U$ preserving the relative metric of the boundary. This means that for any two points $x', x'' \in \mathrm{fr}\, V$, we have $\rho_{\mathrm{fr}\, U,U}(f(x'), f(x'')) = \rho_{\mathrm{fr}\, V,V}(x', x'')$. Then V can be isometrically mapped onto U. In other words, U could be determined in $\mathcal{A}_0$ up to an additional isometric transformation of $\mathbb{R}^n$.

Denote by $\mathcal{A}_1 = \mathcal{A}_1(n)$ the class of all bounded domains U in $\mathbb{R}^n$, $n \geq 2$, whose closures are n-dimensional C^1-manifolds with boundary $\mathrm{fr}\, U$. Using Definition 2.1, we may now restate Theorems 2.1 and 2.2 as follows.

Each bounded convex domain in $\mathbb{R}^n$, $n \geq 2$, is uniquely determined in the class $\mathcal{A}(n)$ by the relative metric on its boundary (Theorem 2.1).

If a domain U belongs to $\mathcal{A}_1(n)$, $n \geq 2$, then it is uniquely determined in the class $\mathcal{A}_1(n)$ by the relative metric on its boundary (Theorem 2.2).

3. Case of unbounded domains

For unbounded domains, the unique determination problem by the relative metrics on their boundaries differs essentially as compared to the case of bounded domains.

Indeed, the half-space

$$\mathbb{R}^n_- = \{(x_1, x_2, \ldots, x_{n-1}, x_n) \in \mathbb{R}^n \mid x_n < 0\},$$

$n \geq 2$ (and consequently, each n-dimensional open half-space of $\mathbb{R}^n$), is not uniquely determined by the relative metric on its boundary, even in the class of all domains whose boundaries are real analytic $(n-1)$-dimensional manifolds without boundary. To see this, let us consider, e.g., the domain

$$U = \{(x_1, x_2, \ldots, x_{n-1}, x_n) \in \mathbb{R}^n \mid x_n < (x_{n-1})^2\}.$$

The boundary of this domain is isometric to the boundary of $\mathbb{R}^n_-$ (in the intrinsic metrics on boundaries). It is not difficult to verify that the mapping $f : \mathrm{fr}\, U \to \mathrm{fr}\, \mathbb{R}^n_-$,

$$f : (x_1, x_2, \ldots, x_{n-2}, x_{n-1}, (x_{n-1})^2) \mapsto (x_1, x_2, \ldots, x_{n-2},$$
$$2^{-1}x_{n-1}\sqrt{1 + 4(x_{n-1})^2} + 4^{-1}\ln(2x_{n-1} + \sqrt{1 + 4(x_{n-1})^2}), 0),$$
$$(x_1, x_2, \ldots, x_{n-1}) \in \mathbb{R}^{n-1},$$

is surjective and isometric in the intrinsic metrics of boundaries of U and $\mathbb{R}^n_-$. Since $\mathbb{R}^n \setminus U$ is convex, we conclude that f is isometric in the relative metrics on boundaries $\mathrm{fr}\, U$ and $\mathrm{fr}\, \mathbb{R}^n_-$. But at the same time these domains are not isometric.

Before starting discussion of results in this Section, we must note that as previously, we consider only domains $U \subset \mathbb{R}^n$, $\neq \mathbb{R}^n$, admitting extension by continuity of inner metrics ρ_U to $\mathrm{cl}\, U$.

The following result is due to A. D. Aleksandrov (see Ref. 4).

Theorem 3.1. *Let U_1 be a strictly convex domain (i.e., a convex domain whose boundary does not contain any linear interval). Assume that U_2 is any domain such that* fr U_1 *and* fr U_2 *are isometric in their relative metrics $\rho_{\mathrm{fr}\, U_1, U_1}$ and $\rho_{\mathrm{fr}\, U_2, U_2}$. Then U_1 and U_2 are isometric.*

Let $\mathcal{B} = \mathcal{B}(n)$ be a class of domains U in $\mathbb{R}^n$ $(n \geq 2)$ with nonempty bounded complements, whose boundaries are connected $(n-1)$-dimensional C^1-manifolds without boundaries. The following result is obtained in Ref. 5.

Theorem 3.2. *Let $n \geq 3$. Then each domain $U \in \mathcal{B}$ is uniquely determined in the class $\mathcal{B}$ by the relative metric on its boundary.*

Remark 3.1. The notion of unique determination of unbounded domains is introduced analogously to Definition 2.1.

Theorem 3.2 is no longer true for $n = 2$ what could be seen in the following example.

Example 3.1. Let U_1 and U_2 be unbounded domains in $\mathbb{R}^2$ with convex bounded complements $\mathbb{R}^2 \setminus U_j$, satisfying $\mathbb{R}^2 \setminus \mathrm{cl}\, U_j \neq \varnothing$ $(j = 1, 2)$ and such that their boundaries fr U_1, fr U_2 are of the same length. Then fr U_1 and fr U_2 are isometric in their intrinsic, hence in relative, metrics. But at the same time, the domains U_1 and U_2 need not be isometric.

Next Theorem proved by A. V. Kuz'minykh,[6] gives a complete solution to the problem of unique determination of convex domains in the class $\mathcal{C} = \mathcal{C}(n)$ of all domains U in $\mathbb{R}^n$, $n \geq 2$, admitting extensions by continuity of their inner metrics to the closures cl U.

Theorem 3.3. *Each convex domain in $\mathbb{R}^n$, $n \geq 2$, different from $\mathbb{R}^n$ and from any open half-space of $\mathbb{R}^n$, is uniquely determined in $\mathcal{C}(n)$ by the relative metric on its boundary.*

Observe that Theorems 2.1, 3.1 are the particular cases of Theorem 3.3.

Let $\Delta(U)$ denote the interior of the convex hull of the complement of U in $\mathbb{R}^n$, $n \geq 2$: $\Delta(U) = \mathrm{int}\,(\mathrm{conv}\,(\mathbb{R}^n \setminus U))$. The following proposition holds (see Ref. 7).

Theorem 3.4. *Assume that the boundary* fr U *of a domain $U \subset \mathbb{R}^n$ $(n \geq 2)$ is a connected $(n-1)$-dimensional C^1-manifold without boundary and satisfies the condition* fr $U \subset \Delta(U)$. *Then U is uniquely determined by the relative metric on its boundary in the class $\mathcal{A}_1^* = \mathcal{A}_1^*(n)$ of all domains in $\mathbb{R}^n$ with smooth boundaries (i.e., domains V whose closures* cl V *are n-dimensional C^1-manifolds with boundary* fr V).

The following result (see Ref. 7) is valid for domains with analytic boundaries, i.e., domains $V \subset \mathbb{R}^n$ $(n \geq 2)$ such that their closures cl V are n-dimensional (real) analytic manifolds with boundary fr V.

Theorem 3.5. *Each domain $U \subset \mathbb{R}^n$ $(n \geq 2)$ with connected analytic boundary satisfying* fr $U \cap \Delta(U) \neq \varnothing$, *is uniquely determined by the relative metric on its boundary in the class $\mathcal{E} = \mathcal{E}(n)$ of all domains in $\mathbb{R}^n$ with analytic boundaries.*

4. Domains with Hausdorff boundary

Domains U we considered above satisfy the condition that the inner metric of U can be extended by continuity onto cl U. But this condition excludes the case of domains whose (Euclidean) boundaries have very simple configuration. For instance, such is an open cube in $\mathbb{R}^3$ "incised" with a half-plane

$$U_1=\{x=(x_1,x_2,x_3)\in\mathbb{R}^3 \mid |x_j|<1, j=1,2,3\}\setminus\{x\in\mathbb{R}^3 \mid x_1\geq 0, x_3=0\}.$$

One more example is the domain

$$U_2=\mathbb{R}^3\setminus\left\{x\in\mathbb{R}^3 \mid \sum_{j=1}^{3}(x_j)^2\leq 1, x_3=0\right\},$$

which is a complement of a disk with respect to $\mathbb{R}^3$.

This drawback could be eliminated if we pass from Euclidean boundaries discussed in previous Sections, to what we call Hausdorff boundaries defined as follows.

Definition 4.1. Let $U\subset\mathbb{R}^n$ $(n\geq 2)$ be a domain and ρ_U its inner metric. Consider a completion of the metric space (U,ρ_U) and let us identify each point $x\in U$ with the point corresponding to x in the completion. Then eliminating U from the completion we obtain a metric space $(\mathrm{fr}_H\, U, \rho_{\mathrm{fr}_H\, U,U})$. The set $\mathrm{fr}_H\, U$ is called *the Hausdorff boundary of the domain* U, while $\rho_{\mathrm{fr}_H\, U,U}$ is called *the relative metric of its Hausdorff boundary.*

Remark 4.1. If a domain U admits an extension by continuity of its inner metric ρ_U onto cl U then the Hausdorff boundary $\mathrm{fr}_H\, U$ of U is identified in a natural way with its Euclidean boundary.

Definition 4.2. Let $\mathcal{U}$ be a class of domains in $\mathbb{R}^n$, $n\geq 2$. We say that *a domain $U\in\mathcal{U}$ is uniquely determined in $\mathcal{U}$ by the relative metric on its Hausdorff boundary* if each domain $V\in\mathcal{U}$, whose Hausdorff boundary is isometric in the relative metrics to the Hausdorff boundary of U, is itself isometric to U (in Euclidean metrics).

Remark 4.2. The isometry of the Hausdorff boundaries of domains U and V in their relative metrics means that there exists a surjective isometric mapping $f:(\mathrm{fr}_H\, U,\rho_{\mathrm{fr}_H\, U,U})\to(\mathrm{fr}_H\, V,\rho_{\mathrm{fr}_H\, V,V})$ of these boundaries.

There are domains U such that U is uniquely determined by the relative metric on its boundary in one class of domains but this is not true in some another, larger class. In this connection, there arises a question about the existence of domains uniquely determined by the relative metric on their boundaries in the class of all domains in $\mathbb{R}^n$. One of possible answers to this question is given by the following proposition which generalizes and strengthens Theorem 2.1 in the case $n = 2$.

Theorem 4.1. *Each bounded convex domain $U \subset \mathbb{R}^2$ is uniquely determined in the class of all domains $V \subset \mathbb{R}^2$ by the relative metric on its Hausdorff boundary (in the sense of Definition* 4.2*).*

The next Theorem (see Ref. 8) gives an example of domains in $\mathbb{R}^n$ of any dimension $n \geq 2$ for which the solution of the unique determination problem appeals necessarily to considering Hausdorff boundaries.

Theorem 4.2. *Each bounded domain $U \subset \mathbb{R}^n$ $(n \geq 2)$ with polyhedral boundary is uniquely determined in the class of all such domains by the relative metric of its Hausdorff boundary.*

By a polyhedral domain (in Theorem 4.2) we mean a domain whose boundary is a polyhedron, i.e., the union of a finitely many cells (possibly of different dimensions), where a cell is defined to be a bounded set which is the intersection of a finite number of closed half-spaces.

Now in conclusion, making use of the notion of Hausdorff boundary, let us give the full statement of A. V. Pogorelov's Theorem on the unique determination of closed convex domains as follows.

Each domain $U \subset \mathbb{R}^3$ whose complement $\mathbb{R}^3 \setminus U$ is a bounded convex set, is uniquely determined in the class of all such domains by the relative metric of its Hausdorff boundary.

Remark 4.3. The case when the complement $\mathbb{R}^3 \setminus U$ of U is a one-dimensional segment was not included in the statement of the Theorem. But since this particular case is trivial, the formulation of Pogorelov's Theorem stated above may be regarded as complete.

In conclusion of the Section, we call reader's attention to very interesting results of recent M. V. Korobkov's works.[9,10] In these papers M.V. Korobkov obtained *the necessary and sufficient conditions for unique determination of a plane domain in the class of all domains by the relative metric on its Hausdorff boundary.*

5. Unique determination of domains (local variant)

In previous Sections we dealt with the problem of the unique determination of domains by the relative metrics of their boundaries. In the present Section we give some results permitting to answer the following natural question. *Is it sufficient for the isometry of domains in Euclidean spaces to assume that their boundaries (Euclidean, Hausdorff) are isometric in their relative metrics in, so to say, some local sense?*

Definition 5.1. Let $\mathcal{M}$ be a class of domains in $\mathbb{R}^n$, $n \geq 2$. We say that *a domain $U \in \mathcal{M}$ is uniquely determined in $\mathcal{M}$ by a condition of local isometry in the relative metrics of Hausdorff boundaries of domains,* if the assumptions 1) $V \in \mathcal{M}$ and 2) $\operatorname{fr}_H V$ is locally isometric in the relative metrics to $\operatorname{fr}_H U$ imply the isometry of U and V (in Euclidean metrics). *The local isometry in the relative metrics of Hausdorff boundaries* $\operatorname{fr}_H U$, $\operatorname{fr}_H V$ *of domains* U, V means that there exists a bijection $f : \operatorname{fr}_H U \to \operatorname{fr}_H V$ *locally isometric in the relative metrics of* $\operatorname{fr}_H U$ *and* $\operatorname{fr}_H V$. In other words, f is such that for each $y \in \operatorname{fr}_H U$, there is $\varepsilon > 0$ satisfying the condition: for each two elements a, b belonging to the ε-neighborhood $Z(y) = \{z \in \operatorname{fr}_H U \mid \rho_{\operatorname{fr}_H U,U}(z,y) < \varepsilon\}$ of y, the equality $\rho_{\operatorname{fr}_H U,U}(a,b) = \rho_{\operatorname{fr}_H V,V}(f(a),f(b))$ holds.

Remark 5.1. The notion of a unique determination of domains by the condition of local isometry in the relative metrics of Euclidean boundaries is introduced analogously. Here we assume that the domains admit extension by continuity of their inner metrics onto the closures of domains.

First results related to the discussed in this Section approach in the study of the unique determination of domains by the condition of local isometry of their boundaries in the relative metrics, were obtained by A. V. Kuz'minykh.[6] One of them is the following assertion.

Theorem 5.1. *Let $U \subset \mathbb{R}^n$ $(n \geq 2)$ be a convex domain such that* $\operatorname{fr} U$ *does not contain any strip, i.e., a convex hull of two parallel $(n-2)$-dimensional planes. Assume that $V \in \mathbb{R}^n$ is a domain admitting extension by continuity of its inner metric ρ_V onto* $\operatorname{cl} V$. *Let $\varepsilon > 0$ be given and suppose that there exists a bijection $f : \operatorname{fr} U \to \operatorname{fr} V$ such that for each $\alpha \in \,]0,\varepsilon[$ and for each points $x', x'' \in \operatorname{fr} U$, the equality $\rho_{\operatorname{fr} U,U}(x',x'') = \alpha$ holds if and only if $\rho_{\operatorname{fr} V,V}(f(x'),f(x'')) = \alpha$. Then f can be extended to an isometry $F : U \to V$ in the Euclidean metrics.*

The next result is obtained by M. K. Borovikova.[11] It contains a complete description of the case where a bounded polygonal domain in $\mathbb{R}^2$ is uniquely determined in the class of all such domains by the condition of local isometry of Hausdorff boundaries in the relative metrics.

Theorem 5.2. *In order that a bounded polygonal domain $U \subset \mathbb{R}^2$ be uniquely determined in the class of all such domains by the condition of local isometry of Hausdorff boundaries in the relative metrics, it is necessary and sufficient that U be convex.*

Observe that the condition "U is bounded" cannot be rejected (see Ref. 11).

We complete this Section with a result by V. A. Aleksandrov,[12] which permits to conclude that the situation in $\mathbb{R}^3$ differs essentially from the case studied by Borovikova in Theorem 5.2 for two-dimensional polyhedral domains. Slightly coarsening, we can formulate Aleksandrov's result as follows:

If a domain $U \subset \mathbb{R}^3$ has the compact polyhedral boundary then it is uniquely determined in the class of all domains in $\mathbb{R}^3$ by the condition of local isometry in the relative metrics of Hausdorff boundaries of domains.

Note that although the boundary of U in the statement of this Theorem is compact, the domain U itself may be unbounded.

6. Unique determination and Riemannian manifolds

Suppose that (X, g) is an n-dimensional connected smooth Riemannian manifold without boundary. Let Y_j, $j = 1, 2$, be two n-dimensional compact connected C^0-submanifolds of X with boundaries ∂Y_j ($\neq \varnothing$) satisfying the following conditions: (1) if $x,\ y \in Y_j$ then

$$\rho_{Y_j}(x, y) = \liminf_{x' \to x, y' \to y; x', y' \in \text{int } Y_j} \{\inf [l(\gamma_{x', y', \text{int } Y_j})]\} < \infty,$$

where $\inf [l(\gamma_{x', y', \text{int } Y_j})]$ is the infimum of the lengths $l(\gamma_{x', y', \text{int } Y_j})$ of the smooth paths $\gamma_{x', y', \text{int } Y_j} : [0, 1] \to \text{int } Y_j$ joining x' and y' in the interior $\text{int } Y_j$ of Y_j, moreover, ρ_{Y_j} is a metric in Y_j; (2) for arbitrary two points $a,\ b \in Y_j$ there exist points $c,\ d \in \partial Y_j$ which may be joint in Y_j by a shortest curve $\gamma : [0, 1] \to Y_j$ in the metric ρ_{Y_j} such that $a,\ b \in \gamma$. Denote also by $\rho_X(x, y)$ (where $x,\ y \in X$) the infimum of the lengths $l(\gamma_{x,y,X})$ of the smooth paths $\gamma_{x,y,X} : [0, 1] \to X$ joining x and y.

Theorem 6.1. *In addition to the above-mentioned conditions, assume that* Y_1 *is strictly convex in the metric* ρ_{Y_1}, *i.e., if* α, $\beta \in Y_1$ *then, for each shortest curve* $\gamma = \gamma_{\alpha,\beta,Y_1} : [0,1] \to Y_1$ *in the metric* ρ_{Y_1} *joining* α *and* β *in* Y_1, *we have* $\gamma(t) \in \operatorname{int} Y_1$, *where* $0 < t < 1$. *Suppose also that the boundaries of* Y_1 *and* Y_2 *are isometric in the metrics* ρ_{Y_j}, $j = 1, 2$, *i.e., there exists a surjection* $f : \partial Y_1 \to \partial Y_2$ *which is an isometry in these metrics. Then* Y_2 *is strictly convex with respect to* ρ_{Y_2}.

Remark 6.1. Assume that the manifold X in Theorem 6.1 has the following property: $\rho_X(x, y) = \rho_Y(x, y)$ for any two points x, y in every n-dimensional compact connected C^0-submanifold $Y \subset X$ with boundary satisfying the condition (1) (where $Y_j = Y$) and strictly convex with respect to the metric ρ_Y. Then (by this Theorem) we obtain the following assertion (A) (on the rigidity, i.e., on the unique determination of the boundary of Y_1): *∂Y_1 and ∂Y_2 are isometric in the metric ρ_X of X.* This assertion is very close to A. D. Aleksandrov's Theorem 3.1: Theorem 3.1 is an immediate consequence of assertion (A) (in the case where the closures of domains U_j, $j = 1, 2$, in Theorem 3.1 are compact C^0-submanifolds in $\mathbb{R}^n$ (with $\partial\{\text{cl } U_j\} = \text{fr } U_j$) satisfying the condition (1)).

Proof. Theorem 6.1 is proved by methods similar to those used in the proof of Theorem 3.1 (see Ref. 4). □

7. On unique determination of conformal type

In the previous Sections we considered the problem of unique determination of domains in Euclidean spaces which has, as may be termed, an isometric type. In other words, we discussed the problem of whether given domains U, V are isometric provided that their boundaries are isometric in some sense.

In the present Section we proceed with further development of that topic discussing the problem of unique determination of conformal type of domains in $\mathbb{R}^n$.

Let us start with details concerning the contents of the Section.

Definition 7.1. Let $U \subset \mathbb{R}^n$ be a domain with Lipschitz boundary (i.e., a domain whose boundary fr U is an $(n-1)$-dimensional Lipschitz manifold without boundary). By *a boundary condenser* $F = \{F_1, F_2\}$ *of the domain* U we mean a couple of disjoint closed subsets F_1, F_2 of fr U, at least one of them being bounded. The sets F_1, F_2 are called *the components of the condenser* F (or *"plates"* of F).

Definition 7.2. *A relative conformal modulus* $M^U(F)$ *of a boundary condenser* F *of a domain* U is an n-modulus $M_n(\Gamma_{F_1,F_2,U})$ of the family $\Gamma_{F_1,F_2,U}$ of all continuous paths $\gamma : [0,1] \to \mathrm{cl}\, U$ connecting the components F_1 and F_2 of the condenser F in the domain U (i.e., we mean paths γ such that $\gamma(0) \in F_1$, $\gamma(1) \in F_2$, and $\gamma(t) \in U$ for $t \in]0,1[$).

Remark 7.1. As to the notion of n-modulus $M_n(\Gamma)$ of a family Γ of continuous curves (paths) in $\mathbb{R}^n$, see, e.g., Ref. 13.

Let $\mathcal{L}_0 = \mathcal{L}_0(n)$ be a subclass of the class $\mathcal{L} = \mathcal{L}(n)$ of all domains $U \subset \mathbb{R}^n$, $n \geq 3$, the closure of each of them being an n-dimensional Lipschitz manifold (with boundary fr $U \neq \varnothing$).

Definition 7.3. We say that *a domain* $U \in \mathcal{L}_0$ *is uniquely determined by the relative conformal moduli of their boundary condensers in the class* $\mathcal{L}_0$ if the following condition is satisfied. Assume that $V \in \mathcal{L}_0$ and there exists a homeomorphism $f : \mathrm{fr}\, V \to \mathrm{fr}\, U$ of fr V onto fr U preserving the relative conformal moduli of boundary condensers: $M^V(F) = M^U(f(F)) = M^U(\{f(F_1), f(F_2)\})$ for each boundary condenser F of V. Then the domain V can be conformally mapped onto U (in other words, U can be determined in the class $\mathcal{L}_0$ up to possibly additional Möbius transformation).

In 2006 (see Refs. 14,15), Kopylov proved the following assertion (about unique determination of convex polyhedral domains by relative conformal moduli of their boundary condensers).

Theorem 7.1. *Assume* $n \geq 4$. *Then each bounded convex polyhedral domain* $U \subset \mathbb{R}^n$ *(i.e., a nonempty bounded intersection of finitely many open n-dimensional half-spaces) is uniquely determined by the relative conformal moduli of their boundary condensers in the class* $\mathcal{P}$ *of all bounded convex polyhedral domains* $V \subset \mathbb{R}^n$. *Moreover, U can be determined in the class* $\mathcal{P}$ *up to possibly additional affine conformal mapping* $F : \mathbb{R}^n \to \mathbb{R}^n$.

It should be observed that Theorem 7.1 (as well as the rest results of our paper) may be considered as one of important applications of notions and ideas of differential geometry in the theory of conformal and (in prospect) quasiconformal mappings for study of their boundary values.

Acknowledgments

This research was carried out with the partial support of the Russian Foundation for Basic Researches (project no. 05–01–00482), the Programm of Support of Leading Scientific School (project no. NSh-8526.2006.1), the Exchange Program between the Russian and Polish Academies of Sciences (2005-2007, project "Stability and regularity of solutions to PDE's system and related problems of the theories of quasiconformal mappings and harmonic fields"), and the Interdisciplinary integration project 2007 of the Siberian Division, Russian Academy of Sciences.

References

1. A.V. Pogorelov, *Extrinsic Geometry of Convex Surfaces* (American Mathematical Society, Providence, 1973).
2. E.P. Sen'kin, Rigidity of convex hypersurfaces, *In: Ukr. geometr. sb., vyp. 12* (Izd-vo KhGU, Kharkov, 1972) 131–152.
3. A.P. Kopylov, On boundary values of mappings which are close to isometric mappings, *Siberian Math. J.* **25** (1985) 438–447.
4. V.A. Aleksandrov, Isometry of domains in $\mathbb{R}^n$ and relative isometry of their boundaries, *Siberian Math. J.* **25** (1985) 339–347.
5. V.A. Aleksandrov, Isometry of domains in $\mathbb{R}^n$ and relative isometry of their boundaries. II, *Siberian Math. J.* **26** (1986) 783–787.
6. A.V. Kuz'minykh, On isometry of domains whose boundaries are isometric in the relative metrics, *Siberian Math. J.* **26** (1986) 380–387.
7. V.A. Aleksandrov, On domains which are uniquely determined by the relative metrics of their boundaries, *In: Trudy Inst. Mat. Akad. Nauk SSSR, Sibirsk. Otd., 7: Research in Geometry and Mathematical Analysis* (Sobolev Institute of Mathematics 1987) 5–19, in Russian.
8. V.A. Aleksandrov, Unique determination of domains with non-Jordan boundaries, *Siberian Math. J.* **30** (1989) 1–8.
9. M.V. Korobkov, On necessary and sufficient conditions for unique determination of plane domains, Preprint/RAS. Sib. Div., Sobolev Institute of Mathematics, Novosibirsk, 2006, no. 174, pp. 27.
10. M.V. Korobkov, On necessary and sufficient conditions for unique determination of plane domains, to appear in *Siberian Math. J.*
11. M.K. Borovikova, On isometry of polygonal domains with boundaries locally isometric in relative metrics, *Siberian Math. J.* **33** (1993) 571–580.
12. V.A. Aleksandrov, On isometry of polyhedral domains with boundaries locally isometric in relative metrics, *Siberian Math. J.* **33** (1992) 177–182.
13. J. Väisälä, *Lectures on n-Dimensional Quasiconformal Mappings* (Springer-Verlag, Berlin-Heidelberg-New York, 1971).
14. A.P. Kopylov, Unique determination of convex polyhedral domains by relative conformal moduli of boundary condensers, *Doklady Mathematics* **74** (2006) 637–639.
15. A.P. Kopylov, Unique determination of domains in Euclidean spaces, *Sovremennaya matematika. Fundamental'nye napravleniya* **22** (2007) 139–167, in Russian.

Differential Geometry and its Applications
Proc. Conf., in Honour of Leonhard Euler, Olomouc, August 2007

Invariance of g-natural metrics on tangent bundles

Oldřich Kowalski

Faculty of Mathematics and Physics, Charles University,
Sokolovská 83, 186 75 *Praha* 8, *Czech Republic*
E-mail: kowalski@karlin.mff.cuni.cz

Masami Sekizawa

Department of Mathematics, Tokyo Gakugei University,
Koganei-shi Nukuikita-machi 4-1-1, *Tokyo* 184-8501, *Japan*
E-mail: sekizawa@u-gakugei.ac.jp

In this paper we prove that each g-natural metric on a tangent bundle TM over a Riemannian manifold (M, g) is invariant with respect to the induced map of a (local) isometry of the base manifold. Then we define natural metrics on unit tangent sphere bundles and prove their homogeneity in case that $(M\ g)$ is a two-point homogeneous space. As a corollary, we re-prove a result from a paper by Musso and Tricerri obtained by another method.

Keywords: Riemannian manifold, tangent bundle, tangent sphere bundle, g-natural metrics, homogeneity.

MS classification: 53C07, 53C20, 53C21, 53C40.

1. Introduction

There are well-known classical examples of "lifted metrics" on the tangent bundle TM over a Riemannian manifold (M, g). Namely, these are the Sasaki metric, the horizontal lift and the vertical lift. As one can see, the classical constructions are examples of "natural transformations of second order". In Ref. 14, the present authors fully classified all (possibly degenerate) naturally lifted metrics "of second order" on TM. They proved that the complete family of such natural metrics (for a fixed base metric) is a module over real functions generated by some generalizations of known classical lifts. Our idea of naturality is closely related to that of A. Nijenhuis, D. B. A. Epstein, P. Stredder and others (see Ref. 13 for the full references). We have used for our purposes the concepts and methods developed by D. Krupka[15,16] and D. Krupka and V. Mikolášová.[17] See also

I. Kolář, P. W. Michor and J. Slovák,[13] pp. 227–280, and D. Krupka and J. Janyška,[18] pp. 160–166 for other presentations of our study and for the concept of naturality in general. We shall use further the name "g-natural metrics" for our natural metrics on TM as it was proposed by M. T. K. Abbassi in Ref. 1.

One of the properties of g-natural lifts of metrics to be expected is the "invariance property" saying that the lifts of (local) isometries are again (local) isometries. After a short survey about our classification we present it in the convenient setting due to M. T. K. Abbassi and M. Sarih.[4] Then we prove that the invariance property really holds. Next we pass over to the tangent sphere bundles. We define g-natural metrics on tangent sphere bundles T_rM of constant radius $r > 0$ as restrictions of g-natural metrics on TM to the hypersurfaces T_rM of TM, and we present an explicit formula belonging to M. T. K. Abbassi. Then we prove that each tangent sphere bundle T_rM equipped with a g-natural metric $\tilde{G}$ over a two-point homogeneous space is a homogeneous space. The local version of this statement follows automatically. As a corollary, we re-prove the result by E. Musso and F. Tricerri in Ref. 19 saying that the homogeneity holds for the Sasaki metric. Another consequence is that, over a space (M, g) of constant curvature, a tangent sphere bundle T_rM with a g-natural metric is locally homogeneous.

2. Lifts of vectors

The tangent bundle TM over a smooth manifold M consists of all pairs (x, u), where x is a point of M and u is a vector from the tangent space M_x of M at x. We denote by p the natural projection of TM to M defined by $p(x, u) = x$.

Let g be a Riemannian metric on the manifold M and ∇ its Levi-Civita connection. Then the tangent space $(TM)_{(x,u)}$ of TM at $(x, u) \in TM$ splits into the horizontal and vertical subspace $H_{(x,u)}$ and $V_{(x,u)}$ with respect to ∇:

$$(TM)_{(x,u)} = H_{(x,u)} \oplus V_{(x,u)}.$$

If a point $(x, u) \in TM$ and a vector $X \in M_x$ are given, then there exists a unique vector $X^h \in H_{(x,u)}$ such that $p_*(X^h) = X$. We call X^h the *horizontal lift* of X to TM at (x, u). The *vertical lift* of X to (x, u) is a unique vector $X^v \in V_{(x,u)}$ such that $X^v(\mathrm{d}f) = Xf$ for all smooth functions f on M. Here we consider a one-form $\mathrm{d}f$ on M as a function on TM defined by $(\mathrm{d}f)(x, u) = uf$ for all $(x, u) \in TM$. The map $X \longmapsto X^h$

is an isomorphism between M_x and $H_{(x,u)}$, and the map $X \longmapsto X^\imath$ is an isomorphism between M_x and $V_{(x,u)}$. In an obvious way we can define horizontal and vertical lifts of vector fields on M. These are uniquely defined vector fields on TM.

3. g-natural metrics

We say that a bundle morphism of the form $\zeta : TM \oplus TM \oplus TM \longrightarrow M \times \mathbb{R}$ is an *F-metric* on M if it is linear in the second and the third argument (and smooth in the first argument). We also say that ζ is symmetric or skew-symmetric if it is symmetric or skew-symmetric with respect to the second and third argument, respectively. (We use here the letter "F" to recall the Finsler geometry.) Any Riemannian metric g on M is a symmetric F-metric which is independent on u. In our special case, letting g be a given Riemannian metric on M, we speak about *natural F-metrics derived from* g which are F-metrics ζ, for a fixed $u \in TM$, whose components $\zeta(u)_{ij} = \zeta(u, \partial/\partial x^i, \partial/\partial x^j)$ with respect to a system of local coordinates $(x^1, x^2, \dots, x^n)$ in M are solutions of the system of differential equations

$$2\sum_{a=1}^{n} g_{ap}\frac{\partial \zeta_{ij}}{\partial g_{aq}} - u^q \frac{\partial \zeta_{ij}}{\partial u^p} = \zeta_{ip}\delta_j^q + \zeta_{pj}\delta_i^q, \quad i,j,p,q = 1,2,\dots,n.$$

We obtain

Theorem 3.1 (Ref. 14). *Let (M,g) be an n-dimensional oriented Riemannian manifold. Then all natural F-metrics ζ on M derived from g are given as follows:*

(1) *For $n = 2$, all symmetric natural F-metrics are of the form*

$$\zeta(u;X,Y) = \alpha(\|u\|^2)g(X,Y) + \beta(\|u\|^2)g(X,u)g(Y,u) + \gamma(\|u\|^2)\{g(X,u)g(Y,Ju) + g(X,Ju)g(Y,u)\},$$

and all skew-symmetric natural F-metrics are of the form

$$\zeta(u;X,Y) = \delta(\|u\|^2)\{g(X,u)g(Y,Ju) - g(X,Ju)g(Y,u)\},$$

where α, β, γ and δ are arbitrary smooth functions of $\|u\|^2 = g(u,u)$ and J is one of the two canonical almost complex structures of (M,g) (for which (M,g,J) is a Kähler manifold).

(2) *For $n = 3$, all symmetric natural F-metrics are of the form*

$$\zeta(u;X,Y) = \alpha(\|u\|^2)g(X,Y) + \beta(\|u\|^2)g(X,u)g(Y,u) \tag{1}$$

and all skew-symmetric natural F-metrics are of the form

$$\zeta(u; X, Y) = \varphi(\|u\|^2)g(X \times Y, u),$$

where α, β *and* φ *are arbitrary smooth functions of* $\|u\|^2 = g(u,u)$, *and* $X \times Y$ *is the usual vector product of* X *and* Y.

(3) *For* $n > 3$, *all natural F-metrics are symmetric and of the form* (1).

I. Kolář, P. W. Michor and J. Slovák have given in Ref. 13, Proposition 33.22 a new and elegant proof of the classification of natural F-metrics on non-oriented Riemannian manifolds.

Theorem 3.2 (Ref. 13). *Let* (M, g) *be an* n*-dimensional non-oriented Riemannian manifold,* $n \geq 1$. *Then all natural F-metrics* ζ *on* M *derived from* g *are symmetric and given by* (1), *where* α *and* β *are arbitrary smooth functions defined on the interval* $(0, \infty)$. *In particular,* $\beta = 0$ *if* $n = 1$.

M. T. K. Abbassi has proved in Ref. 1 explicitly that *all basic functions from Theorems 3.1 and 3.2 can be prolonged to smooth functions on the set of all non-negative real numbers.* This result found many applications in the techniques used for the thorough investigation of g-natural metrics by M. T. K. Abbassi.

For a given Riemannian metric g on M, we define the *classical lifts of F-metrics from* M *to* TM, *with respect to* g, as follows:

(a) The *Sasaki lift* $\zeta^{s,g}$ of a symmetric F-metric ζ with respect to g is defined by

$$\zeta^{s,g}_{(x,u)}(X^h, Y^h) = \zeta_x(u; X, Y), \qquad \zeta^{s,g}_{(x,u)}(X^h, Y^v) = 0,$$
$$\zeta^{s,g}_{(x,u)}(X^v, Y^h) = 0, \qquad \zeta^{s,g}_{(x,u)}(X^v, Y^v) = \zeta_x(u; X, Y)$$

for all $X, Y \in M_x$.

(b) The *horizontal lift* $\zeta^{h,g}$ of an arbitrary F-metric ζ with respect to g is defined by

$$\zeta^{h,g}_{(x,u)}(X^h, Y^h) = 0, \qquad \zeta^{h,g}_{(x,u)}(X^h, Y^v) = \zeta_x(u; Y, X),$$
$$\zeta^{h,g}_{(x,u)}(X^v, Y^h) = \zeta_x(u; X, Y), \qquad \zeta^{h,g}_{(x,u)}(X^v, Y^v) = 0$$

for all $X, Y \in M_x$.

(c) The *vertical lift* ζ^v of a symmetric F-metric ζ with respect to g is defined by

$$\zeta^{v}_{(x,u)}(X^h, Y^h) = \zeta_x(u; X, Y), \qquad \zeta^{v}_{(x,u)}(X^h, Y^v) = 0,$$
$$\zeta^{v}_{(x,u)}(X^v, Y^h) = 0, \qquad \zeta^{v}_{(x,u)}(X^v, Y^v) = 0$$

for all $X, Y \in M_x$.

Obviously, the vertical lift does not depend on the choice of g. We note that $\zeta^{s,g}$, $\zeta^{h,g}$ and ζ^v are (not necessarily regular) pseudo-Riemannian metrics on TM. If we take $\zeta = g$, then $\zeta^{s,g}$, $\zeta^{h,g}$ and ζ^v are just the classical lifts g^s, g^h and g^v, respectively.

Then we can present all metrics on TM which come from a second order natural transformation of a given Riemannian manifold (M, g) in the following form:

Theorem 3.3 (Ref. 14). *Let g be a Riemannian metric on an n-dimensional (oriented or non-oriented) smooth manifold M, $n \geq 2$, and let G be a metric (possibly degenerate) pseudo-Riemannian metric on the tangent bundle TM which comes from a second order natural transformation of g. Then there are natural F-metrics ζ_1, ζ_2 and ζ_3 derived from g, where ζ_1 and ζ_3 are symmetric, such that*

$$G = {\zeta_1}^{s,g} + {\zeta_2}^{h,g} + {\zeta_3}^{v}.$$

Moreover, all natural F-metrics derived from g are given by Theorem 3.1.

If $n = 2$, then the family of all natural metrics G on TM depends on 10 arbitrary functions of one variable, for $n = 3$ it depends on seven arbitrary functions of one variable, and for $n > 3$ on six arbitrary functions of one variable.

M. T. K. Abbassi in Ref. 1 started to call natural metrics on tangent bundles as "g-natural metrics". For the purpose of this paper, we shall use the alternative formulas belonging to Abbassi and Sarih, with a little modification.

Theorem 3.4 (Refs. 1,4). *Let (M, g) be an n-dimensional (non-oriented) Riemannian manifold and G be a g-natural metric on the tangent bundle TM. Then there are real valued functions α_i and β_i, $i = 1, 2, 3$, defined on $[0, \infty)$ such that*

$$\begin{aligned} G_{(x,u)}(X^h, Y^h) &= (\alpha_1 + \alpha_3)(\|u\|^2) g_x(X, Y) \\ &\quad + (\beta_1 + \beta_3)(\|u\|^2) g_x(X, u) g_x(Y, u), \\ G_{(x,u)}(X^h, Y^v) &= \alpha_2(\|u\|^2) g_x(X, Y) + \beta_2(\|u\|^2) g_x(X, u) g_x(Y, u), \\ G_{(x,u)}(X^v, Y^h) &= \alpha_2(\|u\|^2) g_x(X, Y) + \beta_2(\|u\|^2) g_x(X, u) g_x(Y, u), \\ G_{(x,u)}(X^v, Y^v) &= \alpha_1(\|u\|^2) g_x(X, Y) + \beta_1(\|u\|^2) g_x(X, u) g_x(Y, u) \end{aligned} \qquad (2)$$

hold at each point $(x, u) \in TM$ for all $u, X, Y \in M_x$, where $\|\cdot\|$ denotes the norm of the vector. For $n = 1$, the same holds with $\beta_i = 0$, $i = 1, 2, 3$.

4. Invariance of g-natural metrics

Let ϕ be a (local) transformation of a manifold M. Then we define a transformation Φ of TM by

$$\Phi(x, u) = (\phi x, \phi_{*x} u)$$

for all $(x, u) \in TM$.

Proposition 4.1. *Let ϕ be a (local) affine transformation of a manifold M with an affine connection ∇ and let Φ be the lift of ϕ to TM defined as above. Then we have*

$$\Phi_*(X^h) = (\phi_* X)^h, \quad \Phi_*(X^v) = (\phi_* X)^v$$

for all $X \in \mathfrak{X}(M)$.

Proof. We use the formula $p \circ \Phi = \phi \circ p$. For all functions f on M we calculate:

$$\begin{aligned}
&(p_{*\Phi(x,u)}(\Phi_{*(x,u)}(X^h_{(x,u)})))f \\
&= X^h_{(x,u)}(f \circ p \circ \Phi) = X^h_{(x,u)}(f \circ \phi \circ p) \\
&= (p_{*(x,u)}(X^h_{(x,u)}))(f \circ \phi) = X_x(f \circ \phi) \\
&= (\phi_{*x} X_x) f.
\end{aligned}$$

Since Φ preserves the horizontal distribution, we have $\Phi_{*(x,u)}(X^h_{(x,u)}) = (\phi_{*x} X_x)^h_{\Phi(x,u)}$ by the definition of the horizontal lift.

Next, we have $(\mathrm{d}f) \circ \Phi = \mathrm{d}(f \circ \phi)$. In fact, since, by the definition, $(\mathrm{d}f)(x, u) = uf$ at $(x, u) \in TM$ holds for all functions f on M, we calculate at every point $(x, u) \in TM$ as

$$\begin{aligned}
((\mathrm{d}f) \circ \Phi)(x, u) &= (\mathrm{d}f)(\phi x, \phi_{*x} u) = (\phi_{*x} u) f \\
&= u(f \circ \phi) = (\mathrm{d}(f \circ \phi))(x, u).
\end{aligned}$$

Hence we obtain

$$\begin{aligned}
&(\Phi_{*(x,u)} X^v_{(x,u)})(\mathrm{d}f) \\
&= X^v_{(x,u)}((\mathrm{d}f) \circ \Phi) = X^v_{(x,u)}(\mathrm{d}(f \circ \phi)) \\
&= X_x(f \circ \phi) = (\phi_{*x} X_x)(f) = (\phi_{*x} X_x)^v_{(\phi x, \phi_{*x} u)}(\mathrm{d}f)
\end{aligned}$$

Since Φ preserves the vertical distribution, we have $\Phi_{*(x,u)}(X^v_{(x,u)}) = (\phi_{*x} X_x)^v_{\Phi(x,u)}$ by the definition of the vertical lift. $\square$

Theorem 4.1. *Let ϕ be a (local) isometry of a Riemannian manifold (M,g). Then every g-natural metric G on the tangent bundle TM over (M,g) is invariant by the lift Φ of ϕ. In other words, Φ is a local isometry of (TM,G) whose projection on (M,g) is ϕ.*

Proof. Let X, Y be vectors from M_x. Then, by Proposition 4.1 and the first formula of (2), we have at any point $(x,u) \in TM$ that

$$\begin{aligned}
&(\Phi^* G)_{(x,u)}(X^h_{(x,u)}, Y^h_{(x,u)}) \\
&= G_{\Phi(x,u)}(\Phi_{*(x,u)} X^h_{(x,u)}, \Phi_{*(x,u)} Y^h_{(x,u)}) \\
&= G_{(\phi x, \phi_{*x} u)}((\phi_{*x} X)^h_{(\phi x, \phi_{*x} u)}, (\phi_{*x} Y)^h_{(\phi x, \phi_{*x} u)}) \\
&= (\alpha_1 + \alpha_3)(\|\phi_{*x} u\|^2) g_{\phi x}(\phi_{*x} X, \phi_{*x} Y) \\
&+ (\beta_1 + \beta_3)(\|\phi_{*x} u\|^2) g_{\phi x}(\phi_{*x} X, \phi_{*x} u) g_{\phi x}(\phi_{*x} Y, \phi_{*x} u).
\end{aligned}$$

Now, since ϕ is an isometry of (M,g), the right-hand side of this formula is just $G_{(x,u)}(X^h_{(x,u)}, Y^h_{(x,u)})$. Thus we have

$$(\Phi^* G)_{(x,u)}(X^h_{(x,u)}, Y^h_{(x,u)}) = G_{(x,u)}(X^h_{(x,u)}, Y^h_{(x,u)})$$

for all $X, Y \in M_x$.

The rest of the assertion is proved by similar calculations. □

Let us remark that the natural projection p of (TM,G) onto (M,g) is *not* a Riemannian submersion, in general.

5. Tangent sphere bundles

Let r be a positive number. Then the *tangent sphere bundle of radius r* over a Riemannian manifold (M,g) is a smooth hypersurface

$$T_r M = \{(x,u) \in TM \mid g_x(u,u) = r^2\}$$

of the tangent bundle TM. For any vector field X tangent to M, the horizontal lift X^h is always tangent to $T_r M$ at each point $(x,u) \in T_r M$. Yet, in general, the vertical lift X^v is not tangent to $T_r M$ at (x,u).

Proposition 5.1. *Let (M,g) be a Riemannian manifold. Then the tangent space of $T_r M$ at a point (x,u) is given by*

$$(T_r M)_{(x,u)} = \{X^h_{(x,u)} + Y^v_{(x,u)} \mid X \in M_x,\ Y \in \{u\}^\perp \subset M_x\},$$

where $\{u\}^\perp$ is the orthogonal complement of the subspace spanned by u in M_x.

Proof. Let α be a curve of T_rM. Then it is written in the form

$$\alpha : t \longmapsto \alpha(t) = (x(t), u(t)).$$

A vertical vector tangent to T_rM at $(x, u_0) \in T_rM$ is a vector tangent to some curve $\alpha : t \longmapsto \alpha(t) = (x, u(t))$ such that $u(0) = u_0$. The velocity of α at $t = 0$ is $\alpha'(0) = (0, u'(0))$. Since $g(u(t), u(t)) = r^2$ is constant, we have $g_x(u'(0), u(0)) = 0$, that is, $u'(0) \in M_x$ is orthogonal to $u_0 = u(0)$. Now the vertical space $V_{(x,u_0)}$ of TM is spanned by the all vertical lifts $X^v_{(x,u_0)}$ of $X \in M_x$, and hence it is linearly isomorphic to M_x. Moreover, $X^v_{(x,u_0)}$ is tangent to T_rM if and only if X is orthogonal to u_0 in M_x. Hence the assertion follows. □

Definition 5.1. Consider a smooth Riemannian manifold (M, g). A *g-natural metric* on the tangent sphere bundle T_rM over (M, g) is restriction $\tilde{G}$ of any g-natural metric G given by (2) on the tangent bundle TM to the hypersurface T_rM of TM.

All g-natural metrics on the tangent sphere bundle T_rM over a Riemannian manifold (M, g) has been characterized by M. T. K. Abbassi and O. Kowalski (preprint, unpublished). We reproduce it here (with the kind agreement of our friend M. T. K. Abbassi) also the short but detailed proof.

Proposition 5.2. *Let $r > 0$ and (M, g) be a Riemannian manifold. For every Riemannian metric $\tilde{G}$ on T_rM induced by a Riemannian g-natural G on TM, there exist four constants a, b, c and d, with $a > 0$, $a(a+c)-b^2 > 0$ and $a(a + c + dr^2) - b^2 > 0$, such that*

$$\tilde{G} = a\,\tilde{g}^s + b\,\tilde{g}^h + c\,\tilde{g}^v + d\,\tilde{k}^v, \tag{3}$$

where k is the natural F-metric on M defined by

$$k(u; X, Y) = g(X, u)g(Y, u)$$

for all $(u, X, Y) \in TM \oplus TM \oplus TM$, and $\tilde{g}^s$, $\tilde{g}^h$, $\tilde{g}^v$ and $\tilde{k}^v$ are considered here as the metrics on T_rM induced by the lifts g^s, g^h, g^v and k^v, respectively.

Proof. Let us fix $(x, u) \in T_rM$ (*i.e.*, $\|u\| = r$). Then, in the formulas (2) we see that each right-hand side reduces to only one term whenever one of the vectors X, Y is orthogonal to u. Due to Proposition 5.1, the induced metric $\tilde{G}_{(x,u)}$ is completely characterized by the identities

$$\begin{aligned}\tilde{G}_{(x,u)}(X_1^h, X_2^h) &= (\alpha_1 + \alpha_3)(r^2) g_x(X_1, X_2) \\ &\quad + (\beta_1 + \beta_3)(r^2) g_x(X_1, u) g_x(X_2, u), \\ \tilde{G}_{(x,u)}(X_1^h, Y_1^v) &= \alpha_2(r^2) g_x(X_1, Y_1), \\ \tilde{G}_{(x,u)}(Y_1^v, Y_2^v) &= \alpha_1(r^2) g_x(Y_1, Y_2)\end{aligned} \tag{4}$$

for all $X_1, X_2 \in M_x$ and $Y_1, Y_2 \in \{u\}^\perp$. In other words, $\tilde{G}_{(x,u)}$ depends on $(\alpha_1 + \alpha_3)(r^2)$, $(\beta_1 + \beta_3)(r^2)$, $\alpha_2(r^2)$ and $\alpha_1(r^2)$. But, for any element of $(x, u) \in T_rM$, the norm of u is a constant equal to r, then $\tilde{G}_{(x,u)}$ depends on four constants. We put $a = \alpha_1(r^2)$, $b = \alpha_2(r^2)$, $c = \alpha_3(r^2)$ and $d = (\beta_1+\beta_3)(r^2)$. Then a, b, c and d are constants, and the formula (3) follows at once from the formulas (4). Now, the fact that G is Riemannian is equivalent to the inequalities in the Proposition. Hence the Proposition follows. □

It is well-known that if (M, g) is a two-point homogeneous space, then each tangent sphere bundle T_rM, $r > 0$, is a homogeneous space with respect to the induced maps of the isometries of (M, g). From Theorem 4.1 we obtain immediately

Proposition 5.3. *Let (M, g) be a two-point homogeneous space, and let the tangent sphere bundle T_rM be equipped with a g-natural metric $\tilde{G}$ over (M, g). Then $(T_rM, \tilde{G})$ is homogeneous as Riemannian manifold.*

As a corollary, we have reproved the special result by E. Musso and F. Tricerri in Ref. 19 saying that, for the induced Sasaki metric $\tilde{g}^s$, the unit tangent sphere bundle $(T_1M, \tilde{g}^s)$ is a homogeneous space.

We can formulate also the local version of Proposition 5.3:

Proposition 5.4. *Let (M, g) be locally isometric to a two-point homogeneous space, and let the unit tangent sphere bundle T_rM be equipped with a g-natural metric $\tilde{G}$ over (M, g). Then the space $(T_rM, \tilde{G})$ is locally homogeneous.*

Hence we obtain

Corollary 5.1. *Let (M, g) be a space of constant sectional curvature, and let the tangent sphere bundle T_rM be equipped with a g-natural metric $\tilde{G}$ over (M, g). Then the space $(T_rM, \tilde{G})$ is locally homogeneous.*

Acknowledgments

The first author was supported by the grant GA ČR 201/05/2707 and by the project MSM 0021620839.

References

1. M.T.K. Abbassi, Note on the classification theorems of g-natural metrics on the tangent bundle of a Riemannian manifolds (M, g), *Comment. Math. Univ. Carolinae,* **45** (4) (2004) 591–596.
2. M.T.K. Abbassi and G. Calvaruso, g-natural contact metrics on unit tangent sphere bundles, *Monatsh. Math.* **151** (2) (2007) 89–109.
3. M.T.K. Abbassi and O. Kowalski, On g-natural metrics with constant scalar curvature on unit tangent sphere bundles, *In: Topics in almost Hermitian geometry and related fields* (World Sci. Publ., Hackensack, NJ, 2005) 1–29.
4. M.T.K. Abbassi and M. Sarih, On some hereditary properties of Riemannian g-natural metrics on tangent bundles of Riemannian manifolds, *Diff. Geom. Appl.* **22** (2005) 19–47.
5. M.T.K. Abbassi and M. Sarih, On natural metrics on tangent bundles of Riemannian manifolds, *Arch. Math. (Brno)* **41** (2005) 71–92.
6. M.T.K. Abbassi and M. Sarih, On Riemannian g-natural metrics of the form $a.g^s + b.g^h + c.g^v$ on the tangent bundle of a Riemannian manifold (M, g), *Mediter. J. Math.* **2** (2005) 19–43.
7. A.L. Besse, *Einstein manifolds* (Springer-Verlag, Berlin-Heidelberg-New York, 1987).
8. D. Blair, When is the tangent sphere bundle locally symmetric?, *In: Geometry and topology* (World Sci. Publishing, Singapore, 1989) 15–30.
9. E. Boeckx and L. Vanhecke, Geometry of the tangent sphere bundle, *In: Proceedings of the Workshop on Recent Topics in Differential Geometry* ((L.A. Cordero and E. García-Río, Eds.) Santiago de Compostela, Spain, 1997) 5–17.
10. E. Boeckx and L. Vanhecke, Curvature homogeneous unit tangent sphere bundles, *Publ. Math. Debrecen* **35** (1998) 389–413.
11. E. Boeckx and L. Vanhecke, Unit tangent sphere bundles and two-point homogeneous spaces, *Periodica Math. Hung.* **36** (1998) 79–95.
12. P. Dombrowski, On the geometry of the tangent bundles, *J. Reine Angew. Math.* **210** (1962) 73–88.
13. I. Kolář, P.W. Michor and J. Slovák, *Natural Operations in Differential Geometry* (Springer-Verlag, Berlin-Heidelberg-New York, 1993).
14. O. Kowalski and M. Sekizawa, Natural transformations of Riemannian metrics on manifolds to metrics on tangent bundles—A classification—, *Bull. Tokyo Gakugei Univ.* **40** (4) (1988) 1–29; http://ir.u-gakugei.ac.jp/handle /2309/36051.
15. D. Krupka, Elementary theory of differential invariants, *Arch. Math.(Brno)* **4** (1978) 207–214.
16. D. Krupka, *Differential invariants* (Lecture Notes, Faculty of Science, Purkyně University, Brno, 1979).

17. D. Krupka and V. Mikolášová, On the uniqueness of some differential invariants: d, [,], ∇, *Czechoslovak Math. J.* **34** (1984) 588–597.
18. D. Krupka and J. Janyška, *Lectures on Differential Invariants* (University J. E. Purkyně in Brno, 1990).
19. E. Musso and F. Tricerri, Riemannian metrics on tangent bundles, *Ann. Mat. Pura Appl. (4)* **150** (1988) 1–20.
20. S. Sasaki, On the differential geometry of tangent bundles, *Tôhoku Math. J.* **10** (1958) 338–354.

Differential Geometry and its Applications
Proc. Conf., in Honour of Leonhard Euler, Olomouc, August 2007

Some variational problems in geometry of submanifolds

Haizhong Li

Department of Mathematical Sciences, Tsinghua University, Beijing, 100084, People's Republic of China
E-mail: hli@math.tsinghua.edu.cn

In this paper, we present a survey of some variational problems in geometry of submanifolds, which includes our recent research results in geometry of r-minimal submanifolds, geometry of *Willmore* submanifolds and variations of some parametric elliptic functional. We also propose some open problems at the end of paper.

Keywords: r-minimal submanifold, Willmore submanifold, r-anisotropic mean curvature, Wulff shape.

MS classification: 53C42, 53A10, 49Q10.

1. Introduction

This paper is to present our recent research results about variational problems in geometry of submanifolds and divides into six sections. Now we describe in more detail some of the contents of this paper, where we refer to the papers for the explanation of notations and undefined terms.

The section 1 is the introduction. The section 2 is preliminaries and is devoted to some formulas and notations of submanifolds in geometry of submanifolds. In section 3 we present the first and second variational formula of the volume of n-dimensional submanifolds in an $(n+p)$-dimensional manifold N^{n+p}. Besides, we give two important results of J. Simons. In section 4, we will recall our recent works in Ref. 4. We introduce the functional

$$J_r = \int_M F_r(S_0, S_2, \cdots, S_r) dv_g$$

where function F_r is a suitable function on M, and introduce the concepts of r-minimal submanifolds and stability. We study the stability of compact r-minimal submanifold in the unit sphere S^{n+p}. In section 5 we give some facts (see Refs. 10,16,17) about variational problems of Willmore functional on submanifolds. In particular, we recall the Euler-Lagrange equation of $W(x)$ for an n-dimensional submanifold in an $(n+p)$-dimensional Riemannian manifold N^{n+p}, give some typical examples of Willmore submanifolds and give integral inequality of Simons' type for n-dimensional closed Willmore submanifolds in S^{n+p}. In section 6 we consider a variation problem concerning certain parametric elliptic functional and collect some results of B. Palmer and He-Li. We state integral formula of Minkowski's type[8] and new characterizations of the Wulff shape.[9,22]

2. Preliminaries

Let(N^{n+p}, h) be an $n+p$-dimensional oriented smooth Riemannian manifold, and $x : M^n \to N^{n+p}$ be an n-dimensional submanifold of (N^{n+p}, h). We will agree on the following index convention:

$$1 \le i, j, k, \cdots \le n; \quad n+1 \le \alpha, \beta, \gamma, \cdots \le n+p; \quad 1 \le A, B, C, \cdots \le n+p$$

Let$\{e_A\}$ be a local orthonormal basis for TN^{n+p} with dual basis $\{\theta_A\}$ such that when restricted to M^n, $\{e_i\}$ is a local orthonormal basis for TM and $\{e_\alpha\}$ is a local orthonormal basis for the normal bundle of $x : M^n \to N^{n+p}$. Let $\{\omega_{AB}\}$ be the connection forms of (N^{n+p}, h), they are characterized by the following structure equations

$$d\omega_A = \sum_B \omega_{AB} \wedge \omega_B, \quad \omega_{AB} + \omega_{BA} = 0 \tag{1}$$

$$\omega_{AB} = \omega_{AC} \wedge \omega_{CB} - \frac{1}{2} \sum_{C,D} \bar{R}_{ABCD} \omega_C \wedge \omega_D \tag{2}$$

where$\bar{R}_{ABCD}$ are the components of the Riemannian curvature tensor of (N^{n+p}, h).

Now we restrict to a neighborhood of $x : M^n \to N^{n+p}$. Let θ_A, θ_{AB} be the restrictions of ω_A, ω_{AB} to M^n, then we have

$$\theta_\alpha = 0 \tag{3}$$

Taking its exterior derivative and by (1) we get

$$\sum_i \theta_{\alpha i} \wedge \theta_i = 0 \tag{4}$$

By Cartan's lemma we have

$$\theta_{i\alpha} = \sum_j h^\alpha_{ij} \theta_j, \quad h^\alpha_{ij} = h^\alpha_{ji}. \tag{5}$$

We can define the second fundamental form B_{ij} and the mean curvature vector $\mathbb{H}$ of x as follows

$$B_{ij} := \sum_{\alpha} h_{ij}^{\alpha} e_{\alpha}, \qquad \mathbb{H} = \frac{1}{n} \sum_{i,\alpha} h_{ii}^{\alpha} := \frac{1}{n} \sum_{\alpha} H^{\alpha} e_{\alpha} \tag{6}$$

Let $S = |B|^2 = \sum_{i,j,\alpha} (h_{ij}^{\alpha})^2$ and $H = |\mathbb{H}|$ be the norm square of B and the mean curvature of $x : M^n \to N^{n+p}$, respectively. If we denote by R_{ijkl}, R_{ij}, R the Riemannian curvature tensor, Ricci curvature and scalar curvature of M, respectively. We have Gauss equations, Ricci equations and Codazzi equations as followings (see Refs. 3,10)

$$R_{ijkl} = \bar{R}_{ijkl} + \sum_{\alpha} (h_{ik}^{\alpha} h_{jl}^{\alpha} - h_{il}^{\alpha} h_{jk}^{\alpha}) \tag{7}$$

$$R_{\alpha\beta ij} = \bar{R}_{\alpha\beta ij} + \sum_{\alpha} (h_{ik}^{\alpha} h_{kj}^{\beta} - h_{jk}^{\alpha} h_{ki}^{\beta}), \tag{8}$$

$$h_{ijk}^{\alpha} - h_{ikj}^{\alpha} = \bar{R}_{\alpha ikj}, \tag{9}$$

where h_{ijk}^{α} is the covariant derivative of h_{ij}^{α}.

In particular, when the ambient space N^{n+p} is a space form $R^{n+p}(c)$, the Gauss equation, Ricci equation and Codazzi equation are (also see Refs. 14, 15)

$$R_{ijkl} = c(\delta_{ik}\delta_{ji} - \delta_{il}\delta_{jk}) + \sum_{\alpha} (h_{ik}^{\alpha} h_{jl}^{\alpha} - h_{il}^{\alpha} h_{jk}^{\alpha}), \tag{10}$$

$$R_{\alpha\beta ij} = \sum_{\alpha} (h_{ik}^{\alpha} h_{kj}^{\beta} - h_{jk}^{\alpha} h_{ki}^{\beta}), \tag{11}$$

$$h_{ijk}^{\alpha} = h_{ikj}^{\alpha}. \tag{12}$$

3. Minimal submanifolds in a Riemannian manifold

Let $x_0 : M \to N^{n+p}$ be an n-dimensional submanifolds in an $(n+p)$-dimensional manifold N^{n+p}, we assume, without loss of generality, that M is compact with (possibly empty) boundary. If otherwise, we will consider the variation with compact support.

Let $x : M \times R \to N^{n+p}$ be a smooth variation of x_0 such that $x(\cdot, t) = x_0$ and $dx_t(TM) = dx_0(TM)$ on ∂M for each small t, where $x_t(p) = x(p, t)$. Let $V(t)$ be the volume functional of $x_t(M)$, i.e.

$$V(t) = \int_M \theta_1 \wedge \cdots \wedge \theta_n. \tag{13}$$

We have (see Refs. 3,21)

$$V'(t) = -n \int_M \sum_\alpha H^\alpha a_\alpha \theta_1 \wedge \cdots \wedge \theta_n, \tag{14}$$

where $\sum_\alpha a_\alpha e_\alpha$ is the normal variation vector field.

Definition 3.1. Let $x_0 : M \to N^{n+p}$ be an n-dimensional submanifold in an $(n+p)$-dimensional manifold N^{n+p}, if

$$\vec{H} = \frac{1}{n} \sum_\alpha H^\alpha e_\alpha \equiv 0, \tag{15}$$

we call M be a minimal submanifold.

Proposition 3.1 (Refs. 3,27). *Let $x_0 : M \to N^{n+p}$ be an n-dimensional submanifold in an $(n+p)$-dimensional manifold N^{n+p}, then we have*

$$V''(0) = -\int_M \sum_\alpha a_\alpha (\Delta^\perp a_\alpha + \sum_\beta \sigma_{\alpha\beta} a_\beta + \sum_{i=1}^{n} \sum_{\beta=n+1}^{n+p} \tilde{R}_{\alpha i \beta j} a_\beta) \theta_1 \wedge \cdots \wedge \theta_n, \tag{16}$$

where $\tilde{R}_{\alpha i\beta j}$ is the components of Riemannian curvature tensor, $\sigma_{\alpha\beta} = \sum_{i,j} h^\alpha_{ij} h^\beta_{ij}$, and $\Delta^\perp a_\alpha$ is the Laplacian of a_α in the normal bundle.

Definition 3.2. Let $x : M \to N^{n+p}$ be an n-dimensional submanifold in N^{n+p}. If for any normal variational vector $W = \sum_\alpha a_\alpha e_\alpha$, we have $V''(0) \geq 0$, we call M is stable.

Theorem 3.1 (J. Simons, Ref. 27). *There exists no any n-dimensional closed stable minimal submanifolds in an $(n+1)$-dimensional unit sphere S^{n+p}.*

Example 3.1 (see Refs. 1,13). Clifford tori:

$$C_{m,n-m} = S^m(\sqrt{\frac{m}{n}}) \times S^{n-m}(\sqrt{\frac{n-m}{n}}) \ , \quad 1 \leq m \leq n-1$$

are minimal hypersurfaces in S^{n+1}.

Theorem 3.2 (Refs. 1,13,27). *Let M be an n-dimensional $(n \geq 2)$ closed minimal submanifold in $(n+p)$-dimensional unit sphere S^{n+p}. Then we have*

$$\int_M S\left(\frac{n}{2-1/p} - S\right) dv \leq 0. \tag{17}$$

In particular, if

$$0 \le S \le \frac{n}{2-1/p}, \tag{18}$$

then either $S \equiv 0$ and M is totally geodesic, or $S \equiv \dfrac{n}{2-1/p}$. In the latter case, either $p=1$ and M is a Clifford torus $C_{m,n-m}$; or $n=2$, $p=2$ and M is the Veronese surface.

4. r-minimal submanifolds in a space form

We recall the definition of the generalized Kronecker symbols. If $i_1, \cdots, i_r$ and $j_1, \cdots, j_r$ are integers between 1 and n, then $\delta^{i_1\cdots i_r}_{j_1\cdots j_r}$ is +1 or −1 according as the i's are distinct and the j's are an even or odd permutation of the i's and is 0 in all other cases.

Let (M,g) be an n-dimensional submanifold in an $(n+p)$-dimensional space form $R^{n+p}(c)$, and B be the second fundamental form of M. Suppose $\{e_i\}$ is a local orthonormal basis for TM with dual basis $\{\theta_i\}$ and $\{e_\alpha\}$ is a local orthonormal basis for the normal bundle of $x: M^n \to R^{n+p}(c)$, (B_{ij}) is the matrix with respect to the frame $\{e_1, \cdots, e_n\}$ on M. Then for any even integer $r \in \{0, 1, \cdots, n-1\}$, we introduce r-th mean curvature function S_r and $(r+1)$-th mean curvature vector field $\vec{S}_{r+1}$ as follows (see Refs. 4,5,25,26:

$$\begin{aligned}
S_r &= \frac{1}{r!}\sum_{\substack{i_1\cdots i_r\\ j_1\cdots j_r}} \delta^{i_1\cdots i_r}_{j_1\cdots j_r}\langle B_{i_1j_1}, B_{i_2j_2}\rangle \cdots \langle B_{i_{r-1}j_{r-1}}, B_{i_rj_r}\rangle\\
&= \frac{1}{r}\sum_{i,j,\alpha} T_{r-1}{}^{\alpha}{}_{ij} h^{\alpha}_{ij},\\
\vec{S}_{r+1} &= \frac{1}{(r+1)!}\sum_{\substack{i_1\cdots i_{r+1}\\ j_1\cdots j_{r+1}}} \delta^{i_1\cdots i_{r+1}}_{j_1\cdots j_{r+1}}\langle B_{i_1j_1}, B_{i_2j_2}\rangle \cdots \langle B_{i_{r-1}j_{r-1}}, B_{i_rj_r}\rangle B_{i_{r+1}j_{r+1}}\\
&= \frac{1}{r+1}\sum_{i,j,\alpha} T_{rj}^{\ i} h^{\alpha}_{ij} e_\alpha,\\
S_r &= \binom{n}{r} H_r, \quad \vec{S}_{r+1} = \binom{n}{r+1}\vec{H}_{r+1},
\end{aligned}$$

where $\binom{n}{r}$ being the binomial coefficient.

Besides, we define the following $(0,2)$-tensor T_r for $r \in \{1, \cdots, n-1\}$:

• If r is even, we set

$$T_r = \frac{1}{r!} \sum_{\substack{i_1\cdots i_r i\\ j_1\cdots j_r j}} \delta^{i_1\cdots i_r i}_{j_1\cdots j_r j} \langle B_{i_1j_1}, B_{i_2j_2}\rangle \cdots \langle B_{i_{r-1}j_{r-1}}, B_{i_rj_r}\rangle \theta_i \otimes \theta_j$$
$$= \sum_{i,j} T_{r}{}^{i}_{j} \theta_i \otimes \theta_j.$$

• If $r-1$ is odd, we set

$$T_{r-1} = \frac{1}{(r-1)!} \sum_{\substack{i_1\cdots i_{r-1} i\\ j_1\cdots j_{r-1} j}} \delta^{i_1\cdots i_{r-1} i}_{j_1\cdots j_{r-1} j} \langle B_{i_1j_1}, B_{i_2j_2}\rangle \cdots \langle B_{i_{r-3}j_{r-3}}, B_{i_{r-2}j_{r-2}}\rangle$$
$$\cdot B_{i_{r-1}j_{r-1}} \theta_i \otimes \theta_j$$
$$= \sum_{i,j,\alpha} T_{r-1}{}^{\alpha}{}_{ij} \theta_i \otimes \theta_j e_\alpha,$$

By convention, we put $H_0 = S_0 = 1$, T_0 =identity.

Let $x : M \to R^{n+p}(c)$ be an n-dimensional compact submanifold in $R^{n+p}(c)$. Assume that r is even and $r \in \{0, 1, \cdots, n-1\}$, , in Ref. 4 the authors introduce a curvature integral J_r for r even and $r \in \{0, 1, \cdots, n-1\}$

$$J_r = \int_M F_r(S_0, S_2, \cdots, S_r) dv$$

where the function F_r are defined inductively by

$$\begin{cases} F_0 = 1 \\ F_r = S_r + \frac{(n-r+1)c}{r-1} F_{r-2}, \quad \text{for } \; 2 \le r \le n-1. \end{cases}$$

Theorem 4.1 (Ref. 4, the first variational formula). *Let M be an n-dimensional compact, possibly with boundary, submanifold in an $(n+p)$-dimensional space form $R^{n+p}(c)$. Assume that r is even and $r \in \{0, 1, \cdots n-1\}$, then we have*

$$J_r{}'(t) = -(r+1) \int_{M_t} \langle \vec{S}_{r+1}, V\rangle dv_{g_t}$$

where V is the variational vector field.

Definition 4.1 (Ref. 4). Let M be an n-dimensional submanifold in an $(n+p)$-dimensional space form $R^{n+p}(c)$. Assume that r is even and $r \in \{0, 1, \cdots, n-1\}$, we call x to be r-minimal if its $(r+1)$-th mean curvature vector $\vec{S}_{r+1}$ vanishes on M.

Theorem 4.2 (Ref. 4, the second variational formula). *Let M be an n-dimensional compact r-minimal submanifold in $R^{n+p}(c)$ for some even integer $r \in \{0, 1, \cdots, n-1\}$, we have*

$$J''(0) = -\int_M \sum_\alpha V_\alpha \{ \frac{1}{r-1} \sum_{\substack{i_{r-1} i_r i \\ j_{r-1} j_r j \\ \beta}} T_{r-2}{}^{i_{r-1} i_r i}_{j_{r-1} j_r j} h_{i_{r-1} j_{r-1}}^{\alpha} h_{i_r j_r}^{\beta} V_{\beta, ij}$$
$$+ \sum_{i,j} T_{r}{}_j^i V_{\alpha, ij} + c \cdot (n-r)(S_r V_\alpha + \sum_{i,j,\beta} T_{r-1}{}_{ij}^{\alpha} h_{ij}^{\beta} V_\beta)$$
$$- (r+1) \sum_{i,j,\beta} T_{r+1}{}_{ij}^{\alpha} h_{ij}^{\beta} V_\beta \} dv$$

Definition 4.2 (Ref. 4). Assume that r is even and $r \in \{0, 1, \cdots, n-1\}$, and let M be an n-dimensional compact r-minimal submanifold in an $(n+p)$-dimensional space form $R^{n+p}(c)$. If $J_r''(0) \geq 0$ for arbitrary variations, we call M to be stable

Remark 4.1. From definition 4.1, we know that concept of 0-minimal submanifolds is the concept of minimal submanifolds.

Theorem 4.3 (Ref. 4). *Assume that r is even and $r \in \{0, 1, \cdots, n-1\}$. If S_r is positive, then there exists no any closed stable r-minimal submanifold in the unit sphere S^{n+p}.*

Remark 4.2. Noting when $r = 0$, i.e. $S_0 = 1$, it is obvious to see that our Theorem 4.3 reduces to J. Simons' Theorem 3.2.

Theorem 4.4 (Ref. 4). *Assume that r is even and $r \in \{1, \cdots, n-1\}$. If $S_r \geq 0$, then any closed stable r-minimal hypersurface in unit sphere S^{n+1} must be a geodesic sphere.*

5. Willmore submanifolds in a Riemannian manifold

Let N^{n+p} be an $(n+p)$-dimensional Riemannian manifold and $x : M \to N^{n+p}$ an isometric immersion of an n-dimensional Riemannian manifold M. Willmore functional is the following non-negative functional:(see Refs. 2,24 or Ref. 28)

$$W(x) := \int_M \rho^n dv = \int_M (S - nH^2)^{\frac{n}{2}} dv.$$

where $S = \sum_{\alpha,i,j} (h_{ij}^\alpha)^2$ and H are respectively the norm square of the second fundamental form and the mean curvature of the immersion x, dv is the volume element of M. An immersion $x : M \to N^{n+p}$ is called Willmore if it is an extremal submanifold of the Willmore functional. The famous Willmore conjecture can be stated in a equivalent way as that

$$W(x) \geq 4\pi^2$$

holds for all immersed tori $x : M \to S^3$ (see Ref. 30).

It is well known (see Refs. 2,24 or Ref. 28) that Willmore functional is invariant under conformal transformations of N^{n+p}. In Refs. 16,24,28, the authors calculated the first variation formula of Euler-Lagrangian equation of $W(x)$ for an n-dimensional submanifold in $R^{n+p}(c)$. In Ref. 6, the authors calculated the second variation formula of $W(x)$ for Willmore submanifolds $x : M^n \to \mathbb{S}^{n+p}$ without umbilic points in terms of Möbius geometry and gave many examples of Willmore submanifolds.

In Ref. 10 the authors calculated the Euler-Lagrangian equation for the critical points of $W(x)$ for the most general case.

Theorem 5.1 (Ref. 10). *The variation of the Willmore functional depends only on the normal component of the variation vector field. A submanifold $x : M \to N^{n+p}$ is a Willmore submanifold if and only if*

$$\begin{aligned}
&\rho^{n-2}\Big[\sum_{i,j,k,\beta} h^{\beta}_{ij}h^{\beta}_{ik}h^{\alpha}_{kj} + \sum_{i,j,\beta} \tilde{R}_{\beta i\alpha j}h^{\beta}_{ij} - \sum_{i,j,\beta} H^{\beta}h^{\beta}_{ij}h^{\alpha}_{ij} \\
&- \sum_{i,\beta} H^{\beta}\tilde{R}_{\beta i\alpha i} - \rho^2 H^{\alpha}\Big] + \sum_{i,j}\Big\{2(\rho^{n-2})_{,i}h^{\alpha}_{ijj} + (\rho^{n-2})_{,i,j}h^{\alpha}_{ij} \\
&+ \rho^{n-2}h^{\alpha}_{ijij}\Big\} - H^{\alpha}\Delta(\rho^{n-2}) - \rho^{n-2}\Delta^{\perp}H^{\alpha} - 2\sum_{i}(\rho^{n-2})_{,i}H^{\alpha}_{,i} = 0,
\end{aligned} \tag{19}$$

$$n+1 \leq \alpha \leq n+p.$$

where $\tilde{R}_{ABCD}$ are the components of the Riemannian curvature tensor of (N^{n+p}, h).

In particular, when $N^{n+p} = R^{n+p}(c)$, we have

Theorem 5.2 (Refs. 16,24). *A submanifold $x : M^n \to R^{n+p}(c)$ is a Willmore submanifold if and only if*

$$\begin{aligned}
&\rho^{n-2}\Big[\sum_{i,j,k,\beta} h^{\beta}_{ij}h^{\beta}_{ik}h^{\alpha}_{kj} - \sum_{i,j,\beta} H^{\beta}h^{\beta}_{ij}h^{\alpha}_{ij} - \rho^2 H^{\alpha}\Big] \\
&+ (n-1)H^{\alpha}\Delta(\rho^{n-2}) + 2(n-1)\sum_{i}(\rho^{n-2})_{,i}H^{\alpha}_{,i} + (n-1)\rho^{n-2}\Delta^{\perp}H^{\alpha} \\
&- \sum_{i,j}(\rho^{n-2})_{,i,j}(nH^{\alpha}\delta_{ij} - h^{\alpha}_{ij}) = 0, \quad n+1 \leq \alpha \leq n+p.
\end{aligned} \tag{20}$$

Example 5.1 (Ref. 29). Every minimal surface in $R^{n+p}(c)$ is Willmore. We note that there are much more abundance of non-minimal Willmore surfaces in $R^{2+p}(c)$, see, e.g. Refs. 20,23, among many others.

Example 5.2 (Refs. 6,24). Every n-dimensional (≥ 3) minimal and Einstein submanifold in $N^{n+p}(c)$ is Willmore.

Example 5.3 (Refs. 6,16). $W_{n_1,\cdots,n_{p+1}} = S^{n_1}(a_1) \times \cdots \times S^{n_{p+1}}(a_{p+1})$ is an n-dimensional Willmore submanifold in $S^{n+p}(1)$, where $n = n_1 + \cdots + n_{p+1}$ and a_i are defined by

$$a_i = \sqrt{\frac{n - n_i}{np}}, \qquad i = 1, \cdots, p+1.$$

Furthermore, $W_{n_1,\cdots,n_{p+1}}$ is a minimal submanifold in $S^{n+p}(1)$ if and only if it is Einstein with

$$n_1 = \cdots = n_{p+1} = \frac{n}{p+1}, a_i = \sqrt{\frac{1}{p+1}}.$$

Example 5.4 (Refs. 6,16,17). Willmore tori:

$$W_{m,n-m} = S^m(\sqrt{\frac{n-m}{n}}) \times S^{n-m}(\sqrt{\frac{m}{n}}), \quad 1 \leq m \leq n-1$$

are Willmore hypersurfaces in $S^{n+1}(1)$.

Example 5.5 (Ref. 11). Define the Lagrangian sphere $\Psi : \mathbb{S}^n \to \mathbb{C}^n$ by

$$\begin{aligned}&\Psi(x_1, \cdots, x_n, x_{n+1}) \\ &= \frac{2\sqrt{\frac{n-1}{2n}}\, e^{i\beta(x_{n+1})}}{[(1+x_{n+1})^{\sqrt{\frac{2n}{n-1}}} + (1-x_{n+1})^{\sqrt{\frac{2n}{n-1}}}]^{\sqrt{\frac{n-1}{2n}}}} \cdot (x_1, \cdots, x_n),\end{aligned}$$

$$\beta(x_{n+1}) = \sqrt{\frac{2(n-1)}{n}} \arctan\Big(\frac{(1+x_{n+1})^{\sqrt{\frac{n}{2(n-1)}}} - (1-x_{n+1})^{\sqrt{\frac{n}{2(n-1)}}}}{(1+x_{n+1})^{\sqrt{\frac{n}{2(n-1)}}} + (1-x_{n+1})^{\sqrt{\frac{n}{2(n-1)}}}}\Big).$$

Then Ψ is a Lagrangian Willmore submanifold. We call Ψ as *the Lagrangian Willmore sphere*. We note that when $n = 2$, Ψ is *Whitney sphere*

Theorem 5.3 (Refs. 16,17). *Let M be an n-dimensional ($n \geq 2$) closed Willmore submanifold in $(n+p)$-dimensional unit sphere S^{n+p}. Then we have*

$$\int_M \rho^n \left(\frac{n}{2 - 1/p} - \rho^2\right) dv \leq 0. \tag{21}$$

In particular, if

$$0 \leq \rho^2 \leq \frac{n}{2-1/p}, \tag{22}$$

then either $\rho^2 \equiv 0$ *and* M *is totally umbilic, or* $\rho^2 \equiv \frac{n}{2-1/p}$. *In the latter case, either* $p = 1$ *and* M *is a Willmore torus* $W_{m,n-m}$*; or* $n = 2$, $p = 2$ *and* M *is the Veronese surface.*

For $n = 2$, the following result was proved by Li[18] (also see Li-Simon[19])

Theorem 5.4 (Refs. 18,19). *Let* M *be a closed Willmore surface in an* $(2+p)$*-dimensional unit sphere* S^{2+p}. *Then we have*

$$\int_M \rho^2 (2 - \frac{3}{2}\rho^2) dv \leq 0. \tag{23}$$

In particular, if

$$0 \leq \rho^2 \leq \frac{4}{3}, \tag{24}$$

then either $\rho^2 = 0$ *and* M *is totally umbilic, or* $\rho^2 = \frac{4}{3}$. *In the latter case,* $p = 2$ *and* M *is the Veronese surface.*

Remark 5.1. In Ref. 21, the authors give some examples of 3-dimensional Lagrangian Willmore submanifolds in S^6. In Ref. 7, the authors study the variational problem of the functional $F(x) = \int_M (S - nH^2) dv$ for submanifold $x: M^n \to S^{n+p}$ and get an integral inequality of Simons' type.

6. Variations of Some Parametric Elliptic Functional

Let $F\colon S^n \to \mathbb{R}^+$ be a smooth function which satisfies the following convexity condition:

$$(D^2 F + F1)_x > 0, \quad \forall x \in S^n, \tag{25}$$

where $D^2 F$ denotes the intrinsic Hessian of F on S^n and 1 denotes the identity on $T_x S^n$, > 0 means that the matrix is positive definite. We consider the map

$$\begin{aligned} \phi\colon S^n &\to \mathbb{R}^{n+1} \\ x &\to F(x)x + (\operatorname{grad}_{S^n} F)_x, \end{aligned}$$

its image $W_F = \phi(S^n)$ is a smooth, convex hypersurface in $\mathbb{R}^{n+1}$ called the Wulff shape of F (see Refs. 8,9,12,22).

Now let $X\colon M \to \mathbb{R}^{n+1}$ be a smooth immersion of a compact, orientable hypersurface without boundary. Let $\nu\colon M \to S^n$ denotes its Gauss map, that is, ν is an unit inner normal vector of M.

Let $A_F = D^2F + F1$, $S_F = -A_F \circ \mathrm{d}\nu$. S_F is called the F-Weingarten operator, and the eigenvalues of S_F are called anisotropic principal curvatures. Let σ_r be the elementary symmetric functions of the anisotropic principal curvatures $\lambda_1, \lambda_2, \cdots, \lambda_n$:

$$\sigma_r = \sum_{i_1<\cdots<i_r} \lambda_{i_1}\cdots\lambda_{i_r} \quad (1 \le r \le n).$$

We set $\sigma_0 = 1$. The r-anisotropic mean curvature M_r is defined by $M_r = \sigma_r/C_n^r$.

In Ref. 8, we obtained the following integral formulas of Minkowski's type for closed hypersurfaces in $\mathbb{R}^{n+1}$.

Theorem 6.1 (Ref. 8). *Let $X\colon M \to \mathbb{R}^{n+1}$ be an n-dimensional closed hypersurface, $F\colon S^n \to \mathbb{R}^+$ be a smooth function which satisfies (25), then we have the following integral formulas of Minkowski type hold:*

$$\int_M (FM_r + M_{r+1}\langle X, \nu\rangle)\mathrm{d}A_X = 0, \quad r = 0, 1, \cdots, n-1. \tag{26}$$

By use of above integral formulas of Minkowski type, we prove the following new characterizations of the Wulff shape:

Theorem 6.2 (Ref. 8). *Let $X\colon M \to \mathbb{R}^{n+1}$ be an n-dimensional closed hypersurface, $F\colon S^n \to \mathbb{R}^+$ be a smooth function which satisfies (6.1), and $M_1 = const$ and $\langle X, \nu\rangle$ has fixed sign, then up to translations and homotheties, $X(M)$ is the Wulff shape.*

Theorem 6.3 (Ref. 8). *Let $X\colon M \to \mathbb{R}^{n+1}$ be an n-dimensional closed hypersurface, $F\colon S^n \to \mathbb{R}^+$ be a smooth function which satisfies (25). If $M_1 = const$ and $M_r = const$ for some r, $2 \le r \le n$, then up to translations and homotheties, $X(M)$ is the Wulff shape.*

Theorem 6.4 (Ref. 8). *Let $X\colon M \to \mathbb{R}^{n+1}$ be an n-dimensional closed convex hypersurface, $F\colon S^n \to \mathbb{R}^+$ be a smooth function which satisfies (6.1). If $\frac{M_r}{M_k} = const$ for some k and r, with $0 \le k < r \le n$, then then up to translations and homotheties, $X(M)$ is the Wulff shape.*

For each r, $0 \le r \le n$, we set

$$\mathscr{A}_r = \int_M F(\nu)\sigma_r dA_X. \tag{27}$$

Suppose σ_r is positive on M, and consider those hypersurfaces which are critical points of the functional $\mathscr{A}_r$ restricted to those hypersurfaces with the same $\mathscr{A}_{r-1}$, where $r \geq 1$. By a standard argument involving Lagrange multipliers, this means we are considering critical points of the functional

$$\mathscr{F}_{r;\Lambda} = \mathscr{A}_r + \Lambda \mathscr{A}_{r-1}, \tag{28}$$

where Λ is a constant. In Ref. 9 the authors show that the Euler-Lagrange equation of $\mathscr{F}_{r;\Lambda}$ is:

$$(r+1)\sigma_{r+1} + \Lambda r \sigma_r = 0. \tag{29}$$

So the critical points are just hypersurfaces with $M_{r+1}/M_r = \text{const.}$

We call a critical immersion X stable if and only if the second variation of $\mathscr{A}_r$ (or equivalently of $\mathscr{F}_{r;\Lambda}$) is non-negative for all variations of X preserving $\mathscr{A}_{r-1}$. We have the following theorems.

Theorem 6.5 (Ref. 9). *For each r, $1 \leq r \leq n-1$, the Wulff shape W_F is a stable critical immersion.*

Theorem 6.6 (Ref. 9). *Suppose $1 \leq r \leq n-1$. Let $X\colon M \to \mathbb{R}^{n+1}$ be a smooth immersion of an oriented closed, stable critical point of $\mathscr{A}_r$ for all variations of X preserving $\mathscr{A}_{r-1}$. Then, up to translations and homotheties, $X(M)$ is the Wulff shape.*

In Ref. 22, B. Palmer considered a variational problem of the functional $\mathscr{A}_0$ restricted to those hypersurfaces preserving the enclosed volume. He studied the first and second variations of the functional $\mathscr{A}_0$ preserving volume and showed that, up to homothety and translation, the only closed, oriented, stable critical point is the Wulff shape.

At the end of this paper, we would like to propose the following open problems:

Problem 1. *Assume that r is even and $r \in \{2, \cdots, n-1\}$. Is any closed stable r-minimal hypersurface in unit sphere S^{n+1} a geodesic sphere?*

Problem 2. *Let $x : M \to S^{n+1}$ be a n-dimensional Willmore hypersurface ρ^2 =constant, is the value of ρ^2 the discrete?*

Problem 3. *Is any topological sphere in R^3 with constant anisotropic mean curvature the Wulff shape?*

Problem 4. *Is any n-dimensional $(n \geq 2)$ closed embedded hypersurface in R^{n+1} with constant anisotropic mean curvature the Wulff shape?*

Acknowledgments

The author would like to thank Professor Oldrich Kowalski, chairman of Section A: Riemannian Geometry and Submanifolds, in "DIFFERENTIAL GEOMETRY AND ITS APPLICATIONS-the 10th International Conference Dedicated to the 300th Birthday of Leonhard Euler August 27 - 31, 2007, Olomouc, Czech Republic", for inviting him to give a plenary talk in Section A. This work is partially supported by grant No. 10531090 of NSFC and by SRFDP.

References

1. S.S. Chern, M. Do Carmo and S. Kobayashi, Minimal submanifolds of a sphere with second fundamental form of constant length, *In: Functional Analysis and Related Fields* ((F.Brower, Ed.) Springer-Verlag, Berlin, 1970) 59–75.
2. B.Y. Chen, Some conformal invariants of submanifolds and their applications, *Boll. Un. Mat. Ital.* **10** (1974) 380–385.
3. S.S. Chern, *Minimal Submanifolds in a Riemannian Manifold (mimeographed)* (University of Kansas, Lawrence, 1968).
4. L.F. Cao and H. Li, r-minimal submanifolds in space forms, *Ann. Global Anal. Geom.* **32** (2007) 311–341.
5. J.F. Grosjean, Upper bounds for the first eigenvalue of the Laplacian on compact submanifolds, *Pacific J. Math.* **206** (2002) 93–112.
6. Z. Guo, H. Li and C.P. Wang, The second variational formula for Willmore submanifolds in S^n, *Results in Math.* **40** (2001) 205–225.
7. Z. Guo and H. Li, A variational problem for submanifolds in a sphere, *Monatshefte Mathematik* **152** (2007) 295–302.
8. Y.J. He and H. Li, Integral formula of Minkowski type and new characterization of the Wulff shape, 2007, to appear in *Acta Math. Sinica*; arXiv:math.DG/0703187.
9. Y.J. He and H. Li, A new variational characterization of the Wulff shape, to appear in *Diff. Geom. Appl.*, 2007.
10. Z.J. Hu and H. Li, Willmore submanifolds in Riemannian Manifolds, 251-275, *In: Proceedings of the Workshop Contem. Geom. and Related Topics* (Belgrand, Yugoslavia, May, 2002) 15-21.
11. Z.J. Hu and H. Li Willmore Lagrangian spheres in the complex Euclidean space C^n, *Ann. Global Anal. Geom.* **25** (2004) 73–98.
12. M. Koiso and B. Palmer, Geometry and stability of surfaces with constant anisotropic mean curvature, *Indiana Univ. Math. J.* **54** (2005) 1817–1852.
13. H.B. Lawson, Local rigidity theorems for minimal hypersurfaces, *Ann. of Math.* **89** (1969) 187–197.
14. H. Li, Hypersurfaces with constant scalar curvature in space forms, *Math. Ann.* **305** (1996) 665–672.
15. H. Li, Global rigidity theorems of hypersurface, *Ark. Math.* **35** (1997) 327–351.

16. H. Li, Willmore submanifolds in a sphere, *Math. Research Letters* **9** (2002) 771–790.
17. H. Li, Willmore hypersurfaces in a sphere, *Asian J. Math.* **5** (2001) 365–378.
18. H. Li , Willmore surfaces in S^n, *Ann. Global Anal. Geom.* **21** (2002) 203–213.
19. H. Li, U. Simon, Quantization of curvature for compact surfcaes in S^n, *Math. Z.* **245** (2003) 201–216.
20. H. Li and L. Vrancken, New examples of Willmore surfaces in S^n, *Ann. Global Anal. Geom.* **23** (2003) 205–225.
21. H. Li and G.X. Wei, Classification of Lagrangian Willmore submanifolds of the nearly Kaehler 6-sphere $S^6(1)$ with constant scalar curvature, *Glasgow Math. J.* **48** (2006) 53–64.
22. B. Palmer, Stability of the wulff shape, *Proc. Amer. Math. soci.* **126** (12) (1998) 3661–3667.
23. U. Pinkall, Hopf tori in S^3, *Invent. Math.* **81** (1985) 379–386.
24. F.J. Pedit and T.J. Willmore, Conformal geometry, *Atti Sem. Mat. Fis. Univ. Modena* **XXXVI** (1988) 237–245.
25. R. Reilly, Variation properties of functions of the mean curvatures for hypersurfaces in space forms, *J. Diff. Geom.* **8** (1973) 465–477.
26. R. Reilly, Applications of the Hessian operator in a Riemannian manifold, *Indiana Univ. math. Journal* **26** (1977) 459–472.
27. J. Simons, Minimal varieties in Riemmannian manifolds, *Ann. Math.*, 2nd Ser., **88** (1) (1968) 62.
28. C.P. Wang , Moebius geometry of submanifolds in S^n, *Manuscripta Math.* **96** (1998) 517–534.
29. J. Weiner, On a problem of Chen, Willmore, et., *Indiana Univ. Math. J.* **27** (1978) 19–35.
30. T.J. Willmore, Notes on embedded surfaces, *Ann. Stiint. Univ. Al. Cuza*, Iasi, Sect. I a Mat.(N.S.) 11B (1965) 493–496.

Differential Geometry and its Applications
Proc. Conf., in Honour of Leonhard Euler, Olomouc, August 2007

Generalized Veronese manifolds and isotropic immersions

S. Maeda

Department of Mathematics, Saga University,
1 Honzyo, Saga 840-8502 Japan
E-mail: smaeda@ms.saga-u.ac.jp

S. Udagawa

School of Medicine, Nihon University,
Itabashi, Tokyo 173-0032, Japan
E-mail: sudagawa@med.nihon-u.ac.jp

In submanifold theory, an n-dimensional Veronese manifold is defined as a minimal immersion of an n-dimensional sphere $S^n(nc/(2n+2))$ of constant sectional curvature $nc/(2n+2)$ into an $(n+(n^2+n-2)/2)$-dimensional sphere $S^{n+(n^2+n-2)/2}(c)$ of constant sectional curvature c. It is well-known that this minimal immersion has parallel second fundamental form (Ref. 7). In this paper, we generalize the notion of Veronese manifolds and call them generalized Veronese manifolds (for the definition of such submanifolds, see Introduction). We survey some characterizations of generalized Veronese manifolds without proof.

Keywords: Space forms, parallel immersions, isotropic immersions, totally umbilic, Veronese manifolds, generalized Veronese manifolds, sectional curvatures, parallel mean curvature vector.

MS classification: 53C40, 53C42.

1. Introduction

We denote by $M^n(c)$ an n-dimensional *space form* of constant sectional curvature c, which is an n-dimensional complete simply connected Riemannian manifold of constant sectional curvature c. Hence $M^n(c)$ is congruent to either a standard sphere $S^n(c)$, a Euclidean space $\mathbb{R}^n$ or a hyperbolic space $H^n(c)$ according as c is positive, zero or negative, respectively.

In submanifold theory, for a Riemannian submanifold (M^n, f) of $\widetilde{M}^{n+p}$ when the second fundamental form of M^n in $\widetilde{M}^{n+p}$ is parallel, $f : M^n \to \widetilde{M}^{n+p}$ is said to be a *parallel immersion* or M^n is called a *parallel submani-*

fold of $\widetilde{M}^{n+p}$. In this paper, a submanifold $(S^n(nc/(2n+2)), f)$ is said to be a *generalized Veronese manifold* if the isometric immersion f is decomposed as:

$$f = f_2 \circ f_1 : S^n(nc/(2n+2)) \xrightarrow{f_1} S^{n+(n^2+n-2)/2}(c) \xrightarrow{f_2} \widetilde{M}^{n+p}(\tilde{c}), \quad (1)$$

where f_1 is a minimal parallel immersion, f_2 is a totally umbilic embedding and $c \geqq \tilde{c}$. Here, the ambient space $\widetilde{M}^{n+p}(\tilde{c})$ in (1) is either $S^{n+p}(\tilde{c})$, $\mathbb{R}^{n+p}$ or $H^{n+p}(\tilde{c})$. By easy computation we see that $H = \sqrt{c - \tilde{c}}$, where H is the mean curvature of f. So we can see that when $c = \tilde{c}$, the submanifold $(S^n(nc/(2n+2)), f)$ is minimal, but when $c > \tilde{c}$, it is not minimal. Also, for $c = \tilde{c}$, we obtain a Veronese manifold (see, e.g. Ref. 7). Note that when $c = \tilde{c}$, the totally umbilic embedding f_2 in (1) is nothing but a totally geodesic embedding. If $c > \tilde{c}$, then the submanifold is not minimal and hence a generalized Veronese manifold is not a Veronese manifold, in general. It is well-known that generalized Veronese manifolds are the only examples of non totally umbilic but parallel immersions of space forms into space forms. These submanifolds have various nice geometric properties. For example, for each geodesic γ on the submanifold $S^n(nc/(2n+2))$ the curve $f \circ \gamma$ is a circle in the ambient space $\widetilde{M}^{n+p}(\tilde{c})$. Moreover, the immersion f given by (1) is $(\sqrt{(2nc/(n+1)) - \tilde{c}}$-)isotropic (for the definition of "isotropic", see section 2). The notion of isotropic immersions plays a key role in this paper.

In this paper, we study Veronese manifolds and generalized Veronese manifolds in terms of isotropic immersions.

The authors would like to express their hearty thanks to Professor Oldrich Kowalski for his valuable suggestion and encouragement during the preparation of this paper.

2. Isotropic immersions

An isometric immersion $f : M \to \widetilde{M}$ of an n-dimensional Riemannian manifold into an $(n+p)$-dimensional Riemannian manifold is said to be $(\lambda(x)$-)*isotropic* at $x \in M$ if $\|\sigma(X,X)\|/\|X\|^2 (= \lambda(x))$ does not depend on the choice of $X (\neq 0) \in T_xM$, where σ is the second fundamental form of the immersion f. If the immersion is $(\lambda(x)$-)isotropic at every point x of M, then the immersion is said to be (λ-) isotropic (see Ref. 11). When $\lambda = 0$ on the submanifold M, this submanifold is totally geodesic in the ambient manifold $\widetilde{M}$. Note that a totally umbilic immersion is isotropic, but not *vice versa* (for details, see Refs. 2,6,14). The notion of isotropic immersions gives us valuable information in some cases. For example, we consider each parallel submanifold (M^n, f) of a standard sphere $S^{n+p}(\tilde{c})$. Then our manifold M^n is a compact Riemannian symmetric space. Moreover, by the classification theorem of parallel submanifolds in $S^{n+p}(\tilde{c})$ we can see that M^n is of rank one if and only if the isometric immersion $f : M^n \to S^{n+p}(\tilde{c})$ is isotropic (see Refs. 4,13). That is, in this case the notion of isotropic immersions distinguishes symmetric spaces of rank one from symmetric spaces of higher rank.

3. Characterizations of Veronese manifolds

We first review the following fundamental lemma (Ref. 11):

Lemma 3.1. *Let f be a $\lambda(>0)$-isotropic immersion of a Riemannian manifold M^n $(n \geqq 2)$ into a Riemannian manifold $\widetilde{M}^{n+p}$. The discriminant $\Delta_x(X, Y)$ at $x \in M$ is defined by $\Delta_x(X, Y) = K(X, Y) - \widetilde{K}(X, Y)$ for a pair (X, Y) of linearly independent vectors, where $K(X, Y)$(resp. $\widetilde{K}(X, Y)$) denotes the sectional curvature of the plane spanned by $X, Y \in T_xM$ for M (resp. $\widetilde{M}$). Suppose that the discriminant Δ_x at $x \in M$ is constant for each pair of vectors (X, Y). Then the following inequalities hold at x :*

$$-\frac{n+2}{2(n-1)}\lambda(x)^2 \leqq \Delta_x \leqq \lambda(x)^2.$$

Moreover,

$$\begin{aligned}
&(i)\ \Delta_x = \lambda(x)^2 \iff f \text{ is umbilic at } x \iff \dim N_x^1 = 1,\\
&(ii)\ \Delta_x = -\{(n+2)/2(n-1)\}\lambda(x)^2 \iff f \text{ is minimal at } x\\
&\qquad \iff \dim N_x^1 = (n(n+1)/2) - 1,\\
&(iii)\ -\{(n+2)/2(n-1)\}\lambda(x)^2 < \Delta_x < \lambda(x)^2 \iff \dim N_x^1 = n(n+1)/2.
\end{aligned}$$

Here, we denote by N_x^1 the first normal space at x, namely $N_x^1 = Span_{\mathbb{R}}\{\sigma(X, Y) : X, Y \in T_xM\}$.

Using Lemma 3.1, we establish the following local theorem which characterizes Veronese manifolds (cf. Refs. 1,8).

Theorem 3.1. *Let f be a λ-isotropic immersion of an $n(\geqq 2)$-dimensional Riemannian manifold M^n of constant sectional curvature k into an $(n+p)$-dimensional Riemannian manifold $\widetilde{M}^{n+p}$ of constant sectional curvature $\tilde{k}$. Suppose that $p \leqq (n^2+n-2)/2$ and $k < \tilde{k}$. Then f is locally equivalent to a (parallel) minimal immersion of $S^n(k)$ into $S^{n+p}(\tilde{k})$, where $\tilde{k} = 2(n+1)k/n$ and $p = (n^2+n-2)/2$. Hence the submanifold (M^n, f) is locally congruent to a Veronese manifold.*

We here recall the following global theorem which characterizes Veronese manifolds. This theorem can be proved by using a well-known differential equation of the second fundamental form of the immersion (Ref. 12 and Ref. 5).

Theorem 3.2. *Let M^n $(n \geqq 2)$ be a connected compact oriented minimal but not totally geodesic submanifold of $S^{n+p}(\tilde{c})$ through a full isometric immersion f. Suppose that every sectional curvature K of M^n satisfies $K \geqq n\tilde{c}/(2n+2)$. Then $K = n\tilde{c}/(2n+2)$, $p = (n^2+n-2)/2$ and f is a parallel immersion. Hence (M^n, f) is a Veronese manifold.*

4. Characterizations of generalized Veronese manifolds

Motivated by Theorem 3.2, we obtained the following theorem (cf. Ref. 10).

Theorem 4.1. *Let M^n be an $n(\geqq 3)$-dimensional connected compact oriented isotropic submanifold whose mean curvature vector is parallel with respect to the normal connection in an $(n+p)$-dimensional space form $\widetilde{M}^{n+p}(\tilde{c})$ of constant sectional curvature $\tilde{c}$ through an isometric immersion f. Suppose that every sectional curvature K of M^n satisfies $K \geqq (n/2(n+1))(\tilde{c}+H^2)$, where H is the mean curvature of the immersion f. Then the immersion f has parallel second fundamental form and the submanifold (M^n, f) is congruent to one of the following:*

(1) *M^n is a congruent to $S^n(K)$ of constant sectional curvature $K = \tilde{c} + H^2$ and f is a totally umbilic embedding;*
(2) *(M^n, f) is a generalized Veronese manifold. That is, M^n is congruent to $S^n(K)$ of constant sectional curvature $K = (n/2(n+1))(\tilde{c}+H^2)$ and f is decomposed as:*
$f = f_2 \circ f_1 : S^n(K) \xrightarrow{f_1} S^{n+(n^2+n-2)/2}(2(n+1)K/n) \xrightarrow{f_2} \widetilde{M}^{n+p}(\tilde{c})$,
where f_1 is a minimal (parallel) immersion and f_2 is a totally umbilic embedding.

When $n = 2$, Theorem 4.1 also holds without the assumption that M^n is isotropic in $\widetilde{M}^{n+p}(\tilde{c})$ (see Ref. 10).

Theorem 4.2. *Let M^2 be a 2-dimensional connected compact oriented submanifold whose mean curvature vector is parallel with respect to the normal connection in a $(2+p)$-dimensional space form $\widetilde{M}^{2+p}(\tilde{c})$ of constant sectional curvature $\tilde{c}$ through an isometric immersion f. Suppose that every sectional curvature K of M^2 satisfies $K \geqq (1/3)(\tilde{c}+H^2)$, where H is the* *mean curvature of f. Then the immersion f has parallel second fundamental form and the submanifold (M^2, f) is congruent to one of the following:*

(1) *M^2 is congruent to $S^2(K)$ of constant sectional curvature $K = \tilde{c} + H^2$ and f is a totally umbilic embedding;*
(2) *(M^2, f) is a generalized Veronese surface. That is, M^2 is congruent to $S^2(K)$ of constant sectional curvature $K = (1/3)(\tilde{c} + H^2)$ and f is decomposed as:*
 $f = f_2 \circ f_1 : \ S^2(K) \xrightarrow{f_1} S^4(3K) \xrightarrow{f_2} \widetilde{M}^{2+p}(\tilde{c})$,
 where f_1 is a minimal (parallel) immersion and f_2 is a totally umbilic embedding.

Remark 4.1.

(a) In the assumption of Theorem 4.1, if we replace the condition that $K \geqq (n/2(n+1))(\tilde{c}+H^2)$ by a stronger condition that $(n/2(n+1))(\tilde{c}+H^2) \leqq K < \tilde{c} + H^2$, then we obtain a corollary which characterizes just an $n(\geqq 3)$-dimensional generalized Veronese manifold. Also in the assumption of Theorem 4.2, replacing "$K \geqq (1/3)(\tilde{c} + H^2)$" by "$(1/3)(\tilde{c} + H^2) \leqq K < \tilde{c} + H^2$", we get a corollary which characterizes just a generalized Veronese surface.
(b) The following proposition shows that Theorems 4.1 and 4.2 are no longer true if we replace the condition that "the mean curvature vector is parallel with respect to the normal connection" by a weaker condition that "the mean curvature is constant" in assumptions of these theorems (Ref. 9).

Proposition 4.1. *For each $n(\geqq 2)$ there exists at least one Riemannian submanifold M^n of $S^{n+p}(\tilde{c})$ satisfying the following five conditions.*

(1) *M^n is a connected compact oriented Riemannian manifold.*
(2) *M^n is isotropic in $S^{n+p}(\tilde{c})$.*
(3) *The mean curvature H of M^n in $S^{n+p}(\tilde{c})$ is nonzero constant.*
(4) *Every sectional curvature K of M^n satisfies*

$$K \geqq \frac{\tilde{c} + H^2}{2} \Big(> \frac{n}{2n+2}(\tilde{c} + H^2)\Big).$$

(5) *The mean curvature vector of M^n in $S^{n+p}(\tilde{c})$ is not parallel with respect to the normal connection, so that in particular M^n is not a parallel submanifold of $S^{n+p}(\tilde{c})$. Hence M^n is neither totally umbilic in $S^{n+p}(\tilde{c})$ nor a generalized Veronese manifold.*

At the end of this paper, in consideration of Theorems 3.2, 4.1 and 4.2 we pose the following open problem:

Problem. If we remove the assumption that the immersion is isotropic in Theorem 4.1, does this theorem hold?

References

1. N. Boumuki, Isotropic immersions with low codimension of space forms into space forms, *Mem. Fac. Sci. Eng. Shimane Univ. Ser. B: Math. Sci.* **37** (2004) 1–4.
2. R. Bryant, Conformal and minimal immersions of compact surfaces into the 4-sphere, *J. Differential Geom.* **17** (1982) 455–473.
3. M. Do Carmo and N. Wallach, Minimal immersions of spheres into spheres, *Ann. Math.* **93** (1971) 43–62.
4. D. Ferus, Immersions with parallel second fundamental form, *Math. Z.* **140** (1974) 87–92.
5. T. Itoh, Addendum to my paper "On Veronese manifolds" *J. Math. Soc. Japan* **30** (1978) 73–74.
6. T. Itoh and K. Ogiue, Isotropic immersions, *J. Differential Geometry* **8** (1973) 305–316.
7. T. Itoh and K. Ogiue, Isotropic immersions and Veronese manifolds, *Trans. Amer. Math. Soc.* **209** (1975) 109–117.
8. S. Maeda, Isotropic immersions with parallel second fundamental form, *Canad. Math. Bull.* **26** (1983) 291–296.
9. S. Maeda, On some isotropic submanifolds in spheres, *Proc. Japan Acad. Ser. A* **77** (2001) 173–175.
10. S. Maeda and S. Udagawa, Characterization of parallel isometric immersions of space forms into space forms in the class of isotropic immersions, to appear in *Canad. J. Math.*
11. B. O'Neill, Isotropic and Kaehler immersions, *Canadian J. Math.* **17** (1965) 905–915.
12. J. Simons, Minimal varieties in riemannian manifolds, *Ann. Math.* **88** (1968) 62–105.
13. M. Takeuchi, Parallel submanifolds of space forms, *In: Manifolds and Lie groups* (Notre Dame, Ind., 1980, in honor of Y. Matsushima, Birkhäuser, Boston, Mass. Progr. Math. 14, 1981) 429–447.
14. K. Tsukada, Isotropic minimal immersions of spheres into spheres, *J. Math. Soc. Japan* **35** (1983) 355–379.

Differential Geometry and its Applications
Proc. Conf., in Honour of Leonhard Euler, Olomouc, August 2007

Integral formulae for foliations on Riemannian manifolds

V. Rovenski
Mathematical Department, University of Haifa,
Haifa, 31905, Israel
E-mail: rovenski@math.haifa.ac.il
http://math.haifa.ac.il/ROVENSKI/rovenski.html

P. Walczak
Katedra Geometrii, Uniwersytet Łódzki,
ul. Banacha 22 Łódź, Poland
E-mail: pawelwal@math.uni.lodz.pl
http://math.uni.lodz.pl/~pawelwal/

We generalize the integral formulae for foliations on space forms by Brito–Langevin–Rosenberg (1991) and by Brito–Naveira (2000). Our integral formulae concern foliations on complete Riemannian manifolds of a finite volume and involve the co-nullity operator, the mixed curvature operator, and their products. To prove the integral formulae for foliations of arbitrary codimension we introduce and study the tangent sphere bundle of a foliation. This is a submanifold of the tangent sphere bundle with the Sasaki metric studied by Borisenko–Yampol'skii (1991), Kowalski–Sekizawa (2000), and others.

Keywords: Riemannian manifold, foliation, higher mean curvatures, mixed curvature, Sasaki metric.

MS classification: 53C12.

1. Introduction

Recent years brought increasing interest in the dynamics and extrinsic geometry of foliations,[13,18] etc.

In what follows (M, g) denotes a compact (or a finite volume) Riemannian manifold with the Levi-Civita connection ∇, and the curvature tensor R; $\mathcal{F} = \{L\}$ a transversely oriented foliation on M, $T\mathcal{F}$ ($T\mathcal{F}^\perp$) its tangent (orthogonal) distribution, $R^{\rm mix}_\xi(y) = R(y,\xi)\,\xi$ ($\xi \in T\mathcal{F},\ y \in T\mathcal{F}^\perp$), the mixed curvature operator, $K^{\rm mix}$ the mixed sectional curvature (of planes containing some $\xi \in T\mathcal{F}$, and $y \in T\mathcal{F}^\perp$), $C_\xi\, y = (\nabla_y\, \xi)^\perp$ the co-nullity operator, A_ξ Weingarten operator (for the unit ξ), $\sigma_i(A_\xi)$ the mean curvatures of A_ξ, i.e. the coefficients of the polynomial $\det(Id + tA_\xi) = \sum_i \sigma_i(A_\xi)\,t^i$. For the codimension 1 foliations case, ξ denotes the unit normal vector field.

The first known *integral formula* for codimension one foliations reads [12]

$$\int_M H\,dV_M = 0, \quad \text{where } H = \sigma_1(A_\xi)/\dim\mathcal{F} \ - \ \text{the mean curvature.} \quad (1)$$

A general integral formula for pairs of complementary orthogonal plane fields on compact manifolds [15] is

$$\int_M (K(\mathcal{F},\mathcal{F}^\perp) + |A_2|^2 - |H_2|^2 - |T_2|^2 + |A_2|^2 - |H_2|^2 - |T_2|^2)\,dV_M = 0 \quad (2)$$

where $K(\mathcal{F},\mathcal{F}^\perp) = \sum_{i,\alpha} g(R(e_i,e_\alpha)e_i,e_\alpha)$, H_1, T_1, A_1 and H_2, T_2, A_2 are the mean curvature vector, the integrability tensor, and the 2-nd fundamental form of $\mathcal{F}$ and $\mathcal{F}^\perp$. In codimension one foliation case Eq. (2) reduces to

$$\int_M [\sigma_2(A_\xi) - \frac{1}{2}\operatorname{Ric}(\xi)]\,dV_M = 0. \quad (3)$$

These formulae are of some interest, they can be useful for the problems: prescribing higher mean curvatures σ_i of a foliation; minimizing functions like volume and energy defined for plane fields on Riemannian manifolds; existence of foliations with all the leaves enjoying a given geometric property such as being totally geodesic, totally umbilical, minimal, etc. (see, among the others,[4,11,16,17] the survey[13] and the bibliography there)

Brito, Langevin and Rosenberg [3] have shown that the integrals of $\sigma_i(A_\xi)$ of a codimension-one foliation $\mathcal{F}$ on a compact space form $M^{n+1}(c)$ do not depend on $\mathcal{F}$: they depend on n, i, c and the volume of M only.

Theorem 1.1 (Brito–Langevin–Rosenberg[3]). *Let $\mathcal{F}$ be a codimension one foliation on a compact space form $M^{n+1}(c)$. Then the integrals of $\sigma_k(A_\xi)$ (ξ is a unit normal to the leaves) do not depend on $\mathcal{F}$:*

$$\int_M \sigma_k(A_\xi)\,dM = \begin{cases} c^{\frac{k}{2}}\binom{n/2}{k/2}\operatorname{vol}(M), & n,k \text{ even}; \\ 0, & n \text{ or } k \text{ odd}. \end{cases} \quad (4)$$

For $i = 1$ and 2, these integral formulae have known generalizations, Eqs. (1), (3), to arbitrary Riemannian manifolds.

As far as totally geodesic foliations of codimension n on compact space forms $M^{n+p}(c)$ concern, Brito and Naveira [5] have shown that total $2k$-dimensional curvatures, $\gamma_{2k}(\mathcal{F}^\perp) = \int_M \gamma_{2k}(\mathcal{F}^\perp_x)\,dV$ do not depend on a foliation, they depend on k, p, n, c and the volume of M only.

Definition 1.1. Given $x \in M$ take orthonormal bases $\{e_i\}_{1\le i\le p}$ of $T\mathcal{F}_x$ and $\{e_\alpha\}_{1\le\alpha\le n}$ of $T\mathcal{F}_x^\perp$, and set $C^i_{\alpha,\beta} = g(C_{e_i}e_\alpha, e_\beta)$. The $2k$-*dimensional curvature* of $\mathcal{F}_x^\perp$ is defined by

$$\begin{aligned}\gamma_{2k}(\mathcal{F}_x^\perp) = &\frac{1}{(2k)!\,2^k}\sum\nolimits_{\boldsymbol{\alpha},\boldsymbol{\beta}} \varepsilon_{\boldsymbol{\alpha},\boldsymbol{\beta}}\Big[\sum\nolimits_{i_1=1}^{p}(C^{i_1}_{\alpha_1,\beta_1}C^{i_1}_{\alpha_2,\beta_2} - C^{i_1}_{\alpha_1,\beta_2}C^{i_1}_{\alpha_2,\beta_1})\times \\ &\ldots\times\sum\nolimits_{i_k=1}^{p}(C^{i_k}_{\alpha_{2k-1},\beta_{2k-1}}C^{i_k}_{\alpha_{2k},\beta_{2k}} - C^{i_k}_{\alpha_{2k-1},\beta_{2k}}C^{i_k}_{\alpha_{2k},\beta_{2k-1}})\Big]\end{aligned} \tag{5}$$

where $\varepsilon_{\boldsymbol{\alpha},\boldsymbol{\beta}} = \varepsilon_{\alpha_1,\ldots\alpha_{2k};\beta_1,\ldots\beta_{2k}}$ is $+1$ (-1) if $\boldsymbol{\alpha} = (\alpha_1,\ldots\alpha_{2k})$ are $2k$ distinct integers and $\boldsymbol{\beta} = (\beta_1,\ldots\beta_{2k})$ are even (odd) permutations of $\alpha_1,\ldots\alpha_{2k}$; otherwise it is 0. The sum is taken over all α_i, β_i between 1 and n.

Theorem 1.2 (Brito–Naveira[5]). *Let $\mathcal{F}^p$ be a totally geodesic foliation on a compact space form $M^{n+p}(c)$ with $c \ge 0$. Then the total $2k$-dimensional curvatures do not depend on $\mathcal{F}$:*

$$\gamma_{2k}(\mathcal{F}^\perp) = \begin{cases} \dfrac{\binom{p+2k-1}{2k}}{\binom{(p+2k-1)/2}{k}}\binom{n/2}{k}c^k\,\mathrm{vol}(M) & \text{if } n \text{ even, } p \text{ odd;}\\ \dfrac{2^{2k}(k!)^2}{(2k)!}\binom{p/2+k-1}{k}\binom{n/2}{k}c^k\,\mathrm{vol}(M) & \text{if } n, p \text{ are even;}\\ 0 \quad \text{if } n \text{ odd.} \end{cases} \tag{6}$$

Examples of geometrically distinct totally geodesic foliations on space forms are abundant, see discussion in Ref. 5. A compact space form $M(c)$ with $c < 0$ does not admit a totally geodesic foliation of dimension $\ne 0$, dim M.[19]

Brito[4] has shown also relations between some differential forms (involving connection and curvature forms) on arbitrary foliated Riemannian manifolds. His results suggest the existence of similar to Eqs. (1), (3) formulae for σ_i's with $i > 2$. Here, we provide such formulae (see Sec. 3): they generalize Eqs. (4), (6) (the constancy of curvature condition is dropped) and relate the integrals of σ_i's with those involving some algebraic invariants obtained from the *co-nullity operator*, the *mixed curvature operator* and their products. We write several such formulae explicitly, on locally symmetric spaces as well as on arbitrary Riemannian manifolds where they involve also covariant derivatives of the mixed curvature operator.

The structure of the present work is as follows: Sec. 2 contains algebraic preliminaries. Sec. 3 presents the main results (Theorems 1, 2, and their corollaries). Secs. 4, 5 contain proofs.

2. Algebraic preliminaries

We define and study the invariants $\sigma_{\boldsymbol{\lambda}}(A_1,\ldots,A_m)$ of a set of matrices that generalize the elementary symmetric functions of a symmetric matrix A.

2.1. *Invariants of a set of square matrices*

Given arbitrary quadratic matrices $A_1, \dots A_m$ of order n and a unit matrix I_n one can consider the determinant $\det(I_n + t_1 A_1 + \dots + t_m A_m)$ and express it as a polynomial of real variables $\mathbf{t} = (t_1, \dots t_m)$. Given $\boldsymbol{\lambda} = (\lambda_1, \dots \lambda_m)$, a sequence of nonnegative integers with $|\boldsymbol{\lambda}| := \lambda_1 + \dots + \lambda_m \leq n$, we shall denote by $\sigma_{\boldsymbol{\lambda}}(A_1, \dots, A_m)$ its coefficient at $\mathbf{t}^{\boldsymbol{\lambda}} = t_1^{\lambda_1} \cdot \dots t_m^{\lambda_m}$:

$$\det(I_n + t_1 A_1 + \dots + t_m A_m) = \sum\nolimits_{|\boldsymbol{\lambda}| \leq n} \sigma_{\boldsymbol{\lambda}}(A_1, \dots A_m)\, \mathbf{t}^{\boldsymbol{\lambda}}. \qquad (7)$$

Certainly, $\sigma_i(A)$ coincides with the i-th elementary symmetric polynomial of the eigenvalues $\{k_j\}$ of a single symmetric matrix A. The next lemma collects properties of these invariants which will be used in the sequel.

Lemma 2.1 (Ref. 14). *For any $\boldsymbol{\lambda} = (\lambda_1, \dots \lambda_m)$, $a, b \in \mathbb{R}$ and any $n \times n$ matrices A_i, A and B one has*

(I) $\sigma_{\boldsymbol{\lambda}}(0, A_2, \dots A_m) = \sigma_{0,\hat{\boldsymbol{\lambda}}}(A_1, A_2, \dots A_m) = \sigma_{\hat{\boldsymbol{\lambda}}}(A_2, \dots A_m)$, *where* $\hat{\boldsymbol{\lambda}} = (\lambda_2, \dots \lambda_m)$,

(II) $\sigma_{\boldsymbol{\lambda}}(A_{s(1)}, \dots A_{s(m)}) = \sigma_{\boldsymbol{\lambda} \circ s}(A_1, \dots A_m)$, *where $s \in S_m$ is a permutation of m elements and* $\boldsymbol{\lambda} \circ s = (\lambda_{s(1)}, \dots \lambda_{s(m)})$,

(III) $\sigma_{\boldsymbol{\lambda}}(I_n, A_2, \dots A_m) = \binom{n-|\hat{\boldsymbol{\lambda}}|}{\lambda_1} \sigma_{\hat{\boldsymbol{\lambda}}}(A_2, \dots A_m)$,

(IV) $\sigma_{\boldsymbol{\lambda}}(A, A, A_3, \dots A_m) = \binom{\lambda_1+\lambda_2}{\lambda_1} \sigma_{(\lambda_1+\lambda_2, \lambda_3, \dots \lambda_m)}(A, A_3, \dots A_m)$,

(V) $\sigma_{1,\hat{\boldsymbol{\lambda}}}(A + B, A_2, \dots A_m) = \sigma_{1,\hat{\boldsymbol{\lambda}}}(A, \dots A_m) + \sigma_{1,\hat{\boldsymbol{\lambda}}}(B, \dots A_m)$ *and*
$\sigma_{\boldsymbol{\lambda}}(aA_1, A_2, \dots A_m) = a^{\lambda_1} \sigma_{\boldsymbol{\lambda}}(A_1, \dots A_m)$.

The invariants defined above can be used in calculation of the determinant of a matrix $B(t)$ expressed as a power series $B(t) = \sum_{i=0}^{\infty} t^i B_i$. The coefficient at t^j in the series of $\det(B(t))$ depends only on the part $\sum_{i \leq j} t^i B_i$ of $B(t)$.

Lemma 2.2. *If $B(t)$ is $n \times n$ matrix given by $B(t) = \sum_{i=0}^{\infty} t^i B_i$, then*

$$\det(B(t)) = 1 + \sum\nolimits_{k=1}^{\infty} \Big(\sum\nolimits_{\boldsymbol{\lambda}, \|\boldsymbol{\lambda}\| = k} \sigma_{\boldsymbol{\lambda}}(B_1, \dots B_k) \Big) t^k, \qquad (8)$$

where $\|\boldsymbol{\lambda}\| = \lambda_1 + 2\lambda_2 + \dots + k\lambda_k$ *for* $\boldsymbol{\lambda} = (\lambda_1, \dots \lambda_k)$.

By the First Fundamental Theorem of Matrix Invariants,[7] all $\sigma_{\boldsymbol{\lambda}}$ are expressed in terms of the traces of the matrices involved and their products. The following Lemma provides a tool for writing down the explicit formulae for all $\sigma_{\boldsymbol{\lambda}}$'s in terms of traces (of given matrices and their products).

Lemma 2.3 (Ref. 14). *For any $n \times n$-matrices A_i, $\boldsymbol{\lambda} = (\lambda_1, \dots \lambda_m)$, k one has*

$$\begin{aligned}\sigma_{(\boldsymbol{\lambda},k)}(A_1, \dots A_m, A_{m+1}) &= \sigma_{\boldsymbol{\lambda}}(A_1, \dots A_m)\, \sigma_k(A_{m+1}) \\ &- \sum\nolimits_{\boldsymbol{\mu}} \sigma_{\boldsymbol{\mu}}(A_1, \dots A_m, A_{m+1}, A_1 A_{m+1}, \dots A_m A_{m+1}),\end{aligned} \tag{9}$$

where the sum ranges over all the sequences $\boldsymbol{\mu} = (\lambda_1 - j_1, \dots \lambda_m - j_m, k - j_1 - \dots - j_m, j_1, \dots j_m)$ with $j_1 \le \lambda_1, \ \dots \ j_m \le \lambda_m, \ 0 < j_1 + \dots + j_m \le k$.

For example, $\sigma_{k,1}(B, C) = \sigma_k(B)\,\mathrm{Tr}(C) - \sigma_{k-1,1}(B, BC)$, etc.

2.2. *Integral relations for $\sigma_{\boldsymbol{\lambda}}$*

Let $T : (\mathbb{R}^p)^h \times \mathbb{R}^n \to \mathbb{R}^n$ be a multilinear map. For each $\xi \in \mathbb{R}^p$ define a linear operator $T_\xi := T(\xi, \dots \xi; \cdot) : \mathbb{R}^n \to \mathbb{R}^n$. In coordinate form $\xi = \sum_i \xi_i \bar{e}^i$ we have $T_\xi = \sum_{\mathbf{i}} T_{i_1,\dots i_p} \xi_{i_1} \cdot \dots \xi_{i_p}$, where $T_{i_1,\dots i_p} = T(\bar{e}^{i_1}, \dots \bar{e}^{i_p}; \cdot)$.

Lemma 2.4. *If h and m are odd, then $\int_{\|\xi\|=1} \sigma_m(T_\xi)\, d\omega_{p-1} = 0$.*

Proof. By Lemma 2.1 (V), $\sigma_m(T_{-\xi}) = \sigma_m((-1)^h T_\xi) = (-1)^{hm} \sigma_m(T_\xi)$. □

Set $I_{\boldsymbol{\lambda}} := \int_{\|\xi\|=1} \xi^{\boldsymbol{\lambda}}\, d\omega_{p-1}$ where $\xi^{\boldsymbol{\lambda}} = \prod_{i \le p} \xi_i^{\lambda_i}$. It is known [10] that $I_{\boldsymbol{\lambda}} = \frac{2}{\Gamma(\frac{p}{2} + \frac{1}{2}\sum_i \lambda_i)} \prod_{i \le p} \frac{1 + (-1)^{\lambda_i}}{2} \Gamma\big(\frac{1+\lambda_i}{2}\big)$, where Γ is Gamma function. For example, $I_{0,\dots 0} = \frac{2\pi^{p/2}}{\Gamma(p/2)} = \mathrm{Vol}(S_1^{p-1})$, $I_{2\lambda_1, 0,\dots 0} = 2\pi^{\frac{p-1}{2}} \frac{\Gamma(1/2 + \lambda_1)}{\Gamma(p/2 + \lambda_1)}$, etc.

Lemma 2.5. *If $h = 1$ then $\int_{\|\xi\|=1} \sigma_{2k-1}(\sum_{i \le p} \xi_i T_i)\, d\omega_{p-1} = 0$ and*

$$\int_{\|\xi\|=1} \sigma_{2k}(\sum\nolimits_{i \le p} \xi_i T_i)\, d\omega_{p-1} = \sum\nolimits_{|\boldsymbol{\lambda}| = k} I_{2\boldsymbol{\lambda}}\, \sigma_{2\boldsymbol{\lambda}}(T_1, \dots T_p). \tag{10}$$

Proof. Using Eq. (7) one obtains for $T_\xi = \sum_{i \le p} \xi_i T_i$

$$\sum\nolimits_j \sigma_j(T_\xi)\, t^j = \det(I_n + \sum\nolimits_{i \le p} (t\,\xi_i) T_i) = \sum\nolimits_j t^j \sum\nolimits_{|\boldsymbol{\lambda}| = j} \sigma_{\boldsymbol{\lambda}}(T_1, \dots T_p)\, \xi^{\boldsymbol{\lambda}}.$$

Hence

$$\sigma_j(T_\xi) = \sum\nolimits_{|\boldsymbol{\lambda}| = j} \sigma_{\boldsymbol{\lambda}}(T_1, \dots T_p)\, \xi^{\boldsymbol{\lambda}}. \tag{11}$$

From this and definition of $I_{2\boldsymbol{\lambda}}$ follows Eq. (10). □

Example 2.1. The cases $k = 1, 2, \dots$ of Eq. (10) read

$$\int_{\|\xi\|=1} \sigma_2(T_\xi)\, d\omega_{p-1} = I_{2,0,\dots 0} \sum\nolimits_{i \le p} \sigma_2(T_i),$$

$$\int_{\|\xi\|=1} \sigma_4(T_\xi)\, d\omega_{p-1} = I_{4,0,\dots 0} \sum\nolimits_{i \le p} \sigma_4(T_i) + I_{2,2,0,\dots 0} \sum\nolimits_{i,j \le p} \sigma_{2,2}(T_i, T_j),$$

and so on, where $I_{2,0,\dots 0} = \frac{\pi^{p/2}}{\Gamma(p/2+1)}$, $I_{4,0\dots 0} = \frac{3\,\pi^{p/2}}{2\Gamma(p/2+2)}$, $I_{2,2,0\dots 0} = \frac{\pi^{p/2}}{2\Gamma(p/2+2)}$. If $T_\xi = \sum_{i,j} \xi_i\,\xi_j T_{ij}$ $(h=2)$, then similarly to Eq. (11) one has

$$\sigma_k(T_\xi) = \sum\nolimits_{|\boldsymbol{\lambda}|=k} \sigma_{(\lambda_{11},\dots\lambda_{ij},\dots\lambda_{pp})}(T_{11},\dots T_{ij},\dots T_{pp}) \prod\nolimits_{i,j\le p} (\xi_i\,\xi_j)^{\lambda_{ij}}, \quad (12)$$

and so on. The cases $k=1,2$ of Eq. (12) read $\sigma_1(T_\xi) = \sum_{i,j} \sigma_1(T_{ij})\,\xi_i\,\xi_j$, $\sigma_2(T_\xi) = \sum_{i,j} \sigma_2(T_{ij})\,\xi_i\,\xi_j + \sum_{i,j,s,l} \sigma_{1,1}(T_{ij}, T_{s\,l})\,\xi_i\,\xi_j\,\xi_s\,\xi_l$. These integrals can we written similarly to Eq. (10). For instance, $\int_{\|\xi\|=1} \sigma_1(T_\xi)\,d\omega_{p-1} = \frac{\pi^{p/2}}{\Gamma(p/2+1)} \sum_{i\le p} \sigma_1(T_{ii})$.

3. Main Results

In this section we generalize theorems by Brito–Langevin–Rosenberg[3] and by Brito–Naveira,[5] mentioned in Introduction. Namely, we drop the restriction that the ambient space has a constant sectional curvature.

Theorem 3.1. *Let $\mathcal{F}$ be a codimension one foliation (with the unit normal ξ) on a complete locally symmetric space M^{n+1} of a finite volume. Assume that $(M, \mathcal{F})$ has bounded geometry, i.e.*

$$\sup_M \|R_\xi\| < \infty, \qquad \sup_M \|A_\xi\| < \infty. \quad (13)$$

Then for any $m > 0$

$$\int_M \sum\nolimits_{\|\boldsymbol{\lambda}\|=m} \sigma_{\boldsymbol{\lambda}}(B_1,\dots,B_m)\,dV_M = 0, \quad (14)$$

where $B_{2k} = \frac{(-1)^k}{(2k)!} R_\xi^k$, $B_{2k+1} = \frac{(-1)^k}{(2k+1)!} R_\xi^k A_\xi$.

Similar formulae can be derived on arbitrary Riemannian manifolds. They are more complicated since they contain terms which depend on covariant derivatives of R_ξ. More precisely, Eq. (14) contains just terms of the form $R_\xi^{(k)}$ with $k \le m-2$, where $R_\xi^{(1)} = \nabla_\xi R_\xi$, $R_\xi^{(2)} = \nabla_\xi \nabla_\xi R_\xi$ and so on.

For few initial values of $m = 1,\dots 5$, Eq. (14) give Eq. (1), Eq. (3) and

$$\int_M \big(\underline{\sigma_3(A_\xi)} - \frac{1}{2}\,\mathrm{Ric}(\xi)\,\mathrm{Tr}(A_\xi) + \frac{1}{3}\,\mathrm{Tr}(R_\xi A_\xi)\big)dV_M = 0, \quad (15)$$

$$\int_M (\underline{\sigma_4(A_\xi)} + \frac{1}{4}\sigma_2(R_\xi) - \frac{1}{6}\sigma_{1,1}(A_\xi, R_\xi A_\xi) + \frac{1}{24}\sigma_1(R_\xi^2) - \frac{1}{2}\sigma_{2,1}(A_\xi, R_\xi))dV_M = 0,$$

$$\int_M \Big(\underline{\sigma_5(A_x)} + \frac{1}{4}\,\sigma_{1,2}(A_x, R_\xi) - \frac{1}{6}\,\sigma_{2,1}(A_x, R_\xi\,A_x) + \frac{1}{12}\,\sigma_{1,1}(R_\xi, R_\xi\,A_x)$$
$$+ \frac{1}{24}\,\sigma_{1,1}(A_x, R_\xi^2) - \frac{1}{2}\,\sigma_{3,1}(A_x, R_\xi) + \frac{1}{5!}\,\mathrm{Tr}(R_\xi^2\,A_x)\Big)\,dV_M = 0. \quad (16)$$

Corollary 3.1. *Let $\mathcal{F}$ be a codimension 1 foliation on a complete Riemannian manifold M^{n+1} of a finite volume with $K^{\rm mix} = c$ and $\sup_M \|A_\xi\| < \infty$. Then, for any $k > 0$, Eqs. (4) are valid.*

Remark 3.1. (a) From (14) one may get obstructions for existence of codimension 1 totally geodesic/umbilical foliations.[14] (b) We apply Gray's tube formulae[6] and work also with foliations of codimension 1 (or vector fields) which admit "good" (in a sense) singularities, see details in Ref. 14. (c) By Corollary 3.1, there are no codimension 1 foliations with $K^{\rm mix} = const \neq 0$ on compact even-dimensional M.

Theorem 3.2. *Let $\mathcal{F}$ be a p-dimensional totally geodesic foliation on a complete Riemannian manifold M^{p+n} of a finite volume with the condition*

$$\nabla_\xi R^{\rm mix}_\xi = 0 \qquad (\xi \in T\mathcal{F}). \tag{17}$$

Then for any $m > 0$ one has

$$\int_M \Big(\int_{\|\xi\|=1} \Big(\sum\nolimits_{\|\lambda\|=m} \sigma_\lambda(B_{\xi,1}, \dots B_{\xi,m})\Big) d\,\omega\Big) dV_M = 0, \tag{18}$$

where $B_{\xi,2k} = \frac{1}{(2k)!}(-R^{\rm mix}_\xi)^k$, $B_{\xi,2k+1} = \frac{1}{(2k+1)!}(-R^{\rm mix}_\xi)^k C_\xi$.

The expressions for $B_{\xi,i}$ $(i > 2)$ in Theorem 3.2 are more complicated without assumption of Eq. (17), since they also contain terms with covariant derivatives of the mixed curvature operator. For example,

$$B_{\xi,3}{=}-\frac{1}{6}(R^{\rm mix}_\xi C_\xi{+}\nabla_\xi R^{\rm mix}_\xi),\ B_{\xi,4}{=}\frac{1}{4!}((R^{\rm mix}_\xi)^2{-}\nabla^2_\xi R^{\rm mix}_\xi{-}2(\nabla_\xi R^{\rm mix}_\xi)\, C_\xi).$$

Note that $\|\lambda\| = \text{odd} \Rightarrow \int_{\|\xi\|=1} \sigma_\lambda\,(B_{\xi,1}(x), \dots B_{\xi,m}(x))\, d\,\omega_{p-1} = 0$.

For few initial values of m, $m = 2, 4$, Eqs. (18) read as follows:

$$\int_M (\int_{\|\xi\|=1} (\underline{\sigma_2(C_\xi)} - \frac{1}{2}\,{\rm Tr}\, R^{\rm mix}_\xi) d\,\omega)\, dV_M = 0, \tag{19}$$

$$\int_M (\int_{\|\xi\|=1} (\underline{\sigma_4(C_\xi)} + \frac{1}{4}\,\sigma_2(R^{\rm mix}_\xi) + \frac{1}{24}\,{\rm Tr}((R^{\rm mix}_\xi)^2) - \frac{1}{24}\,{\rm Tr}(\nabla^2_\xi R^{\rm mix}_\xi)$$
$$-\frac{1}{12}\,{\rm Tr}((\nabla_\xi R^{\rm mix}_\xi) C_\xi) - \frac{1}{6}\,\sigma_{1,1}(C_\xi, R^{\rm mix}_\xi C_\xi)) d\,\omega)\, dV_M = 0. \tag{20}$$

One may reduce the integrals to ones over M, see Sec. 2.2. The Eq. (20) is new. The Eq. (19) reduces to Eq. (2).

Definition 3.1. The *total k-th mean curvature* is the integral $\sigma_k(\mathcal{F}^\perp) = \int_M \big(\int_{\|\xi\|=1} \sigma_k(C_\xi)\, d\,\omega\big) dV_M$. Note that $\sigma_{2k+1}(\mathcal{F}^\perp) = 0$.

Corollary 3.2. *Let $\mathcal{F}^p$ be a totally geodesic foliation on a complete M^{p+n} of a finite volume with $K^{\rm mix}=c$. Then, for any $k>0$,*

$$\sigma_{2k}(\mathcal{F}^\perp)=\begin{cases}\binom{n/2}{k}\frac{2\pi^{p/2}}{\Gamma(p/2)}\,c^k\,\mathrm{vol}(M), & n \text{ even},\\ 0, & n \text{ odd}.\end{cases}\tag{21}$$

Remark 3.2. (a) From $K^{\rm mix}=c$ follows that $R^{\rm mix}_\xi y=c\,y$, and hence Eq. (17) holds. (b) For $p=1$, Eq. (21) is equivalent to Eq. (6). For $p>1$ one may show that $\gamma_{2k}(\mathcal{F}^\perp_x)\neq\sigma_{2k}(\mathcal{F}^\perp_x)$. For $p=1$ the equality $\gamma_{2k}(\mathcal{F}^\perp_x)=\sigma_{2k}(\mathcal{F}^\perp_x)$ holds, moreover, Eq. (21) and Eq. (6) coincide when $p=1$. (c) By Corollary 3.2, there are no totally geodesic foliations $\mathcal{F}^p$ $(p>0)$ with $K^{\rm mix}=const\neq 0$ on a complete M^{p+n} of a finite volume for n odd.

4. Proof of Theorem 3.1

Let ξ be a smooth unit vector field on a complete oriented Riemannian manifold (M^{n+1},g) of class C^∞ with bounded geometry due (13). For simplicity, one may assume that ξ is orthogonal to a codimension-1 foliation $\mathcal{F}$, $\dim\mathcal{F}=n$. We call $\exp_x$, as usual, the *exponential map* at x: $\exp_x: T_xM\to M$ of M. Let $\gamma_x: t\mapsto \exp_x(t\,\xi)$ be the unique geodesic in M with $\gamma_x(0)=x$ and $\gamma'_x(0)=\xi$. Choose a positively oriented orthonormal frame $(e^1,\dots e^n)$ of $\xi^\perp_x$ $(=T_x\mathcal{F})$ and extend it by parallel translation to the frame $(E^1_x,\dots E^n_x)$ of vector fields along γ_x. Denote also by E^{n+1}_x the parallel field along γ_x with $E^{n+1}_x(0)=\xi_x$. For any $i\le n$, denote by $Y^i_x(t)$ the Jacobi field along γ_x satisfying $Y^i_x(0)=e^i$ and $Y^{i\,\prime}_x(0)=A_x(e^i)$, where A_x is the Weingarten operator (of a leaf at x) relative to ξ_x. Denote by R_ξ the curvature operator $X\mapsto R(X,\xi)\xi$ in $\xi^\perp$ $(=T\mathcal{F})$ and by $R_x(t)$ the matrix with entries $\langle R(E^i_x(t),E^{n+1}_x(t))E^{n+1}_x(t),E^i_x(t)\rangle$ ("*Jacobi operator*"). Denote also by $Y_x(t)$ the $n\times n$ matrix consisting of the scalar products $\langle Y^i_x(t),E^j_x(t)\rangle$ ("*Jacobi tensor*"). Then $Y_x(0)=I_n$ and $Y'_x(0)=A_x$.

Consider the maps $\{\phi_t: M\to M,\ t\in(-\varepsilon,\varepsilon)\}$ defined by $\phi_t(x)=\exp_x(t\,\xi)$. It is known (see, for instance,[8] Lemma 3.1.17) that

$$|d\phi_t(x)|=\det Y_x(t).\tag{22}$$

In view of (13), there exists $\varepsilon>0$ such that for all $x\in M$ there are no focal points of a leaf L_x along $\gamma_x(t)$, $t\in[0,\varepsilon)$. Hence, $\{\phi_t: M\to M\}_{t\in(-\varepsilon,\varepsilon)}$ are diffeomorphisms. Assume that (M,g) is locally symmetric $(\nabla R=0)$ and consider the Jacobian $|d\phi_t(x)|$ of ϕ_t at a point x of M. For short, write $R_\xi:=R_x(0)$. The Jacobi equation $Y''_x=-R_x(t)Y_x$ implies that $Y^{(2m)}_x(0)=(-R_\xi)^m$, $Y^{(2m+1)}_x(0)=(-R_\xi)^m A_x$, $m=0,1,2,\dots$ Hence, our Jacobi tensor $Y_x = \sum_{m=0}^{\infty} Y_x^{(m)}(0)\frac{t^m}{m!}$ has the form

$$Y_x(t) = I_n + tA_x - \frac{t^2}{2!}R_\xi - \frac{t^3}{3!}R_\xi A_x + \frac{t^4}{4!}R_\xi^2 + \dots \tag{23}$$

Certainly, the radius of convergence of the series in Eq. (23) is uniformly bounded from below on M (by $1/\|R\| > 0$). Therefore – by Lebesque Dominated Convergence Theorem – its integration together with Exchange Variable Theorem yield the equality for arbitrary t small enough

$$\mathrm{vol}(M) = \int_M \det\big(I_n + tA_x - \frac{t^2}{2!}R_\xi - \frac{t^3}{3!}R_\xi A_x + \frac{t^4}{4!}R_\xi^2 + \dots\big)\, dV_M. \tag{24}$$

Formula (24) together with Lemma 2.2 imply (14).

5. Proof of Theorem 3.2

First, we recall the basic facts about tangent bundles.[2,9] Let TM be the *tangent bundle* of M with the projection $\pi : TM \to M$. Split TTM (the tangent bundle of TM) into the sum $TTM = V \oplus H$ of vertical and horizontal subbundles. For any $\xi \in (TM)_x$ the subspaces $V_{(x,\xi)}$, $H_{(x,\xi)}$ can be identified with $(TM)_x$ *via* $\pi_* : TTM \to TM$ and the connection map $K : TTM \to TM$. The linear map K is determined by the condition

$$K(Z_*\xi) = \nabla_\xi Z \quad (\text{for all } \xi \in TM \text{ and any vector field } Z \text{ on } M).$$

A *Sasaki metric* $\bar g$ on TM is the unique Riemannian structure for which V and H are orthogonal and all the maps $\pi_{*|H} : H_{(x,\xi)} \to (TM)_x$ and $K_{|V} : V_{(x,\xi)} \to (TM)_x$ are isometries. The *unit tangent sphere bundle* over (M, g) is $T_1M = \{(x,\xi) \in TM : \ |\xi| = 1\}$ (the hypersurface of TM).

Let $\mathcal{F} = \{L\}$ be a transversely oriented foliation on M. We call $T\mathcal{F} = \{(x,\xi) \in TM : x \in M,\ \xi \in (TL)_x\}$ the *tangent bundle of* $\mathcal{F}$, and $\bar{\mathcal{F}} = \{TL\}$ the *lift foliation on* $T\mathcal{F}$. The metric $\bar g$ in $T\mathcal{F}$ is the restriction of the Sasaki metric of TM.

Definition 5.1. We call $T_1\mathcal{F} = \{(x,\xi) \in T\mathcal{F} : \ |\xi| = 1\}$ the *tangent sphere bundle of radius* 1 *of* $\mathcal{F}$, a hypersurface in $T\mathcal{F}$ consisting of all pairs (x,ξ), where $x \in M$ and $\xi \in TL$ is a vector of the length 1. We call $\widetilde{\mathcal{F}} = \{T_1L\}$ the *lift foliation on* $T_1\mathcal{F}$.

The canonical vertical vector field $\bar\xi$ is outward normal to $T_1\mathcal{F}$ in $T\mathcal{F}$ at each point $(x,\xi) \in T_1\mathcal{F}$. The *tangential lift* of y is a vector field $\tilde y^t$ tangent to $T_1\mathcal{F}$ and defined by $\tilde y^t = \bar y^V - \bar g(\bar y^V, \bar\xi)\,\bar\xi$. The hypersurface $T_1\mathcal{F} \subset T\mathcal{F}$ is endowed with the Riemannian metric $\tilde g$

$$\tilde g(\tilde z^H, \tilde y^H) = \bar g(\bar z^H, \bar y^H), \ \ \tilde g(\tilde z^H, \tilde y^t) = 0, \ \ \tilde g(\tilde z^t, \tilde y^t) = \bar g(\bar z^V, \bar y^V) - \bar g(\bar z^V, \bar\xi)\,\bar g(\bar y^V, \bar\xi).$$

The natural projection $\tilde{\pi} : T_1\mathcal{F} \to M$ is the Riemannian submersion with totally geodesic fibers S_1^{p-1}, the spheres of a radius 1. Hence the volume $Vol(T_1\mathcal{F})$ is the product of $Vol(S_1^{p-1})$ and $Vol(M)$, see Ref. 1.

Lemma 5.1. *Let $\mathcal{F} = \{L^p\}$ be a totally geodesic foliation on M. Then $\tilde{\mathcal{F}} = \{T_1L\}$ (the $(2p-1)$-dimensional lift foliation on $T_1\mathcal{F}$) is totally geodesic.*

Proof. follows from the fact that TL is totally geodesic in TM if and only if L is a totally geodesic submanifold in M, see Ref. 2. □

The tangent bundle to $T_1\mathcal{F}$ is orthogonally decomposed into sum of *horizontal* and *vertical* subbundles: $T(T_1\mathcal{F}) = \widetilde{V} \oplus \widetilde{H}$. In view of isomorphism $\tilde{\pi}_* : \widetilde{H} \to TM$, the almost product structure $TM = T\mathcal{F} \oplus T\mathcal{F}^\perp$ induces the orthogonal decomposition of $\widetilde{H}$ into sum of subbundles: $\widetilde{H} = \widetilde{H}^\top \oplus \widetilde{H}^\perp$.

Definition 5.2. The *characteristic vector field* $N \subset \widetilde{H}^\top$ (velocity of the geodesic flow on $T_1\mathcal{F}$) at a point $\widetilde{x} = (x, \xi)$ is defined by $\widetilde{\pi}_*(N) = \xi$. The *characteristic Weingarten* and *Jacobi operators* on $T_1\mathcal{F}$ are defined by $\tilde{A}_N\, \tilde{y} = \widetilde{\nabla}_{\tilde{y}} N - \langle \widetilde{\nabla}_{\tilde{y}} N, N\rangle N$ and $\tilde{R}_N : \tilde{y} \to \tilde{R}(\tilde{y}, N)N$.

For totally geodesic $\mathcal{F}$, the restriction of N on any leaf T_1L generates the geodesic flow on this leaf, and the distribution $\widetilde{H}^\perp$ is $\widetilde{\nabla}$-parallel along N.

Lemma 5.2. *Let $\mathcal{F}$ be a totally geodesic foliation on (M, g), $\tilde{y}$ a vector field on $T_1\mathcal{F}$ and $y = \pi_*(\tilde{y})$ the projection on M. Then*

$$\tilde{A}_N\, \tilde{y} = \widetilde{\nabla}_{\tilde{y}} N \subset \widetilde{V} \oplus \widetilde{H}^\top \qquad (\tilde{y} \subset \widetilde{V} \oplus \widetilde{H}^\top), \tag{25}$$

$$\tilde{A}_N\, \tilde{y} = \widetilde{\nabla}_{\tilde{y}}\, N = \widetilde{(\nabla_y \xi)^\perp} \subset \widetilde{H}^\perp \qquad (\tilde{y} \subset \widetilde{H}^\perp), \tag{26}$$

$$\tilde{R}_N\, \tilde{y} \subset \widetilde{V} \oplus \widetilde{H}^\top \qquad (\tilde{y} \subset \widetilde{V} \oplus \widetilde{H}^\top), \tag{27}$$

$$\tilde{R}_N\, \tilde{y} = \widetilde{R_\xi^{\,\mathrm{mix}} y} \subset \widetilde{H}^\perp \quad (\tilde{y} \subset \widetilde{H}^\perp, \quad \xi = \widetilde{\pi}_*(N)). \tag{28}$$

Proof of Theorem 3.2. By Lemma 5.2, the Weingarten and Jacobi operators, $\tilde{A}_N$, $\tilde{R}_N$, and hence the *Jacobi operator* $\tilde{Y}_N$ have 2×2 block view:

$$\tilde{A}_N = \begin{pmatrix} \tilde{A}_N^L & * \\ 0 & C_\xi \end{pmatrix}, \ \tilde{R}_N = \begin{pmatrix} \tilde{R}_N^L & 0 \\ 0 & R_\xi^{\mathrm{mix}} \end{pmatrix}, \ \tilde{Y}_N(t) = \begin{pmatrix} Y_N^L(t) & * \\ 0 & Y_\xi^\perp(t) \end{pmatrix}$$

where $\tilde{A}_N^L$ and $\tilde{R}_N^L$ are the restrictions of $\tilde{A}_N$ and $\tilde{R}_N$ on T_1L. By Schur identity, we have $\det \tilde{Y}_N(t) = \det Y_N^L(t)\, \det Y_\xi^\perp(t)$. Recall that N generates the geodesic flow on each T_1L and $\det Y_N^L(t)$ presents the volume element in L. Hence, by Liouville's theorem, $\det Y_N^L(t) = 1$ for all t. Finally, we have $\det \tilde{Y}_N(t) = \det Y_\xi^\perp(t)$ for all t. Given $\xi \in (TL)_x$, choose a positive oriented orthonormal frame $(e_1, \dots e_n)$ of $(TL^\perp)_x$ and extend it by parallel translation to the frame $(E_1, \dots E_n)$ of vector fields along the geodesic $\gamma : t \mapsto \exp_x(t\,\xi)$. Hence for the Jacobi tensor $Y_\xi^\perp(t)$ we get

$$\begin{aligned} Y_\xi^\perp(t) &= I_n + tC_\xi - \tfrac{t^2}{2!} R_\xi^{\,\mathrm{mix}} - \tfrac{t^3}{3!}(R_\xi^{\,\mathrm{mix}} C_\xi + \nabla_\xi R_\xi^{\,\mathrm{mix}}) \\ &+ \tfrac{t^4}{4!}((R_\xi^{\,\mathrm{mix}})^2 - \nabla_\xi^{(2)} R_\xi^{\,\mathrm{mix}} - 2\nabla_\xi R_\xi^{\,\mathrm{mix}} C_\xi) + \dots \end{aligned} \tag{29}$$

The diffeomorphisms $\phi_t(x) = \exp_x(t\,\tilde{N}(x))$ of $T_1\mathcal{F}$ are defined for all $t \in \mathbb{R}$ (since the leaves of $\mathcal{F}$ are complete). The Jacobian of ϕ_t satisfies [14] $|d\phi_t(x)| = \det \tilde{Y}_{\tilde{N}}(t)$. The rest of proof is similar to codimension 1 case. □

Proof of Corollary 3.2. By the proof of Theorem 3.2 we have

$$\sum\nolimits_{m \le n} z^m \sigma_m(\mathcal{F}^\perp) = \mathrm{vol}(T_1\mathcal{F})\,(1 \pm |c|\, z^2)^{n/2}.$$

For n odd we obtain $c = 0$ (the case $c \neq 0$ leads to a contradiction). □

References

1. M. Berger, *A Panoramic View of Riemannian Geometry* (Springer, 2002).
2. A. Borisenko and A. Yampolskii, Riemannian geometry of fiber bundles, *Russian Math. Surveys* **46** (6) (1991) 55–106.
3. F. Brito and R. Langevin and H. Rosenberg, Integrales de courbure sur des varietes feuilletees, *J. Diff. Geom.* **16** (1981) 19–50.
4. F. Brito, Une obstruction geometrique a l'existence de feuilletages de codimension 1 totalement geodesiques, *J. Diff. Geom.* **16** (1981) 675–684.
5. F. Brito and A. Naveira, Total extrinsic curvature of certain distributions on closed spaces of constant curvature, *Ann. Global Anal. Geom.* **18** (2000) 371–383.
6. A. Gray, *Tubes* (Birkhauser, Boston, 2nd edition, 2004).
7. G.G. Gurevich, *Foundations of the Theory of Algebraic Invariants* (Noordhof, Groningen, 1964).
8. W. Klingenberg, *Riemannian Geometry* (Walter de Gruyter, Berlin, 1995).
9. O. Kowalskii and M. Sekizawa, On Tangent Sphere Bundles with Small or Large Constant Radius, *Global Anal. Geom.* **18** (2000) 207–219.
10. A.P. Prudnikov, Yu.A. Brychkov and O.I. Marichev, *Integrals and Series* (Vol. 3, Gordon & Breach Sci. Publ., New York, 1990).
11. A. Ranjan, Structural equations and an integral formula for foliated manifolds, *Geom. Dedicata* **20** (1986) 85–91.
12. G. Reeb, Sur la courbure moyenne des variétés intégrales d'une équation de Pfaff $\omega = 0$, *C. R. Acad. Sci. Paris* **231** (1950) 101–102.
13. V. Rovenski, *Foliations on Riemannian Manifolds and Submanifolds* (Birkhauser, Boston, 1998).
14. V. Rovenski and P. Walczak, Integral formulae on foliated symmetric spaces, Preprint, *University of Lodz, Fac. Math. Comp. Sci.* **2007/13** (2007) 1–27.

15. P. Walczak, An integral formula for a Riemannian manifold with two orthogonal complementary distributions, *Colloq. Math.* **58** (1990) 243–252.
16. P. Walczak and F. Brito, On the energy of unit vector fields with isolated singularities, *Ann. Pol. Math.* **73** (3) (2000) 269–274.
17. P. Walczak and P. Schweitzer, Prescribing mean curvature vectors for foliations, *Illinois J. Math.* **48** (1) (2004) 21–35.
18. P. Walczak, *Dynamics of foliations, groups and pseudogroups* (Birkhauser, Basel, 2004).
19. A. Zeghib, Feuilletages géodésiques des variétés localement symétriques, *Topology* **36** (1997) 805–828.

PART 2

Geometric structures

Differential Geometry and its Applications
Proc. Conf., in Honour of Leonhard Euler, Olomouc, August 2007

The prequantization of $T_k^1\mathbb{R}^n$

Adara M. Blaga

Department of Mathematics and Computer Sciences, West University from Timişoara, Timişoara, România
E-mail: adara@math.uvt.ro
www.uvt.ro

The aim of this paper is to give a prequantization of the manifold $T_k^1\mathbb{R}^n$.

Keywords: Prequantization, k-symplectic manifold.

MS classification: 53D05, 53D50.

1. Introduction

Prequantization is the first step in the geometric quantization procedure. Two different ways of performing the prequantization of a symplectic manifold (M, ω) were described by Kostant[1] and respectively, Souriau.[2] The first one consists in finding a complex line bundle with a connection such that its curvature is $\frac{i}{\hbar}\omega$ and the second one refferes to determine a principal S^1-bundle $\pi : Y \longrightarrow M$ with an S^1-invariant 1-form α such that $d\alpha = \pi^*\omega$ and $\int_{fibre} \alpha = 2\pi\hbar$ (where $\hbar$ stands for the Plank's constant).

Using the prequantization of the standard k-symplectic manifold $((T_k^1)^*\mathbb{R}^n, \omega_i, V)_{1\leq i\leq k}$ (see Ref. 3), we will define a prequantization of the k-symplectic manifold $(T_k^1\mathbb{R}^n, (\omega_L)_i, V_L)_{1\leq i\leq k}$ associated to a regular Lagrangian $L : T_k^1\mathbb{R}^n \longrightarrow \mathbb{R}$ and point out an irreducible representation of it.

2. Hamiltonian vector fields on a k-symplectic manifold

Let M be an $n + nk$ dimensional differential manifold.

Definition 2.1 (Ref. 4). A k-symplectic structure on M is a family

$$(\omega_i, V)_{1\leq i\leq k}$$

where ω_i, $1 \leq i \leq k$ are closed 2-forms on M and V is an nk dimensional integrable distribution on M, such that $\omega_i \mid_{V\times V}= 0,\ 1 \leq i \leq k$.

We call $(M, \omega_i, V)_{1\leq i\leq k}$ a k-symplectic manifold.

The standard example of k-symplectic manifold is provided by the Whitney sum of k copies of $T^*\mathbb{R}^n$, i.e. $(T^1_k)^*\mathbb{R}^n$, together with the 2-forms ω_i, $1 \leq i \leq k$ and the distribution V, given in a system of local coordinates (q^α, p^i_α), $1 \leq \alpha \leq n$, $1 \leq i \leq k$, by:

$$\omega_i = dq^\alpha \wedge dp^i_\alpha;$$

$$V = \langle \frac{\partial}{\partial p^i_\alpha} \rangle \tag{1}$$

(the Einstein convention is used whenever a lower and and an equal upper indices appear).

Recall that the manifold of one jets of maps from an n dimensional manifold Q to $\mathbb{R}^k$ with target at $0 \in \mathbb{R}^k$, $J^1(Q, \mathbb{R}^k)_0$, can be canonically identified with the Whitney sum $(T^1_k)^*Q$ of k copies of T^*Q, that is

$$\begin{array}{ccc} J^1(Q,\mathbb{R}^k)_0 & \longrightarrow & T^*Q \oplus \cdots \oplus T^*Q, \\ j^1_{q,0}\sigma & \equiv & (\alpha_1(q), \ldots, \alpha_k(q)) \end{array}$$

where $\alpha_i(q) = d(\pi_i \circ \sigma)(q)$ and $\pi_i : \mathbb{R}^k \longrightarrow \mathbb{R}$ are the canonical projections, $1 \leq i \leq k$.

The Darboux's theorem states the following:

Theorem 2.1. *Around any point of a k-symplectic manifold $(M, \omega_i, V)_{1\leq i\leq k}$, there exists a system of local coordinates (q^α, p^i_α), $1 \leq \alpha \leq n$, $1 \leq i \leq k$, such that:*

$$\omega_i = dq^\alpha \wedge dp^i_\alpha;$$

$$V = \langle \frac{\partial}{\partial p^i_\alpha} \rangle.$$

For $H \in C^\infty(M, \mathbb{R})$ a Hamiltonian on M, we shall obtain the Hamiltonian vector fields

$$X^i_H = (X^i_{1H}, ..., X^i_{kH}),\ 1 \leq i \leq k,$$

as the solutions of the Hamilton's equations (see Ref. 5):

$$\sum_{j=1}^{k} i_{X^i_{jH}} \omega_j = dH,\ 1 \leq i \leq k, \tag{2}$$

of the form:

$$X^i_{jH} = \frac{\partial H}{\partial p^j_\alpha} \cdot \frac{\partial}{\partial q^\alpha} - \delta^i_j \frac{\partial H}{\partial q^\alpha} \cdot \frac{\partial}{\partial p^j_\alpha}, \; 1 \leq i, j \leq k. \qquad (3)$$

Consider $(q^\alpha, v^\alpha_i)_{1\leq\alpha\leq n, 1\leq i\leq k}$ a system of local coordinates on $T^1_k\mathbb{R}^n$, where $T^1_k\mathbb{R}^n$ is the Whitney sum of k copies of $T\mathbb{R}^n$ and let $L \in C^\infty(T^1_k\mathbb{R}^n, \mathbb{R})$ be a hyperregular Lagrangian on $T^1_k\mathbb{R}^n$. Then the Legendre transformation $TL : T^1_k\mathbb{R}^n \longrightarrow (T^1_k)^*\mathbb{R}^n$ defined by

$$(TL(v_{1q}, ..., v_{kq}))^i(w_q) := \frac{d}{ds}|_{s=0} \; L(v_{1q}, ..., v_{iq}+sw_q, ..., v_{kq}), \forall 1 \leq i \leq k, \quad (4)$$

is a diffeomorphism. Using this transformation, we can transfer the k-symplectic structure $(\omega_i, V)_{1\leq i\leq k}$ from $(T^1_k)^*\mathbb{R}^n$ on $T^1_k\mathbb{R}^n$, defining (see Ref. 6):

(1) $(\omega_L)_i = (TL)^*\omega_i$
(2) $V_L = (TL)^{-1}_* V$

We saw that in a weaker case [i.e. when the Lagrangian is only regular (see Ref. 6)] the family

$$(T^1_k\mathbb{R}^n, (\omega_L)_i, V_L)_{1\leq i\leq k}, \qquad (5)$$

is a k-symplectic manifold. Therefore, it happens in our case, too.

3. Poisson structures on the k-symplectic manifold $(T^1_k\mathbb{R}^n, (\omega_L)_i, V_L)_{1\leq i\leq k}$ associated to the Lagrangian L

Let $f, g \in C^\infty((T^1_k)^*\mathbb{R}^n, \mathbb{R})$ and denote by:

$$X^i_f = (X^i_{1f}, ..., X^i_{kf}), \; 1 \leq i \leq k,$$

and

$$X^i_g = (X^i_{1g}, ..., X^i_{kg}), \; 1 \leq i \leq k,$$

the corresponding Hamiltonian vector fields.

Definition 3.1 (Ref. 3). The Poisson bracket of f and g is the smooth, real-valued function defined on $(T^1_k)^*\mathbb{R}^n$:

$$\{f, g\} = \sum_{i=1}^k \omega_i(X^i_{if}, X^i_{ig}). \qquad (6)$$

Using the Legendre transformation associated to the hyperregular Lagrangian L, we can now define a Poisson structure on $T_k^1\mathbb{R}^n$, as follows:

$$\{f,g\}_L = (TL)^*\{(TL)^{*-1}f,(TL)^{*-1}g\}, \tag{7}$$

for any $f, g \in C^\infty(T_k^1\mathbb{R}^n, \mathbb{R})$.

We will verify only the Jacobi identity:

Proposition 3.1. *The Poisson bracket 7 satisfies the Jacobi identity:*

$$\{f,\{g,h\}_L\}_L + \{g,\{h,f\}_L\}_L + \{h,\{f,g\}_L\}_L = 0.$$

Proof. Indeed, let $f, g, h \in C^\infty(T_k^1\mathbb{R}^n, \mathbb{R})$. Then:

$$\begin{aligned}\{f,\{g,h\}_L\}_L &= \{f,(TL)^*(\{(TL)^{*-1}g,(TL)^{*-1}h\})\}_L\\ &= (TL)^*(\{(TL)^{*-1}f,(TL)^{*-1}((TL)^*(\{(TL)^{*-1}g,(TL)^{*-1}h\}))\}_L)\\ &= (TL)^*(\{(TL)^{*-1}f,\{(TL)^{*-1}g,(TL)^{*-1}h\}\}_L)\\ &= \{(TL)^{*-1}f,\{(TL)^{*-1}g,(TL)^{*-1}h\}\}_L \circ TL.\end{aligned}$$

Analogous, we will obtain:

$$\{g,\{h,f\}_L\}_L = \{(TL)^{*-1}g,\{(TL)^{*-1}h,(TL)^{*-1}f\}\}_L \circ TL$$

and

$$\{h,\{f,g\}_L\}_L = \{(TL)^{*-1}h,\{(TL)^{*-1}f,(TL)^{*-1}g\}\}_L \circ TL.$$

Adding the three relations and taking into account that the Poisson bracket $\{,\}$ satisfies the Jacobi identity, we obtain that $\{,\}_L$ satisfies the same identity. □

The proposition above allows us to state:

Proposition 3.2. $(C^\infty(T_k^1\mathbb{R}^n,\mathbb{R}),\{,\}_L)$ *is an infinite dimensional Lie algebra.*

Proof. The axioms of a Lie algebra can now be easily checked. □

4. The prequantization of the k-symplectic manifold $(T_k^1\mathbb{R}^n, (\omega_L)_i, V_L)_{1\leq i\leq k}$ associated to the Lagrangian L

Let $((T_k^1)^*\mathbb{R}^n, \omega_i, V)_{1\leq i\leq k}$ be the standard k-symplectic manifold, the k-symplectic structure being given by:

$$\omega_i = dq^\alpha \wedge dp^i_\alpha;$$

$$V = \langle \frac{\partial}{\partial p^i_\alpha} \rangle$$

for $1 \leq \alpha \leq n$, $1 \leq i \leq k$.

Definition 4.1 (Ref. 3). A prequantization of the standard k-symplectic manifold $((T_k^1)^*\mathbb{R}^n, \omega_i, V)_{1\leq i\leq k}$ is a correspondence which assigns to each function $f \in C^\infty((T_k^1)^*\mathbb{R}^n, \mathbb{R})$ a self-adjoint operator δ_f on a Hilbert space H, such that the Dirac conditions are satisfied:

(1) (D_1) $\delta_{f+g} = \delta_f + \delta_g$;
(2) (D_2) $\delta_{\lambda f} = \lambda\delta_f$;
(3) (D_3) $\delta_{id_{(T_k^1)^*\mathbb{R}^n}} = id_H$;
(4) (D_4) $[\delta_f, \delta_g] = i\hbar\delta_{\{f,g\}}$,

for any $f, g \in C^\infty((T_k^1)^*\mathbb{R}^n, \mathbb{R})$.

Proposition 4.1 (Ref. 3). *The pair (H, δ), where:*

(1) $H = L^2((T_k^1)^*\mathbb{R}^n, \mathbb{C})$;
(2) $\delta : f \in C^\infty((T_k^1)^*\mathbb{R}^n, \mathbb{R}) \longmapsto \delta_f$ *and*

- $\delta_f = -i\hbar X_f - \sum_{j=1}^k \theta_j(X^j_{jf}) + f$;
- $\theta_j = p^j_\alpha dq^\alpha$,

for $1 \leq j \leq k$, *is a prequantization of the k-symplectic manifold* $((T_k^1)^*\mathbb{R}^n, \omega_i, V)_{1\leq i\leq k}$.

Using the Legendre transformation associated to the hyperregular Lagrangian L, we can now define a prequantization of the manifold 5

Proposition 4.2. *The pair (H, δ), where:*

(1) $H = L^2(T_k^1\mathbb{R}^n, \mathbb{C})$;
(2) $\delta : f \in C^\infty(T_k^1\mathbb{R}^n, \mathbb{R}) \longmapsto \delta_f$ *and*

- $\delta_f = -i\hbar X_f - \sum_{j=1}^k (\theta_L)_j(X^j_{jf}) + f$;
- $(\theta_L)_j = \dfrac{\partial L}{\partial v^\alpha_j} dq^\alpha$

for $1 \leq j \leq k$, *is a prequantization of the k-symplectic manifold 5.*

References

1. B. Kostant, *Quantization and unitary reprezentations* (Lectures in modern analysis and applications III, LNM 170, Springer Verlag, 1970).
2. J.M. Souriau, *Structure des Systèmes dynamiques* (Dunod, Paris, 1970).
3. M. Puta, S. Chirici and E. Merino, On the prequantization of $(T_1^k)^*\mathbb{R}^n$ *Bull. Math. Soc. Sc. Math. Roumanie* **44** (2001) 277–284.
4. A. Awane, k-symplectic structures, *Math. Phys.* **33** (1992) 4046–4052.
5. M. Puta, Some remarks on k-symplectic manifolds, *Tensor* **47** (1988) 109–115.
6. F. Munteanu, A. Rey and M. Salgado, The Günther's formalism in classical field theory: momentum map and reduction, *Math. Phys.* **45** (2004) 1730–1751.

Differential Geometry and its Applications
Proc. Conf., in Honour of Leonhard Euler, Olomouc, August 2007

Extension of connections

Miroslav Doupovec
Department of Mathematics, Brno University of Technology, FSI VUT Brno, Technická 2, 616 69 Brno, Czech Republic
E-mail: doupovec@fme.vutbr.cz

Włodzimierz M. Mikulski
Institute of Mathematics, Jagiellonian University, Reymonta 4, Kraków, Poland
E-mail: mikulski@im.uj.edu.pl

We generalize the well known Ehresmann prolongation of a connection $\Gamma : Y \to J^1Y$ in the following way: Given a projectable classical linear connection on Y, we introduce an extension of Γ into an r-th order holonomic connection $Y \to J^rY$. Then we describe all second and third order holonomic extensions of Γ. We also study symmetrizations of semiholonomic and nonholonomic jets.

Keywords: Ehresmann prolongation, holonomic jets, higher order connections.

MS classification: 58A05, 58A20, 58A32.

1. Introduction

In general, an r-th order holonomic connection on a fibered manifold $Y \to M$ in the sense of C. Ehresmann is a smooth section $Y \to J^rY$ of the r-th holonomic jet prolongation of Y. If we replace J^rY by semiholonomic prolongation $\overline{J}^rY$ or by nonholonomic prolongation $\widetilde{J}^rY$, then we obtain the concept of semiholonomic or nonholonomic connection, respectively. Clearly, for $r = 1$ we have a general connection $\Gamma : Y \to J^1Y$, which can be also interpreted as the lifting map $Y \times_M TM \to TY$.

Our starting point is the classical Ehresmann prolongation, which transforms Γ into an r-th order semiholonomic connection. However, the fundamental role in mathematical physics and in differential geometry is played by classical holonomic jets and connections. That is why we study extensions of Γ into higher order holonomic connections. Up till now, the only

---

This paper is in the final form and no version of it will be published elsewhere.

known geometric construction of this type is of the form

$$C^{(2)}(\Gamma * \Gamma) : Y \to J^2Y, \tag{1}$$

where $\Gamma * \Gamma : Y \to \overline{J}^2 Y$ is the second order Ehresmann prolongation of Γ and $C^{(2)} : \overline{J}^2 Y \to J^2 Y$ is the well known symmetrization of second order semiholonomic jets. This simple geometric construction leads us to the idea that the holonomic extension of connections is closely related with the problem of symmetrization of jets.

The aim of this paper is to study both these problems in the systematic way. Our research is based on the results from Ref. 2, where we introduced an r-th order holonomic extension of Γ and also symmetrizations $\overline{J}^r Y \to J^r Y$ and $\widetilde{J}^r Y \to J^r Y$ for any r. We also clarified that for $r > 2$, such geometric constructions depend in an essential way on an auxiliary projectable classical linear connection on Y. The structure of the paper is as follows. In Section 2 we recall some concepts which we need in the sequel. Section 3 contains the survey of some of our recent results from Ref. 2. In Section 4 we describe all second and third order holonomic extensions of a connection Γ. Finally, in Section 5 we present some applications of higher order connections. In particular, we introduce new geometric constructions of connections on higher order principal prolongations.

It is well known that the classical theory of higher order jets and connections was established by C. Ehresmann.[3,4] Next, an important contribution in this research has been made by P. Libermann,[13] J. Pradines,[15] I. Kolář[7,8] and G. Virsik.[16] Other recent results can be found e.g. in Refs. 1,10,12. Moreover, the basic concepts from the theory of connections can be formulated by means of jets. Using such a point of view, the monograph Ref. 9 can serve as an introduction to the systematic theory of jets and connections.

In what follows our geometric constructions will be reflected as natural transformations or natural differential operators in the sense of the book Ref. 9. We denote by $\mathcal{FM}_{m,n}$ the category of fibered manifolds and fiber respecting mappings with m-dimensional bases, n-dimensional fibres and local fibered diffeomorphisms and by $\mathcal{PB}_m(G)$ the category of principal G-bundles with m-dimensional bases and their local principal G-bundle isomorphisms with the identity isomorphisms of G. All manifolds and maps are assumed to be infinitely differentiable.

2. Ehresmann prolongation

First we recall the definition of higher order jet prolongations of a fibered manifold $Y \to M$. The classical r-th holonomic prolongation $J^r Y$ is the bundle of all r-jets of local sections of Y. Using induction, we can define the r-th nonholonomic prolongation $\widetilde{J}^r Y$ by

$$\widetilde{J}^1 Y = J^1 Y, \quad \widetilde{J}^r Y = J^1(\widetilde{J}^{r-1} Y \to M).$$

Then we have the canonical inclusion $J^r Y \subset \widetilde{J}^r Y$ determined by $j^r_x s \mapsto j^1_x(u \mapsto j^{r-1}_u s)$ for every local section s of Y, $u \in M$. Further, write $\overline{J}^1 Y = J^1 Y$ and assume we have defined $\overline{J}^{r-1} Y \subset \widetilde{J}^{r-1} Y$ such that the restriction of $\beta_{\widetilde{J}^{r-2}Y} : \widetilde{J}^{r-1} Y \to \widetilde{J}^{r-2} Y$ maps $\overline{J}^{r-1} Y$ into $\overline{J}^{r-2} Y$, where $\beta_Y : J^1 Y \to Y$ means the projection. Then we can define the r-th semiholonomic prolongation of Y by

$$\overline{J}^r Y = \{U \in J^1 \overline{J}^{r-1} Y; \beta_{\overline{J}^{r-1}Y}(U) = J^1 \beta_{\widetilde{J}^{r-2}Y}(U) \in \overline{J}^{r-1} Y\}.$$

Clearly, for $r > 2$ we have $J^r Y \subset \overline{J}^r Y \subset \widetilde{J}^r Y$. Moreover, we write

$$\overline{J}^{r,r-1} Y := \{U \in \overline{J}^r Y; \beta_{\overline{J}^{r-1}Y}(U) \in J^{r-1} Y\} \tag{2}$$

for the subspace of $\overline{J}^r Y$ such that the underlying $(r-1)$-jets are holonomic. Denoting by (x^i, y^p) the local coordinates on $Y \to M$, the induced coordinates on $\overline{J}^r Y$ are $(x^i, y^p, y^p_i, y^p_{ij}, \ldots, y^p_{i_1 \ldots i_r})$. Then $\overline{J}^{r,r-1} Y$ is characterized by the full symmetry of $y^p_{i_1 \ldots i_s}$ for all $2 \le s \le r-1$ and $J^r Y$ by the full symmetry in all subscripts.

The product of two connections $\Gamma_1 : Y \to \widetilde{J}^r Y$ and $\Gamma_2 : Y \to \widetilde{J}^s Y$ is an $(r+s)$-th order nonholonomic connection $\Gamma_1 * \Gamma_2 : Y \to \widetilde{J}^{r+s} Y$ defined by

$$\Gamma_1 * \Gamma_2 := \widetilde{J}^s \Gamma_1 \circ \Gamma_2.$$

Definition 2.1. Given a connection $\Gamma : Y \to J^1 Y$, its *Ehresmann prolongation* is the r-th order connection $\Gamma^{(r-1)} : Y \to \widetilde{J}^r Y$ defined by

$$\Gamma^{(1)} := \Gamma * \Gamma, \quad \Gamma^{(r-1)} := \Gamma^{(r-2)} * \Gamma.$$

By Ref. 5, $\Gamma^{(r-1)}$ is semiholonomic and G. Virsik[16] proved that $\Gamma^{(r-1)}$ is holonomic if and only if Γ is curvature free. One evaluates directly that if $y^p_i = \Gamma^p_i(x, y)$ is the coordinate expression of Γ, then its Ehresmann prolongation $\Gamma * \Gamma : Y \to \overline{J}^2 Y$ has equations

$$y^p_i = \Gamma^p_i, \quad y^p_{ij} = \frac{\partial \Gamma^p_i}{\partial x^j} + \frac{\partial \Gamma^p_i}{\partial y^q} \Gamma^q_j.$$

3. Holonomic extension of connections and symmetrization of jets

(a) Holonomic extension of connections.

Definition 3.1. An r-th order holonomic connection $\overline{\Gamma} : Y \to J^rY$ is called a *holonomic extension* of $\Gamma : Y \to J^1Y$, if $\pi^r_1 \circ \overline{\Gamma} = \Gamma$, where $\pi^r_s : J^rY \to J^sY$ the jet projection.

Example 3.1. Consider Ehresmann prolongation $\Gamma^{(r-1)} : Y \to \overline{J}^rY$ of $\Gamma : Y \to J^1Y$. Applying an arbitrary natural transformation $C : \overline{J}^rY \to J^rY$, we obtain a holonomic connection

$$C \circ \Gamma^{(r-1)} : Y \to J^rY, \tag{3}$$

which is the holonomic extension of Γ. Clearly, (1) is a special case of (3) for the well known symmetrization $C = C^{(2)}$ of second order semi-holonomic jets. Obviously, $C^{(2)} : \overline{J}^2Y \to J^2Y$ has the coordinate form $(x^i, y^p, y^p_i, y^p_{ij}) \mapsto (x^i, y^p, y^p_i, y^p_{ij} + y^p_{ji})$. We remark that all natural transformations $C : \overline{J}^rY \to J^rY$ will be described in Theorem 3.2 below.

Theorem 3.1 (Ref. 2). *All $\mathcal{FM}_{m,n}$-natural operators A transforming connections $\Gamma : Y \to J^1Y$ into r-th order holonomic connections $A(\Gamma) : Y \to J^rY$ are of the form*

(1) For $m = 1$ and arbitrary natural r, $A(\Gamma) = \Gamma^{(r-1)} : Y \to \overline{J}^rY = J^rY$.
(2) For $r = 1$ and $m \geq 2$, $A(\Gamma) = \Gamma : Y \to J^1Y$.
(3) For $r = 2$ and $m \geq 2$, $A(\Gamma) = C^{(2)}(\Gamma * \Gamma) : Y \to J^2Y$.
(4) For $r \geq 3$ and $m \geq 2$ there is no $\mathcal{FM}_{m,n}$-natural operator $\Gamma \mapsto A(\Gamma)$ in question.

This means that to construct a holonomic extension of Γ for $r \geq 3$ and $m \geq 2$, it is unavoidable to use some additional geometric object. That is why we construct an r-th order holonomic extension of Γ for any r by means of an auxiliary classical linear connection on Y. We recall that a classical linear connection $\nabla : TY \to J^1(TY \to Y)$ on $p : Y \to M$ is called projectable, if there is an underlying connection $\underline{\nabla}$ on M such that $\underline{\nabla} \circ Tp(y) = J^1_xTp \circ \nabla(y)$ for every $y \in Y_x$. Let Γ be a connection on $Y \to M$ and let ∇ be a projectable classical linear connection on Y with the underlying connection $\underline{\nabla}$ on M. Let $y \in Y_x$, $x \in M$.

Lemma 3.1 (Ref. 2). *(1) There is a normal coordinate system (U, Φ) of ∇ with centre y covering a normal coordinate system $(\underline{U}, \underline{\Phi})$ of $\underline{\nabla}$ with centre x such that $J^1\Phi(\Gamma(y)) = j^1_0(0)$.*

(2) If (U, Ψ) is another normal coordinate system of ∇ with centre y covering a normal coordinate system $(\underline{U}, \underline{\Psi})$ of $\underline{\nabla}$ with centre x such that $J^1\Psi(\Gamma(y)) = j_0^1(0)$, then there exists $A \in GL(\mathbf{R}^m)$ and $B \in GL(\mathbf{R}^n)$ such that $\Psi = (A \times B) \circ \Phi$ on some neighbourhood of y.

Example 3.2. Given a general connection Γ on $Y \to M$ and a projectable classical linear connection ∇ on Y, we define an r-th order holonomic connection $A^r(\Gamma, \nabla) : Y \to J^rY$ by

$$A^r(\Gamma, \nabla)(y) := J^r(\Phi^{-1})(j_0^r(0)) \tag{4}$$

where Φ is as in Lemma 3.1(1). By Lemma 3.1(2), the definition of $A^r(\Gamma, \nabla)(y)$ is independent of the choice of Φ. Clearly, $A^r(\Gamma, \nabla)$ is an r-th order holonomic extension of Γ by means of ∇ and the correspondence $A^r : (\Gamma, \nabla) \mapsto A^r(\Gamma, \nabla)$ is an $\mathcal{FM}_{m,n}$-natural operator.

One can show easily that $A^r(\Gamma, \nabla)$ can be also constructed by the composition (3) from Ehresmann prolongation $\Gamma^{(r-1)} : Y \to \overline{J}^rY$ of Γ. Indeed, if $s^r(\nabla) : \widetilde{J}^rY \to J^rY$ is the natural transformation from Example 3.5, then

$$A^r(\Gamma, \nabla) = s^r(\nabla) \circ \Gamma^{(r-1)}.$$

Remark 3.1. From Theorem 3.1 it follows that for $m \geq 2$ and $r \geq 3$ the connection $A^r(\Gamma, \nabla)$ given by (4) depends on ∇ in an essential way. Moreover, Theorem 3.1 also yields that to define an r-th order holonomic extension of $\Gamma : Y \to J^1Y$ for $m \geq 2$ and $r \geq 3$, the use of a projectable classical linear connection ∇ on Y is unavoidable.

Now we introduce an extension of s-th order holonomic connections into r-th order ones, where $s \leq r$.

Example 3.3. Let $\Theta : Y \to J^sY$ be a holonomic s-th order connection on $Y \to M$ and let ∇ be a projectable classical linear connection on Y with the underlying connection $\underline{\nabla}$ on M. Let $y \in Y_x$, $x \in M$ and denote by $\Gamma := \pi_1^s \circ \Theta : Y \to J^1Y$ the first order underlying connection. We define an r-th order holonomic connection $B_s^r(\Theta, \nabla) : Y \to J^rY$ by

$$B_s^r(\Theta, \nabla)(y) := J^r(\Phi^{-1})(j_0^r(\sigma))$$

where Φ is as in Lemma 3.1(1) and $\sigma : \mathbf{R}^m \to \mathbf{R}^n$ is the unique polynomial of degree $\leq s$ such that $J^s\Phi(\Theta(y)) = j_0^s\sigma$. Using Lemma 3.1(2), one evaluates easily that the definition of $B_s^r(\Theta, \nabla)(y)$ is independent of the choice of Φ. Clearly, we have $\pi_s^r \circ B_s^r(\Theta, \nabla) = \Theta$, where $\pi_s^r : J^rY \to J^sY$ is the jet projection. Because of the canonical character of the construction, the correspondence $B^r_s : (\Theta, \nabla) \mapsto B^r_s(\Theta, \nabla)$ is an $\mathcal{FM}_{m,n}$-natural operator. Obviously, we have $B^r_1 = A^r$, where A^r is the operator from Example 3.2.

(b) Symmetrization of jets

By Example 3.1, symmetrization of semiholonomic jets can be used to construct holonomic extension of connections. However, the general idea of symmetrization plays an important role in the theory of jets and also in many areas of mathematical physics. That is why we pay an attention to the problems of symmetrization of semiholonomic and nonholonomic jets. First we recall that I. Kolář[6] defined a symmetrization

$$C^{(r)} : \overline{J}^{r,r-1}Y \to J^rY$$

of special semiholonomic jets (2). Clearly, for $r = 2$ we have $\overline{J}^{2,1}Y = \overline{J}^2Y$, so that $C^{(r)}$ generalizes the well known symmetrization $C^{(2)}$ from Example 3.1. In Ref. 2 we have presented another geometric construction of $C^{(r)}$ and we have proved

Proposition 3.1. *Any $\mathcal{FM}_{m,n}$-natural transformation $A : \overline{J}^{r,r-1}Y \to J^rY$ is $C^{(r)}$.*

Obviously, $C^{(r)}$ has the coordinate expression

$$(x^i, y^p, y^p_i, \dots y^p_{i_1 \dots i_{r-1}}, y^p_{i_1 \dots i_r}) \mapsto (x^i, y^p, y^p_i, \dots y^p_{i_1 \dots i_{r-1}}, y^p_{(i_1 \dots i_r)}).$$

Theorem 3.2 (Ref. 2). *All $\mathcal{FM}_{m,n}$-natural transformations $C : \overline{J}^rY \to J^rY$ are of the form*

(1) For $m = 1$ and arbitrary natural r, $C = id : \overline{J}^rY = J^rY \to J^rY$.
(2) For $r = 1$ and $m \geq 2$, $C = id : \overline{J}^1Y = J^1Y \to J^1Y$.
(3) For $r = 2$ and $m \geq 2$, $C = C^{(2)} : \overline{J}^2Y \to J^2Y$.
(4) For $r \geq 3$ and $m \geq 2$ there is no $\mathcal{FM}_{m,n}$-natural transformation C in question.

Proposition 3.2 (Ref. 2). *For $r \geq 2$ and $m \geq 2$ there is no natural transformation $\widetilde{J}^rY \to J^rY$.*

In Ref. 2 we defined symmetrizations $c^{(r)}(\nabla) : \overline{J}^rY \to J^rY$ and $s^r(\nabla) : \widetilde{J}^rY \to J^rY$ for any r depending on a projectable classical linear connection ∇ on Y.

Example 3.4. Let ∇ be a projectable classical linear connection on Y covering $\underline{\nabla}$ on M. We define a natural transformation $c^{(r)}(\nabla) : \overline{J}^rY \to J^rY$ as follows. Take $\sigma \in (\overline{J}^rY)_y$, $y \in Y_x$, $x \in M$ and let $\underline{\sigma} \in (J^1Y)_y$ be the underlying element of σ. Similarly to Lemma 3.1, denote by Φ a ∇-normal coordinate system on Y with centre y covering a $\underline{\nabla}$-normal coordinate system $\underline{\Phi}$ such that $J^1\Phi(\underline{\sigma}) = j_0^1(0)$. We define

$$c^{(r)}(\nabla)(\sigma) := J^r\Phi^{-1}(I^{-1}(\oplus_{k=1}^r \mathrm{Sym}_k(\overline{I}(\overline{J}^r\Phi(\sigma))))), \tag{5}$$

where $\overline{I} : (\overline{J}^r\mathbf{R}^{m,n})_{(0,0)} \to \oplus_{k=1}^r \otimes^k T_0^*\mathbf{R}^m \otimes V_{(0,0)}\mathbf{R}^{m,n}$ and $I : (J^r\mathbf{R}^{m,n})_{(0,0)} \to \oplus_{k=1}^r S^kT_0^*\mathbf{R}^m \otimes V_{(0,0)}\mathbf{R}^{m,n}$ are the standard identifications and $\mathrm{Sym}_k : \otimes^kT_0^*\mathbf{R}^m \otimes V_{(0,0)}\mathbf{R}^{m,n} \to S^kT_0^*\mathbf{R}^m \otimes V_{(0,0)}\mathbf{R}^{m,n}$ are given by the symmetrizations. If Φ_1 is another ∇-normal coordinate system with centre y satisfying $J^1\Phi_1(\underline{\sigma}) = j_0^1(0)$, then near y we have $\Phi_1 = A \circ \Phi$ for some $A \in \mathrm{GL}(m) \times \mathrm{GL}(n)$ (see also Lemma 3.1(2)). Clearly, $\overline{I}$ and I are $\mathrm{GL}(m) \times \mathrm{GL}(n)$-invariant. One can verify easily that the right hand sides of (5) for Φ and Φ_1 coincide. Then the definition of $c^{(r)}(\nabla)(\sigma)$ is correct (it is independent of the choice of Φ).

Example 3.5. Given a projectable classical linear connection ∇ on Y covering $\underline{\nabla}$ on M, we define a natural transformation $s^r(\nabla) : \widetilde{J}^rY \to J^rY$ as follows. Let $\sigma \in (\widetilde{J}^rY)_y$, $y \in Y_x$, $x \in M$ and let $\underline{\sigma} \in (J^1Y)_y$ be the underlying element of σ. Taking the same Φ as in Example 3.4, we put

$$s^r(\nabla)(\sigma) := J^r\Phi^{-1}(j_0^r(0)) \ . \tag{6}$$

Quite analogously to Example 3.4 we show that the definition of $s^r(\nabla)(\sigma)$ is correct.

Remark 3.2. By Theorem 3.2 and Proposition 3.2, to define symmetrizations $\overline{J}^rY \to J^rY$ or $\widetilde{J}^rY \to J^rY$ for $m \geq 2$ and $r \geq 3$, the use of an auxiliary projectable classical linear connection ∇ on Y is unavoidable.

4. Classification of holonomic extensions

(a) Classification of second order holonomic extensions

From Theorem 3.1 it follows directly that for $m \geq 2$, the operator $\Gamma \mapsto C^{(2)}(\Gamma * \Gamma)$ from Example 3.1 is the unique $\mathcal{FM}_{m,n}$-natural operator transforming general connections $\Gamma : Y \to J^1Y$ into second order holonomic connections $Y \to J^2Y$.

It is well known that $J^rY \to J^{r-1}Y$ is an affine bundle with the associated vector bundle $S^rT^*M \otimes VY$. Thus, we have an $\mathcal{FM}_{m,n}$-natural operator Q sending general connections Γ on $Y \to M$ and projectable classical linear connections ∇ on Y into sections

$$Q(\Gamma, \nabla) := A^2(\Gamma, \nabla) - C^{(2)}(\Gamma * \Gamma) : Y \to S^2T^*M \otimes VY,$$

where A^2 is the operator from Example 3.2 for $r = 2$. In Remark 4.1 below we show another (more classical) interpretation of the operator $Q(\Gamma, \nabla)$.

Theorem 4.1 (Ref. 2). *All $\mathcal{FM}_{m,n}$-natural operators A transforming general connections $\Gamma : Y \to J^1Y$ and projectable torsion free classical linear connections ∇ on Y into second order holonomic connections $A(\Gamma, \nabla) : Y \to J^2Y$ are of the form*

$$A(\Gamma, \nabla) = C^{(2)}(\Gamma * \Gamma) + tQ(\Gamma, \nabla), \quad t \in \mathbf{R}. \tag{7}$$

(b) Classification of third order holonomic extensions

From now on we denote the connection (7) by $A^{\langle t \rangle}(\Gamma, \nabla)$, i.e.

$$A^{\langle t \rangle}(\Gamma, \nabla) := C^{(2)}(\Gamma * \Gamma) + tQ(\Gamma, \nabla).$$

Let B_2^3 be the operator from Example 3.3 transforming second order holonomic connections $\Theta : Y \to J^2Y$ and projectable classical linear connections ∇ on Y into third order holonomic connections $B_2^3(\Theta, \nabla) : Y \to J^3Y$. Then for $\Theta = A^{\langle t \rangle}(\Gamma, \nabla)$ we have third order holonomic connections

$$B^{\langle t \rangle}(\Gamma, \nabla) := B_2^3(A^{\langle t \rangle}(\Gamma, \nabla), \nabla) : Y \to J^3Y, \quad t \in \mathbf{R},$$

which project onto $A^{\langle t \rangle}(\Gamma, \nabla)$ via the jet projection $\pi_2^3 : J^3Y \to J^2Y$.

Theorem 4.2. *All $\mathcal{FM}_{m,n}$-natural operators A transforming general connections $\Gamma : Y \to J^1Y$ and projectable torsion free classical linear connections ∇ on Y into third order holonomic connections $A(\Gamma, \nabla) : Y \to J^3Y$ are of the form*

$$A(\Gamma, \nabla) = B^{\langle t \rangle}(\Gamma, \nabla) + \Delta(\Gamma, \nabla)$$

for some $t \in \mathbf{R}$ and some section $\Delta(\Gamma, \nabla) : Y \to S^3T^*M \otimes VY$.

Proof. Any such $A(\Gamma, \nabla) : Y \to J^3Y$ projects onto $\pi_2^3 \circ B(\Gamma, \nabla) : Y \to J^2Y$, which is exactly $A^{\langle t \rangle}(\Gamma, \nabla)$ for some $t \in \mathbf{R}$ because of Theorem 4.1. Then our assertion follows directly from the fact that $J^3Y \to J^2Y$ is an affine bundle with the associated vector bundle $S^3T^*M \otimes VY$. □

In the rest of this section we describe explicitly all Δ's in question. We show that the vector space of all such Δ's is of dimension 3 for $n \geq 2$ and of dimension 2 for $n = 1$. Roughly speaking, we prove that all $\mathcal{FM}_{m,n}$-natural operators A transforming general connections $\Gamma : Y \to J^1Y$ and projectable torsion free classical linear connections ∇ on Y into third order holonomic connections $A(\Gamma, \nabla) : Y \to J^3Y$ form the one parameter family of three-(resp. two-)dimensional affine spaces if $n \geq 2$ (resp. if $n = 1$). So it suffices to describe all $\mathcal{FM}_{m,n}$-natural operators Δ transforming connections Γ and ∇ as above into sections $\Delta(\Gamma,\nabla) : Y \to S^3T^*M \otimes VY$. We have the following general example of such operators Δ.

Example 4.1. Let $H : J^2(J^1(\mathbf{R}^{m,n} \to \mathbf{R}^m) \to \mathbf{R}^{m,n})_{j^1_0(0)} \to S^3T^*_0\mathbf{R}^m \otimes V_{(0,0)}\mathbf{R}^{m,n}$ (the fibre over $j^1_0(0)$ with respect to the projection $\pi^2_0 : J^2Y \to Y$) be a $\mathrm{GL}(m) \times \mathrm{GL}(n)$-equivariant map (we note that $\mathrm{GL}(m) \times \mathrm{GL}(n)$ preserve $j^1_0(0)$ and then the invariance has sense). Let Γ be a general connection on $Y \to M$ and ∇ be a projectable classical linear connection on Y with the underlying classical linear connection $\underline{\nabla}$ on M. Define $\Delta^H(\Gamma,\nabla) : Y \to S^3T^*M \otimes VY$ as follows

$$\Delta^H(\Gamma,\nabla)(y) := S^3T^*\underline{\Phi}^{-1} \otimes V\Phi^{-1}(H(j^2_{(0,0)}(\Phi_*\Gamma))) \in S^3T^*_xM \otimes V_yY, \quad (8)$$

$y \in Y_x$, $x \in M$ where Φ is a normal fiber coordinate system of ∇ with centre y with the underlying normal coordinate system $\underline{\Phi}$ with centre x such that $J^1\Phi(\Gamma(y)) = j^1_0(0)$. If Φ_1 is another such normal coordinate system, then $\Phi_1 = D \circ \Phi$ for some $D \in \mathrm{GL}(m) \times \mathrm{GL}(n)$. So, because of the $\mathrm{GL}(m) \times \mathrm{GL}(n)$-invariance of H we see that (8) does not depend on the choice of Φ in question. The correspondence $\Delta^H : (\Gamma,\nabla) \to \Delta^H(\Gamma,\nabla)$ is an $\mathcal{FM}_{m,n}$-natural operator.

We prove (see also Proposition 4.3) below

Proposition 4.1. *Any $\mathcal{FM}_{m,n}$-natural operator Δ transforming general connections $\Gamma : Y \to J^1Y$ and torsion free projectable classical linear connections ∇ on Y into sections $\Delta(\Gamma,\nabla) : Y \to S^3T^*M \otimes VY$ is of the form*

$$\Delta(\Gamma,\nabla) = \Delta^H(\Gamma,\nabla)$$

for some $GL(m) \times GL(n)$-equivariant map H from Example 4.1.

Proof. Clearly, Δ is determined by the values

$$\langle(\Delta(\Gamma,\nabla)(0,0), u \odot u \odot u \otimes \omega\rangle \in \mathbf{R} \quad (9)$$

for all connections $\Gamma : \mathbf{R}^{m,n} \to J^1(\mathbf{R}^{m,n} \to \mathbf{R}^m)$, all torsion free projectable classical linear connections ∇ on $\mathbf{R}^{m,n}$, all $u \in T_0\mathbf{R}^m$ and all $\omega \in T^*_0\mathbf{R}^n$. Using respective normal coordinates with centre zero we can assume additionally that $\Gamma(0,0) = j^1_0(0)$, the Christoffel symbols of ∇ are $\nabla^k_{ij}(0) = 0$, $u = u_o = \frac{\partial}{\partial x^1}_{|0}$ and $\omega = d_0y^1$. Because of non linear Peetre theorem (see

Corollary 19.8 in Ref. 9) we can assume that the coefficients of Γ are polynomials in x, y of (fixed) degree $\leq K$ for arbitrary large $K \in \mathbf{N}$,

$$\Gamma = \sum_{i=1}^{m} dx^i \otimes \frac{\partial}{\partial x^i} + \sum_{i=1}^{m}\sum_{j=1}^{n}\sum_{|\alpha|+|\beta|\leq K} \Gamma^j_{i\alpha\beta} x^\alpha y^\beta dx^i \otimes \frac{\partial}{\partial y^j}.$$

Applying the invariance of Δ with respect to the homotheties $(tx^1, \dots, tx^m, ty^1, \dots, ty^n)$ and using homogeneous function theorem from Ref. 9 we deduce that (9) is the linear combination of $\nabla^l_{ij;s}(0,0)$ and of the polynomial in the derivatives of Γ of order ≤ 2. So Δ is determined by the (well-defined) values

$$H(j^2_{(0,0)}\Gamma) := \Delta(\nabla^o, \Gamma)(0,0) \tag{10}$$

for all connections $\Gamma : \mathbf{R}^{m,n} \to J^1\mathbf{R}^{m,n}$ with $\Gamma(0,0) = j^1_0(0)$ (here ∇^o is the usual flat connection on $\mathbf{R}^{m,n}$) together with the values (9) with the trivial connection Γ^o on $\mathbf{R}^{m,n} \to \mathbf{R}^m$ instead of Γ. The values (9) with the trivial connection Γ^o instead of Γ under the above assumptions are linear combinations of $\nabla^k_{ij;s}(0)$. Then using the invariance of Δ with respect to the homotheties $(t_1x^1, \dots, t_mx^m, \tau_1y^1, \dots, \tau_ny^n)$ we see that the last values are proportional to $\nabla^{m+1}_{11;1}(0)$ (we have the assumptions $u = u_o = \frac{\partial}{\partial x^1}_{|0}$ and $\omega = d_0y^1$), and then are zero, as $\nabla^{m+1}_{11;1}(0,0) = 0$ in normal coordinates (see Lemma 4.1 below). By the $\mathrm{GL}(m) \times \mathrm{GL}(n)$-invariance of Δ, the map $H : J^2(J^1(\mathbf{R}^{m,n} \to \mathbf{R}^m) \to \mathbf{R}^{m,n})_{j^1_0(0)} \to S^3T^*_0\mathbf{R}^m \otimes V_{(0,0)}\mathbf{R}^{m,n}$ given by (10) is $\mathrm{GL}(m) \times \mathrm{GL}(n)$-equivariant. Clearly $\Delta = \Delta^H$ because both Δ and Δ^H determine the same values (10). □

Lemma 4.1. *Let ∇ be a classical linear connection on $\mathbf{R}^k$ such that the identity map $id_{\mathbf{R}^k}$ is its normal coordinate system with centre zero. Then $\nabla^l_{11;1}(0) = 0$ for $l = 1, \dots, k$.*

Proof. Obviously, the straight lines passing through 0 are geodesics of ∇. Using the well known equations for geodesics we get

$$\sum_{i,j=1}^{k} \nabla^l_{ij}(x)x^ix^j = 0, \quad l = 1, \dots, k.$$

Then our assertion follows easily from the last equality. Namely, we pass to formal Taylor series at 0 of booth sides of the above equality (for $l = 1, \dots, k$) and then consider the coefficients at $(x^1)^3$. □

So it remains to describe all $\mathrm{GL}(m) \times \mathrm{GL}(n)$-equivariant maps

$$H : J^2(J^1(\mathbf{R}^{m,n} \to \mathbf{R}^n) \to \mathbf{R}^{m,n})_{j^1_0(0)} \to S^3T^*_0\mathbf{R}^m \otimes V_{(0,0)}\mathbf{R}^{m,n}. \tag{11}$$

Obviously, any element $\tau \in J^2(J^1(\mathbf{R}^{m,n} \to \mathbf{R}^m) \to \mathbf{R}^{m,n})_{j^1_0(0)}$ is of the form $\tau = j^2_{(0,0)}\Gamma$ for a general connection

$$\begin{aligned}\Gamma = & \sum_{i=1}^{m} dx^i \otimes \frac{\partial}{\partial x^i} + \sum_{i,j=1}^{m}\sum_{k=1}^{n} a^k_{i,j} x^i dx^j \otimes \frac{\partial}{\partial y^k} \\ & + \sum_{j=1}^{m}\sum_{k,l=1}^{n} b^k_{l,j} y^l dx^j \otimes \frac{\partial}{\partial y^k} + \sum_{l_1,l_2=1}^{n}\sum_{j=1}^{m} c^k_{l_1,l_2,j} y^{l_1} y^{l_2} dx^j \otimes \frac{\partial}{\partial y^k} \\ & + \sum_{i,j=1}^{m}\sum_{k,l=1}^{n} e^k_{i,l,j} x^i y^l dx^j \otimes \frac{\partial}{\partial y^k} + \sum_{i_1,i_2,j=1}^{m}\sum_{k=1}^{n} f^k_{i_1,i_2,j} x^{i_1} x^{i_2} dx^j \otimes \frac{\partial}{\partial y^k}\end{aligned} \tag{12}$$

on $\mathbf{R}^{m,n}$ for some (uniquely determined by τ) real numbers $a^k_{i,j}$, $b^k_{j,l}$, $c^k_{l_1,l_2,j}$, $e^k_{i,l,j}$, $f^k_{i_1,i_2,j}$ with $c^k_{l_1,l_2,j} = c^k_{l_2,l_1,j}$ and $f^k_{i_1,i_2,j} = f^k_{i_2,i_1,j}$. That is why, we have

$$J^2(J^1(\mathbf{R}^{m,n} \to \mathbf{R}^n) \to \mathbf{R}^{m,n})_{j^1_0(0)} = W_1 \oplus W_2 \oplus W_3 \oplus W_4 \oplus W_5$$

modulo the obvious isomorphism of $\mathrm{GL}(m) \times \mathrm{GL}(n)$-modules, where $W_1 = \mathbf{R}^{m*} \otimes \mathbf{R}^{m*} \otimes \mathbf{R}^n$, $W_2 = \mathbf{R}^{n*} \otimes \mathbf{R}^{m*} \otimes \mathbf{R}^n$, $W_3 = S^2\mathbf{R}^{n*} \otimes \mathbf{R}^{m*} \otimes \mathbf{R}^n$, $W_4 = \mathbf{R}^{m*} \otimes \mathbf{R}^{n*} \otimes \mathbf{R}^{m*} \otimes \mathbf{R}^n$ and $W_5 = S^2\mathbf{R}^{m*} \otimes \mathbf{R}^{m*} \otimes \mathbf{R}^n$. Quite similarly,

$$S^3T^*_0\mathbf{R}^m \otimes V_{(0,0)}\mathbf{R}^{m,n} = S^3\mathbf{R}^{m*} \otimes \mathbf{R}^n$$

modulo the obvious isomorphism of $\mathrm{GL}(m) \times \mathrm{GL}(n)$-modules. Thus it remains to describe all $\mathrm{GL}(m) \times \mathrm{GL}(n)$-equivariant maps

$$H : W_1 \oplus W_2 \oplus W_3 \oplus W_4 \oplus W_5 \to S^3\mathbf{R}^{m*} \otimes \mathbf{R}^n. \tag{13}$$

We have the following examples of $\mathrm{GL}(m) \times \mathrm{GL}(n)$-maps (13).

Example 4.2. Let $\tau = (\tau_1, ..., \tau_5) \in W_1 \oplus ... \oplus W_5$. We define $H_i(\tau)$ as follows:

(a) We symmetrize τ_5

$$H_1(\tau) := \mathrm{Sym}(\tau_5) \in S^3\mathbf{R}^{m*} \otimes \mathbf{R}^n.$$

(b) Consider $\tau_1 \otimes \tau_2 \in \mathbf{R}^{m*} \otimes \mathbf{R}^{m*} \otimes \mathbf{R}^n \otimes \mathbf{R}^{n*} \otimes \mathbf{R}^{m*} \otimes \mathbf{R}^n$. Applying two obvious contractions C^3_4 and C^6_4, we obtain $C^3_4(\tau_1 \otimes \tau_2) \in \mathbf{R}^{m^*} \otimes \mathbf{R}^{m*} \otimes \mathbf{R}^{m*} \otimes \mathbf{R}^n$ and $C^6_4(\tau_1 \otimes \tau_2) \in \mathbf{R}^{m*} \otimes \mathbf{R}^{m*} \otimes \mathbf{R}^n \otimes \mathbf{R}^{m*} = \mathbf{R}^{m*} \otimes \mathbf{R}^{m*} \otimes \mathbf{R}^{m*} \otimes \mathbf{R}^n$ (modulo obvious isomorphism of $\mathrm{GL}(m) \times \mathrm{GL}(n)$-modules). Then the symmetrization yields

$$H_2(\tau) := \mathrm{Sym}(C^3_4(\tau_1 \otimes \tau_2)) \in S^3\mathbf{R}^{m*} \otimes \mathbf{R}^n$$

and

$$H_3(\tau) := \mathrm{Sym}(C^6_4(\tau_1 \otimes \tau_2)) \in S^3\mathbf{R}^{m*} \otimes \mathbf{R}^n.$$

Proposition 4.2. *(a) If $n \geq 2$, then any $GL(m) \times GL(n)$-map $H : W_1 \oplus \cdots \oplus W_5 \to S^3\mathbf{R}^{m*} \otimes \mathbf{R}^n$ is a linear combination of H_1, H_2 and H_3 (described in Example 4.2) with real coefficients.*

(b) If $n = 1$, then any $GL(m) \times GL(n)$-map $H : W_1 \oplus \cdots \oplus W_5 \to S^3\mathbf{R}^{m*} \otimes \mathbf{R}^n$ is a linear combination of H_1 and H_2 with real coefficients.

Proof. It is easy to see that if $n \geq 2$, then H_1, H_2 and H_3 are linearly independent in the vector space of all $\mathrm{GL}(m) \times \mathrm{GL}(n)$-maps (13). If $n = 1$, then H_1 and H_2 are also linearly independent. So by the dimension argument and our identifications, it suffices to show that the vector space of all $\mathrm{GL}(m) \times \mathrm{GL}(n)$-maps (11) is of dimension ≤ 3 for $n \geq 2$ and of dimension ≤ 2 for $n = 1$.

Consider a $\mathrm{GL}(m) \times \mathrm{GL}(n)$-map C as in (11). Because of the $\mathrm{GL}(m) \times \mathrm{GL}(n)$-invariance, H is determined by the values

$$\left\langle H(j^2_{(0,0)}\Gamma), u_o \odot u_o \odot u_o \otimes d_0 y^1 \right\rangle \in \mathbf{R} \tag{14}$$

for all Γ as in (12), where $u_o = \frac{\partial}{\partial x^1}_{|0}$. Using equivariance of H with respect to

$$(x^1, tx^2, \ldots, tx^m, y^1, \ldots, y^n) \in \mathrm{GL}(m) \times \mathrm{GL}(n)$$

for $t \neq 0$ and putting $t \to 0$, we see that (14) does not depend on many coordinates of Γ. Consequently we see that H is determined by the values (14) for Γ of the form

$$\begin{aligned}\Gamma = &\sum_{i=1}^m dx^i \otimes \frac{\partial}{\partial x^i} + \sum_{k=1}^n a^k x^1 dx^1 \otimes \frac{\partial}{\partial y^k} + \sum_{k,l=1}^n b^k_l y^l dx^1 \otimes \frac{\partial}{\partial y^k} \\ &+ \sum_{l_1,l_2,k=1}^n c^k_{l_1,l_2} y^{l_1} y^{l_2} dx^1 \otimes \frac{\partial}{\partial y^k} + \sum_{l,k=1}^n e^k_l y^l x^1 dx^1 \otimes \frac{\partial}{\partial y^k} \\ &+ \sum_{k=1}^n f^k (x^1)^2 dx^1 \otimes \frac{\partial}{\partial y^k}\end{aligned} \tag{15}$$

for all real numbers a^k, b^k_l, $c^k_{l_1,l_2}$, e^k_l, f^k with $c^k_{l_1,l_2} = c^k_{l_2,l_1}$. Using equivariance of H with respect to the homotheties $t\mathrm{id}_{\mathbf{R}^{m,n}} \in \mathrm{GL}(m) \times \mathrm{GL}(n)$ for $t \neq 0$ and homogeneous function theorem we see that the left hand side of (14) for Γ as in (15) is the linear combination of $c^k_{l_1,l_2}$, e^k_l, f^k and of the combination of monomials in a^k, b^k_l of degree 2. Next, applying equivariance of H with respect to $\varphi_{t,\tau} = \frac{1}{t}\mathrm{id}_{\mathbf{R}^m} \times \tau\mathrm{id}_{\mathbf{R}^n} \in \mathrm{GL}(m) \times \mathrm{GL}(n)$ for $t, \tau \neq 0$

we get the homogeneous condition

$$\left\langle H(j^2_{(0,0)}((\varphi_{t,\tau})_*\Gamma)), u_o \odot u_o \odot u_o \otimes d_O y^1 \right\rangle$$
$$= t^3\tau \left\langle H(j^2_{(0,0)}\Gamma), u_o \odot u_o \odot u_o \otimes d_o y^1 \right\rangle ,$$

where Γ is as in (15). This yields

$$\left\langle H(j^2_{(0,0)}\Gamma), u_o \odot u_o \odot u_o \otimes d_O y^1 \right\rangle = \sum_{k=1}^{n} \alpha_k f^k + \sum_{k_1,k_2,l_2=1}^{n} \beta^{l_2}_{k_1,k_2} a^{k_1} b^{k_2}_{l_2} \quad (16)$$

for Γ as in (15) for some (determined by H) real numbers α^k, $\beta^{l_2}_{k_1,k_2}$. Using invariance of H with respect to $(x^1,\dots,x^m,y^1,ty^2,\dots,ty^n) \in \mathrm{GL}(m) \times \mathrm{GL}(n)$ for $t \neq 0$ we deduce easily

$$\left\langle H(j^2_{(0,0}\Gamma), u_o \odot u_o \odot u_o \otimes d_O y^1 \right\rangle = \alpha f^1 + \beta a^1 b^1_1 + \sum_{k=2}^{n} \beta^k a^k b^1_k + \sum_{k=2}^{n} \gamma^k a^1 b^k_k \quad (17)$$

for some (determined by H) real numbers α, β, β^k, γ^k, where Γ is as in (15). Therefore if $n = 1$, then the vector space of all H as in (11) is of dimension ≤ 2.

Assume now $n \geq 2$. Using equivariance of H with respect to permutations of coordinates $y^2,\dots,y^n$ (they belong to $\mathrm{GL}(m) \times \mathrm{GL}(n)$) we deduce easily $\beta^2 = \dots = \beta^n$, $\gamma^2 = \dots = \gamma^n$. From (17) we have

$$\left\langle H(j^2_{(0,0)}\Gamma), u_o \odot u_o \odot u_o \otimes d_O y^1 \right\rangle = 0 \quad (18)$$

for $\Gamma = \sum_{i=1}^m dx^i \otimes \frac{\partial}{\partial x^i} + x^1 dx^1 \otimes \frac{\partial}{\partial y^1} + y^2 dx^1 \otimes \frac{\partial}{\partial y^1}$. Applying equivariance of H with respect to $(x^1,\dots,x^m,y^1,y^2+y^1,y^3,\dots,y^n) \in \mathrm{GL}(m) \times \mathrm{GL}(n)$ we have (18) for $\Gamma = \sum_{i=1}^m dx^i \otimes \frac{\partial}{\partial x^i} + x^1 dx^1 \otimes (\frac{\partial}{\partial y^1} + \frac{\partial}{\partial y^2}) + (y^2 - y^1) dx^1 \otimes (\frac{\partial}{\partial y^1} + \frac{\partial}{\partial y^2})$. But from (16) it follows that the left hand side of (18) is equal to $(-\beta + *)$, where $*$ is a linear combination of β^2 and γ^2. Then $(-\beta + *) = 0$. That is why the vector space of all H as in (11) is of dimension ≤ 3. □

Remark 4.1. Let (Γ, ∇) be as in Theorem 4.1 or 4.2. We have the tangent valued 1-form $\Gamma : Y \to T^*Y \otimes VY$ (the horizontal projection of Γ). Its covariant derivative is a tensor field $\nabla\Gamma : Y \to \otimes^2 T^*Y \otimes VY$, where $\nabla\Gamma(U,W) = (\nabla_U \Gamma)(W)$ for vector fields U, W. Composing symmetrization of $\nabla\Gamma$ with the horizontal lifting map $h : Y \to T^*M \otimes TY$ of Γ, we have tensor field $\widetilde{Q}(\Gamma,\nabla) : Y \to S^2 T^*M \otimes VY$. Because of the dimension argument, the operator Q from Theorem 4.1 is proportional to $\widetilde{Q}$.

Write $\nabla^2\Gamma = \nabla(\nabla\Gamma) : Y \to \otimes^3 T^*Y \otimes VY$ for the second covariant derivative. Moreover, we have tensor fields $\nabla\Gamma(\nabla\Gamma, .) : Y \to \otimes^3 T^*Y \otimes VY$

and $C^1_1(\nabla\Gamma)\otimes\nabla\Gamma : Y \to \otimes^3 T^*Y\otimes VY$, where the first tensor is the substitution of $\nabla\Gamma$ into the first position of $\nabla\Gamma$ and C^1_1 denotes the obvious contraction. Composing symmetrization of these tensor fields with the horizontal lifting map h, we define tensor fields $E_i(\Gamma,\nabla) : Y \to S^3T^*M\otimes VY$, $i = 1, 2, 3$. Thus we have the corresponding $\mathcal{FM}_{m,n}$-natural operators E_i transforming Γ and ∇ into tensor fields $E_i(\Gamma,\nabla) : Y \to S^3T^*M\otimes VY$. One can standardly show that the operators E_1, E_2, E_3 are linearly independent for $n \geq 2$. If $n = 1$, then E_1 and E_2 are linearly independent (E_2 and E_3 are proportional).

By Propositions 4.1, 4.2 and the dimension argument we have

Proposition 4.3. *If $n \geq 2$, then any $\mathcal{FM}_{m,n}$-natural operator Δ of the type as in Proposition 4.1 is a linear combination of E_1, E_2, E_3 with real coefficients. If $n = 1$, then such operators are linear combinations of E_1 and E_2.*

5. Some applications of higher order connections

Let $P \to M$ be a principal bundle with structure group G. Then we have a right action of G on J^rP defined by $(j^r_x s(y))g = j^r_x(s(y)g)$, where $s : M \to P$ is a local section, $x, y \in M$, $g \in G$. This enables us to define G-invariant r-th order holonomic connections $\Gamma : P \to J^rP$. Clearly, for $r = 1$ we obtain the well known concept of a principal connection on $P \to M$. We recall that the r-th principal prolongation W^r_mP of P is defined as the space of all r-jets $j^r_{(0,e)}\varphi$ of local trivializations $\varphi : \mathbf{R}^m \times G \to P$, where $e \in G$ is the unit. By Ref. 9, $W^r_mP \to M$ is a principal bundle with the structure group $W^r_mG := J^r_{(0,e)}(\mathbf{R}^m \times G, \mathbf{R}^m \times G)_{(0,-)}$ and the fibered manifold $W^r_mP \to M$ coincides with the fibered product $W^r_mP = P^rM \times_M J^rP$, where $P^rM := \mathrm{inv}J^r_0(\mathbf{R}^m, M)$ is the r-th order frame bundle of M. It is well known that the bundle functor W^r_m plays a fundamental role in the theory of gauge natural operators and in mathematical physics.

Example 5.1. Let Γ be a principal connection on $P \to M$ and Λ be a classical linear connection on M. We have a reduction of principal bundles

$$\mu_\Gamma : P^1M \times_M P \to P^1M \times_M J^1P = W^1_mP, \quad \mu_\Gamma(l,p) = (l,\Gamma(p)) \qquad (19)$$

with the obvious group monomorphism $\mathrm{GL}(m) \times G \to W^1_mG$, cf. Ref. 9. Using right translations, the product connection $\Lambda \times \Gamma$ on $P^1M \times_M P$ can be extended into a principal connection $p(\Gamma,\Lambda)$ on $W^1_mP \to M$. By 17.2. and 16.6. of Ref. 9, reduction μ_Γ can be used in the coordinate description of geometric object fields. Moreover, connection $p(\Gamma, \Lambda)$ plays an important role in the geometric constructions on principal prolongations of principal bundles, see Ref. 10.

Example 5.2. Suppose we have a G-invariant r-th order holonomic connection $D(\Gamma, \Lambda) : P \to J^r P$ depending on a principal connection $\Gamma : P \to J^1 P$ and a classical linear connection Λ on M. Then we can generalize (19) to the reduction

$$\mu_D : P^r M \times_M P \to P^r M \times_M J^r P = W^r_m P,\ \mu_D(l,p) = (l, D(\Gamma, \Lambda)(p)), \quad (20)$$

which can be used for the geometric description of higher order geometric object fields. Moreover, if Λ is a torsion free classical linear connection on M, then we can define its exponential prolongation $\Lambda^{\exp_r} : TM \to J^r TM$, which can be identified with a principal connection on $P^r M$, see e.g. Ref. 14. On $P^r M \times_M P$ we have a product connection $\Lambda^{\exp_r} \times \Gamma$, which can be extended via right translations of $W^r_m G$ to a principal connection

$$p_D(\Gamma, \Lambda) : W^r_m P \to J^1(W^r_m P \to M), \quad (21)$$

which generalizes $p(\Gamma, \Lambda)$ from Example 5.1. Clearly, for $r = 1$ and $D(\Gamma, \Lambda) = \Gamma$, (21) is exactly $p(\Gamma, \Lambda)$.

Example 5.3. We recall that a torsion free classical linear connection Λ on M and a principal connection Γ on $P \to M$ induce a G-invariant projectable classical linear connection $N_P(\Gamma, \Lambda)$ on P, see Ref. 9. Using our operator A^r from Example 3.2, we can write $D(\Gamma, \Lambda) = A^r(\Gamma, N_P(\Gamma, \Lambda))$. Then (21) yields a connection $p^r(\Gamma, \Lambda)$ on $W^r_m P \to M$, which generalizes $p(\Gamma, \Lambda)$.

Example 5.4. One can replace $D(\Gamma, \Lambda)$ in Example 5.3 by an arbitrary $\mathcal{PB}_m(G)$-gauge natural operator $\widetilde{D}$ transforming couples (Γ, Λ) consisting of principal connections Γ on $P \to M$ and classical linear connections Λ on M into G-invariant r-th order holonomic connections $\widetilde{D}(\Gamma, \Lambda) : P \to J^r P$. In the case $r = 2$ or $r = 3$, many examples of such $\widetilde{D}$ can be constructed from operators A classified in Theorems 4.1 and 4.2. In fact, if $Y = P \to M$ is a principal G-bundle, Γ is a principal connection on $P \to M$ and Λ is a torsion free classical linear connection on M, then all operators A from Theorem 4.1 and 4.2 give G-invariant connections $\widetilde{D}(\Gamma, \Lambda) := A(\Gamma, N_P(\Gamma, \Lambda))$. Connections $\widetilde{D}(\Gamma, \Lambda)$ are in fact G-invariant, because the right translations of G on P are in particular $\mathcal{FM}_{m,n}$-maps, such translations preserve Γ (as Γ are principal), they preserve $\nabla = N_P(\Gamma, \Lambda)$ (as $N_P(\Gamma, \Lambda)$ are G-invariant) and they also preserve A (as A are $\mathcal{FM}_{m,n}$-natural). This yields other generalizations of $p(\Gamma, \Lambda)$ to $W^2_m P \to M$ and to $W^3_m P \to M$.

Acknowledgments

The first author was supported by a grant of the GA ČR No 201/05/0523.

References

1. A. Cabras and I. Kolář, Second order connections on some functional bundles, *Arch. Math. (Brno)* **35** (1999) 347–365.
2. M. Doupovec and W.M. Mikulski, Holonomic extension of connections and symmetrization of jets, *Rep. Math. Phys.* **60** (2) (2007) 299–316.
3. C. Ehresmann, Extension du calcul des jets aux jets non holonomes, *CRAS Paris* **239** (1954) 1762–1764.
4. C. Ehresmann, Sur les connexions d' ordre supérieur, *In: Atti del V. Cong. del' Unione Mat. Italiana* (1955, Roma Cremonese, 1956) 344–346.
5. I. Kolář, On the torsion of spaces with connection, *Czech. Math. J.* **21** (1971) 124–136.
6. I. Kolář, The contact of spaces with connection, *J. Diff. Geom.* **7** (1972) 563–570.
7. I. Kolář, Higher order absolute differentiation with respect to generalized connections, *Differential Geometry, Banach Center Publications* **12** (1984) 153–162.
8. I. Kolář, A general point of view to nonholonomic jet bundles, *Cahiers Topol. Géom. Diff.* **44** (2003) 149–160.
9. I. Kolář, P.W. Michor and J. Slovák, *Natural Operations in Differential Geometry* (Springer–Verlag, 1993).
10. I. Kolář, G. Virsik, Connections in first principal prolongations, *Rend. Circ. Mat. Palermo*, Serie II, Suppl. **43** (1996) 163–171.
11. D. Krupka, J. Janyška, *Lectures on Differential Invariants* (Folia Facultatis Scientiarum Naturalium Universitatis Purkynianae Brunensis, Mathematica, University J. E. Purkyně, Brno, 1990).
12. M. Kureš, On the symmetrisation of nonholonomic jets, *Math. Proc. of the Royal Irish Academy* **105A** (2005) 93–106.
13. P. Libermann, Introduction to the theory of semi-holonomic jets, *Arch. Math.* (Brno) **33** (1997) 173–189.
14. W.M. Mikulski, Higher order linear connections from first order ones, *Arch. Math. (Brno)* **43** (4) (2007) 285–288.
15. J. Pradines, Fibrés vectoriels doubles symmétriques et jets holonomes d' ordre 2, *CRAS Paris* **278** (1974) 1557–1560.
16. G. Virsik, On the holonomity of higher order connections, *Cahiers Topol. Géom. Diff.* **12** (1971) 197–212.

Differential Geometry and its Applications
Proc. Conf., in Honour of Leonhard Euler, Olomouc, August 2007

Conformally-projectively flat statistical structures on tangent bundles over statistical manifolds

Izumi Hasegawa

Department of Mathematics, Hokkaido University of Education, Sapporo, 002-8502, Japan
E-mail: hasegawa@sap.hokkyodai.ac.jp

Kazunari Yamauchi

Department of Mathematics, Asahikawa Medical College, Asahikawa, 078-8510, Japan
E-mail: yamauchi@asahikawa-med.ac.jp

Let $(M,\, h,\, \nabla)$ be a statistical manifold and TM the tangent bundle over M. If TM with "horizontal lift" statistical structure $(h^H,\, C^H)$ is conformally-projectively flat, then $(M,\, h,\, \nabla)$ is of constant curvature. In particular, if ∇ is non-flat, then ∇ is the Levi-Civita connection of h.

Keywords: Statistical structure, tangent bundle, horizontal lift, conformally-projectively flat.

MS classification: 53A15, 53B05.

1. Introduction

Let M be a differentiable manifold, h a semi-Riemannian metric on M, and ∇ a symmetric affine connection on M. If the covariant derivative ∇h is symmetric, then ∇ is said to be *compatible* to h. A pair (h, ∇) of semi-Riemannian metric with compatible affine connection is called a *statistical structure* on M. A manifold M together with a statistical structure is called a *statistical manifold*.

On a statistical manifold (M, h, ∇), the symmetric tensor field $C := \nabla h$ is called the *cubic form* of M; $C = (S_{kji}) = S_{kji} dx^k \otimes dx^j \otimes dx^i$, where $S_{kji} := \nabla_k h_{ji}$ and (x^h) is the local coordinate system of M.

The tensor field S of type $(1,2)$ is defined by $h(S(X,Y),Z) := C(X,Y,Z) = (\nabla_X h)(Y,Z)$ for any X, Y, Z. S is called the *skewness operator* of (M, h, ∇); $S = ({S_{ji}}^h) = {S_{ji}}^h dx^j \otimes dx^i \otimes \frac{\partial}{\partial x^h}$, where ${S_{ji}}^h := S_{jia} h^{ah}$ and

$(h^{ji}) := (h_{ji})^{-1}$. Then we have $\nabla = \nabla^0 - \frac{1}{2}S$, i.e., $\Gamma_{ji}{}^h = (\overset{0}{\Gamma})_{ji}{}^h - \frac{1}{2}S_{ji}{}^h$, where ∇^0 denotes the Levi-Civita connection of h and $\Gamma_{ji}{}^h$ (resp. $(\overset{0}{\Gamma})_{ji}{}^h$) are the coefficients of ∇ (resp. ∇^0).

For every statistical manifold (M, h, ∇), there exists a naturally associated symmetric trilinear form C called the cubic form. Conversely let (M, h, C) be a semi-Riemannian manifold with symmetric trilinear form C. Define a tensor field S of type (1, 2) by $h(S(X,Y),Z) := C(X,Y,Z)$, and an affine connection ∇ by $\nabla := \nabla^0 - \frac{1}{2}S$. Then ∇ is symmetric and satisfies $\nabla h = C$. Hence the triplet (M, h, ∇) becomes a statistical manifold. We call this symmetric affine connection a *compatible connection* with respect to $(h,\ C)$. Thus equipping a statistical structure (h, ∇) is equivalent to equipping a pair (h, C) consisting of a semi-Riemannian metric h and a symmetric trilinear form C. Therefore (h, C) is also called a statistical structure.

Let (M, h, ∇) be an n-dimensional statistical manifold. The curvature tensor K of (M, h, ∇) is defined by the curvature tensor of ∇, i.e.,

$$K_{kji}{}^h := \partial_k \Gamma_{ji}{}^h - \partial_j \Gamma_{ki}{}^h + \Gamma_{ji}{}^a \Gamma_{ka}{}^h - \Gamma_{ki}{}^a \Gamma_{ja}{}^h.$$

The curvature tensor satisfies

$$\begin{aligned} &K_{kji}{}^h = -K_{jki}{}^h, \\ &K_{kji}{}^h + K_{jik}{}^h + K_{ikj}{}^h = 0, \\ &K_{kjih} + K_{jhik} + K_{hkij} = 0, \end{aligned}$$

where $K_{kjih} := K_{kji}{}^a h_{ah}$.

The Ricci tensor of (M, h, ∇) is defined by $R_{ji} := K_{aji}{}^a$. In statistical geometry, the Ricci tensor is not necessarily to be symmetric. The scalar curvature of (M, h, ∇) is defined by $\rho := R_{ba}h^{ba}$. (M, h, ∇) is called an *Einstein* statistical manifold if ρ is constant and $R_{ji} = \frac{\rho}{n}h_{ji}$. (M, h, ∇) is said to be *of constant curvature* if ρ is constant and $K_{kji}{}^h = \frac{\rho}{n(n-1)}(\delta_k^h h_{ji} - \delta_j^h h_{ki})$. (M, h, ∇) is said to be *locally flat* if $K = 0$.

In Riemannian geometry we have the following:

Theorem 1.1 (Ref. 11). *Let $(M,\ h)$ be a semi-Riemannian manifold of dimension $n(\geq 2)$. Then $(TM,\ h^C)$ is conformally flat if and only if $(M,\ h)$ is of constant curvature.*

J. Inoguchi[9] suggested to generalize this theorem from the view point of statistical geometry. We can construct many statistical structure on the tangent bundle TM over statistical manifold $(M,\ h,\ \nabla)$. For example, the pair (h^C, ∇^C) of complete lift of h and complete lift of ∇ is a statistical structure on TM. In our previous paper[4] we proved the following:

Theorem 1.2. *Let (M, h, ∇) be a statistical manifold of dimension $n(\geq 2)$. Then (TM, h^C, ∇^C) is conformally-projectively flat if and only if (M, h, ∇) is of constant curvature. In particular, if (TM, h^C, ∇^C) is ± 1-conformally flat, then (M, h, ∇) and (TM, h^C, ∇^C) are locally flat.*

In this paper we prove the following:

Theorem 1.3. *Let (M, h, ∇) be a statistical manifold of dimension $n(\geq 2)$. If (TM, h^H, C^H) is conformally-projectively flat, then (M, h, ∇) is of constant curvature. In particular, if M is non-flat, then (h, ∇) is semi-Riemannian structure (i.e., ∇ is the Levi-Civita connection of h).*

Therefore, if ∇ is not the Levi-Civita connection of h, then (M, h, ∇) is locally flat.

Corollary 1.1. *Let (M, h, ∇) be a statistical manifold of dimension $n(\geq 2)$. If (TM, h^H, C^H) is $\lambda(\neq 0)$-conformally flat, then (M, h, ∇) and (TM, h^H, C^H) are locally flat.*

The following corollary coincides with Theorem 1.1 since the complete lift h^C is coincides with horizontal lift h^H with respect to Levi-Civita connection of h.

Corollary 1.2. *Let (M, h) be a semi-Riemannian manifold of dimension $n(\geq 2)$. Then (TM, h^H) is conformally flat if and only if (M, h) is of constant curvature.*

2. Conformal-projective equivalence

Let (h, ∇) and $(\widetilde{h}, \widetilde{\nabla})$ be two statistical structures on M. We recall the notion of λ-conformality formulated by Kurose.[6]

Definition 2.1. For any fixed real number λ, two statistical structures (h, ∇) and $(\widetilde{h}, \widetilde{\nabla})$ are said to be *λ-conformally equivalent* if there exists a function φ on M satisfying:

$$\widetilde{h}_{ji} = \exp(\varphi) h_{ji} \tag{1}$$

and

$$\widetilde{\Gamma}_{ji}{}^h = \Gamma_{ji}{}^h - \frac{1+\lambda}{2} h_{ji}\varphi^h + \frac{1-\lambda}{2}(\delta_j^h \varphi_i + \delta_i^h \varphi_j), \tag{2}$$

i. e.,

$$\widetilde{S}_{kji} = \exp(\varphi)\left\{S_{kji} + \lambda(h_{kj}\varphi_i + h_{ji}\varphi_k + h_{ik}\varphi_j)\right\},$$

where $\varphi_i := \partial_i\varphi$ and $\varphi^h := h^{ha}\varphi_a$. A statistical manifold (M, h, ∇) is said to be λ-*conformally flat* if (M, h, ∇) is λ-conformally equivalent to a flat statistical manifold in a neighbourhood of an arbitrary point of M.

Remark 2.1 (Remarks of Definition 2.1).
(1) If $\lambda = 0$, then (h, ∇) and $(\widetilde{h}, \widetilde{\nabla})$ are conformally equivalent in statistical geometry. In case of Riemannian geometry, (2) comes from (1).
(2) If $\lambda = -1$, then ∇ and $\widetilde{\nabla}$ are projectively equivalent.
(3) If $\lambda = 1$, then (h, ∇) and $(\widetilde{h}, \widetilde{\nabla})$ are said to be dual-projectively equivalent (cf. Ref. 5).

H. Matsuzoe[8] established the notion of conformal-projective equivalence relation and gave a necessary and sufficient condition for a conformally-projectively flat statistical manifold to be realized by a nondegenerate centroaffine immersion of codimension 2.

Definition 2.2. Two statistical structures (h, ∇) and $(\widetilde{h}, \widetilde{\nabla})$ on M are said to be *conformally-projectively equivalent* if there exist two functions φ and ψ on M satisfying:

$$\widetilde{h}_{ji} = \exp(\varphi + \psi)h_{ji} \tag{3}$$

and

$$\widetilde{\Gamma}_{ji}{}^h = \Gamma_{ji}{}^h - h_{ji}\varphi^h + \delta_j^h\psi_i + \delta_i^h\psi_j, \tag{4}$$

i. e.,

$$\widetilde{S}_{kji} = \exp(\varphi + \psi)\left\{S_{kji} + h_{kj}(\varphi_i - \psi_i) + h_{ji}(\varphi_k - \psi_k) + h_{ik}(\varphi_j - \psi_j)\right\}.$$

A statistical manifold (M, h, ∇) is said to be *conformally-projectively flat* if (M, h, ∇) is conformally-projectively equivalent to a flat statistical manifold in a neighbourhood of an arbitrary point of M.

Remark 2.2 (Remarks of Definition 2.2).
(1) If $\varphi = \psi$, then (h, ∇) and $(\widetilde{h}, \widetilde{\nabla})$ are conformally equivalent in statistical geometry.
(2) If φ is constant, then ∇ and $\widetilde{\nabla}$ are projectively equivalent.
(3) If ψ is constant, then (h, ∇) and $(\widetilde{h}, \widetilde{\nabla})$ are dual-projectively equivalent.

Lemma 2.1. *If two statistical structures* (h, ∇) *and* $(\widetilde{h}, \widetilde{\nabla})$ *on* M *are* λ-*conformally equivalent, then* (h, ∇) *and* $(\widetilde{h}, \widetilde{\nabla})$ *are conformally-projectively equivalent.*

3. Lift metrics and connections on TM (cf. Ref. 12)

Let M be a differentiable manifold with an affine connection ∇, and ${\Gamma_{ji}}^h$ the coefficients of ∇, i.e., ${\Gamma_{ji}}^a\partial_a := \nabla_{\partial_j}\partial_i$, where $\partial_h = \frac{\partial}{\partial x^h}$ and (x^h) is the local coordinates of M. We define a local frame $\{E_i,\ E_{\bar{i}}\}$ of TM as follows:

$$E_i := \partial_i - y^b{\Gamma_{ib}}^a\partial_{\bar{a}} \quad \text{and} \quad E_{\bar{i}} := \partial_{\bar{i}}, \tag{5}$$

where $(x^h,\ y^h)$ is the induced coordinates of TM and $\partial_{\bar{i}} := \frac{\partial}{\partial y^i}$. This frame $\{E_i,\ E_{\bar{i}}\}$ is called the *adapted frame* of TM with respect to ∇. Then $\{dx^h,\ \delta y^h\}$ is the dual frame of $\{E_i,\ E_{\bar{i}}\}$, where $\delta y^h := dy^h + y^b{\Gamma_{ab}}^h dx^a$. Then, by straightforward calculation, we have the following

Lemma 3.1. *The Lie brackets of the adapted frame of* TM *satisfy the following identities:*
(1) $[E_j,\ E_i] = y^b{K_{ijb}}^a E_{\bar{a}}$, (2) $[E_j,\ E_{\bar{i}}] = {\Gamma_{ji}}^a E_{\bar{a}}$, (3) $[E_{\bar{j}},\ E_{\bar{i}}] = 0$,
where $K = ({K_{kji}}^h)$ *denotes the curvature tensor of* $(M,\ \nabla)$ *defined by* ${K_{kji}}^h := \partial_k{\Gamma_{ji}}^h - \partial_j{\Gamma_{ki}}^h + {\Gamma_{ji}}^a{\Gamma_{ka}}^h - {\Gamma_{ki}}^a{\Gamma_{ja}}^h$.

Let h be a semi-Riemannian metric with the components h_{ji} on (M, ∇), i.e., $h = h_{ji}dx^j \otimes dx^i$. The horizontal lift metric h^H of h with respect to ∇ is defined as follows:

$$h^H = h_{ji}(\delta y^j \otimes dx^i + dx^j \otimes \delta y^i). \tag{6}$$

Let $X = X^a\partial_a$ be a vector field on M. The vertical lift X^V and the horizontal lift X^H of X are defined as follows:

$$X^H := X^a E_a \quad \text{and} \quad X^V := X^a E_{\bar{a}}. \tag{7}$$

Let (M, h, ∇) be a statistical manifold. We have the following lift tensors on TM:

$$C^H = S_{kji}(\delta y^k \otimes dx^j \otimes dx^i + dx^k \otimes \delta y^j \otimes dx^i + dx^k \otimes dx^j \otimes \delta y^i). \tag{8}$$

$$S^H = {S_{ji}}^h(\delta y^j \otimes dx^i \otimes E_{\bar{h}} + dx^j \otimes \delta y^i \otimes E_{\bar{h}} + dx^j \otimes dx^i \otimes E_h). \tag{9}$$

Then we have the following:

$$\begin{aligned} &h^H(S^H(\widetilde{X})\widetilde{Y}, \widetilde{Z}) \\ &= C^H(\widetilde{X}, \widetilde{Y}, \widetilde{Z}) \\ &= S_{kji}(\widetilde{X}^{\bar{k}}\widetilde{Y}^j\widetilde{Z}^i + \widetilde{X}^k\widetilde{Y}^{\bar{j}}\widetilde{Z}^i + \widetilde{X}^k\widetilde{Y}^j\widetilde{Z}^{\bar{i}}), \end{aligned} \tag{10}$$

for any $\widetilde{X} = \widetilde{X}^a\partial_a + \widetilde{X}^{\bar{a}}\partial_{\bar{a}}$, $\widetilde{Y} = \widetilde{Y}^a\partial_a + \widetilde{Y}^{\bar{a}}\partial_{\bar{a}}$, $\widetilde{Z} = \widetilde{Z}^a\partial_a + \widetilde{Z}^{\bar{a}}\partial_{\bar{a}} \in \mathfrak{T}^1_0(TM)$. Therefore (h^H, C^H) is a statistical structure on TM.

Lemma 3.2. *Let ∇^H be the compatible connection with respect to (h^H, C^H) on TM. Then we have*

$$\begin{aligned} &\nabla^H_{E_j} E_i = (\Gamma_{ji}{}^a + \frac{1}{2}S_{ji}{}^a)E_a - y^b K^a{}_{ibj} E_{\bar{a}}, \quad \nabla^H_{E_{\bar{j}}} E_{\bar{i}} = 0, \\ &\nabla^H_{E_j} E_{\bar{i}} = (\Gamma_{ji}{}^a - \frac{1}{2}S_{ji}{}^a)E_{\bar{a}}, \quad \nabla^H_{E_{\bar{j}}} E_i = -\frac{1}{2}S_{ji}{}^a E_{\bar{a}}. \end{aligned} \tag{11}$$

We call this structure $(h^H,\ \nabla^H)$ the *horizontal lift statistical structure.*

Remark 3.1. This connection ∇^H is different from the usual horizontal lift connection of ∇. This ∇^H is symmetric connection on TM and coincides with the usual horizontal lift connection if and only if $S = 0$ and $K = 0$.

4. Conformal-projective flatness of TM with $(h^H,\ C^H)$

Theorem 1.3. *Let $(M,\ h,\ \nabla)$ be a statistical manifold of dimension $n(\geq 2)$. If $(TM,\ h^H,\ C^H)$ is conformally-projectively flat, then $(M,\ h,\ \nabla)$ is of constant curvature. In particular, if M is non-flat, then $(h,\ \nabla)$ is a semi-Riemannian structure (i.e., ∇ is the Levi-Civita connection of h).*

Proof. There exists a flat statistical structure $(\widetilde{h},\ \widetilde{\nabla})$ on TM which is conformally-projectively equivalent to $(h^H,\ C^H)$. Then, using Lemma 3.2, we have

$$\begin{aligned} &\widetilde{\nabla}_{E_j} E_i = (\Gamma_{ji}{}^a + \frac{1}{2}S_{ji}{}^a + \delta^a_j\widetilde{\psi}_i + \delta^a_i\widetilde{\psi}_j)E_a - y^b K^a{}_{ibj} E_{\bar{a}}, \\ &\widetilde{\nabla}_{E_j} E_{\bar{i}} = (-h_{ji}\widetilde{\varphi}^a + \delta^a_j\widetilde{\psi}_{\bar{i}})E_a + (\Gamma_{ji}{}^a - \frac{1}{2}S_{ji}{}^a - h_{ji}\widetilde{\varphi}^{\bar{a}} + \delta^a_i\widetilde{\psi}_j)E_{\bar{a}}, \\ &\widetilde{\nabla}_{E_{\bar{j}}} E_i = (-h_{ji}\widetilde{\varphi}^a + \delta^a_i\widetilde{\psi}_{\bar{j}})E_a + (-\frac{1}{2}S_{ji}{}^a - h_{ji}\widetilde{\varphi}^{\bar{a}} + \delta^a_j\widetilde{\psi}_i)E_{\bar{a}}, \\ &\widetilde{\nabla}_{E_{\bar{j}}} E_{\bar{i}} = \left(\delta^a_j\widetilde{\psi}_{\bar{i}} + \delta^a_i\widetilde{\psi}_{\bar{j}}\right)E_{\bar{a}}. \end{aligned} \tag{12}$$

Let $\widetilde{K}$ be the curvature tensor of $\widetilde{\nabla}$. Using (12), we obtain (13), $\cdots$, (18).

$$\begin{aligned}0 &= \widetilde{K}(E_k,\ E_j)E_i \\ &= \Big\{K_{kji}{}^a + \frac{1}{2}(\nabla_k S_{ji}{}^a - \nabla_j S_{ki}{}^a) + \frac{1}{4}(S_{ji}{}^b S_{kb}{}^a - S_{ki}{}^b S_{jb}{}^a) \\ &+ \delta_j^a \alpha_{ki} - \delta_k^a \alpha_{ji}\Big\} E_a + \Big\{y^b(\nabla_j K^a{}_{ibk} - \nabla_k K^a{}_{ibj}) \\ &- \frac{1}{2} y^b (S_{ji}{}^c K^a{}_{cbk} - S_{ki}{}^c K^a{}_{cbj} + S_{jc}{}^a K^c{}_{ibk} \\ &- S_{kc}{}^a K^c{}_{ibj} + S_{ic}{}^a K_{kjb}{}^c)\Big\} E_{\bar{a}}.\end{aligned} \tag{13}$$

$$\begin{aligned}0 &= \widetilde{K}(E_k,\ E_j)E_{\bar{i}} \\ &= \Big\{\delta_j^a \beta_{ki} - \delta_k^a \beta_{ji} - h_{ji} B_k{}^a + h_{ki} B_j{}^a + (\widetilde{\psi}_A \widetilde{\varphi}^A)(\delta_j^a h_{ki} - \delta_k^a h_{ji})\Big\} E_a \\ &+ \Big\{K_{kji}{}^a - h_{ji} A_k{}^a + h_{ki} A_j{}^a \\ &- \frac{1}{2}(\nabla_k S_{ji}{}^a - \nabla_j S_{ki}{}^a) + \frac{1}{4}(S_{ji}{}^b S_{kb}{}^a - S_{ki}{}^b S_{jb}{}^a)\Big\} E_{\bar{a}}.\end{aligned} \tag{14}$$

$$\begin{aligned}0 &= \widetilde{K}(E_k,\ E_{\bar{j}})E_i \\ &= \Big\{ - \delta_k^a \beta_{ij} - h_{ji} B_k{}^a - (\widetilde{\psi}_A \widetilde{\varphi}^A) \delta_k^a h_{ji}\Big\} E_a \\ &+ \Big\{K^a{}_{ijk} - \frac{1}{2}\nabla_k S_{ji}{}^a - \frac{1}{4}(S_{ji}{}^b S_{kb}{}^a + S_{ki}{}^b S_{jb}{}^a) + \delta_j^a \alpha_{ki} - h_{ji} A_k{}^a\Big\} E_{\bar{a}}.\end{aligned} \tag{15}$$

$$\begin{aligned}0 &= \widetilde{K}(E_{\bar{k}},\ E_{\bar{j}})E_i \\ &= \Big\{h_{ki}(\partial_{\bar{j}} \widetilde{\varphi}^a - \widetilde{\varphi}_{\bar{j}} \widetilde{\varphi}^a) - h_{ji}(\partial_{\bar{k}} \widetilde{\varphi}^a - \widetilde{\varphi}_{\bar{k}} \widetilde{\varphi}^a)\Big\} E_a \\ &+ \Big\{\delta_j^a \beta_{ik} - \delta_k^a \beta_{ij} - h_{ji} C_k{}^a + h_{ki} C_j{}^a + (\widetilde{\psi}_A \widetilde{\varphi}^A)(\delta_j^a h_{ki} - \delta_k^a h_{ji})\Big\} E_{\bar{a}}.\end{aligned} \tag{16}$$

$$\begin{aligned}0 &= \widetilde{K}(E_k,\ E_{\bar{j}})E_{\bar{i}} \\ &= \Big\{h_{ki}(\partial_{\bar{j}} \widetilde{\varphi}^a - \widetilde{\varphi}_{\bar{j}} \widetilde{\varphi}^a) - \delta_k^a (\partial_{\bar{j}} \widetilde{\psi}_{\bar{i}} - \widetilde{\psi}_{\bar{j}} \widetilde{\psi}_{\bar{i}})\Big\} E_a \\ &+ \Big\{\delta_j^a \beta_{ki} + h_{ki} C_j{}^a + (\widetilde{\psi}_A \widetilde{\varphi}^A) \delta_j^a h_{ki}\Big\} E_{\bar{a}}.\end{aligned} \tag{17}$$

$$0 = \widetilde{K}(E_{\bar{k}},\ E_{\bar{j}})E_{\bar{i}} = \Big\{\delta_j^a (\partial_{\bar{k}} \widetilde{\psi}_{\bar{i}} - \widetilde{\psi}_{\bar{k}} \widetilde{\psi}_{\bar{i}}) - \delta_k^a (\partial_{\bar{j}} \widetilde{\psi}_{\bar{i}} - \widetilde{\psi}_{\bar{j}} \widetilde{\psi}_{\bar{i}})\Big\} E_{\bar{a}}. \tag{18}$$

Here we put as follows:

$$
\begin{aligned}
\alpha_{ji} :=& E_j\widetilde{\psi}_i - \widetilde{\psi}_j\widetilde{\psi}_i - \Gamma_{ji}{}^c\widetilde{\psi}_c - \frac{1}{2}S_{ji}{}^c\widetilde{\psi}_c + y^b K^c{}_{ibj}\widetilde{\psi}_{\bar{c}}.\\
\beta_{ji} :=& E_j\widetilde{\psi}_{\bar{i}} - \widetilde{\psi}_j\widetilde{\psi}_{\bar{i}} - \Gamma_{ji}{}^c\widetilde{\psi}_{\bar{c}} + \frac{1}{2}S_{ji}{}^c\widetilde{\psi}_{\bar{c}}\\
=& \partial_{\bar{i}}\widetilde{\psi}_j - \widetilde{\psi}_j\widetilde{\psi}_{\bar{i}} + \frac{1}{2}S_{ji}{}^c\widetilde{\psi}_{\bar{c}}.\\
A_j{}^h :=& E_j\widetilde{\varphi}^{\bar{h}} - \widetilde{\varphi}_j\widetilde{\varphi}^{\bar{h}} + \Gamma_{jc}{}^h\widetilde{\varphi}^{\bar{c}} - \frac{1}{2}S_{jc}{}^h\widetilde{\varphi}^{\bar{c}} - y^b K^h{}_{cbj}\widetilde{\varphi}^c.\\
B_j{}^h :=& E_j\widetilde{\varphi}^h - \widetilde{\varphi}_j\widetilde{\varphi}^h + \Gamma_{jc}{}^h\widetilde{\varphi}^c + \frac{1}{2}S_{jc}{}^h\widetilde{\varphi}^c.\\
C_j{}^h :=& \partial_{\bar{j}}\widetilde{\varphi}^{\bar{h}} - \widetilde{\varphi}_{\bar{j}}\widetilde{\varphi}^{\bar{h}} - \frac{1}{2}S_{jc}{}^h\widetilde{\varphi}^c.
\end{aligned}
\tag{19}
$$

From the definition of α_{ji}, we have

$$
\begin{aligned}
\alpha_{ji} - \alpha_{ij} &= E_j\widetilde{\psi}_i - E_i\widetilde{\psi}_j - y^b\big(K^a{}_{jbi}\widetilde{\psi}_{\bar{a}} - K^a{}_{ibj}\widetilde{\psi}_{\bar{a}}\big)\\
&= y^b\big(K_{ijb}{}^a\widetilde{\psi}_{\bar{a}} - K^a{}_{jbi}\widetilde{\psi}_{\bar{a}} + K^a{}_{ibj}\widetilde{\psi}_{\bar{a}}\big)\\
&= 0.
\end{aligned}
\tag{20}
$$

We put $A_{ji} := A_j{}^a h_{ai}$. Then we have also

$$
A_{ji} - A_{ij} = y^b\big(K^a{}_{ibj} - K^a{}_{jbi} - K_{jib}{}^a\big)\widetilde{\varphi}_{\bar{a}} = 0. \tag{21}
$$

From (15), we have

$$
K_{hijk} = \frac{1}{2}\nabla_k S_{jih} - \frac{1}{4}\big(3S_{ji}{}^a S_{kha} + S_{ki}{}^a S_{jha}\big) + A_{kh}h_{ji} - \alpha_{ki}h_{jh}, \tag{22}
$$

from which we have

$$
\begin{aligned}
0 &= K_{hijk} + K_{ihjk}\\
&= \nabla_k S_{jih} - \big(S_{ji}{}^a S_{kha} + S_{ki}{}^a S_{jha}\big)\\
&\quad + \big(A_{kh} - \alpha_{kh}\big)h_{ji} + \big(A_{ki} - \alpha_{ki}\big)h_{jh}.
\end{aligned}
\tag{23}
$$

Applying the first Bianchi identity to (22), we get

$$
\begin{aligned}
0 &= K_{hijk} + K_{ijhk} + K_{jhik}\\
&= \big(A_{kh} - \alpha_{kh}\big)h_{ji} + \big(A_{ki} - \alpha_{ki}\big)h_{jh} + \big(A_{kj} - \alpha_{kj}\big)h_{ih}\\
&\quad + \frac{3}{2}\nabla_k S_{jih} - \big(S_{ji}{}^a S_{kha} + S_{jh}{}^a S_{kia} + S_{ih}{}^a S_{kja}\big).
\end{aligned}
\tag{24}
$$

Comparing (23) with (24), we have

$$
\begin{aligned}
\frac{1}{2}\nabla_k S_{jih} &= S_{ji}{}^a S_{kha} - \big(A_{kh} - \alpha_{kh}\big)h_{ji}\\
&= S_{jh}{}^a S_{kia} - \big(A_{ki} - \alpha_{ki}\big)h_{jh}\\
&= S_{ih}{}^a S_{kja} - \big(A_{kj} - \alpha_{kj}\big)h_{ih},
\end{aligned}
\tag{25}
$$

from which we get

$$(n-1)(A_{ji}-\alpha_{ji}) = S_{ji}{}^{a}S_{ab}{}^{b} - S_{jb}{}^{a}S_{ia}{}^{b} \tag{26}$$

and

$$A_{ji}-\alpha_{ji} = (A_{a}{}^{a}-\alpha_{a}{}^{a})h_{ji} - (S_{ji}{}^{a}S_{ab}{}^{b} - S_{jb}{}^{a}S_{ia}{}^{b}). \tag{27}$$

Using (26) and (27), we have

$$n(A_{ji}-\alpha_{ji}) = (A_{a}{}^{a}-\alpha_{a}{}^{a})h_{ji} \tag{28}$$

and

$$S_{ji}{}^{a}S_{ab}{}^{b} - S_{jb}{}^{a}S_{ia}{}^{b} = \frac{n-1}{n}(A_{a}{}^{a}-\alpha_{a}{}^{a})h_{ji}. \tag{29}$$

From (13) we have

$$\begin{aligned} K_{kjih} &= -\frac{1}{2}(\nabla_k S_{jih} - \nabla_j S_{kih}) + \frac{1}{4}(S_{ji}{}^{a}S_{kha} - S_{ki}{}^{a}S_{jha}) \\ &\quad + \alpha_{ji}h_{kh} - \alpha_{ki}h_{jh}. \end{aligned} \tag{30}$$

On the other hand, using (21) and (22), we have

$$\begin{aligned} K_{kjih} &= K_{khij} - K_{jhik} \\ &= -\frac{1}{2}(\nabla_k S_{jih} - \nabla_j S_{kih}) + \frac{1}{4}(S_{ji}{}^{a}S_{kha} - S_{ki}{}^{a}S_{jha}) \\ &\quad + \alpha_{kh}h_{ji} - \alpha_{jh}h_{ki}, \end{aligned} \tag{31}$$

Comparing (30) with (31), we get

$$\alpha_{ji}h_{kh} - \alpha_{ki}h_{jh} = \alpha_{kh}h_{ji} - \alpha_{jh}h_{ki},$$

from which we have

$$\alpha_{ji} = \frac{1}{n}\alpha_{a}{}^{a}h_{ji}. \tag{32}$$

Substituting (32) into (30), we get

$$\begin{aligned} K_{kjih} =& \frac{1}{n}\alpha_{a}{}^{a}(h_{kh}h_{ji} - h_{ki}h_{jh}) - \frac{1}{2}(\nabla_k S_{jih} - \nabla_j S_{kih}) \\ &+ \frac{1}{4}(S_{ji}{}^{a}S_{kha} - S_{ki}{}^{a}S_{jha}). \end{aligned} \tag{33}$$

Substituting (33) into (25), we obtain

$$A_{ji} = \frac{1}{n}A_{a}{}^{a}h_{ji}. \tag{34}$$

Next, from (14), we have

$$\begin{aligned} K_{kjih} &= A_{kh}h_{ji} - A_{jh}h_{ki} + \frac{1}{2}(\nabla_k S_{jih} - \nabla_j S_{kih}) \\ &\quad - \frac{3}{4}(S_{ji}{}^a S_{kha} - S_{ki}{}^a S_{jha}). \end{aligned} \tag{35}$$

Substituting (34) into (35), we get

$$\begin{aligned} K_{kjih} &= \frac{1}{n}A_a{}^a(h_{kh}h_{ji} - h_{ki}h_{jh}) + \frac{1}{2}(\nabla_k S_{jih} - \nabla_j S_{kih}) \\ &\quad - \frac{3}{4}(S_{ji}{}^a S_{kha} - S_{ki}{}^a S_{jha}). \end{aligned} \tag{36}$$

From (34) and (36), we obtain

$$2K_{kjih} = \frac{1}{n}(A_a{}^a + \alpha_a{}^a)(h_{kh}h_{ji} - h_{ki}h_{jh}) - \frac{1}{2}(S_{ji}{}^a S_{kha} - S_{ki}{}^a S_{jha}), \tag{37}$$

from which

$$K_{kjih} = -K_{kjhi}. \tag{38}$$

Using (33), (36) and (38), we obtain

$$\begin{aligned} 0 &= K_{kjih} + K_{kjhi} \\ &= -\nabla_k S_{jih} + \nabla_j S_{kih} \\ &= \frac{1}{n}(A_a{}^a - \alpha_a{}^a)(h_{kh}h_{ji} - h_{ki}h_{jh}) - S_{ji}{}^a S_{kha} + S_{ki}{}^a S_{jha}, \end{aligned}$$

from which

$$S_{ji}{}^a S_{kha} - S_{ki}{}^a S_{jha} = \frac{1}{n}(A_a{}^a - \alpha_a{}^a)(h_{kh}h_{ji} - h_{ki}h_{jh}). \tag{39}$$

Substituting (39) into (37), we obtain

$$K_{kjih} = \frac{1}{4n}(A_a{}^a + 3\alpha_a{}^a)(h_{kh}h_{ji} - h_{ki}h_{jh}),$$

from which

$$K_{kjih} = \frac{\rho}{n(n-1)}(h_{kh}h_{ji} - h_{ki}h_{jh}) \tag{40}$$

and

$$\rho = \frac{n-1}{4}(A_a{}^a + 3\alpha_a{}^a), \tag{41}$$

where ρ denotes the scalar curvature of K.

Lastly we prove that ρ is constant. From (13) we have

$$\begin{aligned} &\nabla_k K^h{}_{ilj} - \nabla_j K^h{}_{ilk} \\ &= -\frac{1}{2}(S_{ji}{}^a K^h{}_{alk} - S_{ki}{}^a K^h{}_{alj} + S_{ja}{}^h K^a{}_{ilk} - S_{ka}{}^h K^a{}_{ilj} + S_{ia}{}^h K_{kjl}{}^a). \end{aligned} \tag{42}$$

Substituting (40) into (42) and contracting k and h, we have

$$(n-2)(\nabla_j\rho)h_{li} + (\nabla_l\rho)h_{ji} = \frac{1}{2}(S_{la}{}^a h_{ji} + S_{ia}{}^a h_{lj} - (n+1)S_{lji}). \quad (43)$$

Transvecting h^{ji} to (43), we get

$$\nabla_l\rho = 0. \quad (44)$$

Therefore ρ is constant Transvecting h^{li} to (43) and using (44), we get

$$\rho S_{ja}{}^a = -2(n-1)\nabla_j\rho = 0, \quad (45)$$

Substituting (44) and (45) into (43), we obtain

$$\rho S_{kji} = 0. \quad (46)$$

Therefore, if $\rho \neq 0$ (i.e., ∇ is non-flat connection), then $\nabla = \nabla^0$. □

Corollary 1.1. *Let $(M,\ h,\ \nabla)$ be a statistical manifold of dimension $n(\geq 2)$. If $(TM,\ h^H,\ C^H)$ is $\lambda(\neq 0)$-conformally flat, then $(M,\ h,\ \nabla)$ and $(TM,\ h^H,\ C^H)$ are locally flat.*

Proof. First, we prove $(M,\ h,\ \nabla)$ is locally flat. By virtue of Theorem 1.3, it is good that we consider only the case $C = 0$. If $\lambda \neq \pm 1, 0$, then, using (16) and (18), there exists a scalar function φ on M such that

$$\alpha_{ji} = \frac{1-\lambda}{2}(\nabla_j\varphi_i - \frac{1-\lambda}{2}\varphi_j\varphi_i) \quad (47)$$

and

$$A_{ji} = \frac{1+\lambda}{2}(\nabla_j\varphi_i - \frac{1+\lambda}{2}\varphi_j\varphi_i). \quad (48)$$

From (26), (47) and (48) we have

$$0 = A_{ji} - \alpha_{ji} = \lambda(\nabla_j\varphi_i - \varphi_j\varphi_i),$$

from which

$$0 = \nabla_k\nabla_j\varphi_i - \nabla_j\nabla_k\varphi_i = -K_{kji}{}^a\varphi_a = -\frac{\rho}{n(n-1)}(\varphi_k h_{ji} - \varphi_j h_{ki}),$$

i.e.,

$$\rho\varphi_j = 0. \quad (49)$$

If $\lambda = \pm 1$ then, using (39) and (41), $\rho = \frac{n-1}{4}A_a{}^a$(resp., $= \frac{3(n-1)}{4}\alpha_a{}^a) = 0$ according as $(TM,\ h^H,\ C^H)$ is 1-conformally flat (resp., (-1)-conformally flat). Therefore $(M,\ h,\ \nabla)$ is locally flat.

Next, let $\widetilde{K}$ be the curvature tensor of ∇^H and $\widetilde{K}_{DCB}{}^A$ the components of $\widetilde{K}$ with respect to the adapted frame. Then we have

$$
\begin{aligned}
\widetilde{K}_{k\bar{j}i}{}^{\bar{h}} &= -\widetilde{K}_{\bar{j}ki}{}^{\bar{h}} = K^h{}_{ijk} - \frac{1}{2}\nabla_k S_{ji}{}^h + \frac{1}{4}\left(S_{ji}{}^a S_{ka}{}^h + S_{ki}{}^a S_{ja}{}^h\right).\\
\widetilde{K}_{kj\bar{i}}{}^{\bar{h}} &= K_{kji}{}^h - \frac{1}{2}\left(\nabla_k S_{ji}{}^h - \nabla_j S_{ki}{}^h\right) + \frac{1}{4}\left(S_{ji}{}^a S_{ka}{}^h - S_{ki}{}^a S_{ja}{}^h\right).\\
\widetilde{K}_{kji}{}^h &= K_{kji}{}^h + \frac{1}{2}\left(\nabla_k S_{ji}{}^h - \nabla_j S_{ki}{}^h\right) + \frac{1}{4}\left(S_{ji}{}^a S_{ka}{}^h - S_{ki}{}^a S_{ja}{}^h\right). \qquad (50)\\
\widetilde{K}_{kji}{}^{\bar{h}} &= y^b\Big\{\left(\nabla_j K^h{}_{ibk} - \nabla_k K^h{}_{ibj}\right)\\
&\quad - \frac{1}{2}\left(S_{ji}{}^c K^h{}_{cbk} - S_{ki}{}^c K^h{}_{cbj} + S_{jc}{}^h K^c{}_{ibk} - S_{kc}{}^h K^c{}_{ibj} + S_{ic}{}^h K_{kjb}{}^c\right)\Big\}.
\end{aligned}
$$

Other components of $\widetilde{K}$ vanish. On the other hand, from (25), we have

$$\nabla_k S_{ji}{}^h = S_{ji}{}^a S_{ka}{}^h = S_{ki}{}^a S_{ja}{}^h = S_{kj}{}^a S_{ia}{}^h,$$

from which

$$\nabla_k S_{ji}{}^h = \frac{1}{2}\left(S_{ji}{}^a S_{ka}{}^h + S_{ki}{}^a S_{ja}{}^h\right) \qquad (51)$$

and

$$\nabla_k S_{ji}{}^h - \nabla_j S_{ki}{}^h = \frac{1}{2}\left(S_{ji}{}^a S_{ka}{}^h - S_{ki}{}^a S_{ja}{}^h\right) = 0. \qquad (52)$$

Therefore we have $\widetilde{K} = 0$. □

Corollary 1.2. *Let* $(M,\ h)$ *be a semi-Riemannian manifold of dimension* $n(\geq 2)$. *Then* $(TM,\ h^H)$ *is conformally flat if and only if* $(M,\ h,\ \nabla)$ *is of constant curvature.*

Proof. Assume that $(M,\ h)$ is of constant curvature. Let $\widetilde{K}$ the Riemannian curvature tensor of h^H and $\widetilde{K}_{DCB}{}^A$ the components of $\widetilde{K}$ with respect to the adapted frame. Then we have

$$\widetilde{K}_{kji}{}^{\bar{h}} = y^a\left(\nabla_k K_{aji}{}^h - \nabla_j K_{aki}{}^h\right) \qquad (53)$$

and

$$\widetilde{K}_{kji}{}^h = \widetilde{K}_{kj\bar{i}}{}^{\bar{h}} = \widetilde{K}_{\bar{k}ji}{}^{\bar{h}} = \widetilde{K}_{k\bar{j}i}{}^{\bar{h}} = K_{kji}{}^h. \qquad (54)$$

Other components of $\widetilde{K}$ varnish. Then the Ricci tensor $\widetilde{R}$ has the following components:

$$\widetilde{R}_{ji} = 2R_{ji}, \quad \widetilde{R}_{\bar{j}i} = \widetilde{R}_{j\bar{i}} = \widetilde{R}_{\bar{j}\bar{i}} = 0. \qquad (55)$$

The scalar curvature $\widetilde{\rho}$ of $(TM,\ h^H)$ vanishes. Therefore, if M is of constant curvature, then the Weyl conformal curvature tensor of $(TM,\ h^H)$ vanishes, i.e., $(TM,\ h^H)$ is conformally flat. □

References

1. S. Amari and H. Nagaoka, *Methods of Information Geometry* (Amer. Math. Soc., Oxford Univ. Press, 2000).
2. H. Furuhata, H. Matsuzoe and H. Urakawa, Open problems in affine differential geometry and related topics, *Interdisciplinary Information Sciences* **4** (2) (1998) 125–127.
3. I. Hasegawa and K. Yamauchi, λ-conformal flatness of tangent bundle with complete lift statistical structure, *Journal of Hokkaido University of Education (Natural Sciences)* **58** (1) (2007) 1–14.
4. I. Hasegawa and K. Yamauchi, Conformal-projective flatness of tangent bundle with complete lift statistical structure, preprint.
5. S. Ivanov, On dual-projectively flat affine connections, *Journal of Geometry* **53** (1995) 89–99.
6. T. Kurose, On the realization of statistical manifolds in affine space, preprint (1991).
7. S. Lauritzen, Statistical manifolds, Differential Geometry in Statistical Inference, *IMS Lecture Notes Monographs, Hayward California,* Series **10** (1987) 96–163.
8. H. Matsuzoe, On realization of conformally-projectively flat statistical manifolds and the divergences, *Hokkaido Math. J.* **27** (1998) 409–421.
9. H. Matsuzoe and J. Inoguchi, Statistical structures on Tangent bundles, *Applied Sciences* **5** (1) (2003) 55–75.
10. C. Murathan and I. Güney, Vertical and complete lifts on statistical manifolds and on the univariate Gaussian manifolds, *Comm. Fac. Sci. Univ. Ankara Ser. A1 Math. Statist.* **42** (1993) 69–76.
11. T.J. Willmore, Riemann extensions and affine differential geometry, *Results in Math.* **13** (1988) 403–408.
12. K. Yano and S. Ishihara, *Tangent and Cotangent Bundles* (Marcel Dekker, 1973).
13. K. Yano and S. Kobayashi, Prolongations of tensor fields and connections to tangent bundles I, II, III, *J. Math. Soc. Japan* **18** (1966) 194–210; 236–246; **19** (1967) 486–488.

Differential Geometry and its Applications
Proc. Conf., in Honour of Leonhard Euler, Olomouc, August 2007

Morphisms of almost product projective geometries

Jaroslav Hrdina and Jan Slovák

Institute of Mathematics and Statistics, Faculty of Science, Masaryk University, Janáčkovo nám. 2a, 662 95 Brno, Czech Republic
E-mail: jaroslavhrdina@gmail.com, slovak@math.muni.cz

We discuss almost product projective geometry and the relations to a distinguished class of curves. Our approach is based on an observation that well known general techniques[2,5,8] apply, and our goal is to illustrate the power of the general parabolic geometry theory on a quite explicit example. Therefore, some rudiments of the general theory are mentioned on the way, too.

Keywords: Planar curves, almost product projective geometry, parabolic geometry, generalized geodetics.

MS classification: 53C10, 53C22, 53A20.

1. An almost product projective structure

Let M be a smooth manifold of dimension $2m$. *An almost product structure* on M is a smooth trace–free affinor J in $\Gamma(T^\star M \otimes TM)$ satisfying $J^2 = \mathrm{id}_{TM}$.

For better understanding, we describe an almost product structure at each tangent space in a fixed basis, i.e. with the help of real matrices:

$$J := \begin{pmatrix} \mathbb{I}_m & 0 \\ 0 & -\mathbb{I}_m \end{pmatrix}.$$

The eigenvalues of J have to be ± 1 and $T_xM = T_x^L M \oplus T_x^R M$, where the subspaces are of the form

$$T^L M := J_+ = \left\{ \begin{pmatrix} c \\ 0 \end{pmatrix} \mid c \in \mathbb{R}^m \right\}, \quad T^R M := J_- = \left\{ \begin{pmatrix} 0 \\ c \end{pmatrix} \mid c \in \mathbb{R}^m \right\}.$$

Thus, we can equivalently define an almost product structure J on M as a reduction of the linear frame bundle $P^1 M$ to the appropriate structure group, i.e. as a G–structure with the structure group L of all automorphisms preserving the affinor J:

$$L := \left\{ \begin{pmatrix} A & 0 \\ 0 & B \end{pmatrix} |A, B \in GL(m, \mathbb{R}) \right\} \cong GL(m, \mathbb{R}) \times GL(m, \mathbb{R}) \subset GL(2m, \mathbb{R}).$$

These G–structures are of infinite type, however each choice of a linear connection ∇ compatible with the affinor J, i.e. $\nabla J = 0$, determines a finite type geometry similar to products of projective structures, which we shall study below.

The difference Υ between two projectively equivalent connections is a smooth one–form given by

$$\hat{\nabla}_\xi \eta = \nabla_\xi \eta + \Upsilon(\xi)\eta + \Upsilon(\eta)\xi.$$

Instead, we shall consider a class of connections parameterized also by all smooth one–forms Υ, but with the transformation rule

$$\begin{aligned} \hat{\nabla}_{\xi^L+\xi^R}(\eta^L + \eta^R) = \nabla_{\xi^L+\xi^R}(\eta^L + \eta^R) + \Upsilon^L(\xi^L)\eta^L + \Upsilon^L(\eta^L)\xi^L \\ + \Upsilon^R(\xi^R)\eta^R + \Upsilon^R(\eta^R)\xi^R, \end{aligned} \tag{1}$$

where the indices at the forms and fields indicate the components in the subbundles T^LM and T^RM, respectively. Clearly such a transformed connection will make J parallel again.

Definition 1.1. Let M be a smooth manifold of dimension $2m$. *An almost product projective structure* on M is a couple $(J, [\nabla])$, where J is an almost product structure, ∇ is a linear connection preserving J, and $[\nabla]$ is the class of connections obtained from ∇ by the transformations given by all smooth one–forms Υ as in (1).

The standard tool for the study of G–structures is the classical prolongation theory. The first step usually provides a class of distinguished connections with minimized (or preferred) torsions. In our case, the torsion

$$T(\xi, \eta) = \nabla_\xi \eta - \nabla_\eta \xi - [\xi, \eta]$$

of ∇ will always involve the obstructions against the integrability of T^LM and T^RM, i.e. the appropriate projections of the antisymmetric term given by the Lie bracket of vector fields.

Our next goal is to identify a class of almost product projective geometries which fit into a wider class of the so called normal parabolic geometries. We shall see, that the appropriate requirement on the torsion of ∇ will be that the integrability obstructions are the only non–zero components.

Example 1.1 (The homogeneous model). Let us consider the homogeneous space $M = G/P$ given as the product of two projective spaces $G_L/P_L \times G_R/P_R$, i.e. $G_L = G_R = SL(n+1,\mathbb{R})$ while $P = P_L \times P_R$, where $P_R = P_L$ is the usual parabolic subgroup corresponding to the block upper triangular matrices of the block sizes $(1,n)$.

Clearly, any product connection on M built of the linear connections ∇^L and ∇^R from the two projective classes on the product components provides a homogeneous example of an almost product projective structure. At the same time, the Maurer–Cartan form on $G = G_L \times G_R$ provides the homogeneous model of the $|1|$-graded parabolic geometries of type (G,P). The Lie algebra of $P \times P$ is a parabolic subalgebra of the real form $\mathfrak{sl}(n+1,\mathbb{R}) \oplus \mathfrak{sl}(n+1,\mathbb{R})$ of the complex algebra $\mathfrak{sl}(n+1,\mathbb{C}) \oplus \mathfrak{sl}(n+1,\mathbb{C})$.

In matrix form, we can illustrate the grading from our example as:

$$\mathfrak{g} = \begin{pmatrix} \mathfrak{g}_0^L & \mathfrak{g}_1^L & 0 & 0 \\ \mathfrak{g}_{-1}^L & \mathfrak{g}_0^L & 0 & 0 \\ 0 & 0 & \mathfrak{g}_0^R & \mathfrak{g}_1^R \\ 0 & 0 & \mathfrak{g}_{-1}^R & \mathfrak{g}_0^R \end{pmatrix}$$

2. Parabolic geometries and Weyl connections

The classical prolongation procedure for G–structures starts with finding a minimal available torsion for connections belonging to the structure on the given manifold M. Our normalization will come from the general theory of parabolic geometries.

As a rule, the $|1|$–graded parabolic geometries are completely given by certain classical G–structures on the underlying manifolds.[1,2] In our case, however, both components of the semisimple Lie algebra belong to the series of exceptions and only the choice of an appropriate class of connections defines the Cartan geometry completely.[1,2]

The normalization of the Cartan geometries is based on cohomological interpretation of the curvature. More explicitly, the normal geometries enjoy co–closed curvature.[1,2]

In our case, the appropriate cohomology is easily computed by the Künneth formula from the classical Kostant's formulae and the computation[10] provides all six irreducible components of the curvature, only two of which are of torsion type. Of course, the integrability obstructions of the bundles T^LM and T^RM are just those two. Therefore, a *normal almost product projective structure* $(M, J, [\nabla])$ has this minimal torsion.

Now, the general theory provides for each normal almost product projective structure $(M, J, [\nabla])$ the construction of the unique principal bundle $\mathcal{G} \to M$ with structure group P, equipped by the normal Cartan connection. Furthermore, there is the distinguished class of the so called *Weyl connections* corresponding to all choices of reductions of the parabolic structure group to its reductive subgroup G_0. All Weyl connections are parametrized just by smooth one–forms and they all share the torsion of the Cartan connection.

The transformation formulae for the Weyl connections are generally given by the Lie bracket in the algebra in question. Of course, this is just the formula (1) we used for the definition of the almost product projective structures. Let us express (1) as:[2]

$$\hat{\nabla}_X Y = \nabla_X Y + [[X, \Upsilon], Y] \tag{2}$$

where we use the frame forms $X, Y : \mathcal{G} \to \mathfrak{g}_{-1}$ of vector fields, and similarly for $\Upsilon : \mathcal{G} \to \mathfrak{g}_1$. Consequently, $[\Upsilon, X]$ is a frame form of an affinor valued in $\mathfrak{g}_0$ and the bracket with Y expresses the action of such an affinor on the differentiated field. According to the general theory, this transformation rule works for all covariant derivatives ∇ with respect to Weyl connections.[2]

The general theory of the $|1|$–graded geometries also provides a formula for the unique normal Cartan connection ω in terms of any chosen Weyl connection and its curvature. Technically, this formula computes the difference between the two connections as the so called Schouten Rho tensor $\mathrm{P} = -\Box^{-1}\partial^* R$, where R is the curvature of the Weyl connection, ∂^* is the Kostant's codifferential, and $\Box^{-1}$ is the inverse of the Kostant's Laplacian.[2]

This observation shows that our normal almost product projective geometries form a category equivalent to normal Cartan connections of type (G, P) with homogeneous model discussed in Example 1.1.

3. *J*–planar curves

Let us remind the notion of planarity with respect to affinors:[8]

Definition 3.1. Let (M, J) be an almost product structure. A smooth curve $c : \mathbb{R} \to M$ is called J–planar with respect to a linear connection ∇ if $\nabla_{\dot{c}}\dot{c} \in \langle \dot{c}, J(\dot{c})\rangle$, where $\dot{c}$ means the tangent velocity field along the curve c and the brackets indicate the linear hulls of the two vectors in the individual tangent spaces.

Next, we observe that there is a nice link between J–planar curves and connections from the class defining an almost product projective structure:

Theorem 3.1. *Let $(M, J, [\nabla])$ be a smooth almost product projective structure on a manifold M A curve c is J–planar with respect to at least one Weyl connection $\bar{\nabla}$ on M if and only if there is a parametrization of c which is a geodesic trajectory of some Weyl connection ∇. Moreover, this happens if and only if c is J–planar with respect to all Weyl connections.*

Proof. First, let us compute the bracket $[[\dot{c}, \Upsilon], \dot{c}]$ appearing in (2). We write $\dot{c} = c^L + c^R$ and similarly for Υ:

$$\left[\left[\begin{pmatrix} 0 & 0 & 0 & 0 \\ c^L & 0 & 0 & 0 \\ 0 & 0 & 0 & 0 \\ 0 & 0 & c^R & 0 \end{pmatrix}, \begin{pmatrix} 0 & \Upsilon^L & 0 & 0 \\ 0 & 0 & 0 & 0 \\ 0 & 0 & 0 & \Upsilon^R \\ 0 & 0 & 0 & 0 \end{pmatrix}\right], \begin{pmatrix} 0 & 0 & 0 & 0 \\ c^L & 0 & 0 & 0 \\ 0 & 0 & 0 & 0 \\ 0 & 0 & c^R & 0 \end{pmatrix}\right] = \begin{pmatrix} 0 & 0 & 0 & 0 \\ c^L\Upsilon^L(c^L) & 0 & 0 & 0 \\ 0 & 0 & 0 & 0 \\ 0 & 0 & \Upsilon^R(c^R)c^R & 0 \end{pmatrix}.$$

We may write shortly $[[\dot{c}, \Upsilon], \dot{c}] = c^L\Upsilon^L(c^L) + \Upsilon^R(c^R)c^R$.

Now, suppose $c : \mathbb{R} \to M$ is a geodetics with respect to a connection ∇ and compute:

$$\begin{aligned}\hat{\nabla}_{\dot{c}}\dot{c} &= \nabla_{\dot{c}}\dot{c} + [[\dot{c}, \Upsilon], \dot{c}] = [[\dot{c}, \Upsilon], \dot{c}] = c^L\Upsilon^L(c^L) + \Upsilon^R(c^R)c^R = \\ &= (\Upsilon^L(c^L) + \Upsilon^R(c^R))\dot{c} + (\Upsilon^L(c^L) - \Upsilon^R(c^R))J(\dot{c}) \in \langle \dot{c}, J(\dot{c})\rangle.\end{aligned}$$

Thus, the geodetics c is J–planar with respect to connection $\hat{\nabla}$, i.e. with respect to all Weyl connections.

On the other hand, let us suppose that $c : \mathbb{R} \to M$ is J–planar with respect to $\bar{\nabla}$, i.e. $\bar{\nabla}_{\dot{c}}\dot{c} = a(\dot{c})\dot{c} + b(\dot{c})J(\dot{c})$ for some functions $a(\dot{c})$ and $b(\dot{c})$ along the curve. We have to find a one form $\Upsilon = \Upsilon^L + \Upsilon^R$ such that the formula for the transformed connection kills all the necessary terms along the curve c. Since there are many such forms Υ, there is a Weyl connection ∇ such that $\nabla_{\dot{c}}\dot{c} = 0$. □

4. Generalized planar curves and mappings

Definition 4.1 (Ref. 5). Let M be a smooth manifold of dimension n. Let A be a smooth l–rank ($l < n$) vector subbundle in $T^*M \otimes TM$, such that the identity affinor $E = id_{TM}$ restricted to T_xM belongs to $A_x \subset T^*M \otimes TM$ at each point $x \in M$. We say that M is equipped by ℓ–dimensional *A–structure*.

An almost product projective structure $(M, J, [\nabla])$ carries the A–structure with $A = \langle E, J\rangle$. Let us remind, that the A–planarity does not depend on the choice of the Weyl connection ∇ in the class in view of Theorem 3.1.

For any tangent vector $X \in T_xM$, we shall write $A(X)$ for the vector subspace

$$A(X) = \{F(X)|\, F \in A_xM\} \subset T_xM,$$

and we call $A(X)$ the *A-hull of the vector X*. In order to work out relations between morphisms of our geometries and planarity, we shall follow our earlier work.[5] We start by quoting a few definitions and results:

Definition 4.2. Let (M, A) be a smooth manifold M equipped with an ℓ–dimensional A–structure. We say that A–structure has

- *generic rank ℓ* if for each $x \in M$ the subset of vectors $(X, Y) \in T_xM \oplus T_xM$, such that the A–hulls $A(X)$ and $A(Y)$ generate a vector subspace $A(X) \oplus A(Y)$ of dimension 2ℓ is open and dense.
- *weak generic rank ℓ* if for each $x \in M$ the subset of vectors

$$\mathcal{V} := \{X \in T_xM|\,\dim A(X) = \ell\}$$

is open and dense in T_xM.

Lemma 4.1. *Every almost product structure (M, J) on a manifold M, $\dim M \geq 2$, has weak generic rank 2.*

Proof. Let as consider that $X \notin \mathcal{V}$, this fact implies that $\exists F \in A : F(X) = 0$, i.e. the vector X has to be an eigenvector of J, i.e. X has to belong to m–dimensional subspace T^LM or T^RM of TM. Finally, the complement $\mathcal{V}$ is open and dense. □

Theorem 4.1 (Ref. 4). *Let (M, A) be a smooth manifold of dimension n with ℓ–dimensional A–structure, such that $\ell \geqq 2\dim M$.*

- *If A_x is an algebra (i.e. for all $f, g \in A_x$, $fg := f \circ g \in A_x$) for all $x \in M$ and A has weak generic rank ℓ, then the structure has generic rank ℓ.*
- *If $A_x \subset T^\star_xM \otimes T_xM$ is an algebra with inversion then A has generic rank ℓ.*

Each almost product structure on a smooth manifold M has generic rank 2 because of lemma and theorem above.

Definition 4.3 (Refs. 5,8). Let M be a smooth manifold equipped with an A–structure and a linear connection ∇.

- A smooth curve C is told to be *A–planar* if there is trajectory $c : \mathbb{R} \to M$ such that $\nabla_{\dot{c}}\dot{c} \in A(\dot{c})$.
- Let $\bar{M}$ be another manifold with a linear connection $\bar{\nabla}$ and B–structure. A diffeomorphism $f : M \to \bar{M}$ is called *(A, B)–planar* f each A–planar curve C on M is mapped onto the B–planar curve $f_\star C$ on M. In the special case, where A is the trivial structure given by $\langle E\rangle$, we talk about B–planar maps.

Theorem 4.2 (Ref. 5). *Let M be a manifold with a linear connection ∇, let N be a manifold of the same dimension with a linear connection $\bar{\nabla}$ and with A-structure of generic rank ℓ, and suppose $dimM \geqq 2\ell$. Then a diffeomorphism $f : M \to N$ is a A-planar if and only if*

$$\mathrm{Sym}(f^*\bar{\nabla} - \nabla) \in f^*(A^{(1)}) \tag{3}$$

where Sym *denotes the symmetrization of the difference of the two connection.*

Theorem 4.3 (Ref. 5). *Let M be a manifold with linear connection ∇ and an A-structure, N be a manifold of the same dimension with a linear connection $\bar{\nabla}$ and B-structure with generic rank ℓ. Then a diffeomorphism $f : M \to N$ is (A, B)-planar if and only if f is B-planar and $A(X) \subset (f^\star(B))(X)$ for all $X \in TM$.*

Theorem 4.4 (Refs. 4,5). *Let (M, A), (M', A') be smooth manifolds of dimension m equipped with A–structure and A′–structure of the same generic rank $\ell \le 2m$ and assume that the A–structure satisfies the property*

$$\forall X \in T_xM,\ \forall F \in A, \exists c_X \mid \dot{c}_X = X,\ \nabla_{\dot{c}_X}\dot{c}_X = \beta(X)F(X), \tag{4}$$

where $\beta(X) \neq 0$. If $f : M \to M'$ is an (A, A')–planar mapping, then f is a morphism of the A–structures, i.e $f^*A' = A$.

Finally, we can apply the above concepts and theorems to our situation:

Theorem 4.5. *Let $f : M \to M'$ be a diffeomorphism between two almost product projective manifolds of dimension at least four. Then f is a morphism of the almost product structures if and only if it preserves the class of unparameterized geodesics of all Weyl connections on M and M'.*

Proof. In view of the series of theorems above and the fact that $f^*A = A$ implies $f^*J = \pm J$ (i.e. f^*J preserves the eigenspaces of J), we only have to prove that an almost product structure (M, J) has the property (4).

Consider $F = aE + bJ \in \langle E, F\rangle$ and $X \in TM$. First we may solve the system of equations:

$$a(X) + b(X) = 2\Upsilon^L(X^L)$$
$$a(X) - b(X) = 2\Upsilon^R(X^R)$$

with respect to Υ. Second, we may to define a new connection $\hat{\nabla}$, where:

$$\hat{\nabla}_Y Z = \nabla_Y Z - [[Y, \Upsilon], Z]$$

and we may find a geodetics c of $\hat{\nabla}$, such that $\dot{c} = X$. Finally, we recognize that c is the requested curve from (4) because:

$$\begin{aligned} \nabla_X X &= \bar{\nabla}_X X - [[X, \Upsilon], X] = [[X, \Upsilon], X] \\ &= (\Upsilon^L(X^L) + \Upsilon^R(X^R))X = a(X)X + b(X)J(X) = F(X). \end{aligned}$$ □

5. Almost complex projective structure

Another possibility of a Lie algebra whose complexification is $\mathfrak{sl}(n, \mathbb{C}) \oplus \mathfrak{sl}(n, \mathbb{C})$ is the complex algebra $\mathfrak{sl}(n, \mathbb{C})$ viewed as a real algebra. The corresponding geometry analogous to the almost product projective geometry is the almost complex projective geometry.

Definition 5.1. Let M be a smooth manifold of dimension $2m$. *An almost complex projective structure* on M is defined by a smooth affinor I on M satisfying $I^2 = -\mathrm{id}_{TM}$ and by a choice of a linear connection ∇ with $\nabla I = 0$.

In this case, the minimal torsion equals the Nijenhuis tensor obstructing the integrability of I and we may use the same technique as above to verify that this is the only component of the torsion of the normal almost complex projective geometry.

There has been a lot of interest on such geometries recently. All the above approach works equally well and we shall come to further discussion on these questions in the context of existing literature elsewhere.

Acknowledgments

Research supported by grants GACR 201/05/H005 and 201/05/2117.

References

1. A. Čap and J. Slovák, Parabolic geometries, Mathematical Surveys and Monographs, AMS Publishing House, to appear (2008).

2. A. Čap and J. Slovák, Weyl structures for parabolic geometries, *Math. Scand.* **93** (2003) 53–90.
3. J. Hrdina, $\mathbb{H}$-planar curves, *In: Diff. Geom. and its Appl.* (Proc. Conf. Prague, Czech Republic, 2004) 265–272.
4. J. Hrdina , Generalized planar curves and quaternionic geometry, Ph.D. thesis, Faculty of science MU Brno, 2007.
5. J. Hrdina and J. Slovák, Generalized planar curves and quaternionic geometry, *Global analysis and geometry* **29** (2006) 349–360.
6. S. Kobayashi, *Transformation groups in differential geometry* (Springer, 1972).
7. I. Kolář, P.W. Michor and J. Slovák, *Natural operations in differential geometry* (Springer, 1993); http://www.emis.de/monographs/KSM.
8. J. Mikeš and N.S. Sinyukov, On quasiplanar mappings of spaces of affine connection, *Sov. Math.* **27** (1983) (1) 63–70.
9. K. Nomizu and S. Kobayashi, *Foundations of Differential Geometry, vol. 1-2* (John Wiley and Sons, 1963).
10. J. Šilhan, Algorithmic computations of Lie algebras cohomologies, *In: Proceedings of the 22nd Winter School "Geometry and Physics"* (Srní, 2002, Rend. Circ. Mat. Palermo (2) Suppl. No. 71, 2003) 191–197; www.math.muni.cz/~silhan.

Differential Geometry and its Applications
Proc. Conf., in Honour of Leonhard Euler, Olomouc, August 2007

Flows of Spin(7)-structures

Spiro Karigiannis

Mathematical Institute, University of Oxford,
24-29 St Giles', Oxford, OX1 3LB, UK
E-mail: karigiannis@maths.ox.ac.uk
www.maths.ox.ac.uk/~karigiannis

We consider flows of Spin(7)-structures. We use local coordinates to describe the torsion tensor of a Spin(7)-structure and derive the evolution equations for a general flow of a Spin(7)-structure Φ on an 8-manifold M. Specifically, we compute the evolution of the metric and the torsion tensor. We also give an explicit description of the decomposition of the space of forms on a manifold with Spin(7)-structure, and derive an analogue of the second Bianchi identity in Spin(7)-geometry. This identity yields an explicit formula for the Ricci tensor and part of the Riemann curvature tensor in terms of the torsion.

Keywords: Spin(7)-structures, geometric flows.

MS classification: 53C44, 53C10.

1. Introduction

This paper discusses general flows of Spin(7)-structures in a manner similar to the author's analogous results for flows of G_2-structures, which were studied in Ref. 7. Many of the calculations are similar in spirit, although more involved, so we often omit proofs. The reader is advised to familiarize themselves with Ref. 7 first.

A general evolution of a Spin(7)-structure is described by a symmetric tensor h and a skew-symmetric tensor X satisfying some further algebraic condition, and it is only h which affects the evolution of the associated Riemannian metric. However, the evolution of the torsion tensor is determined by both h and X.

In Section 2, we review Spin(7)-structures, the decomposition of the space of forms, and the torsion tensor of a Spin(7)-structure. In Section 3 we compute the evolution equations for the metric and the torsion tensor for a general flow of Spin(7)-structures. In Section 4, we apply our evolution equations to derive a Bianchi-type identity in Spin(7)-geometry. This leads to an explicit formula for the Ricci tensor of a general Spin(7)-structure in terms of the torsion. An Appendix collects various identities in Spin(7)-geometry.

The notation used in this paper is identical to that of Ref. 7. Throughout this paper, M is a (not necessarily compact) smooth manifold of dimension 8 which admits a Spin(7)-structure.

2. Manifolds with Spin(7)-structure

In this section we review the concept of a Spin(7)-structure on a manifold M and the associated decompositions of the space of forms. More details about Spin(7)-structures can be found, for example, in Refs. 1,3–5. We also describe explicitly the torsion tensor associated to a Spin(7)-structure.

Consider an 8-manifold M with a Spin(7) structure Φ. The existence of such a structure is a *topological condition.* The space of 4-forms Φ on M which determine a Spin(7)-structure is a subbundle $\mathcal{A}$ of the bundle Ω^4 of 4-forms on M, called the bundle of *admissible* 4-forms. This is *not* a vector subbundle, and unlike the G_2 case, it is not even an open subbundle.

A Spin(7)-structure Φ determines a Riemannian metric g_Φ and an orientation in a non-linear fashion. Details can be found in Ref. 5, although that paper uses a different orientation convention (see also Ref. 6.) We will not have need for the explicit formula for the metric.

The metric g_Φ and orientation (determined by the volume form) determine a Hodge star operator $*$, and the 4-form Φ is *self-dual.* That is, $*\Phi = \Phi$. The metric also determines the Levi-Civita connection ∇, and the manifold (M, Φ) is called a Spin(7) manifold if $\nabla\Phi = 0$. This is a nonlinear partial differential equation for Φ, since ∇ depends on g which depends non-linearly on Φ. Such manifolds (where Φ is parallel) have Riemannian holonomy $\mathrm{Hol}_g(M)$ contained in the group $\mathrm{Spin}(7) \subset \mathrm{SO}(8)$. A parallel Spin(7)-structure is also called *torsion-free.*

2.1. *Decomposition of the space of forms*

The existence of a Spin(7)-structure Φ on M (with no condition on $\nabla\Phi$) determines a decomposition of the spaces of differential forms on M into irreducible Spin(7) representations. We will see explicitly that the spaces Ω^2, Ω^3, and Ω^4 decompose as

$$\Omega^2 = \Omega^2_7 \oplus \Omega^2_{21} \qquad\qquad \Omega^3 = \Omega^3_8 \oplus \Omega^3_{48}$$
$$\Omega^4 = \Omega^4_1 \oplus \Omega^4_7 \oplus \Omega^4_{27} \oplus \Omega^4_{35}$$

where Ω^k_l has (pointwise) dimension l and this decomposition is orthogonal with respect to the metric g. For $k = 2$ and $k = 3$, the explicit descriptions are as follows:

$$\Omega^2_7 = \{\beta \in \Omega^2;\ *(\Phi \wedge \beta) = -3\beta\} \qquad \Omega^2_{21} = \{\beta \in \Omega^2;\ *(\Phi \wedge \beta) = \beta\} \tag{1}$$

$$\Omega^3_8 = \{X \lrcorner \Phi;\ X \in \Gamma(TM)\} \qquad \Omega^3_{48} = \{\gamma \in \Omega^3;\ \gamma \wedge \Phi = 0\} \tag{2}$$

For $k > 4$, we have $\Omega^k_l = *\Omega^{8-k}_l$.

We need these decompositions in local coordinates. The following proposition is easy to verify.

Proposition 2.1. *Let β_{ij} be a 2-form, γ_{ijk} a 3-form, and X^l a vector field. Then*

$$\beta_{ij} \in \Omega^2_7 \Leftrightarrow \beta_{ab}g^{ap}g^{bq}\Phi_{pqij} = -6\,\beta_{ij} \quad \beta_{ij} \in \Omega^2_{21} \Leftrightarrow \beta_{ab}g^{ap}g^{bq}\Phi_{pqij} = 2\,\beta_{ij}$$

$$\gamma_{ijk} \in \Omega^3_8 \Leftrightarrow \gamma_{ijk} = X^l\Phi_{ijkl} \qquad \gamma_{ijk} \in \Omega^3_{48} \Leftrightarrow \gamma_{ijk}g^{ia}g^{jb}g^{kc}\Phi_{abcd} = 0$$

and the projection operators π_7 and π_{21} on Ω^2 are given by

$$\pi_7(\beta)_{ij} = \frac{1}{4}\,\beta_{ij} - \frac{1}{8}\,\beta_{ab}g^{ap}g^{bq}\Phi_{pqij} \tag{3}$$

$$\pi_{21}(\beta)_{ij} = \frac{3}{4}\,\beta_{ij} + \frac{1}{8}\,\beta_{ab}g^{ap}g^{bq}\Phi_{pqij} \tag{4}$$

Remark 2.1. One can show using Proposition 2.1 and Lemma A.1 that if $\beta_{ij} \in \Omega^2_{21}$,

$$\beta_{ab}g^{bl}\Phi_{lpqr} = \beta_{pi}g^{ij}\Phi_{jqra} + \beta_{qi}g^{ij}\Phi_{jrpa} + \beta_{ri}g^{ij}\Phi_{jpqa}$$

which can then be used to show that Ω^2_{21} is a Lie algebra with respect to the commutator of matrices:

$$[\beta, \mu]_{ij} = \beta_{il}g^{lm}\mu_{mj} - \mu_{il}g^{lm}\beta_{mj}$$

In fact, $\Omega^2_{21} \cong \mathfrak{so}(7)$, the Lie algebra of Spin(7).

The decomposition of the space Ω^4 of 4-forms can be understood by considering the infinitesimal action of $\mathrm{GL}(8,\mathbb{R})$ on Φ. Let $A = A^i_l \in \mathfrak{gl}(8,\mathbb{R})$. Hence $e^{At} \in \mathrm{GL}(8,\mathbb{R})$, and we have

$$e^{At}\cdot\Phi = \frac{1}{24}\,\Phi_{ijkl}\,(e^{At}dx^i) \wedge (e^{At}dx^j) \wedge (e^{At}dx^k) \wedge (e^{At}dx^l)$$

Differentiating with respect to t and setting $t = 0$, we obtain:

$$\left.\frac{d}{dt}\right|_{t=0} (e^{At}\cdot\Phi) = A^m_i\, dx^i \wedge \left(\frac{\partial}{\partial x^m} \lrcorner \Phi\right)$$

Now let $A_i^m = g^{mj} A_{ij}$, and decompose $A_{ij} = S_{ij} + C_{ij}$ into symmetric and skew-symmetric parts, where $S_{ij} = \frac{1}{2}(A_{ij} + A_{ji})$ and $C_{ij} = \frac{1}{2}(A_{ij} - A_{ji})$. We have a map

$$D : \mathfrak{gl}(8, \mathbb{R}) \to \Omega^4$$
$$D : A \mapsto \left.\frac{d}{dt}\right|_{t=0} (e^{At} \cdot \Phi) = A_{ij} g^{jm}\, dx^i \wedge \left(\frac{\partial}{\partial x^m} \lrcorner \Phi\right)$$

Proposition 2.2. *The kernel of D is isomorphic to the subspace Ω^2_{21}. It is also isomorphic to the Lie algebra $\mathfrak{so}(7)$ of the Lie group* Spin(7) *which is the subgroup of* GL(8, ℝ) *which preserves Φ.*

Proof. Since we are defining Spin(7) to be the group preserving Φ, the kernel of D is isomorphic to $\mathfrak{so}(7)$ by definition. To show explicitly that this is isomorphic to Ω^2_{21}, suppose that C_{ij} is in Ω^2_{21}. Then $D(C)$ is

$$\frac{1}{24}\left(C_i^m \Phi_{mjkl} + C_j^m \Phi_{imkl} + C_k^m \Phi_{ijml} + C_l^m \Phi_{ijkm}\right) dx^i \wedge dx^j \wedge dx^k \wedge dx^l$$

From Proposition 2.1, we have $C_{ij} = \frac{1}{2}\, C_{ab} g^{ap} g^{bq} \Phi_{pqij}$. Using this together with the final equation of Lemma A.1, one can compute that

$$C_i^m \Phi_{mjkl} + C_j^m \Phi_{imkl} + C_k^m \Phi_{ijml} + C_l^m \Phi_{ijkm} = \\ -3\left(C_i^m \Phi_{mjkl} + C_j^m \Phi_{imkl} + C_k^m \Phi_{ijml} + C_l^m \Phi_{ijkm}\right)$$

and hence $D(C) = 0$. Thus Ω^2_{21} is in the kernel of D. We show below that D restricted to Ω^2_7 or to $S^2(T)$ is injective. This completes the proof. □

By counting dimensions, we must have $\Omega^4_7 = D(\Omega^2_7)$ and also $\Omega^4_1 \oplus \Omega^4_{35} = D(S^2)$. We now proceed to establish these explicitly. The proofs of the next two propositions are very similar to analogous results in Ref. 7 and are left to the reader.

Proposition 2.3. *Suppose that A_{ij} is a tensor. Consider the 4-form $D(A)$ given by*

$$D(A) = A_{ij} g^{jm}\, dx^i \wedge \left(\frac{\partial}{\partial x^m} \lrcorner \Phi\right)$$

or equivalently

$$D(A)_{ijkl} = A_{im} g^{mn} \Phi_{njkl} + A_{jm} g^{mn} \Phi_{inkl} \\ + A_{km} g^{mn} \Phi_{ijnl} + A_{lm} g^{mn} \Phi_{ijkn} \tag{5}$$

Then the Hodge star of $D(A)$ is

$$*D(A) = D(\bar{A}) = \bar{A}_{ij} g^{jm}\, dx^i \wedge \left(\frac{\partial}{\partial x^m} \lrcorner \Phi\right)$$

where $\bar{A}_{ij} = \frac{1}{4}\operatorname{Tr}_g(A) g_{ij} - A_{ji}$. *That is, as a matrix,* $\bar{A} = \frac{1}{4}\operatorname{Tr}_g(A) I - A^T$.

Proposition 2.4. *Suppose A_{ij} and B_{ij} are two tensors. Let $D(A)$ and $D(B)$ be their corresponding forms in Ω^4. We write $A_{ij} = \frac{1}{8}\operatorname{Tr}_g(A) g_{ij} + (A_0)_{ij} + (A_7)_{ij}$, where A_0 is the symmetric traceless component of A, and A_7 is the component in Ω^2_7. Similarly for B. (We can assume they have no Ω^2_{21} component since that is in the kernel of D.) Then we have*

$$g_\Phi(D(A), D(B)) = \frac{7}{2}\operatorname{Tr}_g(A)\operatorname{Tr}_g(B) + 4\operatorname{Tr}_g(A_0 B_0) - 16\operatorname{Tr}_g(A_7 B_7)$$

where $A_0 B_0$ and $A_7 B_7$ mean matrix multiplication.

Corollary 2.1. *The map $D : \mathfrak{gl}(8, \mathbb{R}) \to \Omega^4$ is injective on $S^2 \oplus \Omega^2_7$. It is therefore an isomorphism onto its image, $\Omega^4_1 \oplus \Omega^4_7 \oplus \Omega^4_{35}$.*

Proof. This follows immediately from Proposition 2.4, since if $D(A) = 0$ and A is pure trace, traceless symmetric, or in Ω^2_7, we see that $A = 0$. □

We still need to understand the space Ω^4_{27}. To do this we give another characterization of the space of 4-forms using the Spin(7)-structure, which may be well-known to experts but has apparently not appeared in print before.

Definition 2.1. We define a Spin(7)-equivariant linear operator Λ_Φ on Ω^4 as follows. Let $\sigma \in \Omega^4$. Use the notation $(\sigma \cdot \Phi)_{ijkl}$ to denote $\sigma_{ijmn} g^{mp} g^{nq} \Phi_{pqkl}$. Then $\Lambda_\Phi(\sigma) \in \Omega^4$ is given by

$$(\Lambda_\Phi(\sigma))_{ijkl} = (\sigma\cdot\Phi)_{ijkl} + (\sigma\cdot\Phi)_{iklj} + (\sigma\cdot\Phi)_{iljk} + (\sigma\cdot\Phi)_{jkil} + (\sigma\cdot\Phi)_{jlki} + (\sigma\cdot\Phi)_{klij}$$

We now explain the motivation for introducing this operator Λ_Φ. If $\sigma \in \Omega^4_{27}$, then $g(\sigma, D(A)) = 0$ for all $A \in \mathfrak{gl}(8, \mathbb{R})$ since $D(A) \in \Omega^4_1 \oplus \Omega^4_7 \oplus \Omega^4_{35}$ and the splitting is orthogonal. Writing this in coordinates using (5) gives

$$\sigma \in \Omega^4_{27} \qquad \Leftrightarrow \qquad \sigma_{cbcd} \Phi_{ijkl} g^{jb} g^{kc} g^{ld} = 0 \quad \text{for all } a, i = 1, \ldots, 8$$

Taking the above expression and contracting it with Φ, and using Lemma A.1, after some laborious calculation one can show that

$$\sigma \in \Omega^4_{27} \qquad \Leftrightarrow \qquad \sigma_{ijkl} = \frac{1}{4}(\Lambda_\Phi(\sigma))_{ijkl}$$

which says that Ω^4_{27} is an eigenspace of Λ_Φ with eigenvalue $+4$. Suppose now that $\sigma = D(A) \in \Omega^4_1 \oplus \Omega^4_7 \oplus \Omega^4_{35}$. Then another brute force calculation using Definition 2.1 and Lemma A.1 shows

$$\Lambda_\Phi(D(A)) = D(6A^T - 6A - 3\,\mathrm{Tr}_g(A)I)$$

and from the above relation it is a simple matter to verify the following characterization of Ω^4.

Proposition 2.5. *The spaces Ω^4_1, Ω^4_7, Ω^4_{27}, and Ω^4_{35} are all eigenspaces of Λ_Φ with distinct eigenvalues. Specifically,*

$$\Omega^4_1 = \{\sigma \in \Omega^4; \Lambda_\Phi(\sigma) = -24\,\sigma\} \qquad \Omega^4_{27} = \{\sigma \in \Omega^4; \Lambda_\Phi(\sigma) = +4\,\sigma\}$$
$$\Omega^4_7 = \{\sigma \in \Omega^4; \Lambda_\Phi(\sigma) = -12\,\sigma\} \qquad \Omega^4_{35} = \{\sigma \in \Omega^4; \Lambda_\Phi(\sigma) = 0\}$$

In addition, we have

$$\Omega^4_1 = \{D(\lambda g); \lambda \in \mathbb{R}\} \quad \Omega^4_7 = \{D(A_7); A_7 \in \Omega^2_7\} \quad \Omega^4_{35} = \{D(A_0); A_0 \in S^2_0\}$$

where S^2_0 is the space of symmetric traceless tensors. Also, Proposition 2.3 shows that

$$\Omega^4_+ = \{\sigma \in \Omega^4; *\sigma = \sigma\} = \Omega^4_1 \oplus \Omega^4_7 \oplus \Omega^4_{27} \quad \Omega^4_- = \{\sigma \in \Omega^4; *\sigma = -\sigma\} = \Omega^4_{35}$$

is the decomposition into self-dual and anti-self dual 4-forms.

Finally, we have the following result, which is also proved using Lemma A.1.

Proposition 2.6. *Let $\sigma \in \Omega^4$. Then if we act on σ by Λ_Φ twice, we have*

$$\Lambda_\Phi(\Lambda_\Phi(\sigma))_{ijkl} = 24\,\sigma_{ijkl} - 16\,\Lambda_\Phi(\sigma)_{ijkl} + 2\,\Phi_{ijmn} g^{mp} g^{nq} \sigma_{pqrs} g^{ra} g^{sb} \Phi_{abkl}$$
$$+ 2\,\Phi_{ikmn} g^{mp} g^{nq} \sigma_{pqrs} g^{ra} g^{sb} \Phi_{ablj} + 2\,\Phi_{ilmn} g^{mp} g^{nq} \sigma_{pqrs} g^{ra} g^{sb} \Phi_{abjk}$$

We will need Propositions 2.5 and 2.6 in Section 2.2 to study the torsion of a Spin(7)-structure.

2.2. *The torsion tensor of a* Spin(7)*-structure*

In order to define the *torsion tensor* T of a Spin(7)-structure Φ, we need to first study the decomposition of $\nabla_X \Phi$ into its components in Ω^4.

Lemma 2.1. *For any vector field X, the 4-form $\nabla_X \Phi$ lies in the subspace Ω^4_7 of Ω^4. Hence $\nabla\Phi$ lies in the space $\Omega^1_8 \otimes \Omega^4_7$, a 56-dimensional space (pointwise).*

Proof. Let $X = \frac{\partial}{\partial x^m}$ and consider the 4-form $\nabla_m \Phi$. Then the second equation of Proposition 5.1 tells us that $\nabla_m \Phi$ is orthogonal to $S^2 \cong \Omega^4_1 \oplus \Omega^4_{35}$, exactly as in the G_2 case as discussed in Ref. 7. However, we need to work harder to show that there is no Ω^4_{27} component.

The essential reason that $\nabla_X \Phi \in \Omega^4_7$ is because of the way that the 4-form Φ determines the metric g. From Ref. 5 (which uses a different orientation convention), we have

$$(u \lrcorner v \lrcorner \Phi) \wedge (w \lrcorner y \lrcorner \Phi) \wedge \Phi = -6\, g(u \wedge v, w \wedge y)\mathsf{vol} + 7\, \Phi(u, v, w, y)\mathsf{vol} \quad (6)$$

Taking ∇_X of this identity gives

$$\begin{aligned}(u \lrcorner v \lrcorner \nabla_X \Phi) \wedge (w \lrcorner y \lrcorner \Phi) \wedge \Phi + (u \lrcorner v \lrcorner \Phi) \wedge (w \lrcorner y \lrcorner \nabla_X \Phi) \wedge \Phi \\ + (u \lrcorner v \lrcorner \Phi) \wedge (w \lrcorner y \lrcorner \Phi) \wedge \nabla_X \Phi = 7\, \nabla_X \Phi(u, v, w, y)\mathsf{vol}\end{aligned}$$

Now since $*\Phi = \Phi$ and $*(\nabla_X \Phi) = \nabla_X(*\Phi) = \nabla_X \Phi$, this can be written as

$$\begin{aligned}g((u \lrcorner v \lrcorner \nabla_X \Phi) \wedge (w \lrcorner y \lrcorner \Phi), \Phi) + g((u \lrcorner v \lrcorner \Phi) \wedge (w \lrcorner y \lrcorner \nabla_X \Phi), \Phi) \\ + g(u \lrcorner v \lrcorner \Phi) \wedge (w \lrcorner y \lrcorner \Phi), \nabla_X \Phi) = 7\, \nabla_X \Phi(u, v, w, y)\end{aligned}$$

We write this expression in coordinates, use Lemma A.1 to simplify the contractions of Φ with itself, and skew-symmetrize the result to obtain

$$\begin{aligned}(\nabla_X \Phi)_{ijkl} = &\frac{1}{12} \left(\Phi_{ijmn} g^{mp} g^{nq} (\nabla_X \Phi)_{pqrs} g^{ra} g^{sb} \Phi_{abkl}\right) \\ &+ \frac{1}{12} \left(\Phi_{ikmn} g^{mp} g^{nq} (\nabla_X \Phi)_{pqrs} g^{ra} g^{sb} \Phi_{ablj}\right) \\ &+ \frac{1}{12} \left(\Phi_{ilmn} g^{mp} g^{nq} (\nabla_X \Phi)_{pqrs} g^{ra} g^{sb} \Phi_{abjk}\right) \\ &- \frac{1}{3} \left((\nabla_X \Phi \cdot \Phi)_{ijkl} + (\nabla_X \Phi \cdot \Phi)_{iklj} + (\nabla_X \Phi \cdot \Phi)_{iljk}\right) \\ &- \frac{1}{3} \left((\nabla_X \Phi \cdot \Phi)_{jkil} + (\nabla_X \Phi \cdot \Phi)_{jlki} + (\nabla_X \Phi \cdot \Phi)_{klij}\right)\end{aligned}$$

using the notation of Definition 2.1. In fact the above expression can also be directly verified using the identities of Lemma A.1 and Proposition 5.1. Now using Definition 2.1 and Proposition 2.6 this becomes

$$(\nabla_X \Phi) = -\frac{1}{3} \Lambda_\Phi(\nabla_X \Phi) - \frac{1}{24} \left(\Lambda_\Phi(\Lambda_\Phi(\nabla_X \Phi)) - 24\, (\nabla_X \Phi) + 16\, \Lambda_\Phi(\nabla_X \Phi)\right)$$

Upon simplification, we have finally succeeded in showing that the basic relation (6) between the metric and the Spin(7)-structure Φ implies

$$\Lambda_\Phi(\Lambda_\Phi(\nabla_X \Phi)) + 8\, \Lambda_\Phi(\nabla_X \Phi) - 48\, (\nabla_X \Phi) = 0 \qquad \text{for any } X \quad (7)$$

Let $\nabla_X \Phi = \sigma_1 + \sigma_7 + \sigma_{27} + \sigma_{35}$ be its decomposition into components, where $\sigma_k \in \Omega^4_k$. Using Proposition 2.5, equation (7) says

$$(336)\,\sigma_1 + (0)\,\sigma_7 + (240)\,\sigma_{27} - (48)\,\sigma_{35} = 0$$

which, by linear independence, shows $\sigma_1 = \sigma_{27} = \sigma_{35} = 0$. Therefore $\nabla_X \Phi \in \Omega^4_7$. □

Remark 2.2. The above result was first proved in Ref. 2 by Fernàndez, using different methods.

Definition 2.2. Lemma 2.1 says that $\nabla\Phi$ can be written as

$$\begin{aligned}\nabla_m \Phi_{ijkl} &= D(T_m)_{ijkl} \\ &= T_{m;ip} g^{pq} \Phi_{qjkl} + T_{m;jp} g^{pq} \Phi_{iqkl} + T_{m;kp} g^{pq} \Phi_{ijql} + T_{m;lp} g^{pq} \Phi_{ijkq}\end{aligned}$$

where for each fixed m, $T_{m;ab}$ is in Ω^2_7. This defines the *torsion tensor* T of the Spin(7)-structure, which is an element of $\Omega^1_8 \otimes \Omega^2_7$.

The following lemma gives an explicit formula for $T_{m;ab}$ in terms of $\nabla\Phi$. This will be used in Section 3.1 to derive the evolution equation for the torsion tensor.

Lemma 2.2. *The torsion tensor $T_{m;\alpha\beta}$ is equal to*

$$T_{m;\alpha\beta} = \frac{1}{96} (\nabla_m \Phi_{\alpha jkl}) \Phi_{\beta bcd} g^{jb} g^{kc} g^{ld} \tag{8}$$

Proof. This is a simple computation using Definition 2.2 and the identities in Lemma A.1. □

We close this section with some remarks about the decomposition of T into irreducible components. One can show that $\Omega^1_8 \otimes \Omega^2_7 \cong \Omega^3 = \Omega^3_8 \oplus \Omega^3_{48}$. Therefore the torsion tensor T is actually a 3-form, with two irreducible components. In fact under this isomorphism T is essentially $\delta\Phi$, which is the content of the following result.

Theorem 2.1 (Fernández, 1986). *The* Spin(7)*-structure corresponding to Φ is torsion-free if and only if $d\Phi = 0$. Since $*\Phi = \Phi$, this is equivalent to $\delta\Phi = 0$.*

Suppose M is simply-connected, as it must be to admit a metric with holonomy exactly equal to Spin(7) in the compact case (see Ref. 4.) Then as in the G_2 case, which is described in Ref. 7, the component of the torsion in Ω^3_8 can always be conformally scaled away, once we have made the Ω^3_{48}

component vanish, without changing that other component. Therefore in principle we can restrict our attention to trying to make the Ω^3_{48} component of the torsion vanish, although it is not clear if this is really a simplification. We will not pursue this here.

3. General flows of Spin(7)-structures

In this section we derive the evolution equations for a general flow $\frac{\partial}{\partial t}\Phi$ of a Spin(7)-structure Φ. Let $A_{ij} = h_{ij} + X_{ij}$, where $h_{ij} \in S^2$ and $X_{ij} \in \Omega^2_7$. Then from the discussion in Section 2.1, a general variation of Φ can be written as $\frac{\partial}{\partial t}\Phi = D(A)$. In coordinates, using (5), this is

$$\begin{aligned}\frac{\partial}{\partial t}\Phi_{ijkl} &= D(A)_{ijkl} \\ &= A_{im}g^{mn}\Phi_{njkl} + A_{jm}g^{mn}\Phi_{inkl} + A_{km}g^{mn}\Phi_{ijnl} + A_{lm}g^{mn}\Phi_{ijkn}\end{aligned} \tag{9}$$

The first thing we need to do is to derive the evolution equations for the metric g and objects related to the metric, specifically the volume form vol and the Christoffel symbols Γ^k_{ij}. We do this using a much simpler argument than that presented in Ref. 7 for the G_2 case. This method works for that case as well.

Proposition 3.1. *The evolution of the metric g_{ij} under the flow (9) is given by*

$$\frac{\partial}{\partial t} g_{ij} = 2\, h_{ij} \tag{10}$$

Proof. We want to know what the first order variation of the metric g_Φ is, given a first order variation $D(A)$ of the Spin(7)-structure Φ. It suffices to consider any path $\Phi(t)$ of Spin(7)-structures that satisfies $\frac{\partial}{\partial t}\big|_{t=0}\Phi(t) = D(A)$. We take

$$\Phi(t) = e^{At}\cdot\Phi = \frac{1}{24}\Phi_{ijkl}\,(e^{At}dx^i)\wedge(e^{At}dx^j)\wedge(e^{At}dx^k)\wedge(e^{At}dx^l)$$

Then if $g = g_{ij}dx^i dx^j$ is the metric of $\Phi = \Phi(0)$, it is easy to see that the metric $g(t)$ of $\Phi(t)$ is

$$g(t) = g_{ij}(e^{At}dx^i)(e^{At}dx^j)$$

Now we differentiate

$$\begin{aligned}\frac{\partial}{\partial t}\bigg|_{t=0} g(t) &= g_{ij}\,(A^i_k dx^k)dx^j + g_{ij}\,dx^i(A^j_l dx^l) = A_{kj}\,dx^k dx^j + A_{li}\,dx^i dx^l \\ &= (A_{ij}+A_{ji})\,dx^i dx^j = 2\,h_{ij}\,dx^i dx^j\end{aligned}$$

since h is the symmetric part of A. This completes the proof. □

Corollary 3.1. *The evolution of the inverse g^{ij} of the metric, the volume form* vol*, and the Christoffel symbols Γ^k_{ij}, under the flow* (9)*, are given by*

$$\frac{\partial}{\partial t} g^{ij} = -2\, h^{ij} \quad \frac{\partial}{\partial t}\mathsf{vol} = \mathrm{Tr}_g(h)\,\mathsf{vol} \quad \frac{\partial}{\partial t}\Gamma^k_{ij} = g^{kl}\left(\nabla_i h_{jl} + \nabla_j h_{il} - \nabla_l h_{ij}\right)$$

Proof. This is a standard result. □

3.1. *Evolution of the torsion tensor*

In this section we derive the evolution equation for the torsion tensor T of Φ under the general flow (9). We begin with the evolution of $\nabla_m \Phi_{ijkl}$.

Lemma 3.1. *The evolution of $\nabla_m \Phi_{ijkl}$ under the flow* (9) *is given by*

$$\begin{aligned}
\frac{\partial}{\partial t}(\nabla_m \Phi_{ijkl}) \;=\; & A_{ip} g^{pq}(\nabla_m \Phi_{qjkl}) + A_{jp} g^{pq}(\nabla_m \Phi_{iqkl}) \\
& + A_{kp} g^{pq}(\nabla_m \Phi_{ijql}) + A_{lp} g^{pq}(\nabla_m \Phi_{ijkq}) + (\nabla_p h_{im}) g^{pq} \Phi_{qjkl} \\
& + (\nabla_p h_{jm}) g^{pq} \Phi_{iqkl} + (\nabla_p h_{km}) g^{pq} \Phi_{ijql} + (\nabla_p h_{lm}) g^{pq} \Phi_{ijkq} \\
& - (\nabla_i h_{pm}) g^{pq} \Phi_{qjkl} - (\nabla_j h_{pm}) g^{pq} \Phi_{iqkl} - (\nabla_k h_{pm}) g^{pq} \Phi_{ijql} \\
& - (\nabla_l h_{pm}) g^{pq} \Phi_{ijkq} + (\nabla_m X_{ip}) g^{pq} \Phi_{qjkl} + (\nabla_m X_{jp}) g^{pq} \Phi_{iqkl} \\
& + (\nabla_m X_{kp}) g^{pq} \Phi_{ijql} + (\nabla_m X_{lp}) g^{pq} \Phi_{ijkq}
\end{aligned}$$

where $A_{ij} = h_{ij} + X_{ij} \in S^2 \oplus \Omega^2_7$.

Proof. Recall that

$$\nabla_m \Phi_{ijkl} = \frac{\partial}{\partial x^m}\Phi_{ijkl} - \Gamma^n_{mi}\Phi_{njkl} - \Gamma^n_{mj}\Phi_{inkl} - \Gamma^n_{mk}\Phi_{ijnl} - \Gamma^n_{ml}\Phi_{ijkn}$$

so if we differentiate this equation with respect to t and simplify, we obtain

$$\begin{aligned}
\frac{\partial}{\partial t}(\nabla_m \Phi_{ijkl}) = \nabla_m \left(\frac{\partial}{\partial t}\Phi_{ijkl}\right) - \left(\frac{\partial}{\partial t}\Gamma^n_{mi}\right)\Phi_{njkl} \\
- \left(\frac{\partial}{\partial t}\Gamma^n_{mj}\right)\Phi_{inkl} - \left(\frac{\partial}{\partial t}\Gamma^n_{mk}\right)\Phi_{ijnl} - \left(\frac{\partial}{\partial t}\Gamma^n_{ml}\right)\Phi_{ijkn}
\end{aligned}$$

Now we substitute (9) and use Corollary 3.1. After we use the product rule on the first term, all the terms involving $\nabla_m h$ cancel in pairs. The result now follows. □

Theorem 3.1. *The evolution of the torsion tensor $T_{m;\alpha\beta}$ under the flow (9) is given by*

$$\frac{\partial}{\partial t}T_{m;\alpha\beta} = A_{\alpha p}g^{pq}T_{m;q\beta} - A_{\beta p}g^{pq}T_{m;q\alpha} + \pi_7(\nabla_\beta h_{\alpha m} - \nabla_\alpha h_{\beta m} + \nabla_m X_{\alpha\beta}) \quad (11)$$

where $A_{ij} = h_{ij} + X_{ij}$ is the element of $S^2 \oplus \Omega^2_7$ corresponding to the flow of Φ, and π_7 denotes the projection onto Ω^2_7 of the tensor skew-symmetric in α, β for fixed m.

Proof. This is a long computation, but is similar in spirit to the analogous result for G$_2$-structures in Ref. 7. We will describe the main steps, and leave the details to the reader. Begin with Lemma 2.2, and differentiate to obtain:

$$\begin{aligned}\frac{\partial}{\partial t}T_{m;\alpha\beta} &= \frac{1}{96}\left(\frac{\partial}{\partial t}\nabla_m \Phi_{\alpha jkl}\right)\Phi_{\beta bcd}g^{jb}g^{kc}g^{ld} \\ &+ \frac{1}{96}(\nabla_m \Phi_{\alpha jkl})\left(\frac{\partial}{\partial t}\Phi_{\beta bcd}\right)g^{jb}g^{kc}g^{ld} - \frac{6}{96}(\nabla_m \Phi_{\alpha jkl})\Phi_{\beta bcd}h^{jb}g^{kc}g^{ld}\end{aligned} \quad (12)$$

where we have used $\frac{\partial}{\partial t}g^{ij} = -2\,h^{ij}$ from Corollary 3.1. Recall that for a tensor B_{ij} we defined

$$D(B)_{ijkl} = B_{ip}g^{pq}\Phi_{qjkl} + B_{jp}g^{pq}\Phi_{iqkl} + B_{kp}g^{pq}\Phi_{ijql} + B_{lp}g^{pq}\Phi_{ijkq}$$

Let us define a similar shorthand notation $D_m(B)_{ijkl}$ to denote

$$B_{ip}g^{pq}\nabla_m \Phi_{qjkl} + B_{jp}g^{pq}\nabla_m \Phi_{iqkl} + B_{kp}g^{pq}\nabla_m \Phi_{ijql} + B_{lp}g^{pq}\nabla_m \Phi_{ijkq}$$

Then Lemma 3.1 says that

$$\frac{\partial}{\partial t}\nabla_m \Phi_{ijkl} = D_m(A) + D(B) \quad (13)$$

where we define

$$B_{\alpha\beta} = \nabla_\beta h_{\alpha m} - \nabla_\alpha h_{\beta m} + \nabla_m X_{\alpha\beta} \quad (14)$$

We also have $\frac{\partial}{\partial t}\Phi = D(A)$. Therefore (12) becomes

$$\begin{aligned}\frac{\partial}{\partial t}T_{m;\alpha\beta} &= \frac{1}{96}D_m(A)_{\alpha jkl}\Phi_{\beta bcd}g^{jb}g^{kc}g^{ld} + \frac{1}{96}D(B)_{\alpha jkl}\Phi_{\beta bcd}g^{jb}g^{kc}g^{ld} \\ &+ \frac{1}{96}(\nabla_m \Phi_{\alpha jkl})D(A)_{\beta bcd}g^{jb}g^{kc}g^{ld} - \frac{6}{96}(\nabla_m \Phi_{\alpha jkl})\Phi_{\beta bcd}h^{jb}g^{kc}g^{ld}\end{aligned} \quad (15)$$

We will break up the computation into several manageable pieces. First, we need the following identity. If $A = h + X \in S^2 \oplus \Omega^2_7$, then:

$$\begin{aligned} D(A)_{ijkl}\Phi_{abcd}g^{kc}g^{ld} &= 4\,h_{ia}g_{jb} - 4\,h_{ib}g_{ja} + 4\,h_{jb}g_{ia} - 4\,h_{ja}g_{ib} \\ &\quad + 2\,\mathrm{Tr}_g(h)(g_{ia}g_{jb} - g_{ib}g_{ja} - \Phi_{ijab}) + 16\,X_{ia}g_{jb} - 16\,X_{ib}g_{ja} \\ &\quad + 16\,X_{jb}g_{ia} - 16\,X_{ja}g_{ib} - 2\,A_{ip}g^{pq}\Phi_{qjab} + 2\,A_{jp}g^{pq}\Phi_{qiab} \\ &\quad + 2\,A_{pa}g^{pq}\Phi_{qbij} - 2\,A_{pb}g^{pq}\Phi_{qaij} \end{aligned} \tag{16}$$

which can be proved using Lemma A.1. Also, it is easy to check that if X and Y are both in Ω^2_7, then

$$X_{ip}g^{pq}Y_{qj}g^{ia}g^{jb}\Phi_{abkl} = X_{kp}g^{pq}Y_{ql} - Y_{kp}g^{pq}X_{ql} \tag{17}$$

which essentially says that the Lie bracket of two elements of Ω^2_7 is always in Ω^2_{21}. Now using the identities (16) and (17), and Lemma A.1 again, along with some patience, one can establish the following four expressions:

$$\begin{aligned} D(A)_{\alpha jkl}\Phi_{\beta bcd}g^{jb}g^{kc}g^{ld} &= 24\,h_{\alpha\beta} + 18\,\mathrm{Tr}_g(h)\,g_{\alpha\beta} + 96\,X_{\alpha\beta} \\ D_m(A)_{\alpha jkl}\Phi_{\beta bcd}g^{jb}g^{kc}g^{ld} &= 48\,(h_{\alpha p}g^{pq}T_{m;q\beta} + h_{\beta p}g^{pq}T_{m;q\alpha}) \\ &\quad + 48\,(\mathrm{Tr}_g(h)\,T_{m;\alpha\beta} - g(X, T_m)g_{\alpha\beta}) \\ (\nabla_m\Phi_{\alpha jkl})D(A)_{\beta bcd}g^{jb}g^{kc}g^{ld} &= 48\,(-h_{\alpha p}g^{pq}T_{m;q\beta} - h_{\beta p}g^{pq}T_{m;q\alpha}) \\ &\quad + 48\,(\mathrm{Tr}_g(h)\,T_{m;\alpha\beta} + g(X, T_m)g_{\alpha\beta}) \\ &\quad + 96\,(X_{\alpha p}g^{pq}T_{m;q\beta} - X_{\beta p}g^{pq}T_{m;q\alpha}) \\ (\nabla_m\Phi_{\alpha jkl})\Phi_{\beta bcd}h^{jb}g^{kc}g^{ld} &= 16\,(-h_{\alpha p}g^{pq}T_{m;q\beta} - h_{\beta p}g^{pq}T_{m;q\alpha}) \\ &\quad + 16\,(\mathrm{Tr}_g(h)\,T_{m;\alpha\beta}) \end{aligned}$$

Now we use the above four expressions to simplify equation (15). We need to substitute B as defined in (14) for A when we use the first of these expressions. After much cancellation and collecting like terms, we are left with exactly (11). □

We remark that, just as in the G_2 case, the terms with ∇h and with ∇X play quite different roles in the evolution of the torsion tensor in equation (11). One hopes that it is possible to choose h and X in terms of T and possibly also ∇T so that the evolution equations have nice properties. In particular we would like the equation to be parabolic transverse to the action of the diffeomorphism group, for short-time existence. Ideally such a flow exists where the L^2-norm $||T||$ of the torsion decreases. These are questions for future research.

4. Bianchi-type identity and curvature formulas

In this section, we apply the evolution equation (11) to derive a Bianchi-type identity for manifolds with Spin(7)-structure. This yields explicit formulas for the Ricci tensor and part of the Riemann curvature tensor in terms of the torsion tensor. As the calculations here are extremely similar to those in Ref. 7, we will be brief.

Proposition 4.1. *The diffeomorphism invariance of the metric g as a function of the 4-form Φ is equivalent to the vanishing of the $\Omega^4_1 \oplus \Omega^4_{35}$ component of $\nabla_Y \Phi$ for any vector field Y. This is the fact which was proved earlier in Lemma 2.1.*

Proof. The proof is identical to the G_2 case. In both cases it is due to the fact that the evolution of the metric g depends only on the symmetric part h of $A = h + X$. Notice that in the Spin(7) case, there is a stronger result that the Ω^4_{27} component of $\nabla_Y \Phi$ also vanishes, Lemma 2.1, which does not follow from here. □

Theorem 4.1. *The diffeomorphism invariance of the torsion tensor T as a function of the 4-form Φ is equivalent to the following identity:*

$$\begin{aligned}\nabla_q T_{p;\alpha\beta} - \nabla_p T_{q;\alpha\beta} &= \frac{1}{4}R_{pq\alpha\beta} - \frac{1}{8}R_{pqij}g^{ia}g^{jb}\Phi_{ij\alpha\beta} \\ &\quad + 2\,T_{q;\alpha m}g^{mn}T_{p;n\beta} - 2\,T_{p;\alpha m}g^{mn}T_{q;n\beta}\end{aligned} \tag{18}$$

Proof. The proof is very similar to the analogous result for G_2-structures described in Ref. 7, and is left to the reader. The identity (18) can also be established directly by using (8), Lemma A.1, and the Ricci identities. □

We now examine some consequences of Theorem 4.1. For i and j fixed, the Riemann curvature tensor R_{ijkl} is skew-symmetric in k and l. Hence we can use the decomposition of Ω^2 to write it as

$$R_{ijkl} = (\pi_7(\text{Riem}))_{ijkl} + (\pi_{21}(\text{Riem}))_{ijkl}$$

where by equation (3), we have

$$(\pi_7(\text{Riem}))_{ijkl} = \frac{1}{4}R_{ijkl} - \frac{1}{8}R_{ijab}g^{ap}g^{bq}\Phi_{pqkl} \tag{19}$$

Therefore the identity (18) says that

$$\begin{aligned}(\pi_7(\text{Riem}))_{pq\alpha\beta} &= \nabla_q T_{p;\alpha\beta} - \nabla_p T_{q;\alpha\beta} \\ &\quad + 2\,(T_{p;\alpha m}g^{mn}T_{q;n\beta} - T_{q;\alpha m}g^{mn}T_{p;n\beta})\end{aligned} \tag{20}$$

Corollary 4.1. *If Φ is torsion-free, then the Riemann curvature tensor $R_{ijkl} \in S^2(\Omega^2)$ actually takes values in $S^2(\Omega^2_{21})$, where $\Omega^2_{21} \cong \mathfrak{so}(7)$, the Lie algebra of* Spin(7).

Proof. Setting $T = 0$ in (20) shows the for fixed i, j, we have $R_{ijkl} \in \Omega^2_{21}$ as a skew-symmetric tensor in k, l. The result now follows from the symmetry $R_{ijkl} = R_{klij}$. □

Remark 4.1. This result is well-known. When $T = 0$, the Riemannian holonomy of the metric g_Φ is contained in the group Spin(7). By the Ambrose-Singer holonomy theorem, the Riemann curvature tensor of the metric is thus an element of $S^2(\mathfrak{so}(7))$.

Lemma 4.1. *Let $Q_{ijkl} = R_{ijab} g^{ap} g^{bq} \Phi_{pqkl}$. Then we have $Q_{ijkl} g^{il} = 0$.*

Proof. This is identical to the G_2 case proved in Ref. 7. □

From Lemma 4.1 and equation (19), we see that the Ricci tensor R_{jk} can be expressed as

$$R_{jk} = R_{ijkl} g^{il} = 4\,(\pi_7(\text{Riem}))_{ijkl} g^{il} \tag{21}$$

Proposition 4.2. *Given a* Spin(7)*-structure Φ with torsion tensor $T_{m;\alpha\beta}$, its associated metric g has Ricci curvature R_{jk} given by*

$$R_{jk} = 4\, g^{il} \nabla_i T_{j;lk} - 4\, \nabla_j \left(g^{il} T_{i;lk}\right) + 8\, T_{j;mk} T_{i;nl} g^{mn} g^{il} - 8\, T_{j;ml} T_{i;nk} g^{mn} g^{il}$$

Proof. This follows immediately from equations (20) and (21). □

This explicit expression for the Ricci tensor R_{jk} in terms of the torsion tensor appears to be new.

Corollary 4.2. *The metric of a torsion-free* Spin(7)*-structure is necessarily Ricci-flat.*

Remark 4.2. This is classical, originally proved by Bonan. Here we see a direct proof of this fact.

Appendix A.

5. Identities in Spin(7)-geometry

In this appendix we collect several identities involving the 4-form Φ of a Spin(7)-structure. They are derived by methods analogous to those for the G_2 case as explained in Ref. 7, so we omit the proofs.

In local coordinates $x^1, x^2, \ldots, x^8$, the 4-form Φ is

$$\Phi = \frac{1}{24}\Phi_{ijkl}\, dx^i \wedge dx^j \wedge dx^k \wedge dx^l$$

where Φ_{ijkl} is totally skew-symmetric. The metric is given by $g_{ij} = g(\frac{\partial}{\partial x^i}, \frac{\partial}{\partial x^j})$.

Lemma A.1. *The following identities hold:*

$$\begin{aligned}
\Phi_{ijkl}\Phi_{abcd}g^{ia}g^{jb}g^{kc}g^{ld} &= 336\\
\Phi_{ijkl}\Phi_{abcd}g^{jb}g^{kc}g^{ld} &= 42g_{ia}\\
\Phi_{ijkl}\Phi_{abcd}g^{kc}g^{ld} &= 6g_{ia}g_{jb} - 6g_{ib}g_{ja} - 4\Phi_{ijab}\\
\Phi_{ijkl}\Phi_{abcd}g^{ld} &= g_{ia}g_{jb}g_{kc} + g_{ib}g_{jc}g_{ka} + g_{ic}g_{ja}g_{kb}\\
&\quad - g_{ia}g_{jc}g_{kb} - g_{ib}g_{ja}g_{kc} - g_{ic}g_{jb}g_{ka}\\
&\quad - g_{ia}\Phi_{jkbc} - g_{ja}\Phi_{kibc} - g_{ka}\Phi_{ijbc}\\
&\quad - g_{ib}\Phi_{jkca} - g_{jb}\Phi_{kica} - g_{kb}\Phi_{ijca}\\
&\quad - g_{ic}\Phi_{jkab} - g_{jc}\Phi_{kiab} - g_{kc}\Phi_{ijab}
\end{aligned}$$

Proposition A.1. *The following identities hold:*

$$\begin{aligned}
(\nabla_m \Phi_{ijkl})\Phi_{abcd}g^{ia}g^{jb}g^{kc}g^{ld} &= 0\\
(\nabla_m \Phi_{ijkl})\Phi_{abcd}g^{jb}g^{kc}g^{ld} &= -\Phi_{ijkl}(\nabla_m \Phi_{abcd})g^{jb}g^{kc}g^{ld}\\
(\nabla_m \Phi_{ijkl})\Phi_{abcd}g^{kc}g^{ld} &= -\Phi_{ijkl}(\nabla_m \Phi_{abcd})g^{kc}g^{ld} - 4\nabla_m \Phi_{ijab}
\end{aligned}$$

References

1. R.L. Bryant and S.M. Salamon, On the Construction of Some Complete Metrics with Exceptional Holonomy, *Duke Math. J.* **58** (1989) 829–850.
2. M. Fernández, A Classification of Riemannian Manifolds with Structure Group Spin(7), *Ann. Mat. Pura Appl. (IV)* **143** (1986) 101–122.
3. D.D. Joyce, Compact 8-Manifolds with Holonomy Spin(7), *Invent. Math.* **123** (1996) 507–552.
4. D.D. Joyce, *Compact Manifolds with Special Holonomy* (Oxford University Press, 2000).
5. S. Karigiannis, Deformations of G_2 and Spin(7)-structures on Manifolds, *Canad. J. Math.* **57** (2005) 1012–1055.
6. S. Karigiannis, Some Notes on G_2 and Spin(7) Geometry; arXiv:math.DG/0608618.
7. S. Karigiannis, Geometric Flows on Manifolds with G_2-structure, I., submitted for publication; arXiv:math/0702077v2.

Differential Geometry and its Applications
Proc. Conf., in Honour of Leonhard Euler, Olomouc, August 2007

Connections on principal prolongations of principal bundles

Ivan Kolář

Institute of Mathematics and Statistics, Faculty of Science, Masaryk University, Janáčkovo nám 2a, 602 00 Brno, Czech Republic
E-mail: kolar@math.muni.cz

We study the principal connections of the r-th principal prolongation W^rP of a principal bundle $P(M,G)$ by using the related Lie algebroids. We deduce that both basic approaches to the concept of torsion are naturally equivalent. We prove that the torsion-free connections on W^rP are in bijection with the reductions of $W^{r+1}P$ to the group $G^1_m \times G$. Special attention is paid to the flow prolongation of connections.

Keywords: Principal prolongation of principal bundle, gauge theories, connection, torsion, Lie algebroid.

MS classification: 53A05, 58A20, 58A32.

Consider a principal bundle $P(M,G)$, $\dim M = m$. Its r-th order principal prolongation W^rP is the bundle of all r-jets $j^r_{(0,e)}\varphi$ of local principal bundle isomorphisms

$$\varphi\colon \mathbb{R}^m \times G \to P\,, \quad 0 \in \mathbb{R}^m,\ e = \text{the unit of } G.$$

This is a principal bundle over M with structure group $W^r_mG := W^r_0(\mathbb{R}^m \times G)$, whose action on W^rP is given by the jet composition.[11] If $G = \{e\}$ is the one-element group, then $M \times \{e\}$ is identified with M and $W^r(M \times \{e\}) = P^rM$ is the r-th order frame bundle of M. The r-th principal prolongation W^rP is a fundamental structure for both the general theory of geometric object fields,[11] and the gauge theories of mathematical physics.[3]

Our main subject are the principal connections on W^rP. So we omit the adjective "principal" as a rule. At a few places (in particular at the beginning of Section 4), where we mention arbitrary connections on an arbitrary fibered manifold Y, we call them explicitly "general connections".

It has been clarified recently that the connections Λ on P^rM are in bijection with the r-th order linear connections $\lambda\colon TM \to J^rTM$ on TM. In Section 2 we point out that in the case of W^rP the role of TM is replaced by the Lie algebroid $LP = TP/G$ of P. Using the flow prolongation of right -invariant vector fields, we identify $J^r(LP \to M)$ with the Lie algebroid LW^rP and prove that the connections Δ on W^rP are in bijection with the linear splittings $\delta\colon TM \to J^rLP$. The torsion of Δ is defined as the covariant exterior differential of the canonical one-form of W^rP, while the torsion of δ is introduced by means of the bracket of LP. In Section 3 we prove that the torsions of Δ and δ are naturally equivalent.

A connection Γ on P and a connection Λ on P^rM determine a connection $\mathcal{W}^r(\Gamma, \Lambda)$ on W^rP by means of the flow prolongation of vector fields. To demonstrate the applicability of the algebroid approach, we express explicitly the torsion of $\mathcal{W}^1(\Gamma, \Lambda)$ in terms of the torsion of Λ and the curvature of Γ in Section 4. For arbitrary r, we then deduce that $\mathcal{W}^r(\Gamma, \Lambda)$ is torsion-free, if and only if Λ is torsion-free and Γ is curvature-free. In Section 5 we prove that, analogously to the case of P^rM, the torsion-free connections on W^rP are identified with certain reductions of $W^{r+1}P$.

From a general point of view, J^r is a fiber product preserving bundle functor. In Section 6 we study an arbitrary functor F of this type and we deduce the algebroid formula for the flow prolongation of the above-mentioned pair of connections Γ and Λ.

All manifolds and maps are assumed to be infinitely differentiable. Unless otherwise specified, we use the terminology and notations from the book.[11]

1. The torsion of connections on W^rP

We write G^r_m for the r-th jet group in dimension m and $T^r_mG = J^r_0(\mathbb{R}^m, G)$. We have $W^rP = P^rM \times_M J^rP$ and

$$W^r_0(\mathbb{R}^m \times G) = W^r_mG = G^r_m \rtimes T^r_mG \tag{1}$$

is the group semidirect product with the group composition

$$(g_1, C_1)(g_2, C_2) = \big(g_1 \circ g_2, (C_1 \circ g_2) \bullet C_2\big)\,, \tag{2}$$

where $\bullet$ denotes the induced group composition in T^r_mG.[11] The first product projection $W^rP \to P^rM$ is a principal bundle morphism with the associated group homomorphism $W^r_mG \to G^r_m$ determined by (1). Write $\mathcal{PB}_m(G)$ for the category of principal G-bundles with m-dimensional bases and principal G-morphisms with local diffeomorphisms as base maps. Let $\bar{P}(\bar{M}, G)$

be another object of $\mathcal{PB}_m(G)$. For every $\mathcal{PB}_m(G)$-morphism $f\colon P \to \bar{P}$ with base map $\underline{f}\colon M \to \bar{M}$, we define

$$W^r f = P^r \underline{f} \times_{\underline{f}} J^r f\colon P^r M \times_M J^r P \to P^r \bar{M} \times_{\bar{M}} J^r \bar{P}\,. \tag{3}$$

Then W^r is a functor from the category $\mathcal{PB}_m(G)$ into $\mathcal{PB}_m(W^r_m G)$.[11]

On $P^r M$, we have the canonical one-form $\varphi_r\colon TP^r M \to \mathbb{R}^m \times \mathfrak{g}^{r-1}_m$. On $W^r P$, we introduce analogously a canonical one-form $\Theta_r\colon TW^r P \to \mathbb{R}^m \times \mathfrak{w}^{r-1}_m G = T_{(0,e_{r-1})} W^{r-1}(\mathbb{R}^m \times G)$, $e_{r-1} =$ the unit of $W^{r-1}_m G$. Consider $u = j^r_{(0,e)}\varphi \in W^r P$ and write $u_1 = \pi^r_{r-1}(u) \in W^{r-1}P$, where π^r_{r-1} is the jet projection. The tangent map

$$\widetilde{u} = T_{(0,e_{r-1})} W^{r-1}\varphi\colon \mathbb{R}^m \times \mathfrak{w}^{r-1}_m G \to T_{u_1} W^{r-1} P \tag{4}$$

is a linear isomorphism depending on u only. For every $Z \in T_u W^r P$, we define

$$\Theta_r(Z) = \widetilde{u}^{-1}\big(T\pi^r_{r-1}(Z)\big)\,.$$

Clearly, the following diagram commutes

$$\begin{array}{ccc} TW^r P & \xrightarrow{\ \Theta_r\ } & \mathbb{R}^m \times \mathfrak{w}^{r-1}_m G \\ \downarrow & & \downarrow \\ TP^r M & \xrightarrow{\ \varphi_r\ } & \mathbb{R}^m \times \mathfrak{g}^{r-1}_m \end{array} \tag{5}$$

For a connection Λ on $P^r M$, Yuen defined its torsion to be the covariant exterior differential $D_\Lambda \varphi_r$. Analogously, in Ref. 13 we introduced

Definition 1.1. The torsion of a connection Δ on $W^r P$ is the covariant exterior differential $D_\Delta \Theta_r$.

2. Another approach to connections on $W^r P$

A linear splitting $TM \to J^r TM$ is said to be a linear r-th order connection on TM. Since P^r is an r-th order bundle functor from the category $\mathcal{M}f_m$ of m-dimensional manifolds and local diffeomorphisms into $\mathcal{PB}_m(G^r_m)$, the flow prolongation $\mathcal{P}^r X$ of every vector field X on M is a right-invariant vector field on $P^r M$.[11] The restriction $\mathcal{P}^r X \mid P^r_x M$ depends on $j^r_x X$ only. This defines an identification

$$I^r_M\colon J^r TM \to LP^r M\,, \tag{6}$$

where $LP^r M$ is the Lie algebroid of $P^r M$.[15] Clearly, for every linear splitting $\lambda\colon TM \to J^r TM$, the rule

$$\Lambda(Z) = I^r_M\big(\lambda(Z)\big)\,, \qquad Z \in TM \tag{7}$$

defines a connection Λ on P^rM. This establishes a bijection $\lambda \to \Lambda$ between linear r-th order connections on TM and connections on P^rM.[8]

Such a bijection can be generalized to the case of W^rP. The role of TM is replaced by the Lie algebroid $LP = TP/G$ of P. Every section $\sigma\colon M \to LP$ is identified with a right-invariant vector field $\bar{\sigma}\colon P \to TP$ such that $\sigma = \bar{\sigma}/G$. Since W^r is a functor from $\mathcal{PB}_m(G)$ into $\mathcal{PB}_m(W^r_mG)$, the flow prolongation $\mathcal{W}^r(\bar{\sigma})$ is a right-invariant vector field on W^rP. The rule

$$I^r_P(j^r_x\sigma) = \big(\mathcal{W}^r(\bar{\sigma}) \mid W^r_xP\big)/W^r_mG \tag{8}$$

defines a bijection $I^r_P\colon J^r(LP \to M) \to LW^rP$. This is the principal bundle form of an identification that was established in the Lie algebroid form in Ref. 14. In the same way as in (7), we obtain

Proposition 2.1. (8) *identifies connections Δ on W^rP with linear splittings*

$$\delta\colon TM \to J^rLP\,. \tag{9}$$

We say that δ is the algebroid form of Δ.

3. The torsions on W^rP and J^rLP

The $(r-1)$-jet at $x \in M$ of the bracket $[X_1, X_2]$ of two vector fields X_1 and X_2 on M depends on the r-jets $j^r_xX_1$ and $j^r_xX_2$. This defines a bilinear morphism

$$[\ ,\ ]_r\colon J^rTM \times_M J^rTM \to J^{r-1}TM\,.$$

For a linear r-th order connection $\lambda\colon TM \to J^rTM$, one defines its torsion $\tau\lambda\colon TM \times_M TM \to J^{r-1}TM$ by

$$(\tau\lambda)(Z_1, Z_2) = \big[\lambda(Z_1), \lambda(Z_2)\big]_r\,, \qquad (Z_1, Z_2) \in TM \times_M TM\,. \tag{10}$$

Our result from Ref. 8 reads: If Λ is the principal bundle form of λ, then the torsion $D_\Lambda\varphi_r$ is naturally identified with the torsion $\tau\lambda$. We are going to deduce the same result for the case of W^rP.

Since the bracket $[\![\ ,\ ]\!]$ of LP is a first order operator, it determines a bilinear morphism

$$[\![\ ,\ ]\!]_r\colon J^rLP \times_M J^rLP \to J^{r-1}LP$$

analogously to $[\ ,\ ]_r$.

Definition 3.1. The torsion of a connection in the algebroid form $\delta\colon TM \to J^rLP$ is the morphism

$$\tau\delta\colon TM \times_M TM \to J^{r-1}LP$$

defined by

$$(\tau\delta)(Z_1, Z_2) = [\![\delta(Z_1), \delta(Z_2)]\!]_r\,, \qquad (Z_1, Z_2) \in TM \times_M TM\,.$$

Clearly, $\tau\delta$ can be viewed as a section of $J^{r-1}LP \otimes \Lambda^2 T^*M$.

Write $U_m^{r-1} = J_0^{r-1}L(\mathbb{R}^m \times G)$. Since $J^{r-1}L$ is an r-th order gauge natural bundle, every $u = j^r_{(0,e)}\varphi \in W^rP$ can be interpreted as a map

$$J^{r-1}L(u) := (J^{r-1}L)(\varphi)\mid_0\colon U_m^{r-1} \to J_x^{r-1}LP\,. \tag{11}$$

Our identification $I_P^{r-1}\colon J^{r-1}LP \to LW^{r-1}P$ is a natural equivalence of functors $J^{r-1}L$ and LW^{r-1}. Write $V_m^{r-1} = L_0W^{r-1}(\mathbb{R}^m \times G) = \mathbb{R}^m \times \mathfrak{w}_m^{r-1}$. Analogously to (11), we construct

$$LW^{r-1}(u) := LW^{r-1}(\varphi)\mid_0\colon V_m^{r-1} \to L_xW^{r-1}P\,. \tag{12}$$

The restriction of $I_{\mathbb{R}^m \times G}^{r-1}$ over $0 \in \mathbb{R}^m$ yields a bijection $\varepsilon\colon U_m^{r-1} \to V_m^{r-1}$. By naturality, the following diagram commutes

$$\begin{array}{ccc} U_m^{r-1} & \xrightarrow{\;J^{r-1}L(u)\;} & J_x^{r-1}LP \\ \Big\downarrow{\scriptstyle\varepsilon} & & \Big\downarrow{\scriptstyle(I_P^{r-1})_x} \\ V_m^{r-1} & \xrightarrow{\;LW^{r-1}(u)\;} & L_xW^{r-1}P \end{array} \tag{13}$$

If we identify $L_xW^{r-1}P$ with $T_{u_1}W^{r-1}P$, $u_1 = \pi^r_{r-1}(u)$, then $LW^{r-1}(u)$ is identified with $\widetilde{u}$ from (4). Since $J^{r-1}LP \otimes \Lambda^2T^*M$ is a fiber bundle associated to W^rP with standard fiber $U_m^{r-1} \otimes \Lambda^2\mathbb{R}^{m*}$ and $\tau\delta$ is a section, we can consider its frame form[11]

$$\{\tau\delta\}\colon W^rP \to U_m^{r-1} \otimes \Lambda^2\mathbb{R}^{m*}\,. \tag{14}$$

On the other hand, $D_\Delta\Theta_r$ is a horizontal 2-form, so that it can be interpreted as a map

$$\{D_\Delta\Theta_r\}\colon W^rP \to V_m^{r-1} \otimes \Lambda^2\mathbb{R}^{m*}\,. \tag{15}$$

Further we can construct

$$\varepsilon \otimes \mathrm{id}_{\Lambda^2\mathbb{R}^{m*}}\colon U_m^{r-1} \otimes \Lambda^2\mathbb{R}^{m*} \to V_m^{r-1} \otimes \Lambda^2\mathbb{R}^{m*}\,.$$

Proposition 3.1. *Under the identifications (14) and (15),*

$$\{D_\Delta \Theta_r\} = \left(\varepsilon \otimes \mathrm{id}_{\Lambda^2\mathbb{R}^{m*}}\right) \circ \frac{1}{2}\{\tau\delta\}\,. \tag{16}$$

Proof. If η_2 is a vector field on W^rP, then $\Theta_r(\eta_2)$ is an $(\mathbb{R}^m \times \mathfrak{w}_m^{r-1}G)$-valued function on W^rP. Thus, if η_1 is another vector field on W^rP, we can consider the derivative $\eta_1\Theta_r(\eta_2)\colon W^rP \to \mathbb{R}^m \times \mathfrak{w}_m^{r-1}G$ of $\Theta_r(\eta_2)$ in the direction of η_1. First we deduce that for every sections σ_1, $\sigma_2\colon M \to LP$ we have

$$\mathcal{W}^r(\bar\sigma_1)\,\Theta_r\big(\mathcal{W}^r(\bar\sigma_2)\big) = \Theta_r\big([\bar{\mathcal{W}}^r\bar\sigma_1, \mathcal{W}^r\bar\sigma_2]\big)\,. \tag{17}$$

Indeed, the rule $\bar\sigma \mapsto \Theta^r(\mathcal{W}^r\bar\sigma)$ is a gauge-natural operator. Hence it commutes with the Lie differentiation.[11] But the Lie derivative of $\bar\sigma_2$ with respect to $\bar\sigma_1$ is the bracket $[\bar\sigma_1, \bar\sigma_2]$.

Consider now $u \in W_x^rP$ and Z_1, $Z_2 \in T_xM$, $\delta(Z_i) = j_x^r\sigma_i$, $i = 1, 2$. Write $u_0\colon \mathbb{R}^m \to T_xM$ for the underlying map of u. If we interpret $\{D_\Delta\Theta_r\}$ as a map $W^rP \times \mathbb{R}^m \times \mathbb{R}^m \to V_m^{r-1}$, we have

$$\{D_\Delta\Theta_r\}\big(u, u_0^{-1}(Z_1), u_0^{-1}(Z_2)\big) = d\Theta_r\big(\mathcal{W}^r\bar\sigma_1(u), \mathcal{W}^r\bar\sigma_2(u)\big)\,.$$

Applying the classical formula for $d\Theta_r$ and (17), we obtain

$$\begin{aligned} 2d\Theta_r(\mathcal{W}^r\bar\sigma_1, \mathcal{W}^r\bar\sigma_2) &= (\mathcal{W}^r\bar\sigma_1)\Theta_r(\mathcal{W}^r\bar\sigma_2) - (\mathcal{W}^r\bar\sigma_2)\Theta_r(\mathcal{W}^r\bar\sigma_1) \\ &\quad - \Theta_r\big([\mathcal{W}\bar\sigma_1, \mathcal{W}\bar\sigma_2]\big) = \Theta_r\big([\mathcal{W}^r\bar\sigma_1, \mathcal{W}^r\bar\sigma_2]\big)\,. \end{aligned}$$

By the commutativity of (13), the last expression corresponds to $[\![\delta(Z_1), \delta(Z_2)]\!]_r$. □

4. The flow prolongation of principal connections

First we recall a general result on the flow prolongation of connections on an arbitrary fibered manifold $Y \to M$. Let Σ be a general connection on Y considered in the lifting form

$$\Sigma\colon Y \times_M TM \to TY\,.$$

Write $\mathcal{FM}_{m,n}$ for the category of fibered manifolds with m-dimensional bases and n-dimensional fibers and their local isomorphisms. Let F be a bundle functor on $\mathcal{FM}_{m,n}$ of the base order r.[11] For every vector field X on M we first construct the Σ-lift $\Sigma X\colon Y \to TY$. The flow prolongation $\mathcal{F}(\Sigma X)$ depends on the r-jets of X. This defines a map

$$\mathcal{F}\Sigma\colon FY \times_M J^rTM \to TFY\,.$$

Let Λ be a principal connection on P^rM and $\lambda\colon TM \to J^rTM$ be the corresponding splitting. Then

$$\mathcal{F}(\Sigma,\Lambda) := \mathcal{F}\Sigma \circ (\mathrm{id}_{FY} \times_M \lambda)\colon FY \times_M TM \to TFY$$

is a general connection on FY, that is called the flow prolongation of Σ with respect to Λ.[11]

If we consider a principal connection Γ on a principal bundle $P(M,G)$ in the role of Σ, then $\mathcal{W}^r(\Gamma,\Lambda)$ is a principal connection on W^rP. The algebroid form $\gamma\colon TM \to LP$ of Γ is a fibered morphism over id_M. Its r-th jet prolongation is a map $J^r\gamma\colon J^rTM \to J^rLP$. The following assertion will be proved in Section 6 in a more general setting.

Proposition 4.1. *The algebroid form of $\mathcal{W}^r(\Gamma,\Lambda)$ is*

$$J^r\gamma \circ \lambda\colon TM \to J^rLP\,. \tag{18}$$

The first application of (18) is the following assertion, that we deduced in a quite different way in Ref. 13. However, we find remarkable that the algebroid approach reduces the proof to a simple direct evaluation.

Proposition 4.2. *$\mathcal{W}^1(\Gamma,\Lambda)$ is torsion-free, iff λ is torsion-free and Γ is curvature-free.*

Proof. By locality, it suffices to discuss the case $P = \mathbb{R}^m \times G$, so that $LP = T\mathbb{R}^m \times \mathfrak{g}$. The sections of LP are pairs (X,ϱ) of a vector field X on $\mathbb{R}^m$ and a map $\varrho\colon \mathbb{R}^m \to \mathfrak{g}$ with the bracket

$$[\![(X_1,\varrho_1),(X_2,\varrho_2)]\!] = \left([X_1,X_2], X_1\varrho_2 - X_2\varrho_1 + [\varrho_1,\varrho_2]_{\mathfrak{g}}\right), \tag{19}$$

where $[X_1,X_2]$ is the bracket of vector fields and $[\ ,\ ]_{\mathfrak{g}}$ is the bracket in $\mathfrak{g}$.

Consider the canonical coordinates x^i on $\mathbb{R}^m$, the induced coordinates y^i on $T\mathbb{R}^m$ and some linear coordinates z^p on $\mathfrak{g}$. Let y^i_j, z^p_i be the induced coordinates on J^1LP. The map $[\![\ ,\ ]\!]_1$ has the coordinate expression

$$\left(y_1^j y_{2j}^i - y_2^j y_{1j}^i,\, y_1^i z_{2i}^p - y_2^i z_{1i}^p + c^p_{qr} z_1^q z_2^r\right), \tag{20}$$

where c^p_{qr} are the structure constants of G. Let $\delta\colon TM \to J^1LP$ be a connection of the form

$$z^p = \Delta^p_i(x)y^i\,, \quad y^i_j = \Delta^i_{kj}(x)y^k\,, \quad z^p_i = \Delta^p_{ji}(x)y^j\,. \tag{21}$$

Then the coordinate expression of $\tau\delta$ is

$$\frac{1}{2}\left(\Delta^k_{ij}, \Delta^p_{ij} + c^p_{qr}\Delta^q_i\Delta^r_j\right) dx^i \wedge dx^j\,. \tag{22}$$

On the other hand, let γ and λ be of the form

$$z^p = \Gamma^p_i(x)y^i\,, \quad y^i_j = \Lambda^i_{kj}(x)y^k\,. \tag{23}$$

Then the coordinate expression of $J^1\gamma$ is

$$z^p_i = \frac{\partial \Gamma^p_j}{\partial x^i}\,y^j + \Gamma^p_j y^j_i\,. \tag{24}$$

Hence $J^1\gamma \circ \lambda$ is of the form $y^i_j = \Lambda^i_{kj}y^k$ and

$$z^p_i = \Big(\frac{\partial \Gamma^p_j}{\partial x^i} + \Gamma^p_k \Lambda^k_{ij}\Big)y^j\,. \tag{25}$$

By (22), the torsion of $J^1\gamma \circ \lambda$ is

$$\frac{1}{2}\Big(\Lambda^k_{ij}, \frac{\partial \Gamma^p_j}{\partial x^i} + \Gamma^p_k \Lambda^k_{ij} + c^p_{qr}\Gamma^q_i \Gamma^r_j\Big)\,dx^i \wedge dx^j\,. \tag{26}$$

The first term in (26) is the torsion of Λ. If it vanishes, the second term coincides with the algebroid expression

$$\frac{1}{2}\Big(\frac{\partial \Gamma^p_j}{\partial x^i} + c^p_{qr}\Gamma^q_i \Gamma^r_j\Big)\,dx^i \wedge dx^j \tag{27}$$

of the curvature of Γ. □

Now it is easy to prove the general result.

Proposition 4.3. *$\mathcal{W}^r(\Gamma, \Lambda)$ is torsion-free, iff Λ is torsion-free and Γ is curvature-free.*

Proof. If $\mathcal{W}^r(\Gamma, \Lambda)$ is torsion-free, then $\mathcal{W}^1(\Gamma, \Lambda)$ is also torsion-free, so that Γ is integrable. Hence there is a local trivialization of P such that $\Gamma^p_i = 0$ identically. Then all non-trivial coefficients of $J^r\gamma$ are also zero and the coordinate expression of $[\![\ ,\ ]\!]_r$ reduces to the case of Λ. So the coordinate expressions of $\tau(J^r\gamma \circ \lambda)$ and $\tau\lambda$ coincide and our assertion follows from Proposition 4.2 in Ref. 8. □

5. Torsion-free connections as reductions

Every $a \in G^1_m$ is a matrix that defines a linear map $l(a)\colon \mathbb{R}^m \to \mathbb{R}^m$. This yields an injection

$$G^1_m \hookrightarrow G^r_m\,, \qquad a \mapsto j^r_0 l(a)\,.$$

In Ref. 5, we deduced that the torsion-free connections on P^rM are in bijection with the reductions of $P^{r+1}M$ to the subgroup $G^1_m \subset G^{r+1}_m$. The

r-jets $j_0^r\widehat{g}$, $g \in G$, of the constant maps $\widehat{g}\colon \mathbb{R}^m \to G$, $x \mapsto g$, define an injection $G \to T_m^r G$. Then $G_m^1 \times G$ is a subgroup of $W_m^r G$. In Ref. 13, we proved that the torsion-free connections on W^1P are in bijection with the reductions of W^2P to $G_m^1 \times G$. We are going to deduce such a result for an arbitrary order r.

For every fibered manifold $Y \to M$, the r-th contact morphism is a map $\psi_r\colon TJ^rY \to VJ^{r-1}Y \approx J^{r-1}(VY \to M)$. In the case of a principal bundle $P(M,G)$, we have $VP = P \times \mathfrak{g}$. Then $J^{r-1}VP = J^{r-1}P \times_M J^{r-1}(M,\mathfrak{g})$. Every frame of $P_x^{r-1}M$ identities $J_x^{r-1}(M,\mathfrak{g})$ with the Lie algebra $\mathfrak{t}_m^{r-1}G$ of $T_m^{r-1}G$. If we modify ψ_r in this way, we obtain a map $\bar{\psi}_r\colon TW^rP \to \mathfrak{t}_m^{r-1}G$. On the other hand, $\mathfrak{w}_m^{r-1}G = \mathfrak{g}_m^{r-1} \times \mathfrak{t}_m^{r-1}G$, so that we have the product projection $\pi\colon \mathbb{R}^m \times \mathfrak{w}_m^{r-1}G \to \mathfrak{t}_m^{r-1}G$. One verifies directly that

$$\bar{\psi}_r = \pi \circ \Theta_r \,. \tag{28}$$

We have $J^1(W^rP) = J^1P^rM \times_M J^1J^rP$. In Ref. 5, we described an injection $P^{r+1}M \hookrightarrow J^1P^rM$. On the other hand, we have the classical inclusion $J^{r+1}P \hookrightarrow J^1J^rP$. This defines an injection

$$i_r\colon W^{r+1}P \to J^1(W^rP)\,. \tag{29}$$

Let $\Gamma\colon P \to J^1P$ be a connection on $P = W^0P$. The rule

$$\varrho(\Gamma)(u,v) = \big(u, \Gamma(v)\big)\,, \qquad (u,v) \in P^1M \times_M P\,,$$

defines a reduction $\varrho(\Gamma)(P^1M \times_M P) \subset W^1P$ to $G_m^1 \times G$.[13] For a connection $\Delta\colon W^rP \to J^1W^rP$, we proceed by the following induction. Let Δ be such that the underlying connection Δ_1 on $W^{r-1}P$ is torsion-free. Hence Δ_1 defines a reduction

$$\varrho(\Delta_1)\colon P^1M \times_M P \to W^rP$$

by the induction hypothesis.

Proposition 5.1. *Δ is torsion-free, iff the values of $\Delta \circ \varrho(\Delta_1)$ lie in $i_r(W^{r+1}P)$. Then we define*

$$\varrho(\Delta) = i_r^{-1} \circ \Delta \circ \varrho(\Delta_1)\colon P^1M \times_M P \to W^{r+1}P\,.$$

Proof. Every $\varrho(\Delta)(u,v)$, $(u,v) \in P^1M \times_M P$, represents a linear m-dimensional subspace S in TW^rP, which is identified with a pair of m-dimensional linear subspaces $S_1 \subset TP^rM$ and $S_2 \subset TJ^rP$. By (5) and (28), $d\Theta_r \mid S$ can be considered as the pair $\big(d\varphi_r \mid S_1, d\bar{\psi}_r \mid S_2\big)$. By Ref. 5,

$d\varphi_r \mid S_1 = 0$ if and only if S_1 corresponds to an element of $P^{r+1}M$. Analogously to Ref. 6, $d\bar{\psi}_r \mid S_2 = 0$ if and only if S_2 corresponds to an element of $J^{r+1}P$. □

Proposition 5.2. *Proposition 5.1 establishes a bijection between the torsion-free connections on W^rP and the reductions of $W^{r+1}P$ to the subgroup $G^1_m \times G \subset W^{r+1}_m G$.*

Proof. On one hand, one verifies directly that $\varrho(\Delta)$ is a reduction to the subgroup $G^1_m \times G$. On the other hand, let $Q\colon P^1M \times_M P \to W^{r+1}P$ be a reduction to the subgroup $G^1_m \times G$. Write $Q_1 = \pi^{r+1}_r \circ Q\colon P^1M \times_M P \to W^rP$. For every $Q_1(u,v) \in W^rP$, $(u,v) \in P^1M \times_M P$, $Q(u,v)$ represents an m-dimensional horizontal subspace of TW^rP. Since our maps are $(G^1_m \times G)$-equivariant, these subspaces are canonically extended into a connection on W^rP. By the proof of Proposition 5.1, this connection is torsion-free. □

6. The case of $W^F P$

The r-th jet prolongation of fibered manifolds is a fiber product preserving bundle functor J^r on the category $\mathcal{FM}_m$ of fibered manifolds with m-dimensional bases and fibered morphisms with local diffeomorphisms as base maps. In Ref. 12 we characterized all these functors in terms of Weil algebras, see also Ref. 9. Every such functor F has finite order. If the base order of F is r, then we have an identification $F = (A, H, t)$, where A is a Weil algebra, $H\colon G^r_m \to \operatorname{Aut} A$ is a group homomorphism and $t\colon \mathbb{D}^r_m \to A$ is an equivariant algebra homomorphism, provided Aut A means the group of all algebra automorphisms of A and $\mathbb{D}^r_m$ is the Weil algebra $J^r_0(\mathbb{R}^m, \mathbb{R})$. In the case of J^r, we have $A = \mathbb{D}^r_m$, so that $\operatorname{Aut} \mathbb{D}^r_m = G^r_m$, $H = \operatorname{id}_{G^r_m}$ and $t = \operatorname{id}_{\mathbb{D}^r_m}$.

Analogously to the case of J^r, every $F = (A, H, t)$ determines a bundle functor W^F on $\mathcal{PB}_m(G)^2$

$$W^F P = P^r M \times_M FP\,, \quad W^F f = P^r \underline{f} \times_{\underline{f}} Ff\,.$$

Similarly to the case of W^r, $W^F(\mathbb{R}^m \times G)$ is a Lie group

$$W^A_H G = G^r_m \rtimes T^A G \tag{30}$$

with the group composition

$$(g_1, C_1)(g_2, C_2) = \left(g_1 \circ g_2, H_G(g_2^{-1})(C_1) \bullet C_2\right), \tag{31}$$

where $\bullet$ denotes the induced group composition in $T^A G$.

Further, $W^F P$ is a principal bundle over M with structure group $W^A_H G$. The values of W^F are in the category $\mathcal{PB}_m(W^A_H G)$.

For every fibered manifold $Y \to M$, t induces a map $t_Y \colon J^r Y \to FY$,

$$t_Y(j^r_x s) = (Fs)(x)\,, \qquad x \in M\,, \tag{32}$$

where s is a local section of Y, which is interpreted as a fibered morphism from the trivial fibered manifold $M \to M$ into Y, so that $Fs \colon M \to FY$. In particular, we have $t_{TM} \colon J^r TM \to FTM$. On the other hand, the anchor map $q \colon LP \to TM$ induces $FLP \to FTM$. In Ref. 10 we deduced, by using the theory of semi-direct products, that the Lie algebroid of $W^F P$ is

$$LW^F P = J^r TM \times_{FTM} FLP\,. \tag{33}$$

For every section $\sigma \colon M \to LP$, the vector field $\bar{\sigma}$ on P induces the flow prolongation $\mathcal{W}^F(\bar{\sigma})$, which is a right-invariant vector field on $W^F P$. To found its algebroid form, we use our general idea of the flow natural transformation of F. According to Ref. 7, see also Ref. 9, for every $Y \to M$ there exists a map

$$\psi^F_Y \colon J^r TM \times_{FTM} F(TY \to M) \to T(FY)$$

with the property that for every projectable vector field η on Y over ξ on M, the flow prolongation $\mathcal{F}\eta$ satisfies

$$\mathcal{F}\eta = \psi^F_Y \circ (j^r\xi \times_{F\xi} F\eta)\,,$$

provided η is considered as a fibered morphism of $TY \to M$ into $TM \to M$. In particular, this yields

Proposition 6.1. *For every section $\sigma \colon M \to LP$ over $X = q \circ \sigma \colon M \to TM$, the flow prolongation $\mathcal{W}^F(\bar{\sigma})$ corresponds to the section*

$$j^r X \times_{FX} F\sigma \colon M \to LW^F P\,, \quad j^r X \colon M \to J^r TM\,,$$
$$F\sigma \colon M \to F(LP \to M)$$

Let Γ be a connection on P and Λ a connection on $P^r M$. Hence the flow prolongation $\mathcal{W}^F(\Gamma, \Lambda)$ is a connection on $W^F P$. The algebroid form $\gamma \colon TM \to LP$ of Γ is a base preserving morphism, so that we can construct $F\gamma \colon FTM \to FLP$. Further, we have $\lambda \colon TM \to J^r TM$. By the very definition of $\mathcal{W}^F(\Gamma, \Lambda)$, we deduce

Proposition 6.2. *The algebroid form of $\mathcal{W}^F(\Gamma, \Lambda)$ is $(\lambda, F\gamma \circ t_{TM} \circ \lambda) \colon TM \to LW^F P$.*

In the case $F = J^r$, we have $t_{TM} = \mathrm{id}_{J^rTM}$, so that Proposition 4.1 is a special case of Proposition 6.2.

Remark 6.1. There is a natural question whether one can define the torsion of connections on an arbitrary principal bundle $W^F P$. The definition of the canonical form on $W^r P$ is essentially based on the fact that W^{r-1} is the underlying functor of W^r of the order $r - 1$. However, Doupovec clarified that the general concept of underlying functors of arbitrary F is rather sophisticated.[1] So it seems to be reasonable to restrict ourselves to the subfunctors $E \subset J^1 F$ with the property that the jet projection $EY \to FY$ is surjective. Then Proposition 2 of Ref. 4 implies that there is a canonical form on $W^F P$ with good properties and the procedures of the present paper can be applied.

Acknowledgments

The author was supported by the Ministry of Education of the Czech Republic under the project MSM 0021622409 and the grant GACR No. 201/05/0523

References

1. M. Doupovec, On the underlying lower order bundle functors, *Czechoslovak Math. J.* **55** (2005) 901–916.
2. M. Doupovec and I. Kolář, Iteration of Fiber Product Preserving Bundle Functors, *Monatsh. Math.* **134** (2001) 39–50.
3. L. Fatibene and M. Francaviglia, *Natural and Gauge Natural Formalism for Classical Fields Theories* (Kluwer, 2003).
4. I. Kolář, Generalized G-structures and G-structures of higher order, *Boll. Un. Math. Ital.*, Suppl. fasc. **3** (1975) 249–256.
5. I. Kolář, Torsion-free connections on higher order frame bundles, *In: New Developments in Differential Geometry* (Proceedings, Kluwer, 1996) 233–241.
6. I. Kolář, On holonomicity criteria in second order geometry, *Beiträge zur Algebra und Geometrie* **39** (1998) 283–290.
7. I. Kolář, On the geometry of fiber product preserving bundle functors, *In: Diff. Geom. and Its Applications* (Proceedings, Silesian University of Opava, 2002) 85–92.
8. I. Kolář, On the torsion of linear higher order connections, *Central European Journal of Mathematics* **3** (2003) 360–366.
9. I. Kolář, Weil Bundles as Generalized Jet Spaces, *In: Handbook of Global Analysis* (Elsevier, 2007) 625–665.
10. I. Kolář and A. Cabras, On the functorial prolongations of principal bundles, *Comment. Math. Univ. Carolin.* **47** (2006) 719–731.

11. I. Kolář, P.W. Michor and J. Slovák, *Natural Operations in Differential Geometry* (Springer Verlag, 1993).
12. I. Kolář and W. Mikulski, On the fiber product preserving bundle functors, *Differential Geometry and Its Applications* **11** (1999) 105–115.
13. I. Kolář and G. Virsik, Connections in first principal prolongations, *Suppl. ai Rendiconti del Circolo Matematico di Palermo*, Serie II **43** (1996) 163–171.
14. A. Kumpera and D. Spencer, *Lie equations I* (Princeton University Press, 1972).
15. K. Mackenzie, *General Theory of Lie Groupoids and Lie Algebroids* (Cambridge University Press, Cambridge 2005).

Differential Geometry and its Applications
Proc. Conf., in Honour of Leonhard Euler, Olomouc, August 2007

The (κ, μ, ν)-contact metric manifolds and their classification in the 3-dimensional case

Th. Koufogiorgos

Department of Mathematics, University of Ioannina,
Ioannina 45100, Greece
E-mail: tkoufog@cc.uoi.gr

M. Markellos and V.J. Papantoniou

Department of Mathematics, University of Patras,
Rion, GR-26500, Greece
E-mail: mark@upatras.gr, bipapant@math.upatras.gr

The (κ, μ, ν)-contact metric manifolds have been recently introduced by the authors (Ref. 11, 2007). In this aspect, we locally classify three dimensional (κ, μ, ν)-contact metric manifolds $M(\eta, \xi, \phi, g)$ which satisfy the condition $\nabla_\xi \tau = 2a\tau\phi$, where a is smooth function on M with $\xi(a) = 0$ and $\tau = \mathcal{L}_\xi g$. Moreover, we consider the same condition with a a constant function.

Keywords: Contact metric manifolds, (κ, μ, ν)-contact metric manifolds.

MS classification: 53C15, 53C25, 53C30.

1. Introduction

Let $M(\eta, \xi, \phi, g)$ be a contact metric manifold. Chern and Hamilton[7] introduced the torsion $\tau = \mathcal{L}_\xi g$, where $\mathcal{L}_\xi$ is the Lie derivative of g with respect to the characteristic vector field ξ, in their study of compact metric three-manifolds. The classification of conformally flat contact metric manifolds is another problem which has been investigated by many researchers. At one hand, in many cases conformally flat contact metric manifolds must have constant sectional curvature.[6,8,15] So, it is natural to ask if there are any conformally flat contact metric structures which are not of constant curvature. To this direction, D. E. Blair (Ref. 1, page 108) constructed examples of conformally flat contact metric three-manifolds which don't have constant sectional curvature. G. Calvaruso[4] proved that Blair's examples satisfy the condition

$$\nabla_\xi \tau = 2a\tau\phi, \tag{1}$$

where a is a smooth function with $\xi(a) = 0$. Here, the composition $(\tau\phi)(X,Y)$ has to be interpreted as $\tau(\phi X, Y)$.

In Ref. 11 the authors proved the existence of a new class of contact metric manifolds: the so called (κ,μ,ν)-contact metric manifolds. Such a manifold M is defined through the condition

$$\begin{aligned} R(X,Y)\xi = \kappa(\eta(Y)X - \eta(X)Y) + \mu(\eta(Y)hX - \eta(X)hY) \\ + \nu(\eta(Y)\phi hX - \eta(X)\phi hY) \end{aligned} \tag{2}$$

for every $X, Y \in \mathcal{X}(M)$ and κ, μ, ν are smooth functions on M. Furthermore, it is shown in Ref. 11 that if $\dim M > 3$, then κ, μ are constants and ν is the zero function. In the same paper, the authors gave a nice geometric interpretation of three dimensional (κ,μ,ν)-contact metric manifolds in terms of harmonic vector fields. More precisely, they proved that the characteristic vector field ξ of a 3-dimensional contact metric manifold M is a harmonic vector field if and only if M is a (κ,μ,ν)-contact metric manifold on an everywhere open and dense subset of M, generalizing a similar result of Perrone.[14]

The paper is organized in the following way. Section 2 contains the presentation of some basic notions about contact manifolds and (κ,μ)-contact metric manifolds.

In Section 3 we set the question of the existence of (κ,μ,ν)-contact metric manifolds where κ,μ,ν are smooth functions independent of the choice of the vector fields X, Y. We state that the answer to the above question is negative for dimension greater than three and positive for dimension equal to three. More precisely, for dimensions greater than three (κ,μ,ν)-contact metric manifolds are reduced to (κ,μ)-contact metric manifolds manifolds or to Sasakian manifolds.

In Section 4 we give a partial classification of 3-dimensional (κ,μ,ν)-contact metric manifolds. Especially, we assume that these manifolds satisfy the condition (1) where a is a smooth function or a constant. In the case which a is a constant, we give a geometrical meaning of (1).

2. Preliminaries

We start by collecting some fundamental notions about contact Riemannian geometry. We refer to Ref. 1 for further details. All manifolds in the present paper are assumed to be connected and smooth.

A $(2n+1)$-dimensional manifold is called *contact manifold* if it admits a global 1-form η(contact form) such that $\eta \wedge (d\eta)^n \neq 0$ everywhere on M. Given η, there exists a unique vector field ξ, called the *characteristic vector field* or the *Reeb vector field*, such that $\eta(\xi) = 1$ and $d\eta(\xi, X) = 0$ for every vector field X on M. It is well known that there also exists a Riemannian metric g and a tensor field ϕ of type (1, 1) such that

$$\phi(\xi) = 0, \qquad \phi^2 = -Id + \eta \otimes \xi, \qquad \eta \circ \phi = 0 \tag{3}$$

$$g(\phi X, \phi Y) = g(X, Y) - \eta(X)\eta(Y) \tag{4}$$

for all vector fields X, Y on M. Moreover, the quadruple (η, ξ, ϕ, g) can be chosen so that $d\eta(X, Y) = g(X, \phi Y)$. The manifold M together with the structure tensors (η, ξ, ϕ, g) is called a *contact metric manifold* and is denoted by $M(\eta, \xi, \phi, g)$. We denote by ∇ the Levi - Civita connection, and by R the corresponding Riemannian curvature tensor field given by

$$R(X, Y) = [\nabla_X, \nabla_Y] - \nabla_{[X,Y]}$$

for all vector fields X, Y on M. Given a contact Riemannian manifold M, we define on M the operators h and τ by

$$hX = \tfrac{1}{2}(\mathcal{L}_\xi \phi)X, \; \tau(X, Y) = (\mathcal{L}_\xi g)(X, Y)$$

where $\mathcal{L}_\xi$ is the Lie differentiation in the direction of ξ. The tensor field h of type (1,1) is self-adjoint and satisfies

$$h\xi = 0, \; trh = trh\phi = 0, \; h\phi = -\phi h. \tag{5}$$

If X is an eigenvector of h corresponding to the eigenvalue λ, then ϕX is also an eigenvector of h corresponding to the eigenvalue $-\lambda$, since h anticommutes with ϕ.

We also have the following formulas for a contact metric manifold:

$$\tau = 2g(\phi\cdot, h\cdot) \tag{6}$$

$$\nabla_\xi \tau = 2g(\phi\cdot, \nabla_\xi h\cdot) \tag{7}$$

Formulas (6) and (7) occur also in Ref. 12.

A contact metric manifold for ξ being a Killing vector field is called a $K-$*contact manifold.* It is well known that a contact metric manifold is $K-$contact if and only if $h = 0$.

A contact Riemannian manifold $M(\eta, \xi, \phi, g)$ is called a Sasakian manifold if and only if

$$R(X, Y)\xi = \eta(Y)X - \eta(X)Y \tag{8}$$

for every $X, Y \in \mathcal{X}(M)$.

Every Sasakian manifold is K-contact, but the converse is true only in the three dimensional case.

The (κ,μ)-nullity distribution of a contact metric manifold $M(\eta,\xi,\phi,g)$ for the pair $(\kappa,\mu) \in \mathbb{R}^2$ is the distribution

$$N(\kappa,\mu): p \to N_p(\kappa,\mu) = \{Z \in T_pM | R(X,Y)Z = \kappa(g(Y,Z)X - g(X,Z)Y) + \mu(g(Y,Z)hX - g(X,Z)hY)\}.$$

for every $X, Y \in T_pM$. So, if the characteristic vector field ξ belongs to the (κ,μ)-nullity distribution, then

$$R(X,Y)\xi = \kappa(\eta(Y)X - \eta(X)Y) + \mu(\eta(Y)hX - \eta(X)hY) \tag{9}$$

and the manifold M is called (κ,μ)-*contact metric manifold.*[2] If κ, μ are non - constant smooth functions on M, the manifold M is called *generalized* (κ,μ)-*contact metric manifold.* It is shown in Ref. 9 that if $\dim M > 3$, then κ and μ are necessarily constants. On the contrary, if $\dim M = 3$ then such generalized manifolds exist.

We mention that the class of (κ,μ)-contact metric manifolds extends the class of Sasakian manifolds ($\kappa = 1$ and relation (8)).

3. (κ,μ,ν)-contact metric manifolds

The following question comes up naturally. Do there exist contact metric manifolds satisfying (2) with κ,μ,ν non-constant smooth functions, independent of the choice of vector fields X, Y? The answer is positive for the 3-dimensional case and in the following we give an example.[11]

Example 3.1. *Consider the 3-dimensional manifold* $M = \{(x,y,z) \in \mathbb{R}^3;\ x > 0, y > 0, z > 0\}$, *where* (x,y,z) *are the cartesian coordinates in* $\mathbb{R}^3$. *We define the following vector fields on* M:

$$e_1 = \frac{\partial}{\partial x}, \qquad e_2 = \frac{\partial}{\partial y}, \qquad e_3 = -\frac{4}{z}e^G G_y \frac{\partial}{\partial x} + \beta \frac{\partial}{\partial y} + e^{\frac{G}{2}} \frac{\partial}{\partial z}$$

where $G = G(y,z) < 0$ *for every* (y,z), *is a solution of the partial differential equation*

$$2G_{yy} + G_y^2 = -ze^{-G},$$

and the function $\beta = \beta(x,y,z)$ *is a solution of the following system of partial differential equations*

$$\beta_x = \frac{4}{zx^2}e^G \quad \text{and} \quad \beta_y = \frac{1}{2z}e^{\frac{G}{2}} - \frac{G_z e^{\frac{G}{2}}}{2} - \frac{4e^G G_y}{xz}.$$

The vector fields e_1, e_2, e_3 are linearly independent at each point of M. We define a Riemannian metric g on M such that $g(e_i, e_j) = \delta_{ij}, i, j = 1, 2, 3$. Let η be the 1-form defined by $\eta(W) = g(W, e_1)$ for every $W \in \mathcal{X}(M)$. Then η is a contact form since $\eta \wedge d\eta \neq 0$ everywhere on M. Let ϕ be the tensor field of type (1,1), defined by $\phi e_1 = 0$, $\phi e_2 = e_3, \phi e_3 = -e_2$. Using the linearity of ϕ, $d\eta$ and g, we easily obtain that $\eta(e_1) = 1$, $d\eta(Z, W) = g(\phi Z, W)$ and $g(\phi Z, \phi W) = g(Z, W) - \eta(Z)\eta(W)$ for every vector fields Z, W on M. Hence $M(\eta, e_1, \phi, g)$ is a contact metric manifold. Computing the Lie brackets of e_1, e_2, e_3 and using the Koszul's formula, we easily prove that $M(\eta, e_1, \phi, g)$ is a (κ, μ, ν)-contact metric manifold with $\kappa = 1 - \frac{4e^{2G}}{z^2 x^4}, \mu = 2(1 + \frac{2e^G}{zx^2}), \nu = -\frac{2}{x}$.

Let $M(\eta, \xi, \phi, g)$ be a $(2n+1)$-dimensional (κ, μ, ν)-contact metric manifold and $B = \{p \in M | \kappa(p) = 1\}$. Then, the set $N = M \setminus B$ is an open subset of M and hence it inherits the contact structure of M i.e. $N(\eta, \xi, \phi, g)$ is a contact metric manifold which satisfies (2) with $\kappa < 1$ everywhere.

On the contrary, for dimensions greater than three the following Theorem is valid.

Theorem 3.1. *Every (κ, μ, ν)-contact metric manifold $M(\eta, \xi, \phi, g)$ of dimension greater than 3 is either a Sasakian manifold or a (κ, μ)-contact metric manifold, i.e. the functions κ, μ are constants and ν is the zero function.*

The proof of this Theorem is given in Ref. 11 and depends largely on the following lemmas.

Lemma 3.1. *On every $(2n+1)$-dimensional (κ, μ, ν)-contact metric manifold $M(\eta, \xi, \phi, g)$ the following relations are satisfied*

$$h^2 = (\kappa - 1)\phi^2, \; \kappa \leq 1 \tag{10}$$

$$\nabla_\xi h = \mu h\phi + \nu h \tag{11}$$

$$\xi(\kappa) = 2\nu(\kappa - 1) \tag{12}$$

Lemma 3.2. *On every $(2n+1)$-dimensional (κ, μ, ν)-contact metric manifold $M(\eta, \xi, \phi, g)$ the following differential equation is satisfied:*

$$\begin{aligned} &\xi(\kappa)[\eta(Y)X - \eta(X)Y] + \xi(\mu)[\eta(Y)hX - \eta(X)hY] \\ &+ \xi(\nu)[\eta(Y)\phi hX - \eta(X)\phi hY] - X(\kappa)\phi^2 Y + Y(\kappa)\phi^2 X \\ &+ X(\mu)hY - Y(\mu)hX + X(\nu)\phi hY - Y(\nu)\phi hX = 0 \end{aligned} \tag{13}$$

Lemma 3.3 (Ref. 9, Lemma 3.4). *For every $p \in N$ there exists an open neighborhood W of p and orthonormal local vector fields $X_i, \phi X_i, \xi, i = 1, 2, \ldots n$, defined on W, such that*

$$hX_i = \lambda X_i,\ h\phi X_i = -\lambda\phi X_i,\ h\xi = 0,\ i = 1, 2, \ldots n \tag{14}$$

where $\lambda = \sqrt{1 - \kappa}$.

4. Classification of 3-dimensional (κ, μ, ν)-contact metric manifolds with $\nabla_\xi \tau = 2a\tau\phi$

The existence of (κ, μ, ν)-contact metric manifolds in dimension 3 as described in Section 3, raises the question of their classification. To this direction, we give some partial answers assuming additionally that they satisfy the condition (1). We mention that for the cases $a = 0$ or $a = 1$, we have a complete classification of 3-dimensional (κ, μ, ν)-contact metric manifolds. For the sake of completeness, we give the corresponding classifications.[11]

Theorem 4.1. *Let $M(\eta, \xi, \phi, g)$ be a complete 3-dimensional (κ, μ, ν)-contact metric manifold with $\nabla_\xi \tau = 0$. Then M is a $(\kappa, 0)$-contact metric manifold with $\kappa =$ constant. In particular, M is either a Sasakian manifold (if $\kappa = 1$) or locally isometric to one of the following Lie groups, equipped with a left invariant metric: $SU(2)$ if $0 < \kappa < 1$, $SL(2, \mathbb{R})$ if $\kappa < 0$ and $E(2)$ if $\kappa = 0$.*

Theorem 4.2. *Let $M(\eta, \xi, \phi, g)$ be a complete 3-dimensional (κ, μ, ν)-contact metric manifold for which $\nabla_\xi \tau = 2\tau\phi$. Then M is a $(\kappa, 2)$-contact metric manifold where κ is a constant. Moreover, M is either a Sasakian manifold (if $\kappa = 1$) or locally isometric to $SL(2, \mathbb{R})$ if $\kappa \neq 1$, equipped with a left invariant metric. In the second case, any two $(\kappa, 2)$-contact metric manifolds are D-invariant under a specific D-homothetic transformation.*

A contact metric manifold $M(\eta, \xi, \phi, g)$ is said to be *homogeneous* if there exists a connected Lie group of isometries acting transitively on M and leaving η invariant. It is said to be *locally homogeneous* if the pseudogroup of local isometries acts transitively on M and leaves η invariant. For more details about homogeneous Riemannian manifolds see Ref. 16.

In Ref. 13 , D. Perrone studied three-dimensional manifolds admitting a homogeneous contact metric structure. He showed that these manifolds are locally isometric to a Lie group G with a left-invariant contact metric structure (η, ξ, ϕ, g). Moreover, all such manifolds satisfy the condition (1) with a

constant (Ball-homogeneous is defined as a Riemannian manifold which has the property that the volume of sufficiently small geodesic spheres or balls depends only on their radius and not on their center). Moreover, G. Calvaruso and D. Perrone[5] proved that a three-dimensional contact metric manifold is locally homogeneous if and only if it is ball-homogeneous and satisfies the condition (1) with a constant. In the sequence, we completely classify 3-dimensional (κ, μ, ν)-contact metric manifolds which additionally satisfy the condition (1) with a constant.

Theorem 4.3. *Let $M(\eta, \xi, \phi, g)$ be a complete 3-dimensional (κ, μ, ν)-contact metric manifold. If M satisfies (1) with $a =$ constant, then M is a (κ, μ)-contact metric manifold. Particularly, M is either a Sasakian manifold ($\kappa = 1$) or locally isometric to one of the following Lie groups with a left invariant metric: $SU(2)$ (or $SO(3)$), $SL(2, \mathbb{R})$ (or $O(1,2)$), $E(2)$ (the group of rigid motions of the Euclidean 2-space), $E(1,1)$ (the group of rigid motions of the Minkowski 2-space).*

Proof. Using (1), (3), (4), (6) and (7), we get

$$(\nabla_\xi h)\phi X = -2ahX$$

for every $X \in \mathcal{X}(M)$. Combining (11) and the last relation, we easily obtain

$$\begin{aligned} -2ahX &= (\nabla_\xi h)\phi X \\ &= \mu h\phi^2 X + \nu h\phi X \end{aligned}$$

or

$$(2a - \mu)hX + \nu h\phi X = 0 \tag{15}$$

for every $X \in \mathcal{X}(M)$. Let $p \in N$, then Lemma 3.3 implies the existence of a local orthonormal basis $\{e, \phi e, \xi\}$ on W which satisfies the relations (14). Applying (15) on e we get

$$0 = (2a - \mu)he + \nu h\phi e = (2a - \mu)\lambda e - \lambda\nu\phi e$$

which gives $\mu = 2a$ and $\nu = 0$ on W. On the other hand, substituting $X = e$ and $Y = \phi e$ in (13), we get that

$$\begin{aligned} e(\kappa) - \lambda e(\mu) - \lambda\phi e(\nu) &= 0 \\ -\phi e(\kappa) - \lambda\phi e(\mu) + \lambda e(\nu) &= 0. \end{aligned} \tag{16}$$

Since the function μ is constant on W and the function ν is the zero function on W, the relations (12) and (16) lead to the constancy of the continuous function κ on every connected component of W. If $M \setminus N \neq \emptyset$, then due to the continuity of the function κ we have that $\kappa = 1$ everywhere on M i.e. M is a Sasakian manifold. If $M \setminus N = \emptyset$ i.e $M = N$, then we have $\mu = 2a, \nu = 0$ and $\kappa = c \neq 1$ everywhere on M, where c is a constant i.e. M is a (κ, μ)-contact metric manifold. From the classification of 3-dimensional (κ, μ)-contact metric manifolds in Ref. 2 it follows that in the non-Sasakian case M is locally isometric to the above Lie groups equipped with a left invariant metric. □

Remark 4.1.

- i) In Ref. 3 , E. Boeckx proved that (κ, μ)-contact metric manifolds are locally homogeneous. As a consequence, 3-dimensional (κ, μ, ν)-contact metric manifolds which satisfy (1) with a constant, are locally homogeneous.
- ii) In Example 3.1, the eigenvalues of h are non-constant smooth functions. In fact, they are $\frac{2e^G}{zx^2}, -\frac{2e^G}{zx^2}$ and 0. This shows that the (κ, μ, ν)-contact metric manifolds are not necessarily locally homogeneous contact Riemannian manifolds[13] or, more generally, curvature homogeneous.

Generalizing Theorem 4.3, we obtain

Theorem 4.4. *Let $M(\eta, \xi, \phi, g)$ be a 3-dimensional (κ, μ, ν)-contact metric manifold for which $\nabla_\xi \tau = 2a\tau\phi$ where a is a smooth function on M. Then, M is either a Sasakian manifold $(\kappa = 1)$ or a generalized (κ, μ)-contact metric manifold with $\kappa < 1$.*

Proof. Using (1), (6) and (7) we obtain that

$$(\nabla_\xi h)\phi X = -2ahX$$

for every $X \in \mathcal{X}(M)$. Combining (11) and the last relation, we easily deduce

$$(2a - \mu)hX + \nu h\phi X = 0 \tag{17}$$

for every $X \in \mathcal{X}(M)$. Let $p \in N$, then Lemma 3.3 implies the existence of a local orthonormal basis $\{e, \phi e, \xi\}$ on W which satisfies the relations (14). Applying (17) on e we get $\mu = 2a$ and $\nu = 0$ on W. If $M \setminus N \neq \emptyset$, then due to the continuity of the function κ we have that $\kappa = 1$ everywhere on M i.e. M is a Sasakian manifold. If $M \setminus N = \emptyset$ i.e $M = N$, then we have $\mu = 2a, \nu = 0$ and $\kappa < 1$ everywhere on M. But, in this case κ, μ are smooth functions on M i.e. M is a generalized (κ, μ)-contact metric manifold with $\kappa < 1$. □

Now, we give an example of a generalized (κ, μ)-contact metric manifold which satisfy the condition (1) with a smooth function.

Example 4.1. We consider the 3-dimensional manifold $M = \{(x_1, x_2, x_3) \in \mathbb{R}^3 | x_3 \neq 0\}$, where (x_1, x_2, x_3) are the standard coordinates in $\mathbb{R}^3$. The vector fields

$$e_1 = \frac{\partial}{\partial x_1}, \quad e_2 = -2x_2x_3\frac{\partial}{\partial x_1} + \frac{2x_1}{x_3^3}\frac{\partial}{\partial x_2} - \frac{1}{x_3^2}\frac{\partial}{\partial x_3}, \quad e_3 = \frac{1}{x_3}\frac{\partial}{\partial x_2}$$

are linearly independent at each point of M. Let g be the Riemannian metric defined by $g(e_i, e_j) = \delta_{ij}, i, j = 1, 2, 3$ and η the dual 1-form to the vector field e_1. We define the tensor field ϕ of type $(1,1)$ by $\phi e_1 = 0, \phi e_2 = e_3, \phi e_3 = -e_2$. Following Ref. 9 , we have that $M(\eta, e_1, \phi, g)$ is a generalized (κ, μ)-contact metric manifold with $\kappa = \frac{x_3^4 - 1}{x_3^4}$ and $\mu = 2(1 - \frac{1}{x_3^2})$. By a straightforward calculation, we deduce that M satisfies the condition:

$$\nabla_\xi \tau = 2(1 - \frac{1}{x_3^2})\tau\phi$$

Remark 4.2. We assume that the condition (1) is satisfied on a 3-dimensional (κ, μ, ν)-contact metric manifold $M(\eta, \xi, \phi, g)$. Then, applying Theorem 4.4 and Theorem 1.1 of Ref. 14 , we deduce that the characteristic vector field $\xi : (M, g) \to (T_1M, g_S)$ determines an harmonic map, where (T_1M, g_S) is the unit tangent sphere bundle equipped with the Sasaki metric g_S.

Proposition 4.1. *Let $M(\eta, \xi, \phi, g)$ be a non-Sasakian 3-dimensional (κ, μ, ν)-contact metric manifold. Suppose that $\nabla_\xi \tau = 2a\tau\phi$ where a is a smooth function which is constant along the geodesic foliation generated by ξ. Then, M is a generalized (κ, μ)-contact metric manifold with $\xi(\mu) = 0$. We denote by $\{\xi, X, \phi X\}$ a local orthonormal frame of eigenvectors of h such that $hX = \lambda X, \lambda = \sqrt{1-\kappa} > 0$. Furthermore, for every $p \in M$ there exists a chart $(U, (x, y, z)), z < 1$, such that the function κ depends only of the variable z and $\mu = 2(1 - \sqrt{1-\kappa})$ or $\mu = 2(1 + \sqrt{1-\kappa})$. In the first case, the following are valid,*

$$\xi = \frac{\partial}{\partial x}, \phi X = \frac{\partial}{\partial y}, X = \alpha\frac{\partial}{\partial x} + b\frac{\partial}{\partial y} + \frac{\partial}{\partial z}.$$

In the second case, the following are valid,

$$\xi = \frac{\partial}{\partial x}, X = \frac{\partial}{\partial y}, \phi X = \alpha_1\frac{\partial}{\partial x} + b_1\frac{\partial}{\partial y} + \frac{\partial}{\partial z},$$

where $\alpha(x, y, z) = -2y + f(z), \alpha_1(x, y, z) = 2y + f(z), b(x, y, z) = b_1(x, y, z) = 2\lambda(z)x - \frac{\lambda'(z)}{2\lambda(z)}y + h(z), \lambda(z) = \sqrt{1 - \kappa(z)} > 0$ and f, h are smooth functions of z.

Proof. By Theorem 4.4, we get that M is a generalized (κ, μ)-contact metric manifold with $\mu = 2a$ and $\kappa < 1$. Since $\xi(a) = 0$, we easily obtain that $\xi(\mu) = 0$. The remaining part of the proposition follows immediately from Ref. 10. □

Acknowledgments

The second author was partially supported by the Greek State Scholarships Foundation (I.K.Y.) and by the C. Caratheóodory grant no.C.161 2007 - 10, University of Patras.

References

1. D.E. Blair, *Riemannian geometry of contact and symplectic manifolds* (Vol. 203, Progress in Math., Birkhauser, Boston, 2002).
2. D.E. Blair, Th. Koufogiorgos and B. Papantoniou, Contact metric manifolds satisfying a nullity condition, *Israel J. Math.* **91** (1995) 189–214.
3. E. Boeckx, A class of locally ϕ-symmetric contact metric spaces, *Arch. Math.* **72** (1999) 466–472.
4. G. Calvaruso, Einstein-like and conformally flat contact metric three-manifolds, *Balkan J. Geom. Appl.* **5** (2000) 17–36.
5. G. Calvaruso and D. Perrone, Torsion and homogeneity on contact metric three-manifolds, *Ann. Mat. Pura Appl.* **178** (2000) 271–285.
6. G. Calvaruso, D. Perrone and L. Vanhecke, Homogeneity on three-dimensional contact metric manifolds, *Israel J. Math.* **114** (1999) 301–321.
7. S.S. Chern and R.S. Hamilton On Riemannian metrics adapted to three-dimensional contact manifolds, *In: Proc. Meet. Max-Planck-Inst. Math.* (Vol.1111, Lect. Notes in Math., Springer-Verlag, Berlin/New York, 1985) 279-308.
8. F. Gouli-Andreou and Ph.J. Xenos, Two classes of conformally flat contact metric 3-manifolds, *J.Geom.* **64** (1999) 80–88.
9. Th. Koufogiorgos and C. Tsichlias, On the existence of a new class of contact metric manifolds, *Canad. Math. Bull.* **43** (2000) 440–447.
10. Th. Koufogiorgos and C. Tsichlias, Generalized (κ, μ)-manifolds with $\xi\mu = 0$, to be appeared in *Tokyo J. Math.*
11. T. Koufogiorgos, M. Markellos and V.J. Papantoniou, The harmonicity of the Reeb vector field on contact metric 3-manifolds, *Pacific J. Math.* **234** (2008) 325–344.
12. D. Perrone, Torsion and critical metrics on contact three-manifolds, *Kodai Math. J.* **13** (1990) 88–100.
13. D. Perrone, Homogeneous contact Riemannian three-manifolds, *Illinois J. Math.* **42** (1998) 243–256.
14. D. Perrone, Harmonic characteristic vector fields on contact metric manifolds, *Bull. Austral. Math.* **67** (2003) 305–315.

15. S. Tanno, Locally symmetric K-contact Riemannian manifolds, *Proc. Japan Acad.* **43** (1967) 581–583.
16. F. Tricerri and L. Vanhecke, *Homogeneous structures on Riemannian manifolds* (Vol. 83, Lect. Note Series, London Math. Soc., Cambridge University Press, 1989).

Differential Geometry and its Applications
Proc. Conf., in Honour of Leonhard Euler, Olomouc, August 2007

Invariants and submanifolds in almost complex geometry

Boris Kruglikov

Institute of Mathematics and Statistics, University of Tromsø,
Tromsø 90-37, Norway
E-mail: kruglikov@math.uit.no

In this paper we describe the algebra of differential invariants for $\mathrm{GL}(n, \mathbb{C})$-structures. This leads to classification of almost complex structures of general positions. The invariants are applied to the existence problem of higher-dimensional pseudoholomorphic submanifolds.

Keywords: Almost complex structure, equivalence, differential invariant, Nijenhuis tensor, pseudoholomorphic submanifold.

MS classification: 32Q60, 53C15, 53A55.

1. Introduction

Let (M, J) be an almost complex manifold, $J^2 = -\mathbf{1}$. In this paper we discuss only local aspects and so suppose $n = \frac{1}{2}\dim M > 1$. In this case the Nijenhuis tensor $N_J(\xi, \eta) = [J\xi, J\eta] - J[\xi, J\eta] - J[J\xi, \eta] - [\xi, \eta]$ (which is a skew-symmetric (2,1)-tensor) is generically non-zero. Vanishing of N_J is equivalent to local integrability of J.[11]

It is known that all differential invariants can be expressed via the jet of the Nijenhuis tensor.[5] In the first part of the paper we describe how this can be used to solve the equivalence problem of $\mathrm{GL}(n, \mathbb{C})$-structures. This problem is void for $n = 1$ and was solved in Ref. 7 for $n = 2$, but we present here a uniform approach via differential invariants suitable for all n.

The differential invariants of an almost complex structure also occur in the problem of establishing pseudoholomorphic (PH) submanifolds. They played the crucial role in the proof of non-existence of PH-submanifolds for generic almost complex (M, J).[6]

In dimension $2n \geq 8$ existence of a single higher-dimensional submanifold already imposes restrictions on the Nijenhuis tensor, so for their existence N_J should be degenerate (though can stay far from being zero). On the other hand existence of 4-dimensional PH-submanifolds for 6-

dimensional (M,J) does not impose identities-restrictions on the tensor N_J (there are open subsets of admissible tensors in the space of all Nijenhuis tensors).

At the second half of the paper we discuss when (M,J) can have PH-foliations and when their number is bounded in non-integrable case.

2. First order invariants of almost complex structures

For $n=1$ the almost complex structures J are complex and possess no local invariants. So this case will not be considered in what follows.

For $n=2$ there are non-integrable structures, but there are no first order differential invariants. To explain this let us note that all such invariants must be derived from the Nijenhuis tensor N_J. In dimension 4 the linear Nijenhuis tensor (a purely tensorial object at a point, i.e. an element of $\Lambda^2 T^* \otimes_{\bar{\mathbb{C}}} T$ with $T = T_x M$, see Ref. 5) is special and the $\mathrm{GL}(2,\mathbb{C}) \subset \mathrm{GL}(4)$ orbit space consists of two points: zero and non-zero tensor N_J.

For non-zero tensor we can talk of the image $\Pi^2 = \mathrm{Im}(N_J)$ which, if we vary the point x, is a two-distribution in M^4, called Nijenhuis tensor characteristic distribution.[7] Provided J is generic, Π^2 is generic as well. In particular, there's the derived rank 3 distribution $\Pi^3 = \partial\Pi^2$. This leads to the fact that there's no second order invariants as well. However, we can associate the second order e-structure $\{\xi_i\}_{i=1}^4$, $\xi_i \in \mathcal{D}_M$, to J as follows:

$$\xi_1 \in C^\infty(\Pi^2),\ \xi_3 \in C^\infty(\Pi^3),\ N_J(\xi_1,\xi_3) = \xi_1,$$
$$\xi_2 = J\xi_1,\ [\xi_1,\xi_2] = \xi_3,\ \xi_4 = J\xi_3.$$

This defines the pair ξ_1,ξ_2 canonically up to ± 1 and the pair ξ_3,ξ_4 absolutely canonically.[7]

When $n=3$ there are moduli in the space of linear Nijenhuis tensors.[6] This is clearly seen from Theorem 7 loc.sit. Indeed the statement means that the space of differential invariants of order 1 is two-dimensional and the constants of the normal forms provide the invariants. In Ref. 2 Bryant arrived independently to the result about dimension 2, observing that codimension of generic orbits w.r.t. the $\mathrm{GL}(3,\mathbb{C}) \subset \mathrm{GL}(6)$ action on the space of Nijenhuis tensors $\Lambda^2 T^* \otimes_{\bar{\mathbb{C}}} T$ is 2 (where $T = T_x M$ is a model tangent space of dimension 6), because the stabilizer is two-dimensional.

Moreover in Ref. 2 some invariants of almost complex 6-dimensional manifolds were introduced. All of them are expressed via a (1,1)-form ω, which is given in coordinates via the components of the Nijenhuis tensor as follows:

$$\omega_i^j = \frac{N_{ik}^l \bar{N}_{jl}^k - N_{jk}^l \bar{N}_{il}^k}{\sqrt{-1}}.$$

Here complex coordinates adapted at the point to J are used (in fact, in Ref. 2 non-holonomic, i.e. frames). Note that this is a real-valued form and it can be written in invariant terms as follows (now we assume all tensors real):

$$\omega(\xi,\eta) = \mathrm{Tr}[N_J(\xi, JN_J(\eta,\cdot)) - N_J(\eta, JN_J(\xi,\cdot))].$$

In particular, $\omega(\xi, J\xi) = 2\,\mathrm{Tr}[N_J(\xi, N_J(\xi,\cdot))]$ is not identically zero.

The form ω is J-compatible: $\omega(J\xi, J\eta) = \omega(\xi,\eta)$ and we can associate the quadric $q(\xi,\eta) = \omega(\xi, J\eta)$, which equals

$$q(\xi,\eta) = \mathrm{Tr}[N_J(\xi, N_J(\eta,\cdot)) + N_J(\eta, N_J(\xi,\cdot))].$$

Indeed, these both 2-tensors are skew-symmetric and symmetric parts of the form $T_{ij} = N_{ik}^l \bar{N}_{jl}^k$ and the pair (S,ω) forms a Hermitian metric provided q (or equivalently ω) is non-degenerate (it can be indefinite).

Let us investigate ω via the normal forms of N_J.[6]

Proposition 2.1. *The (1,1)-form ω is degenerate precisely in the following cases in terms of differential invariants from classification theorem 7 of Ref. 6:*

NDG_1: $\lambda = \pm 1$, $\varphi = 0, \pi$.
NDG_2: $\varphi = 0, \pi$.
NDG_3: $\psi = \pm\frac{\pi}{4} \pm \frac{\pi}{2} = \pm\varphi \pm \pi$.
NDG_4, DG_1 (4): Never.
DG_1 (1-3,5), DG_2 (1-2): Always.

This is a straightforward tedious calculation. It shows generic non-degeneracy of the 2-form ω. In Refs. 2,12 global implications of non-degeneracy are discussed. The above local aspects show which open strata of Ref. 6 contain non-degenerate forms ω, and this yields topological restrictions on global realization of such ω.

The canonical G_2-invariant almost complex structure J on S^6 corresponds to NDG.3 $\varphi = 0, \psi = \frac{\pi}{2}$, so in this case ω is non-degenerate. Also note that when the form ω is degenerate, then M possesses a canonical distribution (kernel), which can be used to construct classification in the case of non-general position.

In dimension $n > 3$ the orbit space of $\mathrm{GL}(n, \mathbb{C})$-action on $\Lambda^2 T^* \otimes_{\bar{\mathbb{C}}} T$ is quite complicated. And indeed the space of invariants is pretty big, as will be discussed below.

3. General background on differential invariants

The equivalence problem of geometric structures on M is usually solved either via differential invariants algebra or by constructing a canonical e-structure. In the first case the algebra can be represented either via some basic invariants and invariant differentiations or via some more differential invariants and Tresse derivatives.

However in the case of geometric structures the number of required differential invariants is smaller and equals $n = \dim M$, provided the restrictions of them to the structure are functionally independent (this is generically so).

Indeed, let π be a bundle of geometric structures (associated with a tensorial bundle over M) and $\mathcal{E}$ a section of it (i.e. a geometric structure of the specified type), which can be represented as the image of a section $j : M \to E_\pi$. Let ρ be the induced action of the pseudogroup $\mathrm{Diff}_{\mathrm{loc}}(M)$ on π and I be a differential invariant. Its restriction to $\mathcal{E}$ is the function $I_\mathcal{E} = j^* I \in C^\infty_{\mathrm{loc}}(M)$.

Given n functionally independent invariants $I^1, \dots, I^n$ we assume their restrictions $I^1_\mathcal{E}, \dots, I^n_\mathcal{E}$ are functionally independent (here and in what follows one can assume local treatment), so that they can be considered as local coordinates. Then one gets local frames $\partial_i = \frac{\partial}{\partial I^i_\mathcal{E}}$ and coframes $\omega^i = dI^i_\mathcal{E}$. Any tensorial field T can be expressed as $T = T^{j_1 \dots j_t}_{i_1 \dots i_s} \partial_{j_1} \otimes \dots \partial_{j_t} \otimes \omega^{i_1} \otimes \dots \omega^{j_s}$ and the coefficients are scalar differential invariants.

Being expressed via $I^i_\mathcal{E}$ they form the complete set of invariant relations for equivalence problem. This is the principle of n-invariants.[1]

Two remarks are of order. First: It is clear in this case that canonical frame field ω^i gives e-structure; otherwise around is also true, so that e-structures approach[4] is equivalent to one with differential invariants. Second: Lifts of the derivations ∂_i are invariant differentiations and coefficients of $dJ|_\mathcal{E} = \frac{DJ}{DI^i_\mathcal{E}} \omega^i$ are exactly Tresse derivatives of a differential invariant J by the basis I^i (see Refs. 9,10). Thus all the discussed approaches are equivalent.

Let us apply this to classification of almost complex structures of general position. This means that π is the bundle of almost complex structures over

M:

$$\pi^{-1}(x) = \{J \in \mathrm{GL}(T_x M) : J^2 = -\mathbf{1}\} \simeq \mathrm{GL}(2n,\mathbb{R})/\,\mathrm{GL}(n,\mathbb{C}) \stackrel{\mathrm{def}}{=} \mathcal{J}(2n).$$

The pseudogroup $G = \mathrm{Diff}_{\mathrm{loc}}(M)$ acts on π. The groupoid of its jets is denoted by $G^l \subset J^l(M,M)$, with natural projections being denoted by $\rho_{l,k} : G^l \to G^k$ and $\rho_l : G^l \to M$. Denote G^l_x the fiber over the point $(x,x) \in G^0 = M \times M$, which is also a sub-groupoid of G called the differential group of order l (we will sometimes omit reference to the point x).

Then G^l_x acts on the fiber of $\pi_k : J^k\pi \to M$ over point x. Moreover denoting $\mathfrak{G}^{l+1}_x = \mathrm{Ker}[\rho_{l-1,l} : G^{l+1}_x \to G^l_x]$ we obtain action of this normal subgroup on the fiber of the bundle $\pi_{l,l-1}$. Since for $l > 0$ the group $\mathfrak{G}^{l+1}$ is abelian, the orbits in $F_l = \pi^{-1}_{l,l-1}(x_{l-1})$ are affine and so the differential invariants can be chosen affine in derivatives of order l. Usually they are non-linear in lower-order derivatives.

4. Equivalence problem for almost complex structures

Let (M,J) be an almost complex manifold. If it is in general position, then as we have noticed above, it is enough to find $2n = \dim M$ differential invariants for local classification. Solution of the equivalence problem depends on n (which we can assume to be > 1).

Theorem 4.1. *The basic scalar differential invariants of J solving the equivalence problem via the described methods can be specified as follows.*

$n = 2$: *There are no differential invariants of order ≤ 2, but in order 3 there are (no less than) 4 differential invariants;*

$n = 3$: *There are precisely 2 differential invariants of order 1 and 4 invariants of order 2;*

$n > 3$: *There are at least $n^2(n-3) > 2n$ differential invariants of order 1.*

Proof. Consider the cases.

$\mathbf{n} = 2$. It is clear from the description in Section 2 that J has no differential invariants of order 1 or 2. To get invariants of order 3 one proceeds as follows: the Maurer-Cartan coefficients c^k_{ij} for the described canonical e-structure ξ_s (defined by the 2-jet of J) are the invariants of order 3: $[\xi_i, \xi_j] = c^k_{ij}\xi_k$. Since $c^k_{12} = \delta^k_3$ and $[\xi_2, \xi_4]$ can be expressed via other brackets from $N_J(\xi_1, \xi_3) = \xi_3$, the number of such differential invariants is 16. Note that the invariants of Ref. 5 §6.1 (canonical 1- and 2-forms on Π^2) can be expressed via c^k_{ij}.

Let us notice that the result can be obtained via pure dimensional count. Indeed, rank of $\rho_{1,0}$ is 16 and that of π is 8. The action of $\mathfrak{G}^1$ on $F_0 = \pi^{-1}(x)$ is transitive (8-dimensional stabilizer). Rank of $\rho_{2,1}$ is 40 and that of $\pi_{1,0}$ is 32. Again the action of $\mathfrak{G}^2$ on F_1 is transitive (8-dimensional stabilizer). Next the rank of $\rho_{3,2}$ is 80 and that of $\pi_{2,1}$ is 80 as well, the corresponding action is transitive. Finally rank of $\rho_{4,3}$ is 140 and that of $\pi_{3,2}$ is 160. The action of $\mathfrak{G}^4$ cannot be transitive. Moreover if we consider the action of G^4 on $\pi_3^{-1}(x)$, it cannot be transitive as well, because even though the action of G^3 has $8+8=16$-dimensional stabilizer, the difference in dimension is $160-140-16=4$. Thus there are at least 4 differential invariants. Note though that there are more (as we explained above), so that the action of G^4 has a large stabilizer.

n $= 3$. We do the dimensional count. Rank of $\rho_{1,0}$ is 36 and that of π is 18. The action of $\mathfrak{G}^1$ on F_0 is transitive (18-dimensional stabilizer). Rank of $\rho_{2,1}$ is 126 and that of $\pi_{1,0}$ is 108. It seems that the stabilizer should be 18-dimensional, but as we explained in Section 2 the action of G^2 has orbits of codimension 2, so the dimension of stabilizer is by two bigger than can be expected. Next rank of $\rho_{3,2}$ is 336 and that of $\pi_{2,1}$ is 378, so that the pure difference of dimensions gives at least $378-336-18\cdot 2-2=4$ differential invariants of order 3. All these invariants can be expressed via the normal forms of Ref. 6.

n > 3. Here the dimensional count can be misleading, so we better calculate codimension of orbits of $\mathrm{GL}(2n)$-action on the space of linear Nijenhuis tensors. The stabilizer of a linear complex structure J_0 on T is $\mathrm{GL}(n,\mathbb{C})$. Since $\dim \mathrm{GL}(n,\mathbb{C}) = 2n^2$ and $\dim \Lambda^2 T^* \otimes_{\bar{\mathbb{C}}} T = n^2(n-1)$ the largest orbits have codimension $n^2(n-3) + \dim \mathrm{St}$, where St is a stabilizer of a generic point. The result follows. □

This solves the equivalence problem for almost complex structures.

5. Existence of almost complex submanifolds

In a private communication M. Gromov asked the following question: how many higher-dimensional PH-submanifolds can an almost complex manifold possess? According to Refs. 3,6 generically there are none.

On the other end, for integrable J there're plenty. What happens in between? This question is quite difficult if PH-submanifolds are isolated, so we treat the case when they come in families, regularly fashioned, namely as PH-foliations (this allows investigation via linearization[8]).

We will consider in details 6-dimensional situation, the general case allows certain generalizations. Let us start with some examples.

Example 5.1. Let $M = \mathbb{C}^3$ with almost complex structure J being given in 2×2 block form $J = \operatorname{diag}(A_1, A_2, A_3)$, where the coefficients of $A_i \in C^\infty(M, \mathcal{J}(2))$ do depend on all 6 coordinates $(x^1, \dots, x^6) \in M$ in a generic way. Then the Nijenhuis tensor N_J is non-degenerate. Indeed, we have for i, j odd: $N_J(\partial_i, \partial_j) \equiv \mathbb{C}\langle\partial_i\rangle \oplus \mathbb{C}\langle\partial_j\rangle$, whence existence of kernel $\xi = \sum_{i \text{ odd}} \alpha_i \partial_i$ ($\alpha_i \in \mathbb{C}$ and multiplication means $\alpha \cdot \eta = \operatorname{Re}\alpha \cdot \eta + \operatorname{Im}\alpha \cdot J\eta$) implies that some of the vectors $N_J(\partial_i, \partial_j)$ have zero components in the above $\mathbb{C}^2$ decomposition.

In other words if we denote the above splitting as $M = V_1 \oplus V_2 \oplus V_3$, then the genericity condition is $N_J(V_i, V_j) \not\subset V_i$ for all $i \neq j$.

Thus we see that it is possible to have 3 transversal PH-foliation of (M^6, J) with J being maximally non-degenerate at each point. Generalization to dimension $2n$ is straightforward.

Example 5.2. The above example can be modified as follows. Let $V_{ij} = V_i \oplus V_j$ and let the almost complex structure have a block form $J = \operatorname{diag}(A, B)$ in the splitting $M = V_1 \oplus V_{23}$, where $A \in C^\infty(M, \mathcal{J}(2))$ and $B \in C^\infty(M, \mathcal{J}(4))$. While A-block is allowed to be arbitrary, the B-block is assumed symmetric in V_1-direction, i.e. independent of (x^1, x^2)-coordinates.

Then any PH-curve $\mathcal{C}^2 \subset V_{23}$ lifts to the 4D PH-submanifold $V_1 \times \mathcal{C}^2 \subset M$. Thus we have an infinite-dimensional family of PH-submanifolds $\mathbb{C} \times \mathcal{C}^2$, all of which intersect by a leaf of the 2D PH-foliation V_1.

Note that this family will persist if we allow the structure to have the form

$$J = \begin{pmatrix} A & B \\ 0 & D \end{pmatrix}$$

in the splitting $M = V_1 \oplus V_{23}$ with the block D projectable along V_1.

Definition 5.1. A family of 4D PH-submanifolds Φ_α intersecting by a PH curve $\mathcal{C}$ is a *pencil* if there exists a 2D PH-foliation in a neighborhood of $\mathcal{C}$ such that the projection along it is a PH-map and each Φ_α is projected to a PH-curve.

In other words in a neighborhood of the curve J is represented by the above upper-triangular block form. Then for such a pencil the tensor N_J is degenerate.

Let us recall basics about degenerations of linear Nijenhuis tensors.[6] Such a tensor can be considered as a $\mathbb{C}$-antilinear map $N_J : \Lambda^2_{\mathbb{C}} T \to T$ of vector spaces of $\dim_{\mathbb{C}} = 3$. So if $N_J \neq 0$ the following situations are possible:

NDG: $\dim_{\mathbb{C}} \operatorname{Im} N_J = 3$ (non-degenerate);
DG$_1$: $\dim_{\mathbb{C}} \operatorname{Im} N_J = 2$ (weakly degenerate);
DG$_2$: $\dim_{\mathbb{C}} \operatorname{Im} N_J = 1$, there is a kernel $V \in \operatorname{Gr}_1^{\mathbb{C}}(T)$, $N_J(V, \cdot) = 0$.

Generically a pencil belongs to DG$_1$ case. However DG$_2$ can be obtained in the two following cases:

1. A is projectable along V_{23} and $B = 0$ or A is constant in the above splitting and B is projectable along V_1. Then if the tensor $N_J \neq 0$, its kernel coincides with V_1.

2. Almost complex structure D on V_{23} is integrable. Then if the tensor $N_J \neq 0$, its kernel is transversal to V_1.

Proposition 5.1. *Let Φ_α be a family of 4-dimensional PH-foliations of 6-dimensional (M, J), intersecting by a common foliation V by PH-curves, such that shifts along V is a symmetry of the family as foliations.*

Let cardinality of indices α be at least 4 and at almost every point x there be 4 leaves Φ_{α_i} with $T_x\Phi_{\alpha_i}/T_xV$ of general position in T_xM/T_xV.

Then the family Φ_α is a pencil: There exists a PH-submersion $\pi : (M^6, J) \to (W^4, \tilde{J})$ with V-fibers.

Before proving this let us discuss the problem how an almost complex structure J is characterized by its PH-submanifolds. This question is non-void even in dimension 4, on which we concentrate. In this case the problem can be reformulated as a PH-analog of plane webs.

Lemma 5.1. *Let Ψ_a be a PH 4-web of almost complex (W^4, J), i.e. there are foliations Ψ_a, $1 \le a \le 4$, by PH-curves, none two of them being tangent anywhere. Then J is determined by Ψ_a up to sign.*

Proof. We will prove a more general statement: Let Ψ_a be a 4-webs of surfaces in W^4 with the same condition of general position at each point. Then there are at most two almost complex structures $\pm J$ making Ψ_a into PH-web.

Indeed, this is the question of linear algebra. We have $T_xW = \Pi_1 \oplus \Pi_2$, where $\Pi_a = T_x\Psi_a$ are 2-dimensional subspaces. Complex structures on Π_1 and Π_2 determine that on T_xW.

Since Π_3 is a complex subspace it is a graph of a complex linear map $F : \Pi_1 \to \Pi_2$. This map is nondegenerate and the complex structure on Π_1 determines that of Π_2. Now using Π_4, which is also a graph, we get a complex automorphism $L : \Pi_1 \to \Pi_1$, not proportional to identity. So no two different (up to sign) complex structures can commute with it. This proves uniqueness.

Let us discuss existence. It is equivalent to the claim the the spectrum of L is purely complex. Necessity is obvious: if $\mathrm{Sp}(L)$ is real simple or L is a Jordan box, no rotation can commute with it. On the other hand, if $\mathrm{Sp}(L) = \{\frac{\lambda \pm i}{\beta}\}$, then $J = \beta L - \lambda I$ is a complex structure on Π_1 and this gives the complex structure on $T_x W$. □

The above problem is equivalent to the following: Given a family of PH-foliations Ψ_a on (W^{2n}, J) and a diffeomorphism $f : W \to W$, mapping them to PH-foliations, how large should be the index set $\{a\}$ to ensure that f is a PH-map or anti-PH: $f^*J = \pm J$. Imposing general position of leaves, making it into PH-web, the modification of the above proof gives the answer $n+2$.

Proof of Proposition 5.1. Shift along transversal V maps transversal foliations Φ_α/V into themselves. Since they are complex PH-lines in TM/TV, Lemma 5.1 implies that the complex structure in quotient tangent spaces is preserved. Thus shifts along V preserve the almost complex structure J in normal direction. Thus the complex structure becomes of the upper-triangular block form and the result follows. □

Let us call pencils from this Proposition 4-pencils, because there are 4 foliations in it (but then it extends to a continuous family).

Remark 5.1. This proposition has certain generalizations to dimensions $2n > 6$, but then one should make more specifications (dimension of PH-foliations in the pencil, their number etc), so we do not discuss it.

6. Criteria of integrability

We present several approaches basing on existence of many PH-submanifolds. It was shown in Ref. 6 that whenever through every point $x \in (M^{2n}, J)$ and every complex [$\dim_{\mathbb{C}} = k$]-dimensional subspace in $T_x M$ passes a PH-submanifold of dimension $2k$ (or PH 2-jet), then J is integrable (k is fixed).

But with this we require an infinite number of PH-submanifolds to ensure integrability. This requirement can be much weakened with the same conclusion. We will specify as above to the case $n = 3$.

1. We can use the pencils of Proposition 5.1 to get another criterion as follows. Consider 5 foliations V_i of M^6 by PH-curves (these always exist locally), $1 \le i \le 5$, none two of which are tangent and none three have complex dependent tangents at almost any point.

Theorem 6.1. *Assume that (M, J) admits 10 PH-foliations Φ_{ij}, such that Φ_{ij} contain both V_i and V_j and are symmetric with respect to shifts along them. Then the structure J is integrable.*

Proof. Indeed in this case (M, J) admits 5 pencils of PH-foliations of dimension 4. Proposition 5.1 applies. In fact in this case the pencils become just as in Example 5.2 (without upper-triangular modification) because for any family $A_a = \{\Phi_{ak}\}_{k\neq a}$ there is a transversal PH-foliation Φ_{ij}, $i, j \neq a$. Thus the tensor N_J is degenerate.

Two pencils can have weak degeneracy along the same complex 2-plane from $\mathrm{Gr}_2(TM, \mathbb{C})$, but then the next two show another weak degeneracy, so that there is a kernel. The last pencil gives a weak degeneracy of N_J, independent of this kernel, whence the Nijenhuis tensor vanishes. □

The hypotheses of the theorem can be modified to have four 4-pencils, each having 3 common PH-foliations with the other pencils, leading to the same conclusion. However this provides the same total amount 10 of PH-foliations.

2. We can skip organizing PH-foliations in pencils and get the same claim, but then the number of foliations should grow.

Let us call family Φ_α of PH-foliations of dimension 4 quadratically non-degenerate if at almost every point $x \in M$ the tangents $T_x\Phi_\alpha \in \mathrm{Gr}_2(T_xM, \mathbb{C}) \simeq \mathbb{C}P^2$ do not belong to any real quadric of codimension 1. Note that any 14 points in $\mathbb{C}P^2$ do belong to a real quadric.

Theorem 6.2. *Let Φ_α be a family of PH-foliations of dimension 4 in (M^6, J), $\alpha = 1, \dots, 15$. If it is quadratically non-degenerate, then J is integrable.*

Proof. If $N_J \neq 0$, then the Grassmanian of 4-planes in T_xM, which are invariants with respect to both J and N_J is a real quadric of codimension 2 in NDG case or codimension 1 in DG cases of Ref. 6 (it can be also empty, then its codimension is 4). But no 15 generic points in $\mathrm{Gr}_2^{\mathbb{C}}(T_xM)$ can belong to a quadric. □

3. We can have some intermediate criteria between approach 1, using fewer number of PH-submanifolds though with some integrability assumptions, and approach 2, using larger number of PH-submanifolds but only genericity conditions. For example, assume we have 14 families of 4D PH-foliations of (M^6, J), which have generic arrangements of tangents at almost every point.

Then we have a field of quadrics $Q_x \subset T_xM$, $x \in M$. If the structure is non-integrable, this field satisfies certain integrability criteria ($\Xi(\pi_*\Theta_H(\Pi)) = 0$ from Ref. 6). This is a binding requirement.

Theorem 6.3. *Let Φ_α be a family of PH-foliations of dimension 4 in (M^6, J), $\alpha = 1, \dots, 14$ with generic arrangements of tangents a.e. If the corresponding family of quadrics Q is non-integrable, then J is integrable.*

Acknowledgments

The author thanks IHES for hospitality during the visit of January 2007, when this work was initiated.

References

1. D. Alekseevskij, V. Lychagin and A. Vinogradov, *Basic ideas and concepts of differential geometry* (Encyclopedia of mathematical sciences 28, Geometry 1, Springer-Verlag, 1991).
2. R. Bryant, On the geometry of almost complex 6-manifolds, *Asian J. Math.* **10** (3) (2006) 561–605.
3. M. Gromov, Pseudo-holomorphic curves in symplectic manifolds, *Invent. Math.* **82** (1985) 307–347.
4. S. Kobayashi, *Transformation groups in Differential geometry* (Springer-Verlag, 1972).
5. B.S. Kruglikov, Nijenhuis tensors and obstructions for pseudoholomorphic mapping constructions, *Math. Notes* **63** (4) (1998) 541–561.
6. B.S. Kruglikov, Non-existence of higher-dimensional pseudoholomorphic submanifolds, *Manuscripta Mathematica* **111** (2003) 51–69.
7. B.S. Kruglikov, Characteristic distributions on 4-dimensional almost complex manifolds, *In: Geometry and Topology of Caustics - Caustics '02* (Banach Center Publications 62, 2004) 173–182.
8. B.S. Kruglikov, Tangent and normal bundles in almost complex geometry, *Diff.Geom.Appl.* **25** (4) (2007) 399–418.

9. B.S. Kruglikov and V.V. Lychagin, Invariants of pseudogroup actions: Homological methods and Finiteness theorem, *Int. J. Geomet. Meth. Mod. Phys.* **3** (5 & 6) (2006) 1131–1165.
10. A. Kumpera, Invariants differentiels d'un pseudogroupe de Lie. I, *J. Differential Geometry* **10** (2) (1975) 289–345; II, ibid. **10** (3) (1975) 347–416.
11. A. Newlander, L. Nirenberg, Complex analytic coordinates in almost-complex manifolds, *Ann. Math.* **65**, ser. 2, issue 3 (1957) 391–404.
12. M. Verbitsky, Hodge theory on nearly Kahler manifolds; arXiv:math.DG/0510618 (2005).

Differential Geometry and its Applications
Proc. Conf., in Honour of Leonhard Euler, Olomouc, August 2007

Hirzebruch signature operator for transitive Lie algebroids

Jan Kubarski

Institute of Mathematics, Technical University of Łódź,
Poland
E-mail: kubarski@p.lodz.pl

The aim of the paper is to construct Hirzebruch signature operator for transitive invariantly oriented Lie algebroids.

Keywords: Lie algebroid, Hirzebruch signature operator, $*$-Hodge operator, elliptic complex, invariantly oriented Lie algebroid.

MS classification: 58A14, 58J10.

1. Signature of Lie algebroids

1.1. *Definition of Lie algebroids, Atiyah sequence*

Lie algebroids appeared as infinitesimal objects of Lie groupoids, principal fibre bundles, vector bundles (Pradines, 1967), TC-foliations and non-closed Lie subgroups (Molino, 1977), Poisson manifolds (Dazord, Coste, Weinstein, 1987), etc. Their algebraic equivalents are known as Lie pseudo-algebras (Herz, 1953) called also further as Lie-Rinehart algebras (Huebschmann, 1990).

A *Lie algebroid* on a manifold M is a triple

$$A = (A, [\![\cdot,\cdot]\!], \#_A)$$

where A is a vector bundle on M, $(\operatorname{Sec} A, [\![\cdot,\cdot]\!])$ is an $\mathbb{R}$-Lie algebra,

$$\#_A : A \to TM$$

is a linear homomorphism (called the *anchor*) of vector bundles and the following Leibniz condition is satisfied

$$[\![\xi, f\cdot\eta]\!] = f\cdot[\![\xi,\eta]\!] - \#_A(\xi)(f)\cdot\eta, \quad f\in C^\infty(M),\ \xi,\eta\in\operatorname{Sec} A.$$

The anchor is bracket-preserving,[12]

$$\#_A \circ [\![\xi,\eta]\!] = [\#_A\circ\xi, \#_A\circ\eta].$$

A Lie algebroid is called *transitive* if $\#_A$ is an epimorphism.

For a transitive Lie algebroid A we have the Atiyah sequence

$$0 \longrightarrow \boldsymbol{g} \hookrightarrow A \xrightarrow{\#_A} TM \longrightarrow 0,$$

$\boldsymbol{g} := \ker \#_A$. The fiber $\boldsymbol{g}_x$ of the bundle $\boldsymbol{g}$ in the point $x \in M$ is the Lie algebra with the commutator operation being

$$[v, w] = [\![\xi, \eta]\!](x), \quad \xi, \eta \in \operatorname{Sec} A, \quad \xi(x) = v, \eta(x) = w, \quad v, w \in \boldsymbol{g}_x.$$

The Lie algebra $\boldsymbol{g}_x$ is called the isotropy Lie algebra of A at $x \in M$. The vector bundle $\boldsymbol{g}$ is a Lie Algebra Bundle (LAB in short), called the *adjoint* of A, the fibres are isomorphic Lie algebras.

TM is a Lie algebroid with $id : TM \to TM$ as the anchor,

$\mathfrak{g}$ -finitely dimensional Lie algebra - is a Lie algebroid over a point $M = \{*\}$.

1.2. *Cohomology algebra, ellipticity of the complex of exterior derivatives* $\{d_A^k\}$

To a Lie algebroid A we associate the cohomology algebra $\mathbf{H}(A)$ defined via the DG-algebra of A-differential forms (with real coefficients) $(\Omega(A), d_A)$, where

$$\Omega(A) = \operatorname{Sec} \bigwedge A^*, \quad \text{- the space of cross-sections of } \bigwedge A^*$$

$$d_A : \Omega^\bullet(A) \to \Omega^{\bullet+1}(A)$$

$$\begin{aligned}(d_A\omega)(\xi_0, ..., \xi_k) &= \sum_{j=0}^{k} (-1)^j \, (\#_A \circ \xi_j)(\omega(\xi_0, ...\hat{\jmath}..., \xi_k)) \\ &\quad + \sum_{i<j} (-1)^{i+j} \, \omega([\![\xi_i, \xi_j]\!], \xi_0, ...\hat{\imath}...\hat{\jmath}..., \xi_k),\end{aligned}$$

$\omega \in \Omega^k(A)$, $\xi_i \in \operatorname{Sec} A$. The operators d_A^k satisfy

$$d_A(\omega \wedge \eta) = d_A\omega \wedge \eta + (-1)^k \omega \wedge d_A\eta,$$

so they are of first order and the symbol of d_A^k is equal to

$$\begin{aligned} S(d_A^k)_{(x,v)} &: \bigwedge^k A_x^* \to \bigwedge^{k+1} A_x^* \\ S(d_A^k)_{(x,v)}(u) &= (v \circ (\#_A)_x) \wedge u, \quad 0 \neq v \in T_x^* M.\end{aligned}$$

In consequence the sequence of symbols

$$\bigwedge^k A_x^* \overset{S(d_A^k)_{(x,v)}}{\longrightarrow} \bigwedge^{k+1} A_x^* \overset{S(d_A^{k+1})_{(x,v)}}{\longrightarrow} \bigwedge^{k+2} A_x^*$$

is exact if and only if A is transitive and then the complex $\{d_A^k\}$ is an elliptic complex.

The exterior derivative d_A introduces the cohomology algebra

$$\mathbf{H}(A) = \mathbf{H}(\Omega(A), d_A).$$

For the trivial Lie algebroid TM - the tangent bundle of the manifold M - the differential d_{TM} is the usual de-Rham differential d_M of differential forms on M whereas, for $L = \mathfrak{g}$ - a Lie algebra $\mathfrak{g}$ - the differential $d_{\mathfrak{g}}$ is the usual Chevalley-Eilenberg differential, $d_{\mathfrak{g}} = \delta_{\mathfrak{g}}$.

1.3. *Invariantly oriented Lie algebroids and signature*

The following theorem describes the class of transitive Lie algebroids (over compact oriented manifold) for which $\mathbf{H}^{\mathrm{top}}(A) \neq 0$.

Theorem 1.1 (Ref. 5). *For each transitive Lie algebroid* $(A, [\![\cdot,\cdot]\!], \#_A)$ *with the Atiyah sequence*

$$0 \to \boldsymbol{g} \to A \overset{\#_A}{\longrightarrow} TM \to 0,$$

over a compact oriented manifold M the following conditions are equivalent ($m = \dim M$, $n = \dim \boldsymbol{g}_{|x}$, *i.e.* $\operatorname{rank} A = m + n$)

(a) $\mathbf{H}^{m+n}(A) \neq 0$,

(b) $\mathbf{H}^{m+n}(A) \cong \mathbb{R}$ *and* $\mathbf{H}(A)$ *is an Poincaré algebra, i.e. the pairing* $\mathbf{H}^j(A) \times \mathbf{H}^{m+n-j}(A) \to \mathbf{H}^{m+n}(A) \cong \mathbb{R}$ *is nondegenerate,* $\mathbf{H}^j(A) \cong (\mathbf{H}^{m+n-j}(A))^*$,

(c) *there exists a global nonsingular cross-section* $\varepsilon \in \operatorname{Sec}(\bigwedge^n \boldsymbol{g})$ *invariant with respect to the adjoint representation* ad_A, *that is, A is the so-called a TUIO-Lie algebroid, see Ref. 3, (shortly, A is invariantly oriented),*

(d) *the vector bundle $\boldsymbol{g}$ is orientable and the modular class of A is trivial,* $\theta_A = 0$.

We recall the definition of the isomorphism $\mathbf{H}^{m+n}(A) \cong \mathbb{R}$ (for invariantly oriented transitive Lie algebroids). In Ref. 3 there is defined (for arbitrary transitive Lie algebroids) the so-called *fibre integral operator*

$$\int_A : \Omega^\bullet(A) \to \Omega_{dR}^{\bullet - n}(M)$$

by the formula

$$\left(\oint_A \omega^k\right)_x (w_1, ..., w_{k-n}) = (-1)^{kn}\, \omega_x (\varepsilon_x, \tilde{w}_1, ..., \tilde{w}_{k-n}), \quad \#_A (\tilde{w}_i) = w_i,$$

where $\varepsilon \in \mathrm{Sec}\,(\bigwedge^n \boldsymbol{g})$ is a nonsingular cross-section. The operator $\oint_A$ commutes with the differentials d_A and d_M if and only if ε is invariant. Then, the fibre integral gives a homomorphism in cohomology

$$\oint_A^{\#} : \mathbf{H}^\bullet (A) \to \mathbf{H}_{dR}^{\bullet - n} (M).$$

Assume in the sequel that a transitive Lie algebroid A over compact oriented manifold M is invariantly oriented and $\varepsilon \in \mathrm{Sec}\,(\bigwedge^n \boldsymbol{g})$ is an invariant cross-section. The scalar Poincaré product

$$\mathcal{P}_A^k : \mathbf{H}^k (A) \times \mathbf{H}^{m+n-k} (A) \to \mathbb{R},$$

$$([\omega], [\eta]) \longmapsto \int_A \omega \wedge \eta \quad \left(:= \int_M \left(\oint_A \omega \wedge \eta \right) \right)$$

is defined and is nondegenerated;[4] in consequence

$$\begin{aligned} &\mathbf{H}^k (A) \cong \mathbf{H}^{m+n-k} (A), \\ &\mathbf{H}^{m+n} (A) \cong \left(\mathbf{H}^0 (M)\right)^* = \mathbb{R}, \\ &\dim \mathbf{H} (M) < \infty, \end{aligned}$$

and we can consider an isomorphism

$$\mathbf{H}^{m+n} (A) \cong \mathbb{R}, \quad [\omega] \longmapsto \int_A \omega.$$

The pairing of A-differential forms

$$\begin{aligned} &\langle\langle \cdot, \cdot \rangle\rangle^k : \Omega^k (A) \times \Omega^{m+n-k} (A) \to \mathbb{R} \\ &\langle\langle \omega, \eta \rangle\rangle^k = \int_M \oint_A \omega \wedge \eta \end{aligned} \tag{1}$$

has the property

$$\langle\langle \omega, \eta \rangle\rangle^k = (-1)^{k(m+n-k)} \langle\langle \eta, \omega \rangle\rangle^{m+n-k}$$

and

$$\langle\langle d_A \omega, \eta \rangle\rangle = (-1)^{k+1} \langle\langle \omega, d_A \eta \rangle\rangle \quad \text{for} \quad \omega \in \Omega^k (A), \ \eta \in \Omega^{m+n-(k+1)} (A).$$

If

$$m + n = 4p$$

then

$$\mathcal{P}_A^{2p} : \mathbf{H}^{2p}(A) \times \mathbf{H}^{2p}(A) \to \mathbb{R}$$

is nondegenerated and symmetric. Therefore its signature is defined and is called the signature of A, and is denoted by

$$\mathrm{Sign}\,(A)\,.$$

The problem is:[4]

- to calculate the signature Sign (A) and give some conditions to the equality Sign $(A) = 0$. There are examples for which Sign $(A) \neq 0$ (this is announced in Ref. 6).

2. $*$- operator and exterior coderivative d_A^*

2.1. *Associated scalar product and $*$-Hodge operator*

Consider

- any Riemannian tensor G_1 in the vector bundle $\boldsymbol{g} = \ker \#_A$ for which ε is the volume tensor (such a tensor exists).
- any Riemannian tensor G_2 on M.

Next, taking an arbitrary connection $\lambda : TM \to A$ in the Lie algebroid A i.e. a splitting of the Atiyah sequence

$$0 \longrightarrow \boldsymbol{g} \hookrightarrow A \underset{\lambda}{\overset{\#_A}{\longrightarrow}} TM \longrightarrow 0,$$

and the horizontal space

$$H = \operatorname{Im} \lambda,$$
$$A = \boldsymbol{g} \bigoplus H$$

we define a Riemannian tensor G (called *scalar product associated* to ε) on $A = \boldsymbol{g} \bigoplus H$ such that $\boldsymbol{g}$ and H are orthogonal, on $\boldsymbol{g}$ we have G_1 but on H we have the pullback $\lambda^* G_2$. The vector bundle A is oriented (since $\boldsymbol{g}$ and M are oriented).

At each point $x \in M$ we consider the scalar product G_x on $A_{|x}$ and the pairing of tensors

$$\langle , \rangle_x^k : \bigwedge^k A_x^* \times \bigwedge^{m+n-k} A_x^* \to \bigwedge^{m+n} A_x^* \overset{\rho_x}{\to} \mathbb{R}$$

where ρ_x is defined via the volume form for G_x.

We can notice that ρ_x is equal to the composition

$$\begin{array}{ccc} \bigwedge^{m+n} A_x^* & \xrightarrow{\rho_x} & \mathbb{R} \\ \downarrow (-1)^{(m+n)n} i_{\varepsilon_x} \searrow \fint_{A_p} & & \nearrow \rho_{G_2 x} \\ \bigwedge^{m} A_x^* & \xrightarrow{\rho_{\lambda x}} \bigwedge^{m} T_x^* M & \end{array}$$

$$i_{\varepsilon_x}\omega_x\left(\left(v_1, ..., v_{k-n}\right)\right) = \omega_x\left(\varepsilon_x, v_1, ..., v_{k-n}\right),$$
$$\rho_{\lambda x}\left(\omega_x\right)\left(w_1, ..., w_{k-n}\right) = \omega_x\left(\varepsilon_x, \lambda_x\left(w_1\right), ..., \lambda_x\left(w_{k-n}\right)\right).$$

Standardly, we can extend the scalar product G_x in A_x to a scalar product $(\cdot,\cdot)_x$ in $\bigwedge A_x^*$. There exists exactly one the so-called $*$-Hodge operator $*_x : \bigwedge^k A_x^* \to \bigwedge^{m+n-k} A_x^*$ such that

$$\langle \alpha_x, \beta_x \rangle = \left(\alpha_x, *_x \beta_x\right),$$

and it is given by

$$*_x\left(e_{i_1}^* \wedge ... \wedge e_{i_k}^*\right) = \operatorname{sgn}\left(j_1, ..., j_{m+n-k}, i_1, .., i_k\right) e_{j_1}^* \wedge ... \wedge e_{j_{m+n-k}}^*$$

where $1 \leq i_1 < ... < i_k \leq m+n$, $1 \leq j_1 < ... < j_{m+n-k} \leq m+n$ and $\{i_1, ..., i_k\} \cap \{j_1, ..., j_{m+n-k}\} = \varnothing$ (for ON positive frame e_i in A_x). Using all points $x \in M$ we obtain two $C^\infty(M)$-linear 2-tensors

$$\langle , \rangle, (,) : \Omega(A) \times \Omega(A) \to C^\infty(M)$$

defined as above point by point. Integrating along M we get $\mathbb{R}$-linear 2-tensors

$$\langle\langle , \rangle\rangle, ((,)) : \Omega(A) \times \Omega(A) \to \mathbb{R}.$$

The first $\langle\langle , \rangle\rangle$ is, clearly, given by (1)

$$\langle\langle \omega, \eta \rangle\rangle = \int_M \langle \omega, \eta \rangle = \int_M \fint_A \omega \wedge \eta.$$

The $*$-Hodge operator $* : \Omega(A) \to \Omega(A)$ is defined point by point

$$*(\omega)(x) = *_x(\omega_x)$$

and we have

$$\langle\langle \alpha, \beta \rangle\rangle = ((\alpha, *\beta)).$$

2.2. *Exterior coderivative*

We define exterior coderivative $d_A^* : \Omega^k(M) \longrightarrow \Omega^{m+n-k}(A)$ by the formula

$$d_A^*(\omega) = (-1)^{k(m+n-k)}(-1)^k * d_A * (\omega), \quad \omega \in \Omega^k(A),$$

where $*$ is the $*$-Hodge operator in $\Omega(A)$. We have

(a)

$$** (\omega) = (-1)^{k(m+n-k)} \cdot \omega, \quad \omega \in \Omega^k(A), \tag{2}$$

(b)

$$((d^*(\omega), \eta)) = ((\omega, d_A(\eta))),$$

i.e. d_A^* is adjoint to d_A with respect to the scalar product $((,))$ in $\Omega(A)$.

3. Laplacian and harmonic differential forms

It enables us to introduce the Laplacian

$$\Delta_A = (d_A + d_A^*)^2 = d_A d_A^* + d_A^* d_A.$$

Clearly

$$\ker \Delta_A = (\operatorname{Im} \Delta_A)^{\perp}$$

(with the respect to the scalar product $((,))$).

Proposition 3.1. *The Laplacian Δ_A is elliptic, self-adjoint and nonnegative operator. In consequence*

$$\Omega(A) = \ker \Delta_A \bigoplus \operatorname{Im} \Delta_A. \tag{3}$$

Proof. The first property follows from the ellipticity of the complex $\{d_A^k\}$ (namely: the symbol of the adjoint operator d_A^{*k} is equal to the minus of the adjoint symbol of d_A^k and to prove the ellipticity of the Laplacian we use Remark 6.34 from Ref. 8), the next two properties are trivial consequence of the definition. The last property (3) can be proved in the same way as the Theorem 55 from Ref. 7 using extension $\operatorname{Sec} \bigwedge A^*$ to the Hilbert Sobolev spaces $H_s(\bigwedge A^*)$, extension Δ to continuous operator $\Delta_s : H_s(\bigwedge A^*) \to H_{s-2}(\bigwedge A^*)$ and the fact that $\ker \Delta_s$ consists only of smooth sections, $\ker \Delta_s = \ker \Delta$. □

A A-differential form $\omega \in \Omega(A)$ is called *harmonic* if $d_A\omega = 0$ i $d_A^*\omega = 0$. Denote the space of harmonic A-differential forms by $\mathcal{H}(A)$ and harmonic k- A-differential forms by $\mathcal{H}^k(A)$. $\mathcal{H}(A)$ is a graded vector space

$$\mathcal{H}(A) = \bigoplus_{k=0}^{m+n} \mathcal{H}^k(A)$$

and

$$\mathcal{H}(A) = \ker \Delta_A.$$

is the eigenspace of the operator Δ corresponding to the zero value of the eigenvalue.

Simple calculations assert that $\ker \Delta^k$ and $\operatorname{Im} d^{k-1}$ are orthogonal, therefore the inclusion

$$\mathcal{H}^k(W) = \ker \Delta^k \hookrightarrow \ker d^k$$

induce a monomorphism

$$\ker \Delta^k \rightarrowtail \mathbf{H}^k(W). \tag{4}$$

Since $\operatorname{Im} \Delta^k \subset Imd_A^{k-1} + \operatorname{Im} d_A^{*(k+1)}$, the inclusion (3) yields $\Omega^k(A) = \ker \Delta_A^k + \operatorname{Im} d_A^{k-1} + \operatorname{Im} d_A^{*(k+1)}$ and easily we can notice that these three subspaces are orthogonal. Therefore

$$\Omega^k(A) = \ker \Delta_A^k \bigoplus \operatorname{Im} d_A^{k-1} \bigoplus \operatorname{Im} d_A^{*(k+1)}$$

and

$$\ker d^k = \ker \Delta^k \bigoplus \operatorname{Im} d^{k-1}$$

which implies the Hodge Theorem for Lie algebroids:

Corollary 3.1. *The monomorphism (4) is an isomorphism*

$$\mathcal{H}^k(W) \cong \ker d^k / \operatorname{Im} d^{k-1} = \mathbf{H}^k(W).$$

It means that in each cohomology class $\alpha \in \mathbf{H}^k(A)$ there is exactly one harmonic A-differential form $\omega \in \mathcal{H}^k(W)$.

Let $\varepsilon_k = (-1)^{k(n+m-k)}$. Simple calculations yields the equality

$$*\Delta_A\omega = \varepsilon_{k-1}\varepsilon_k(-1)^{n+m+1}\Delta_A * \omega, \quad \omega \in \Omega^k(A),$$

therefore

$$*\left[\mathcal{H}^k(W)\right] \subset \mathcal{H}^{m+n-k}(W),$$

and (thanks (2))

$$* : \mathcal{H}^k(W) \to \mathcal{H}^{m+n-k}(W)$$

is an isomorphism. In consequence we obtain (independently on Ref. 4) the Duality Theorem

$$\mathbf{H}^k(A) \simeq \mathbf{H}^{m+n-k}(A).$$

We restrict the scalar product $((\cdot,\cdot)) : \Omega^k(A) \times \Omega^k(A) \to \mathbb{R}$ to the space of harmonic A-differential forms

$$((\cdot,\cdot)) : \mathcal{H}^k(A) \times \mathcal{H}^k(A) \to \mathbb{R},$$

and we restrict the tensor $\langle\langle \cdot,\cdot \rangle\rangle : \Omega^k(A) \times \Omega^{m+n-k}(A) \to \mathbb{R}$ to harmonic A-differential forms

$$\mathcal{B}^k = \langle\langle \cdot,\cdot \rangle\rangle : \mathcal{H}^k(A) \times \mathcal{H}^{m+n-k}(A) \to \mathbb{R}.$$

Using the isomorphism $\mathcal{H}^k(A) \cong \mathbf{H}^k(A)$ we see that

$$\mathcal{B}^k = \mathcal{P}^k_A,$$

therefore, if $m+n=4p$ then

$$\text{Sign}\,(A) = \text{Sign}\,\mathcal{B}^{2p}.$$

4. Hirzebruch signature operator

Assume $m+n = 4p$. Considering the direct sum $\mathcal{H}^{2p}(A) = \mathcal{H}^{2p}_+(A) \bigoplus \mathcal{H}^{2p}_-(A)$, where

$$\mathcal{H}^{2p}_\pm(W) = \left\{\omega \in \mathcal{H}^{2p}(W);\ *\omega = \pm\omega\right\},$$

and noticing that $\mathcal{B}^{2p}$ is positive on $\mathcal{H}^{2p}_+(A)$, and is negative on $\mathcal{H}^{2p}_-(A)$ we see that

$$\text{Sign}\,\left(\mathcal{B}^{2p}\right) = \dim_{\mathbb{R}} \mathcal{H}^{2p}_+(W) - \dim_{\mathbb{R}} \mathcal{H}^{2p}_-(W).$$

To construction the Hirzebruch signature operator the fundamental role is played by an auxiliary operator

$$\tau : \Omega(A) \to \Omega(A)$$

defined by

$$\tau^k\left(\omega^k\right) = \tilde{\varepsilon}_k * \left(\omega^k\right),\quad \tilde{\varepsilon}_k \in \{-1,+1\},\quad \omega^k \in \Omega^k(A),$$

fulfilling the properties

i) $\tau \circ \tau = Id$,
ii) $d_A^* = -\tau \circ d_A \circ \tau$,
iii) $\tau^{2p} = *$.

Lemma 4.1. *The operator τ fulfills axioms i)-iii) if and only if* $\tilde{\varepsilon}_k = (-1)^{\frac{k(k+1)}{2}}(-1)^p$.

Proof. Easy calculations. □

We put

$$\Omega(A)_+ = \{\omega \in \Omega(A);\ \tau\omega = +\omega\},$$
$$\Omega(A)_- = \{\omega \in \Omega(A);\ \tau\omega = -\omega\},$$

The spaces $\Omega(A)_+$ and $\Omega(A)_-$ are eigenspaces of τ corresponding to the eigenvalues $+1$ i -1 and are spaces of cross-sections of suitable vector bundles.

We notice that

$$(d_A + d_A^*)\left[\Omega(A)_+\right] \subset \Omega(A)_- .$$

Definition 4.1. The operator

$$(D_A)_+ = d_A + d_A^* : \Omega(A)_+ \to \Omega(A)_-$$

is called the *Hirzebruch operator* (or the *signature operator*) for the Lie algebroid A.

Clearly

$$(D_A)_+^* = d_A + d_A^* : \Omega(A)_- \to \Omega(A)_+ .$$

Theorem 4.1.

$$\operatorname{Sign} A = \operatorname{Ind}(D_A)_+ = \dim_{\mathbb{R}} \ker\left((D_A)_+\right) - \dim_{\mathbb{R}} \ker\left((D_A)_+^*\right).$$

Proof. It is sufficient to prove that $\operatorname{Ind}(D_A)_+ = \operatorname{Sign}\left(\mathcal{B}^{2p}\right)$. The proof is analogous to the classical case.[9] Firstly, we notice that subspaces $\mathcal{H}^s(A) + \mathcal{H}^{m+n-s}(A)$ are τ-stable and for $s = 0, 1, ..., 2p-1$

$$\begin{aligned} \varphi_\pm : \mathcal{H}^s(A) &\to \left(\mathcal{H}^s(A) + \mathcal{H}^{m+n-s}(A)\right)_\pm \\ \omega &\longmapsto \frac{1}{2}(\omega \pm \tau\omega) \end{aligned}$$

is an isomorphism of real vector spaces. Secondly $\Omega^{2p}(A)_\pm \cap \mathcal{H}^{2p}(A) = \mathcal{H}_\pm^{2p}(A)$, the space $\Omega^s(A) + \Omega^{m+n-s}(A)$ is τ-stable and

$$\Omega(A) = \bigoplus_{s=0}^{2p-1}\left(\Omega^s(A) + \Omega^{m+n-s}(A)\right) \bigoplus \Omega^{2p}(A),$$

therefore

$$\Omega(A)_+ = \bigoplus_{s=0}^{2p-1} \left(\Omega^s(A) + \Omega^{m+n-s}(A)\right)_+ \bigoplus \Omega^{2p}(A)_+ .$$

Thirdly, (a)

$$\begin{aligned}
&\ker (D_A)_+ \\
&= \Omega(A)_+ \cap \ker\left(d_A + d_A^* : \Omega(A) \to \Omega(A)\right) \\
&= \Omega(A)_+ \cap \mathcal{H}(A) \\
&= \bigoplus_{s=0}^{2p-1} \left(\Omega^s(A) + \Omega^{m+n-s}(A)\right)_+ \bigoplus \Omega^{2p}(A)_+ \\
&\cap \bigoplus_{s=0}^{2p-1} \left(\mathcal{H}^s(A) + \mathcal{H}^{m+n-s}(A)\right) \oplus \mathcal{H}^n(A) \\
&= \bigoplus_{s=0}^{2p-1} \left(\mathcal{H}^s(A) + \mathcal{H}^{m+n-s}(A)\right)_+ \bigoplus \mathcal{H}^{2p}(A)_+
\end{aligned}$$

(b)

$$\begin{aligned}
&\dim\ker (D_A)_+ - \dim\ker (D_A)_+^* \\
&= \sum_{s=0}^{2p-1} \dim_{\mathbb{R}} \left(\mathcal{H}^s(A) + \mathcal{H}^{m+n-s}(A)\right)_+ + \dim_{\mathbb{R}} \mathcal{H}_+^{2p}(A) \\
&- \sum_{s=0}^{2p-1} \dim_{\mathbb{R}} \left(\mathcal{H}^s(A) + \mathcal{H}^{m+n-s}(A)\right)_- - \dim_{\mathbb{R}} \mathcal{H}_-^{2p}(A) \\
&= \dim_{\mathbb{R}} \mathcal{H}_+^{2p}(A) - \dim_{\mathbb{R}} \mathcal{H}_-^{2p}(A) \\
&= \operatorname{Sign}(\mathcal{B}^n) .
\end{aligned}$$

□

Thanks the above Theorem, we can use the Atiyah-Singer formula for calculating the signature of A.

References

1. B. Balcerzak, J. Kubarski and W. Walas, Primary characteristic homomorphism of pairs of Lie algebroids and Mackenzie algebroid, *In: Lie Algebroids and Related Topics in Differential Geometry* (Banach Center Publicationes, Volume 54, IMPAN Warszawa, 2001) 71–97.
2. J.C. Herz, Pseudo-algèbres de Lie, *C.R.Acad.Sci.Paris* **236** (1953), I, 1935–1937 and II, 2289–2291.
3. J. Kubarski, Fibre integral in regular Lie algebroids, *In: New Developments in Differential Geometry* (Budapest 1996, KLUWER ACADEMIC PUBLISHERS, Dordrecht, 1999, Proceedings of the Conference on Differential Geometry, Budapest, Hungary, 27-30 July 1996) 173–202.
4. J. Kubarski, Poincaré duality for transitive unimodular invariantly oriented Lie algebroids, *Topology and Its Applications* **121** (2002) 333–355.

5. J. Kubarski and A. Mishchenko, Nondegenerate cohomology pairing for transitive Lie algebroids, characterization, *Central European Journal of Mathematics* **2** (5), pp. 1–45, (2004) 663–707.
6. J. Kubarski and A. Mishchenko, On signature of transitive unimodular Lie algebroids, *Doklady Mathematical Sciences* **68** (5/1) (2003) 166–169.
7. G. Luke and A.S. Mishchenko, *Vector Bundles and Their Applications* (Kluwer Academic Publishers, 1998).
8. F.W. Warner, *Foundations of differentiable manifolds and Lie groups* (Scott, Foresman and Company, Gleniew, Illinois, London, 1971).
9. Y. Yu, *Index Theorem and Heat Equation Method* (World Scientific, Nankai Tracts in Mathematics, Vol. 2, 2000).

Differential Geometry and its Applications
Proc. Conf., in Honour of Leonhard Euler, Olomouc, August 2007

Quadratic Nambu-Poisson structures

Nobutada Nakanishi

Department of Mathematics, Gifu Keizai University,
5-50 Kitagata, Ogaki-city, Gifu, 503-8550, Japan
E-mail: nakanisi@gifu-keizai.ac.jp

Using the result of the classification of linear Nambu-Poisson structures (Dufour-Zung[2]), we will classify quadratic Nambu-Poisson structures on a vector space. This is equivalent to the classification of quadratic integrable differential forms. Our method to get quadratic integrable differential forms is available for higher degree homogeneous integrable differential forms.

Keywords: Nambu-Poisson structures, integrable differential forms, Koszul operator.

MS classification: 53D17

1. Introduction

It is an interesting problem to classify homogeneous foliations. As is well-known,[5,7] a Nambu-Poisson structure is, in general, a singular foliation. (See the next section for the precise definition of Nambu-Poisson structures.), In 1999, J.-P.Dufour and N.T.Zung[2] gave the classification of linear Nambu-Poisson structures. This result can be seen as the first step of the classification of homogeneous foliations.

As the next step, here we will classify quadratic Nambu-Poisson structures. Quadratic Nambu-Poisson structures on a vector space are those Nambu-Poisson structures where the Nambu-Poisson brackets of coordinates functions are homogeneous quadratic polynomials. To classify them, we use the result of linear Nambu-Poisson structures and Koszul operator.[4]

Let η be a quadratic Nambu-Poisson structure of order q on a vector space of dimension n, and let Ω be a constant volume form. Then the interior product $i_\eta\Omega$ is a quadratic integrable $(n-q)$- form.

A p-form ω is called *integrable* if ω satisfies the following two conditions: For any $(p-1)$-vector A,

(1) $i_A\omega \wedge \omega = 0$,

(2) $i_A\omega \wedge\ d\omega = 0$.

Hence the classification of homogeneous integrable differential forms is equivalent to that of homogeneous Nambu-Poisson structures. (See Proposition 2.2.) C. Camacho and A.L. Neto[1] considered integrable 1-forms. By using our method, we can extend their results.

2. Nambu-Poisson structures and integrable differential forms

Let M be a smooth n-dimensional manifold and $C^\infty(M)$ the algebra of real-valued C^∞-functions on M. We denote by $\Gamma(\Lambda^q TM)$ the space of sections from M to $\Lambda^q TM$. And each element of $\Gamma(\Lambda^q TM)$ is simply called *q-vector*. Then each q-vector η defines a bracket of functions $f_i \in C^\infty(M)$ by

$$\{f_1, ..., f_q\} = \eta(df_1, ..., df_q).$$

This bracket also defines a vector fields $X_{f_1,..,f_{q-1}}$ by

$$X_{f_1,...,f_{q-1}}(g) = \{f_1, ..., f_{q-1}, g\}, \quad g \in C^\infty(M).$$

This vector field is called a Hamiltonian vector field.

Definition 2.1. An element η of $\Gamma(\Lambda^q TM)$, $q \geq 3$, is called a Nambu-Poisson structure of order q if η satisfies

$$\mathcal{L}_{X_{f_1,...,f_{n-1}}}\eta = 0,$$

for any Hamiltonian vector field $X_{f_1,...,f_{n-1}}$.

This definition was proposed by L.Takhtajan[7] in 1994. If $q = 2$, this is nothing but the definition of usual Poisson structure. In studying the geometry of Nambu-Poisson manifolds, the following theorem, which is called "local structure theorem" is fundamental. Let $\eta(x) \neq 0$, $x \in M$. Then η is said to be *regular* at x, and x is called *a regular point.*

Theorem 2.1 (Refs. 3,5). *If η is a Nambu-Poisson structure of order $q \geq 3$, then for any regular point x, there exists a coordinate neighborhood U with local coordinates $(x_1, ..., x_q, x_{q+1}, ..., x_n)$ around x such that*

$$\eta = \frac{\partial}{\partial x_1} \wedge \cdots \wedge \frac{\partial}{\partial x_q}$$

on U, and vice versa.

Remark 2.1. Let M_0 be the set of regular points of η. Then it is easy to see that η is a Nambu-Poisson structure on M if and only if η is a Nambu-Poisson structure on M_0.

Let Ω be a volume form on M and $\eta \in \Gamma(\Lambda^q TM)$. Then the interior product $i_\eta\Omega$ is $(n-q)$-form. The following definition of *integrable differential forms* is due to J.-P.Dufour and N.T.Zung.[2]

Definition 2.2. A p-form ω on M is called integrable if it satisfies for any $A \in \Gamma(\Lambda^{p-1}TM)$,

(1) $i_A\omega \wedge \omega = 0$,

(2) $i_A\omega \wedge d\omega = 0$.

Remark 2.2. The above definition does not depend on the choice of volume forms.

Around a regular point of ω, the above definition can be rewritten as follows:

Proposition 2.1 (Ref. 6). *ω is (locally) integrable if and only if*

(1) ω is locally decomposable,

(2) There exists a local 1-form ψ such that $d\omega = \psi \wedge \omega$.

Using Theorem 2.1, we can get the following.

Proposition 2.2 (Refs. 2,6). *Let $\eta \in \Gamma(\Lambda^q TM)$. Put $\omega = i_\eta\Omega$. If $q \geq 3$, then ω is integrable if and only if η is Nambu-Poisson structure. If $q = 2$, and rank $\eta(x) \leq 2$ for any $x \in M$, then ω is integrable if and only if η is Poisson structure.*

3. Koszul Operator

Let Ω be a volume form on an n-dimensional manifold M. Then Ω induces a linear isomorphism Φ from $\Gamma(\Lambda^q TM)$ to the space of p-forms by $\Phi(\eta) = i_\eta\Omega$, where $p = n - q$. In this section, let us recall the following operator D_Ω due to Koszul.[4]

Definition 3.1. Koszul operator D_Ω is defined by

$$D_\Omega = \Phi^{-1} \circ d \circ \Phi.$$

By definition, it is clear that D_Ω maps $\Gamma(\Lambda^q TM)$ into $\Gamma(\Lambda^{q-1}TM)$. It is clear that $D_\Omega \circ D_\Omega = 0$. Moreover, with respect to the Schouten bracket, the following result is due to Koszul.

Proposition 3.1 (Refs. 2,4). *For any $A \in \Gamma(\Lambda^a TM)$ and $B \in \Gamma(\Lambda^b TM)$, we have*

$$D_\Omega(A \wedge B) = (-1)^b[A, B] + A \wedge (D_\Omega B) + (-1)^b(D_\Omega A) \wedge B.$$

4. Results

Throughout this section, we assume that $M = \mathbb{R}^n$. As a volume form on $\mathbb{R}^n$, we adopt the standard volume form $\Omega = dx_1 \wedge dx_2 \wedge \cdots \wedge dx_n$.

Proposition 4.1 (See also Refs. 2,6). *Let $q \geq 4$. Let $\eta \in \Gamma(\Lambda^q T\mathbb{R}^n)$ be a quadratic Nambu-Poisson structure. Then $D_\Omega\eta$ is a linear Nambu-Poisson structure of order $(q-1)$. In particular, if $q = 3$, then $D_\Omega\eta$ is a linear Poisson structure of rank $D_\Omega\eta \leq 2$.*

Proof. Put $\omega = i_\Omega\eta$. Since ω is integrable, it is easy to see that $d\omega$ is also integrable (see Proposition 2.2). Moreover since $D_\Omega\eta = \Phi^{-1}(d\omega)$, $D_\Omega\eta$ is a linear Nambu-Poisson structure. If $q = 3$, we can put $\eta = \frac{\partial}{\partial x_1} \wedge \frac{\partial}{\partial x_2} \wedge \frac{\partial}{\partial x_3}$ and $\Omega = f dx_1 \wedge \cdots \wedge dx_n$, for some $f \in C^\infty(\mathbb{R}^n)$. Then we have

$$D_\Omega\eta = \frac{1}{f}\Big(\frac{\partial f}{\partial x_1}\frac{\partial}{\partial x_2} \wedge \frac{\partial}{\partial x_3} + \frac{\partial f}{\partial x_2}\frac{\partial}{\partial 3} \wedge \frac{\partial}{\partial x_1} + \frac{\partial f}{\partial x_3}\frac{\partial}{\partial x_1} \wedge \frac{\partial}{\partial x_2}\Big).$$

Direct calculation shows that $D_\Omega\eta \wedge D_\Omega\eta = 0$, and $[D_\Omega\eta, D_\Omega\eta] = 0$. Hence $D_\Omega\eta$ is a linear Poisson structure of rank $D_\Omega\eta \leq 2$. □

We explain the decomposition of $\eta \in \Gamma(\Lambda^q T\mathbb{R}^n)$, where η is quadratic. η is not necessarily Nambu-Poisson structure. And we generalize the result of Ping Xu and Z.J.Liu.[8]

Put $D_\Omega\eta = \hat{\pi}$, and $\hat{I} = \sum_{i=1}^n x_i \frac{\partial}{\partial x_i}$. Then $\hat{\pi}$ is linear and order $(q-1)$, and moreover we get $[\hat{\pi}, \hat{I}] = (q-2)\hat{\pi}, D_\Omega(\hat{\pi}) = 0$, and $D_\Omega\hat{I} = n$. By Proposition 3.1, we have

$$\begin{aligned} D_\Omega(\eta - \frac{1}{n-q+2}\hat{\pi} \wedge \hat{I}) &= \hat{\pi} - \frac{1}{n-q+2} D_\Omega(\hat{\pi} \wedge \hat{I}) \\ &= \hat{\pi} - \frac{1}{n-q+2}\{-[\hat{\pi}, \hat{I}] - D_\Omega(\hat{\pi}) \wedge \hat{I} + \hat{\pi} \wedge D_\Omega\hat{I}\} \\ &= 0. \end{aligned}$$

Hence $\Phi(\eta - \frac{1}{n-q+2}\hat{\pi} \wedge \hat{I})$ is closed $(n-q)$-form on $\mathbb{R}^n$, so that we can find an $(n-q-1)$-form θ such that $\Phi(\eta_\theta) = d\theta$, where $\eta_\theta = \eta - \frac{1}{n-q+2}\hat{\pi} \wedge \hat{I}$. Thus we proved the following.

Proposition 4.2. *Let $\eta \in \Gamma(\Lambda^q T\mathbb{R}^n)$ be a quadratic q-vector Put $p = n-q$. Then η has the following decomposition:*

$$\eta = \eta_\theta + \frac{1}{p+2}\hat{\pi} \wedge \hat{I},$$

where θ is a $(p-1)$-form which satisfies $\Phi(\eta_\theta) = d\theta$.

Put $\omega = \Phi(\eta) = i_\eta \Omega$, and $\alpha = \Phi(\hat{\pi} \wedge \hat{I})$. By Proposition 4.2, we have the decomposition of p-form ω:

$$(*) \qquad \omega = d\theta + \frac{1}{p+2}\alpha.$$

Using this decomposition, we can obtain a criterion for η to be a Nambu-Poisson structure. We resume it in the following theorem.

Theorem 4.1. *Let $\eta \in \Gamma(\Lambda^q T\mathbb{R}^n)$ be a quadratic q-vector. According to the decomposition (*), η is a Nambu-Poisson structure of order q if and only if ω satisfies the following two conditions:*

For any $A \in \Gamma(\Lambda^{p-1} T\mathbb{R}^n)$,

(1) $i_A(d\theta + \frac{1}{p+2}\alpha) \wedge (d\theta + \frac{1}{p+2}\alpha) = 0,$

(2) $i_A(d\theta + \frac{1}{p+2}\alpha) \wedge d\alpha = 0.$

If η is a quadratic Nambu-Poisson structure, then $D_\Omega \eta = \hat{\pi}$ is a linear Nambu-Poisson structure. If $q \geq 4$, we can make use of the result of Dufour-Zung.[2] In fact, $\hat{\pi}$ is one of two types in Theorem 6.4.4.[2] In particular, if $q = 3$, $\hat{\pi}$ is a linear Poisson structure with rank≤ 2, and we know all such Poisson structures (see Ref. 2 for example). So in our notations, α is already known, and we have only to calculate $d\theta$ to satisfy (1) and (2) of the above theorem.

Next we show that two quadratic Nambu-Poisson structures are not isomorphic to each other if the corresponding linear Nambu-Poisson (or linear Poisson) structures through Koszul operators are not isomorphic to each other.

Theorem 4.2. *Let η_1 and η_2 be two quadratic Nambu-Poisson structures on $\mathbb{R}^n$. Assume that $\phi_* \eta_1 = \eta_2$ for some diffeomorphism ϕ on $\mathbb{R}^n$, which satisfies $\phi(0) = 0$. Define Φ_1 and Φ_2 by $\Phi_1(\eta_1) = i_{\eta_1}\Omega_1$, and $\Phi_2(\eta_2) = i_{\eta_2}\Omega_2$, where $\phi^* \Omega_2 = \Omega_1$. Then it holds that $D_{\Omega_2} \circ \phi_* = \phi_* \circ D_{\Omega_1}$.*

Proof. By the same manner as Ping Xu and Z.J.Liu,[8] we can easily prove that ϕ must be a linear isomorphism. Then we have

$$\begin{aligned}\phi^*(\Phi_2(\phi_*\eta_1)) &= \phi^*(i_{\eta_2}\Omega_2)) \\ &= i_{(\phi^{-1})_*\eta_2}\phi^*\Omega_2 \\ &= i_{\eta_1}\Omega_1 \\ &= \Phi_1(\eta_1).\end{aligned}$$

Hence $\Phi_2 \circ \phi_* = (\phi^{-1})^* \circ \Phi_1$. This leads us directly to

$$D_{\Omega_2} \circ \phi_* = \phi_* \circ D_{\Omega_1}.$$

□

Remark 4.1. Put $\hat{\pi_1} = D_{\Omega_1}\eta_1$ and $\hat{\pi_2} = D_{\Omega_2}\eta_2$. If $\phi_*\eta_1 = \eta_2$, then we have $\phi_*\hat{\pi_1} = \hat{\pi_2}$ by Theorem 4.2. In other words, if $\hat{\pi_1}$ is not isomorphic to $\hat{\pi_2}$, then η_1 is not isomorphic to η_2.

5. Examples

In this section, we determine quadratic Nambu-Poisson structures η of order 3 defined on $\mathbb{R}^4$ and $\mathbb{R}^5$. We treat here only the cases of "Type I tensors", which are the terminology due to Dufour-Zung.[2] In the following subsections 5.1 and 5.2, four linear Poisson structures $\hat{\pi}$ are not isomorphic to each other by Ref. 2. Thus by Theorem 4.2, four quadratic Nambu-Poisson structures η obtained here are not isomorphic to each other.

5.1. *Quadratic Nambu-Poisson structures η of order 3 defined on $\mathbb{R}^4$*

If η is a quadratic Nambu-Poisson structure of order 3 on $\mathbb{R}^4$,by Proposition 4.1, $\hat{\pi} = D_\Omega\eta$ is a linear Poisson structure of rank$\leq$ 2. There are essentially following 4 cases for $\hat{\pi}$. Here ∂_i means $\frac{\partial}{\partial x_i}$.

(a) $\hat{\pi} = x_1\partial_2 \wedge \partial_3 + x_2\partial_3 \wedge \partial_1 + x_3\partial_1 \wedge \partial_2$.

(b) $\hat{\pi} = x_1\partial_2 \wedge \partial_3 + x_2\partial_3 \wedge \partial_1 + x_4\partial_1 \wedge \partial_2$.

(c) $\hat{\pi} = x_1\partial_2 \wedge \partial_3 + x_4\partial_3 \wedge \partial_1$.

(d) $\hat{\pi} = x_4\partial_2 \wedge \partial_3$.

Since $p = 1$, the condition (1) of Theorem 4.1 is void. So we have only to check the condition (2) of Theorem 4.1. Note that θ is a cubic polynomial. We must find a cubic polynomial which satisfies $i_A(d\theta + \frac{1}{3}\alpha) \wedge d\alpha = 0$ for any vector field A. Results are as follows. A, B, C, D, E are arbitrary constants.

Case(a). $\hat{\pi} = x_1\partial_2 \wedge \partial_3 + x_2\partial_3 \wedge \partial_1 + x_3\partial_1 \wedge \partial_2$.

$$\begin{aligned}\eta &= \{A(x_1^2 + x_2^2 + x_3^2) + Bx_4^2\}\partial_1 \wedge \partial_2 \wedge \partial_3 \\ &\quad - (2A-1)(x_3x_4\partial_1 \wedge \partial_2 \wedge \partial_4 - x_2x_4\partial_1 \wedge \partial_3 \wedge \partial_4 \\ &\quad + x_1x_4\partial_2 \wedge \partial_3 \wedge \partial_4). \\ \omega &= (2A-1)(x_1x_4dx_1 + x_2x_4dx_2 + x_3x_4dx_3) \\ &\quad + \{A(x_1^2 - x_2^2 + x_3^3) + Bx_4^2\}dx_4.\end{aligned}$$

Case (b). $\hat{\pi} = x_1\partial_2 \wedge \partial_3 + x_2\partial_3 \wedge \partial_1 + x_4\partial_1 \wedge \partial_2$.

$$\begin{aligned}\eta &= \{A(x_1^2 + x_2^2) + (4A-1)x_3x_4 + Bx_4^2\}\partial_1 \wedge \partial_2 \wedge \partial_3 \\ &\quad - (2A-1)(x_4^2\partial_1 \wedge \partial_2 \wedge \partial_4 - x_2x_4\partial_1 \wedge \partial_3 \wedge \partial_4 \\ &\quad + x_1x_4\partial_2 \wedge \partial_3 \wedge \partial_4). \\ \omega &= (2A-1)(x_1x_4dx_1 + x_2x_4dx_2 + x_4^2dx_3) \\ &\quad + \{A(x_1^2 + x_2^2) + (4A-1)x_3x_4 + Bx_4^2\}dx_4.\end{aligned}$$

Case (c). $\hat{\pi} = x_1\partial_2 \wedge \partial_3 + x_4\partial_3 \wedge \partial_1$.

$$\begin{aligned}\eta &= \{Ax_1^2 + (4A-1)x_2x_4 + Bx_4^2\}\partial_1 \wedge \partial_2 \wedge \partial_3 \\ &\quad + (2A-1)(x_4^2\partial_1 \wedge \partial_3 \wedge \partial_4 - x_1x_4\partial_2 \wedge \partial_3 \wedge \partial_4). \\ \omega &= (2A-1)(x_1x_4dx_1 + x_4^2dx_2) \\ &\quad + \{Ax_1^2 + (4A-1)x_2x_4 + Bx_2^4\}dx_4.\end{aligned}$$

Case (d). $\hat{\pi} = x_4\partial_2 \wedge \partial_3$.

$$\begin{aligned}\eta &= \{Ax_1^2 + (2C+1)x_1x_4 + Ex_4^2\}\partial_1 \wedge \partial_2 \wedge \partial_3 \\ &\quad - (Bx_1^2 + 2Ax_1x_4 + Cx_4^2)\partial_2 \wedge \partial_3 \wedge \partial_4. \\ \omega &= (Bx_1^2 + 2Ax_1x_4 + Cx_4^2)dx_1 \\ &\quad + \{Ax_1^2 + (2C+1)x_1x_4 + Ex_4^2\}dx_4.\end{aligned}$$

5.2. *Quadratic Nambu-Poisson structures η of order 3 defined on $\mathbb{R}^5$*

In this case, η is a quadratic Nambu-Poisson structure of order 3 on $\mathbb{R}^5$, so $\hat{\pi} = D_\Omega\eta$ is a linear Poisson structure of rank≤ 2. θ is a cubic 1-form, and α is a quadratic 2-form. There are also following 4 cases.

(a) $\hat{\pi} = x_1\partial_2 \wedge \partial_3 + x_2\partial_3 \wedge \partial_1 + x_3\partial_1 \wedge \partial_2$.

(b) $\hat{\pi} = x_1\partial_2 \wedge \partial_3 + x_2\partial_3 \wedge \partial_1 + x_4\partial_1 \wedge \partial_2$.

(c) $\hat{\pi} = x_1\partial_2 \wedge \partial_3 + x_4\partial_3 \wedge \partial_1 + x_5\partial_1 \wedge \partial_2$.

(d) $\hat{\pi} = x_4\partial_2 \wedge \partial_3 + x_5\partial_3 \wedge \partial_1$.

Since $p = 2$, we first determine $d\theta$ which satisfies the condition (2) of Theorem 4.1. We easily know that this $d\theta$ also satisfies the condition (1) of Theorem 4.1. Throughout this subsection, $k(x_4, x_5) = c_1x_4^2 + c_2x_4x_5 + c_3x_5^2$ denotes an arbitrary quadratic polynomial of two variables x_4 and x_5. A, B, C, D also denote arbitrary constants.

Case (a). $\hat{\pi} = x_1\partial_2 \wedge \partial_3 + x_2\partial_3 \wedge \partial_1 + x_3\partial_1 \wedge \partial_2$.

$$\begin{aligned}
\eta = {} & \Big\{\Big(\frac{C-B+1}{2}\Big)(x_1^2+x_2^2+x_3^2) + k(x_4,x_5)\Big\}\partial_1 \wedge \partial_2 \wedge \partial_3 \\
& + (Ax_4 + Bx_5)(x_1\partial_2 \wedge \partial_3 - x_2\partial_1 \wedge \partial_3 + x_3\partial_1 \wedge \partial_2) \wedge \partial_5 \\
& - (Cx_4 + Dx_5)(x_1\partial_2 \wedge \partial_3 - x_2\partial_1 \wedge \partial_3 + x_3\partial_1 \wedge \partial_2) \wedge \partial_4. \\
\omega = {} & \Big\{\Big(\frac{C-B+1}{2}\Big)(x_1^2+x_2^2+x_3^2) + k(x_4,x_5)\Big\}dx_4 \wedge dx_5 \\
& + (Ax_4 + Bx_5)(x_1dx_1 + x_2dx_2 + x_3dx_3) \wedge dx_4 \\
& + (Cx_4 + Dx_5)(x_1dx_1 + x_2dx_2 + x_3dx_3) \wedge dx_5.
\end{aligned}$$

Case (b). $\hat{\pi} = x_1\partial_2 \wedge \partial_3 + x_2\partial_3 \wedge \partial_1 + x_4\partial_1 \wedge \partial_2$.

$$\begin{aligned}
\eta = {} & \Big\{\Big(\frac{1-B-C}{2}\Big)(x_1^2+x_2^2) + (1-B-2C)x_3x_4 \\
& - Dx_3x_5 + k(x_4,x_5)\Big\}\partial_1 \wedge \partial_2 \wedge \partial_3 \\
& + (Ax_4 + Bx_5)(x_1\partial_2 \wedge \partial_3 - x_2\partial_1 \wedge \partial_3 + x_4\partial_1 \wedge \partial_2) \wedge \partial_5 \\
& + (Cx_4 + Dx_5)(x_1\partial_2 \wedge \partial_3 - x_2\partial_1 \wedge \partial_3 + x_4\partial_1 \wedge \partial_2) \wedge \partial_4. \\
\omega = {} & \Big\{\Big(\frac{1-B-C}{2}\Big)(x_1^2+x_2^2) + (1-B-2C)x_3x_4 \\
& - Dx_3x_5 + k(x_4,x_5)\Big\}dx_4 \wedge dx_5 \\
& + (Ax_4 + Bx_5)(x_1dx_1 + x_2dx_2 + x_4dx_3) \wedge dx_4 \\
& - (Cx_4 + Dx_5)(x_1dx_1 + x_2dx_2 + x_4dx_3) \wedge dx_5.
\end{aligned}$$

Case (c). $\hat{\pi} = x_1\partial_2 \wedge \partial_5 + x_4\partial_3 \wedge \partial_1 + x_5\partial_1 \wedge \partial_2$.

$$\begin{aligned}\eta = \Big\{\Big(\frac{1-B-C}{2}\Big)x_1^2 + (1-B-2C)x_2x_4 - Dx_2x_5 \\ + (1-2B-C)x_3x_5 - Ax_3x_4 + k(x_4,x_5)\Big\}\partial_1 \wedge \partial_2 \wedge \partial_3 \\ + (Ax_4 + Bx_5)(x_1\partial_2 \wedge \partial_3 - x_4\partial_1 \wedge \partial_3 + x_5\partial_1 \wedge \partial_2) \wedge \partial_5 \\ - (Cx_4 + Dx_5)(x_1\partial_2 \wedge \partial_3 - x_4\partial_1 \wedge \partial_3 + x_5\partial_1 \wedge \partial_2) \wedge \partial_4.\end{aligned}$$

$$\begin{aligned}\omega = \Big\{\Big(\frac{1-B-C}{2}\Big)x_1^2 + (1-B-2C)x_2x_4 - Dx_2x_5 \\ + (1-2B-C)x_3x_5 - Ax_3x_4 + k(x_4,x_5)\Big\}dx_4 \wedge dx_5 \\ + (Ax_4 + Bx_5)(x_1dx_1 + x_4dx_2 + x_5dx_3) \wedge dx_4 \\ - (Cx_4 + Dx_5)(x_1dx_1 + x_4dx_2 + x_5dx_3) \wedge dx_5.\end{aligned}$$

Case (d). $\hat{\pi} = x_4\partial_2 \wedge \partial_3 + x_5\partial_3 \wedge \partial_1$.

$$\begin{aligned}\eta = \{(1-B+2C)x_1x_4 + (1-2B+C)x_2x_5 - Ax_2x_4 \\ + Dx_1x_5 + k(x_4,x_5)\}\partial_1 \wedge \partial_2 \wedge \partial_3 \\ + (Ax_4 + Bx_5)(x_4\partial_2 - x_5\partial_1)\partial_3 \wedge \partial_5 \\ - (Cx_4 + Dx_5)(x_4\partial_2 - x_5\partial_1)\partial_3 \wedge \partial_4.\end{aligned}$$

$$\begin{aligned}\omega = \{(1-B+2C)x_1x_4 + (1-2B+C)x_2x_5 - Ax_2x_4 + Dx_1x_5 \\ + k(x_4,x_5)\}dx_4 \wedge dx_5 + (Ax_4 + Bx_5)(x_4dx_1 + x_5dx_2) \wedge dx_4 \\ + (Cx_4 + Dx_5)(x_4dx_1 + x_5dx_2) \wedge dx_5.\end{aligned}$$

References

1. C. Camacho and A.L. Neto, The topology of integrable differential forms near a singularity, *I. H. E. S. Publ. Math.* (55) (1982) 5–35.
2. J.-P. Dufour and N.T. Zung, Linearization of Nambu structures, *Compositio Math.* **117** (1) (1999) 77–98.
3. P. Gautheron, Some remarks concerning Nambu mechanics, *Lett. Math. Phys.* **37** (1) (1996) 103–116.
4. J.-L. Koszul, Crochet de Schouten-Nijenhuis et cohomologie, *Astérisque*, (1985) 257–271.
5. N. Nakanishi, On Nambu-Poisson manifolds, *Rev. Math. Phys.* **10** (4) (1998) 499-510.
6. N. Nakanishi, Nambu-Poisson tensors on Lie groups, *Banach Center Publ.* **51** (2000) 243–249.
7. L. Takhtajan, On foundation of the generalized Nambu mechanics, *Comm. Math. Phys.* **160** (1994) 295–315.
8. P. Xu and Z.-J. Liu, On quadratic Poisson structures, *Lett. Math. Phys.* **26** (1) (1992) 33–42.

Differential Geometry and its Applications
Proc. Conf., in Honour of Leonhard Euler, Olomouc, August 2007

Bochner-Kaehler metrics and connections of Ricci type

Martin Panák

Department of Algebra and Geometry, Masaryk University,
Janáčkovo nám. 2a, 602 00, Brno, Czech Republic
E-mail: naca@math.muni.cz

Lorenz J. Schwachhöfer

Universität Dortmund,
Vogelpothsweg 87, 44221 Dortmund, Germany
E-mail: lschwach@math.uni-dortmund.de

We apply the results from Ref. 3 about special symplectic geometries to the case of Bochner-Kaehler metrics. We obtain a (local) classification of these based on the orbit types of the adjoint action in $\mathfrak{su}(n,1)$. The relation between Sasaki and Bochner-Kaehler metrics in cone and transversal metrics constructions is discussed. The connection of the special symplectic and Weyl connections is outlined. The duality between certain classes of the Ricci-type and Bochner-Kaehler metrics is shown.

Keywords: Bochner-Kähler metric, Sasaki metric, Ricci type connection, Weyl structure.

MS classification: 53D05 53C55.

1. Bochner-Kaehler metrics

The curvature tensor of the Levi-Civita connection of a Kaehler metric g decomposes (under the action of $\mathfrak{u}(n)$) into its Ricci and Bochner part. The metric is said to be Bochner-Kaehler, iff the Bochner part of its curvature tensor vanishes.

A remarkable relationship was revealed among following types of geometric structures in the article Ref. 3: manifolds with a connection of Ricci type, manifolds with a connection with the special symplectic holonomy, pseudo-Riemannian Bochner-Kähler structures, manifolds with a Bochner-bi-Lagrangian connection. All these geometric objects are instantons of the

This article is a short version of the article Ref. 6, where you can find all the proofs. The results are uniquely published in these proceedings.

same construction, and they are called special symplectic geometries. The word "symplectic" comes from the fact that they all carry a symplectic connection; special stands for the common special type of the curvature of the connection: let (M, ω) be a symplectic manifold. Then the curvature of the special symplectic geometries is of the form

$$R_h(X,Y) = 2\omega(X,Y)h + X \circ (hY) - Y \circ (hX), \tag{1}$$

where $\mathfrak{h} \subset \mathfrak{sp}(n,\mathbb{R})$ (or $\mathfrak{sp}(n,\mathbb{C})$) is a Lie algebra, $h \in \mathfrak{h}$, $\circ : S^2(TM) \to \mathfrak{h}$, is an $\mathfrak{h}$-equivariant product with special properties.[3] For Bochner-Kaehler structures the special form of curvature translates as follows: let (M, g, J, ω) be a Kaehler structure on a manifold M. That is J is the orthogonal complex structure which is parallel with respect to the Levi-Civita connection of g, and the Kaehler form ω is defined by $\omega(x,y) = g(x, Jy)$. The Kaehler structure is Bochner-Kaehler iff the curvature of the Levi-civita connection of the metric has the above form, where $\mathfrak{h} = \mathfrak{u}(n)$, we get then

$$\begin{aligned} R_\rho(X,Y) = {} & 2g(X,JY)\rho + 2g(X,\rho Y)J + (\rho Y \wedge JX) \\ & - (\rho X \wedge JY) + (X \wedge J\rho Y) - (Y \wedge J\rho X), \end{aligned} \tag{2}$$

where $(X \wedge Y)Z = g(X,Z)Y - g(Y,Z)X$.

2. Kaehler and Sasaki manifolds

The following considerations are motivated by the lecture given by Krzysztof Galicki at Winter School of Geometry and Physics, Srní, Czech republic, 2004.[1]

Sasaki metric. One of the possible (equivalent) definition of the *Sasakian manifold* C is that it is a Riemannian manifold with the metric g, on which there exists a unit length Killing vector field ξ such that the curvature tensor R of the Levi-Civita connection ∇ of g satisfies:

$$R(X,\xi)Y = g(\xi,Y)X - g(X,Y)\xi.$$

The one form λ dual to ξ defines a contact distribution $\mathfrak{D} = \{X \in TC | \lambda(X) = g(X,\xi) = 0\}$. The vector field ξ is called the characteristic vector field of the contact distribution $\mathfrak{D}$.

Transversal Kaehler metric. Further consider J defined by $J(X) = -\nabla_X \xi$. It is an automorphism of the tangent bundle TC and its restriction to $\mathfrak{D}$ gives rise to a complex structure J on $\mathfrak{D}$. Then $(\nabla_X J)(Y) = 0$ for $X, Y \in \mathfrak{D}$ and thus there is a so called *transversal Kaehler structure* on $\mathfrak{D}$. The Kaehler structure then factorizes to the set of leaves of the foliation generated by ξ (the characteristic foliation) if this is locally an orbifold. See Ref. 1 for details.

Conversely, the transversal Kaehler structure on a compact distribution $\mathfrak{D}$ (given as $\mathfrak{D} = \{X|\lambda(X) = 0, \lambda \in \Omega^1(C)\}$) on a manifold C translates to a Sasakian structure on C: given a metric $g_{\mathfrak{D}}$ on $\mathfrak{D}$ with a parallel complex structure J on $\mathfrak{D}$, and a transversal symmetry ξ of $\mathfrak{D}$, one extends $g_{\mathfrak{D}}$ to the whole of TC with $g(X,Y) = g_{\mathfrak{D}}(X,Y)$ for X, $Y \in \mathfrak{D}$, $g(\xi,\xi) = 1$ and $g(\xi, X) = 0$.

Cone metric. On the other hand, a manifold C is Sasakian if and only if the "cone metric" $(t^2 \cdot g + (dt)^2)$ on $C \times \mathbb{R}_+$ is Kaehler, where the complex structure J' on the cone is the extension of J

$$J'(\xi) = t\partial_t, \quad J'(\partial_t) = -\frac{1}{t}\xi, \tag{3}$$

$\xi \in \mathfrak{X}(\widehat{C})$ being the lift of the characteristic vector field ξ on C.

Following the ideas from Ref. 1, there arise questions what happens if we require the transversal metric to be Bochner-Kaehler. What special has to be the Sasaki metric on C? Will then the "cone metric" be Bochner-Kaehler? We give the answer:

Theorem 2.1. *Let M be a $2n$-dimensional (real) manifold with Bochner-Kaehler metric g, J be the corresponding complex structure. Then there exists a Sasakian manifold C with the dimension $2n+1$, such that the set of leaves of the characteristic foliation (together with the structure induced from C) is isomorphic to M. Moreover $\widehat{C} = C \times \mathbb{R}_+$ with the complex structure defined by (3) is Bochner Kaehler if and only if M is locally isomorphic (as the Kaehler structure) to the complex projective space $\mathbb{C}P^n$. The cone $\widehat{C}$ is then a flat manifold.*

3. General construction

Let us quickly review the construction from Ref. 3, which gives rise to all special symplectic geometries. All manifolds with special symplectic connection are locally isomorphic to the factor manifold of the oriented projectivization of the cone $\mathcal{C} = Ad_G x \subset \mathfrak{g}$, where x is an appropriate element in the parabolic 2-gradable Lie algebra $\mathfrak{g}$, where we factor along the flow of the convenient vector field. The special symplectic connection is then induced on the factor from one of the components of the Cartan form on $\mathfrak{g}$, which decomposes due to the 2-grading.

Some of the standard notions from the theory of contact structures are used without definitions. The reader can find them and all the proofs of the theorems stated in this section, in Ref. 3.

Symplectic algebra as subalgebra of a 2-graded algebra. Let V be a vector space (either real or complex) with a symplectic form ω. Let $\mathfrak{h} \subset \mathfrak{sp}(V,\omega) = \{h \in \mathrm{End}(V) | \omega(x,y) + \omega(x,hy) = 0 \text{ for all } x, y \in V\}$ such, that there exists an $\mathfrak{h}$-equivariant map $\circ : S^2(V) \to \mathfrak{h}$ and an $\mathrm{ad}_{\mathfrak{h}}$-invariant inner product $(,)$ which satisfy the following identities:

$$(h, x \circ y) = \omega(hx, y) = \omega(hy, x)$$
$$(x \circ y)z - (x \circ z)y = 2\,\omega(y,z)x - \omega(x,y)z + \omega(x,z)y,$$

for all $x, y, z \in V$ and $h \in \mathfrak{h}$. All such $\mathfrak{h}$s were classified in Ref. 3.

Then there exists a unique simple Lie algebra $\mathfrak{g}$ with a 2-grading of the parabolic type, that is

$$\mathfrak{g} = \mathfrak{g}^{-2} \oplus \mathfrak{g}^{-1} \oplus \mathfrak{g}^{0} \oplus \mathfrak{g}^{1} \oplus \mathfrak{g}^{2},$$

where $\mathfrak{g}^{-2}$ and $\mathfrak{g}^{2}$ are one-dimensional. The grading corresponds to $\mathfrak{h}$ in the following sense:

$$\mathfrak{g}^{ev} := \mathfrak{g}^{-2} \oplus \mathfrak{g}^{0} \oplus \mathfrak{g}^{2} \cong \mathfrak{sl}_{\alpha_0} \oplus \mathfrak{h} \text{ and } \mathfrak{g}^{odd} := \mathfrak{g}^{-1} \oplus \mathfrak{g}^{1} \cong \mathbb{F}^2 \otimes V \text{ as a } \mathfrak{g}^{ev}\text{-module,}$$

where $\mathfrak{g}^{2}$, resp. $\mathfrak{g}^{-2}$, are root spaces of a long root α_0, resp $-\alpha_0$, and $\mathfrak{sl}_{\alpha_0}$ is the Lie algebra isomorphic to $\mathfrak{sl}(2,\mathbb{F})$ generated by the root spaces and the corresponding coroot H_{α_0} which lies in $\mathfrak{g}^0$. We will also write $\mathfrak{p} = \mathfrak{g}^0 \oplus \mathfrak{g}^1 \oplus \mathfrak{g}^2$ for the parabolic subalgebra of $\mathfrak{g}$ and $\mathfrak{p}_0 := \mathfrak{h} \oplus \mathfrak{g}^1 \oplus \mathfrak{g}^2$. Let further P and P_0 be corresponding connected subgroups of G.

Further we fix a non-zero $\mathbb{F}$-bilinear area form $a \in \Lambda^2(\mathbb{F}^2)^*$. There is a canonical $\mathfrak{sl}(2,\mathbb{F})$-equivariant isomorphism

$$S^2(\mathbb{F}^2) \longrightarrow \mathfrak{sl}(2,\mathbb{F}), \quad (ef)\cdot g := a(e,g)f + a(f,g)e \text{ for all } e, f, g \in \mathbb{F}^2, \tag{4}$$

and under this isomorphism, the Lie bracket on $\mathfrak{sl}(2,\mathbb{F})$ is given by

$$[ef, gh] = a(e,g)fh + a(e,h)fg + a(f,g)eh + a(f,h)eg. \tag{5}$$

Thus, if we fix a basis $e_+, e_- \in \mathbb{F}^2$ with $a(e_+, e_-) = 1$, then we have the identifications

$$H_{\alpha_0} = -e_+e_-, \quad \mathfrak{g}^{\pm 2} = \mathbb{F}e_{\pm}^2, \quad \mathfrak{g}^{\pm 1} = e_{\pm} \otimes V.$$

The cone in 2-gradable algebra and its projectivization. Using the Cartan-Killing form (up to the multiple) we identify $\mathfrak{g}$ and $\mathfrak{g}^*$, and we define the root cone $\hat{\mathcal{C}}$ and its (oriented) projectivization $\mathcal{C}$ as follows:

$$\hat{\mathcal{C}} := \mathrm{G} \cdot e_+^2 \subset \mathfrak{g} \cong \mathfrak{g}^*, \qquad \mathcal{C} := p(\hat{\mathcal{C}}) \subset \mathbb{P}^o(\mathfrak{g}) \cong \mathbb{P}^o(\mathfrak{g}^*),$$

where $\mathbb{P}^o(\mathfrak{g})$ is the set of *oriented* lines in $\mathfrak{g}$, i.e. $\mathbb{P}^o \cong S^d$ if $\mathbb{F} = \mathbb{R}$, and $\mathbb{P}^o \cong \mathbb{CP}^d$ if $\mathbb{F} = \mathbb{C}$, where $d = \dim \mathfrak{g} - 1$, and where $p : \mathfrak{g}\backslash 0 \to \mathbb{P}^o(\mathfrak{g})$ is the principal $\mathbb{R}^+$-bundle ($\mathbb{C}^*$-bundle, respectively) defined by the canonical projection. Thus, the restriction $p : \hat{\mathcal{C}} \to \mathcal{C}$ is a principal bundle as well.

Contact structure on the projectivized cone. Being a coadjoint orbit, $\hat{\mathcal{C}}$ carries a canonical G-invariant symplectic structure Ω. Moreover, the *Euler vector field* defined by

$$E_0 \in \mathfrak{X}(\hat{\mathcal{C}}), \qquad (E_0)_v := v$$

generates the principal action of p and satisfies $\mathfrak{L}_{E_0}(\Omega) = \Omega$, so that the distribution

$$\mathcal{D} = dp(E_0^{\perp_\Omega}) \subset T\mathcal{C} \tag{6}$$

yields a G-invariant contact distribution on $\mathcal{C}$, see Ref. 3, Proposition 3.2.

The cone as homogeneous space. Let $\lambda = \iota_{E_0}(\Omega)$. Then we define the bundle $\mathcal{R}$:

$$\mathfrak{R} := \{(\lambda, \hat{\xi}) \in \hat{\mathcal{C}} \times T\hat{\mathcal{C}} \subset T^*\mathcal{C} \times T\hat{\mathcal{C}} \mid \lambda(dp(\hat{\xi})) = 1\}.$$

Let P and P_0 be subgroups of G corresponding to the subalgebras $\mathfrak{p}$ and $\mathfrak{p}_0$ of $\mathfrak{g}$.

Lemma 3.1. *As homogeneous spaces, we have $\mathcal{C} = G/P$, $\hat{\mathcal{C}} = G/P_0$, and $\mathfrak{R} = G/H$.*

Transversal symmetry defines the geometry. For each $a \in \mathfrak{g}$ we define the vector fields $a^* \in \mathfrak{X}(\mathcal{C})$ and $\hat{a}^* \in \mathfrak{X}(\hat{\mathcal{C}})$ corresponding to the infinitesimal action of a, i.e.

$$(a^*)_{[v]} := \left.\frac{d}{dt}\right|_{t=0} (\exp(ta) \cdot [v]) \quad \text{and} \quad (\hat{a}^*)_v := \left.\frac{d}{dt}\right|_{t=0} (\exp(ta) \cdot v). \tag{7}$$

Note that a^* is a contact symmetry (with respect to the canonical contact distribution on $\hat{\mathcal{C}}$), and $\hat{a}^*$ is its Hamiltonian lift. Let

$$\hat{\mathcal{C}}_a := \{\lambda \in \hat{\mathcal{C}} \mid \lambda(a^*) \in \mathbb{R}^+ (\in \mathbb{C}^*, \text{ respectively})\} \text{ and } \mathcal{C}_a := p(\hat{\mathcal{C}}_a) \subset \mathcal{C}, \tag{8}$$

so that $p : \hat{\mathcal{C}}_a \to \mathcal{C}_a$ is a principal $\mathbb{R}^+$-bundle ($\mathbb{C}^*$-bundle, respectively) and the restriction of a^* to $\mathcal{C}_a$ is a positively transversal contact symmetry. Then there exists a unique section λ of the bundle $p : \hat{\mathcal{C}}_a \to \mathcal{C}_a$ such that $\lambda(a^*) = 1$ and therefore, we obtain the section

$$\sigma_a : \mathcal{C}_a \longrightarrow \mathfrak{R} = \mathrm{G/H}, \qquad \sigma_a(u) := (\lambda(u), \hat{a}^*(u)) \in \mathfrak{R}. \tag{9}$$

Let $\pi : \mathrm{G} \to \mathrm{G}/\mathrm{H} = \mathfrak{R}$ be the canonical projection, and let $\Gamma_a := \pi^{-1}(\sigma_a(\mathcal{C}_a)) \subset \mathrm{G}$. The restriction $\pi : \Gamma_a \to \sigma_a(\mathcal{C}_a) \cong \mathcal{C}_a$ is then a principal H-bundle.

Theorem 3.1. *Let $a \in \mathfrak{g}$ be such that $\mathcal{C}_a \subset \mathcal{C}$ from (8) is non-empty, define $a^* \in \mathfrak{X}(\mathcal{C})$ and $\hat{a}^* \in \mathfrak{X}(\hat{\mathcal{C}})$ as in (7), and let $\pi : \Gamma_a \to \mathcal{C}_a$ with $\Gamma_a \subset G$ be the principal H-bundle from above. Then there are functions $\rho : \Gamma_a \to \mathfrak{h}$, $u : \Gamma_a \to V$, $f : \Gamma_a \to \mathbb{F}$ such that*

$$\mathrm{Ad}_{g^{-1}}(a) = \frac{1}{2}e_-^2 + \rho + e_+ \otimes u + \frac{1}{2}fe_+^2 \tag{10}$$

for all $g \in \Gamma_a$.

The restriction of the $\mu_{\mathfrak{h}} + \mu_{-1} + \mu_{-2}$ part of the Maurer-Cartan form to Γ_a yields a pointwise linear isomorphism $T\Gamma_a \to \mathfrak{h} \oplus \mathfrak{g}^{-1} \oplus \mathfrak{g}^{-2}$, and we can further decompose it as

$$\mu_{\mathfrak{h}} + \mu_{-1} + \mu_{-2} = -2\kappa \left(\frac{1}{2}e_-^2 + \rho\right) + e_- \otimes \theta + \eta,$$

$\kappa \in \Omega^1(\Gamma_a)$, $\theta \in \Omega^1(\Gamma_a) \otimes V$, $\eta \in \Omega^1(\Gamma_a) \otimes \mathfrak{h}$.

Theorem 3.2. *Let $a \in \mathfrak{g}$ and $\mathcal{C}_a \subset \mathcal{C}$ as before. Let $U \subset \mathcal{C}_a$ be a regular open subset , i.e. the local quotient $M_U := T_a^{loc}\backslash U$ is a manifold, where*

$$T_a := \exp(\mathbb{F}a) \subset G.$$

Let $\omega \in \Omega^2(M)$ be the unique symplectic form on M_U, such that $\pi^*(\omega) = -2d(E_0 \lrcorner \Omega)$. Then M_U carries a canonical special symplectic connection associated to $\mathfrak{g}$, and the (local) principal T_a-bundle $\pi : U \to M$ admits a connection $\kappa \in \Omega^1(U)$ whose curvature is given by $d\kappa = \pi^*(\omega)$.

Conversely, any manifold with special symplectic connection comes in this way (locally). Namely we have the following theorem (Theorem B from Ref. 3):

Theorem 3.3. *Let (M, ω) be a symplectic manifold with a special symplectic connection of class C^4, and let $\mathfrak{g}$ be the Lie algebra associated to the special symplectic condition as described at the beginning of this section.*

i) Then there is a principal $\hat{T}$-bundle $\hat{M} \to M$, where $\hat{T}$ is a one dimensional Lie group which is not necessarily connected, and this bundle carries a principal connection with curvature ω.

ii) Let $T \subset \hat{T}$ be the identity component. Then there is an $a \in \mathfrak{g}$ such that $T \equiv T_a \subset G$, and a T_a-equivariant local diffeomorphism $\hat{\iota} : \hat{M} \to C_a$ which for each sufficiently small open subset $V \subset \hat{M}$ induces a connection preserving diffeomorphism $\iota : T^{loc}\mathcal{V} \to T^{loc}U = M_U$, where $U := \hat{\iota}(V) \subset C_a$ and M_U carries the connection from 3.2.

4. Construction of Bochner-Kaehler metrics

A little of linear algebra. Let V be a complex $(n+1)$-dimensional vector space, h a hermitian form of the signature $(n, 1)$ on V.

Example 4.1. We consider the standard hermitian form on the complex space $\mathbb{C}^{n+1}$ of signature $(n, 1)$ ($h(\mathbf{x}, \mathbf{y}) = \sum_{i=1}^{n} x_i \overline{y}_i - x_{n+1}\overline{y}_{n+1}$, for $\mathbf{x}$, $\mathbf{y} \in C^{n+1}$. Then as a matrix algebra $\mathfrak{su}(n, 1)$ can be written as

$$\mathfrak{su}(n,1) = \left\{ \left(\begin{array}{c|c} A & \mathbf{v} \\ \hline \mathbf{v}^* & -\operatorname{tr} A \end{array} \right), A \in \mathfrak{u}(n), \mathbf{v} = \begin{pmatrix} v_1 \\ \vdots \\ v_n \end{pmatrix} \in \mathbb{C}^n \right\}.$$

The elements of the bundle Γ_a from the general construction in the previous section, can be described with the structure functions from (3.1) as follows:

$$\begin{pmatrix} \rho - \frac{1}{n+2}(\operatorname{tr}\rho)\mathbb{I}_n & u & u \\ -u^* & -\frac{1}{2}(\operatorname{tr}\rho - i(f+1)) & \frac{i}{2}(f-1) \\ u^* & \frac{i}{2}(1-f) & -\frac{1}{2}(\operatorname{tr}\rho + i(f+1)) \end{pmatrix}. \tag{11}$$

The grading of $\mathfrak{su}(n,1) = \mathfrak{g}^{-2} \oplus \mathfrak{g}^{-1} \oplus \mathfrak{g}^0 \oplus \mathfrak{g}^1 \oplus \mathfrak{g}^2 \equiv \mathbb{R}e_-^2 \oplus e_- \otimes V \oplus (\mathfrak{h} \oplus \mathbb{R}e_+e_-) \oplus e_+ \otimes V \oplus \mathbb{R}e_+^2$ is given as follows (all the matrices are $(n+1) \times (n+1)$ ones):

$$e_\pm^2 = \left(\begin{array}{c|cc} \mathbf{O} & \begin{matrix} 0 \\ \vdots \\ 0 \end{matrix} & \begin{matrix} 0 \\ \vdots \\ 0 \end{matrix} \\ \hline 0 \dots 0 & i & \pm i \\ 0 \dots 0 & \mp i & -i \end{array} \right), \quad \mathfrak{h} = \left\{ \left(\begin{array}{c|cc} A & \begin{matrix} 0 \\ \vdots \\ 0 \end{matrix} & \begin{matrix} 0 \\ \vdots \\ 0 \end{matrix} \\ \hline 0 \dots 0 & -\frac{1}{2}\operatorname{tr} A & 0 \\ 0 \dots 0 & 0 & -\frac{1}{2}\operatorname{tr} A \end{array} \right), A \in \mathfrak{u}(n-1) \right\},$$

$$e_+e_- = \left(\begin{array}{c|cc} \mathbf{O} & \begin{matrix} 0 \\ \vdots \\ 0 \end{matrix} & \begin{matrix} 0 \\ \vdots \\ 0 \end{matrix} \\ \hline 0 \dots 0 & 0 & 1 \\ 0 \dots 0 & 1 & 0 \end{array} \right), \quad \mathfrak{g}^{-1} = \left\{ \left(\begin{array}{c|cc} \mathbf{O} & \mathbf{v} & \pm\mathbf{v} \\ \hline -\mathbf{v}^* & 0 & 1 \\ \pm\mathbf{v}^* & 1 & 0 \end{array} \right), \mathbf{v} = \begin{pmatrix} v_1 \\ \vdots \\ v_{n-1} \end{pmatrix} \in \mathbb{C}^{n-1} \right\}.$$

The action of the algebra $\mathfrak{h} = \mathfrak{u}(n-1)$ on V is then given as the adjoint matrix action and one easily computes that for $\rho \in \mathfrak{u}(n-1)$ there is

$$\rho \cdot u = \rho u + \frac{1}{2}\operatorname{tr}(\rho) u \tag{12}$$

The hermitian form h is uniquely determined either by its real part g (real valued symmetric bilinear form on V) or by its imaginary part ω, the antisymmetric real valued form on V. ($\omega(x,y) = g(x, Jy)$, where J is the complex structure on V).

Lemma 4.1. *There is a $U(n,1)$-equivariant map $m : V \to \mathfrak{u}(n,1) : x \mapsto x \wedge Jx$, where*

$$(x \wedge Jx)z = g(x,z)Jx - g(Jx,z)x,$$

and the $U(n,1)$ action on $\mathfrak{u}(n,1)$ is given by ad *representation.*

Remark 4.1.

i) The image of the morphism m are from the definition rank one morphisms.
ii) The value of the morphism $x \wedge Jx$ on a vector z is actually $i\overline{h(x,z)}x$, but we stick to write it in the form $g(x,z)Jx - g(Jx,z)x$, which comes from the morphism $x \wedge y$: $(x \wedge y)z = g(x,z)y - g(y,z)x$.

Lemma 4.2. *The morphism m is not injective: $x \wedge Jx$ and $y \wedge Jy$ determine the same element in $\mathfrak{u}(n,1)$ if and only if $x = e^{ik}y$, $k \in \mathbb{R}$, x, $y \in V$.*

Lemma 4.3. *A morphism $x \wedge Jx$ lies in $\mathfrak{su}(n,1)$ iff $g(x,x) = 0$.*

Let us follow the construction for $G = SU(n,1)$. There is $\widehat{C} = Ad_{SU(n,1)}(x) \subset \mathfrak{su}(n,1)$, where x is some maximal root element. The roots in $\mathfrak{su}(n,1)$ have the same length and any rank 1 matrix in $\mathfrak{su}(n,1)$ is a maximal root element.

Lemma 4.4. *The cone $C \subset \mathfrak{su}(n,1)$ is isomorphic to $\mathbb{C}^n \setminus \{0,\ldots,0\}$ and the projectivized cone $C \subset \mathbb{P}_o(\mathfrak{su}(n,1))$ is isomorphic to S^{2n-1}.*

The module structure of the cone. Being a homogeneous space (see (3.1)), the cone C is a $SU(n,1)$, resp. $\mathfrak{su}(n,1)$-module.

Let us notice, that the identification of the cone C with the sphere S^{n+1} is subject to the choice of the standard hermitian form on $\mathbb{C}^{n+1}$, respectively to the choice of the normal base with respect to it. We have to have this in mind when considering different block forms of the matrices in $\mathfrak{su}(n,1)$.

The proof of the lemma (4.6) shows that if a matrix of the form $\begin{pmatrix} A & 0 \\ 0 & -\operatorname{tr}(A) \end{pmatrix}$ is in $\mathfrak{su}(n,1)$, then $A \in \mathfrak{u}(n)$ (with respect to some orthonormal basis of the given hermitian form).

The one-parametric subgroup generated by A are then matrices of the form

$\begin{pmatrix} G & 0 \\ 0 & \frac{1}{\det(G)} \end{pmatrix}$, where $G \in U(n)$. As we have seen above, a point in our sphere S^{2n-1} corresponds to a class of null vectors in $\mathbb{C}^{n+1}$ which differ by a complex multiple. According to the lemma 4.1 the group $SU(n,1)$ acts on these classes in a standard way and we get

$$G \cdot \mathbf{x} \sim \begin{pmatrix} G & 0 \\ 0 & \frac{1}{\det(G)} \end{pmatrix} (\mathbf{x}, 1) = \left(G\mathbf{x}, \frac{1}{\det(G)} \right) \sim \det(G) G\mathbf{x} \tag{13}$$

For the action of the matrices in $\mathfrak{u}(n)$ of the above form on the tangent bundle of the sphere we get then:

$$A \cdot \mathbf{x} = (\operatorname{tr}(A)E + A)\mathbf{x}, \tag{14}$$

which is in accordance with the action of $\mathfrak{u}(n-1)$ on $\mathbb{C}^{n-1}$ from (12), if we represent A in the matrix form from (11).

Thus we get in fact the structure of $U(n)$ and $\mathfrak{u}(n)$-module on C regarded as a sphere S^{2n-1}.

The action of the whole group $SU(n,1)$ or the whole algebra $\mathfrak{su}(n,1)$ respectively is then non-linear (as a $\mathbb{C}^n \to \mathbb{C}^n$ mapping).

Lemma 4.5. *The canonical symplectic form on $\widehat{C} \subset \mathfrak{g}^*$ corresponds under our identification to the standard symplectic form on $\mathbb{C}^n$, the one form $\lambda = \iota_{E_0}(\Omega)$ is then $\lambda = z_i\, d\overline{z_i} + \overline{z_i}\, dz_i$ in the complex coordinates z_i on $\mathbb{C}^n$.*

Up to now, the construction was common for all the Bochner-Kaehler geometries. The choice of a transversal symmetry of the canonical contact distribution actually determines the geometry. Let $A \in \mathfrak{su}(n,1)$ and let us consider the vector field $\xi_v = \frac{\partial}{\partial t}|_0 \operatorname{Ad}(\exp(tA))v$ on $\widehat{C}$. This vector field is a contact symmetry with respect to the distribution $\widehat{\mathfrak{D}}$ on $\widehat{C}$ and thus it determines a section of $\widehat{C} \to C$ that is a contact form on C (the identification of $\mathfrak{g}$ and $\mathfrak{g}^*$ gives an identification of $C \subset \mathfrak{su}(n,1)$ and $C^* \subset \mathfrak{su}(n,1)^*$).

The section $\lambda : C \to \widehat{C}$ is given by the equation $\lambda(\xi) = 1$. The image of C in $\widehat{C} = \mathbb{C}^n - \{0\}$ is then a hyperplane, which we will call Σ_A. The tangent space of Σ_A is then characterized by $T_v(C) = \{X \in \mathbb{C}^n | \omega(X, iA \cdot v) = 0\}$

and consequently there is

$$\mathfrak{D}_v = \{X_v | g(X, iv) = 0 = g(X, iA \cdot v)\}. \tag{15}$$

The projectivized cone C and the CR-sphere. According to the lemma (3.1), the projectivized cone C is a homogeneous space $SU(n,1)/P$, where P is a parabolic subgroup of $SU(n,1)$, corresponding to the subalgebra $\mathfrak{u}(n) \oplus \mathbb{C}^{n-1} \oplus e_+^2$ of $\mathfrak{su}(n,1)$, see (4.1). As we have seen in the previous lemma, it is isomorphic to the sphere S^{2n-1}. The adjoint action of $SU(n,1)$ on $\widehat{C}$ corresponds to the standard action of $SU(n,1)$ on the null-vectors (with respect to the standard hermitian form of the signature $(n,1)$) in $\mathbb{C}^{n+1}$, and thus as a homogeneous space it is exactly the CR-sphere (see Ref. 4).

4.1. *Classification of Bochner-Kaehler metrics*

All Bochner-Kaehler manifolds come from the mentioned construction for the Lie algebra $\mathfrak{su}(n,1)$ and the resulting manifolds are isomorphic if we take in the course of construction matrices A_0 lying on the same adjoint orbit of $SU(n,1)$ in $\mathfrak{su}(n,1)$. Thus we can classify the Bochner-Kaehler manifolds according to which orbit of the action induces the given manifold.

There are four types of orbits of the adjoint action of $GL(n+1,\mathbb{C})$ on $\mathfrak{su}(n,1)$. Three types of these orbits are $SU(n,1)$ orbits as well, the fourth one splits into two $SU(n,1)$ orbits. We describe the orbit types according to the Jordan blocks of the matrices in the orbits.

Lemma 4.6. *There are five types of orbits of the adjoint action of the $SU(n,1)$ on $\mathfrak{su}(n,1)$. If we represent a morphism in the $\mathfrak{su}(n,1)$ with a matrix A, than the orbit types look as follows:*

1. *The matrix is diagonizable and its eigenvalues are purely imaginary.*
2. *The eigenvalues of matrices in the orbit are pure imaginary and there is just one Jordan block of the dimension 2 (there are n eigenvectors). There exists an eigenvector e and a root vector f, both in the block, such that*

   2a. $(e, f) = i$.
   2b. $(e, f) = -i$.
3. *The eigenvalues of matrices in the orbit are pure imaginary and there is just one Jordan block of the dimension 3 (there are $n-1$ eigenvectors).*
4. *There are $n-1$ pure imaginary eigenvalues corresponding to $n-1$ eigenvectors and two eigenvalues $\lambda = \lambda_1 + i\lambda_2$ and $\mu = \mu_1 + i\mu_2$.*

Characteristic polynomial of the metrics. The characteristic polynomial determines all invariants of the adjoint orbit of the matrix $A \in \mathfrak{g}^*$ Thus the different types of adjoint orbits correspond to different types of characteristic polynomials (distinguished according to their roots) and we get invariants of the equivalent classes of Bochner-Kaehler metric.

The characteristic polynomial p_A of the matrices in Γ_A is

$$\det\begin{pmatrix} \rho - \frac{1}{n+2}(\operatorname{tr}\rho)\mathbb{I}_n & u & u \\ -u^* & -\frac{1}{2}(\operatorname{tr}\rho - i(f+1)) & \frac{i}{2}(f-1) \\ u^* & \frac{i}{2}(1-f) & -\frac{1}{2}(\operatorname{tr}\rho + i(f+1)) \end{pmatrix} - t\mathbb{I}_{n+2}$$

$$= \det(\rho - \frac{1}{n+2}(\operatorname{tr}\rho) - t\mathbb{I}_n)(t^2 + (\operatorname{tr}\rho)t + f + \frac{1}{4}(\operatorname{tr}\rho)^2)$$

$$+ u^* \operatorname{Cof}(\rho - \frac{1}{n+2}(\operatorname{tr}\rho) - t\mathbb{I}_n)u,$$

where $\operatorname{Cof}(X)$ means the cofactor matrix of X.

This is in accordance with the Bryant's result (see Ref. 2)† on the orbits of the diagonalizable matrices with pure imaginary eigenvalues.

4.2. *One "nice" type of Bochner-Kaehler metrics*

We are now going to describe the first one of the five mentioned types of Bochner-Kaehler metrics in more detail. Namely let us investigate those metrics which come from the construction if we take in the course of it the matrix generating transversal symmetry to be diagonalizable with all eigenvalues pure imaginary (or zero). These are the matrices which acts as linear morphism ($U(n)$-morphism actually) on the sphere S^{2n-1} regarded as a projectivized cone C.

Any diagonalizable matrix $\widehat{A} \in \mathfrak{su}(n,1)$ can be written in the form $\operatorname{diag}(i\lambda_1, \ldots, i\lambda_n, -i\sum_{i=1}^{n} \lambda_i)$, where $A = \operatorname{diag}(i\lambda_1, \ldots, i\lambda_n)$ is in $\mathfrak{u}(n)$.

Then according to (14), A acts on the tangent bundle of the spere:

$$A \cdot (x_1, \ldots, x_n) = (i(\lambda_1 + \sigma)x_1, \ldots, i(\lambda_n + \sigma)x_n) \in TC, \quad \sigma = \sum_{i=1}^{n} \lambda_i.$$

The action of a matrix A on C^n thus corresponds to the multiplication with the matrix A', where $A' = A + \operatorname{tr}(A)I$, I being the identity matrix.

---

*Let $\mathfrak{g} \subset \operatorname{End}(V)$ be an irreducible representation of the Lie algebra g. Let $\varphi : \mathfrak{g} \to \mathbb{R}(t)$, such that $\varphi(\operatorname{Ad}_g x) = \varphi(x)$ for all $x, g \in \mathfrak{g}$ and $\operatorname{gr}(\varphi(x)) \leq n$ Then $\varphi(x)$ is a constant multiple of the characteristic polynomial of x.

†The functions ρ, u, f from 3 2 correspond to Bryant's functions S, T, U as follows: $\rho = iS$, $u = -T$, $U = -f$.

Let $A_0 \in \mathfrak{su}(n,1)$, $A_0 = \begin{pmatrix} A_0 & 0 \\ 0 & -\operatorname{tr}(A_0) \end{pmatrix}$ be of the above diagonal form such that $\xi_0 v = A_0 \cdot v$ defines a transversal symmetry on C_0, a non-empty open subset of C, that is $\xi_0 v \notin \mathfrak{D}_v$ on C_0. This symmetry then defines a section λ of the line bundle $\widehat{C} \to C$. Then $\lambda(C) = \Sigma_A \subset \widehat{C}$,

$$\Sigma_A = \{x \in \mathbb{C}^n | \sum_{i=1}^{n} (\lambda_i + \sigma)|x_i|^2 = 1\}. \tag{16}$$

Consider U a regular open subset of C with respect to ξ_0, that is there is a submersion $\pi_U : U \to M_U$ onto some manifold M_U, the set of leaves of the foliation generated on U by ξ. The whole of C can be covered by regular subsets. We write $M_U = U/T$.

Our goal is to determine the Bochner-Kaehler connection on M_U which is induced there according to the general construction of special symplectic geometries. The Bochner-Kaehler connections are with one-to-one correspondence with the Bochner-Kaehler metrics which are further in one-to-one correspondence with the pair consisting of the fundamental form of the Kaehler structure and the complex structure on M_U.

There is the unique symplectic (which turns to the fundamental Kaehler one with the complex structure on M_U) form ω_U on M_U such that the pull-back of this form to $U \times \mathbb{R} \subset \widehat{C} \subset \mathbb{C}^n$ is the canonical symplectic form on $\mathbb{C}^n$ (this is the form that comes with the above identifications from the Cartan-Killing form on $\mathfrak{su}(n,1)$).

Example 4.2. The complex projective space $\mathbb{C}P^n$ comes from our construction for $\mathfrak{g} = \mathfrak{su}(n+1,1)$ and $A = \operatorname{diag}(-\frac{i}{2(n+2)}, \dots, -\frac{i}{2(n+2)}, \frac{i(n+1)}{2(n+2)})$.

The generating metric on the contact distribution. The Bochner-Kaehler structure on the sphere S^{2n-1} lifts to the structure on Σ_A. The canonical complex structure on the contact distribution on C, that is canonical CR-distribution on the sphere lifts to the complex structure on the distribution $\mathfrak{D}_\Sigma \subset T\Sigma_A$, that is to the complex structure on the contact distribution on the section Σ_A of $\widehat{C} \to C$:

$$J_M(X) = JX - (X, A_0 p)p - \frac{(X,p)}{|p|^2} Jp, \tag{17}$$

for vectors X, $Y \in T_p\Sigma$, and $|p|^2 = (p,p)$. On $\mathfrak{D}_\Sigma$ we get the metric

$$g(X,Y) = \omega(J_M X, Y) = (J_M X, JY) = (X,Y) - \frac{(X,p)(Y,p)}{|p|^2}. \tag{18}$$

This metric factors to the metric on $M_U \cong T\Sigma$, which is, according to the construction, Bochner-Kaehler. This can be confirmed also with the direct computation. We can view the metric (18) as a degenerated metric on the whole Σ. The corresponding Levi-Civita connection is then

$$(\nabla_X Y)_p = \nabla^0_X Y + g(X,Y)\eta + (X, A_0 JY)p, \tag{19}$$

where $X, Y, Z \in \mathfrak{X}(\Sigma_A)$ ∇^0 is a flat connection in $\mathbb{C}^n$, $\eta = A_o Jp - |A_0 p|^2 p$, $p \in \Sigma_A$. For the Levi-Civita connection of the metric on M_U we have then

$$\overline{\nabla_X Y} = \nabla_{\overline{X}} \overline{Y} - \frac{1}{2}\omega(X,Y)\xi_0 + \alpha(X)\overline{J_M Y} + \alpha(Y)\overline{J_M X},$$

where $X, Y \in \mathfrak{X}(M)$, $\overline{X}$ is a lift of a vector field X on $\mathfrak{X}(M)$ to a vector field on Σ_A ($\overline{X} \in \mathfrak{D}_\Sigma$), and $\alpha(X) := g(\overline{X}, \xi_0) = (\overline{X}, \xi_0)$.

The Bochner-Kaehler form of the curvature of the metric. Let us define the mapping $\rho : TM_U \to TM_U$ as

$$\overline{\rho X} := \nabla_{\xi_0}\overline{X} = \nabla_{\overline{X}}\xi_0 = A_0\overline{X} + g(\overline{X}, \xi_0)\eta + g(\overline{X}, A_0^2 Jp)p. \tag{20}$$

With the help of the map ρ (which is in $\mathfrak{u}(n)$), we can express the curvature of the Levi-Civita connection on M_U in the Bochner-Kaehler form (2).

Proposition 4.1. *Let M_U be a Bochner-Kaehler manifold which comes from the general construction in Ref. 3 for $\mathfrak{h} = \mathfrak{u}(n-1)$, $\mathfrak{g} = \mathfrak{su}(n,1)$, and $A_0 \in \mathfrak{su}(n,1)$. Then the curvature R of the Bochner-Kaehler metric on M_U is given by*

$$R = R_{\frac{1}{2}\rho + \frac{1}{4}|A_0' p|^2 J}$$

Example 4.3.

For the complex projective space we get the following characteristic polynomial $p(t)$ for the class of the Bochner-Kaehler metrics with the constant holomorphic curvature equal 1:

$$p(t) = \left(t - \frac{i}{2(n+2)}\right)^{n+1}\left(t - \frac{i(n+1)}{2(n+2)}\right),$$

which corresponds to the Bryant's one (see section 4.1.1. in Ref. 2).

5. Bochner-Kähler and Ricci-type connections duality

In this section we describe the duality between the manifolds with the Bochner-Kähler metrics of type 1. (see 4.6) and Ricci flat connections.

Recall the general construction from the section 3. So far we were interested in the case with $\mathfrak{g}_1 = \mathfrak{su}(n,1)$. If we consider the construction for the parabolic 2-gradable algebra$\mathfrak{g}_2 := \mathfrak{sp}(n, \mathbb{R})$, we get a manifold with the connection of Ricci type. Recall the two standard embeddings of $\mathfrak{u}(n+1)$, first into $\mathfrak{su}(n+1,1)$ (that was described in the previous section), second into $\mathfrak{sp}(n+1, \mathbb{R})$.

Theorem 5.1. *Consider the action of the Lie algebras* $\mathfrak{g}_1 = \mathfrak{su}(n+1,1)$ *and* $\mathfrak{g}_2 = \mathfrak{sp}(n+1, \mathbb{R})$ *on the projectivized cones* $\widehat{C}_1$, $\widehat{C}_2$. *Then the following are equivalent*

i) For $a_i \in \mathfrak{g}_i$ *the actions of* $T_{a_i} \subset G_i$ *on* $\widehat{C}_i$ *are conjugate for i=1,2.*
ii) $a_i \in \mathfrak{u}(n+1)$, *where* $\mathfrak{u}(n+1) \subset \mathfrak{g}_i$ *for* $i = 1, 2$ *via the two standard embeddings.*

Thus we get the following theorem:

Theorem 5.2.

i) Let (M, ω, ∇) *be a symplectic manifold with a connection of Ricci type, and suppose that the corresponding element* $A \in \mathfrak{sp}(n+1, \mathbb{R})$ *from 3.3 is conjugate to an element of* $\mathfrak{u}(n+1) \subset \mathfrak{sp}(n+1, \mathbb{R})$. *Then M carries a canonical Bochner-Kähler metric whose Kähler form is given by* ω.
ii) Conversely, let (M, J, ω) *be a Bochner-Kähler metric such that the element* $a \in \mathfrak{su}(n+1,1)$ *from 3.3 is conjugate to an element of* $\mathfrak{u}(n+1) \subset \mathfrak{su}(n+1,1)$. *Then* (M, ω) *carries a canonical connection of Ricci-type.*

Acknowledgments

Both authors were supported by the Schwerpunktprogramm Globale Differentialgeometrie of the Deutsche Forschungsgesellschaft, the first one also by the grant nr. 201/05/P088 of the Grant academy of the Czech republic.

References

1. Ch.P. Boyer and K. Galicki Sasakian Geometry, Hypersurface Singularities, and Einstein Metrics, *In: Geometry and Physics* (lecture notes, Srni, Czech Republic, in January of 2004, 30); arXiv:math.DG/0405256.
2. R. Bryant, Bochner-Kähler metrics *J.AMS* **14** (3) (2001) 623–715.
3. M. Cahen and L.J. Schwachhoefer, Special symplectic connections; arXiv: math.DG/0402221.
4. A. Čap and J. Slovák, Parabolic geometries, to apear quite soon.
5. S. Kobayashi and K. Nomizu, *Foundations of Differential Geometry, I* (Interscience publishers, 1963).
6. M. Panák and L. Swachhöfer, Bochner-Kaehler metrics and connections of Ricci type; arXiv:0710.0164.

Differential Geometry and its Applications
Proc. Conf., in Honour of Leonhard Euler, Olomouc, August 2007

Special n-forms on a $2n$-dimensional vector space

Jiří Vanžura

Mathematical Institute, Academy of Sciences of the Czech Republic, branch Brno, Žižkova 22, 616 62 Brno, Czech Republic
E-mail: vanzura@ipm.cz

The configuration of regular 3-forms in dimension 6 is generalized to n-forms in dimension $2n$. The algebras of complex, paracomplex, and dual numbers are systematically used. The automorphism groups of all forms are determined.

Keywords: $2n$-dimensional vector space, n-form, automorphism group of an n-form, 2-dimensional algebra.

MS classification: 15A75.

In the last decade there has arisen interest in exterior forms of higher degree, first of all in forms of degree 3 (see Ref. 2 and Ref. 3). It is known (see Ref. 4 and Ref. 1) that on a 6-dimensional real vector space there are exactly three types (= orbits) of regular 3-forms, and that these types are closely related with 2-dimensional unital algebras. In this note we show that these forms can be generalized to higher dimensions. Namely, using 2-dimensional unital algebras we construct n-forms on a $2n$-dimensional real vector space, and investigate their properties.

We shall consider all three 2-dimensional unital, associative and commutative real algebras, namely

$$\begin{aligned}
&\mathbb{C} = [1, i], && i^2 = -1 && \text{algebra of complex numbers,}\\
&\mathbb{D} = [1, d], && d^2 = 1 && \text{algebra of paracomplex numbers,}\\
&\mathbb{E} = [1, e], && e^2 = 0 && \text{algebra of dual numbers.}
\end{aligned}$$

Let V be a $2n$-dimensional real vector space, $n \geq 3$. On this vector space we shall consider consider three endomorphisms J, D, and E, respectively.

We shall assume that they satisfy

$$J^2 = -I \quad \text{(complex structure)}$$
$$D^2 = I, \ \dim \operatorname{Ker}(D - I) = \dim \operatorname{Ker}(D + I) = n \quad \text{(product structure)}$$
$$E^2 = 0, \ \dim \operatorname{im} E = \dim \operatorname{Ker} E = n \quad \text{(tangent structure)}.$$

If V is endowed with a complex structure J (resp. product structure D, resp. tangent structure E), we can introduce on V a structure of a $\mathbb{C}$-module (i. e. complex vector space) (resp. $\mathbb{D}$-module, resp. $\mathbb{E}$-module) in the following way

$$(a+bi)v = av+bJv \,(\text{resp. } (a+bd)v = av+bDv, \text{resp. } (a+be)v = av+bEv).$$

On the other hand if V carries a structure of a $\mathbb{C}$-module (resp. $\mathbb{D}$-module, resp. $\mathbb{E}$-module), we can introduce on V a complex structure J (resp. product structure D, resp. tangent structure E) by the formula

$$Jv = iv, \quad (\text{resp. } Dv = dv, \quad \text{resp. } Ev = ev).$$

Writing V we shall always consider V as a real vector space. If we want to consider V as a $\mathbb{C}$-module (resp. $\mathbb{D}$-module, resp. as a $\mathbb{E}$-module), we shall write $(V, \mathbb{C})$ (resp. $(V, \mathbb{D})$, resp. $(V, \mathbb{E})$). If there is no danger of confusion, we shall very often abbreviate $W = (V, \mathbb{C})$ (resp. $W = (V, \mathbb{D})$, resp. $W = (V, \mathbb{E})$). Using the above assumptions on J (resp. D, resp. E), we can easily see that W is an n-dimensional free $\mathbb{C}$-module (resp. $\mathbb{D}$-module, resp. $\mathbb{E}$-module).

We shall consider the group

$$GL^*(V;\mathbb{C}) = \{A \in GL(V;\mathbb{R}); AJ = JA \text{ or } AJ = -JA\}.$$

We define also

$$GL^+(V;\mathbb{C}) = \{A \in GL(V;\mathbb{R}); AJ = JA\} \quad \text{and}$$
$$GL^-(V;\mathbb{C}) = \{A \in GL(V;\mathbb{R}); AJ = -JA\}.$$

We have $GL^*(V;\mathbb{C}) = GL^+(V;\mathbb{C}) \cup GL^-(V;\mathbb{C})$, and

$$GL^+(V;\mathbb{C}) \cdot GL^+(V;\mathbb{C}) = GL^+(V;\mathbb{C}),$$
$$GL^+(V;\mathbb{C}) \cdot GL^-(V;\mathbb{C}) = GL^-(V;\mathbb{C}) \cdot GL^+(V;\mathbb{C}) = GL^-(V;\mathbb{C})$$
$$GL^-(V;\mathbb{C}) \cdot GL^-(V;\mathbb{C}) = GL^+(V;\mathbb{C})$$

Along the same lines we introduce $GL^*(V;\mathbb{D})$ and $GL^*(V;\mathbb{E})$.

Let us consider a $\mathbb{C}$-module $(V, \mathbb{C})$ (resp. $\mathbb{D}$-module $(V, \mathbb{D})$, resp. $\mathbb{E}$-module $(V, \mathbb{E})$). A real form of this module is an n-dimensional real subspace $V_0 \subset V$ such that

$$V_0 + iV_0 = V \quad (\text{resp. } V_0 + dV_0 = V, \quad \text{resp. } V_0 + eV_0 = V).$$

Because the module W is free, it is easy to see that in all these three cases a real form exists.

For a k-form ω on V, $k \geq 2$, we can define a homomorphism

$$V \to \Lambda^{k-1}V^*, \quad v \mapsto \iota_v \omega = \omega(v, \cdot, \ldots, \cdot).$$

The form ω is called *regular* (or *multisymplectic*) if the above homomorphism is a monomorphism.

Next for a k-form ω on V, $k \geq 3$, we can consider all endomorphisms A of V satisfying

$$\omega(Av_1, v_2, \ldots, v_k) = \omega(v_1, Av_2, \ldots, v_k) = \cdots = \omega(v_1, v_2, \ldots, Av_k).$$

It is easy to see that such endomorphisms constitute a unital associative real algebra. This algebra is commutative (which can be very easily proved). We shall denote it by the symbol $\mathcal{A}_\omega$.

If A is an endomorphism of V and ω is a k-form, we define a k-form $A^*\omega$ in the following way

$$(A^*\omega)(v_1, \ldots, v_k) = \omega(Av_1, \ldots, Av_k)$$

Next we define the derivation $\mathcal{D}_A\omega \in \Lambda^k V^*$ by the formula

$$(\mathcal{D}_A\omega)(v_1, \ldots, v_k) = \sum_{i=1}^{k} \omega(v_1, \ldots, v_{i-1}, Av_i, v_{i+1}, \ldots, v_k).$$

1. Forms of the complex type

W is here an n-dimensional complex vector space, and $e_1, \ldots, e_n$ is its basis.

Lemma 1.1. *Let $\theta \neq 0$ be a complex n-form. Then $\mathcal{A}_\theta = [I, J] = \{cI; c \in \mathbb{C}\}$.*

Proof. Obviously $J \in \mathcal{A}_\theta$. If $A \in \mathcal{A}_\theta$, then $AJ = JA$, which means that A is a complex endomorphism. Let $A \in \mathcal{A}_\theta$. Then we have

$$\theta(e_1, Ae_1, e_3, \ldots, e_n) = \theta(e_1, e_1, Ae_3, \ldots, e_n) = 0.$$

Let us write $Ae_i = \sum_{j=1}^{n} a_{ij}e_j$. From the above equality it follows that $a_{12} = 0$. Along the same lines we can easily prove $a_{ij} = 0$ for $i \neq j$. Consequently, we have $Ae_i = a_{ii}e_i$ for $i = 1, \ldots, n$. From the equalities

$$\theta(Ae_1, e_2, \ldots, e_n) = \theta(e_1, Ae_2, \ldots, e_n) = \cdots = \theta(e_1, e_2, \ldots, Ae_n)$$

we get $a_{11} = a_{22} = \cdots = a_{nn}$. Therefore we have $A = cI$, where $c \in \mathbb{C}$. This finishes the proof. $\square$

Let $\theta \neq 0$ be a complex n-form on W. Then we define real n-forms ω_- and $\tilde{\omega}_-$ on V by the formula

$$\theta = \tilde{\omega}_- + i\omega_-.$$

Lemma 1.2. *The automorphism J belongs to the both algebras $\mathcal{A}_{\omega_-}$ and $\mathcal{A}_{\tilde{\omega}_-}$.*

Proof. We have

$$\begin{aligned}
&\tilde{\omega}_-(Jv_1, v_2, \ldots, v_n) + i\omega_-(Jv_1, v_2, \ldots, v_n) \\
&\quad = \theta(Jv_1, v_2, \ldots, v_n) = i\theta(v_1, v_2, \ldots, v_n) = \theta(v_1, Jv_2, \ldots, v_n) \\
&\quad = \tilde{\omega}_-(v_1, Jv_2, \ldots, v_n) + i\omega_-(v_1, Jv_2, \ldots, v_n).
\end{aligned}$$ $\square$

Further we have

$$\begin{aligned}
&-\omega_-(v_1, v_2, \ldots, v_n) + i\tilde{\omega}_-(v_1, v_2, \ldots, v_n) \\
&\quad = i[\tilde{\omega}_-(v_1, v_2, \ldots, v_n) + i\omega_-(v_1, v_2, \ldots, v_n)] = i\theta(v_1, v_2, \ldots, v_n) \\
&\quad = \theta(Jv_1, v_2, \ldots, v_n) = \tilde{\omega}_-(Jv_1, v_2, \ldots, v_n) + i\omega_-(Jv_1, v_2, \ldots, v_n)
\end{aligned}$$

Hence we get

$$\begin{aligned}
\tilde{\omega}_-(v_1, v_2, \ldots, v_n) &= \omega_-(Jv_1, v_2, \ldots, v_n) \text{ and} \\
\omega_-(v_1, v_2, \ldots, v_n) &= -\tilde{\omega}_-(Jv_1, v_2, \ldots, v_n).
\end{aligned}$$

This result can be reformulated in the following way.

Lemma 1.3. $\tilde{\omega}_- = \frac{1}{n}\mathcal{D}_J\omega_-, \quad \omega_- = -\frac{1}{n}\mathcal{D}_J\tilde{\omega}_-.$

Lemma 1.4. *The forms ω_- and $\tilde{\omega}_-$ are regular.*

Proof. Let us assume that $\iota_v\omega_- = 0$. We have then

$$\begin{aligned}
(\iota_v\tilde{\omega}_-)(v_1, v_2, \ldots, v_{n-1}) &= \tilde{\omega}_-(v, v_1, v_2, \ldots, v_{n-1}) \\
&= \omega_-(Jv, v_1, v_2, \ldots, v_{n-1}) = \omega_-(v, Jv_1, v_2, \ldots, v_{n-1}) \\
&= (\iota_v\omega_-)(Jv_1, v_2, \ldots, v_{n-1}) = 0,
\end{aligned}$$

which proves that $\iota_v\tilde{\omega}_- = 0$. Consequently $\iota_v\theta = 0$, and this implies that $v = 0$. We have thus proved that the form ω_- is regular. Expressing $\tilde{\omega}_-$ using ω_- and J, we find easily that $\tilde{\omega}_-$ is also regular. $\square$

Proposition 1.1. $\mathcal{A}_{\omega_-} = \mathcal{A}_{\tilde{\omega}_-} = [I, J]$.

Proof. Let $A \in \mathcal{A}_{\omega_-}$. We have

$$\begin{aligned}\tilde{\omega}_-(Av_1, v_2, \ldots, v_n) &= (1/n)(\mathcal{D}_J\omega_-)(Av_1, v_2, \ldots, v_n)\\ &= \omega_-(JAv_1, v_2, \ldots, v_n) = \omega_-(AJv_1, v_2, \ldots, v_n)\\ &= \omega_-(Jv_1, Av_2, \ldots, v_n) = (1/n)(\mathcal{D}_J\omega_-)(v_1, Av_2, \ldots, v_n)\\ &= \tilde{\omega}_-(v_1, Av_2, \ldots, v_n),\end{aligned}$$

which shows that $\mathcal{A}_{\omega_-} \subset \mathcal{A}_{\tilde{\omega}_-}$. The converse inclusion can be proved in a similar way. For $A \in \mathcal{A}_{\omega_-}$ we have

$$\begin{aligned}\theta(Av_1, v_2, \ldots, v_n) &= \tilde{\omega}_-(Av_1, v_2, \ldots, v_n) + i\omega_-(Av_1, v_2, \ldots, v_n)\\ &= \tilde{\omega}_-(v_1, Av_2, \ldots, v_n) + i\omega_-(v_1, Av_2, \ldots, v_n) = \theta(v_1, Av_2, \ldots, v_n).\end{aligned}$$

According to Lemma 1.1 we have $\mathcal{A}_\omega \subset [I, J]$. The converse inclusion is obvious. □

Let us consider now an automorphism $A \in \mathrm{Aut}(\omega_-)$. We have

$$\begin{aligned}\omega_-(AJA^{-1}v_1, v_2, v_3, \ldots, v_n) &= \omega_-(JA^{-1}v_1, A^{-1}v_2, A^{-1}v_3, \ldots, A^{-1}v_n)\\ &= \omega_-(A^{-1}v_1, JA^{-1}v_2, A^{-1}v_3, \ldots, A^{-1}v_n) = \omega_-(v_1, AJA^{-1}v_2, v_3, \ldots, v_n).\end{aligned}$$

This shows that $AJA^{-1} \in \mathcal{A}_{\omega_-}$. Consequently, there are $a, b \in \mathbb{R}$ such that $AJA^{-1} = aI + bJ$. Squaring this identity we get

$$-I = (a^2 - b^2)I + 2abJ.$$

Obviously there must be $b \neq 0$. Consequently we have $a = 0$, and then $b = \pm 1$. This means that we have $AJA^{-1} = \pm J$ or equivalently $AJ = \pm JA$. We have thus proved the following lemma.

Lemma 1.5. *Every automorphism of ω_- is a complex linear or complex antilinear mapping.*

We define

$$\mathrm{Aut}^+(\omega_-) = \mathrm{Aut}(\omega_-) \cap GL^+(V; J), \quad \mathrm{Aut}^-(\omega_-) = \mathrm{Aut}(\omega_-) \cap GL^-(V; J).$$

We have $\mathrm{Aut}(\omega_-) = \mathrm{Aut}^+(\omega_-) \cup \mathrm{Aut}^-(\omega_-)$.

Lemma 1.6. *$Aut^+(\omega_-) = Aut^+(\tilde{\omega}_-)$.*

Proof. Let $A \in \mathrm{Aut}^+(\omega_-)$. Then we have

$$\begin{aligned}\tilde{\omega}_-(Av_1, Av_2, \ldots, Av_n) &= \omega_-(JAv_1, Av_2, \ldots, Av_n)\\ &= \omega_-(AJv_1, Av_2, \ldots, Av_n) = \omega_-(Jv_1, v_2, \ldots, v_n) = \tilde{\omega}_-(v_1, v_2, \ldots, v_n).\end{aligned}$$

This shows that $\mathrm{Aut}^+(\omega_-) \subset \mathrm{Aut}^+(\tilde{\omega}_-)$. The converse inclusion can be proved in the same way. □

If $A \in \mathrm{Aut}^+(\omega_-)$, then $A \in \mathrm{Aut}(\tilde{\omega}_-)$ and we find easily

$$A^*\theta = \det_{\mathbb{C}} A \cdot \theta, \quad A^*\theta = A^*\tilde{\omega}_- + iA^*\omega_- = \tilde{\omega}_- + i\omega_- = \theta.$$

We have thus shown that if $A \in \mathrm{Aut}^+(\omega_-)$, then $\det_{\mathbb{C}} A = 1$. Conversely, it can be easily seen that if $A \in GL^+(V;\mathbb{C})$ and $\det_{\mathbb{C}} A = 1$, then $A \in \mathrm{Aut}^+(\omega_-)$.

Proposition 1.2. *An automorphism $A \in GL^+(V;\mathbb{C})$ belongs to $Aut^+(\omega_-)$ if and only if $\det_{\mathbb{C}} A = 1$. Consequently $Aut^+(\omega_-) = SL(V;\mathbb{C}) \cong SL(n;\mathbb{C})$.*

Our next aim is to investigate the set $\mathrm{Aut}^-(\omega_-)$. First of all we must see that his set is not empty. We shall start with the following lemma.

Lemma 1.7. *Let $A \in Aut(\omega_-)$. Then $A \in Aut^-(\omega_-)$ if and only if one of the following two equivalent conditions is satisfied.*

(i) $A^*\tilde{\omega}_- = -\tilde{\omega}_-$,
(ii) $A^*\theta = -\bar{\theta}$.

Proof. Let us assume first that $A \in \mathrm{Aut}^-(\omega_-)$. Then we have

$$(A^*\tilde{\omega}_-)(v_1, v_2, \ldots, v_n) = \tilde{\omega}_-(Av_1, Av_2, \ldots, Av_n) = \omega_-(JAv_1, Av_2, \ldots, Av_n)$$
$$= -\omega_-(AJv_1, Av_2, \ldots, Av_n) = -\omega_-(Jv_1, v_2, \ldots, v_n) = -\tilde{\omega}_-(v_1, v_2, \ldots, v_n).$$

On the other hand, let us suppose that $A^*\tilde{\omega}_- = -\tilde{\omega}_-$. Then we have

$$\begin{aligned}
\tilde{\omega}_-(Av_1, Av_2, \ldots, Av_n) &= -\tilde{\omega}_-(v_1, v_2, \ldots, v_n) \\
\omega_-(JAv_1, Av_2, \ldots, Av_n) &= -\omega_-(Jv_1, v_2, \ldots, v_n) \\
\omega_-(A^{-1}JAv_1, v_2, \ldots, v_n) &= \omega_-(-Jv_1, v_2, \ldots, v_n).
\end{aligned}$$

Because the form ω_- is regular, the last equality implies that $A^{-1}JA = -J$, which shows that $A \in \mathrm{Aut}^-(\omega_-)$.

If $A^*\tilde{\omega}_- = -\tilde{\omega}_-$, then we have

$$A^*\theta = A^*\tilde{\omega}_- + iA^*\omega = -\tilde{\omega}_- + i\omega = -\bar{\theta}.$$

The converse direction is now obvious. This finishes the proof. □

Now we shall consider the complex conjugate $\bar{W}$ of W. We recall that $\bar{W} = W$, and multiplication by a complex number c in $\bar{W}$ is defined by the formula $c * v = \bar{c}v$. We have

$$\bar{\theta}(c * v_1, v_2, \ldots, v_n) = \bar{\theta}(\bar{c}v_1, v_2, \ldots, v_n) = c\bar{\theta}(v_1, v_2, \ldots, v_n),$$

which shows that $\bar{\theta}$ is a complex n-form on the complex vector space $\bar{W}$.

Because W and $\bar{W}$ are n-dimensional vector spaces, there exists an isomorphism $B : W \to \bar{W}$ such that $B^*\bar{\theta} = \theta$.

Lemma 1.8. *$A \in Aut^-(\omega_-)$ if and only if* $\det_{\mathbb{C}}(AB) = -1$.

Proof. $A \in \mathrm{Aut}^-(\omega_-)$ if and only if $A^*\theta = -\bar{\theta}$, which is equivalent to

$$A^*\theta = -\bar{\theta}$$
$$B^*A^*\theta = -B^*\bar{\theta}$$
$$(AB)^*\theta = -\theta.$$

The last equality holds if and only if $\det_{\mathbb{C}}(AB) = -1$. □

From this lemma we can easily see that $\mathrm{Aut}^-(\omega_-) \neq \emptyset$. Now it is obvious that

$$\mathrm{Aut}^+(\omega_-) \cdot \mathrm{Aut}^+(\omega_-) = \mathrm{Aut}^+(\omega_-),$$
$$\mathrm{Aut}^+(\omega_-) \cdot \mathrm{Aut}^-(\omega_-) = \mathrm{Aut}^-(\omega_-) \cdot \mathrm{Aut}^+(\omega_-) = \mathrm{Aut}^-(\omega_-),$$
$$\mathrm{Aut}^-(\omega_-) \cdot \mathrm{Aut}^-(\omega_-) = \mathrm{Aut}^+(\omega_-).$$

Summarizing, we have the following proposition.

Proposition 1.3. *The automorphism group $Aut(\omega_-)$ consists of two connected components $Aut^+(\omega_-)$ and $Aut^-(\omega_-)$, where $Aut^+(\omega_-)$ is the connected component of the unit. The group $Aut^+(\omega_-) = SL(V;\mathbb{C}) \cong SL(3;\mathbb{C})$. Moreover* $\dim_{\mathbb{R}} Aut(\omega_-) = 2(n^2-1)$.

2. Forms of the product type

It is obvious that the elements $\rho = (1/2)(1+d)$ and $\sigma = (1/2)(1-d)$ form a basis of $\mathbb{D}$ and that we have

$$\rho^2 = \rho, \quad \rho\sigma = 0, \quad \sigma^2 = \sigma.$$

Let us choose a real form V_0 of V. It is easy to see that every element $v \in V$ can be uniquely expressed in the form

$$v = \rho x + \sigma y, \quad \text{where } x, y \in V_0.$$

Lemma 2.1. *Vectors $w_1 = \rho u_1 + \sigma v_1, \ldots, w_k = \rho u_k + \sigma v_k$ are linearly independent in the $\mathbb{D}$-module W if and only if the vectors $u_1, \ldots, u_k$ are linearly independent in the vector space V_0 and the vectors $v_1, \ldots, v_k$ are linearly independent in the same vector space V_0. If the vectors $u_1, \ldots, u_k$ are linearly independent in V_0 and the vectors $v_1, \ldots, v_k$ are linearly independent in V_0, then the vectors $w_1, \ldots, w_k$ can be completed to a basis of W.*

Proof. Let us write

$$c_i = a_i\rho + b_i\sigma, \quad w_i = \rho u_i + \sigma v_i, \quad i = 1, \ldots, k.$$

We have

$$c_1 w_1 + \cdots + c_k w_k = \rho(a_1 u_1 + \cdots + a_k u_k) + \sigma(b_1 v_1 + \cdots + b_k v_k).$$

We can easily see that the vectors $w_1, \ldots, w_k$ are linearly independent in the module W if and only if the vectors $u_1, \ldots, u_n$ are linearly independent in V_0 and the vectors $v_1, \ldots, v_n$ are linearly independent in V_0. Moreover, it is obvious that if the vectors $w_1, \ldots, w_k$ are linearly independent, they can be completed to a basis. □

Lemma 2.2. *Let θ be a $\mathbb{D}$-multilinear n-form on W with values in the algebra $\mathbb{D}$. Then $\mathcal{A}_\theta = [I, D] = \{cI; c \in \mathbb{D}\}$.*

Proof. The proof proceeds along the same lines as the proof of Lemma 1.1. □

We take now a non-zero n-form θ on the $\mathbb{D}$-module W. We introduce real valued $\mathbb{R}$-multilinear n-forms ω_+ and $\tilde{\omega}_+$ on V by the formula

$$\theta = \tilde{\omega}_+ + e\omega_+.$$

Lemma 2.3. *The automorphism D belongs to the both algebras $\mathcal{A}_{\omega_+}$ and $\mathcal{A}_{\tilde{\omega}_+}$.*

Proof. This proof is the same as the proof of Lemma 1.2. □

Lemma 2.4. *The n-forms ω_+ and $\tilde{\omega}_+$ satisfy the relations $\tilde{\omega}_+ = \frac{1}{n}\mathcal{D}_D\omega_+$ and $\omega_+ = \frac{1}{n}\mathcal{D}_D\tilde{\omega}_+$.*

Proof. Here it suffices to proceed in the same way as in the proof of Lemma 1.3. □

Lemma 2.5. *The forms ω_+ and $\tilde{\omega}_+$ are regular.*

Proof. The proof is same as the proof of Lemma 1.4. □

Proposition 2.1. $\mathcal{A}_{\omega_+} = \mathcal{A}_{\tilde{\omega}_+} = [I, D]$.

Proof. The proof follows the lines of the proof of Proposition 1.1. □

Let us assume now that $A \in \mathrm{Aut}(\omega_+)$. Then we have

$$\begin{aligned}\omega_+(ADA^{-1}v_1, v_2, \dots, v_n) &= \omega_+(DA^{-1}v_1, A^{-1}v_2, \dots, A^{-1}v_n)\\ &= \omega_+(A^{-1}v_1, DA^{-1}v_2, \dots, A^{-1}v_n) = \omega_+(v_1, ADA^{-1}v_2, \dots, v_n),\end{aligned}$$

which shows that $ADA^{-1} \in \mathcal{A}_{\omega_+}$. This implies that there are $a, b \in \mathbb{R}$ such that $ADA^{-1} = aI + bD$. Taking the second power of this equality, we get

$$I = (a^2 + b^2)I + 2abD.$$

Obviously, there must be either $a = 0$ or $b = 0$. Let us assume first that $b = 0$. Then we get $I = a^2 I$, which implies $a = \pm 1$. In this situation we have the following two possibilities.

$$\begin{aligned} ADA^{-1} &= I & \qquad ADA^{-1} &= -I\\ AD &= A & AD &= -A\\ D &= I & D &= -I \end{aligned}$$

But according to the assumptions concerning D neither $D = I$ nor $D = -I$ is possible. Consequently there must be $a = 0$ and $b = \pm 1$. Then $ADA^{-1} = \pm D$, and we have the following two possibilities.

$$\begin{aligned} ADA^{-1} &= D & \qquad ADA^{-1} &= -D\\ AD &= DA & AD &= -DA \end{aligned}$$

We have thus proved the following lemma.

Lemma 2.6. *Every automorphism of ω_+ is a paracomplex linear or paracomplex antilinear mapping.*

We denote

$$\begin{aligned}\mathrm{Aut}^+(\omega_+) &= \{A \in \mathrm{Aut}(\omega_+); AD = DA\},\\ \mathrm{Aut}^-(\omega_+) &= \{A \in \mathrm{Aut}(\omega_+); AD = -DA\}.\end{aligned}$$

We have obviously $\mathrm{Aut}(\omega_+) = \mathrm{Aut}^+(\omega_+) \cup \mathrm{Aut}^-(\omega_+)$.

$$\begin{gathered}\mathrm{Aut}^+(\omega_+) \cdot \mathrm{Aut}^+(\omega_+) = \mathrm{Aut}^+(\omega_+),\\ \mathrm{Aut}^+(\omega_+) \cdot \mathrm{Aut}^-(\omega_+) = \mathrm{Aut}^-(\omega_+) \cdot \mathrm{Aut}^+(\omega_+) = \mathrm{Aut}^-(\omega_+),\\ \mathrm{Aut}^-(\omega_+) \cdot \mathrm{Aut}^-(\omega_+) = \mathrm{Aut}^+(\omega_+).\end{gathered}$$

Lemma 2.7. $Aut^+(\omega_+) = Aut^+(\tilde{\omega}_+)$.

Proof. The proof follows the lines of the proof of Proposition 1.6. □

Proposition 2.2. *An automorphism* $A \in GL^+(V;\mathbb{D})$ *belongs to* $Aut^+(\omega_+)$ *if and only if* $\det_{\mathbb{D}} A = 1$. *Consequently* $Aut^+(\omega_+) = SL(V;\mathbb{D})$.

Proof. Here we proceed as in the proof of Proposition 1.2. □

Every endomorphism Q of the $\mathbb{D}$-module $(V, \mathbb{D})$ can be uniquely expressed in the form $Q = \rho R + \sigma S$, where R and S are real endomorphisms, i. e. endomorphisms satisfying $RV_0 \subset V_0$ and $SV_0 \subset V_0$. Then it is obvious that

$$\begin{aligned}\det_{\mathbb{D}} Q &= \rho \det_{\mathbb{R}}(R|V_0) + \sigma \det_{\mathbb{R}}(S|V_0)\\ &= \frac{1}{2}(\det_{\mathbb{R}}(R|V_0) + \det_{\mathbb{R}}(S|V_0)) + d\frac{1}{2}(\det_{\mathbb{R}}(R|V_0) - \det_{\mathbb{R}}(S|V_0)).\end{aligned}$$

We can see that $\det_{\mathbb{D}} Q = 1$ if and only if $\det_{\mathbb{R}}(R|V_0) = \det_{\mathbb{R}}(S|V_0) = 1$. Now we get easily the following proposition.

Proposition 2.3. *The group* $Aut^+(\omega_+) = SL(V;\mathbb{D})$ *is isomorphic with the group* $SL(V_0) \times SL(V_0) \cong SL(n;\mathbb{R}) \times SL(n;\mathbb{R})$, *and consequently is connected.*

In the algebra $\mathbb{D}$ of paracomplex numbers we can introduce conjugation by the standard formula $\overline{a + db} = a - db$. This conjugation has moreless the same properties as the conjugation of complex numbers. If W is a $\mathbb{D}$-module, we can introduce the conjugate $\mathbb{D}$-module $\bar{W}$ by setting $\bar{W}$ and $c * v = \bar{c}v$. If W is an n-dimensional free $\mathbb{D}$-module, then $\bar{W}$ is also an n-dimensional free $\mathbb{D}$-module. Consequently, there exist a $\mathbb{D}$-module isomorphism $B : W \to \bar{W}$ such that $B^*\bar{\theta} = \theta$. Now, along the same lines as in the complex case, we get the following two lemmas.

Lemma 2.8. *Let* $A \in Aut(\omega_+)$. *Then* $A \in Aut^-(\omega_+)$ *if and only if one of the following two equivalent conditions is satisfied.*

(i) $A^*\tilde{\omega}_+ = -\tilde{\omega}_+$,
(ii) $A^*\theta = -\bar{\theta}$.

Lemma 2.9. $A \in Aut^-(\omega_+)$ *if and only if* $\det_{\mathbb{D}}(AB) = -1$.

Now we can easily see that there is

$$\begin{aligned} \mathrm{Aut}^+(\omega_+)\cdot\mathrm{Aut}^+(\omega_+) &= \mathrm{Aut}^+(\omega_+),\\ \mathrm{Aut}^+(\omega_+)\cdot\mathrm{Aut}^-(\omega_+) = \mathrm{Aut}^-(\omega_+)\cdot\mathrm{Aut}^+(\omega_+) &= \mathrm{Aut}^-(\omega_-),\\ \mathrm{Aut}^-(\omega_+)\cdot\mathrm{Aut}^-(\omega_+) &= \mathrm{Aut}^+(\omega_+). \end{aligned}$$

Summarizing, we obtain the following proposition.

Proposition 2.4. *The automorphism group $Aut(\omega_+)$ consists of two connected components $Aut^+(\omega_+)$ and $Aut^-(\omega_+)$, where $Aut^+(\omega_+)$ is the connected component of the unit. The group $Aut^+(\omega_+) = SL(V_0)\times SL(V_0)\cong SL(n;\mathbb{R})\times SL(n;\mathbb{R})$. Moreover $\dim_{\mathbb{R}} Aut(\omega_+) = 2(n^2-1)$.*

According to our assumptions concerning the automorphism D of V we can write

$$V = V_+\oplus V_-, \text{ where } V_+ = \{v\in V; Dv = v\},\quad V_- = \{v\in V; Dv = -v\}.$$

Lemma 2.10. *Let $v_+\in V_+$, $v_-\in V_-$, and $v_3,\dots,v_n\in V$. Then $\omega_+(v_+,v_-,v_3,\dots,v_n) = 0$.*

Proof. We get

$$\begin{aligned} &\omega_+(v_+,v_-,v_3,\dots,v_n) = \omega_+(Dv_+,v_-,v_3,\dots,v_n) =\\ &= \omega_+(v_+,Dv_-,v_3,\dots,v_n) = -\omega_+(v_+,v_-,v_3,\dots,v_n), \end{aligned}$$

which shows that $\omega_+(v_+,v_-,v_3,\dots,v_n) = 0$. □

If $v_1 = v_{1+}+v_{1-},\dots,v_n = v_{n+}+v_{n-}$, then we obviously have

$$\omega_+(v_1,\dots,v_n) = \omega_+(v_{1+},\dots,v_{n+}) + \omega_+(v_{1-},\dots,v_{n-}).$$

Moreover it is easy to see that $\omega_+|V_+$ and $\omega_+|V_-$ are regular forms. (Otherwise the form ω_+ would be singular.) Let $\pi_+ : V\to V_+$ and $\pi_- : V\to V_-$ denote the projections. We get the following proposition.

Proposition 2.5. *For the form ω_+ we have $\omega_+ = \pi_+^*(\omega_+|V_+)+\pi_-^*(\omega_+|V_-)$.*

3. Forms of the tangent type

In this section we shall consider an n-dimensional $\mathbb{E}$-module $(V,\mathbb{E}) = W$. First we introduce the mapping

$$\rho : \mathbb{E}\to\mathbb{E}/(e)\cong\mathbb{R},$$

which is projection onto the quotient by the ideal (e). Now we are going to prove the following lemma.

Lemma 3.1. *The elements* $w_1, \dots, w_k$ *are linearly independent in* W *if and only if the vectors* $Ew_1, \dots, Ew_k$ *are linearly independent in* V. *The elements* $w_1, \dots, w_n$ *constitute a basis of* W *if and only if the vectors* $Ew_1, \dots, Ew_n$ *are linearly independent in* V.

Proof. Let us assume that the elements $w_1, \dots, w_k$ are linearly independent in W. Let $a_1, \dots, a_k$ be real numbers such that $a_1Ew_1+\cdots+a_kEw_k = 0$. Then there is $E(a_1w_1 + \cdots + a_kw_k) = 0$, which means that we can find $w \in W$ such that

$$a_1w_1 + \cdots + a_kw_k = ew$$
$$ea_1w_1 + \cdots + ea_kw_k = 0$$

Because the elements $w_1, \dots, w_k$ are linearly independent in W, we have $a_1e = \cdots = a_ke = 0$, which implies $a_1 = \cdots = a_k = 0$. We have thus proved that the vectors $Ew_1, \dots, Ew_k$ are linearly independent in V.

On the other hand, let us suppose that the vectors $Ew_1, \dots, Ew_k$ are linearly independent in V. Let $c_1, \dots, c_k \in \mathbb{E}$ be such that $c_1w_1 + \cdots + c_kw_k = 0$. Writing $c_i = c_i' + ec_i''$, $i = 1, \dots, k$, we get from the last equality $c_1'Ew_1 \cdots + c_k'Ew_k = 0$. This implies that we have $c_1' = \cdots = c_k' = 0$. Consequently, we get

$$e(c_1''w_1 + \cdots + c_k''w_k) = 0$$
$$c_1''Ew_1 + \cdots + c_k''Ew_k = 0,$$

which again implies $c_1'' = \cdots = c_k'' = 0$. We have thus proved that $w_1, \dots, w_k$ are linearly independent in the $\mathbb{E}$-module W.

If $w_1, \dots, w_n$ is a basis of W, then the elements $w_1, \dots, w_n$ are linearly independent in W, and consequently the vectors $Ew_1, \dots, Ew_n$ are linearly independent in V. Conversely, let us assume that the vectors $Ew_1, \dots, Ew_n$ are linearly independent in V. Then $w_1, \dots, w_n$ are linearly independent in W. Finally, let $w \in W$. Then we can find uniquely determined $c_1', \dots, c_n' \in \mathbb{R}$ such that

$$Ew = c_1'Ew_1 + \cdots + c_n'Ew_n.$$

Consequently, $E(w - c_1'w_1 - \cdots - c_n'w_n) = 0$, and there is $w'' \in W$ such that $w - c_1'Ew_1 - \cdots - c_n'Ew_n = Ew''$. We can find uniquely determined $c_1'', \dots, c_n'' \in \mathbb{R}$ such that $Ew'' = c_1''Ew_1 + \cdots + c_n''Ew_n$. We can see that

$$w = (c_1' + ec_1'')w_1 + \cdots + (c_n' + ec_n'')w_n$$

which proves that $w_1, \ldots, w_n$ is a basis of the $\mathbb{E}$-module W. □

Let θ be a non-zero n-form on W. Then, proceeding as in the proofs of Lemma 1.1 1.1we get easily the following lemma.

Lemma 3.2. *Let $\theta \neq 0$ be an n-form on the n-dimensional $\mathbb{E}$-module W. Then $\mathcal{A}_\theta = [I, E] = \{cI; c \in \mathbb{E}\}$.*

We introduce real valued n-forms ω_0 and $\tilde{\omega}_0$ by the formula

$$\theta = \tilde{\omega}_0 + e\omega_0.$$

Lemma 3.3. *The automorphism E belongs to the both algebras $\mathcal{A}_{\omega_0}$ and $\mathcal{A}_{\tilde{\omega}_0}$.*

Proof. The proof follows the lines of the proof of Lemma 1.2. □

Lemma 3.4. *The n-forms ω_0 and $\tilde{\omega}_0$ satisfy the relation $\tilde{\omega}_0 = \frac{1}{n}\mathcal{D}_E\omega_0$ and $\mathcal{D}_E\tilde{\omega}_0 = 0$.*

Proof.

$$\theta(Ew_1, w_2, \ldots, w_n) = \tilde{\omega}_0(Ew_1, w_2, \ldots, w_n) + e\omega_0(Ew_1, w_2, \ldots, w_n),$$
$$\theta(Ew_1, w_2, \ldots, w_n) = e\theta(w_1, w_2, \ldots, w_n) = e\tilde{\omega}_0(w_1, w_2, \ldots, w_n),$$

which shows that $\tilde{\omega}_0(w_1, w_2, \ldots, w_n) = \omega_0(Ew_1, w_2, \ldots, w_n)$ and $\tilde{\omega}_0(Ew_1, w_2, \ldots, w_n) = 0$. In other words $\tilde{\omega}_0 = (1/n)\mathcal{D}_E\omega_0$ and $\mathcal{D}_E\tilde{\omega}_0 = 0$. □

Lemma 3.5. *The form ω_0 is regular.*

Proof. This lemma can be proved in the same way as Lemma 1.4 □

Lemma 3.6. $\mathcal{A}_{\omega_0} = [I, E]$, $[I, E] \subset \mathcal{A}_{\tilde{\omega}_0}$.

Proof. Let $A \in \mathcal{A}_\theta$. Then we have

$$\tilde{\omega}_0(Av_1, v_2, \ldots, v_n) + e\omega_0(Av_1, v_2, \ldots, v_n) = \theta(Av_1, v_2, \ldots, v_n)$$
$$= \theta(v_1, Av_2, \ldots, v_n) = \tilde{\omega}_0(v_1, Av_2, \ldots, v_n) + e\omega_0(v_1, Av_2, \ldots, v_n),$$

which shows that $\mathcal{A}_\theta \subset \mathcal{A}_{\omega_0}$ and $\mathcal{A}_\theta \subset \mathcal{A}_{\tilde{\omega}_0}$. Next, let us assume that $A \in \mathcal{A}_{\omega_0}$. Then according to Lemma 3.3 there is $AE = EA$. We have then

$$\tilde{\omega}_0(Av_1, v_2, \ldots, v_n) = \omega_0(EAv_1, v_2, \ldots, v_n) = \omega_0(AEv_1, v_2, \ldots, v_n)$$
$$= \omega_0(Ev_1, Av_2, \ldots, v_n) = \tilde{\omega}_0(v_1, Av_2, \ldots, v_n),$$

which shows that $A \in \mathcal{A}_{\tilde{\omega}_0}$. Consequently $A \in \mathcal{A}_\theta$, and we have $\mathcal{A}_{\omega_0} = \mathcal{A}_\theta = [I, E]$. □

In the same way as in the complex case we can prove that if $A \in \mathrm{Aut}(\omega_0)$, then $AEA^{-1} \in \mathcal{A}_{\omega_0}$. Consequently there are real numbers a, b such that $AEA^{-1} = aI + bE$. Taking the square of this relation we get $0 = a^2 I + 2abE$ and this implies $a = 0$. We obtain

Lemma 3.7. *Every automorphism $A \in Aut(\omega_0)$ satisfies the relation $AE = \kappa EA$ with $\kappa \in \mathbb{R}^* = \mathbb{R} - \{0\}$.*

Lemma 3.8. *Every automorphism $A \in Aut(\omega_0)$ is a conformal automorphism of the form $\tilde{\omega}_0$. More precisely, if $A \in Aut(\omega_0)$ satisfies $AE = \kappa EA$, then $A^*\tilde{\omega}_0 = (1/\kappa)\tilde{\omega}_0$.*

Proof. We have

$$\begin{aligned}(A^*\tilde{\omega}_0)(v_1, v_2, \ldots, v_n) &= \tilde{\omega}_0(Av_1, Av_2, \ldots, Av_n)\\ &= \omega_0(EAv_1, Av_2, \ldots, Av_n) = (1/\kappa)\omega_0(AEv_1, Av_2, \ldots, Av_n)\\ &= (1/\kappa)\omega_0(Ev_1, v_2, \ldots, v_n) = (1/\kappa)\tilde{\omega}_0(v_1, v_2, \ldots, v_n).\end{aligned}$$

□

Lemma 3.9. *For every $\kappa \in \mathbb{R}^*$ there exists an automorphism $A \in Aut(\omega_0)$ such that $AE = \kappa EA$.*

Proof. We choose a basis $\beta_1, \ldots, \beta_n$ of W^* such that $\theta = \beta_1 \wedge \cdots \wedge \beta_n$ and the corresponding dual basis $e_1, \ldots, e_n$ of W. Then $e_1, \ldots, e_n, Ee_1, \ldots, Ee_n$ is a basis of the vector space V. We take the dual basis $\alpha_1, \ldots, \alpha_n, \alpha_{n+1}, \ldots, \alpha_{2n}$ of the vector space V^*. We find easily that

$$\beta_1 = \alpha_1 + e\alpha_{n+1}, \ldots, \beta_n = \alpha_n + e\alpha_{2n}.$$

Now we can see that

$$\theta = \alpha_1 \wedge \cdots \wedge \alpha_n + e \sum_{i=1}^{n} \alpha_1 \wedge \cdots \wedge \alpha_{i-1} \wedge \alpha_{i+n} \wedge \alpha_{i+1} \wedge \cdots \wedge \alpha_n.$$

We define now an automorphism of V by the following formulas.

$$\begin{aligned}&A^*\alpha_1 = \frac{1}{\kappa}\alpha_1, A^*\alpha_2 = \alpha_2, \ldots, A^*\alpha_n = \alpha_n,\\ &A^*\alpha_{n+1} = \alpha_{n+1}, A^*\alpha_{n+2} = \kappa\alpha_{n+2}, \ldots, A^*\alpha_{2n} = \kappa\alpha_{2n}.\end{aligned}$$

Now it is obvious that $A^*\omega_0 = \omega_0$ and $A^*\tilde{\omega}_0 = \frac{1}{\kappa}\tilde{\omega}_0$. Hence we have $AE = \kappa EA$. □

From the above considerations we get easily the following lemma.

Lemma 3.10. *The n-form $\tilde{\omega}_0$ is decomposable.*

We can now define an epimorphism $K : \mathrm{Aut}(\omega_0) \to \mathbb{R}^*$. If $A \in \mathrm{Aut}(\omega_0)$, then there is a unique $\kappa \in \mathbb{R}^*$ such that $AE = \kappa EA$. We set $K(A) = \kappa$. Using Lemma 3.9 we obtain the short exact sequence

$$1 - \mathrm{Ker}K \to \mathrm{Aut}(\omega_0) \xrightarrow{K} \mathbb{R}^* \to 1$$

If $A \in \mathrm{Ker}K$, then $AE = EA$, A is an $\mathbb{E}$-linear automorphism, $A^*\omega_0 = \omega_0$ and $A^*\tilde{\omega}_0 = \tilde{\omega}_0$. Consequently,

$$A^*\theta = A^*\tilde{\omega}_0 + eA^*\omega_0 = \tilde{\omega}_0 + e\omega_0 = \theta.$$

Hence we can see that $\mathrm{Ker}K = SL(V;\mathbb{E}) \cong SL(n,\mathbb{E})$. The above exact sequence can now be written in the form

$$1 \to SL(V;\mathbb{E}) \to \mathrm{Aut}(\omega_0) \xrightarrow{K} \mathbb{R}^* \to 1.$$

Introducing the subsets

$$\mathrm{Aut}^+(\omega) = \{A \in \mathrm{Aut}(\omega); K(A) > 0\}, \mathrm{Aut}^-(\omega) = \{A \in \mathrm{Aut}(\omega); K(A) < 0\},$$

we have an exact sequence

$$1 \to SL(V;\mathbb{E}) \to \mathrm{Aut}^+(\omega_0) \xrightarrow{K} \mathbb{R}^+ \to 1.$$

Lemma 3.11. *The group $Aut(\omega_0)$ is a semidirect product $SL(V;\mathbb{E}) \ltimes \mathbb{R}^*$. Analogously, the group $Aut^+(\omega_0)$ is a semidirect product $SL(V;\mathbb{E}) \ltimes \mathbb{R}^+$.*

Proof. In the first case it suffices to find a splitting $\sigma : \mathbb{R}^* \to \mathrm{Aut}(\omega_0)$. We use the same bases as in Lemma 3.9. To $\kappa \in \mathbb{R}^*$ we assign an automorphism $\sigma(\kappa)$ defined by the formulas

$$\begin{array}{llll} \sigma(\kappa)e_1 = \frac{1}{\kappa}e_1, & \sigma(\kappa)e_2 = e_2, & \dots, & \sigma(\kappa)e_n = e_n, \\ \sigma(\kappa)e_{n+1} = e_{n+1}, & \sigma(\kappa)e_{n+2} = \kappa e_{n+2}, & \dots, & \sigma(\kappa)e_{2n} = \kappa e_{2n}. \end{array}$$

It can be immediately seen that σ is a splitting. It is also obvious that $\sigma(\mathbb{R}^+) \subset \mathrm{Aut}^+(\omega_0)$, which means that the same splitting can be used also in the second case. □

We shall now investigate the group $SL(V;\mathbb{E})$. Let $A \in GL(V;\mathbb{E})$. Because A is $\mathbb{E}$-linear, it preserves the subspace $\mathrm{im}\,A$, and consequently induces an automorphism $\hat{A}$ of the quotient $\hat{V} = V/\,\mathrm{im}\,E$. We have the projection $\pi : V \to \hat{V}$, which satisfies $\pi(av) = \rho(a)\pi(v)$. This projection induces also a projection

$$\Lambda^n \pi : \Lambda^n_{\mathbb{E}} V \to \Lambda^n_{\mathbb{R}} \hat{V}, \quad v_1 \wedge_{\mathbb{E}} \cdots \wedge_{\mathbb{E}} v_n \mapsto \pi(v_1) \wedge_{\mathbb{R}} \cdots \wedge_{\mathbb{R}} \pi(v_n).$$

We have

$$\begin{aligned}
&A_*(e_1 \wedge_{\mathbb{E}} \cdots \wedge_{\mathbb{E}} e_n) = \det_{\mathbb{E}} A \cdot e_1 \wedge_{\mathbb{E}} \cdots \wedge_{\mathbb{E}} e_n,\\
&\hat{A}_*(\pi(e_1) \wedge_{\mathbb{R}} \cdots \wedge_{\mathbb{R}} \pi(e_n)) = \det_{\mathbb{R}} \hat{A} \cdot \pi(e_1) \wedge_{\mathbb{R}} \cdots \wedge_{\mathbb{R}} \pi(e_n),\\
&\hat{A}_*(\pi(e_1) \wedge_{\mathbb{R}} \cdots \wedge_{\mathbb{R}} \pi(e_n)) = \hat{A}_*(\Lambda^n \pi)(e_1 \wedge_{\mathbb{E}} \cdots \wedge_{\mathbb{E}} e_n)\\
&\quad = (\Lambda^n \pi) A_*(e_1 \wedge_{\mathbb{E}} \cdots \wedge_{\mathbb{E}} e_n) = (\Lambda^n \pi)(\det_{\mathbb{E}} A \cdot e_1 \wedge_{\mathbb{E}} \cdots \wedge_{\mathbb{E}} e_n)\\
&\quad = \rho(\det_{\mathbb{E}} A) \cdot (\Lambda^n \pi)(e_1 \wedge_{\mathbb{E}} \cdots \wedge_{\mathbb{E}} e_n) = \rho(\det_{\mathbb{E}} A) \cdot \pi(e_1) \wedge_{\mathbb{R}} \cdots \wedge_{\mathbb{R}} \pi(e_n).
\end{aligned}$$

We have thus proved the formula

$$\rho(\det_{\mathbb{E}} A) = \det_{\mathbb{R}} \hat{A}.$$

We denote Q the homomorphism assigning to an automorphism $A \in GL(V;\mathbb{E})$ the induced automorphism $\hat{A} \in GL(\hat{V})$. It is easy to see that we get a short exact sequence

$$1 \to \mathrm{Ker} Q \to GL(V;\mathbb{E}) \xrightarrow{Q} GL(\hat{V}) \to 1.$$

Let us denote first

$$\mathcal{B} = \{B \in gl(V;\mathbb{E}); BV \subset \operatorname{im} E\}.$$

If $B \in \mathcal{B}$, then $B(\operatorname{im} E) = 0$. Namely, if $v \in \operatorname{im} E$, then $v = ev'$ for some $v' \in V$. Then $Bv = B(ev') = eB(v') = 0$. Consequently, if $B, B' \in \mathcal{B}$, then $BB' = 0$.

Every $A \in \mathrm{Ker} Q$ can be expressed in the form $A = I + B$, where $B \in \mathcal{B}$. On the other hand, every endomorphism of the form $I + B$ with $B \in \mathcal{B}$ is an automorphism. Namely,

$$(I + B)(I - B) = I - B + B - BB = I.$$

Moreover $A = I + B$ obviously belongs to $\mathrm{Ker} Q$. If $A = I + B$ and $A' = I + B'$ we have

$$(I + A)(I + B') = I + B + B' + BB' = I + B + B'.$$

Hence we can see that $\mathrm{Ker} Q$ is an abelian group isomorphic with $\mathcal{B} \cong \mathbb{R}^{n^2}$.

Lemma 3.12. *The group $GL(V;\mathbb{E})$ is a semidirect product $\mathcal{B} \ltimes GL(\hat{V})$.*

Proof. We use again the same bases as in the proof of lemma 3.9. Obviously the classes $[e_1], \ldots, [e_n]$ constitute a basis of the vector space $\hat{V}$. Any automorphism $\varphi \in GL(\hat{V})$ can be expressed in the form

$$\varphi[e_i] = \sum_{j=1}^{n} a_{ij}[e_j],$$

where a_{ij} are real numbers. We define an automorphism $A \in GL(V; \mathbb{E})$ by the formulas

$$Ae_i = \sum_{j=1}^{n} a_{ij} e_j.$$

Setting $\sigma(\varphi) = A$, we get a splitting $\sigma : GL(\hat{V}) \to GL(V; \mathbb{E})$. □

Let us remind that if $v_1, v_2 \in \operatorname{im} E$, then $\theta(v_1, v_2, \ldots, v_n) = 0$. Namely, we have $v_1 = ev_1'$ and $v_2 = ev_2'$, and we get

$$\theta(v_1, v_2, v_3, \ldots, v_n) = \theta(ev_1', ev_2', v_3, \ldots, v_n) = e^2\theta(v_1', v_2', v_3, \ldots, v_n) = 0.$$

Let $A = I + B$ be again an element of Ker$Q$. We obtain

$$\begin{aligned}(A^*\theta)(v_1, v_2, \ldots, v_n) &= \theta(v_1 + Bv_1, v_2 + Bv_2, \ldots, v_n + Bv_n)\\ &= \theta(v_1, v_2, \ldots, v_n) + \sum_{i=1}^{n} \theta(v_1, \ldots, v_{i-1}, Bv_i, v_{i+1}, \ldots, v_n)\\ &= \theta(v_1, v_2, \ldots, v_n) + \operatorname{tr}(B)\theta(v_1, v_2, \ldots, v_n) = (1 + \operatorname{tr}(B))\theta(v_1, v_2, \ldots, v_n).\end{aligned}$$

We have thus proved that if $A \in \operatorname{Ker}Q$, then $\det_{\mathbb{E}} A = 1 + \operatorname{tr}(B)$.

Using the formula $\rho(\det_{\mathbb{E}} A) = \det_{\mathbb{R}} \hat{A}$ we get another short exact sequence

$$1 \to \operatorname{Ker}q \to SL(V; \mathbb{E}) \xrightarrow{q} SL(\hat{V}) \to 1,$$

where $q = Q|SL(V; \mathbb{E})$. Applying the last determinant formula we find easily that

$$\operatorname{Ker}q = \mathcal{B}_0 = \{B \in \mathcal{B}; \operatorname{tr}(B) = 0\} \cong \mathbb{R}^{n^2-1}.$$

Lemma 3.13. *The group $SL(V; \mathbb{E})$ is a semidirect product $\mathcal{B}_0 \ltimes SL(\hat{V})$.*

Proof. The proof follows the same lines as the proof of Prop. 3.12. □

Summarizing we have the following proposition.

Proposition 3.1. *The automorphism group $Aut(\omega_0)$ consists of two connected components $Aut^+(\omega_0)$ and $Aut^-(\omega_0)$, where $Aut^+(\omega_0)$ is the connected component of the unit. Moreover* $\dim_{\mathbb{R}} Aut(\omega_0) = 2n^2 - 1$.

Acknowledgments

The author was supported by the Academy of Sciences of the Czech Republic, Institutional Research Plan No. AV0Z10190503 and by the Grant Agency of the Academy of Sciences of the Czech Republic, Grant No. A100190701.

References

1. J. Bureš and J. Vanžura, Unified Treatment of Multisymplectic 3-Forms in Dimension 6; arXiv:math/0405101.
2. N. Hitchin, The geometry of three-forms in six dimensions, *J. Diff. Geom.* **55** (3) (2000) 547–576.
3. N. Hitchin, Stable forms and special metrics, *In: Global differential geometry: the mathematical legacy of Alfred Gray* (Bilbao, 2000, Contemp. Math. 288, Amer. Math. Soc., Providence, RI, 2001) 70–89.
4. D.Z. Djoković, Classification of trivectors of an eight-dimensional real vector space, *Linear and Multilinear Algebra* **13** (1) (1983) 3–39.

Differential Geometry and its Applications
Proc. Conf., in Honour of Leonhard Euler, Olomouc, August 2007

Symmetries of almost Grassmannian geometries

Lenka Zalabová
Masaryk University, Eduard Čech Center for Algebra and Geometry,
Janáčkovo n. 2a, 662 95 Brno, Czech Republic
E-mail: zalabova@math.muni.cz

We study symmetries of almost Grassmannian and almost quaternionic structures. We generalize the classical definition for locally symmetric spaces and we discuss the existence of symmetries on the homogeneous models. We also conclude some observations on the general curved geometries.

Keywords: Cartan geometries, parabolic geometries, almost Grassmannian structures, almost quaternionic structures, symmetric spaces.

MS classification: 53C15, 53C05, 53A40, 53C35.

The aim of this paper is to introduce and discuss the symmetries for the almost Grassmannian and almost quaternionic structures. These two geometries belong into the class of so called $|1|$–graded parabolic geometries and we use the language of Cartan and parabolic geometries, which allows us to generalize the classical concepts. For the class of $|1|$–graded geometries, the symmetries are defined in the same intuitive way as in the affine geometry.

At the same time, much of the classical theory of affine symmetries extends. In particular, the existence of a symmetry at a point kills the torsion of the geometry at this point. In view of the nice general theory of parabolic geometries, this already proves the local flatness of the symmetric geometries for most cases of almost Grassmannian geometries.

There are also some more interesting types of almost Grassmannian and almost quaternionic geometries, which can carry some symmetry in the point with nonzero curvature. We show, that there can be at most one symmetry in such point.

1. Definitions and basic facts

Throughout the paper we use the standard notation and concepts.[6,12] We recall some of the basic facts and notation related to the parabolic geometries below.

1.1. *Cartan and parabolic geometries*

A *Cartan geometry* of type (G, P) on a smooth manifold M is a principal P–bundle $p : \mathcal{G} \to M$ with a *Cartan connection* $\omega \in \Omega^1(\mathcal{G}, \mathfrak{g})$, where the 1–form ω is an absolute parallelism which is P–equivariant and reproduces fundamental vector fields. The simplest examples are so called *homogeneous models*, which are the P–bundles $G \to G/P$ endowed with the (left) Maurer Cartan form ω_G.

A *parabolic geometry* is a Cartan geometry $(\mathcal{G} \to M, \omega)$ of type (G, P), where P is a parabolic subgroup of a semisimple Lie group G. The algebra $\mathfrak{g}$ of the group G is equipped (up to the choice of Levi factor $\mathfrak{g}_0$ in $\mathfrak{p}$) with a grading of the form $\mathfrak{g} = \mathfrak{g}_{-k} \oplus \cdots \oplus \mathfrak{g}_0 \oplus \cdots \oplus \mathfrak{g}_k$ such that the algebra $\mathfrak{p}$ of P is exactly $\mathfrak{p} = \mathfrak{g}_0 \oplus \cdots \oplus \mathfrak{g}_k$. We suppose that the gradation is fixed. We also have the subgroup $G_0 \subset P$ with Lie algebra $\mathfrak{g}_0$ of elements such that their adjoint action preserves the gradation. Remark, that P is exactly the group of all elements such that their adjoint action preserves *induced filtration* given as $\mathfrak{g} = \mathfrak{g}^{-k} \supset \mathfrak{g}^{-k+1} \supset \cdots \supset \mathfrak{g}^k = \mathfrak{g}_k$, where $\mathfrak{g}^i = \mathfrak{g}_i \oplus \cdots \oplus \mathfrak{g}_k$. If the length of the gradation of $\mathfrak{g}$ is k, then the geometry is called $|k|$*–graded*. We are mainly interested in $|1|$–graded geometries and we formulate most of the facts for them.

1.2. *Almost Grassmannian structures*

We take $G = Sl(p+q, \mathbb{R})$. This group acts on $\mathbb{R}^{p+q} \simeq \mathbb{R}^p \oplus \mathbb{R}^q$ and the parabolic subgroup P is exactly the stabilizer of $\mathbb{R}^p$. The reductive subgroup G_0 is of the form $S(Gl(p, \mathbb{R}) \times Gl(q, \mathbb{R}))$. We obtain $\mathfrak{g} = \mathfrak{sl}(p+q, \mathbb{R})$ and we get gradation of the form $\mathfrak{g}_{-1} \simeq L(\mathbb{R}^p, \mathbb{R}^q)$, $\mathfrak{g}_0 \simeq \mathfrak{s}(\mathfrak{gl}(p, \mathbb{R}) \oplus \mathfrak{gl}(q, \mathbb{R}))$ and $\mathfrak{g}_1 \simeq L(\mathbb{R}^q, \mathbb{R}^p)$. Thus, this choice of a Lie group corresponds to a $|1|$–graded geometry and its homogeneous model is the Grassmannian of p-dimensional subspaces of $\mathbb{R}^{p+q}$.

Remind that these geometries are given (as G_0-structures) by the identification of the tangent bundle $TM \simeq \mathcal{G}_0 \times_{G_0} \mathfrak{g}_{-1}$ with tensor product of two bundles $TM \simeq E \otimes F^*$ together with the preferred trivialization of $\wedge^q E \otimes \wedge^p F$. The rank q bundle $E \to M$ and the rank p bundle $F \to M$ are of the form of the associated vector bundles $E \simeq \mathcal{G}_0 \times_{G_0} \mathbb{R}^q$ and $F \simeq \mathcal{G}_0 \times_{G_0} \mathbb{R}^p$ for the standard actions of G_0.

1.3. *The curvature*

The curvature of $|1|$-graded geometry can be described by the *curvature function* $\kappa : \mathcal{G} \to \wedge^2 \mathfrak{g}_{-1}^* \otimes \mathfrak{g}$, where

$$\kappa(u)(X, Y) = [X, Y] - \omega([\omega^{-1}(X), \omega^{-1}(Y)](u)).$$

It is valued in the cochains for the second cohomology $H^2(\mathfrak{g}_{-1}, \mathfrak{g})$. This group can be also computed as the homology of the codifferential

$$\partial^* : \wedge^{k+1} \mathfrak{g}_{-1}^* \otimes \mathfrak{g} \to \wedge^k \mathfrak{g}_{-1}^* \otimes \mathfrak{g}.$$

The parabolic geometry is called *normal* if the curvature satisfies $\partial^* \circ \kappa = 0$. If the geometry is normal, we can define the *harmonic part of curvature*, $\kappa_H : \mathcal{G} \to H^2(\mathfrak{g}_{-1}, \mathfrak{g})$, as the composition of the curvature and the projection to the second cohomology group.

Thanks to the gradation of $\mathfrak{g}$, there are several decompositions of the curvature of the $|1|$-graded geometry. One of the possibilities is the decomposition into *homogeneous components*, which is of the form $\kappa = \kappa^{(1)} + \kappa^{(2)} + \kappa^{(3)}$, where $\kappa^{(i)}(u)(X, Y) \in \mathfrak{g}_{p+q+i}$ for all $X \in \mathfrak{g}_p, Y \in \mathfrak{g}_q$ and $u \in \mathcal{G}$. The parabolic geometry is called *regular* if the curvature function κ satisfies $\kappa^{(r)} = 0$ for all $r \leq 0$ and clearly, $|1|$-graded geometries are always regular. The crucial structural description of the curvature of parabolic geometry is provided by the following theorem:[14]

Theorem 1.1. *The curvature κ of regular normal geometry vanishes if and only if its harmonic part κ_H vanishes. Moreover, if all homogeneous components of κ of degrees less than j vanish identically and there is no cohomology $H_j^2(\mathfrak{g}_-, \mathfrak{g})$, then also the curvature component of degree j vanishes.*

Another possibility is the decomposition of the curvature according to the values $\kappa = \kappa_{-1} + \kappa_0 + \kappa_1$ and in an arbitrary frame u we have $\kappa_j(u) \in \mathfrak{g}_{-1} \wedge \mathfrak{g}_{-1} \to \mathfrak{g}_j$. The component κ_{-1} valued in $\mathfrak{g}_{-1}$ is the *torsion*. An important feature of $|1|$-graded geometries is that κ_i is equal to $\kappa^{(i+2)}$ for $i = -1, 0, 1$.

Notice that the Maurer–Cartan equations imply that the curvature of homogeneous model is zero. It can be proved that if the curvature of a Cartan geometry of type (G, P) vanishes, then the geometry is locally isomorphic with the homogeneous model $(G \to G/P, \omega_G)$.[11] Cartan geometry is called *locally flat* if the curvature κ vanishes. Homogeneous models are sometimes called *flat* models.

1.4. *Curvatures of almost quaternionic and almost Grassmannian geometries*

Next, we come back to the almost Grassmannian and almost quaternionic geometries which are special cases of $|1|$–graded geometries and we provide the complete list with their non-zero components of the harmonic curvature:

- *almost Grassmannian structures,* $\mathfrak{g} = \mathfrak{sl}(p+q, \mathbb{R})$, $\mathfrak{g}_0 = \mathfrak{s}(\mathfrak{gl}(p, \mathbb{R}) \times \mathfrak{gl}(q, \mathbb{R}))$:

  $p = 1$, $q = 2$; $p = 2$, $q = 1$: the projective structures dim $= 2$, one curvature of homogeneity 3

  $p = 1$, $q > 2$; $p > 2$, $q = 1$: the projective structures dim > 2, one curvature of homogeneity 2

  $p = 2$, $q = 2$: dim $= 4$, two curvatures of homogeneity 2

  $p = 2$, $q > 2$; $p > 2$, $q = 2$: dim $= pq$, one torsion, one curvature of homogeneity 2

  $p > 2$, $q > 2$: dim $= pq$, two torsions

- *almost quaternionic structures,* $\mathfrak{g} = \mathfrak{sl}(p+1, \mathbb{H})$:

  $p = 1$: the almost quaternionic geometries, dim $= 4$, two curvatures of homogeneity 2

  $p > 1$: the almost quaternionic geometries, dim $= 4p$, one torsion, one curvature of homogeneity 2

- *the geometries modeled on quaternionic Grassmannians*:

  two torsions

1.5. *Automorphisms and symmetries*

We shall deal with the automorphisms of Cartan geometries. These are P–bundle morphisms, which preserve Cartan connection on the geometry. They have to preserve the structure given by the existence of the Cartan connection.

It is well known that all automorphisms of the homogeneous model $(G \to G/P, \omega_G)$ of a Cartan geometry are exactly the left multiplications by elements of G.[11]

We define symmetry on the $|1|$–graded parabolic geometry in the following way:

Definition 1.1. Let $(\mathcal{G} \to M, \omega)$ be a $|1|$–graded geometry. The *symmetry at the point* x is a locally defined diffeomorphism s_x on M such that:

- $s_x(x) = x$
- $T_x s_x|_{T_x M} = -\,\mathrm{id}_{T_x M}$
- s_x is covered by an automorphism of the Cartan geometry.

The geometry is called *(locally) symmetric* if there is a symmetry at each point $x \in M$.

Thus the definition of the symmetries of $|1|$–graded geometries follows completely the classical intuitive idea.

2. Homogeneous models

Homogeneous models are simplest candidates for symmetric geometries. We ask, whether homogeneous models of almost Grassmannian geometries and almost quaternionic geometries are symmetric or not. It can be proved:[15,17]

Proposition 2.1. *All symmetries of homogeneous models of $|1|$–graded geometries at the origin are exactly left multiplications by elements $g \in P$ such that $g = g_0 \exp Z$ satisfying $\mathrm{Ad}_{g_0}(X) = -X$ for all $X \in \mathfrak{g}_{-1}$ and $Z \in \mathfrak{g}_1$ is arbitrary.*

Thus we are looking for these elements in the models for the latter two geometries.

2.1. *Almost Grassmannian structures*

We have $\mathfrak{g} = \mathfrak{sl}(p+q,\mathbb{R})$ and we first take $G = Sl(p+q,\mathbb{R})$, see section 1.2. The algebra $\mathfrak{g}$ consists of block elements $\left(\begin{smallmatrix} X & Y \\ Z & W \end{smallmatrix}\right)$ with block size p and q where $tr(X) + tr(W) = 0$ and elements of $\mathfrak{g}_{-1}$ are those with X, Y, W vanishing.

The adjoint action of some $\left(\begin{smallmatrix} S & 0 \\ 0 & T \end{smallmatrix}\right) \in G_0$ on $\left(\begin{smallmatrix} 0 & 0 \\ V & 0 \end{smallmatrix}\right) \in \mathfrak{g}_{-1}$ is TVS^{-1} and we look for S and T such that $TV = -VS$ for all $V \in L(\mathbb{R}^p,\mathbb{R}^q)$. The properties of matrix multiplication give that T and S are diagonal, elements on the diagonal of T are equal, elements on the diagonal of S are equal and elements on the diagonal of T are equal to minus elements from S. The condition on determinant gives that only the elements $\left(\begin{smallmatrix} E & 0 \\ 0 & -E \end{smallmatrix}\right)$ and $\left(\begin{smallmatrix} -E & 0 \\ 0 & E \end{smallmatrix}\right)$ satisfy all latter restrictions.

We have to discuss the dependence on p and q to resolve whether some of these two elements belong to $Sl(p+q,\mathbb{R})$ and give a symmetry. We get some symmetry if at least one of p and q is even. If only p is even, then all symmetries are given by elements $\left(\begin{smallmatrix} -E & X \\ 0 & E \end{smallmatrix}\right)$ for all $X \in L(\mathbb{R}^q,\mathbb{R}^p)$.

If only q is even, then all symmetries are given by elements $\left(\begin{smallmatrix} E & X \\ 0 & -E \end{smallmatrix}\right)$ for all $X \in L(\mathbb{R}^q, \mathbb{R}^p)$. If both sizes are even, then all latter elements give symmetries. If p and q are odd, then there is no symmetry.

The situation where p and q are both even is exactly the situation in non–effective models. In this special case, the geometry also has nontrivial kernel. This is of the form $\{\left(\begin{smallmatrix} E & 0 \\ 0 & E \end{smallmatrix}\right), \left(\begin{smallmatrix} -E & 0 \\ 0 & -E \end{smallmatrix}\right)\}$. Clearly, the second element belongs to the group $Sl(p+q, \mathbb{R})$ if and only if p and q are both even (or odd, but this case is not interesting). In this case, there are two different elements giving the same symmetry and these two elements differ by the multiplication by $-E$. We can take $G = PSl(p+q, \mathbb{R})$ instead of $Sl(p+q, \mathbb{R})$ to get effective geometry. With this choice, each symmetry is given by exactly one class represented by some of the above elements.

2.2. *Almost quaternionic structures*

Now we consider almost quaternionic structures, we have $\mathfrak{g} = \mathfrak{sl}(m+1, \mathbb{H})$. There are again two interesting choices of the groups. We can choose $G = Sl(m+1, \mathbb{H})$ with the canonical action on $\mathbb{H}^{m+1}$. The parabolic subgroup P is the stabilizer of the quaternionic line spanned by the first basis vector in $\mathbb{H}^{m+1}$. Then $G_0 = \{\left(\begin{smallmatrix} a & 0 \\ 0 & A \end{smallmatrix}\right) \mid |a|^4 \det_{\mathbb{R}} A = 1\}$.

Next, we can take $G = PGl(m+1, \mathbb{H})$, the quotient of all invertible quaternionic linear endomorphisms by the subgroup of real multiples of identity. Let P be the (factor of the) stabilizer of the quaternionic line spanned by the first basis vector. The subgroup G_0 consists of classes in P of block diagonal matrices which are represented by matrices of the form $\left(\begin{smallmatrix} a & 0 \\ 0 & A \end{smallmatrix}\right)$ such that $0 \neq a \in \mathbb{H}$ and $A \in Gl(m, \mathbb{H})$.

We have $\mathfrak{g}_{-1} = \{\left(\begin{smallmatrix} 0 & 0 \\ X & 0 \end{smallmatrix}\right) \mid X \in \mathbb{H}^m\}$ and we look for elements $\left(\begin{smallmatrix} q & 0 \\ 0 & B \end{smallmatrix}\right)$ such that $BX = -Xq$ for each X. Again, such an element must be diagonal and the elements on the diagonal of B are equal to $-q$. Suppose that $q = a+bi+cj+dk$. If we choose $X = \left(\begin{smallmatrix} i \\ 0 \end{smallmatrix}\right)$ we get $(-a-bi-cj-dk)i = -i(a+bi+cj+dk)$, thus $-ai+b+ck-dj = -ai+b-ck+dj$ and so $c = d = 0$. Then the choice $X = \left(\begin{smallmatrix} j \\ 0 \end{smallmatrix}\right)$ gives that q has to be real. We again get the element $\left(\begin{smallmatrix} 1 & 0 \\ 0 & -E \end{smallmatrix}\right)$.

In the case of $PGl(m+1, \mathbb{H})$, this element clearly represents the class giving a symmetry. In the case of $Sl(m+1, \mathbb{H})$ it should again depend on the dimension of the manifold. But the real dimension equals to $4m$ and also in this case, the symmetry is well defined. All elements giving symmetries look like $\left(\begin{smallmatrix} 1 & W \\ 0 & -E \end{smallmatrix}\right)$ for all $W \in \mathbb{H}^{m*}$.

Thus we see that homogeneous models are mostly symmetric. If not, we can change the group G to get symmetric geometry.

3. Torsion restrictions

We are mainly interested in almost Grassmannian and almost quaternionic structures and we will formulate most facts only for them. We shall follow the general well known results on all $|1|$–graded geometries.[15,17]

We know that curved geometries could be symmetric because the homogeneous models are. We look for restrictions on the curvature of these geometries. The following theorem plays a crucial role for us.

Theorem 3.1. *Symmetric $|1|$–graded parabolic geometries are torsion free.*

The Theorem 1.1 and the whole theory on harmonic curvature of parabolic geometries give as a simple consequence the following corollary.

Theorem 3.2. *The following symmetric normal geometries have to be locally flat:*

- *almost Grassmannian geometries such that $p > 2$ and $q > 2$*
- *geometries modeled on quaternionic Grassmannians (but not the almost quaternionic ones).*

The crucial point is that the components in harmonic curvatures are only of homogeneity one, see 1.4. Similar argument applies for geometries where the only available homogeneity is three and we get the following Theorem.

Theorem 3.3. *Symmetric normal projective geometries of dimension 2 are locally flat.*

There are also some more interesting cases, which can carry some symmetry in a point with nonzero curvature. We are interested in curved versions of:

- projective geometries of dim > 2
- almost quaternionic geometries
- almost Grassmannian structures such that $p = 2$ or $q = 2$.

These geometries have homogeneous component of curvature of degree 2 and we concentrate on it.

4. Further curvature restrictions

We use the theory of Weyl structures to study more interesting cases of almost Grassmannian and almost quaternionic geometries. We introduce Weyl structures[6] only in the $|1|$–graded case.

4.1. *Weyl structures*

Remind that there is the *underlying bundle* $\mathcal{G}_0 := \mathcal{G}/\exp \mathfrak{g}_1$ for each parabolic geometry, which is the principal bundle $p_0 : \mathcal{G}_0 \to M$ with structure group G_0. At the same time we get the principal bundle $\pi : \mathcal{G} \to \mathcal{G}_0$ with structure group $P_+ = \exp \mathfrak{g}_1$. The *Weyl structure* σ for parabolic geometry is a global smooth G_0–equivariant section of the projection $\pi : \mathcal{G} \to \mathcal{G}_0$. There exists some Weyl structure $\sigma : \mathcal{G}_0 \to \mathcal{G}$ on an arbitrary parabolic geometry, and for arbitrary two Weyl structures σ and $\hat{\sigma}$, there is exactly one G_0–equivariant mapping $\Upsilon : \mathcal{G}_0 \to \mathfrak{g}_1$ such that

$$\hat{\sigma}(u) = \sigma(u) \cdot \exp \Upsilon(u)$$

for all $u \in \mathcal{G}_0$. The equivariancy allows to extend Υ to P–equivariant mapping $\mathcal{G} \to \mathfrak{g}_1$ and in fact, it is a 1–form on M. Weyl structures form an affine space modeled over the vector space of all 1–forms and in this sense we can write $\hat{\sigma} = \sigma + \Upsilon$.

The choice of the Weyl structure σ defines the decomposition of G_0–equivariant 1–form $\sigma^*\omega \in \Omega^1(\mathcal{G}_0, \mathfrak{g})$ such that $\sigma^*\omega = \sigma^*\omega_{-1} + \sigma^*\omega_0 + \sigma^*\omega_1$. The part $\sigma^*\omega_0 \in \Omega^1(\mathcal{G}_0, \mathfrak{g}_0)$ defines the principal connection on $p_0 : \mathcal{G}_0 \to M$, the *Weyl connection.* For each representation $\lambda : G_0 \to Gl(V)$ we get the induced Weyl connection ∇^σ on $\mathcal{G}_0 \times_{G_0} V$ and for arbitrary two Weyl structures σ and $\hat{\sigma} = \sigma \cdot \exp \Upsilon$ we get explicit formula for the change of corresponding connection ∇^σ and $\nabla^{\hat{\sigma}}$. It can be nicely written in the language of tractor calculi in the following way.[4,6] We have

$$\nabla^{\hat{\sigma}}_\xi(s) = \nabla^{\sigma}_\xi(s) + \{\xi, \Upsilon\} \bullet s$$

for $\xi \in \mathfrak{X}(M)$ and $s \in \Gamma(\mathcal{G}_0 \times_{G_0} V)$. Here $\{\ ,\ \}$ is the algebraic bracket of a vector field with a 1–form, which becomes an endomorphism on TM and $\bullet$ is the algebraic action derived from λ.

For each automorphism φ of the geometry, there is the pullback $\varphi^*\sigma$ of the Weyl structure σ. It is again Weyl structure and there is exactly one Υ such that $\varphi^*\sigma = \sigma + \Upsilon$. In addition, it respects the affine structure, i.e. $\varphi^*(\sigma + \Upsilon) = \varphi^*\sigma + \varphi^*\Upsilon$.

Finally we remind the existence of the *normal Weyl structure* at u. It is the only G_0–equivariant section $\sigma_u : \mathcal{G}_0 \to \mathcal{G}$ satisfying

$$\sigma_u \circ \pi \circ \mathrm{Fl}_1^{\omega^{-1}(X)}(u) = \mathrm{Fl}_1^{\omega^{-1}(X)}(u).$$

We remark that the pullback of normal Weyl structure is again normal Weyl structure.

4.2. *The Weyl curvature*

The curvature of each symmetric geometry is of the form $\kappa = \kappa^0 : \mathcal{G} \to \wedge^2 \mathfrak{g}_{-1}^* \otimes \mathfrak{g}^0$, because these geometries are torsion free. If we choose some Weyl structure σ, we can take the decomposition $\sigma^*\kappa^0 = \sigma^*\kappa_0 + \sigma^*\kappa_1$. The part

$$\sigma^*\kappa_0 : \mathcal{G}_0 \to \wedge^2 \mathfrak{g}_{-1}^* \otimes \mathfrak{g}_0$$

does not change, if we change the Weyl structure, because it is the lowest part of decomposition. This part is called *Weyl curvature* and is usually denoted by W. It is the most interesting part of the curvature an we will focus on it.

Next, we summarize here some facts on Weyl structures and Weyl curvature of symmetric $|1|$–graded geometries.[16,17]

Proposition 4.1. *Let $(\mathcal{G} \to M, \omega)$ be a $|1|$–graded geometry. Suppose, that there is a symmetry s_x in $x \in M$ and let φ be some covering of s_x. Remark, the following properties do not depend on the choice of covering φ of s_x.*

(1) There is a Weyl structure σ such that $\varphi^*\sigma = \sigma$ in the point x
(2) There is exactly one normal Weyl structure σ such that $\varphi^*\sigma = \sigma$ over some neighborhood of x.
(3) There exists a Weyl connection ∇^σ such that $\nabla^\sigma W = 0$ in x. The connection corresponds to the fixed Weyl structure in x.
(4) Assume there are two different symmetries in x on a $|1|$–graded geometry. Then

$$\{\xi, \Upsilon\} \bullet W = 0$$

holds in x for any $\xi \in \mathfrak{X}(M)$ and one fixed 1–form Υ. Here Υ is a suitable multiple of the difference between 'fixed' Weyl structures of the symmetries.

4.3. *The main results*

We use the fourth property of Proposition 4.1 to prove the following restriction on the Weyl curvature and thus the whole curvature of geometries.

Theorem 4.1. *Suppose that there exist two different symmetries in x on an almost Grassmannian structure of the type $(2, q)$ and suppose that the latter Υ has maximal rank. Then W vanishes in x and thus the whole curvature vanishes in x.*

If there are two such different symmetries in each point, then the geometry is locally flat.

Proof. We write up to a choice of the frame the expression $\{\xi, \Upsilon\} \bullet W = 0$ in our concrete geometry. The idea is to choose ξ such that $\{\xi, \Upsilon\}$ corresponds to a reasonable element of $\mathfrak{g}$.

If we write $\Upsilon = \left(\begin{smallmatrix}1&0&\cdots&0\\0&1&\cdots&0\end{smallmatrix}\right)$ and take $\xi = \left(\begin{smallmatrix}1&0&\cdots&0\\0&1&\cdots&0\end{smallmatrix}\right)^T$, we get the decomposition of $TM = U \oplus V$ such that U and V are eigenspaces with eigenvalues 1 and 2. Clearly, the bracket corresponds to the element

$$\left[\left(\begin{array}{cc|cccc} 0&0&0&0&\dots&0\\ 0&0&0&0&\dots&0\\ \hline 1&0&0&0&\dots&0\\ 0&1&0&0&\dots&0\\ \vdots&\vdots&\vdots&\vdots&\ddots&\vdots\\ 0&0&0&0&\dots&0 \end{array}\right), \left(\begin{array}{cc|cccc} 0&0&1&0&\dots&0\\ 0&0&0&1&\dots&0\\ \hline 0&0&0&0&\dots&0\\ 0&0&0&0&\dots&0\\ \vdots&\vdots&\vdots&\vdots&\ddots&\vdots\\ 0&0&0&0&\dots&0 \end{array}\right)\right] = \left(\begin{array}{cc|cccc} -1&0&0&0&\dots&0\\ 0&-1&0&0&\dots&0\\ \hline 0&0&1&0&\dots&0\\ 0&0&0&1&\dots&0\\ \vdots&\vdots&\vdots&\vdots&\ddots&\vdots\\ 0&0&0&0&\dots&0 \end{array}\right)$$

and the action of the bracket on the vector is

$$\left[\left(\begin{array}{cc|cccc} -1&0&0&0&\dots&0\\ 0&-1&0&0&\dots&0\\ \hline 0&0&1&0&\dots&0\\ 0&0&0&1&\dots&0\\ \vdots&\vdots&\vdots&\vdots&\ddots&\vdots\\ 0&0&0&0&\dots&0 \end{array}\right), \left(\begin{array}{cc|cccc} 0&0&0&0&\dots&0\\ 0&0&0&0&\dots&0\\ \hline a&b&0&0&\dots&0\\ c&d&0&0&\dots&0\\ \vdots&\vdots&\vdots&\vdots&\ddots&\vdots\\ e&f&0&0&\dots&0 \end{array}\right)\right] = \left(\begin{array}{cc|cccc} 0&0&0&0&\dots&0\\ 0&0&0&0&\dots&0\\ \hline 2a&2b&0&0&\dots&0\\ 2c&2d&0&0&\dots&0\\ \vdots&\vdots&\vdots&\vdots&\ddots&\vdots\\ e&f&0&0&\dots&0 \end{array}\right)$$

Similarly, T^*M decomposes into eigenspaces with eigenvalues -2 and -1.

Weyl curvature W lives in a component of $T^*M \wedge T^*M \otimes TM \otimes T^*M$, which splits according to the decomposition of TM. The action of the bracket on each component is an integer from $\{-1, -2, -3, -4, -5\}$. Then W has to vanish. In this case, the whole curvature vanishes. □

Remark, that this argument does not work in the case of singular Υ. There is no ξ such that all the eigenvalues of the bracket are non-zero.

Similar arguments and computations work in the other cases.

Theorem 4.2.

(1) Suppose that there exist two different symmetries in x on an almost Grassmannian structure of the type $(p, 2)$ and suppose that the latter Υ has maximal rank. Then W vanishes in x.

(2) Suppose that there exist two different symmetries in x on an almost quaternionic structure of the type $(2, q)$. Then W vanishes in x.

Acknowledgments

Discussions with Jan Slovák and Boris Doubrov were very useful during the work on this paper. This research has been supported by the grant GACR 201/05/H005.

References

1. T.N. Bailey and M.G. Eastwood, Complex Paraconformal Manifolds - their Differential Geometry and Twistor Theory, *Forum Mathematikum* **3** (1991) 61–103.
2. A. Čap, Two constructions with parabolic geometries, *In: Proceedings of the 25th Winter School on Geometry and Physics* (Srni, 2005, Rend. Circ. Mat. Palermo Suppl. ser. II.).
3. A. Čap and R. Gover, Tractor Bundles for Irreducible Parabolic Geometries, *In: Global analysis and harmonic analysis* (SMF, Séminaires et Congrès, n.4, 2000) 129–154.
4. A. Čap and R. Gover, Tractor Calculi for Parabolic Geometries, *Trans. Amer. Math. Soc.* **354** (2002) 1511–1548.
5. A. Čap and H. Schichl Parabolic geometries and canonical Cartan connection, *Hokkaido Math. J.* **29** (2000) 453–505.
6. A. Čap and J. Slovák, Weyl Structures for Parabolic Geometries, *Math. Scand.* **93** (2003) 53–90.
7. A. Čap, J. Slovák, Parabolic Geometries, Mathematical Surveys and Monographs, AMS Publishing House, to appear.
8. S. Kobayashi and K. Nomizu, *Foundations of Differential Geometry. Vol II* (John Wiley & Sons, New York-London-Sydney, 1969, pp. 470).
9. I. Kolář, P.W. Michor and J. Slovák, *Natural Operations in Differential Geometry* (Springer-Verlag, 1993, pp. 434).
10. F. Podesta, A Class of Symmetric Spaces, *Bulletin de la S.M.F.* **117** (3) (1989) 343–360.
11. R.W. Sharpe, *Differential geometry: Cartan's generalization of Klein's Erlangen program* (Graduate Texts in Mathematics 166, Springer-Verlag 1997).
12. J. Slovák, Parabolic geometries, Research Lecture Notes, Part of DrSc-dissertation, Masaryk University, 1997, pp. 70, IGA Preprint 97/11, University of Adelaide.
13. J. Šilhan, Cohomology of Lie algebras, algorithm for computation available at http://bart.math.muni.cz/ silhan/lie/.
14. K. Yamaguchi, Differential systems associated with simple graded Lie algebras, *Advanced Studies in Pure Mathematics* **22** (1993) 413–494.
15. L. Zalabová, Remarks on Symmetries of Parabolic Geomeries, *Arch. Math.* **42** (Supplement) 357–368.
16. L. Zalabová, Symmetries of Parabolic Geometries, submitted to Differential Geometry and Its Aplications.
17. L. Zalabová, Symmetries of Parabolic Geometries, Ph.D. thesis, 2007
18. V. Žádník, Generalised Geodesics, Ph.D. thesis, 2003, pp. 65.

PART 3

Global analysis

Differential Geometry and its Applications
Proc. Conf., in Honour of Leonhard Euler, Olomouc, August 2007

Zeta regularized integral and Fourier expansion of functions on an infinite dimensional torus

Akira Asada

Sinsyu University, 3-6-21 Nogami, Takarazuka, 665-0022 Japan
E-mail: asada-a@poporo.ne.jp

Let $\{H, G\}$ be a pair of a Hilbert space and a Schatten class operator G such that $\zeta(G, s) = \mathrm{tr}G^s$ is holomorphic at $s = 0$. Then regularization of integrals on suitable subsets of $H^\sharp$, an extension of H obtained to add a 1-dimensional space, is defined by using $\zeta(G, s)$. In this paper, we apply this regularized integral to periodic functions of $H^\sharp$, and show practical computations of Fourier expansions of periodic functions. Results also show there exists de Rham type cohomology having the Poincaré duality exists on suitable infinite dimensional torus.

Keywords: Zeta regularization, regularized infinite dimensional integral, infinite dimensional torus.

MS classification: 58B25, 42B99, 58J52.

1. Introduction

To overcome the difficulty of divergence in the calculus on a Hilbert space H, we have proposed to consider the pair $\{H, G\}$, where G is a (positive) Schatten class operator on H, such that its ζ-function $\zeta(G, s) = \mathrm{tr}G^s$ is holomorphic at $s = 0$.[1,2] By using G, we introduce the Sobolev k-norm $\|x\|_k$, $x \in \mathcal{D}(G^{-k})$, by

$$\|x\|_k = \|G^{-k/2}x\| = (G^{-k}x, x).$$

Sobolev space constructed by this norm and $\mathcal{D}(G^{-k})$ is denoted by W^k. The complete ortho-normal basis of H is fixed to be $\{e_1, e_2, \ldots\}$; $Ge_n = \mu_n e_n$, $\mu_1 \geq \mu_2 \geq \cdots > 0$. Selection of $e_1, e_2, \ldots$ gives an additional structure to $\{H, G\}$. But we do not discuss on this point. We set $\nu = \zeta(G, 0)$ and d the location of the first pole of $\zeta(G, s)$. Then $e_\infty = \sum_{n=1}^{\infty} \mu_n^{d/2} e_n \in W^l$, $l < 0$ but not belongs to H. We set

$$H^\sharp = H \oplus \mathbb{K}e_\infty \subset \bigcap_{l<0} W^l.$$

Here $\mathbb{K}$ is either of $\mathbb{R}$ or $\mathbb{C}$, according to H is a real Hilbert space or complex Hilbert space. Then we can define regularized infinite dimensional integral $\int_{\mathcal{D}_{\mathbf{a},\mathbf{b}}} f(x) : d^\infty x :$, if f is Frechét differentiable on

$$\mathcal{D}_{\mathbf{a},\mathbf{b}} = \{\sum_{n=1}^{\infty} x_n e_n \in H^\sharp | a_n \leq x_n \leq b\}, \quad \mathbf{a} = \sum_{n=1}^{\infty} x_n e_n, \ \mathbf{b} = \sum_{n=1}^{\infty} b_n e_n \in H^\sharp.$$

We also define regularized infinite product $: \prod_{n=1}^{i} nfty x_n :$ by

$$: \prod_{n=1}^{\infty} x_n := \prod_{n=1}^{\infty} x_n^{\mu_n^s}|_{s=0}.$$

Then we have

$$\int_{\mathcal{D}_{\mathbf{a},\mathbf{b}}} \prod_{n=1}^{\infty} f_n(x_n) : d^\infty x :=: \prod_{n=1}^{\infty} \int_{a_n}^{b_n} f_n(x_n) dx_n : .$$

By this formula, we can compute $\int_{\mathcal{D}_{0,\mu^{d/2}}} (\prod_n f_n(x_n))(\prod_n g_n(x_n)) : d^\infty x :$, where $0 = (0, 0, \ldots)$, $\mu^{d/2} = (\mu_1^{d/2}.\mu_n^{d/2}, \ldots)$ and $f_n(x_n)$, $g_n(x_n)$, $n = 1, 2, \ldots$, are either of

$$\sin(2m_n \pi \mu_n^{-d/2} x_n), \quad \cos(2m_n \pi \mu_n^{-d/2} x_n).$$

$f(x)$, $g(x)$ are periodic functions with the period generated by $\mu_n^{d/2} e_n$, $n = 1, 2, \ldots$. The abelian group generated by these periods is denoted by $\mathbb{Z}^\infty$. It is a subgroup of H. While in $H^\sharp$, the group of periods should be $\hat{\mathbb{Z}}^\infty = \mathbb{Z}^\infty \oplus \mathbb{Z} e_\infty$. We set

$$\mathbb{T}^\infty = H/\mathbb{Z}^\infty, \quad \hat{\mathbb{T}}^\infty = H^\sharp/\hat{\mathbb{T}}^\infty.$$

By definitions, we have $\hat{\mathbb{T}}^\infty = \mathbb{T}^\infty \times S^1$.

On H, $f = 0$ unless $f_n = 1$ except finite factors $f_{n_1}, \ldots, f_{n_m}$ of f. While on $H^\sharp$, $f \neq 0$, if $\lim_{n\to\infty} m_n = m_\infty$ exists, and except finite factors of f, $f_n = \sin(2m_n \pi \mu_n^{-d/2} x_n)$, or $f_n = \cos(2m_n \pi \mu_n^{-d/2} x_n)$. Fourier expansion of Frechét differentiable functions on $\mathbb{T}^\infty$ is possible by using finite products of trigonometric functions. On th other hand functions on $\hat{\mathbb{T}}^\infty$, comes from $S^1 = \mathbb{R} e_\infty / \mathbb{Z} e_\infty$ are expansed by using infinite products of trigonometric functions. So Fourier expansion of Frechét differentiable functions on $\hat{\mathbb{T}}^\infty$ are possible and computed by regularized infinite dimensional integral (cf. Ref. 4).

As an application of Fourier expansion of functions on $\hat{\mathbb{T}}^\infty$, we can determine eigenvalues and eigenfunctions of the periodic boundary value problem of regularized Laplacian.[3,5] These results also suggest existence of de Rham type cohomology of $\hat{\mathbb{T}}^\infty$ having Poincaré duality. This is possible, but since it needs rigorous definitions of $(\infty - p)$-forms and study of their algebraic and analytic properties. So we omit its precise description in this paper.

2. Hilbert space equipped with a Schatten class operator

Let H be a Hilbert space over $\mathbb{K}$, $\mathbb{K}$ is either of $\mathbb{R}$ or $\mathbb{C}$. G a (positive) Schatten class operator (cf. Ref. 11), such that its ζ-function $\zeta(G,s) = \mathrm{tr}G^s$ is holomorphic at $s = 0$. Typical example of such pair is $H = L^2(X,E)$, X a compact Riemannian manifold, E a vector bundle over X, and G is the Green operator of D, a (positive) non degenerate selfadjoint elliptic (pseudo) differential operator acting on the sections of E.[9] In the rest, we set

(a) $\nu = \zeta(G,0)$: the regularized dimension of H.
(b) d: The location of the first pole of $\zeta(G,s)$.

Let $\mu_1 \geq \mu_2 \geq \cdots > 0$ be eigenvalues of G with normalized eigenfunctions $e_1, e_2, \ldots$; $Ge_n = \mu_n e_n$. We fix the complete orthonormal basis of H to be $e_1, e_2, \ldots$. Note that selecting $e_1, e_2, \ldots$ gives an additional structure to $\{H, G\}$. But at this stage, we can not obtain any additional informations from this selection.

The k-th power G^k of G is defined by $G^k e_n = \mu_n^k e_n$. It does not depend on the choice of eigenvectors, but if $k < 0$, it does not defined on H. The domain of G^k is denoted by $\mathcal{D}(G^k)$.

Let (x,y) and $\|x\|$ be inner product and norm of H. Then we define the Sobolev k-inner product and norm by

$$(x,y)_k = (G^{-k/2}x, G^{-k/2}y), \quad \|x\|_k = \sqrt{(x,x)_k} = \|G^{-k/2}x\|. \tag{1}$$

The Sobolev space constructed by this Sobolev norm and elements of $\mathcal{D}(G^{k/2})$ is denoted by W^k ($W^0 = H$). The complete orthonormal basis of W^k is given by $e_{1,k}, e_{2,k}, \ldots$; $e_{n,k} = \mu_n^{k/2} e_n = G^{k/2} e_n$.

As sets, $W^k \subset W^l$, $l < k$. We set $\cap_{l<k} W^l = W^{k-0}$, $\cap_{l<0} W^l = H^-$. We say a series $x_1, x_2, \ldots$ of elements of W^{k-0} tends to $x \in W^{k-0}$, if $\lim_{n\to\infty} \|x - x_n\|_l = 0$, for some $l < k$.

By definition, $e_{\infty,k} = \sum_{n=1}^{\infty} \mu_n^{d/2} e_{n,k}$ belongs to W^{k-0} but not belongs to W^k. We define

$$W^{k,\sharp} = W^k \oplus \mathbb{K}e_{\infty,k} \subset W^{k-0}, \quad H^\sharp = H \oplus \mathbb{K}e_\infty \subset H^-. \tag{2}$$

$W^{k,\sharp}$ is not a Hilbert space. But to define an inner product $\langle x, y\rangle_K$, $x = x_f + te_{\infty,k}$, $x_f \in W^k$, by

$$\langle x, y\rangle_k = \lim_{s\downarrow 0}(x_f + t\sqrt{s}G^{s/2}e_{\infty,k}, y_f + u\sqrt{s}G^{s/2}e_{\infty,k})_k,$$

$W^{k,sharp}$ has a Hilbert space structure such that $W^k \perp \mathbb{K}e_{\infty,k}$ and $\|e_{\infty,k}\| = \sqrt{c}$, $c = \mathrm{Res}_{s=d}\zeta(G,s)$. This space is denoted by $H^{\natural}$.

The polar coordinate of H is given by

$$x_1 = r\cos\theta_1,\ x_2 = r\sin\theta_1\cos\theta_2, \dots, x_n = r\sin\theta_1\cdots\sin\theta_{n-1}\cos\theta_n, \dots,$$

$r = \|x\|$.[2] This polar coordinate has no longitude. Latitude variables $\theta_1, \theta_2, \dots$ need to satisfy

$$\lim_{n\to\infty}\sin\theta_1\cdots\sin\theta_n = 0.$$

Removing this constraint, we obtain a new variable $t_\infty = \pm r\prod_{n=1}^{\infty}\sin\theta_n$. Denoting this extended space of H by $\hat{H}$, there is a map $\rho : H^{\natural} \cong \hat{H}$.

3. Regularized infinite product and regularized determinant

Let $x_1, x_2, \dots$ be a series having an Agmon angle θ. Then we define the regularized infinite product $:\prod_{n=1}^{\infty} x_n :=: \prod_{n=1}^{\infty} x_n :_{G,\theta}$ by

$$:\prod_{n=1}^{\infty} x_n :_{G,\theta} = \prod_{n=1}^{\infty} x_n^{\mu_n^s}|_{s=0}, \quad x_n^{\mu_n^s} = |x_n|^{\mu_n^s}\mathrm{e}^{\mu_n^s \mathrm{Arg}x_n}, \qquad (3)$$

$$\theta < \mathrm{Arg}x_n < \theta + 2\pi.$$

By definition, we have

$$:\prod_{n=1}^{\infty}\mu_n := \prod_{n=1}^{\infty}\mu_n^{\mu_n^s}|_{s=0} = \mathrm{e}^{\sum_{n=1}^{\infty}\log\mu_n\mu_n^s}|_{s=0} = \mathrm{e}^{\zeta'(G,0)}.$$

Hence $:\prod_{n-1}^{\infty}\mu_n :$ is the Ray-Singer determinant of G.

If $x = \sum_{n=1}^{\infty} x_n e_n \in H^{\sharp}$, then we have $x_n = x_{n,f} + t\mu_n^{d/2}$. If $t \neq 0$, fixing $\mathrm{Arg}t = \phi$, we can choose $\mathrm{Arg}x_n$ to be $\lim_{n\to\infty}\mathrm{Arg}x_n = \phi$. Hence $x_1, x_2, \dots$ has an Agmon angle, and we have

$$:\prod_{n=1}^{\infty} x_n := t^{\nu}(\mathrm{det}G)^{d/2}\prod_{n=1}^{\infty}(1 + \frac{x_n}{t\mu_n^{d/2}})^{\mu_n^s}|_{s=0}.$$

Hence $:\prod_{n=1}^{\infty} x_n :$ is defined on dense subset of $H^{\sharp}$. It is a single valued function if and only if ν is an integer.

Let I_x; $x = (x_1, x_2, \ldots)$, be a scaling operator defined by $I_x e_n = x_n e_n$. If none of $x_1, x_2, \ldots$ is equal to 0 and $x_1, x_2, \ldots$ has an Agmon angle, we can determine $\log x_n$, simultaneously. Hence $\log I_x = I_{\log x}$; $\log x = (\log x_1, \log x_2, \ldots)$ is well defined. Then we have

$$:\prod_{n=1}^{\infty} x_n := \mathrm{e}^{\mathrm{tr}(G^s \log I_x)}|_{s=0}.$$

In general, if a densely defined linear operator T has a logarithm $\log T = s$; $\mathrm{e}^S = T$, then we define regularized determinant $\det_G T$ of T with respect to G by

$$\det_G T = \mathrm{e}^{\mathrm{tr}(G^s S)}|_{s=0}. \tag{4}$$

Note that $\mathrm{tr}(G^s S)|_{s=0}$ is the renormalized trace of S in the sense of Paycha.[6,10]

If G is the Green operator of an elliptic operator D, then $\det_G D$ is the Ray-Singer determinant of D. By definitions, we have

$$\det_G(T^m) = (\det_G T)^m, \quad \det_G(tT) = t^{\nu}\det_G T. \tag{5}$$

Since $\log T$ is not unique, $\det_G T$ is not unique in general. For example, if G is the Green operator of $D = \not{D}^2$, $\not{D}$ is the Dirac operator, then

$$\det_G \not{D} = \det \not{D} = (-1)^{\nu_-}\sqrt{\det D},$$

where $\nu_- = \dfrac{\zeta(D,0) - \eta(\not{D})}{2}$. Since there are two different expressions of -1; $\mathrm{e}^{\pi i}$ and $\mathrm{e}^{-\pi i}$, $(-1)^{\nu_-}$ is not unique unless ν_- is not an integer. Therefore $\det \not{D}$ is unique if and only if ν_- is an integer.

In general, we do not have $\det_G(T_1 T_2) = \det_G T_1 \det_G T_2$. But if Campbell-Haussdorff formula is hold for T_1^t, T_2^t and $G^s S_2 = S_2 G^s$, $S_2 = \log T_2$, $s, t \in \mathbb{R}$, then we have $\det_G(T_1 T_2) = \det_G T_1 \det_G T_2$.

We have $\det_G(PTP^{-1}) = \det_{P^{-1}GP} T$. But do not have $\det_G(PTP^{-1}) = \det G T$, in general. For example, if

$$\begin{cases} Ge_{2n-1} = \frac{1}{n} e_{2n-1}, \\ Ge_{2n} = \frac{1}{n+1} e_{2n}, \end{cases} \quad \begin{cases} Te_{2n-1} = 3e_{2n-1}, \\ Te_{2n} = 2e_{2n}, \end{cases} \quad \begin{cases} Pe_{2n-1} = e_{2n}, \\ Pe_{2n} = e_{2n-1}, \end{cases}$$

then we have

$$\det_G T = 3^{\zeta(s)} 2^{\zeta(s)-1}|_{s=0} = \frac{1}{2\sqrt{6}}, \quad \det_G(PTP^{-1}) = 2^{\zeta(s)} 3^{\zeta(s)-1} = \frac{1}{3\sqrt{6}}.$$

Hence they are different.

4. Special values of higher derivatives of $\zeta(G, s)$ and regularized determinant

$\{G^t | t \geq 0\}$ is a 1-parameter semigroup of bounded operators on H. But since $\lim_{n\to\infty} \mu_n = 0$, we have $\lim_{t\to 0} \|G^t - I\| = 1$, so $\lim_{t\to 0} G^t$ converges to I only by the strong topology of operators. The generating operator of this semigroup is $\log G$; $\log G e_n = \log \mu_n e_n$:

$$\log G = \lim_{t\to 0} \frac{G^t - I}{t},$$

where the limit is taken as the inductive topology of the strong topology of operators on W^l, $l > 0$. $\log G$ is defined on $H^+ = \cup_{l>0} W^l$. We can regard $\{G^t | t \in \mathbb{R}\}$ as the 1-parameter group of operators on $W^\infty = \bigcap_l W^l$.

We define the operator $\mathrm{e}^{t(\log G)^k G^m}$ by

$$\mathrm{e}^{t(\log G)^k G^m} e_n = \mathrm{e}^{t(\log \mu_n)^k \mu_n^m} e_n.$$

If G is the Green operator of D, then $\mathrm{e}^{tG^{-m}}$ is the heat kernel of D^m.

By definition, if $k \geq 1$, we have

$$\mathrm{e}^{t(\log G)^k G^m} = G^{t(\log G)^{k-1} G^m}.$$

Theorem 4.1. *If $\zeta(G, s)$ is holomorphic at $s = m$, then we have*

$$\det{}_G \mathrm{e}^{t(\log G)^k G^m} = \mathrm{e}^{t\zeta^{(k)}(G,m)}. \tag{6}$$

Proof. Since $\dfrac{d}{ds}\mathrm{tr} G^{m+s} = \mathrm{tr}\left(\dfrac{d}{ds} G^{m+s}\right)$, we have

$$\begin{aligned}
\mathrm{e}^{\zeta^{(k)}(G,m+s)}|_{s=0} &= \mathrm{e}^{t\frac{d^k}{ds^k}\mathrm{tr} G^{m+s}}|_{s=0} = \mathrm{e}^{t\mathrm{tr}(\frac{d^k}{ds^k} G^{m+s})}|_{s=0} \\
&= \mathrm{e}^{t\mathrm{tr}((\log G)^k G^{m+s}}|_{s=0} = \det{}_G \mathrm{e}^{t(\log G)^k G^m}.
\end{aligned}$$

Hence we have Theorem. □

For example, we have

$$\det{}_G \mathrm{e}^{tG^m} = \mathrm{e}^{t\zeta(G,m)}.$$

Hence if G is the Green operator of D and $\zeta(G, -m) = 0$, regularized determinant of the heat kernel of D^m does not depend on t.

If $k \geq 1$, we can rewrite (6) as follows;

$$\det{}_G G^{t(\log G)^{k-1} G^m} = \mathrm{e}^{t\zeta^{(k)}(G,m)}. \tag{7}$$

By (6), we have

$$:\prod_{n=1}^{\infty}\mu_n^{t(\log\mu_n)^k\mu_n^m}(1+x_n):=\mathrm{e}^{t\zeta^{(k)}(G,m)}\prod_{n=1}^{\infty}(1+x_n),$$

if $\zeta(G,s)$ is holomorphic at $s=m$ and $\sum_{n=1}^{\infty}|x_n|<\infty$.

5. Regularized infinite dimensional integral

Let $\mathbf{a}=(a_1,a_2,\ldots)$ and $\mathbf{b}=(b_1,b_2,\ldots)$, $a_n<b_n, n=1,2,\ldots$ be $a=\sum_n a_ne_n, b=\sum_n b_ne_n\in H^\sharp$. Then we set

$$\mathcal{D}_{\mathbf{a},\mathbf{b}}=\{\sum_{n=1}^{\infty}x_ne_n\in H^\sharp|a_n\le x_n\le b_n\}.$$

We also set

$$\mathcal{D}_{\mathbf{a},\mathbf{b}}^N=\{\sum_{n=1}^{N}x_ne_n\in\mathbb{R}^N|a_n\le x_n\le b_n\}.$$

Let $*=(*_1,*_2,\ldots)$, $\sum_n *_n\in\mathcal{D}_{\mathbf{a},\mathbf{b}}$ and $*^N=(*_{N+1},*_{N+2},\ldots)$, and let f be a Frechét differentiable function on $\mathcal{D}_{\mathbf{a},\mathbf{b}}$. Then we define regularized integral $\int_{\mathcal{D}_{\mathbf{a},\mathbf{b}}}f(x):d^\infty x:$ of f on $\mathcal{D}_{\mathbf{a},\mathbf{b}}$ by

$$\int_{\mathcal{D}_{\mathbf{a},\mathbf{b}}}f(x):d^\infty x:=\lim_{N\to\infty}\int_{\mathcal{D}_{\mathbf{a},\mathbf{b}}^n}f(x_1,\ldots,x_N,*^N)d(x_1^{\mu_1^s})\cdots d(x_N^{\mu_N^s})|_{s=0}.\quad(8)$$

It is shown this integral does not depend on the choice of $*$, if it exists.

$\mathcal{D}_{\mathbf{a},\mathbf{b}}$ is also defined for $a_n=-\infty$ or $b_n=\infty$. Regularized integrals on such domains are similarly defined.

Regularized integral is also defined if $a,b\in W^{k,\sharp}$. Such integral may be regarded as the integral on $G^{k/2}\mathcal{D}_{\mathbf{a},\mathbf{b}}$. In general, if I_x is a scaling operator, and $I_x^\sharp f(y)=f(I_x^\sharp y)$, where f is a function on $I_x\mathcal{D}$, we have

$$\int_{\mathcal{D}}I_x^\sharp f(y):d^\infty y:=\int_{I_x\mathcal{D}}|\mathrm{det}_G I_x|^{-1}f(\xi):d^\infty\xi:,\quad \xi=I_xy.\qquad(9)$$

If G is the Green operator of D, taking $I_x=G^{1/2}$, we obtain

$$\int_{W^{1/2,\sharp}}\mathrm{e}^{-\pi(x,Dx)}:d^\infty x:=\frac{1}{\sqrt{\mathrm{det}D}}.$$

If $f(x) = \prod_{n=1}^{\infty} f_n(x_n)$ and $\displaystyle\int_{a_n}^{b_n} f_n(x_n)dx_n = c_n$, $c_n \neq 0, \pm\infty$, then by using scaling transformation I_c, $c = (c_1, c_2, \ldots)$, we have

$$\int_{\mathcal{D}_{\mathbf{a},\mathbf{b}}} \prod_{n=1}^{\infty} f_n(x_n) : d^{\infty}x :=: \prod_{n=1}^{\infty} \int_{a_n}^{b_n} f_n(x_n)dx_n : . \tag{10}$$

6. Periodic functions on $H^{\sharp}$

We denote $\mathbb{Z}^{\infty}$ the abelian group generated by $\mu_n^{d/2} e_n$, $n = 1, 2, \ldots$. It is a closed subgroup of H, but not closed as a subgroup of $H^{\sharp}$. The closure $\hat{\mathbb{Z}}^{\infty}$ of $\mathbb{Z}^{\infty}$ in $H^{\sharp}$ is

$$\hat{\mathbb{Z}}^{\infty} = \mathbb{Z}^{\infty} \oplus \mathbb{Z}e_{\infty}. \tag{11}$$

$\mathbb{Z}^{\infty}$ and $\hat{\mathbb{Z}}^{\infty}$ are discrete subgroups of H and $H^{\sharp}$, respectively. We can take $\mathcal{D}_{0,\mu^{d/2}}$, $0 = (0, 0, \ldots)$ and $\mu^{d/2} = (\mu_1^{d/2}, \mu_2^{d/2}, \ldots)$ as a fundamental domain of $\hat{\mathbb{Z}}^{\infty}$. Then we can identify a periodic function f; $f(x) = f(x + \mathbf{n})$, $\mathbf{n} \in \hat{\mathbb{Z}}^{\infty}$ and a function on f such that

$$f(x)|_{x_n=0} = f(x)|_{x_n=\mu_n^{d/2}}, \quad n = 1, 2, \ldots. \tag{12}$$

We also set

$$\mathbb{T}^{\infty} = H/\mathbb{Z}^{\infty}, \quad \hat{\mathbb{T}}^{\infty} = H^{\sharp}/\hat{\mathbb{Z}}^{\infty}. \tag{13}$$

By definitions, we have

$$\hat{\mathbb{T}}^{\infty} = \mathbb{T}^{\infty} \times S^1, \quad S^1 = \mathbb{R}e_{\infty}/\mathbb{Z}e_{\infty}. \tag{14}$$

The projections from $\hat{\mathbb{T}}^{\infty}$ to $\mathbb{T}^{\infty}$ and S^1 are denoted by p_0 and p_{∞}, respectively.

Let $C_b^1(\hat{\mathbb{T}}^{\infty})$ and $C_b^1(\mathbb{T}^{\infty})$ be the space of Frechét differentiable C^1-class functions on $\hat{\mathbb{T}}^{\infty}$) and $\mathbb{T}^{\infty}$). Then $p_0^*(C_b^1(\mathbb{T}^{\infty})) \otimes p_{\infty}^*(C^1(S^1))$ is dense in $C_b^1(\hat{\mathbb{T}}^{\infty})$.

Taking $\mathcal{D}_{0,\mu^{d/2}}$ as the fundamental domain, functions of $p_0^*(C_b^1(\mathbb{T}^{\infty}))$ are expansed by finite products of trigonometric functions

$$s_{m_n} = \sin(2m_n\pi\mu_n^{-d/2}x_n), \quad c_{m_n} = \cos(2m_n\pi\mu_n^{-d/2}x_n). \tag{15}$$

On the other hand, since the variable x_{∞} of $\mathbb{R}e_{\infty}$ is $\sum_{n=1}^{\infty} \mu_n^{-d/2} x_n$, (complex valued) functions of $C^1(S^1)$ is expansed by $\exp(2m\pi \sum_{n=1}^{\infty} \mu_n^{-d/2} x_n)$. Hence functions of $p_{\infty}^*(C^1(S^1))$ are expansed by infinite products

$$f_I(x) = \prod_{n=1}^{\infty} *_{m_n}, \quad I = (m_{1,*}, m_{2,*}, \ldots),$$

where $*$ is either of s or c. On $H^\sharp$, $f_I = 0$ unless $\lim_{n\to\infty} m_n = m_\infty$ exists and except finite factors, $* = s$, or $* = c$. We say such functions to be elementary trigonometric functions on $H^\sharp$.

Theorem 6.1. *Functions in $C_b^1(\hat{\mathbb{T}}^\infty)$ is expansed by elementary trigonometric functions on $H^\sharp$.*

7. Fourier expansions of functions on $\hat{\mathbb{T}}^\infty$

By theorem 6.1, if $f \in C_b^1(\hat{\mathbb{T}} - \infty)$, and the fundamental domain of $\hat{\mathbb{T}}^\infty$ is taken to be $\mathcal{D}_{0,\mu^{d/2}}$, then we have

$$f(x) = \sum_I c_I f_I(x). \tag{16}$$

Regarding f to be the periodic function with the period $\hat{\mathbb{Z}}$, this gives the Fourier expansion of periodic functions on $H^\sharp$.

Regarding f_I to be a function on $\mathcal{D}_{0,\mu^{d/2}}$, we have by (10)

$$\int_{\mathcal{D}_{0,\mu^{d/2}}} f_I f_J : d^\infty x : = 0, \quad I \neq J. \tag{17}$$

$$\int_{\mathcal{D}_{0,\mu^{d/2}}} f_I^2 : d^\infty x : = \epsilon_I, \tag{18}$$

$$\epsilon_I = \begin{cases} \frac{1}{2^k}(\det G)^{d/2}, & m_n = 0 \text{ except } n \in \{n_1, \ldots, n_k\}, \\ \frac{1}{2^{\nu-k}}(\det G)^{d/2}, & m_n \neq 0 \text{ except } n \in \{n_1, \ldots, n_k\}. \end{cases} \tag{19}$$

Hence we have

$$c_I = \frac{1}{\epsilon_I} \int_{\mathcal{D}_{0,\mu^{d/2}}} f(x) f_I(x) : d^\infty x : . \tag{20}$$

Therefore we obtain

Theorem 7.1. *If f is a C^1-class Frechét differentiable function on $H^\sharp$ having the period (12), then we have*

$$f(x) = \sum_I \frac{1}{\epsilon_I} \int_{\mathcal{D}_{0,\mu^{d/2}}} f(x) f_I(x) : d^\infty x : f_I(x). \tag{21}$$

As an application, we have

$$\int_{\mathcal{D}_{0,\mu^{d/2}}} |f(x)|^2 : d^\infty x := \sum_I \epsilon_I |c_I|^2 > 0, \tag{22}$$

if f is a C^1-class Frechét differentiable function and $f \neq 0$. Therefore we can define regularized inner product : (f,g) : of $f, g \in C_b^1(\hat{\mathbb{T}}^\infty)$ by

$$: (f,g) := \int_{\mathcal{D}_{0,\mu^{d/2}}} f\bar{g} : d^\infty x : . \tag{23}$$

The Hilbert space obtained from $C_b^1(\hat{\mathbb{T}}^\infty)$ and this inner product is denoted by $L^2(\hat{\mathbb{T}}^\infty)$. Then we have

$$L^2(\hat{\mathbb{T}}^\infty) \cong \{\sum_I c_I f_I | \sum_I \epsilon_I |c_I|^2 < \infty\}. \tag{24}$$

8. Some applications of Fourier expansions

Let $\Delta = \sum_{n=1}^{\infty} \frac{\partial^2}{\partial x_n^2}$ be the Laplacian of H. Δ can not act on $r(x) = \|x\|$. To avoid this difficulty, we have introduced regularized Laplacian : Δ : by

$$: \Delta : f = \sum_{n=1}^{\infty} \mu_n^{2s} \frac{\partial^2 f}{\partial x_n^2}|_{s=0}. \tag{25}$$

For example, we have

$$: \Delta : r(x) = \frac{\nu - 1}{r(x)^2}.$$

We consider the following boundary value problem of : Δ :.

$$f(x)|_{x_n=0} = f(x)|_{x_n=\mu_n^{d/2}}, \quad \frac{\partial f(x)}{\partial x_n}|_{x_n=0} = \frac{\partial f(x)}{\partial x_n}|_{x_n=\mu_n^{d/2}}. \quad n = 1, 2, \ldots$$

If $f(x) = \prod_{n=1}^{\infty} f_n(x_n)$ is an eigenfunction of this boundary value problem, it must be

$$f_n(x_n) = A \sin(2m_n \pi \mu_n^{-d/2} x_n) + B \cos(2m_n \pi \mu_n^{-d/2} x_n).$$

Hence an elementary trigonometric function f_I may be an eigenfunction of the above boundary value problem of : Δ :.

Let $I = (m_{1,*}, m_{2,*}, \ldots)$. Then f_I is an eigenfunction of : Δ belonging to $-\sum_{k=1}^{N} m_{n_k}^2 \mu_n^{-d}$, if $m_{n,*} = 0_{n,c}$, except $n \in \{n_1, \ldots, n_k\}$. While if $\lim_{n\to\infty} m_n = m_\infty \geq 1$, $f_I(x)$ becomes an eigenfunction of : Δ : provided $\zeta(G, s)$ is holomorphic at $s = -d$. If $\zeta(G, -d)$ exists, we have

$$: \Delta : f_I(x) = -(m_\infty^2 \zeta(G, -d) + \sum_{k=1}^{N} (m_{n_k}^2 - m_\infty^2 \mu_{n_k}^{-d}) f_I(x).$$

Since elementary trigonometric functions spans $L^2(\hat{\mathbb{T}}^\infty)$, these eigenfunctions and eigenvalues exhaust the eigenvalues and eigenfunctions of the above boundary condition of : Δ :.

We may consider

$$\int_{\mathcal{D}_{0,\mu^{d/2}}} 1 : d^\infty x :=: \prod_{n=1}^{\infty} \mu_n^{d/2} := (\det G)^{d/2},$$

to be the regularized volume of $\hat{\mathbb{T}}^\infty$. This shows existence of regularized volume form of $\hat{\mathbb{T}}^\infty$. By using regularized volume form, we can define $(\infty - p)$-forms on $\hat{\mathbb{T}}^\infty$ and de Rham type cohomology of $\hat{\mathbb{T}}^\infty$ having the Poincaré duality. Since the commutation rule of a p-form ϕ^p and an $(\infty - q)$-form $\psi^{\infty-q}$ should be

$$\phi^p \wedge \psi^{\infty-q} = (-1)^{p(\nu-q)} \psi^{\infty-q} \wedge \phi^p,$$

such cohomology can not defined unless ν is not an integer, as a real cohomology group. Note that to define

$$\phi^p \wedge \psi^{\infty-q} = \mathrm{e}^{p(\nu-q)\pi \mathrm{i}} \psi^{\infty-q} \wedge \phi^p, \quad \psi^{\infty-q} \wedge \phi^p = \mathrm{e}^{-p(\nu-q)\pi \mathrm{i}} \phi^p \wedge \psi^{\infty-q},$$

complex coefficients de Rham type cohomology of $\hat{\mathbb{T}}^\infty$ is well defined (cf. Refs. 7,8). But we omit details.

References

1. A. Asada, Regularized calculus: An application of zeta-regularization to infinite dimensional geometry and analysis, *Int. J. Geom. Meth. Mod. Phys.* **1** (2004) 107–157.
2. A. Asada, Regularized volume form of the sphere of a Hilbert space with the determinant bundle, *In: Differential Geometry and Its Applications* (Proc. DGA2004, (J. Bureš, O. Kowalski, D. Krupka and J. Slovák, Eds.) matfyzpress, Prague, 2005) 397–409.
3. A. Asada, Fractional calculus and regularized residue of infinite dimensional space, *In: Mathematical Methods in Engineering* ((K. Tas, J.A. Tenreiro Machado and D. Baleanu, Eds.) Springer, 2007) 3–11.
4. A. Asada, Regularized integral and Fourier expansion of functions on infinite dimensional tori, preprint.
5. A. Asada and N. Tanabe, Regularization of differential operators of a Hilbert space and meanings of zeta-regularization, *Rev. Bull. Cal. Math. Soc.* **11** (2003) 45–52.
6. A. Cardona, C. Ducourtioux and S. Paycha, From tracial anomalies to anomalies in quantum filed theory, *Commun. Math. Phys.* **242** (2002) 31–65.
7. A. Connes, Entire cyclic cohomology of Banach algebra and characters of θ-summable Fredholm module, *K-theory* **1** (1988) 519–548.

8. J. Cuntz, Cyclic Theory, Bivariant K-theory and the Chern-Connes Character, *In: Cyclic Cohomology in Non-Commutative Geometry* (EMS 121, Springer, 2001) 1–71.
9. P. Gilkey, The residue of global η function at the origin, *Adv. Math.* **40** (1981) 290–307.
10. S. Paycha, Renormalized trace as a looking glass into infinite dimensional geometry, *Infin. Dim. Anal. Quantum Prob. Relat. Top.* **4** (2001) 221–266.
11. B. Simon, *Trace Ideals and Their Applications* (Cambridge, 1979).

Differential Geometry and its Applications
Proc. Conf., in Honour of Leonhard Euler, Olomouc, August 2007

On the direction independence of two remarkable Finsler tensors

S. Bácsó and Z. Szilasi

Institute of Informatics, University of Debrecen,
Debrecen, Hungary
E-mail: bacsos@inf.unideb.hu, szzoltan13@t-online.hu
www.inf.unideb.hu

Finsler manifolds some of whose characteristic tensors are direction independent provide stimulation for current research. In this paper we show that the direction independence of the Landsberg and the stretch tensor implies the vanishing of these tensors.

Keywords: Finsler manifolds, Landsberg tensor, stretch tensor, direction independence.

MS classification: 53B40.

1. Basic constructions

Throughout this paper, M will be an n-dimensional smooth manifold. $C^\infty(M)$ denotes the ring of real-valued smooth functions on M. T_pM is the tangent space to M at $p \in M$ $TM := \bigcup_{p\in M} T_pM$ is the tangent bundle of M, $\tau : TM \to M$ is the natural projection. $\overset{\circ}{T}M$ denotes the open subset of the nonzero tangent vectors to M, $\overset{\circ}{\tau} := \tau \restriction \overset{\circ}{T}M$. $\mathfrak{X}(M)$ is the $C^\infty(M)$-module of (smooth) vector fields on M. Capitals $X, Y, \ldots$ will denote vector fields on M, while, usually, Greek letters $\xi, \eta, \zeta, \ldots$ will stand for vector fields on TM. i_ξ is the substitution operator induced by $\xi \in \mathfrak{X}(TM)$, d denotes the operator of the exterior derivative.

All of our considerations will be purely of local character, so we may assume without loss of generality that our base manifold M admits a global coordinate system $\left(u^i\right)_{i=1}^n$; this assumption simplifies a little the notation. Then

$$\left(x^i, y^i\right)_{i=1}^n; \quad x^i := u^i \circ \tau,\ y^i := du^i$$

is a coordinate system for TM. These coordinate systems yield the bases

$$\left(\frac{\partial}{\partial u^i}\right)_{i=1}^n \text{ and } \left(\frac{\partial}{\partial x^i}, \frac{\partial}{\partial y^i}\right)_{i=1}^n =: \left(\partial_i, \dot{\partial}_i\right)_{i=1}^n$$

of $\mathfrak{X}(M)$ and $\mathfrak{X}(TM)$, respectively.

Let $T_s^r M := \bigcup_{p\in M} T_s^r(T_pM)$ be the bundle of type $\binom{r}{s}$ tensors over M, and let $\tau_s^r : T_s^r M \longrightarrow M$ be the natural projection. Following Z. I. Szabó,[9] by a type $\binom{r}{s}$ *Finsler tensor field* over M we mean a smooth map

$$\widetilde{A} : \overset{\circ}{T}M \longrightarrow T_s^r M \text{ such that } \tau_s^r \circ \widetilde{A} = \overset{\circ}{\tau}.$$

These tensor fields form a $C^\infty(\overset{\circ}{T}M)$-module, which will be denoted by $\mathscr{T}_s^r\left(\overset{\circ}{\tau}\right)$. In particular, $\mathfrak{X}\left(\overset{\circ}{\tau}\right) := \mathscr{T}_0^1\left(\overset{\circ}{\tau}\right)$ is the module of *Finsler vector fields*, and $\mathfrak{X}^*\left(\overset{\circ}{\tau}\right)$ is its dual. In what follows, Finsler tensor fields will simply be mentioned as *tensors*, or, for obvious reasons, *tensors along the projection* $\overset{\circ}{\tau}$. Evidently, the construction also works on the whole TM, and leads to the $C^\infty(TM)$-modules $\mathscr{T}_s^r(\tau)$. Via restrictions, $\mathscr{T}_s^r(\tau)$ may be interpreted as a submodule of $\mathscr{T}_s^r\left(\overset{\circ}{\tau}\right)$; we shall use this harmless inclusion in what follows.

If X is a vector field on M, then $\widehat{X} := X \circ \tau$ is a Finsler vector field, called a *basic vector field* along τ. In particular, $\left(\frac{\widehat{\partial}}{\partial u^i}\right)_{i=1}^n$ is a base for the module $\mathfrak{X}(\tau)$. Besides the class of basic vector fields, a distinguished role is played by the *canonical Finsler vector field* $\delta := 1_{TM}$, called classically *support element.* In coordinates, $\delta = y^i\left(\frac{\widehat{\partial}}{\partial u^i}\right)$ (with sum convention in force).

We have a canonical $C^\infty(TM)$-linear injection $\mathbf{i} : \mathfrak{X}(\tau) \longrightarrow \mathfrak{X}(TM)$ and a surjection $\mathbf{j} : \mathfrak{X}(TM) \longrightarrow \mathfrak{X}(\tau)$ such that

$$\mathbf{i}\left(\frac{\widehat{\partial}}{\partial u^i}\right) = \frac{\partial}{\partial y^i};\ \mathbf{j}\left(\frac{\partial}{\partial x^i}\right) = \frac{\widehat{\partial}}{\partial u^i},\ \mathbf{j}\left(\frac{\partial}{\partial y^i}\right) = 0;\ i \in \{1, \ldots, n\}.$$

(For an intrinsic construction of $\mathbf{i}$ and $\mathbf{j}$, see e.g. Ref. 10. $\mathfrak{X}^v(TM) := \mathbf{i}(\mathfrak{X}(\tau))$ is the module of *vertical vector fields* on TM, $X^v := \mathbf{i}\left(\widehat{X}\right)$ is the *vertical lift* of $X \in \mathfrak{X}(M)$. $C := \mathbf{i}(\delta)$ is called the *Liouville vector field.* In coordinates, $C = y^i\frac{\partial}{\partial y^i}$. $\mathbf{J} := \mathbf{i} \circ \mathbf{j}$ is said to be the *vertical endomorphism* of $\mathfrak{X}(TM)$. It follows at once that

$$Im(\mathbf{J}) = Ker(\mathbf{J}) = \mathfrak{X}^v(TM), \quad \mathbf{J}^2 = 0.$$

We define the $d_{\mathbf{J}}$*-differential* of a smooth function f on TM as the *one-form* $d_{\mathbf{J}}f := df \circ \mathbf{J}$ *on* TM. In coordinates, $d_{\mathbf{J}}f = \frac{\partial f}{\partial y^i}dx^i$.

The formalism can go on. Let $\widetilde{X}$ be a Finsler vector field over M. We define a tensor derivation $\nabla^v_{\widetilde{X}}$ on the algebra of Finsler tensor fields by the following requirements:

(i) *On functions,* $\nabla^v_{\widetilde{X}} f := \left(\mathbf{i}\widetilde{X}\right) f \ ; \ f \in C^\infty(TM)$.

(ii) *On Finsler vector fields,* $\nabla^v_{\widetilde{X}} \widetilde{Y} := \mathbf{j}\left[\mathbf{i}\widetilde{X}, \eta\right]$, *where* $\eta \in \mathfrak{X}(TM)$ *is such that* $\mathbf{j}(\eta) = \widetilde{Y}$.

(iii) *If* $\widetilde{A} \in \mathcal{T}^r_s(\tau)$, then $\nabla^v_{\widetilde{X}} \widetilde{A}$ *is given by the* product rule.

$\nabla^v_{\widetilde{X}}$ is called the *(canonical) v-covariant derivative* with respect to $\widetilde{X}$. In coordinates: if $\widetilde{X} = \xi^i \widehat{\frac{\partial}{\partial u^i}}$, $\widetilde{Y} = \eta^i \widehat{\frac{\partial}{\partial u^i}}$, then

$$\nabla^v_{\widetilde{X}} f = \xi^i \frac{\partial f}{\partial y^i}, \quad \nabla^v_{\widetilde{X}} \widetilde{Y} = \xi^i \frac{\partial \eta^j}{\partial y^i} \frac{\partial}{\partial y^j}.$$

We see that $\nabla^v_{\widetilde{X}} \widetilde{Y}$ is well-defined: it does not depend on the choice of the vector field η. We have, in particular, $\nabla^v_{\widehat{X}} \widehat{Y} = 0$ for any vector fields X,Y on M.

As a final step, we define the *vertical differential* of a type $\binom{r}{s}$ Finsler tensor field $\widetilde{A}$ as the $\binom{r}{s+1}$ tensor $\nabla^v \widetilde{A}$ which 'collects all the v-covariant derivatives' of $\widetilde{A}$. For simplicity, if $r = s = 1$, then

$$\nabla^v \widetilde{A}\left(\widetilde{\alpha}, \widetilde{Y}, \widetilde{X}\right) := \left(\nabla^v_{\widetilde{X}} \widetilde{A}\right)\left(\widetilde{\alpha}, \widetilde{Y}\right) \overset{(iii)}{=}$$
$$\left(\mathbf{i}\widetilde{X}\right) \widetilde{A}\left(\widetilde{\alpha}, \widetilde{Y}\right) - \widetilde{A}\left(\nabla^v_{\widetilde{X}} \widetilde{\alpha}, \widetilde{Y}\right) - \widetilde{A}\left(\widetilde{\alpha}, \nabla^v_{\widetilde{X}} \widetilde{Y}\right)$$

for all $\widetilde{X}, \widetilde{Y} \in \mathfrak{X}(\tau)$ and $\widetilde{\alpha} \in \mathfrak{X}^*(\tau)$. More generally, if the components of an $\binom{r}{s}$ tensor $\widetilde{A}$ are

$$\widetilde{A}^{i_1 \dots i_r}_{i_1 \dots j_s} := \widetilde{A}\left(\widehat{du^{i_1}}, \dots, \widehat{du^{i_r}}, \widehat{\frac{\partial}{\partial u^{j_1}}}, \dots, \widehat{\frac{\partial}{\partial u^{j_s}}}\right) \left(\widehat{du^i} := du^i \circ \tau^0_1\right),$$

then the components of $\nabla^v \widetilde{A}$ are $\dot{\partial}_j \widetilde{A}^{i_1 \dots i_r}_{i_1 \dots j_s}$; these functions will be denoted by $\widetilde{A}^{i_1 \dots i_r}_{i_1 \dots j_s \cdot j}$. We recognize that in components 'vertical differentiation reduces to partial differentiation with respect to the fibre coordinates'. Notice that $\nabla^v f$ and $d_{\mathbf{J}} f$ are related by

$$d_{\mathbf{J}} f = \nabla^v f \circ \mathbf{j} \ , \ f \in C^\infty(TM) \ .$$

2. Finsler functions. The h-Berwald derivative

By a *Finsler function* we mean a continuous function $F : TM \longrightarrow [0, \infty[$, satisfying the three conditions:

(i) *F is smooth* on $\overset{\circ}{T}M$.
(ii) *F is positive-homogeneous of degree 1, i.e., $F(\lambda v) = \lambda F(v)$ for all $\lambda \in \mathbb{R}_+^*$ and $v \in TM$.*
(iii) *The* metric tensor $g := \frac{1}{2}\nabla^v\nabla^v F^2$ *is pointwise non-degenerate on* $\overset{\circ}{T}M$.

A manifold endowed with a Finsler function is said to be a *Finsler manifold*. Quite surprisingly, under these conditions the metric tensor g is positive definite, see Ref. 6. The components $g_{ij} := g\left(\frac{\widehat{\partial}}{\partial u^i}, \frac{\widehat{\partial}}{\partial u^j}\right)$ of g are just the functions $\frac{1}{2}\dot{\partial}_i\dot{\partial}_j F^2$.

In the remainder of the paper, (M,F) will be a Finsler manifold.

We show that the Finsler function F and the metric tensor g are related by

$$g(\delta, \delta) = F^2. \tag{1}$$

Indeed, by the homogeneity of F, we have $CF^2 = 2F^2$, and it can easily be checked that $\nabla^v \delta = 1_{\mathfrak{X}(\tau)}$. Thus

$$\begin{aligned} g(\delta,\delta) &= \frac{1}{2}\nabla^v\nabla^v F^2(\delta,\delta) = \frac{1}{2}\nabla^v_\delta\left(\nabla^v F^2\right)(\delta) = \frac{1}{2}\left(C\left(CF^2\right) - \nabla^v F^2(\delta)\right) \\ &= \frac{1}{2}\left(4F^2 - 2F^2\right) = F^2. \end{aligned}$$

In the Finslerian case the canonical vector field δ has a dual 1-form δ^* along $\overset{\circ}{\tau}$ given by

$$\delta^*\left(\widetilde{X}\right) := g\left(\widetilde{X}, \widetilde{Y}\right) \; ; \widetilde{X} \in \mathfrak{X}^*\left(\overset{\circ}{\tau}\right). \tag{2}$$

Then $\delta^*(\delta) = F^2$ by (1). The components of δ^* can be obtained from the components of δ by index lowering:

$$y_i := \delta^*\left(\frac{\widehat{\partial}}{\partial u^i}\right) = g\left(\frac{\widehat{\partial}}{\partial u^i}, y^j\frac{\widehat{\partial}}{\partial u^j}\right) = g_{ij}y^j \; , \; i \in \{1, \ldots, n\}.$$

It follows immediately that

$$y_i y^i = F^2. \tag{3}$$

By the *Cartan tensor* of (M, F) we mean the type $\binom{0}{3}$ Finsler tensor $\mathscr{C}_\flat := \frac{1}{2}\nabla^v g$. Its components are

$$C_{ijk} := \mathscr{C}_\flat\left(\frac{\widehat{\partial}}{\partial u^i}, \frac{\widehat{\partial}}{\partial u^j}, \frac{\widehat{\partial}}{\partial u^k}\right) = \frac{1}{2}\dot{\partial}_k g_{ij} = \frac{1}{4}\dot{\partial}_k\dot{\partial}_j\dot{\partial}_i F^2,$$

thus $\mathscr{C}_\flat$ is totally symmetric. Raising an index, we get the *vector-valued Cartan tensor* $\mathscr{C}$, metrically equivalent to $\mathscr{C}_\flat$. More pedantically, $\mathscr{C}$ is defined by

$$g\left(\mathscr{C}\left(\widetilde{X}, \widetilde{Y}\right), \widetilde{Z}\right) = \mathscr{C}_\flat\left(\widetilde{X}, \widetilde{Y}, \widetilde{Z}\right);\ \widetilde{X}, \widetilde{Y}, \widetilde{Z} \in \mathfrak{X}\left(\mathring{\tau}\right),$$

so its components are

$$g^{ir}C_{jkr} =: C^i_{jk}\ ;\ \ (g^{ij}) := (g_{ij})^{-1}.$$

It is a fundamental fact, that F determines a unique spray $S : TM \longrightarrow TTM$ via the Euler-Lagrange equation

$$i_S dd_{\mathbf{J}} F^2 = -dF^2.$$

S is called the *canonical spray* of the Finsler manifold. In coordinates, $S = y^i\partial_i - 2G^i\dot{\partial}_i$, where

$$G^i = \frac{1}{4}g^{ij}\left(\frac{\partial^2 F^2}{\partial x^r \partial y^j}y^r - \frac{\partial F^2}{\partial x^j}\right)\ ,\ i \in \{1, \ldots, n\}.$$

The *spray coefficients* G^i are of class C^1 on TM, smooth on $\mathring{T}M$ and are positively homogeneous of degree 2. The canonical spray determines the *canonical Ehresmann connection* $\mathscr{H} : \mathfrak{X}\left(\mathring{\tau}\right) \longrightarrow \mathfrak{X}(TM)$ of (M, F) by *Crampin's construction*[4]

$$\widehat{X} \in \mathfrak{X}\left(\mathring{\tau}\right) \longmapsto X^h := \mathscr{H}\left(\widehat{X}\right) := \tfrac{1}{2}\left(X^c + [X^v, S]\right),\ X \in \mathfrak{X}(M)$$

(X^c denotes the complete lift of X). X^h is called the *horizontal lift* of X. The horizontal lifts of the coordinate vector fields $\frac{\partial}{\partial u^j}$ take the form

$$\left(\frac{\partial}{\partial u^j}\right)^h = \frac{\partial}{\partial x^j} - \frac{\partial G^i}{\partial y^j}\frac{\partial}{\partial y^i},\ j \in \{1, \ldots, n\};$$

the functions $G^i_j := \dot{\partial}_j G^i$ are said to be the *Christoffel symbols* of $\mathscr{H}$.

The *horizontal* and the *vertical projector* associated to $\mathscr{H}$ are $\mathbf{h} := \mathscr{H} \circ \mathbf{j}$ and $\mathbf{v} = 1_{\mathfrak{X}\left(\mathring{T}M\right)} - \mathbf{h}$, respectively. Following Berwald's terminology,[3] we call the type $\binom{1}{2}$ Finsler tensor $\mathbf{R}$ defined by

$$\mathbf{i}\mathbf{R}\left(\widehat{X},\widehat{Y}\right) := -\mathbf{v}\left[X^h, X^v\right] \; ; \; X,Y \in \mathfrak{X}(M)$$

the *fundamental affine curvature* of the Finsler manifold. To be in harmony with Matsumoto's conventions,[8] we define the components R^i_{jk} of $\mathbf{R}$ by $R^i_{jk}\frac{\widehat{\partial}}{\partial u^i} = \mathbf{R}\left(\frac{\widehat{\partial}}{\partial u^k}, \frac{\widehat{\partial}}{\partial u^j}\right)$. If $G^i_{jk} := \dot{\partial}_k G^i_j$, then

$$R^i_{jk} = \left(\frac{\partial}{\partial u^k}\right)^h G^i_j - \left(\frac{\partial}{\partial u^j}\right)^h G^i_k = \partial_k G^i_j - \partial_j G^i_k + G^r_j G^i_{rk} - G^r_k G^i_{rj}. \quad (4)$$

In the spirit of Berwald's above mentioned paper, by the *affine curvature tensor* of (M,F) we mean the type $\binom{1}{3}$ tensor $\mathbf{H} := \nabla^v \mathbf{R}$. The components of $\mathbf{H}$ are determined by $H^i_{jkl}\frac{\widehat{\partial}}{\partial u^i} := \mathbf{H}\left(\frac{\widehat{\partial}}{\partial u^l}, \frac{\widehat{\partial}}{\partial u^k}\right)\frac{\widehat{\partial}}{\partial u^j}$. Obviously,

$$H^i_{jkl} = \dot{\partial}_j R^i_{kl} = R^i_{kl\cdot j}, \quad (5)$$

We define a further important tensor, the *Berwald curvature* $\mathbf{B}$, by

$$\mathbf{i}\mathbf{B}\left(\widehat{X},\widehat{Y}\right)\widehat{Z} := \left[\left[X^v, Y^h\right], Z^v\right] \; ; \; X,Y,Z \in \mathfrak{X}(M).$$

Its components are $G^i_{jkl} := \dot{\partial}_l G^i_{jk}$.

Following the above scheme, we construct a further tensor derivation on the algebra of Finsler tensors, depending on the canonical Ehresmann connection. Let $\widetilde{X} \in \mathfrak{X}\left(\overset{\circ}{\tau}\right)$. Define the operator $\nabla^h_{\widetilde{X}}$

(i) *on functions by* $\nabla^h_{\widetilde{X}} f := (\mathscr{H}\widetilde{X})f$, $f \in C^\infty(\overset{\circ}{T}M)$;

(ii) *on Finsler vector fields by* $\mathbf{i}\nabla^h_{\widetilde{X}}\widetilde{Y} := \mathbf{v}\left[\mathscr{H}\widetilde{X}, \mathbf{i}\widetilde{Y}\right]$;

(iii) *on type* $\binom{r}{s}$ *tensors by the* product rule.

$\nabla^h_{\widetilde{X}}$ is said to be the *h-Berwald derivative* with respect to $\widetilde{X}$. Its Christoffel symbols are just the functions G^i_{jk}, i.e., we have

$$\nabla^h_{\frac{\widehat{\partial}}{\partial u^k}} \frac{\widehat{\partial}}{\partial u^j} =: G^i_{jk}\frac{\widehat{\partial}}{\partial u^i}.$$

After this the *h-Berwald differential* ∇^h can be defined in the same way as the vertical differential ∇^v (formally: replace the canonical injection $\mathbf{i}$ by the surjection $\mathcal{H}$). As for the index gymnastics, if the components of $\widetilde{A}$ are $\widetilde{A}^{i_1\ldots i_r}_{j_1\ldots j_s}$, then the components of $\nabla^h\widetilde{A}$ will be denoted by $\widetilde{A}^{i_1\ldots i_r}_{j_1\ldots j_s;j}$. These functions are much more complicated than the components of $\nabla^v\widetilde{A}$. As an illustration, we calculate the components of the h-Berwald differential of the metric tensor g:

$$g_{ij;k} := \nabla^h g\left(\frac{\widehat{\partial}}{\partial u^i}, \frac{\widehat{\partial}}{\partial u^j}, \frac{\widehat{\partial}}{\partial u^k}\right) := \left(\nabla^h_{\frac{\widehat{\partial}}{\partial u^k}} g\right)\left(\frac{\widehat{\partial}}{\partial u^i}, \frac{\widehat{\partial}}{\partial u^j}\right)$$
$$= \left(\frac{\widehat{\partial}}{\partial u^k}\right)^h g_{ij} - g\left(\nabla^h_{\frac{\widehat{\partial}}{\partial u^k}} \frac{\widehat{\partial}}{\partial u^i}, \frac{\widehat{\partial}}{\partial u^j}\right) - g\left(\frac{\widehat{\partial}}{\partial u^i}, \nabla^h_{\frac{\widehat{\partial}}{\partial u^k}} \frac{\widehat{\partial}}{\partial u^j}\right)$$
$$= \left(\frac{\partial}{\partial u^k}\right)^h g_{ij} - G^r_{ik} g_{rj} - G^r_{jk} g_{ir}.$$

3. Landsberg tensor depending only on the position

By the *Landsberg tensor* of a Finsler manifold (M, F) we mean the type $\binom{0}{3}$ tensor

$$\mathbf{P} := -\tfrac{1}{2}\nabla^h g$$

along $\overset{\circ}{\tau}$. Its components are

$$P_{ijk} = -\tfrac{1}{2} g_{ij;k} \ ,$$

where the functions $g_{ij;k}$ have just been calculated. The Landsberg tensor and the Cartan tensor $\mathscr{C}_\flat$ are related by

$$\mathbf{P} = \nabla^h_\delta \mathscr{C}_\flat. \tag{6}$$

In components,

$$P_{ijk} = C_{ijk;l} y^l, \tag{7}$$

which may easily be shown. A coordinate-free proof of (6) needs a little more effort, see Ref. 10, section 3.11.

Now we are in a position to show that the property $\nabla^v \mathbf{P} = 0$ implies a drastic consequence.

Proposition 3.1. *If the Landsberg tensor of a Finsler manifold depends only on the position, then it vanishes identically.*

Proof. Keeping the notation introduced above, suppose that $\dot{\partial}_l P_{ijk} = P_{ijk \cdot l} = 0$. Then differentiation of relation (7) with respect to $\dot{\partial}_l$ leads to

$$C_{ijk;l} + C_{ijk;r \cdot l} y^r = 0. \tag{8}$$

Now we use the Ricci identity (Ref. 8, 2.5.5) for the h-Berwald derivative and the vertical derivative. Then we obtain

$$C_{ijk;r\cdot l} - C_{ijk\cdot l;r} = -C_{sjk}G^s_{ilr} - C_{isk}G^s_{jlr} - C_{ijs}G^s_{klr} \tag{9}$$

(recall that $G^i_{jkl} = \dot{\partial}_l G^i_{jk}$ are the components of the Berwald tensor). Since the functions G^i_{jk} are positively homogeneous of degree 0, we have $G^i_{jkl}y^l = 0$. Thus, transvection of (9) with y^r leads to

$$C_{ijk;r\cdot l}y^r = C_{ijk\cdot l;r}y^r.$$

Hence (8) takes the form

$$C_{ijk;l} + C_{ijk\cdot l;r}y^r = 0. \tag{10}$$

Interchanging indices k and l, we obtain

$$C_{ijl;k} + C_{ijl\cdot k;r}y^r = 0. \tag{11}$$

Since $C_{ijk\cdot l} = C_{ijl\cdot k}$, if we subtract (11) from (10) we find that

$$C_{ijk;l} - C_{ijl;k} = 0. \tag{12}$$

But transvection of (12) with y^l yields

$$C_{ijk;l}y^l = 0, \tag{13}$$

since $C_{ijl}y^l = \frac{1}{2}\frac{\partial g_{ij}}{\partial y^l}y^l = 0$ by the 0^+-homogenity of the functions g_{ij}, and by the commutation of contractions and covariant derivatives.
Relations (13) and (7) imply our assertion $\mathbf{P} = 0$. □

4. Stretch tensor depending only on the position

Inspired by a manuscript of L. Kozma,[5] we define the *stretch tensor* Σ of a Finsler manifold (M, F) by

$$\frac{1}{2}\Sigma\left(\widetilde{X},\widetilde{Y},\widetilde{Z},\widetilde{U}\right) := \nabla^h\mathbf{P}\left(\widetilde{X},\widetilde{Y},\widetilde{Z},\widetilde{U}\right) - \nabla^h\mathbf{P}\left(\widetilde{X},\widetilde{Y},\widetilde{U},\widetilde{Z}\right), \tag{14}$$

where $\mathbf{P}$ is the Landsberg tensor discussed above, and $\widetilde{X},\widetilde{Y},\widetilde{Z},\widetilde{U}$ are arbitrary Finsler vector fields. Since the components of $\nabla^h\mathbf{P}$ are

$$P_{ijk;l} = \left(\frac{\widehat{\partial}}{\partial u^l}\right)^h P_{ijk} - G^r_{il}P_{rjk} - G^r_{jl}P_{irk} - G^r_{kl}P_{ijr},$$

it follows that the components of Σ are

$$\Sigma_{ijkl} = 2\left(P_{ijk;l} - P_{ijl;k}\right). \tag{15}$$

This is just the formula obtained by M. Matsumoto for the stretch tensor in Ref. 7. Notice that the stretch tensor was discovered by L. Berwald.[1] He also found an important relation between the affine curvature and the stretch tensor, which may be formulated as follows:

$$\Sigma\left(\widetilde{X},\widetilde{Y},\widetilde{Z},\widetilde{U}\right) = -\delta^{*}\left(\nabla^{v}\mathbf{H}\left(\widetilde{Z},\widetilde{U},\widetilde{X},\widetilde{Y}\right)\right), \widetilde{X},\widetilde{Y},\widetilde{Z},\widetilde{U} \in \mathfrak{X}\left(\overset{\circ}{\tau}\right). \quad (16)$$

In terms of tensor components, (16) leads to

$$\Sigma_{ijkl} = -y_r \dot{\partial}_j H^r_{ikl}, \quad (17)$$

this is just formula (14) of Berwald's paper.[2] From (5) and (17) it follows that we also have

$$\Sigma_{ijkl} = -y_r \dot{\partial}_j \dot{\partial}_i R^r_{kl}. \quad (18)$$

Relations (16) and (18) imply that the stretch tensor vanishes, if the fundamental affine curvature, or, equivalently, the affine curvature tensor of the Finsler manifold depends only on the position. Now we shall show that this conclusion is also true, if Σ itself has this property.

Proposition 4.1. *If the stretch tensor of a Finsler manifold depends only on the position, then it vanishes identically.*

Proof. We use the same tactics as in the previous proof. By our condition,

$$\Sigma_{ijkl\cdot m} = \dot{\partial}_m \Sigma_{ijkl} = 0\ ,$$

so from (15) we get for the Landsberg tensor

$$P_{ijk;l\cdot m} - P_{ijl;k\cdot m} = 0. \quad (19)$$

Using the Ricci identity for $P_{ijk;l\cdot m}$ we get

$$P_{ijk;l\cdot m} = P_{ijk\cdot m;l} - P_{rjk}G^r_{iml} - P_{irk}G^r_{jml} - P_{ijr}G^r_{kml}. \quad (20)$$

Transvection of (20) with y^i leads to

$$P_{ijk;l\cdot m}y^i = P_{ijk\cdot m;l}y^i \quad (21)$$

because of $P_{ijk}y^i \overset{(7)}{=} C_{ijk;l}y^iy^l = 0$. In the same way we obtain

$$P_{ijl;k\cdot m}y^i = P_{ijl\cdot m;k}y^i. \quad (22)$$

Relations (19), (21) and (22) imply that

$$P_{ijk\cdot m;l}y^i = P_{ijl\cdot m;k}y^i. \quad (23)$$

On the other hand, from the identity $P_{rjk}y^r = 0$ we obtain by repeated covariant differentiation

$$0 = (P_{rjk}y^r)_{\cdot i;l} = P_{ijk;l} + P_{rjk\cdot i;l}y^r \ .$$

Interchanging indices k and l, we get

$$P_{ijl;k} + P_{rjl\cdot i;k}y^r = 0 \ .$$

(23) and the last two relations imply that $P_{ijk;l} - P_{ijl;k} = 0$. Hence, by (15), $\Sigma_{ijkl} = 0$. □

Acknowledgments

The first author was supported by National Science Research Foundation OTKA No. T48878.

References

1. L. Berwald, Über Parallelübertragung in Räumen mit allgemeiner Massbestimmung, *Jber. Deutsch Math.-Verein* **34** (1926) 213–220.
2. L. Berwald, Parallelübertragung in allgemeinen Räumen, *Atti. Congr. Intern. Mat. Bologna* **4** (1928) 263–270.
3. L. Berwald, Ueber Finslersche und Cartansche Geometrie IV, *Ann. Math.* **48** (1947) 755–781.
4. M. Crampin, On horizontal distribution on the tangent bundle of a differentiable manifold, *J. London Math. Soc. (2)* **3** (1971) 178–182.
5. L. Kozma, Unpublished manuscript.
6. R.L. Lovas, A note on Finsler-Minkowski norms, *Houston J. Math.* **33** (2007) 701–707.
7. M. Matsumoto, On the stretch curvature of a Finsler space and certain open problems, *J. Nat. Acad. Math. India* **11** (1997) 22–32.
8. M. Matsumoto, Finsler Geometry in the 20th-Century, *In: Handbook of Finsler Geometry* ((P. Antonelli, Ed.) Kluwer Academic Publishers, Dordrecht, 2003).
9. Z.I. Szabó, Über Zusammenhänge vom Finsler-Typ, *Publ. Math. Debrecen* **27** (1980) 77–88.
10. J. Szilasi, A Setting for Spray and Finsler Geometry, *In: Handbook of Finsler Geometry* ((P. Antonelli, Ed.) Kluwer Academic Publishers, Dordrecht, 2003).

Differential Geometry and its Applications
Proc. Conf., in Honour of Leonhard Euler, Olomouc, August 2007

Lie systems and integrability conditions of differential equations and some of its applications

J.F. Cariñena and J. de Lucas

Departamento de Física Teórica, Facultad de Ciencias, Universidad de Zaragoza, 50009 Zaragoza, Spain
E-mail: jfc@unizar.es, dlucas@unizar.es

The geometric theory of Lie systems is used to establish integrability conditions for several systems of differential equations, in particular some Riccati equations and Ermakov systems. Many different integrability criteria in the literature will be analysed from this new perspective, and some applications in physics will be given.

Keywords: Integrability criteria, Riccati, Ermakov systems, Milne–Pinney equation, superposition rules, Lie systems.

MS classification: 34A26.

1. Introduction

Non-autonomous systems of first-order and second-order differential equations appear in many places in physics. For instance, Hamilton equations are systems of first-order differential equations, while Euler–Lagrange equations for regular Lagrangians are systems of second-order differential equations. A system of second-order differential equations in n variables of the form $\ddot{x}^i = F^i(x, \dot{x}, t)$, with $i = 1, \ldots, n$, is related with a system of first-order equations in $2n$ variables:

$$\begin{cases} \dot{x}^i = v^i \\ \dot{v}^i = F^i(x, v, t) \end{cases}, \qquad i = 1, \ldots n\,. \tag{1}$$

Therefore, it is enough to restrict ourselves to study systems of first-order differential equations. From the geometric viewpoint a system

$$\dot{x}^i = X^i(x, t)\,, \qquad i = 1, \ldots, n\,, \tag{2}$$

is associated with the t-dependent vector field $X = X^i(x, t)\,\partial/\partial x^i$ whose integral curves are determined by the solutions of the system.

Unfortunately, there is no general method for solving such equations. Relevant questions about integrability are how to find a particular solution (determined by $x(0) = x_0$), or a r-parameter family of solutions, or even the general solution (a n-parameter family of solutions). Finally, when is it possible to find and how to determine a superposition rule for solutions?

We shall understand that to find a solution means to reduce the problem to carry out some quadratures. For instance, the general solution of the inhomogeneous linear differential equation $dx/dt = b_0(t) + b_1(t)x$ can be found with two quadratures and it is given by

$$x(t) = \exp\left(\int_0^t b_1(s)\,ds\right) \times \left(x_0 + \int_0^t b_0(t')\exp\left(-\int_0^{t'} b_1(s)\,ds\right)dt'\right) .$$

Actually, when the systems we are dealing with are linear, there is a *linear superposition principle* allowing us to find the general solution as a linear combination of n particular solutions. For instance, for the harmonic oscillator with a t-dependent angular frequency $\omega(t)$:

$$\ddot{x} = -\omega^2(t)\,x \Longleftrightarrow \begin{cases} \dot{x} = v \\ \dot{v} = -\omega^2(t)\,x \end{cases} ,$$

whose solutions are the integral curves of the t-dependent vector field $X = v\partial/\partial x - \omega^2(t)x\,\partial/\partial v$, if we know a particular solution, the general solution can be found by means of one quadrature and if we know two particular solutions, x_1 and x_2, the general solution is a linear combination (no quadrature is needed) $x(t) = k_1\,x_1(t) + k_2\,x_2(t)$.

There are systems whose general solution can be written as a nonlinear function of some particular solutions. For instance, for Riccati equation: if a particular solution is known, the general solution is obtained by two quadratures, if two particular solutions are known the problem reduces to one quadrature and, finally, when three particular solutions are known, x_1, x_2 and x_3, the general solution can be found from the cross-ratio relation

$$\frac{x - x_1}{x - x_2} : \frac{x_3 - x_1}{x_3 - x_2} = k ,$$

which provides us a nonlinear superposition rule.[1]

There also exist cases in which we can superpose solutions of one system for finding solutions of another one. We have seen one example: the general solution of the inhomogeneous linear equation can be written as $x(t) = x_1(t) + C\,x_0(t)$, where $x_0(t)$ is a solution of the associated homogeneous equation and $x_1(t)$ is a particular solution of the inhomogeneous linear one.

Milne-Pinney equation[2] $\ddot{x} = -\omega^2(t)x + k/x^3$ is usually studied together with the time-dependent harmonic oscillator $\ddot{y}+\omega^2(t)y = 0$ and the system is called Ermakov system. Pinney showed in a short paper[2] that the general solution of the first equation can be written as a nonlinear superposition of two solutions of the associated harmonic oscillator. All these properties can be better understood in the framework of Lie systems, conveniently extended in some cases to include systems of second-order differential equations. These systems have a lot of applications not only in mathematics but also in many different branches of classical and quantum physics.

Let us look for systems admitting a (maybe nonlinear) superposition rule. The main result was given by Lie:[3,4]

Theorem 1.1. *Given (2) a necessary and sufficient condition for the existence of a function* $\Phi : \mathbb{R}^{n(m+1)} \to \mathbb{R}^n$ *such that the general solution is* $x = \Phi(x_{(1)}, \ldots, x_{(m)}; k_1, \ldots, k_n)$, *with* $\{x_{(a)} \mid a = 1, \ldots, m\}$ *being a set of particular solutions of the system and* $k_1, \ldots, k_n$, *are* n *arbitrary constants, is that the system can be written as*

$$\frac{dx^i}{dt} = Z^1(t)\xi^i_1(x) + \cdots + Z^r(t)\xi^i_r(x),$$

where $Z^1, \ldots, Z^r$, *are* r *functions depending only on* t *and* ξ^i_α, $\alpha = 1, \ldots, r$, *are functions of* $x = (x^1, \ldots, x^n)$, *such that the* r *vector fields in* $\mathbb{R}^n$ *given by* $X_\alpha \equiv \sum_{i=1}^{n} \xi^i_\alpha(x^1, \ldots, x^n)\partial/\partial x^i$, $\alpha = 1, \ldots, r$, *close on a real finite-dimensional Lie algebra, i.e. the* X_α *are l.i. and there are* r^3 *real numbers,* $c_{\alpha\beta}{}^\gamma$, *such that* $[X_\alpha, X_\beta] = \sum_{\gamma=1}^{r} c_{\alpha\beta}{}^\gamma X_\gamma$. *The number* r *satisfies* $r \leq mn$.

The condition in the Theorem 1.1 is that $X(x,t)$ can be written as $X(x,t) = \sum_{\alpha=1}^{r} Z^\alpha(t)X_\alpha(x)$, with X_α as mentioned above.

Non-autonomous systems corresponding to such t-dependent vector fields will be called Lie systems. One instance of Lie system is the Riccati equation[1]

$$\frac{dx(t)}{dt} = b_2(t)\,x^2(t) + b_1(t)\,x(t) + b_0(t)\ , \tag{3}$$

for which $m = 3$ and the superposition principle comes from the relation

$$\frac{x - x_1}{x - x_2} : \frac{x_3 - x_1}{x_3 - x_2} = k \implies x = \frac{k\,x_1(x_3 - x_2) + x_2(x_1 - x_3)}{k\,(x_3 - x_2) + (x_1 - x_3)}\ .$$

The associated Lie algebra is generated by X_0, X_1 and X_2 given by

$$X_0 = \frac{\partial}{\partial x}\,, \quad X_1 = x\,\frac{\partial}{\partial x}\,, \quad X_2 = x^2\,\frac{\partial}{\partial x}\,,$$

which close on a $\mathfrak{sl}(2,\mathbb{R})$ 3-dimensional real Lie algebra, because

$$[X_0, X_1] = X_0\,, \quad [X_0, X_2] = 2X_1\,, \quad [X_1, X_2] = X_2\,. \tag{4}$$

The time-dependent harmonic oscillator is also an example of physical relevance. It is described by a Hamiltonian

$$H = \frac{1}{2}\,\frac{p^2}{m(t)} + \frac{1}{2}\,m(t)\omega^2(t)x^2\,,$$

which gives rise to the dynamics defined by the t-dependent vector field

$$X(x,p,t) = \frac{1}{m(t)}\,p\,\frac{\partial}{\partial x} - m(t)\omega^2(t)\,x\frac{\partial}{\partial p}\,.$$

If we consider the set of vector fields

$$X_0 = p\frac{\partial}{\partial x}, \quad X_1 = \frac{1}{2}\left(x\frac{\partial}{\partial x} - p\frac{\partial}{\partial p}\right), \quad X_2 = -x\frac{\partial}{\partial p}\,,$$

which close on a $\mathfrak{sl}(2,\mathbb{R})$ Lie algebra with the same commutation relations as (4), the corresponding t-dependent vector field X can be written as a linear combination $X(\cdot,t) = m(t)\omega^2(t)\,X_2(\cdot) + (1/m(t))\,X_0(\cdot)$, i.e. it is a linear combination with t-dependent coefficients $X(\cdot,t) = \sum_{\alpha=0}^{2} b_\alpha(t)X_\alpha(\cdot)$ with $b_0(t) = 1/m(t)$, $b_1(t) = 0$ and $b_2(t) = m(t)\omega^2(t)$.

The prototype of Lie system is the time-dependent right-invariant vector fields in a Lie group G. Let $\{a_1, \ldots, a_r\}$ denote a basis of T_eG. A right-invariant vector field X^R is one such that $X^R(g) = R_{g*e}X_e$. Define X^R_α by $X^R_\alpha(g) = R_{g*e}a_\alpha$. The t–dependent right-invariant vector field

$$\bar{X}(g,t) = -\sum_{\alpha=1}^{r} b_\alpha(t)X^R_\alpha(g)\ .$$

defines a Lie system in G whose integral curves are solutions of the system $\dot{g} = -\sum_{\alpha=1}^{r} b_\alpha(t)\,X^R_\alpha(g)$, and when applying $R_{g^{-1}}$ to both sides we see that $g(t)$ satisfies

$$R_{g^{-1}(t)*g(t)}\dot{g}(t) \;= -\sum_{\alpha=1}^{r} b_\alpha(t)a_\alpha \in T_eG\ . \tag{5}$$

Let H be a closed subgroup of G and consider the homogeneous space $M = G/H$. Then, G can be seen as a principal bundle over G/H: $(G,\tau,G/H)$.

The X^R_α are τ-projectable on the corresponding fundamental vector fields of the left-action $\lambda : (g, g'H) \in G \times M \to (gg'H) \in M$ given by $-X_\alpha = -X_{a_\alpha}$ with $\tau_{*g} X^R_\alpha(g) = -X_\alpha(gH)$, the projected vector field in M will be $X(x,t) = \sum_{\alpha=1}^r b_\alpha(t) X_\alpha(x)$, and its integral curves are the solutions of the system of differential equations: $\dot{x} = \sum_{\alpha=1}^r b_\alpha(t) X_\alpha(x)$. The solution of this last system starting from x_0 is $x(t) = \lambda(g(t), x_0)$, with $g(t)$ being the solution of (5) such that $g(0) = e$. This means that solving such Lie system in G we are simultaneously solving the corresponding problems in all its homogeneous spaces.

2. SODE Lie systems

A system of second order differential equations can be studied through the corresponding system of first-order differential equations as indicated in (1). We call SODE Lie systems those for which the associated first-order one is a Lie system, i.e. it can be written as a linear combination with t-dependent coefficients of vector fields closing on a finite-dimensional real Lie algebra. An example is the 1-dimensional harmonic oscillator with time-dependent frequency, but the same is true for the 2-dimensional isotropic harmonic oscillator with time-dependent frequency, with an associated vector field

$$X = v_1 \frac{\partial}{\partial x_1} - \omega^2(t) x_1 \frac{\partial}{\partial v_1} + v_2 \frac{\partial}{\partial x_2} - \omega^2(t) x_2 \frac{\partial}{\partial v_2} ,$$

which is a linear combination $X = X_2 - \omega^2(t) X_1$ with

$$X_1 = x_1 \frac{\partial}{\partial v_1} + x_2 \frac{\partial}{\partial v_2} , \qquad X_2 = v_1 \frac{\partial}{\partial x_1} + v_2 \frac{\partial}{\partial x_2} ,$$

and then they close once again on a Lie algebra isomorphic to $\mathfrak{sl}(2,\mathbb{R})$:

$$[X_1, X_2] = 2\, X_3 , \quad [X_1, X_3] = -X_1 , \quad [X_2, X_3] = X_2 , \tag{6}$$

with

$$X_3 = \frac{1}{2} \left(x_1 \frac{\partial}{\partial x_1} - v_1 \frac{\partial}{\partial v_1} + x_2 \frac{\partial}{\partial x_2} - v_2 \frac{\partial}{\partial v_2} \right) .$$

The search for the superposition rule for a Lie system consists on looking for enough number of first integrals, independent of the time-dependent coefficients, in an extended space in which we consider several replicas of the given vector field.

The 2-dimensional case admits an invariant F given by the first integral $F(x_1, x_2, v_1, v_2) = x_1 v_2 - x_2 v_1$, which can be seen as a partial superposition rule.[5] Actually, if $x_1(t)$ is a solution of the first equation, then we obtain for each real number k the first-order differential equation for the variable x_2, $x_1(t)\,dx_2/dt = k + \dot{x}_1(t)x_2$, from where x_2 can be found to be given by $x_2(t) = k'x_1(t) + k\,x_1(t)\int^t x_1^{-2}(\zeta)\,d\zeta$. With three copies of the same harmonic oscillator, i.e. X_1 and X_2 given by

$$X_1 = v_1\frac{\partial}{\partial x_1} + v_2\frac{\partial}{\partial x_2} + v\frac{\partial}{\partial x}\,, \qquad X_2 = x_1\frac{\partial}{\partial v_1} + x_2\frac{\partial}{\partial v_2} + x\frac{\partial}{\partial v}\,,$$

there exist two independent first integrals $F_1(x_1, x_2, x, v_1, v_2, v) = xv_1 - x_1 v$ and $F_2(x_1, x_2, x, v_1, v_2, v) = xv_2 - x_2 v$, from where we obtain the expected superposition rule:

$$x = k_1\,x_1 + k_2\,x_2\,, \qquad v = k_1\,v_1 + k_2\,v_2\,.$$

Another interesting non-linear example is the Pinney equation, the second order non-linear differential equation:

$$\ddot{x} = -\omega^2(t)x + \frac{k}{x^3}\,,$$

where k is a constant, with associated t-dependent vector field

$$X = v\frac{\partial}{\partial x} + \left(-\omega^2(t)x + \frac{k}{x^3}\right)\frac{\partial}{\partial v}\,,$$

which is a Lie system because it can be written as $X = L_2 - \omega^2(t)L_1$, where $L_1 := x\,\partial/\partial v$ and $L_2 = (k/x^3)\,\partial/\partial v + v\,\partial/\partial x$ generate a three-dimensional real Lie algebra isomorphic to $\mathfrak{sl}(2,\mathbb{R})$ with nonzero defining relations similar to (6) with $L_3 = (1/2)\,(x\,\partial/\partial x - v\,\partial/\partial v)$.

Note that this isotonic oscillator shares with the harmonic one the property of having a period independent of the energy, i.e. they are isochronous, and in the quantum case they have a equispaced spectrum. The fact that they have the same associated Lie algebra means that they can be solved simultaneously in the group $SL(2,\mathbb{R})$ by the same equation

$$R_{g^{-1}*g}\dot{g} = \omega^2(t)\,a_1 - a_2\,, \quad g(0) = e\,.$$

3. Ermakov systems

We can consider the generalised Ermakov system given by:

$$\begin{cases} \ddot{x} = \dfrac{1}{x^3}f(y/x) - \omega^2(t)x \\ \ddot{y} = \dfrac{1}{y^3}g(y/x) - \omega^2(t)y \end{cases}$$

which when $f(u) = k$ and $g(u) = 0$ reduces to the Ermakov system.

This system is described by the t-dependent vector field

$$X = v_x \frac{\partial}{\partial x} + v_y \frac{\partial}{\partial v_y} + \left(-\omega^2(t)x + \frac{1}{x^3} f(y/x)\right)\frac{\partial}{\partial v_x} + \left(-\omega^2(t)y + \frac{1}{y^3} g(y/x)\right)\frac{\partial}{\partial v_y},$$

which can be written as a linear combination $X = N_2 - \omega^2(t)\, N_1$, where N_1 and N_2 are the vector fields

$$N_1 = x\frac{\partial}{\partial v_x} + y\frac{\partial}{\partial v_y}, \quad N_2 = v_x\frac{\partial}{\partial x} + \frac{1}{x^3} f(y/x)\frac{\partial}{\partial v_x} + v_y\frac{\partial}{\partial y} + \frac{1}{y^3} g(y/x)\frac{\partial}{\partial v_y},$$

that generate a three-dimensional real Lie algebra isomorphic to $\mathfrak{sl}(2,\mathbb{R})$ with a third generator

$$N_3 = \frac{1}{2}\left(x\frac{\partial}{\partial x} - v_x\frac{\partial}{\partial v_x} + y\frac{\partial}{\partial y} - v_y\frac{\partial}{\partial v_y}\right).$$

There exists a first integral for the motion, $F : \mathbb{R}^4 \to \mathbb{R}$, for any $\omega^2(t)$, which satisfies $N_i F = 0$ for $i = 1, \ldots, 3$, but as $[N_1, N_2] = 2N_3$ it is enough to impose $N_1 F = N_2 F = 0$. The condition $N_1 F = 0$, implies that there exists a function $\bar{F} : \mathbb{R}^3 \to \mathbb{R}$ such that $F(x, y, v_x, v_y) = \bar{F}(x, y, \xi = xv_y - yv_x)$. Then using the method of the the characteristics in condition $N_2 F = 0$, we can obtain the first integral:

$$F(x, y, v_x, v_y) = \frac{1}{2}(xv_y - yv_x)^2 + \int^{x/y}\left[-\frac{1}{u^3} f\left(\frac{1}{u}\right) + u\, g\left(\frac{1}{u}\right)\right] du\,.$$

For the Ermakov system with $f(1/u) = k$ and $g(1/u) = 0$ we obtain the known Ermakov invariant

$$F(x, y, v_x, v_y) = \frac{k}{2}\left(\frac{y}{x}\right)^2 + \frac{1}{2}(xv_y - yv_x)^2$$

We can now consider a system made up by a Pinney equation with two associated harmonic oscillator equations, with associated t-dependent vector field

$$X = v_x\frac{\partial}{\partial x} + v_y\frac{\partial}{\partial y} + v_z\frac{\partial}{\partial z} + \frac{k}{y^3}\frac{\partial}{\partial v_y} - \omega^2(t)\left(x\frac{\partial}{\partial v_x} + y\frac{\partial}{\partial v_y} + z\frac{\partial}{\partial v_z}\right)$$

which can be expressed as $X = N_2 - \omega^2(t)N_1$ where N_1 and N_2 are:

$$N_1 = y\frac{\partial}{\partial v_y} + x\frac{\partial}{\partial v_x} + z\frac{\partial}{\partial v_z}, \quad N_2 = v_y\frac{\partial}{\partial y} + \frac{1}{y^3}\frac{\partial}{\partial v_y} + v_x\frac{\partial}{\partial x} + v_z\frac{\partial}{\partial z},$$

These vector fields generate a three-dimensional real Lie algebra isomorphic to $\mathfrak{sl}(2,\mathbb{R})$ with the vector field N_3 given by

$$N_3 = \frac{1}{2}\left(x\frac{\partial}{\partial x} - v_x\frac{\partial}{\partial v_x} + y\frac{\partial}{\partial y} - v_y\frac{\partial}{\partial v_y} + z\frac{\partial}{\partial z} - v_z\frac{\partial}{\partial v_z}\right).$$

In this case there exist three first integrals for the distribution generated by these fundamental vector fields: The Ermakov invariant I_1 of the subsystem involving variables x and y, the Ermakov invariant I_2 of the subsystem involving variables y and z, and finally, the Wronskian W of the subsystem involving variables x and z. They are given by $W = xv_z - zv_x$,

$$I_1 = \frac{1}{2}\left((yv_x - xv_y)^2 + k\left(\frac{x}{y}\right)^2\right), \quad I_2 = \frac{1}{2}\left((yv_z - zv_y)^2 + k\left(\frac{z}{y}\right)^2\right).$$

In terms of these three integrals we can obtain an explicit expression of y in terms of x, z and the integrals I_1, I_2, W:

$$y = \frac{\sqrt{2}}{W}\left(I_2x^2 + I_1z^2 \pm \sqrt{4I_1I_2 - cW^2}xz\right)^{1/2}.$$

This can be interpreted as saying that there is a superposition rule allowing us to express the general solution of the Pinney equation in terms of two independent solutions of the corresponding harmonic oscillator with time-dependent frequency.

4. The reduction method and integrability criteria

Given an equation (5) on a Lie group, it may happen that the only non-vanishing coefficients are those corresponding to a subalgebra $\mathfrak{h}$ of $\mathfrak{g}$ and then the equation reduces to a simpler equation on a subgroup, involving less coordinates. An important result is that if we know a particular solution of the problem associated in a homogeneous space, the original solution reduces to one on the isotopy subgroup.

One can show that there is an action of the group $\mathcal{G}$ of curves in G on the set of right-invariant Lie systems in G (see e.g. Ref. 6 for a geometric justification), and we can take advantage of such an action for transforming a given Lie system into another simpler one.

So, if $g(t)$ is a solution of the given Lie system and we choose a curve $g'(t)$ in the group G, and define a curve $\overline{g}(t)$ by $\overline{g}(t) = g'(t)g(t)$, then the new curve in G, $\overline{g}(t)$, determines a new Lie system. Indeed,

$$R_{\overline{g}(t)^{-1}*\overline{g}(t)}(\dot{\overline{g}}(t)) = R_{g'^{-1}(t)*g'(t)}(\dot{g}'(t)) - \sum_{\alpha=1}^{r} b_\alpha(t)\mathrm{Ad}\,(g'(t))\mathrm{a}_\alpha\,,$$

which is similar to the original one, with a different right-hand side. Therefore, the aim is to choose the curve $g'(t)$ in such a way that the new equation be simpler. For instance, we can choose a subgroup H and look for a choice of $g'(t)$ such that the right hand side lies in T_eH, and hence $\overline{g}(t) \in H$ for all

t. This can be done when we know a solution of the associated Lie system in G/H allows us to reduce the problem to one in the subgroup H.[7]

Theorem 4.1. *Each solution of (5) on the group G can be written in the form $g(t) = g_1(t)\,h(t)$, where $g_1(t)$ is a curve on G projecting onto a solution $\tilde{g}_1(t)$ for the left action λ of G on the homogeneous space G/H and $h(t)$ is a solution of an equation but for the subgroup H, given explicitly by*

$$(R_{h^{-1}*h}\dot{h})(t) = -\mathrm{Ad}(g_1^{-1}(t))\left(\sum_{\alpha=1}^{r} b_\alpha(t)\mathrm{a}_\alpha + (R_{g_1^{-1}*g_1}\dot{g}_1)(t)\right) \in T_eH\,.$$

This fact is very important because one can show that Lie systems associated with solvable Lie algebras are solvable by quadratures and therefore, given a Lie system with an arbitrary G having a solvable subgroup, we should look for a possible transformation from the original system to one which reduces to the subalgebra and therefore integrable by quadratures.

By the last Theorem there always exists a curve in G that transforms the initial Lie system into a new one related with solvable a Lie subgroup of G. Nevertheless, it can be difficult to find out a solution of the equation in M that determines this transformation. Then, to be able to obtain one is more interesting to suppose also that this transformation is a curve in a certain subset of G, i.e. a one-dimensional subgroup. It would be easier to obtain a transformation but it may be that such a transformation does not exist. In summary: *The conditions for the existence of such a transformation of a certain form are integrability conditions for the system.*

We could choose for showing this assertion a particular example: Riccati equation. One can find in the literature a lot of integrability criteria for Riccati equation,[8–10] all of them particular examples of the above method.[11] We can also consider other equivalent examples as the Pinney equation, the (generalized) Ermakov system or more relevant examples in Physics, for instance, time-dependent harmonic oscillators. The results obtained for one system are valid for the other; they are essentially conditions for the equation in the group, and all are examples of Lie systems associated with the same Lie group: $SL(2,\mathbb{R})$. Consider, for instance, the Riccati equation (3). The group $SL(2,\mathbb{R})$ contains the affine group (either the one generated by X_0 and X_1 or the one generated by X_1 and X_2), which is SOLVABLE. Therefore, a transformation from the given equation to one of this subgroup allows us to express the general solution in terms of quadratures. This happens when we know a particular solution x_1 of the given equation:

$x = x_1 + z$, what corresponds to choose

$$\bar{g}(t) = \begin{pmatrix} 1 & -x_1 \\ 0 & 1 \end{pmatrix}$$

reduces the equation to $dz/dt = (2\, b_2\, x_1 + b_1)z + b_2\, z^2$. The reduction by the knowledge of two or three quadratures has also been studied from this perspective and similarly for the Strelchenya criterion.[10]

Each Riccati equation can be considered as a curve in $\mathbb{R}^3$ and we can transform every function in $\mathbb{R}$, $x(t)$, under an element of the group $\mathcal{G}$ of smooth $SL(2,\mathbb{R})$-valued curves $\mathrm{Map}(\mathbb{R}, SL(2,\mathbb{R}))$, as follows:

$$\begin{aligned} \Theta(A, x(t)) &= \tfrac{\alpha(t)x(t)+\beta(t)}{\gamma(t)x(t)+\delta(t)}\,, \quad \text{if } x(t) \neq -\tfrac{\delta(t)}{\gamma(t)}\,, \\ \Theta(A, \infty) &= \alpha(t)/\gamma(t)\,, \quad \Theta(A, -\delta(t)/\gamma(t)) = \infty\,, \\ &\text{when } A = \begin{pmatrix} \alpha(t) & \beta(t) \\ \gamma(t) & \delta(t) \end{pmatrix} \in \mathcal{G}\,. \end{aligned}$$

The image $x'(t) = \Theta(\bar{A}(t), x(t))$ of a curve $x(t)$ solution of the given Riccati equation satisfies a new Riccati equation with the coefficients b'_2, b'_1, b'_0:

$$\begin{aligned} b'_2 &= \bar{\delta}^2\, b_2 - \bar{\delta}\bar{\gamma}\, b_1 + \bar{\gamma}^2\, b_0 + \bar{\gamma}\dot{\bar{\delta}} - \bar{\delta}\dot{\bar{\gamma}}\,, \\ b'_1 &= -2\,\bar{\beta}\bar{\delta}\, b_2 + (\bar{\alpha}\bar{\delta} + \bar{\beta}\bar{\gamma})\, b_1 - 2\,\bar{\alpha}\bar{\gamma}\, b_0 + \bar{\delta}\dot{\bar{\alpha}} - \bar{\alpha}\dot{\bar{\delta}} + \bar{\beta}\dot{\bar{\gamma}} - \bar{\gamma}\dot{\bar{\beta}}\,, \\ b'_0 &= \bar{\beta}^2\, b_2 - \bar{\alpha}\bar{\beta}\, b_1 + \bar{\alpha}^2\, b_0 + \bar{\alpha}\dot{\bar{\beta}} - \bar{\beta}\dot{\bar{\alpha}}\,. \end{aligned}$$

This expression defines an affine action of the group $\mathcal{G}$ on the set of Riccati equations or analogous Lie systems.

Lie systems in $SL(2,\mathbb{R})$ defined by a constant curve, $\mathrm{a}(t) = \sum_{\alpha=0}^{2} c_\alpha \mathrm{a}_\alpha$, are integrable and the same happens for curves of the form $\mathrm{a}(t) = D(t)\left(\sum_{\alpha=0}^{2} c_\alpha \mathrm{a}_\alpha\right)$, where D is an arbitrary function, because a time reparametrisation reduces the problem to the previous one, i.e. the system is essentially a Lie system on a one-dimensional Lie group.

We can prove the following theorem which is valid for both Riccati equation and any other Lie system with Lie group $SL(2,\mathbb{R})$:

Theorem 4.2. *The necessary and sufficient conditions for the existence of a transformation:* $y' = G(t)y$, *i.e.* $\bar{A}(t) = \begin{pmatrix} \alpha(t) & 0 \\ 0 & \alpha^{-1}(t) \end{pmatrix}$, *relating the Riccati equation (3) with (for* $b_0 b_2 \neq 0$, *with an integrable one given by*

$$\frac{dy'}{dt} = D(t)(c_0 + c_1 y' + c_2 y'^2)\,, \tag{7}$$

where c_i *are real numbers,* $c_i \in \mathbb{R}$, *is that* $c_0 c_2 \neq 0$, *and:*

$$D^2(t) c_0 c_2 = b_0(t) b_2(t), \qquad \sqrt{\frac{c_0 c_2}{b_0(t) b_2(t)}}\left(b_1(t) + \frac{1}{2}\left(\frac{\dot{b}_2(t)}{b_2(t)} - \frac{\dot{b}_0(t)}{b_0(t)}\right)\right) = c_1.$$

The unique transformation is then $y' = (b_2(t)c_0)^{1/2}(b_0(t)c_2)^{-1/2}\, y$.

As a consequence, given (7) if there are constants K, L such that

$$\sqrt{\frac{L}{b_0(t)b_2(t)}}\left(b_1(t)+\frac{1}{2}\left(\frac{\dot b_2(t)}{b_2(t)}-\frac{\dot b_0(t)}{b_0(t)}\right)\right)=K$$

then there exists a time-dependent linear change of variables transforming the given equation into the solvable Riccati equation (7) with $c_1 = K, c_0\, c_2 = L$ and $D(t)$ is given as above.

The existence of such constant K can be considered a sufficient condition for integrability of the given Riccati equation or the corresponding Milne–Pinney equation.

References

1. J.F. Cariñena and A. Ramos, Integrability of the Riccati equation from a group theoretical viewpoint, *Int. J. Mod. Phys. A* **14** (1999) 1935–1951.
2. E. Pinney, The nonlinear differential equation $y'' + p(xy + cy^{-3} = 0$, *Proc. Am. Math. Soc.* **1** (1950) 681.
3. S. Lie and G. Scheffers, *Vorlesungen über continuierliche Gruppen mit geometrischen und anderen Anwendungen* (Edited and revised by G. Scheffers, Teubner, Leipzig, 1893).
4. J.F. Cariñena, J. Grabowski and G. Marmo, *Lie–Scheffers systems: a geometric approach* (Bibliopolis, Napoli, 2000).
5. J.F. Cariñena, J. Grabowski and G. Marmo, Superposition rules, Lie theorem and partial differential equations, *Rep. Math. Phys.* **60** (2007) 237–58; arXiv:math-ph/0610013.
6. J.F. Cariñena and A. Ramos, Lie systems and Connections in fibre bundles: Applications in Quantum Mechanics, *In: 9th Int. Conf. Diff. Geom and Appl.* (2004, (J. Bureš *et al.* Eds.) Matfyzpress, Prague, 2005) 437–452.
7. J.F. Cariñena, J. Grabowski and A. Ramos, Reduction of time-dependent systems admitting a superposition principle, *Acta Appl. Math.* **66** (2001) 67–87.
8. E. Kamke, *Differentialgleichungen: Lösungsmethoden und Lösungen* (Akademische Verlagsgeselischaft, Leipzig, 1959).
9. G.M. Murphy, *Ordinary differential equations and their solutions* (Van Nostrand, New York, 1960).
10. V.M. Strelchenya, A new case of integrability of the general Riccati equation and its application to relaxation problems, *J. Phys. A: Math. Gen.* **24** (1991) 44965–4967.
11. J.F. Cariñena, A. Ramos and J. de Lucas, A geometric approach to integrability conditions for Riccati equations, *Electron. J. Diff. Eqns.* **122** (2007) 1–14.

Differential Geometry and its Applications
Proc. Conf., in Honour of Leonhard Euler, Olomouc, August 2007

Transverse Poisson structures: The subregular and minimal orbits

P.A. Damianou

Department of Mathematics and Statistics, University of Cyprus,
P.O. Box 20537, 1678 Nicosia, Cyprus
E-mail: damianou@ucy.ac.cy

H. Sabourin and P. Vanhaecke

Laboratoire de Mathematiques, UMR 6086 du CNRS, Université de Poitiers,
86962 Futuroscope Chasseneuil Cedex France
E-mail: herve.sabourin@math.univ-poitiers.fr, pol.vanhaecke@math.univ-poitiers.fr

We study the transverse Poisson structures to adjoint orbits in complex simple Lie algebras with special emphasis on the subregular and minimal orbits.

Keywords: Poisson manifolds, coadjoint orbits, simple Lie algebras, singularities.

MS classification: 53D17, 17B10, 14J17.

1. Some general results

In this first section we prove some general results on the transverse Poisson structures to coadjoint orbits. In the final two sections we specialize to the two extreme and most interesting cases, i.e. the subregular and minimal orbits. With the exception of the results on the minimal orbit which are new we only give the statements of the theorems and we refer to Ref. 5 for detailed proofs.

1.1. *Transverse Poisson Structures to adjoint orbits*

The definition of the transverse Poisson structure to a symplectic leaf in a Poisson manifold goes back to A.Weinstein.[8] It is given in the following splitting theorem:

Theorem 1.1 (A. Weinstein, 1983). *Let x_0 be a point in a Poisson manifold M. Then near x_0, M is isomorphic to a product $S \times N$ where*

S is the symplectic leaf of M, passing through x_0 and N is a submanifold of M transverse to S at x_0, which inherits a Poisson structure from M vanishing at x_0. This Poisson structure on N is called the transverse Poisson structure at x_0.

When M is the dual $\mathfrak{g}^*$ of a complex Lie algebra $\mathfrak{g}$, equipped with its standard Lie-Poisson structure, we know that the symplectic leaf through $\mu \in \mathfrak{g}^*$ is the co-adjoint orbit $\mathbf{G} \cdot \mu$ of the adjoint Lie group $\mathbf{G}$ of $\mathfrak{g}$. In this case, a natural transverse slice to $\mathbf{G} \cdot \mu$ is obtained in the following way: we choose any complement $\mathfrak{n}$ to the centralizer $\mathfrak{g}(\mu)$ of μ in $\mathfrak{g}$ and we take N to be the affine subspace $\mu + \mathfrak{n}^\perp$ of $\mathfrak{g}^*$. Since $\mathfrak{g}(\mu)^\perp = \mathrm{ad}^*_{\mathfrak{g}}\, \mu$ we have

$$T_\mu(\mathfrak{g}^*) = T_\mu(\mathbf{G} \cdot \mu) \oplus T_\mu(N),$$

so that N is indeed a transverse slice to $\mathbf{G} \cdot \mu$ at μ. Furthermore, defining on $\mathfrak{n}^\perp$ any system of linear coordinates $(q_1, \ldots, q_k)$, and using the explicit formula for Dirac reduction, one can write down explicit formulas for the Poisson matrix $\Lambda_N := (\{q_i, q_j\}_N)_{1 \le i,j \le k}$ of the transverse Poisson structure, from which it follows easily that the coefficients of Λ_N are actually rational functions in $(q_1, \ldots, q_k)$. As a corollary, in the Lie-Poisson case, the transverse Poisson structure is always rational.

One immediately wonders in which cases the Poisson structure on N is polynomial; more precisely, for which Lie algebras $\mathfrak{g}$, for which co-adjoint orbits, and for which complements $\mathfrak{n}$.

In 1989 P. A. Damianou[4] made the conjecture that in gl_n, for a specific choice of slice (orthogonal to the orbit with respect to the Killing form) the transverse Poisson structure is always polynomial. The conjecture was verified for all nilpotent orbits of gl_n, for $n \le 7$ and was proved for some special cases i.e. subregular and minimal orbits. In 2002 R. Cushman and M. Roberts[3] proved that there exists for any nilpotent adjoint orbit of a semi-simple Lie algebra a special choice of a complement $\mathfrak{n}$ such that the corresponding transverse Poisson structure is polynomial. In 2005 H. Sabourin in Ref. 6 gave a more general class of complements where the transverse structure is polynomial, using in an essential way the machinery of semi-simple Lie algebras. In this paper the transverse slice is always chosen to lie in the class of complements prescribed by Sabourin.

1.2. *Reduction to nilpotent orbits*

It turns out that the transverse Poisson structure to any adjoint orbit $\mathbf{G} \cdot x$ of a semi-simple (or reductive) algebra $\mathfrak{g}$ is essentially determined by the transverse Poisson structure to the underlying nilpotent orbit $\mathbf{G}(s) \cdot e$ defined by its Jordan decomposition $x = s+e$. In fact we have the following result:[5]

Theorem 1.2. *Let $x \in \mathfrak{g}$ be any element, $\mathbf{G} \cdot x$ its adjoint orbit and $x = s+e$ its Jordan-Chevalley decomposition. Given any complement $\mathfrak{n}_e$ of $\mathfrak{g}(x)$ in $\mathfrak{g}(s)$ and putting $\mathfrak{n} := \mathfrak{n}_s \oplus \mathfrak{n}_e$, where $\mathfrak{n}_s = \mathfrak{g}(s)^\perp$, the parallel affine spaces $N_x := x + \mathfrak{n}^\perp$ and $N := e + \mathfrak{n}^\perp$ are respectively transverse slices to the adjoint orbit $\mathbf{G} \cdot x$ in $\mathfrak{g}$ and to the nilpotent orbit $\mathbf{G}(s) \cdot e$ in $\mathfrak{g}(s)$. The translation which sends N_x to N realises an isomorphism between the transverse Poisson structure on N_x and N. The Poisson structure on both transverse slices is given by the same Poisson matrix in terms of the same affine coordinates restricted to the corresponding transverse slice.*

1.3. *The polynomial and quasi-homogeneous character of the transverse Poisson structure*

The next general result is that the transverse Poisson structure is a quasi-homogeneous Poisson structure of degree -2 with respect to a set of weights that arise from the representation theory of the corresponding simple Lie algebra. We begin with the definition of quasi-homogeneous Poisson structure; see e.g. Ref. 1.

Let $\nu = (\nu_1, \ldots, \nu_d)$ be non-negative integers. A polynomial $P \in \mathbf{C}[x_1, \ldots, x_d]$ is said to be *quasi-homogeneous* (relative to ν) if for some integer κ,

$$\forall t \in \mathbf{C}, P(t^{\nu_1}x_1, \ldots, t^{\nu_d}x_d) = t^\kappa P(x_1, \ldots, x_d).$$

The integer κ is then called the *quasi-degree* of P, denoted by $\varpi(P)$. Similarly, a polynomial Poisson structure $\{\cdot\,,\cdot\}$ on $\mathbf{C}[x_1, \ldots, x_d]$ is said to be *quasi-homogeneous* (relative to ν) if there exists $\kappa \in \mathbf{Z}$ such that, for every quasi-homogeneous polynomials F and G, their Poisson bracket $\{F, G\}$ is quasi-homogeneous of degree

$$\varpi(\{F, G\}) = \varpi(F) + \varpi(G) + \kappa;$$

Before stating the result, we need some notions from the representation theory of simple Lie algebras. First, we choose a Cartan subalgebra $\mathfrak{h}$ of the semi-simple Lie algebra $\mathfrak{g}$, with corresponding root system $\Delta(\mathfrak{h})$, from which a basis $\Pi(\mathfrak{h})$ of simple roots is selected. Let $\mathcal{O}$ be a nilpotent orbit. According to the Jacobson-Morosov-Kostant correspondence, there exists a canonical triple (h, e, f) of elements of $\mathfrak{g}$, associated with $\mathcal{O}$ and completely determined by the following properties:

(1) (h, e, f) is a $\mathfrak{sl}_2$-triple, i.e., $[h, e] = 2e$, $[h, f] = -2f$ and $[e, f] = h$;
(2) h is the characteristic of $\mathcal{O}$, i.e., $h \in \mathfrak{h}$ and $\alpha(h) \in \{0, 1, 2\}$ for every simple root $\alpha \in \Pi(\mathfrak{h})$.
(3) $\mathcal{O} = \mathbf{G} \cdot e$.

The triple (h, e, f) leads to two decompositions of $\mathfrak{g}$:

(1) A decomposition of $\mathfrak{g}$ into eigenspaces relative to ad_h. Each eigenvalue being an integer we have

$$\mathfrak{g} = \bigoplus_{i \in \mathbf{Z}} \mathfrak{g}(i),$$

where $\mathfrak{g}(i)$ is the eigenspace of ad_h that corresponds to the eigenvalue i. For example, $e \in \mathfrak{g}(2)$ and $f \in \mathfrak{g}(-2)$.

(2) Let $\mathfrak{s}$ be the Lie subalgebra of $\mathfrak{g}$ isomorphic to $\mathfrak{sl}_2$, which is generated by h, e and f. The Lie algebra $\mathfrak{g}$ is an $\mathfrak{s}$-module, hence it decomposes as

$$\mathfrak{g} = \bigoplus_{j=1}^{k} V_{n_j},$$

where each V_{n_j} is a simple $\mathfrak{s}$-module, with $n_j + 1 = \dim V_{n_j}$ and with ad_h-weights $n_j, n_j - 2, n_j - 4, \ldots, -n_j$.

Let $\mathcal{N}_h$ be the set of ad_h-invariant complements to $\mathfrak{g}(e)$.

Theorem 1.3. *Let g be a semi-simple Lie algebra, let $\mathcal{O}$ be a nilpotent orbit of $\mathfrak{g}$ with canonical triple (h, e, f), and let $\mathfrak{n}$ be in $\mathcal{N}_h$. The transverse Poisson structure on $N := e + \mathfrak{n}^{\perp}$ is a polynomial Poisson structure that is quasi-homogeneous of degree -2, relative to the quasi-degrees $n_1 + 2, \ldots, n_k + 2$, where $n_1, \ldots, n_k$ denote the highest weights of $\mathfrak{g}$ as an $\mathfrak{s}$-module.*

A transverse Poisson structure given by theorem 1.3 will be called an adjoint transverse Poisson structure, or ATP-structure.

2. The subregular case

We will give an explicit description of the Transverse Poisson structure in the case of the subregular orbit $\mathcal{O}_{sr} \subset \mathfrak{g}$, where $\mathfrak{g}$ is a semi-simple Lie algebra. Recall that an element Z in $\mathfrak{g}$ is subregular if $\dim \mathfrak{g}(Z) = Rk(\mathfrak{g}) + 2$. In this case, the generic rank of the ATP-structure on N is 2 and we know

$\dim N - 2$ independent Casimirs, namely the basic Ad-invariant functions on $\mathfrak{g}$, restricted to N. It follows that the ATP-structure is the determinantal structure (also called Nambu structure), determined by these Casimirs, up to multiplication by a function. What is much less trivial to show is that this function is actually just a non-zero *constant*. For this we will use Brieskorn's theory of simple singularities.

2.1. *Invariant functions and Casimirs*

Let $\mathcal{O}_{sr} = \mathbf{G} \cdot e$, be a subregular orbit in the semi-simple Lie algebra $\mathfrak{g}$ of rank ℓ, let (h, e, f) be the corresponding canonical $\mathfrak{sl}_2$-triple and consider the transverse slice $N := e + \mathfrak{n}^{\perp}$ to $\mathbf{G} \cdot e$, where $\mathfrak{n}$ is an ad_h-invariant complement to $\mathfrak{g}(e)$. We know that the transverse structure on N, equipped with the linear coordinates $q_1, \dots, q_k$, is a quasi-homogeneous polynomial Poisson structure of generic rank 2. Let $S(\mathfrak{g}^*)^{\mathbf{G}}$ be the algebra of Ad-invariant polynomial functions on $\mathfrak{g}$. By a classical theorem due to Chevalley, $S(\mathfrak{g}^*)^{\mathbf{G}}$ is a polynomial algebra, generated by ℓ homogeneous polynomials $(G_1, \dots, G_\ell)$, whose degree $d_i := \deg(G_i) = m_i + 1$, where $m_1, \dots, m_\ell$ are the exponents of $\mathfrak{g}$. These functions are Casimirs of the Lie-Poisson structure on $\mathfrak{g}$.

If we denote by χ_i the restriction of G_i to the transverse slice N, then it follows that these functions are independent Casimirs of the transverse Poisson structure.

2.2. *Simple singularities*

Let $\mathfrak{h}$ be a Cartan subalgebra of $\mathfrak{g}$. The Weyl group $\mathcal{W}$ acts on $\mathfrak{h}$ and the algebra $S(\mathfrak{g}^*)^{\mathbf{G}}$ of Ad-invariant polynomial functions on $\mathfrak{g}$ is isomorphic to $S(\mathfrak{h}^*)^{\mathcal{W}}$, the algebra of $\mathcal{W}$-invariant polynomial functions on $\mathfrak{h}^*$. The inclusion homomorphism $S(\mathfrak{g}^*)^{\mathbf{G}} \hookrightarrow S(\mathfrak{g}^*)$, is dual to a morphism $\mathfrak{g} \to \mathfrak{h}/\mathcal{W}$, called the *adjoint quotient*. Concretely, the adjoint quotient is given by

$$\begin{aligned} G : \mathfrak{g} &\to \mathbf{C}^\ell \\ x &\mapsto (G_1(x), G_2(x), \dots, G_\ell(x)). \end{aligned}$$

The zero-fiber $G^{-1}(0)$ of G is exactly the nilpotent variety $\mathcal{N}$ of $\mathfrak{g}$. We are interested in $N \cap \mathcal{N} = N \cap G^{-1}(0) = \chi^{-1}(0)$, which is an affine surface with an isolated, simple singularity.

Up to conjugacy, there are five types of finite subgroups of $\mathbf{SL}_2 = \mathbf{SL}_2(\mathbf{C})$, the cyclic, dihedral and three exceptional types, denoted by

$\mathcal{C}_p, \mathcal{D}_p, \mathcal{T}, \mathcal{O}$ and $\mathcal{I}$. Given such a subgroup $\mathbf{F}$, one looks at the corresponding ring of invariant polynomials $\mathbf{C}[u,v]^{\mathbf{F}}$. In each of the five cases, $\mathbf{C}[u,v]^{\mathbf{F}}$ is generated by three fundamental polynomials X, Y, Z, subject to only one relation $R(X,Y,Z) = 0$, hence the quotient space $\mathbf{C}^2/\mathbf{F}$ can be identified, as an affine surface, with the singular surface in $\mathbf{C}^3$, defined by $R = 0$. The origin is its only singular point; it is called a *(homogeneous) simple singularity.* The exceptional divisor of the minimal resolution of $\mathbf{C}^2/\mathbf{F}$ is a finite set of projective lines. If two of these lines meet, then they meet in a single point, and transversally. Moreover, the intersection pattern of these lines forms a graph that coincides with one of the simply laced Dynkin diagrams of type A_ℓ, D_ℓ, E_6, E_7 or E_8. This type is called the type of the singularity.

For the other simple Lie algebras (of type B_ℓ, C_ℓ, F_4 or G_2), there exists a similar correspondence. By definition, an *(inhomogeneous) simple singularity* of type Δ is a couple (V, Γ) consisting of a homogeneous simple singularity $V = \mathbf{C}^2/\mathbf{F}$ and a group $\Gamma = \mathbf{F}'/\mathbf{F}$ of automorphisms of V.

We can now state the following extension of a theorem of Brieskorn, which is due to Slodowy[7]

Theorem 2.1. *Let $\mathfrak{g}$ be a simple complex Lie algebra, with Dynkin diagram of type Δ. Let $\mathcal{O}_{sr} = \mathbf{G}\cdot e$ be the subregular orbit and $N = e + \mathfrak{n}^\perp$ a transverse slice to $\mathbf{G}\cdot e$. The surface $N \cap \mathcal{N} = \chi^{-1}(0)$ has a (homogeneous or inhomogeneous) simple singularity of type Δ.*

2.3. *The determinantal Poisson structure*

In terms of linear coordinates $q_1, q_2, \ldots, q_{\ell+2}$ on $\mathbf{C}^{\ell+2}$, the formula

$$\{f,g\}_{det} := \frac{df \wedge dg \wedge d\chi_1 \wedge \cdots \wedge d\chi_\ell}{dq_1 \wedge dq_2 \wedge \cdots \wedge dq_{\ell+2}}$$

defines a Poisson bracket on $\mathbf{C}^{\ell+2}$ with Casimirs $\chi_1, \ldots, \chi_\ell$.

In our case it means that we have two polynomial Poisson structures on the transverse slice N which have $\chi_1, \ldots, \chi_\ell$ as Casimirs on $N \cong \mathbf{C}^{\ell+2}$, namely the transverse Poisson structure and the determinantal structure, constructed by using these Casimirs. The fact that both structures have the same quasi-degree -2, combined with the fact that the singularity of $\chi^{-1}(0)$ is isolated yields the following result :

Theorem 2.2. *Let $\mathcal{O}_{sr}$ be the subregular nilpotent adjoint orbit of a complex semi-simple Lie algebra $\mathfrak{g}$ and let (h,e,f) be the canonical triple, associated to $\mathcal{O}_{sr}$. Let $N = e + \mathfrak{n}^\perp$ be a transverse slice to $\mathcal{O}_{sr}$, where $\mathfrak{n}$ is* *an* ad_h*-invariant complementary subspace to* $\mathfrak{g}(e)$. *Let* $\{\cdot,\cdot\}_N$ *and* $\{\cdot,\cdot\}_{det}$ *denote respectively the transverse Poisson structure and the determinantal structure on* N. *Then* $\{\cdot,\cdot\}_N = c\{\cdot,\cdot\}_{det}$ *for some* $c \in \mathbf{C}^*$.

2.4. *Reduction to a* 3×3 *Poisson matrix*

Let $\mathcal{O}_{sr}$ be the subregular nilpotent adjoint orbit of a complex semi-simple Lie algebra $\mathfrak{g}$ of rank ℓ.

Our goal now is to show that, in well-chosen coordinates, the transverse Poisson structure $\{\cdot,\cdot\}_N$ on N is essentially given by a 3×3 skew-symmetric matrix, closely related to the polynomial that defines the singularity. The non-Poisson part of the following theorem is due to Brieskorn in the case of ADE singularities and was extended later to the other types of simple Lie algebras by Slodowy.[7] Brieskorn's theorem says that the map $\chi : N \to \mathbf{C}^\ell$, which is the restriction of the adjoint quotient to the slice N, is a semi-universal deformation of the singular surface $N \cap \mathcal{N}$. Using these results and the determinantal formula we obtain the following result:

Theorem 2.3. *After possibly relabeling the coordinates* q_i *and the Casimirs* χ_i, *the* $\ell + 2$ *functions*

$$\chi_i, 1 \le i \le \ell - 1, \quad \textit{and} \quad q_\ell, q_{\ell+1}, q_{\ell+2}$$

form a system of coordinates on the affine space N. *The Poisson matrix of the transverse Poisson structure on* N *takes, in terms of these coordinates, the form*

$$\Lambda_N = \begin{pmatrix} 0 & 0 \\ 0 & \Omega \end{pmatrix},$$

where

$$\Omega = \begin{pmatrix} 0 & \dfrac{\partial \chi_\ell}{\partial q_{\ell+2}} & -\dfrac{\partial \chi_\ell}{\partial q_{\ell+1}} \\ -\dfrac{\partial \chi_\ell}{\partial q_{\ell+2}} & 0 & \dfrac{\partial \chi_\ell}{\partial q_\ell} \\ \dfrac{\partial \chi_\ell}{\partial q_{\ell+1}} & -\dfrac{\partial \chi_\ell}{\partial q_\ell} & 0 \end{pmatrix}.$$

It has the polynomial χ_ℓ *as Casimir, which reduces to the polynomial which defines the singularity, when setting* $\chi_j = 0$ *for* $j = 1, 2, \ldots, \ell - 1$.

3. The minimal orbit

In this section we consider the transverse Poisson structure to the minimal orbit $\mathcal{O}_m$ in an arbitrary semi-simple Lie algebra $\mathfrak{g}$, whose Killing form will be denoted by $\langle \cdot \mid \cdot \rangle$. This orbit is the nilpotent orbit of minimal dimension (besides the trivial orbit $\{0\}$). It is unique and is generated by a root vector E_m, associated to a highest root, with respect to a fixed Cartan subalgebra $\mathfrak{h}$ and a choice of simple roots. Let (H_m, E_m, F_m) denote the canonical triplet, associated to $\mathcal{O}_m$ and let $\mathfrak{g}^{E_m}$ denote the centralizer of E_m in $\mathfrak{g}$.

3.1. *Properties of the minimal orbit*

We first list the properties of the minimal orbit that we will use (see Ref. 2 for proofs). The Lie algebra $\mathfrak{g}$ decomposes in eigenspaces, relatively to ad_{H_m}, as follows:

$$\mathfrak{g} = \mathfrak{g}(-2) \oplus \mathfrak{g}(-1) \oplus \mathfrak{g}(0) \oplus \mathfrak{g}(1) \oplus \mathfrak{g}(2).$$

This decomposition has the following properties.

(F1) $\mathfrak{g}(2) = \mathbf{C}.E_m$;
(F2) $\mathfrak{g}(-2) = \mathbf{C}.F_m$;
(F3) $\mathfrak{g}(0)$ is a reductive subalgebra of $\mathfrak{g}$ and $\mathfrak{g}(0) = \mathfrak{g}^{E_m}(0) \oplus \mathbf{C}.H_m$, where $\mathfrak{g}^{E_m}(0) := \mathfrak{g}^{E_m} \cap \mathfrak{g}(0)$;
(F4) $\mathfrak{g}^{E_m}(0) \perp \mathbf{C}.H_m$;
(F5) $\mathfrak{g}^{E_m}(0)$ is reductive, $\mathfrak{g}^{E_m}(0) = \mathbf{Z}_e \oplus \mathfrak{m}_e$ where $\mathbf{Z}_e \subset \mathfrak{h}$ is the center of $\mathfrak{g}^{E_m}(0)$ and $\mathfrak{m}_e$ its semi-simple part;
(F6) Let $\mathfrak{n}_e^+ := \mathfrak{m}_e \cap \mathfrak{n}^+ = \langle X_\alpha,\ \alpha(H_m) = 0 \rangle$, let $\mathfrak{n}_e^-$ denote its opposite and let $\mathfrak{h}_e := \mathfrak{m}_e \cap \mathfrak{h}$. Then $\mathfrak{m}_e = \mathfrak{n}_e^- \oplus \mathfrak{h}_e \oplus \mathfrak{n}_e^+$ and $\mathfrak{h} = \mathfrak{h}_e \oplus \mathbf{Z}_e \oplus \mathbf{C}.H_m$.
(F7) Let K_f be the skew-symmetric bilinear form defined by $K_f(X,Y) := \langle F_m | [X,Y] \rangle$. The space $\mathfrak{g}(1)$, equipped with K_f, is a symplectic space of dimension $2s$. Moreover, we can choose a basis $(Z_1, \ldots, Z_s, Z_{s+1}, \ldots, Z_{2s})$ such that each vector Z_i is a root vector X_{α_i}, associated to a positive root, and

$$[Z_i, Z_j] = 0 = [Z_{i+s}, Z_{j+s}], \qquad [Z_i, Z_{2s+1-j}] = \delta_{ij} E_m,$$

for all i, j with $1 \leq i, j \leq s$;
(F8) The same result as in (F7) holds for the space $\mathfrak{g}(-1)$ equipped with the bilinear form $K_e(X,Y) = \langle E_m | [X,Y] \rangle$. The corresponding basis, defined by the same properties as in (F7), will be denoted by $X_1, \ldots, X_{2s}$.

3.2. *The ATP-structure associated to $\mathcal{O}_m$*

Since the centralizer $\mathfrak{g}^{E_m}$ is given by

$$\mathfrak{g}^{E_m} = \mathfrak{g}^{E_m}(0) \oplus \mathfrak{g}(1) \oplus \mathfrak{g}(2),$$

we have a decomposition $\mathfrak{g} = \mathfrak{g}^{E_m} \oplus \mathfrak{n}$, where $\mathfrak{n} = \mathfrak{g}(-2) \oplus \mathfrak{g}(-1) \oplus \mathbf{C}.H_m$. It is clear that the Lie subalgebra $\mathfrak{n}$ of $\mathfrak{g}$ is ad_{H_m}-invariant, has dimension $2s+2$ and that its orthogonal is given by $\mathfrak{n}^\perp = \mathfrak{g}(-2) \oplus \mathfrak{g}(-1) \oplus \mathfrak{g}^{E_m}(0)$. Thus, we choose $N_m := E_m + \mathfrak{n}^\perp$ as transverse slice to $\mathcal{O}_m$. The Poisson matrix of the corresponding ATP-Poisson structure on N_m is given, at $n \in N_m$, by the Dirac formula

$$\Lambda_m(n) = A(n) + B(n)C^{-1}(n)B(n)^T,$$

where the matrices $A(n), B(n), C(n)$ are constructed as follows. Let $Z_{2s+2}, \dots, Z_{2s+p+1}$ be a basis of $\mathfrak{g}^{E_m}(0)$ and let $Z_{2s+1} := E_m$, so that $Z_1, \dots, Z_{2s+p+1}$ is a basis of $\mathfrak{g}^{E_m}$. Also, set $X_{2s+1} := \frac{1}{2}H_m$ and $X_{2s+2} := F_m$. Then, $X_1, \dots, X_{2s+2}$ is a basis of $\mathfrak{n}$. In terms of these bases, the matrices $A(n), B(n)$ and $C(n)$ are given by

$$A(n)_{ij} = \langle n | [Z_i, Z_j] \rangle, \quad B(n)_{ik} = \langle n | [Z_i, X_k] \rangle, \quad C(n)_{kl} = \langle n | [X_k, X_l] \rangle,$$

where $1 \le i, j \le 2s + p + 1$ and $1 \le k, l \le 2s + 2$. In terms of the $2q \times 2q$-matrices J_q, defined by

$$J_q = \begin{pmatrix} 0 & & \cdots & & 0 & 1 \\ & & & & \iddots & 0 \\ \vdots & & \iddots & 1 & & \\ & & -1 & \iddots & & \vdots \\ 0 & \iddots & & & & \\ -1 & 0 & & \cdots & & 0 \end{pmatrix},$$

we have:

$$C(n) = \begin{pmatrix} J_s & 0 \\ 0 & -J_2 \end{pmatrix} \qquad \text{and} \qquad C^{-1}(n) = -C(n).$$

Proposition 3.1. *The ATP-structure of the minimal orbit $\mathcal{O}_m$ is the sum of two Poisson structures $\Lambda_m = \mathcal{A} + \mathcal{Q}$, where*

1 *$\mathcal{A}$ is a linear Poisson structure, isomorphic to the Lie-Poisson structure on the dual of the Lie algebra $\mathfrak{g}^{E_m}$;*

2 *$\mathcal{Q}$ is a quadratic Poisson bracket, whose Poisson matrix at $n \in N_m$ is given by $Q(n) := B(n)C^{-1}(n)B(n)^T$. Moreover, its generic rank is* $\dim \mathcal{O}_m - 2$.

Proof. The first statement is clear because the matrix $A(n)$ in the Dirac formula is the matrix of the Lie-Poisson structure on $(\mathfrak{g}^{E_m})^*$. Both matrices A and $\Lambda_m = A+Q$ are Poisson. Moreover, since $C(n)$ is constant (independent of $n \in N_m$), all entries of Q are quadratic polynomials. Since Q is the highest degree term of the Poisson matrix Λ, it is also a Poisson matrix. Therefore $\mathcal{A}$ and $\mathcal{Q}$ are compatible Poisson structures. We show that the rank of $\mathcal{Q}$ is $\dim \mathcal{O}_m - 2$. To do this let n be any element in N_m. We will restrict our attention now to the matrix $B(n)$. From the definitions, we get

$$[Z_i, F_m] = [Z_i, H_m] = 0, \qquad [Z_i, X_k] \in \mathfrak{g}(-1),$$

for i, k such that $2s+2 \le i \le 2s+p+1$ and $1 \le k \le 2s$. It implies that $B(n)_{ik} = 0$ for the latter values of i and k. Thus, the last p rows of $B(n)$ are zero,

$$B(n) = \begin{pmatrix} D(n) \\ 0 \end{pmatrix},$$

where $D(n)$ is the $(2s+1) \times (2s+2)$-matrix, whose entries are given by

$$D(n)_{ik} = \langle n \mid [Z_i, X_k] \rangle,$$

where $1 \le i \le 2s+1$ and $1 \le k \le 2s+2$. If $1 \le i \le 2s$ then $[Z_i, X_{2s+2}] \in \mathfrak{g}(-1)$ and $[E_m, X_{2s+2}] = H_m$. Using (F4), it follows that the last column of the matrix $D(n)$ is zero,

$$D(n) = \left(D'(n)\ 0\right).$$

Thus, for $n \in E_m + \mathfrak{n}^\perp$, we have

$$Q(n) = \begin{pmatrix} D(n)C^{-1}(n)D(n)^T & 0 \\ 0 & 0 \end{pmatrix} = \begin{pmatrix} D'(n)C'(n)D'(n)^T & 0 \\ 0 & 0 \end{pmatrix},$$

where $C'(n)$ is the submatrix of $C^{-1}(n)$, obtained by removing its last column and its last row. This implies that

$$\mathrm{Rk}(Q(n)) = \mathrm{Rk}(D'(n)C'(n)D'(n)^T) \le 2s+1,$$

for all $n \in E_m + \mathfrak{n}^\perp$. Since Q is skew-symmetric, its rank is at most $2s = \dim \mathcal{O}_m - 2$. We show that there exists a point $n \in N_m$ where the rank of Q is $2s$. Recall that the symplectic vector space $\mathfrak{g}(1)$ is generated by root vectors, $Z_i = X_{\alpha_i}$, for $1 \le i \le 2s$. So, we can define $P_i := \mathfrak{n}^\perp \cap H_{\alpha_i}^\perp$, for $1 \le i \le 2s$, which are hyperplanes of $\mathfrak{n}^\perp$. Let P denote their union. Let $n \in \mathfrak{n}^\perp \backslash P + E_m$. Then

1. If $1 \le i \ne k \le 2s$ then $[Z_i, X_k] \subset \mathfrak{n}_e^- \oplus \mathfrak{n}_e^+ \subset V$, so $D'(n)_{ik} = \langle n \mid [Z_i, X_k] \rangle = 0$;

2. For all i with $1 \leq i \leq 2s$, $D'(n)_{ii} = \langle n \mid H_{\alpha_i} \rangle \neq 0$;
3. For all k with $1 \leq k \leq 2s+1$, $[Z_{2s+1}, X_k] \in \mathfrak{g}(1) \oplus \mathfrak{g}(2)$, so that $D'(n)_{2s+1,k} = 0$.

Consider the submatrix C'' of C', and similarly D'' of D', obtained by removing its last row and its last column. Then the upper left $2s \times 2s$-minor of $D'(n)C'(n)D'(n)^T$ is $\det\left(D''(n)C''(n)D''(n)^T\right)$, which is non-zero, This proves that $\mathrm{Rk}(Q(n)) = 2s = \dim \mathcal{O}_m - 2$, so that $\mathrm{Rk}(\mathcal{Q}) = \dim \mathcal{O}_m - 2$.□

References

1. M. Adler, P. Van Moerbeke and P. Vanhaecke, *Algebraic integrability, Painlevé geometry and Lie algebra* (Ergebnisse der Mathematik und ihrer grenzgebiete 47, Springer-Verlag, Berlin Heidelberg, 2004).
2. D.H. Collingwood and W. Mc Govern, *Nilpotent orbits in semisimple Lie algebras* (Van Nostrand Reinhold, Mathematics series, 1993).
3. R. Cushman and M. Roberts, Poisson structures transverse to coadjoint orbits, *Bull.Sci.Math.* **126** (2002) 525–534.
4. P.A. Damianou, Nonlinear Poisson brackets, Ph.D. Dissertation, University of Arizona, 1989.
5. P.A. Damianou, H. Sabourin and P. Vanhaecke, Transverse Poisson structures to adjoint orbits in semi-simple Lie algebras, *Pac. J. of Math.* **232** (2007) 111–139.
6. H. Sabourin, Sur la structure transverse à une orbite nilpotente adjointe, *Canad.J.Math* **57** (2005) 750–770.
7. P. Slodowy, *Simple singularities and simple algebraic groups* (Lect.Notes in Math. 815, Springer Verlag, Berlin 1980).
8. A. Weinstein, Local structure of Poisson manifolds, *J. Diffential Geom.* **18** (1983) 523–557.

Differential Geometry and its Applications
Proc. Conf., in Honour of Leonhard Euler, Olomouc, August 2007

On first-order differential invariants of the non-conjugate subgroups of the Poincaré group $P(1,4)$

V.M. Fedorchuk

Institute of Mathematics, Pedagogical Academy,
Podchorążych 2, 30-084, Kraków, Poland;
Pidstryhach Institute of Applied Problems of Mechanics and Mathematics
of the National Ukrainian Academy of Sciences,
Naukova Street 3b, Lviv, 79601, Ukraine
E-mail: vasfed@gmail.com

V.I. Fedorchuk

Pidstryhach Institute of Applied Problems of Mechanics and Mathematics
of the National Ukrainian Academy of Sciences,
Naukova Street 3b, Lviv, 79601, Ukraine
E-mail: volfed@gmail.com

The number of non-equivalent functional bases of the first-order differential invariants for all non-conjugate subgroups of the group $P(1,4)$ is established. All these bases have been classified according to their dimensions. Some application of the results obtained are considered.

Keywords: Poincaré group $P(1,4)$, Lie algebra of the group $P(1,4)$, non-conjugate subgroups of the group, first-order differential invariants, non-equivalent functional bases.

MS classification: 17B05, 17B81.

1. Introduction

It is well known that the differential invariants of Lie groups of point transformations play an important role in geometry, group analysis of differential equations, theoretical and mathematical physics, gas dynamics, etc. (see, for example, Refs. 1–23).

The development of theoretical and mathematical physics required various extensions of the four-dimensional Minkowski space $M(1,3)$ and, correspondingly, various extensions of a Poincaré group $P(1,3)$. The natural extension of this group is a generalized Poincaré group $P(1,4)$. The group $P(1,4)$ is a group of rotations and translations of the five-dimensional

Minkowski space $M(1,4)$. This group has many applications in theoretical and mathematical physics (see, for example, Refs. 24–26). The group $P(1,4)$ has many subgroups used in theoretical physics.[27–31] Among these subgroups, there are the Poincaré group $P(1,3)$ (the symmetry group of relativistic physics) and the extended Galilei group $\widetilde{G}(1,3)$ (see also Ref. 32) (the symmetry group of non-relativistic physics). Therefore, the results obtained with the help of the subgroup of the group $P(1,4)$ will be useful in relativistic and non-relativistic physics.

The articles Refs. 33–39 are devoted to the construction and the classification of functional bases of the first-order differential invariants of the splitting and non-splitting subgroups of the group $P(1,4)$. It should be noted that these results were obtained for the splitting and non-splitting subgroups separately.

The purpose of this paper is to present some results obtained, which would be valid to all non-conjugate subgroups of the group $P(1,4)$. We plan to find the number of the non-equivalent functional bases of the first-order differential invariants for all non-conjugate subgroups of the group $P(1,4)$. All these bases will be classified according to their dimensions.

In order to present some of the results obtained, we have to consider the Lie algebra of the group $P(1,4)$.

2. The Lie algebra of the group $P(1,4)$ and its non-conjugate subalgebras

The Lie algebra of the group $P(1,4)$ is given by the 15 basis elements $M_{\mu\nu} = -M_{\nu\mu}$ $(\mu,\nu = 0,1,2,3,4)$ and P'_μ $(\mu = 0,1,2,3,4)$, satisfying the commutation relations

$$[P'_\mu, P'_\nu] = 0, \qquad [M'_{\mu\nu}, P'_\sigma] = g_{\mu\sigma}P'_\nu - g_{\nu\sigma}P'_\mu,$$

$$[M'_{\mu\nu}, M'_{\rho\sigma}] = g_{\mu\rho}M'_{\nu\sigma} + g_{\nu\sigma}M'_{\mu\rho} - g_{\nu\rho}M'_{\mu\sigma} - g_{\mu\sigma}M'_{\nu\rho},$$

where $g_{00} = -g_{11} = -g_{22} = -g_{33} = -g_{44} = 1$, $g_{\mu\nu} = 0$, if $\mu \neq \nu$. Here and in what follows, $M'_{\mu\nu} = iM_{\mu\nu}$.

We consider the following representation of the Lie algebra of the group $P(1,4)$:

$$P'_0 = \frac{\partial}{\partial x_0}, \quad P'_1 = -\frac{\partial}{\partial x_1}, \quad P'_2 = -\frac{\partial}{\partial x_2}, \quad P'_3 = -\frac{\partial}{\partial x_3},$$

$$P'_4 = -\frac{\partial}{\partial x_4}, \quad M'_{\mu\nu} = -\left(x_\mu P'_\nu - x_\nu P'_\mu\right).$$

In order to study the non-conjugate subalgebras of the Lie algebra of the group $P(1,4)$, we used the method proposed in Ref. 40. These subalgebras have been described in Refs. 27–31. They have been divided by splitting and non-splitting ones.

Splitting subalgebras $P_{i,a}$ of the Lie algebra of the group $P(1,4)$ can be written in the following form:

$$P_{i,a} = F_i \overset{\circ}{+} N_{ia},$$

where F_i are subalgebras of the Lie algebra of the group $O(1,4)$, N_{ia} are subalgebras of the Lie algebra of the translations group $T(5) \subset P(1,4)$.

Non-splitting subalgebras $\widetilde{P}_{j,k}$ are subalgebras, for which the basis can be choosen in the form:

$$\widetilde{B}_k = B_k + \sum_i c_{ki} X_i, \qquad \sum_j d_{rj} X_j,$$

where c_{ki} and d_{rj} are fixed real constants (not equal zero simultaneously). B_k are bases of subalgebras of the Lie algebra of the group $O(1,4)$, X_i are bases of subalgebras of the Lie algebra of the group $T(5)$.

One of the important consequences of the study of the non-conjugate subalgebras of the Lie algebra of the group $P(1,4)$ is that the Lie algebra of the group $P(1,4)$ contains, as subalgebras, the Lie algebras of the following important for theoretical and mathematical physics groups: $P(1,3)$, $\widetilde{G}(1,3)$ (see also Ref. 32), $O(1,4)$, $O(4)$, $E(4)$, etc.

Further, instead of $M'_{\mu\nu} = -M'_{\nu\mu}$ $(\mu, \nu = 0,1,2,3,4)$ and P'_μ $(\mu = 0,1,2,3,4)$, we will use the following basis elements:

$$G = M'_{40}, \quad L_1 = M'_{32}, \quad L_2 = -M'_{31}, \quad L_3 = M'_{21},$$

$$P_a = M'_{4a} - M'_{a0}, \quad C_a = M'_{4a} + M'_{a0}, \quad (a = 1,2,3),$$

$$X_0 = \frac{1}{2}\left(P'_0 - P'_4\right), \quad X_k = P'_k \quad (k = 1,2,3), \quad X_4 = \frac{1}{2}\left(P'_0 + P'_4\right).$$

The Lie algebra of the group $\widetilde{G}(1,3)$ is generated by the following basis elements:

$$L_1,\ L_2,\ L_3,\ P_1,\ P_2,\ P_3,\ X_0,\ X_1,\ X_2,\ X_3,\ X_4.$$

3. The non-equivalent functional bases of first-order differential invariants of non-conjugate subgroups of the group $P(1,4)$

For all non-conjugate subgroups of the group $P(1,4)$, the functional bases of the first-order differential invariants have been constructed. In the construction of these bases, it has turned out that different non-conjugate subalgebras of the Lie algebra of the group $P(1,4)$ may have the same ones. Consequently, there is no one-to-one correspondence between the non-conjugate subalgebras of the Lie algebra of the group $P(1,4)$ and their respective functional bases of the first-order differential invariants. Moreover, some of the functional bases (which are of the same dimension) may be equivalent. Our aim is to obtain non-equivalent functional bases only. Let $\{J_1^{(1)}, J_2^{(1)}, ..., J_t^{(1)}\}$ and $\{J_1^{(2)}, J_2^{(2)}, ..., J_t^{(2)}\}$ be the functional bases of the first-order differential invariants, which correspond to the subalgebras L^1 and L^2 of the Lie algebra of the group $P(1,4)$.

Definition 3.1. We say that the functional bases $\{J_1^{(1)}, J_2^{(1)}, ..., J_t^{(1)}\}$ and $\{J_1^{(2)}, J_2^{(2)}, ..., J_t^{(2)}\}$ be equivalent if there exist smooth functions $f_1, f_2, ..., f_t$ and $g_1, g_2, ..., g_t$ such that

$$\begin{aligned} J_1^{(2)} &= f_1(J_1^{(1)}, J_2^{(1)}, ..., J_t^{(1)}) \\ J_2^{(2)} &= f_2(J_1^{(1)}, J_2^{(1)}, ..., J_t^{(1)}) \\ &\dots\dots\dots \\ J_t^{(2)} &= f_t(J_1^{(1)}, J_2^{(1)}, ..., J_t^{(1)}) \end{aligned} \quad \text{and} \quad \begin{aligned} J_1^{(1)} &= g_1(J_1^{(2)}, J_2^{(2)}, ..., J_t^{(2)}) \\ J_2^{(1)} &= g_2(J_1^{(2)}, J_2^{(2)}, ..., J_t^{(2)}) \\ &\dots\dots\dots \\ J_t^{(1)} &= g_t(J_1^{(2)}, J_2^{(2)}, ..., J_t^{(2)}). \end{aligned}$$

Lemma 3.1. *Two functional bases* $\{J_1^{(1)}, J_2^{(1)}, ..., J_t^{(1)}\}$ *and* $\{J_1^{(2)}, J_2^{(2)}, ..., J_t^{(2)}\}$ *are equivalent if and only if they satisfy the following conditions:*

$$\begin{aligned} \widetilde{X}_1^{(1)} J_1^{(2)} = 0, \widetilde{X}_1^{(1)} J_2^{(2)} = 0, ..., \widetilde{X}_{r_1}^{(1)} J_t^{(2)} = 0 \\ \widetilde{X}_1^{(2)} J_1^{(1)} = 0, \widetilde{X}_1^{(2)} J_2^{(1)} = 0, ..., \widetilde{X}_{r_2}^{(2)} J_t^{(1)} = 0 \end{aligned} \qquad (*)$$

where $\{\widetilde{X}_1^{(1)}, \widetilde{X}_2^{(1)}, ..., \widetilde{X}_{r_1}^{(1)}\}$, $\{\widetilde{X}_1^{(2)}, \widetilde{X}_2^{(2)}, ..., \widetilde{X}_{r_2}^{(2)}\}$ *are the first-prolonged bases operators of the Lie subalgebra* L^1 *and* L^2*, respectively;* r_1, r_2 *are the dimensions of the subalgebras* L^1 *and* L^2.

Proof. The Proof of this Lemma can be found in Ref. 39. □

Proposition 3.1. *There exist 498 non-equivalent functional bases of the first-order differential invariants for the non-conjugate subgroups of the group* $P(1,4)$.

Proof. The list of the non-conjugate (the conjugation was considered under the group $P(1,4)$) subalgebras of the Lie algebra of the group $P(1,4)$ contains 555 ones.[41]

As resulting from the calculation of the general ranks of matrices, which contain coordinates of the one-prolonged basis elements of the subalgebras of the Lie algebra considered, and using the theorem on number of invariants of the Lie group of the point transformations (see, for example, Refs 6,8) we make sure that all of the non-conjugate subalgebras of the Lie algebra of the group $P(1,4)$ have the functional bases of the first-order differential invariants. Thus, there are 555 functional bases of the first-order differential invariants. Among them, there are equivalent ones. Equivalent functional bases can only be among those, which have the same dimensions.

Let L^1 be a non-conjugate subalgebra of the Lie algebra of the group $P(1,4)$, which has a t-dimensional functional basis of the first-order differential invariants $\{J_1^{(1)}, J_2^{(1)}, ..., J_t^{(1)}\}$. To find the bases, which are equivalent to $\{J_1^{(1)}, J_2^{(1)}, ..., J_t^{(1)}\}$, we use the Lemma. Let $\{J_1^{(2)}, J_2^{(2)}, ..., J_t^{(2)}\}$ be the t-dimensional functional basis of the first-order differential invariants of the other non-conjugate subalgebra L^2. Based on the Lemma, if these functional bases satisfy the conditions (*), then, the considered bases are equivalent. Otherwise the considered bases are not equivalent. In the analogous manner, we check whether or not other t-dimensional functional bases of the first-order differential invariants are equivalent to the $\{J_1^{(1)}, J_2^{(1)}, ..., J_t^{(1)}\}$. In this way, we obtain all t-dimensional functional bases, which are equivalent to $\{J_1^{(1)}, J_2^{(1)}, ..., J_t^{(1)}\}$.

In the analogous manner, we construct classes of the equivalent functional bases of other dimensions.

The direct check provides 498 non-equivalent functional bases of the first-order differential invariants for the non-conjugate subgroups of the group $P(1,4)$. The Proposition is proved. □

It is impossible to present here all non-equivalent functional bases. Therefore, below we only give a short review of the results obtained.

Let N be a number of non-equivalent functional bases invariant under non-conjugate subgroups of the group $P(1,4)$,

N_s be a number of non-equivalent functional bases invariant under splitting non-conjugate subgroups of the group $P(1,4)$,

N_{ns} be a number of non-equivalent functional bases invariant under non-splitting non-conjugate subgroups of the group $P(1,4)$.

3.1. *One-dimensional non-equivalent functional bases*

There are no one-dimensional non-equivalent functional bases.

3.2. *Two-dimensional non-equivalent functional bases*

There exists one two-dimensional non-equivalent functional basis. In this case $N_{ns} = 0$. Thus, $N = N_s = 1$. The functional basis is invariant under the non-galilean subalgebras of the Lie algebra of the group $P(1,4)$.

Below, for some of the non-conjugate subalgebras of the Lie algebra of the group $P(1,4)$, we write their basis elements and the corresponding functional bases of differential invariants.

$\langle G,\ P_1,\ P_2,\ P_3,\ X_0,\ X_1,\ X_2,\ X_3,\ X_4\rangle$,

$\langle L_3 + eG,\ P_1,\ P_2,\ P_3,\ X_0,\ X_1,\ X_2,\ X_3,\ X_4,\ e > 0\rangle$,

$\langle G,\ L_3,\ P_1,\ P_2,\ P_3,\ X_0,\ X_1,\ X_2,\ X_3,\ X_4\rangle$,

$\langle G,\ L_1,\ L_2,\ L_3,\ P_1,\ P_2,\ P_3,\ X_0,\ X_1,\ X_2,\ X_3,\ X_4\rangle$,

$\langle G,\ C_1,\ C_2,\ C_3,\ L_1,\ L_2,\ L_3,\ P_1,\ P_2,\ P_3,\ X_0,\ X_1,\ X_2,\ X_3,\ X_4\rangle$,

$J_1 = u, \quad J_2 = u_0^2 - u_1^2 - u_2^2 - u_3^2 - u_4^2\ ;$

$$u_\mu \equiv \frac{\partial u}{\partial x_\mu}, \quad \mu = 0,1,2,3,4.$$

3.3. *Three-dimensional non-equivalent functional bases*

There exist $N = 7$ non-equivalent three-dimensional functional bases. In this case $N_s = 7$, $N_{ns} = 4$. Thus, $N_s + N_{ns} - N = 4$ functional bases are invariant under splitting and non-splitting subalgebras. Let us give an example:

$\langle G,\ P_1,\ P_2,\ X_0,\ X_1,\ X_2,\ X_3,\ X_4\rangle$,

$\langle L_3 + eG,\ P_1,\ P_2,\ X_0,\ X_1,\ X_2,\ X_3,\ X_4,\ e > 0\rangle$,

$\langle G,\ L_3,\ P_1,\ P_2,\ X_0,\ X_1,\ X_2,\ X_3,\ X_4\rangle$,

$\langle G + a_3X_3,\ L_3 + d_3X_3,\ P_1,\ P_2,\ X_0,\ X_1,\ X_2,\ X_4,\ a_3 < 0,\ d_3 < 0\rangle$,

$\langle G,\ L_3 + d_3X_3,\ P_1,\ P_2,\ X_0,\ X_1,\ X_2,\ X_4,\ d_3 < 0\rangle$,

$J_1 = u, \quad J_2 = u_3, \quad J_3 = u_0^2 - u_1^2 - u_2^2 - u_4^2\ .$

It should be noted that among the three-dimensional functional bases there is one basis invariant under the subalgebras of the Lie algebra of the group $\widetilde{G}(1,3)$ and 6 bases invariant under the non-galilean subalgebras.

3.4. *Four-dimensional non-equivalent functional bases*

There exist $N = 25$ non-equivalent four-dimensional functional bases. In this case $N_s = 18$, $N_{ns} = 10$. Thus, $N_s + N_{ns} - N = 3$ functional bases are invariant under splitting and non-splitting subalgebras. Let us give an example:

$\langle P_1,\ P_2,\ P_3,\ X_1,\ X_2,\ X_3,\ X_4\rangle,$

$\langle L_3 - P_3,\ P_1,\ P_2,\ X_1,\ X_2,\ X_3,\ X_4\rangle,$

$\langle L_3,\ P_1,\ P_2,\ P_3,\ X_1,\ X_2,\ X_3,\ X_4\rangle,$

$\langle L_1,\ L_2,\ L_3,\ P_1,\ P_2,\ P_3,\ X_1,\ X_2,\ X_3,\ X_4\rangle,$

$\langle L_3 + d_3X_3,\ P_1,\ P_2,\ P_3,\ X_1,\ X_2,\ X_4,\ d_3 < 0\rangle,$

$J_1 = x_0 + x_4, \quad J_2 = u, \quad J_3 = u_0 - u_4, \quad J_4 = u_0^2 - u_1^2 - u_2^2 - u_3^2 - u_4^2\ .$

It should be noted that among the four-dimensional functional bases there are 7 bases invariant under the subalgebras of the Lie algebra of the group $\widetilde{G}(1,3)$ and 18 bases invariant under the non-galilean subalgebras.

3.5. *Five-dimensional non-equivalent functional bases*

There exist $N = 63$ non-equivalent five-dimensional functional bases. In this case $N_s = 37$, $N_{ns} = 28$. Thus, $N_s + N_{ns} - N = 2$ functional bases are invariant under splitting and non-splitting subalgebras. Let us give an example:

$\langle P_1,\ P_2,\ X_1,\ X_2,\ X_3,\ X_4\rangle,$

$\langle L_3,\ P_1,\ P_2,\ X_1,\ X_2,\ X_3,\ X_4\rangle,$

$\langle L_3 + d_3X_3,\ P_1,\ P_2,\ X_1,\ X_2,\ X_4,\ d_3 < 0\rangle,$

$J_1 = x_0 + x_4, \quad J_2 = u, \quad J_3 = u_3, \quad J_4 = u_0 - u_4,$

$J_5 = u_0^2 - u_1^2 - u_2^2 - u_4^2\ .$

It should be noted that among five-dimensional functional bases there are 22 bases invariant under the subalgebras of the Lie algebra of the group $\widetilde{G}(1,3)$ and 41 bases invariant under the non-galilean subalgebras.

3.6. *Six-dimensional non-equivalent functional bases*

There exist $N = 108$ non-equivalent six-dimensional functional bases. In this case $N_s = 51$, $N_{ns} = 57$. Thus, $N_s + N_{ns} - N = 0$ functional bases are invariant under splitting and non-splitting subalgebras.

Let us give an example of the functional basis, which is invariant under the non-splitting non-galilean subalgebras:

$$\langle L_3 + dG + \alpha_3 X_3,\ P_3,\ X_1,\ X_2,\ X_4,\ d > 0,\ \alpha_3 < 0\rangle,$$

$$J_1 = u, \quad J_2 = \frac{x_0 + x_4}{u_0 - u_4}, \quad J_3 = d \arctan \frac{u_2}{u_1} - \ln(x_0 + x_4),$$

$$J_4 = dx_3 - \alpha_3 \ln(x_0 + x_4) + du_3 \frac{x_0 + x_4}{u_0 - u_4}, \quad J_5 = u_1^2 + u_2^2, \quad J_6 = u_0^2 - u_3^2 - u_4^2 .$$

It should be noted that among the six-dimensional functional bases there are 53 bases invariant under the subalgebras of the Lie algebra of the group $\widetilde{G}(1,3)$ and 55 bases invariant under the non-galilean subalgebras.

3.7. *Seven-dimensional non-equivalent functional bases*

There exist $N = 136$ non-equivalent seven-dimensional functional bases. In this case $N_s = 58$, $N_{ns} = 78$. Thus, $N_s + N_{ns} - N = 0$ functional bases are invariant under splitting and non-splitting subalgebras.

Let us give an example of the functional basis, which is invariant under the splitting non-galilean subalgebras:

$$\langle G,\ X_1,\ X_2,\ X_3\rangle,$$

$$J_1 = (x_0^2 - x_4^2)^{1/2}, \quad J_2 = u, \quad J_3 = (x_0 + x_4)(u_0 + u_4), \quad J_4 = u_1,$$

$$J_5 = u_2, \quad J_6 = u_3, \quad J_7 = u_0^2 - u_4^2 .$$

It should be noted that among the seven-dimensional functional bases there are 75 bases invariant under the subalgebras of the Lie algebra of the group $\widetilde{G}(1,3)$ and 61 bases invariant under the non-galilean subalgebras.

3.8. *Eight-dimensional non-equivalent functional bases*

There exist $N = 89$ non-equivalent eight-dimensional functional bases. In this case $N_s = 40$, $N_{ns} = 49$. Thus, $N_s + N_{ns} - N = 0$ functional bases are invariant under splitting and non-splitting subalgebras.

Let us give an example of the functional basis, which is invariant under the splitting galilean subalgebras:

$\langle L_3,\ X_3,\ X_4\rangle$,

$$J_1 = x_0 + x_4, \quad J_2 = (x_1^2 + x_2^2)^{1/2}, \quad J_3 = u, \quad J_4 = x_1 u_2 - x_2 u_1,$$

$$J_5 = u_0, \quad J_6 = u_3, \quad J_7 = u_4, \quad J_8 = u_1^2 + u_2^2\ .$$

It should be noted that among the eight-dimensional functional bases there are 52 bases invariant under the subalgebras of the Lie algebra of the group $\widetilde{G}(1,3)$ and 37 bases invariant under the non-galilean subalgebras.

3.9. *Nine-dimensional non-equivalent functional bases*

There exist $N = 49$ non-equivalent nine-dimensional functional bases. In this case $N_s = 21$, $N_{ns} = 28$. Thus, $N_s + N_{ns} - N = 0$ functional bases are invariant under splitting and non-splitting subalgebras.

Let us give an example of the functional basis, which is invariant under the non-splitting galilean subalgebras:

$\langle P_3 + X_2,\ X_1\rangle$,

$$J_1 = x_0 + x_4, \quad J_2 = \frac{x_3}{x_0 + x_4} + x_2, \quad J_3 = (x_0^2 - x_3^2 - x_4^2)^{1/2},$$

$$J_4 = u, \quad J_5 = (x_0 + x_4)u_3 + (u_0 - u_4)x_3, \quad J_6 = u_1, \quad J_7 = u_2,$$

$$J_8 = u_0 - u_4, \quad J_9 = u_0^2 - u_3^2 - u_4^2\ .$$

It should be noted that among the nine-dimensional functional bases there are 30 bases invariant under the subalgebras of the Lie algebra of the group $\widetilde{G}(1,3)$ and 19 bases invariant under the non-galilean subalgebras.

3.10. *Ten-dimensional non-equivalent functional bases*

There exist $N = 20$ non-equivalent ten-dimensional functional bases. In this case $N_s = 10$, $N_{ns} = 10$. Thus, $N_s + N_{ns} - N = 0$ functional bases are invariant under splitting and non-splitting subalgebras.

Let us give an example of the functional basis, which is invariant under the non-splitting non-galilean subalgebras:

$\langle P_3 + C_3 + eL_3 + \alpha(X_0 + X_4),\ e > 2, \alpha < 0\rangle$,

$$J_1 = (x_1^2 + x_2^2)^{1/2}, \quad J_2 = (x_3^2 + x_4^2)^{1/2}, \quad J_3 = ex_0 - \alpha \arctan\frac{x_1}{x_2},$$

$$J_4 = u, \quad J_5 = x_1 u_2 - x_2 u_1, \quad J_6 = x_3 u_4 - x_4 u_3,$$

$$J_7 = 2x_0 + \alpha \arctan \frac{u_4}{u_3}, \quad J_8 = u_0, \quad J_9 = u_1^2 + u_2^2, \quad J_{10} = u_3^2 + u_4^2 \,.$$

It should be noted that among the ten-dimensional functional bases there are 12 bases invariant under the subalgebras of the Lie algebra of the group $\widetilde{G}(1,3)$ and 8 bases invariant under the non-galilean subalgebras.

4. On some applications of the results obtained

It is well known (see, for example, Refs. 4,6,8,24–26,41) that differential equations with non-trivial symmetry groups play an important role in theoretical and mathematical physics, mechanics, gas dynamics etc. Therefore, the construction and investigation of equations of this type are important from the physical and mathematical points of view. In particular, the results obtained can be used in order to construct the first-order differential equations in the space $M(1,4) \times R(u)$, which are invariant under the non-conjugate subgroups of the group $P(1,4)$. Here, and in what follows, $R(u)$ is the number axis of the dependent variable u.

Indeed, (see, for example, Refs. 1,6,8), in many cases these equations can be written in the following form:

$$F(J_1, J_2, ..., J_t) = 0,$$

where F is an arbitrary smooth function of its arguments, $\{J_1, J_2, ..., J_t\}$ are functional bases of the first-order differential invariants of the corresponding non-conjugate subalgebras of the Lie algebra of the group $P(1,4)$. In this way, we have constructed 498 classes of the first-order differential equations in the space $M(1,4) \times R(u)$ with non-trivial symmetry.

Now, let us present some of the obtained classes of the first-order differential equations, which are invariant under some important for theoretical and mathematical physics subalgebras of the Lie algebra of the group $P(1,4)$. Below, for each of the such subalgebras we write their basis elements and corresponding classes of the first-order differential equations in the space $M(1,4) \times R(u)$.

(1) $\langle L_1, \; L_2, \; L_3, \; P_1 + C_1, \; P_2 + C_2, \; P_3 + C_3 \rangle (\cong O(4))$,

$$F\left(x_0, \; (x_1^2 + x_2^2 + x_3^2 + x_4^2)^{1/2}, \; u, \; x_1u_1 + x_2u_2 + x_3u_3 + x_4u_4, \; u_0, \; u_1^2 + u_2^2 + u_3^2 + u_4^2\right) = 0 \;;$$

(2) $\langle L_1, \; L_2, \; L_3, \; P_1{+}C_1, \; P_2{+}C_2, \; P_3{+}C_3, \; X_1, \; X_2, \; X_3, \; X_0{-}X_4 \rangle (\cong E(4))$,

$\langle L_1 + \frac{1}{2}(P_1 + C_1), \; L_2 + \frac{1}{2}(P_2 + C_2), \; L_3 + \frac{1}{2}(P_3 + C_3), \; X_1, \; X_2,$

$X_3,\ X_0 - X_4\rangle,$

$\langle L_1 + \frac{1}{2}(P_1 + C_1),\ L_2 + \frac{1}{2}(P_2 + C_2),\ L_3 + \frac{1}{2}(P_3 + C_3),$

$L_3 - \frac{1}{2}(P_3 + C_3),\ X_1,\ X_2,\ X_3,\ X_0 - X_4\rangle,$

$F\left(x_0,\ u,\ u_0,\ u_1^2 + u_2^2 + u_3^2 + u_4^2\right) = 0\ ;$

(3) $\langle L_1,\ L_2,\ L_3,\ P_1 - C_1,\ P_2 - C_2,\ P_3 - C_3,\ X_1,\ X_2,\ X_3,$

$X_0 + X_4\rangle (\cong P(1,3)).$

$F\left(x_4,\ u,\ u_4,\ u_0^2 - u_1^2 - u_2^2 - u_3^2\right) = 0\ ;$

(4) $\langle G,\ C_1,\ C_2,\ C_3,\ L_1,\ L_2,\ L_3,\ P_1,\ P_2,\ P_3\rangle (\cong O(1,4)),$

$F\left((x_0^2 - x_1^2 - x_2^2 - x_3^2 - x_4^2)^{1/2},\ u,\ x_0u_0 + x_1u_1 + x_2u_2 + x_3u_3 + x_4u_4,\right.$

$\left. u_0^2 - u_1^2 - u_2^2 - u_3^2 - u_4^2\right) = 0\ ;$

(5) $\langle L_1,\ L_2,\ L_3,\ P_1,\ P_2,\ P_3,\ X_0,\ X_1,\ X_2,\ X_3,\ X_4\rangle (\cong \widetilde{G}(1,3)),$

$\langle P_1,\ P_2,\ P_3,\ X_0,\ X_1,\ X_2,\ X_3,\ X_4\rangle,$

$\langle L_3 - P_3,\ P_1,\ P_2,\ X_0,\ X_1,\ X_2,\ X_3,\ X_4\rangle,$

$\langle L_3,\ P_1,\ P_2,\ P_3,\ X_0,\ X_1,\ X_2,\ X_3,\ X_4\rangle,$

$\langle L_3 - X_0,\ P_1,\ P_2,\ P_3,\ X_1,\ X_2,\ X_3,\ X_4\rangle,$

$\langle P_1,\ P_2,\ P_3 + X_0,\ L_3 + \beta X_0,\ X_1,\ X_2,\ X_3,\ X_4,\ \beta < 0\rangle,$

$F\left(u,\ u_0 - u_4,\ u_0^2 - u_1^2 - u_2^2 - u_3^2 - u_4^2\right) = 0\ ;$

(6) $\langle G,\ C_1,\ C_2,\ C_3,\ L_1,\ L_2,\ L_3,\ P_1,\ P_2,\ P_3,\ X_0,\ X_1,\ X_2,$

$X_3,\ X_4\rangle (\cong P(1,4)),$

$\langle G,\ P_1,\ P_2,\ P_3,\ X_0,\ X_1,\ X_2,\ X_3,\ X_4\rangle,$

$\langle L_3 + eG,\ P_1,\ P_2,\ P_3,\ X_0,\ X_1,\ X_2,\ X_3,\ X_4,\ e > 0\rangle,$

$\langle G,\ L_3,\ P_1,\ P_2,\ P_3,\ X_0,\ X_1,\ X_2,\ X_3,\ X_4\rangle,$

$\langle G,\ L_1,\ L_2,\ L_3,\ P_1,\ P_2,\ P_3,\ X_0,\ X_1,\ X_2,\ X_3,\ X_4\rangle,$

$F\left(u,\ u_0^2 - u_1^2 - u_2^2 - u_3^2 - u_4^2\right) = 0\ .$

Since the Lie algebra of the group $P(1,4)$ contains, as subalgebras, the Lie algebra of the Poincaré group $P(1,3)$ and the Lie algebra of the extended Galilei group $\widetilde{G}(1,3)$ (see also Ref. 32), the obtained differential equations can be used in relativistic and non-relativistic physics.

References

1. S. Lie, Über Differentialinvarianten, *Math. Ann.* **24** (1) (1884) 537–578.
2. S. Lie, *Vorlesungen über continuierliche gruppen* (Teubner, Leipzig, 1893).
3. A. Tresse, Sur les invariants differentiels des groupes continus de transformations, *Acta math.* **18** (1894) 1–88.
4. E. Vessiot, Sur l'integration des sistem differentiels qui admittent des groupes continus de transformations, *Acta math.* **28** (1904) 307–349.
5. A. Kumpera, Invariants différentiels d'un pseudogroupe de Lie, Géométrie differentielle, *In: Lecture Notes in Math.* (V.392, Springer, Berlin, 1974) 121–162.
6. L.V. Ovsiannikov, *Group Analysis of Differential Equations* (Academic Press, New York, 1982).
7. E.I. Chakyrov, Differential invariants of certain extentions of the Galilei group, *Dinamika Sploshn. Sredy* (69) (1985) 123–149.
8. P.J. Olver, *Applications of Lie Groups to Differential Equations* (Springer-Verlag, New York, 1986).
9. W.I. Fushchych and I.A. Yehorchenko, Differential Invariants for Galilei Algebra, *Dokl. Acad. Nauk. Ukr. SSR, Ser.A* (4) (1989) 19–34.
10. W.I. Fushchych and I.A. Yehorchenko, Differential Invariants for Poincaré Algebra and Conformal Algebra, *Dokl. Acad. Nauk. Ukr. SSR, Ser.A* (5) (1989) 46–53.
11. D. Krupka and J. Janyška, *Lectures on Differential Invariants* (Brno University, Czech Republic, 1990).
12. P.J. Olver, Differential invariants and invariant differential equations, *Lie Groups and their Appl.* **1** (1994) 177–192.
13. P.J. Olver, Differential invariants. Geometric and algebraic structures in differential equations, *Acta Applicandae Math.* **41** (1-3) (1995) 271–284.
14. D.Q. Chao and D. Krupka, 3rd order differential invariants of coframes, Preprint 10/1997, Slezská univerzita v Opavě, Matematický ústav v Opavě, 1997.
15. X. Xiaoping, Differential invariants of classical groups, *Duke Math. J.* **94** (3) (1998) 543–572.
16. G.M. Beffa and P.J. Olver, Differential Invariants for parametrized projective surfaces, *Comm. Anal. Geom.* **7** (4) (1999) 807–839.
17. J. Šeděnková, Differential invariants of the meric tensor, *In: Proceedings of the Seminar on Differential Geometry, Mathematical Publications* (V. 2, Silesian University in Opava, Opava, 2000) 145–158.
18. Y. Nutku and M.B. Sheftel, Differential Invariants and group foliation for the complex Monge-Ampère equation, *J. Phys. A: Math. Gen.* **34** (2001) 137–156.
19. R.O. Popovych and V.M. Boyko, Differential Invariants and Application to Riccati-Type Systems, *In: Proc. of the Fourth Internat. Conf. Symmetry in Nonlinear Mathematical Physics* (Ukr. Math. Congr. Dedicated to 200th An-

niversary of Mykhailo Ostrograds'kyi, 9–15 July 2001, Kyiv, Ukraine, Proc. of Inst. of Math. of NAS of Ukraine, Kyiv, 43, Part 1, 2002) 184–193.

20. A.P. Chupakhin, Differential invariants: theorem of commutativity *Communications in Nonlinear Science and Numerical Simulation* **9** Issue 1, (2004) 25–33.
21. S.V. Golovin, Applications of the differential invariants of infinite dimensional groups in hydrodynamics *Communications in Nonlinear Science and Numerical Simulation* **9** Issue 1, (2004) 35–51.
22. N.H. Ibragimov, M. Torrisi and A. Valenti, Differential invariants of nonlinear equations $\nu_{tt} = f(x, \nu_x)\nu_{xx} + g(x, \nu_x)$, *Communications in Nonlinear Science and Numerical Simulation* **9** Issue 1, (2004) 69–80.
23. R. Tracina, Invariants of a family of nonlinear wave equations, *Communications in Nonlinear Science and Numerical Simulation* **9** Issue 1, (2004) 127–133.
24. W.I. Fushchych, Representations of full inhomogeneous de Sitter group and equations in five-dimensional approach. I, *Teoret. i mat. fizika* **4** (3) (1970) 360–367.
25. V.G. Kadyshevsky, New approach to theory electromagnetic interactions, *Fizika elementar. chastitz. i atomn. yadra* **11** (1) (1980) 5–39.
26. W.I. Fushchych and A.G. Nikitin, *Symmetries of Equations of Quantum Mechanics* (Allerton Press Inc., New York, 1994).
27. V.M. Fedorchuk, Continuous subgroups of the inhomogeneous de Sitter group $P(1,4)$, Preprint, N 78.18, Inst. Matemat. Acad. Nauk Ukr. SSR, Kyiv, 1978.
28. V.M. Fedorchuk, Splitting subalgebras of the Lie algebra of the generalized Poincaré group $P(1,4)$, *Ukr. Mat. Zh.* **31** (6) (1979) 717–722.
29. V.M. Fedorchuk and W.I. Fushchych, On subgroups of the generalized Poincaré group, *In: Proceedings of the International Seminar on Group Theoretical Methods in Physics* (V.1, Nauka, Moscow, 1980) 61–66.
30. V.M. Fedorchuk, Nonsplitting subalgebras of the Lie algebra of the generalized Poincaré group $P(1,4)$, *Ukr. Mat. Zh.* **33** (5) (1981) 696–700.
31. W.I. Fushchich, A.F. Barannik, L.F. Barannik and V.M. Fedorchuk, Continuous subgroups of the Poincaré group $P(1,4)$, *J. Phys. A: Math. Gen.* **18** (14) (1985) 2893–2899.
32. W.I. Fushchich and A.G. Nikitin, Reduction of the representations of the generalized Poincaré algebra by the Galilei algebra, *J. Phys.A: Math. and Gen.* **13** (7) (1980) 2319–2330.
33. V.M. Fedorchuk and V.I. Fedorchuk, Differential invariants of first-order of splitting subgroups of the generalized Poincaré group $P(1,4)$, *Mat. Metody i Fiz.- Mekh. Polya* **44** (1) (2001) 16–21.
34. V.M. Fedorchuk and V.I. Fedorchuk, On first-order differential invariants for splitting subgroups of the generalized Poincaré group $P(1,4)$, *Dopov. Nats. Akad. Nauk Ukrainy* (5) (2002) 36–42.
35. V.M. Fedorchuk and V.I. Fedorchuk, On Differential Invariants of First - and Second-Order of the Splitting Subgroups of the Generalized Poincaré Group $P(1,4)$ *In: Proc. 4th Internat. Conf. Symmetry in Nonlinear Mathematical Physics* (Ukr. Math. Congr, Dedicated to 200th Anniversary of Mykhailo

Ostrohrads'kyi, 9–15 July, 2001, Kyiv, Ukraine, Proc. of Inst. of Math. of NAS of Ukraine, Kyiv, 43, Part 1, 2002) 140–144.

36. V.M. Fedorchuk and V.I. Fedorchuk, On the Differential First- Order Invariants of the Non-Splitting Subgroups of the Poincare group $P(1,4)$, *In: Proc. of Inst. of Math. of NAS of Ukraine* (Kyiv, 50, Part 1, 2004) 85–91.
37. V.M. Fedorchuk and V.I. Fedorchuk, On the differential first-order invariants for the non-splitting subgroups of the generalized Poincare group $P(1,4)$, *Annales Academiae Paedagogicae Cracoviensis Studia Mathematica* **4** (2004) 65–74.
38. V.M. Fedorchuk and V.I. Fedorchuk, On functional bases of first-order differential invariants of continuous subgroups of the Poincaré group $P(1,4)$, *Mat. Metody i Fiz. - Mekh. Polya* **48** (4) (2005) 51–58.
39. V.M. Fedorchuk and V.I. Fedorchuk, First-Order Differential Invariants of the splitting subgroups of the Poincaré group $P(1,4)$, *Universitatis Iagellonicae Acta Mathematica* **44** (2006) 21–30.
40. J. Patera, P. Winternitz and H. Zassenhaus, Continuous subgroups of the fundamental groups of physics. I. General method and the Poincaré group, *J. Math. Phys.* **16** (8) (1975) 1597–1614.
41. W. Fushchych, L. Barannyk and A. Barannyk, *Subgroup analysis of the Galilei and Poincaré groups and reductions of nonlinear equations* (Naukova Dumka, Kiev, 1991).

Differential Geometry and its Applications
Proc. Conf., in Honour of Leonhard Euler, Olomouc, August 2007

On the flag curvature of the dual of Finsler metric

Dragos Hrimiuc

Department of Mathematics, University of Alberta,
Edmonton, Canada, T6G 2G1
E-mail: dhrimiuc@ualberta.ca

We define the flag curvature of the dual of a Finsler metric and study its relationship with the flag curvature of the initial Finsler metric via Legendre transformation.

Keywords: Finsler metrics, nonlinear connection, Legendre transformation, flag curvature.

MS classification: 53B40, 53C60, 53Z05.

1. Introduction

The flag curvature of a Finsler metric, an extension of the sectional curvature in Riemannian geometry, is a significant invariant in Finsler geometry that captured the attention of an important number of researchers. I would like merely to mention the recent work of Bao and Robles,[1] Shen,[2] Bao, Robles and Shen,[3] Bryant.[4] It is very known that some problems written on the tangent bundle can be simplified if they are re-written by using the machinery of the dual. For example the geodesics of a Finsler metric (the integral curves of a certain second order vector field) can be viewed as the projections of the geodesics of the dual metric (integral curves of a Hamiltonian vector field) on the base manifold. There are other important cases when the metric could have a complicated format or is implicitly given, while its dual could be simpler. Some of such metrics will be presented in the next section. The flag curvature for the dual metric is introduced in the last section and its connection with the curvature of initial metric is given by using the Legendre transformation.

Dedicated to Professor Radu Miron on the occasion of his 80th birthday

2. The dual of a Finsler metric

Let M be a smooth manifold and (TM,τ), (T^*M,τ^*) its tangent and cotangent bundles. Let $A \subset TM$ be an open set and $A_x := A \cap T_xM$, $x \in M$.

The *fiber derivative* of a differentiable function $f : A \to \mathbb{R}$ at $y \in A_x$ in the direction of $u \in T_xM$ is the mapping $\mathbb{D}f : A \to T^*M$, $\mathbb{D}f(y)u = \langle \mathbb{D}f(y), u\rangle := \frac{d}{dt}f(y+tu)|_{t=0}$.

The *second fibre derivative* of a twice differentiable function $f : A \to \mathbb{R}$ at $y \in A_x$ is the mapping $\mathbb{D}^2 f : A \to L^2_{sym}(TM, \mathbb{R})$ given by

$$\mathbb{D}^2 f(y)(u,v) = \frac{\partial^2 f}{\partial s \partial t}(y + su + tv))|_{s=t=0},\ u, v \in T_xM,\ x \in M$$

Notice that for each $y \in A_x$ we have the linear mapping $\mathbb{D}^2 f(y) : T_xM \to T_x^*M$, $\langle\ \mathbb{D}^2 f(y)u, v\rangle = \langle\ \mathbb{D}^2 f(y)v, u\rangle = \mathbb{D}^2 f(y)(u,v)$, $u, v \in T_xM$, $x \in M$.

Definition 2.1. A *Finsler metric* on TM is a continuous function $F : TM \to [0,\infty)$, C^∞ on the slit tangent bundle $\overset{\circ}{TM} := TM\backslash\{0\}$ that is a also a Minkowski norm on each fibre i.e.

(a) $F(\lambda y) = \lambda F(y)$, $\lambda > 0, y \in T_xM$
(b) For each $y \in T_xM$, $\mathbb{D}^2\left(\frac{1}{2}F^2\right)(y)$ is an inner product on T_xM, $x \in M$.

If set $L := \frac{1}{2}F^2$, then the following formulae are direct consequences of the above definition:[5]

$$F^2(y) = \left\langle \mathbb{D}^2(L)(y)y, y\right\rangle,\ y \in T_xM,\ x \in M \tag{1}$$

$$\langle \mathbb{D}F(y), u\rangle \leq F(u),\ y, u \in T_xM, x \in M \tag{2}$$

$$F(u+v) \leq F(u) + F(v),\ u, v \in T_xM,\ x \in M \tag{3}$$

Let F be a Finsler metric on TM and $\partial\Omega_{F,x} := \{y \in T_xM; F(y) = 1\}$ the sphere (indicatrix) of the metric at each $x \in M$.

Definition 2.2. The *dual* of the Finsler metric F is is the function $F^* : T^*M \to \mathbb{R}$ defined by

$$F^*(p) = \sup_{u \in \partial\Omega_{F,x}} \langle p, u\rangle = \sup_{u \neq 0} \frac{\langle p, u\rangle}{u},\ p \in T_x^*M,\ x \in M$$

Definition 2.3. The Legendre transformation generated by F, denoted $\mathcal{L}_F$, is the following bundle map (not necessarily linear) over the identity:

$$\mathcal{L}_F : \overset{\circ}{TM} \to \overset{\circ}{T^*M}, \mathcal{L}_F = \mathbb{D}(L) = \mathbb{D}\left(\frac{1}{2}F^2\right)$$

Notice that

$$\mathcal{L}_F(y) = \mathbb{D}(L)(y) = \mathbb{D}^2(L)(y)\,y,\ \ F^2(y) = \langle \mathcal{L}_F(y), y\rangle \tag{4}$$

and from (2) we get:

$$\langle \mathcal{L}_F(y), u\rangle \leq F(y)F(u) \tag{5}$$

By using (1)-(5) we can prove the following result:

Proposition 2.1. *(a)* $\mathcal{L}_F : \overset{\circ}{TM} \to \overset{\circ}{T^*M}$ *is a smooth 1-positive homogeneous diffeomorphism*
(b) $F^* = F \circ \mathcal{L}_F^{-1}$
(c) F^* *is a Finsler metric on* T^*M.

Notice that $\mathcal{L}_F$ can be extended to a continuous mapping on TM by taking $\mathcal{L}_F(0) = 0$.

Remark 2.1. The following equations hold:

$$F^*(p) = F(y),\ p = \mathcal{L}_F(y) \tag{6}$$

$$\mathcal{L}_F^{-1} = \mathbb{D}(H),\ H := \frac{1}{2}F^{*2} \tag{7}$$

We also mention that $\mathcal{L}_F$ is a 1-positive homogeneous diffeomorphism that preserves the norms F and F^* and $\mathcal{L}_F$ is linear if and only if F (or F^*) is Riemannian.

Theorem 2.1. *Let* F *be a Finsler metric on* TM *,* w *a smooth vector field on* M*, such that* $F(w) < 1$ *and let* $\beta = \mathcal{L}_F(w)$*. The following properties hold:*
(a) $\widetilde{F}(y) = F(y) - \langle \beta, y\rangle$ *is a Finsler metric on* TM
(b) $F^*(p) = \widetilde{F}^*(p - F^*(p)\beta)$
(c) $\widetilde{F}^*(p) = F^*\left(p + \widetilde{F}^*(p)\beta\right)$.

Proof. (a) The proof is quite similar to that given for Randers metrics[5]
(b) We have:

$$\begin{aligned}\mathcal{L}_{\widetilde{F}}(y) &= \widetilde{F}(y)\mathbb{D}\widetilde{F}(y) = (F(y) - \langle \beta, y\rangle)\left(\frac{\mathcal{L}_F(y)}{F(y)} - \beta\right)\\ &= (F^*(p) - \langle \beta \circ \mathcal{L}_{F^*}, p\rangle)\left(\frac{p}{F^*(p)} - \beta\right)\\ &= \left(1 - \frac{\langle \beta \circ \mathcal{L}_{F^*}, p\rangle}{F^*(p)}\right)p + (-F^*(p) + \langle \beta \circ \mathcal{L}_{F^*}, p\rangle)\beta\end{aligned}$$

If $\widetilde{p} := \mathcal{L}_{\widetilde{F}}(y)$ then

$$\widetilde{F}^*(\widetilde{p}) = \widetilde{F}(y) = F(y) - \langle \beta, y \rangle = F^*(p) - \langle \beta \circ \mathcal{L}_{F^*}, p \rangle$$

Hence

$$\widetilde{p} = \frac{\widetilde{F}^*(\widetilde{p})}{F^*(p)} p - \widetilde{F}^*(\widetilde{p})\,\beta = \widetilde{F}^*(\widetilde{p}) \left(\frac{p}{F^*(p)} - \beta \right)$$

and therefore

$$\widetilde{F}^*(\widetilde{p}) = \widetilde{F}^*(\widetilde{p})\, \widetilde{F}^* \left(\frac{p}{F^*(p)} - \beta \right)$$

and (b) follows. Now (c) can be obtained from (b). □

We remark that generally $\widetilde{F}^*$ can not be found in terms of F^* and β. However, when F is Riemannian we have:[6]

Corollary 2.1. *Let $F(y) = |y|$ be a Riemannian metric on M and $\widetilde{F}(y) = |y| - \langle \mathcal{L}_{|\cdot|}(w), y \rangle$, $|w| < 1$ (a Randers metric). Then*

$$\widetilde{F}^*(p) = \frac{\sqrt{\langle w, p \rangle^2 + (1 - |w|^2)\,|p|^{*2}}}{1 - |w|^2} + \frac{\langle w, p \rangle}{1 - |w|^2} \tag{8}$$

where $|\cdot|^*$ is the dual of the Riemannian norm $|\cdot|$.

Proof. Use equation (c) from the above proposition. □

Remark 2.2. If F is a Finsler metric on TM and w is a smooth vector field on M, such that $F(w) < 1$ we can also define $\widehat{F}(y) = F(y) - \langle \mathcal{L}_F(y), w \rangle$. $\widetilde{F} \equiv \widehat{F}$ if and only if F is Riemannian. It is still an open question if $\widehat{F}$ is or is not a Finsler metric.

Let F is a Finsler metric on TM and w is a smooth vector field on M, such that $F(w) < 1$. Then, every one of the following equations

$$F(y) = \overline{F}(y - F(y)w) \tag{9}$$

$$\overline{F}(y) = F(y + \overline{F}(y)w) \tag{10}$$

defines a new Finsler metric $\overline{F}$ on TM.[2,7] This Finsler metric is related to so called Zermello navigation problem used by Shen,[2] Bao, Robles and Shen.[3] Notice the similarity between these equations and those of Theorem 2.1. We remark that $\overline{F}$ can not be generally obtained in terms of F . However, we have:

Theorem 2.2. *The dual of $\overline{F}$ is*

$$\overline{F}^{*}(p) = F^{*}(p) - \langle p, w\rangle \,, \; p \in T_x^* M, \;\; w \in T_x M, \;\; x \in M. \tag{11}$$

Proof. Let $\varphi : TM \to TM$, $\varphi(y) = y - F(y)w$, $y, w \in T_x M$; then $\overline{F} = F \circ \varphi^{-1}$.Thus

$$\overline{F}^{*} = \overline{F} \circ \mathcal{L}_{\overline{F}}^{-1} = F \circ \left(\mathcal{L}_{\overline{F}} \circ \varphi\right)^{-1}$$

and also

$$\mathcal{L}_{\overline{F}}\left(\varphi(y)\right) = \mathcal{L}_F(y) \circ \left(\mathbb{D}\varphi\right)^{-1}$$

On the other hand

$$\mathbb{D}\varphi(y) = id - w \otimes \mathbb{D}F(y) \;\; \text{and} \;\; \left(\mathbb{D}\varphi(y)\right)^{-1} = id + \frac{1}{1 - B(y)} w \otimes \mathbb{D}F(y)$$

where $B(y) := \mathbb{D}F\left(y\right) w = \frac{1}{F(y)} \mathcal{L}_F(y) w$. Hence

$$\mathcal{L}_{\overline{F}}\left(\varphi(y)\right) = \mathcal{L}_F(y) \circ \left[id + \frac{1}{1 - B(y)} w \otimes \mathbb{D}F(y)\right] = \frac{1}{1 - B\left(y\right)} \mathcal{L}_F(y)$$

and

$$\left(1 - B(y)\right) y = \left(\mathcal{L}_{\overline{F}} \circ \varphi\right)^{-1} \circ \mathcal{L}_F(y)$$

Therefore

$$\overline{F}^{*}\left(\mathcal{L}_F(y)\right) = F \circ \left(\mathcal{L}_{\overline{F}} \circ \varphi\right)^{-1} \circ \mathcal{L}_F(y) = F\left((1 - B(y))y\right) = F\left(y\right) - \mathcal{L}_F(y) w$$

and (11) follows. □

3. Finsler metrics and flag curvature

Let (F, M) be a Finsler space. The canonical connection(nonlinear) associated to (F, M) is the almost product structure

$$N = -L_\xi J \tag{12}$$

where J is the canonical tangent structure of TM and ξ is the second order vector field associated to F, i.e. the solution of the equation

$$i_\xi \omega = dL \tag{13}$$

with $\omega = -dd_J\left(L\right)$ and $L = \frac{1}{2} F^2$. This connection produces the standard splitting

$$TTM = VTM \oplus HTM \tag{14}$$

where $VTM = \mathrm{Ker}\,(id + N)$ is the vertical bundle and $HTM = \mathrm{Ker}\,(id - N)$ is the horizontal bundle. The horizontal and vertical projectors induced by N are $h = \frac{1}{2}\,(id + N)$ and $v = \frac{1}{2}\,(id - N)$. If $[h, h]$ denotes the Nijenhuis bracket, the curvature of N defined as

$$\Omega = -\frac{1}{2}\,[h, h] \tag{15}$$

gives the obstruction against the integrability of HTM.

For each $y \in T_xM$, $x \in M$ let us consider

$$(\cdot)_y^v : T_xM \to V_yTM,\ (\cdot)_y^h : T_xM \to H_yTM \tag{16}$$

the vertical and respectively the horizontal lift mappings and let $C = (y)_y^v$ be the Liouville vector field. For a more detailed description of the above concepts please see Miron,[8] Szilasi.[9]

Now we are ready to define an essential concept:

Definition 3.1. The Riemannian curvature tensor at $y \in T_xM$ is given by

$$R_y = -i_\xi \Omega \circ (\cdot)_y^h \tag{17}$$

We mention that the Riemannian curvature tensor is defined here globally, in terms of the curvature tensor of the nonlinear connection and not in terms of Chern's linear connection as in Bao.[5]

The Finsler metric F induces a metric tensor $\overset{\vee}{g}$ on the vertical bundle VTM defined by the following equation:

$$\overset{\vee}{g} \circ (J \times J) = -\omega_F \circ (J \times id) \tag{18}$$

Definition 3.2. Let $y \in T_xM$, $x \in M$. The flag curvature at $u \in T_xM$ is defined by

$$K_y(u) = \frac{\overset{\vee}{g}\left(R_y(u), \overset{\vee}{u}\right)}{\overset{\vee}{g}(C, C)\,\overset{\vee}{g}\left(\overset{\vee}{u}, \overset{\vee}{u}\right) - \left[\overset{\vee}{g}\left(C, \overset{\vee}{u}\right)\right]^2} \tag{19}$$

where $\overset{\vee}{u} := (u)_y^v$.

Notice that the way we define the flag curvature is slightly different from the classical one. This definition seems to be more convenient for our approach that intensively makes use of Legendre transformation.

Let $X_H := (\mathcal{L}_F)_* \xi$ and $J^* := (\mathcal{L}_F)_* J$ be the push-forward of ξ an respectively J by $\mathcal{L}_F$. X_H is the Hamiltonian vector field associated to the

Hamiltonian $H = \frac{1}{2}F^{*2}$ and J^* is a tangent structure on T^*M that will be called the dual of J.

Proposition 3.1. *$N^* = -L_{X_H}J^*$ is a connection (nonlinear) on T^*M.*

Proof. It is enough to verify the that $J^*N^* = J^*$ and $N^*J^* = -J^*$. These equalities are obtained from those verified by N by using Legendre transformation. □

The connection defined above is called the *dual* of N. See also Ref. 6. Let

$$TT^*M = VT^*M \oplus HT^*M \tag{20}$$

the splitting generated by N^*.

Theorem 3.1. *(a) $(\mathcal{L}_F)_*(HTM) = HT^*M$ and $(\mathcal{L}_F)_*(VTM) = VT^*M$ (b) The horizontal and vertical projectors h^* and respectively v^* induced by the splitting(20)are the duals of h and v i.e. $h^* = (\mathcal{L}_F)_* h = (\mathcal{L}_F)_* \circ h \circ (\mathcal{L}_F)_*^{-1}$ and $v^* = (\mathcal{L}_F)_* v = (\mathcal{L}_F)_* \circ v \circ (\mathcal{L}_F)_*^{-1}$.*

Notice that based on part (a) of the above theorem we obtain that the duals (the push forward by Legendre transformation) of the vertical and horizontal lift mappings (16) at $p = \mathcal{L}_F(y), y \in T_xM$ are:

$$(\cdot)_p^{*v} : T_x^*M \to V_pT^*M,\ (\cdot)_p^{*h} : T_x^*M \to H_pT^*M \tag{21}$$

i.e., $(u^*)_p^{*v} = (\mathcal{L}_F)_* \left(\mathcal{L}_F^{-1}(u^*)\right)_y^v$ and $(u^*)_p^{*h} = (\mathcal{L}_F)_* \left(\mathcal{L}_F^{-1}(u^*)\right)_y^h$, $u^* \in T_x^*M$.

Theorem 3.2. *(a) The dual of Ω is just the curvature of N^**

$$\Omega^* = (\mathcal{L}_F)_* \Omega = -\frac{1}{2}[h^*, h^*] \tag{22}$$

(b) The dual of the Riemannian curvature tensor (17) of N at $y \in T_xM$ is the Riemannian curvature tensor of N^* at $p = \mathcal{L}_F(y) \in T_x^*M$ i.e.

$$R_p^* = (\mathcal{L}_F)_* R_y = (\mathcal{L}_F)_* \circ R_y \circ \mathcal{L}_F^{-1} = -i_{X_H}\Omega^* \circ (\cdot)_p^{*h} \tag{23}$$

Proof. (a) $-\frac{1}{2}[h^*, h^*] = -\frac{1}{2}[(\mathcal{L}_F)_*(h), (\mathcal{L}_F)_*(h)] = (\mathcal{L}_F)_*\left(-\frac{1}{2}[h, h]\right) = (\mathcal{L}_F)_* \Omega = \Omega^*$.

(b) $R_p^*(u^*) = (\mathcal{L}_F)_* (R_y \mathcal{L}_F^{-1}(u^*)) = (\mathcal{L}_F)_* (-i_\xi \Omega (\mathcal{L}_F^{-1}(u^*))_y^h = -i_{X_H}\Omega^*((u^*)_p^{*h})$. □

We mention that the dual of the metric tensor $\overset{\vee}{g}$ is

$$\overset{\vee}{g}^* = (\mathcal{L}_F)_* \overset{\vee}{g} \tag{24}$$

and we have

$$\overset{\vee}{g}^* \circ (J^* \times J^*) = \omega \circ (id \times J^*) \tag{25}$$

with ω the canonical two form on TT^*M.

Definition 3.3. Let $p \in T_x^*M$, $x \in M$. The co-flag curvature tensor at $u^* \in T_x^*M$ is defined by

$$K_p^*(u^*) = \frac{\overset{\vee}{g}^* \left(R_p^*(u^*), \overset{\vee}{u}^* \right)}{\overset{\vee}{g}^* (C^*, C^*) \overset{\vee}{g}^* \left(\overset{\vee}{u}^*, \overset{\vee}{u}^* \right) - \left[\overset{\vee}{g}^* \left(C^*, \overset{\vee}{u}^* \right) \right]^2}$$

where $\overset{\vee}{u}^* := (u^*)_p^{*v}$ and $C^* := (\mathcal{L}_F)_* (C)$.

By using Theorem 3.2 and (19), (24) we cam prove the following

Theorem 3.3. *If $y, u \in T_xM$ then $K^*_{\mathcal{L}_F(y)}(\mathcal{L}_F(u)) = K_y(u)$ i.e. the flag curvature is $\mathcal{L}_F$ invariant.*

Corollary 3.1. *(F, M) is of scalar curvature(constant) if and only if (F^*, M) is of scalar(constant) curvature.*

References

1. D. Bao and C. Robles, Ricci and Flag Curvatures in Finsler Geometry, A sampler of Riemann-Finsler Geometry, *MSRIP* **50** (2004) 198–256.
2. Z. Shen, Finsler metric with $K = 0$ and $S = 0$, *Canadian J. Math.* **55** (2003) 112–132.
3. D. Bao , C. Robles and Z. Shen, Zermelo navigation on Riemannian manifolds, *J. Diff. Geometry* **66** (2004) 391–449.
4. R.L. Bryant, Some remarks on Finsler manifolds with constant flag curvature, *Houston J. Math.* **28** (2002) 221–262.
5. D. Bao, S.S. Chern and Z. Shen, *An Introduction to Riemann-Finsler Geometry* (Springer, 2000).
6. D. Hrimiuc and H. Shimada, On the L-duality between Lagrange and Hamilton manifolds, *Nonlinear World* **3** (1996) 613–641.

7. D. Hrimiuc, On the affine deformation of a Minkowski norm, *Periodica Mathematica Hungarica* **48** (2004) 49–60.
8. R. Miron, D. Hrimiuc, H. Shimada and S.V. Sabau, *The Geometry of Hamilton and Lagrange Spaces* (Kluwer, 2001).
9. J. Szilasi, A setting for a spray and Finsler geometry, *In: Handbook of Finsler Geometry* (Vol 2, (P.L. Antonelli, Ed.) Kluwer Academic Publishers, 2003) 1185–1426.

Differential Geometry and its Applications
Proc. Conf., in Honour of Leonhard Euler, Olomouc, August 2007

A note on the height of the canonical Stiefel–Whitney classes of the oriented Grassmann manifolds

Július Korbaš

Department of Algebra, Geometry, and Mathematical Education, Faculty of Mathematics, Physics, and Informatics, Comenius University, Bratislava 4, SK-842 48 Bratislava 4, Slovakia
or Mathematical Institute, Slovak Academy of Sciences, Štefánikova 49, SK-814 73 Bratislava 1, Slovakia
E-mail: korbas@fmph.uniba.sk

We derive a functional upper bound $\tilde{\kappa}_i(n,k)$ for the $\mathbb{Z}_2$-height of the ith Stiefel–Whitney class of the canonical oriented k-plane bundle over the oriented Grassmann manifold $\tilde{G}_{n,k}$ of oriented k-dimensional vector subspaces in Euclidean n-space. It turns out that $\tilde{\kappa}_2(n,k)$ is for many pairs (n,k) better than the upper bound implied by Dutta and Khare's (2002) results on the height of the second Stiefel–Whitney class of the canonical k-plane bundle over the Grassmann manifold $G_{n,k}$ of k-dimensional vector subspaces in Euclidean n-space. In addition to this, $\tilde{\kappa}_2(n,2)$ always coincides with the exact value of the height (well known for $\tilde{G}_{n,2}$). We also prove that $\tilde{\kappa}_2(2^{2t},3)$ $(t \geq 2)$ implies an upper bound which differs from the exact value of the height of the second Stiefel–Whitney class by at most one.

Keywords: Oriented Grassmann manifold, Stiefel–Whitney class, cup product, height.

MS classification: 57R19, 53C30, 55M30, 57R20, 57T15.

1. Introduction and statement of the main results

The oriented Grassmann manifold $\tilde{G}_{n,k} \cong SO(n)/SO(k) \times SO(n-k)$ $(2k \leq n)$ consists of oriented k-dimensional vector subspaces in $\mathbb{R}^n$; its dimension is $k(n-k)$. We shall suppose that $k \geq 2$, to exclude the spheres $\tilde{G}_{n,1} \cong S^{n-1}$ from our considerations.

As is well known, the manifold $\tilde{G}_{n,k}$ is a universal double covering space of the Grassmann manifold $G_{n,k} \cong O(n)/O(k) \times O(n-k)$, the latter consisting of all k-dimensional vector subspaces in $\mathbb{R}^n$. Let $p : \tilde{G}_{n,k} \to G_{n,k}$ be the obvious covering projection. For the tangent bundles we have $T\tilde{G}_{n,k} \cong p^*(TG_{n,k})$.

As is usual for smooth manifolds, by the ith Stiefel–Whitney class of $\tilde{G}_{n,k}$ we understand the ith Stiefel–Whitney class of its tangent bundle, hence we have $w_i(\tilde{G}_{n,k}) = w_i(T\tilde{G}_{n,k}) \in H^i(\tilde{G}_{n,k};\mathbb{Z}_2)$; at the same time we have $w_i(G_{n,k}) = w_i(TG_{n,k}) \in H^i(G_{n,k};\mathbb{Z}_2)$.

Let $\tilde{\gamma}_{n,k}$ (briefly $\tilde{\gamma}$) be the canonical oriented k-plane bundle over $\tilde{G}_{n,k}$, and let $\gamma_{n,k}$ (briefly γ) be the canonical k-plane bundle over $G_{n,k}$ (see, e.g., Ref. 9). By the ith *canonical* Stiefel–Whitney class of $\tilde{G}_{n,k}$ (resp. $G_{n,k}$) we mean the ith Stiefel–Whitney class of the canonical bundle, hence the class $w_i(\tilde{\gamma}_{n,k})$, briefly denoted by $\tilde{w}_i$ when n and k are clear from the context (resp. $w_i(\gamma_{n,k})$, briefly denoted by w_i when n and k are clear from the context).

Given a topological space X, we recall that the $\mathbb{Z}_2$-height (briefly height) of a cohomology class $y \in H^*(X;\mathbb{Z}_2)$ is defined to be

$$\sup\{t; y^t \neq 0 \in H^*(X;\mathbb{Z}_2)\},$$

and we denote it by height(y). For instance, height$(w_1(\gamma_{n,k}))$ was calculated by R. Stong in Ref. 10, J. Lörinc in Ref. 8, extending earlier results of S. Ilori and D. Ajayi (Ref. 5), calculated the height of the first Stiefel–Whitney characteristic class of any flag manifold $O(n_1+\cdots+n_q)/O(n_1)\times\cdots\times O(n_q)$, and S. Dutta and Khare in Ref. 4 calculated height$(w_2(\gamma_{n,k}))$.

For some purposes, it is desirable to know the height of $w_i(\tilde{\gamma}_{n,k})$. Our interest in this invariant stems from the intention to derive results on the cup-length and Lyusternik-Shnirel'man category of the manifolds $\tilde{G}_{n,k}$, proceeding in the spirit of Ref. 7.

To calculate height$(w_2(\tilde{\gamma}_{n,2}))$, it is enough to use the knowledge of the cohomology algebra $H^*(G_{n,2};\mathbb{Z}_2)$ (see, e.g., Ref. 2), the Gysin exact sequence associated with the covering projection mentioned above (see, e.g., S12 in Ref. 9), and elementary considerations. So one readily verifies (but it is also well known) that

$$\text{height}(w_2(\tilde{\gamma}_{n,2})) = \begin{cases} \frac{n-3}{2} & \text{if } n \text{ is odd}, \\ \frac{n-2}{2} & \text{if } n \text{ is even}. \end{cases} \tag{1}$$

For $k \geq 3$, as in contrast to $k = 2$, the algebra $H^*(\tilde{G}_{n,k};\mathbb{Z}_2)$ is in general unknown. In spite of this drawback, again the knowledge of the algebra $H^*(G_{n,k};\mathbb{Z}_2)$ and the corresponding Gysin exact sequence make it possible — *in theory* — to calculate the height of any $w_i(\tilde{\gamma}_{n,k})$. But *in practice* the matter is different: the computations are quite complicated, and with growing n and k they become unmanageable, even when one uses a computer.

In view of this, it would be desirable to have some explicit formulae, hence something like (1). But in general no formulae for height$(w_i(\tilde{\gamma}_{n,k}))$ with $k \geq 3$ are available up to now. As a step on the way to desired formulae, it seems reasonable to look for some good general estimates for height$(w_i(\tilde{\gamma}_{n,k}))$. Having this in mind, we mainly derive here some upper bounds.

We first define two functions: $\tilde{\kappa}_i(n,k)$ for positive integers i, n, k, and $\kappa_2(n,k)$ for positive integers n and k such that $6 \leq 2k \leq n$. We put

$$\tilde{\kappa}_i(n,k) = \begin{cases} \lfloor \frac{(k-1)(n-k)}{i} \rfloor & \text{if } n \text{ is odd,} \\ \lfloor \frac{(k-1)(n-k+1)}{i} \rfloor & \text{if } n \text{ is even,} \end{cases}$$

where $\lfloor a \rfloor$ denotes the integer part of $a \in \mathbb{R}$, and

$$\kappa_2(n,3) = \begin{cases} 2^s - 1 & \text{if } n = 2^s + 1, \\ 2^s & \text{if } n = 2^s + 2, \\ 2^s + 2^{p+1} - 2 & \text{if } n = 2^s + 2^p + 1, s > p \geq 1, \\ 2^s + 2^{p+1} - 1 & \text{if } n = 2^s + 2^p + t + 1, s > p \geq 1. \\ & \quad 1 \leq t \leq 2^p - 1; \end{cases}$$

$$\kappa_2(n,4) = \begin{cases} 2^s - 1 & \text{if } n = 2^s + 1, \\ 2^{s+1} - 4 & \text{if } n = 2^s + 2, \\ 2^{s-1} - 4 & \text{if } n = 2^s + 3, \\ 2^{s+1} - 1 & \text{if } 2^s + 4 \leq n \leq 2^{s+1}; \end{cases}$$

$$\kappa_2(n,k) = \begin{cases} 2^s - 1 & \text{if } n = 2^s + 1, k \geq 5, \\ 2^{s+1} - 1 & \text{if } 2^s + 2 \leq n \leq 2^{s+1}, k \geq 5. \end{cases}$$

We note that height$(w_2(\gamma_{n,k})) = \kappa_2(n,k)$ by Ref. 4.

We now state the following upper bounds for height$(w_i(\tilde{\gamma}_{n,k}))$.

Proposition 1.1. *For the oriented Grassmann manifold $\tilde{G}_{n,k}$ with $4 \leq 2k \leq n$ we have*

$$\text{height}(w_i(\tilde{\gamma}_{n,k})) \leq \tilde{\kappa}_i(n,k)$$

for $i = 2, \ldots, k$.

In some situations, this proposition gives interesting upper bounds. Indeed, one can see immediately (from Proposition 1.1 and the formula (1)) that height$(w_2(\tilde{\gamma}_{n,2})) = \tilde{\kappa}_2(n,2)$. In addition to this, the following corollary exhibits an infinite family of manifolds for which $\tilde{\kappa}$ leads to an upper bound differing from the exact value by at most one.

Corollary 1.1. *For the oriented Grassmann manifolds* $\tilde{G}_{2^{2t},3}$, $t \geq 2$, *we have*

$$2^{2t} - 4 \leq \text{height}(w_2(\tilde{\gamma}_{2^{2t},3})) \leq 2^{2t} - 3.$$

For the second canonical Stiefel–Whitney class, the upper bound for its height given by Proposition 1.1 can sometimes be improved by the following.

Proposition 1.2. *For the oriented Grassmann manifold* $\tilde{G}_{n,k}$, $6 \leq 2k \leq n$, *we have*

$$\text{height}(w_2(\tilde{\gamma}_{n,k})) \leq \min\{\tilde{\kappa}_2(n,k), \kappa_2(n,k)\}.$$

Applications of our results to calculations of the cup-length and Lyusternik-Shnirel'man category of the oriented Grassmann manifolds will be presented in a forthcoming paper.

2. Proofs of the results

In this section, we prove our results stated in Sec. 1. Before passing to the proofs, we recall (see, e.g., S12 in Ref. 9) that besides the compact CW-complexes $\tilde{G}_{n,k}$ we also have their limit space $\tilde{G}_{\infty,k}$; this infinite CW-complex can be identified with the classifying space $BSO(k)$. Let $\tilde{\gamma}_{\infty,k}$ denote the canonical k-plane bundle over $\tilde{G}_{\infty,k}$. Recall that the cohomology algebra $H^*(\tilde{G}_{\infty,k};\mathbb{Z}_2)$ is a $\mathbb{Z}_2$-polynomial algebra; we can write

$$H^*(\tilde{G}_{\infty,k};\mathbb{Z}_2) = \mathbb{Z}_2[w_2(\tilde{\gamma}_{\infty,k}), \ldots, w_k(\tilde{\gamma}_{\infty,k})].$$

As is well known (or a proof can be obtained by mimicking the corresponding part of the proof of Lemma 3.1 in Ref. 1), the restriction homomorphism $j^* : H^*(\tilde{G}_{\infty,k};\mathbb{Z}_2) \to H^*(\tilde{G}_{n,k};\mathbb{Z}_2)$, induced by the obvious "inclusion" $j : \tilde{G}_{n,k} \to \tilde{G}_{\infty,k}$, is an isomorphism in dimensions $\leq n-k-1$ and a monomorphism in dimension $n-k$. As a consequence (we have $j^*(w_i(\tilde{\gamma}_{\infty,k})) = \tilde{w}_i$), the classes $\tilde{w}_2, \ldots, \tilde{w}_k \in H^*(\tilde{G}_{n,k};\mathbb{Z}_2)$ do not obey any relations in dimensions $\leq n-k$.

Proof of Proposition 1.1. Let us suppose that n is odd. By an obvious adjustment of 3.6.2 of Ref. 1 or by Theorem 1.1 in Ref. 6, the Stiefel–Whitney class $w_i(\tilde{G}_{n,k})$ $(i \leq k \leq n-k)$ can now be expressed uniquely as

$$w_i(\tilde{G}_{n,k}) = \tilde{w}_i + Q_i(\tilde{w}_2, \ldots, \tilde{w}_{i-1}),$$

where Q_i is a $\mathbb{Z}_2$-polynomial.

An easy induction shows that

$$\tilde{w}_i = w_i(\tilde{G}_{n,k}) + P_i(w_2(\tilde{G}_{n,k}), \dots, w_{i-1}(\tilde{G}_{n,k})),$$

where P_i is a polynomial for $i = 2, \dots, k$. Indeed, we have $\tilde{w}_2 = w_2(\tilde{G}_{n,k})$ (it is enough to use the formula for $w_2(G_{n,k})$ from Ref. 1 together with the fact that $p^*(\gamma) = \tilde{\gamma}$ and $T\tilde{G}_{n,k} \cong p^*(TG_{n,k})$). By what we have said above, one has

$$\tilde{w}_j = w_j(\tilde{G}_{n,k}) + \text{a polynomial in } \tilde{w}_2, \dots, \tilde{w}_{j-1}$$

for $j \geq 2$. The induction hypothesis then implies that

$$\tilde{w}_j = w_j(\tilde{G}_{n,k}) + P_j(w_2(\tilde{G}_{n,k}), \dots, w_{j-1}(\tilde{G}_{n,k}))$$

for some polynomial P_j.

So the algebra $H^*(\tilde{G}_{n,k}; \mathbb{Z}_2)$ is in dimensions $\leq n - k - 1$ generated by the canonical Stiefel–Whitney classes $\tilde{w}_2, \dots, \tilde{w}_k$, or — thanks to the fact that n is odd — alternatively by the Stiefel–Whitney classes $w_2(\tilde{G}_{n,k}), \dots, w_k(\tilde{G}_{n,k})$.

We need to prove that $\tilde{w}_i^{1+\lfloor\frac{(k-1)(n-k)}{i}\rfloor} = 0$. Let us suppose that the contrary is true. Then, by Poincaré duality, there is a cohomology class

$$y \in H^{k(n-k)-i-i\lfloor\frac{(k-1)(n-k)}{i}\rfloor}(\tilde{G}_{n,k}; \mathbb{Z}_2)$$

such that the cup product

$$\tilde{w}_i^{1+\lfloor\frac{(k-1)(n-k)}{i}\rfloor} y \in H^{k(n-k)}(\tilde{G}_{n,k}; \mathbb{Z}_2) \cong \mathbb{Z}_2$$

is nonzero. One readily verifies that the number $k(n-k) - i - i\lfloor\frac{(k-1)(n-k)}{i}\rfloor$ does not exceed $n-k-1$. Hence the cohomology class y can be expressed in terms of the Stiefel–Whitney classes $w_2(\tilde{G}_{n,k}), \dots, w_k(\tilde{G}_{n,k})$, and one sees that the value of the product

$$\tilde{w}_i^{1+\lfloor\frac{(k-1)(n-k)}{i}\rfloor} y \in H^{k(n-k)}(\tilde{G}_{n,k}; \mathbb{Z}_2)$$

at the fundamental $\mathbb{Z}_2$-homology class of the manifold $\tilde{G}_{n,k}$ is nothing but a Stiefel–Whitney number of $\tilde{G}_{n,k}$. But each $\tilde{G}_{n,k}$ is bordant to zero (this is well known; if needed, consult, e.g., Theorem 31.1 of Ref. 3 keeping in mind that the involution on $\tilde{G}_{n,k}$ reversing the orientation of oriented k-dimensional vector subspaces in $\mathbb{R}^n$ has no fixed point). Therefore, by the Pontrjagin-Thom theorem (see, e.g., Ref. 9) all Stiefel–Whitney numbers of $\tilde{G}_{n,k}$ vanish. As a result, also the product

$$\tilde{w}_i^{1+\lfloor\frac{(k-1)(n-k)}{i}\rfloor} y \in H^{k(n-k)}(\tilde{G}_{n,k}; \mathbb{Z}_2)$$

must be zero. This contradiction proves that we really have $\tilde{w}_i^{1+\lfloor\frac{(k-1)(n-k)}{i}\rfloor} = 0$, and therefore the height of $\tilde{w}_i$ is at most $\lfloor\frac{(k-1)(n-k)}{i}\rfloor$, as claimed.

Now let n be even. We have the obvious "inclusion" $j : \tilde{G}_{n,k} \to \tilde{G}_{n+1,k}$, and for the pullback bundles we obtain $j^*(\tilde{\gamma}_{n+1,k}) = \tilde{\gamma}_{n,k}$. Hence the cohomology homomorphism induced by j maps any canonical Stiefel–Whitney class to the corresponding canonical Stiefel–Whitney class,

$$j^*(w_i(\tilde{\gamma}_{n+1,k})) = w_i(\tilde{\gamma}_{n,k}).$$

More precisely, since $j^* : H^*(\tilde{G}_{n+1,k};\mathbb{Z}_2) \to H^*(\tilde{G}_{n+1,k};\mathbb{Z}_2)$ is a homomorphism of algebras, we have

$$j^*(w_i(\tilde{\gamma}_{n+1,k})^a) = w_i(\tilde{\gamma}_{n,k})^a$$

for any a. So we see that if $w_i(\tilde{\gamma}_{n,k})^a \neq 0$, then also $w_i(\tilde{\gamma}_{n+1,k})^a \neq 0$. Therefore we always have

$$\text{height}(w_i(\tilde{\gamma}_{n,k})) \leq \text{height}(w_i(\tilde{\gamma}_{n+1,k})).$$

Since $n+1$ is odd, by what we have proved we obtain that

$$\text{height}(w_i(\tilde{\gamma}_{n,k})) \leq \text{height}(w_i(\tilde{\gamma}_{n+1,k})) \leq \frac{(k-1)(n-k+1)}{i},$$

and the proof of Proposition 1.1 is complete. □

Proof of Corollary 1.1. It is known (see p. 104 in Ref. 10) that the cup product

$$w_1^{2^{2t}-1} w_2^{2^{2t}-4} \in H^{3\cdot 2^{2t}-9}(G_{2^{2t},3};\mathbb{Z}_2) \cong \mathbb{Z}_2$$

does not vanish. Since (see p. 103 in Ref. 10) $\text{height}(w_1) = 2^{2t}-1$, this means that the class $w_2^{2^{2t}-4}$ is not a multiple of w_1. So the standard Gysin sequence argument implies that we have $\tilde{w}_2^{2^{2t}-4} \neq 0$, and therefore

$$\text{height}(\tilde{w}_2) \geq 2^{2t}-4,$$

as claimed.

In addition to this, we know (see Proposition 1.1) that

$$\text{height}(\tilde{w}_2) \leq 2^{2t}-2.$$

But $\tilde{w}_2^{2^{2t}-2} = 0$, which means that

$$\text{height}(\tilde{w}_2) \leq 2^{2t}-3,$$

as claimed.

Indeed, let us suppose that $\tilde{w}_2^{2^{2t}-2} \neq 0$. Then by Poincaré duality there exists a cohomology class $z \in H^{2^{2t}-5}(\tilde{G}_{2^{2t},3};\mathbb{Z}_2)$ such that the cup product

$$\tilde{w}_2^{2^{2t}-2} z \in H^{\mathrm{top}}(\tilde{G}_{2^{2t},3};\mathbb{Z}_2) \cong \mathbb{Z}_2$$

does not vanish. By what we have said above, since $2^{2t}-5 < 2^{2t}-4$, the element z must be decomposable in the sense that $z = \tilde{w}_2^{a_1}\tilde{w}_3^{b_1} + \ldots \tilde{w}_2^{a_r}\tilde{w}_3^{b_r}$ for some $a_1, \ldots, a_r, b_1, \ldots, b_r$. We conclude that there are some numbers a and b such that the element

$$\tilde{w}_2^{2^{2t}-2+a}\tilde{w}_3^{b} \in H^{\mathrm{top}}(\tilde{G}_{2^{2t},3};\mathbb{Z}_2)$$

does not vanish.

Since $\mathrm{height}(\tilde{w}_2) \leq 2^{2t}-2$, it must be $a = 0$, and b must be a positive integer. In other words, we have, for some $b > 0$, a nonzero element

$$\tilde{w}_2^{2^{2t}-2}\tilde{w}_3^{b} \in H^{3(2^{2t}-3)}(\tilde{G}_{2^{2t},3};\mathbb{Z}_2) \cong \mathbb{Z}_2.$$

By counting dimensions we obtain that $2^{2t+1}-4+3b = 3\cdot 2^{2t}-9$, hence $2^{2t}-5 = 3b$. But the latter is impossible: indeed, we have $2^{2t}-5 = 2^{2(t+1)}-5-3\cdot 2^{2t}$, hence if 3 does not divide $2^{2t}-5$, then 3 does not divide $2^{2(t+1)}-5$; of course, 3 does not divide 2^4-5. This shows that $\tilde{w}_2^{2^{2t}-2} = 0$, and the proof of the corollary is complete. □

Proof of Proposition 1.2. By Ref. 4, we know that

$$\mathrm{height}(w_2(\gamma_{n,k})) = \kappa_2(n,k).$$

The fact that $p^*(w_2(\gamma^a_{n,k})) = w_2(\tilde{\gamma}^a_{n,k})$ implies that we always have

$$\mathrm{height}(w_2(\tilde{\gamma}_{n,k})) \leq \mathrm{height}(w_2(\gamma_{n,k})) = \kappa_2(n,k).$$

At the same time, by Proposition 1.1 we have

$$\mathrm{height}(w_2(\tilde{\gamma}_{n,k})) \leq \tilde{\kappa}_2(n,k).$$

One readily verifies that the bound $\tilde{\kappa}_2(n,k)$ is for many pairs (n,k) better than $\kappa_2(n,k)$; but it is not so for all pairs (n,k). That is why we always take the minimum of the two upper bounds.

Proposition 1.2 is proved. □

Acknowledgments

The author was supported in part by two grants of VEGA (Slovakia). He thanks Tibor Macko for his comments.

References

1. V. Bartík and J. Korbaš, Stiefel–Whitney characteristic classes and parallelizability of Grassmann manifolds, *Rend. Circ. Mat. Palermo (2)* **33** (Suppl. 6) (1984) 19–29.
2. A. Borel, La cohomologie mod 2 de certains espaces homogènes, *Comment. Math. Helvetici* **27** (1953) 165–193.
3. P. Conner, *Differentiable Periodic Maps* (2nd edition, LNM 738, Springer, Berlin, 1979).
4. S. Dutta and S.S. Khare, On second Stiefel–Whitney class of Grassmann manifolds and cuplength, *J. Indian Math. Soc.* **69** (2002) 237–251.
5. S. Ilori and D. Ajayi, The height of the first Stiefel–Whitney class of the real flag manifolds, *Indian J. Pure Appl. Math.* **31** (2000) 621–624.
6. J. Korbaš, Some partial formulae for Stiefel–Whitney classes of Grassmannians, *Czechoslovak Math. J.* **36** (111) (1986) 535–540.
7. J. Korbaš, Bounds for the cup-length of Poincaré spaces and their applications, *Topology Appl.* **153** (2006) 2976–2986.
8. J. Lörinc, The height of the first Stiefel–Whitney class of any nonorientable real flag manifold, *Math. Slovaca* **53** (2003) 91–95.
9. J. Milnor and J. Stasheff, *Characteristic Classes* (Princeton University Press, Princeton, N. J., 1974).
10. R.E. Stong, Cup products in Grassmannians, *Topology Appl.* **13** (1982) 103–113.

Differential Geometry and its Applications
Proc. Conf., in Honour of Leonhard Euler, Olomouc, August 2007

Differential invariants of velocities and higher order Grassmann bundles

D. Krupka and Z. Urban
Department of Algebra and Geometry, Palacký University, Olomouc, Czech Republic
E-mail: krupka@inf.upol.cz, urban@inf.upol.cz
http://globanal.upol.cz

In this paper we present the theory of higher order velocities and their scalar differential invariants. We consider a natural action of a differential group on manifolds of higher order velocities, and study properties of its orbits (contact elements) and orbit spaces (higher order Grassmann bundles). We show that this action defines on a manifold of regular velocities the structure of a principal bundle with structure group the differential group. The bundle projection is then naturally interpreted as the basis of scalar invariants of higher order velocities. We give a recurrence formula for differential invariants. Explicit description of the basis is given for velocities of order ≤ 3. Analogous methods can be applied to the problem of finding bases of differential invariants of different geometric objects.

Keywords: Jet, velocity, differential invariant, Grassmann bundle.

MS classification: 53A55, 58A20, 58A32.

1. Introduction

This paper represents a shortened version of a plenary lecture, delivered by the first author at the conference *Differential Geometry and its Applications*, Olomouc, Czech Republic, August 2007. Its aim is to describe the structure of scalar differential invariants of (higher order) regular velocities.

By the r-th *differential group* L^r_n of $\mathbf{R}^n$ we mean the Lie group of invertible r-jets with source and target at the origin $0 \in \mathbf{R}^n$. Let Y be a manifold of dimension $n+m$. By the (r,n)-*velocity* with values in Y we mean an r-jet with source at $0 \in \mathbf{R}^n$ and target in Y; we denote the set of velocities by T^r_nY. L^r_n acts naturally on T^r_nY, the action being induced by the composition of jets. A real-valued function, constant on orbits of this action, is said to be a *scalar differential invariant*. Every scalar differential invariant is completely described, in the well-known sense, by means of the quotient projection of $T_n^r Y$ onto the quotient space $T_n^r Y/L_n^r$.

Our analysis of the group action of L_n^r on $T_n^r Y$ shows that the manifold of *regular* velocities $\mathrm{imm}\, T_n^r Y \subset T_n^r Y$ is a *principal* L_n^r*-bundle* over $\mathrm{imm}\, T_n^r Y/L_n^r$. We call $\mathrm{imm}\, T_n^r Y/L_n^r$ the *Grassmann prolongation* of Y, and the bundle projection the *basis* of scalar differential invariants of (r, n)-velocities. We find explicit chart description for the quotient projections of regular velocities of order ≤ 3.

The concepts, used in this paper, follow basic ideas of the Ehresmann's theory of jets and contact elements[1] (for generalities see also Ref. 3 and Ref. 5). It should be pointed out that the concept of a higher order Grassmann fibrations has also been considered in a different framework by Olver.[8] The exposition, used in this paper, is based on Grigore, D. Krupka, and M. Krupka.[2,4,6,7] Explicit analysis of the differential invariants of order $r \leq 3$ is taken from Ref. 9; the proofs can be found in Ref. 6 and Ref. 9.

2. Differential invariants

Recall that if G is a group acting on two sets P and Q on the left, then a mapping $f : P \to Q$ is said to be G*-equivariant*, if $f(g \cdot p) = g \cdot f(p)$ for all $g \in G$ and $p \in P$. G-equivariant mappings are also called *invariants* of the group G. If Q is the real line $\mathbf{R}$, endowed with the trivial action of G, an equivariant mapping $f : P \to \mathbf{R}$ is called a *scalar invariant.* These concepts can be applied to the case when G is a differential group. By the r-th *differential group* L_n^r of $\mathbf{R}^n$ we mean the Lie group of invertible r-jets with source and target at the origin $0 \in \mathbf{R}^n$. Recall that L_n^r as the set consists of invertible r-jets $J_0^r\alpha$ of diffeomorphisms α of open sets $U \subset \mathbf{R}^n$, containing the origin $0 \in \mathbf{R}^n$, such that $\alpha(0) = 0$. The multiplication $L_n^r \ni (J_0^r\alpha, J_0^r\beta) \to J_0^r\alpha \cdot J_0^r\beta = J_0^r(\alpha \circ \beta) \in L_n^r$ is given by the composition of jets, and defines the structure of a Lie group on L_n^r. Clearly, L_n^1 is canonically identified with the general linear group $GL_n(\mathbf{R})$, and the canonical jet projection of L_n^r onto L_n^1 is a Lie group homomorphism. Denoting by K_n^r its *kernel*, we can represent L_n^r as the *semi-direct product* of L_n^1 and K_n^r, $L_n^r = L_n^1 \times_s K_n^r$; K_n^r is a *nilpotent* normal subgroup of L_n^r.

If P and Q are two left L_n^r-manifolds, then a mapping $f : P \to Q$ is said to be a *differential invariant*, if for all $J_0^r\alpha \in L_n^r$ and $p \in P$,

$$f(J_0^r\alpha \cdot p) = J_0^r\alpha \cdot f(p). \tag{1}$$

If $\pi : P \to P/L_n^r$ is the quotient projection and we take for Q the real line $\mathbf{R}$ with *trivial* action of the group L_n^r, we have a commutative diagram

$$\begin{array}{ccc} P & \xrightarrow{f} & \mathbf{R} \\ {\scriptstyle\pi}\downarrow & \nearrow_{f_0} & \\ P/L_n^r & & \end{array} \tag{2}$$

defining a mapping $f_0 : P/L_n^r \to \mathbf{R}$. The decomposition

$$f = f_0 \circ \pi \tag{3}$$

then allows us to express the scalar differential invariant f in terms of π and f_0. If, moreover, the orbit space P/L_n^r has a manifold structure such that π is a submersion, then f_0 is differentiable if and only if f is differentiable. We call π the *basis of differential invariants* on the left L_n^r-manifold P.

Note that if Q is a left L_n^1-manifold, e.g. a vector space of tensors over $\mathbf{R}^n$, endowed with the tensor action of the group L_n^1, and $f : P \to Q$ is a differential invariant, then the restriction of the group action of L_n^r to K_n^r yields the condition $f(J_0^r\alpha \cdot p) = f(p)$. This observation indicates the meaning of the semi-direct product structure of L_n^r for computation of differential invariants (the *orbit reduction method*): Each differential invariant is expressible as the composite of a scalar invariant of the group K_n^r and an invariant of the general linear group L_n^1.

3. Velocities

From now on, $m, n \geq 1$ and $r \geq 0$ are integers. $J^r_{(x,y)}(X, Y)$ denotes the set of r-jets with source x in a manifold X and target y in a manifold Y. The composition of jets is denoted $\circ$.

Let Y be a manifold of dimension $m + n$. By an n-velocity of order r at a point $y \in Y$ we mean an r-jet $P \in J^r_{(0,y)}(\mathbf{R}^n, Y)$, $P = J_0^r\zeta$. We denote

$$T_n^r Y = \bigcup_{y \in Y} J^r_{(0,y)}(\mathbf{R}^n, Y), \tag{4}$$

and define surjective mappings $\tau_n^{r,s} : T_n^r Y \to T_n^s Y$, where $0 \leq s \leq r$, by $\tau_n^{r,s}(J_0^r\zeta) = J_0^s\zeta$. The set $T_n^r Y$ is endowed with a right action of the group L_n^r, defined by the jet composition

$$T_n^r Y \times L_n^r \ni (P, A) \to P \circ A \in T_n^r Y. \tag{5}$$

For every chart (V, ψ), $\psi = (y^K)$, on Y we set $V_n^r = (\tau_n^{r,0})^{-1}(V)$, and $\psi_n^r = (y^K, y^K_{i_1}, y^K_{i_1 i_2}, \ldots, y^K_{i_1 i_2 \ldots i_r})$, where $1 \leq K \leq n + m$, $1 \leq i_1 \leq i_2 \leq \ldots \leq i_r \leq n$, and for every $P \in V_n^r$, $P = J_0^r\zeta$, $y^K_{i_1 i_2 \ldots i_l}(P) =$

$D_{i_1} D_{i_2} \ldots D_{i_l}(y^K \zeta)(0)$. Note that this formula can also be written in a different way. Denote by tr_ξ the *translation* of $\mathbf{R}^{n+m}$, expressed by the equation $\mathrm{tr}_\xi(x) = x - \xi$. Writing tr_ξ^K for the components of the translation, we can write $y^K_{i_1 i_2 \ldots i_l}(P) = D_{i_1} D_{i_2} \ldots D_{i_l}(\mathrm{tr}^K_{\psi\zeta(0)} \psi\zeta)(0)$.

In the following lemma we denote $L^r_{n,m} = J^r_{(0,0)}(\mathbf{R}^n, \mathbf{R}^m)$.

Lemma 3.1. *There exists one and only one smooth structure on $T^r_n Y$ such that for every chart (V, ψ), the pair (V^r_n, ψ^r_n) is a chart on $T^r_n Y$. The dimension of $T^r_n Y$ is $N = (n+m)\binom{n+r}{n}$. In this smooth structure $T^r_n Y$ is a smooth fibration with fiber $L^r_{n,m}$, base Y, and projection $\tau^{r,0}_n$. The group action (5) is smooth.*

The set $T^r_n Y$ endowed with the smooth structure and with the group action, defined by Lemma 3.1, is called the *manifold of n-velocities of order r* over Y. The chart (V^r_n, ψ^r_n) on $T^r_n Y$ is said to be *associated* with the chart (V, ψ).

The group action (5) can be easily determined in the canonical coordinates on L^r_n and a chart (V, ψ) on Y. Using the associated chart (V^r_n, ψ^r_n), we get the equations

$$\bar{y}^K = y^K, \quad \bar{y}^K_{i_1 i_2 \ldots i_l} = \sum_{p=1}^{l} y^K_{j_1 j_2 \ldots j_p} \sum_{(I_1, I_2, \ldots, I_p)} a^{j_1}_{I_1} a^{j_2}_{I_2} \ldots a^{j_p}_{I_p}, \tag{6}$$

where the second sum is extended to all partitions $(I_1, I_2, \ldots, I_p)$ of the set $\{i_1, i_2, \ldots, i_l\}$.

Let γ be a smooth mapping of an open set $U \subset \mathbf{R}^n$ into Y. Then for any $t \in U$, the mapping $x \to \gamma \circ \mathrm{tr}_{-t}(x)$ is defined on a neighbourhood of the origin $0 \in \mathbf{R}^n$ so the mapping $U \ni t \to (T^r_0 \gamma)(t) = J^r_0(\gamma \circ \mathrm{tr}_{-t}) \in T^r_n Y$ is defined. This mapping is called the *r-prolongation* of γ. Its chart expression is given by

$$y^K_{i_1 i_2 \ldots i_l} \circ T^r_0 \gamma(t) = D_{i_1} D_{i_2} \ldots D_{i_l}(y^K \gamma)(t). \tag{7}$$

Let $P \in T^r_n Y$, $P = J^r_0 \zeta$. A representative ζ of P defines the tangent mapping $T_0 T^{r-1}_0 \zeta$, which sends a tangent vector $\xi \in T_0 \mathbf{R}^n$ to the tangent vector $T_0 T^{r-1}_0 \zeta \cdot \xi$ of $T^{r-1}_n Y$ at $\tau^{r,r-1}_n(P) = J^{r-1}_0 \zeta$. If $\xi = \xi^i (\partial/\partial t^i)_0$, then by (7),

$$T_0 T^{r-1}_0 \zeta \cdot \xi = \xi^i d_i(P), \tag{8}$$

where

$$d_i(P) = \sum_{l=0}^{r-1} \sum_{i_1 \le i_2 \le \ldots \le i_l} y^K_{i_1 i_2 \ldots i_l i}(P) \left(\frac{\partial}{\partial y^K_{i_1 i_2 \ldots i_l}} \right)_{J^{r-1}_0 \zeta} \tag{9}$$

defines the *i-th formal derivative morphism* $T_n^r Y \ni P \to d_i(P) \in TT_n^{r-1}Y$ over $T_n^{r-1}Y$. The tangent vectors (9) are defined independently of the chart. Note that $\partial/\partial y^K_{i_1 i_2 \ldots i_l}$ are understood as tangent vectors to $T_n^{r-1}Y$; (9) is not a vector field.

The *i-th formal derivative* of a function $f : V_n^{r-1} \to \mathbf{R}$ is defined by

$$d_i f = \sum_{l=0}^{r-1} \sum_{i_1 \le i_2 \le \ldots \le i_l} y^K_{i_1 i_2 \ldots i_l i} \frac{\partial f}{\partial y^K_{i_1 i_2 \ldots i_l}}. \tag{10}$$

Since by (7), $d_i f \circ T_0^r \gamma = D_i(f \circ T_0^{r-1}\gamma)$, we have $d_i d_j f = d_j d_i f$. Note that if we take $f = y^L_{i_1 i_2 \ldots i_l}$ in (10), we get

$$d_i y^L_{i_1 i_2 \ldots i_l} = y^L_{i_1 i_2 \ldots i_l i}. \tag{11}$$

We give explicit transformation formulas between the associated charts on $T_n^r Y$. Suppose we have the transformation equations between two charts (V, ψ) to $(\bar{V}, \bar{\psi})$

Our aim now will be to derive explicit transformation formulas between the induced charts on $T_n^r Y$. Let us write the transformation equations from (V, ψ) and $(\bar{V}, \bar{\psi})$, $\bar{y}^K = F^K(y^L)$. We wish to determine the functions $F^K_{i_1}, F^K_{i_1 i_2}, \ldots, F^K_{i_1 i_2 \ldots i_r}$, defining induced transformations $\bar{y}^K_{i_1 i_2 \ldots i_l} = F^K_{i_1 i_2 \ldots i_l}(y^L, y^L_{j_1}, y^L_{j_1 j_2}, \ldots, y^L_{j_1 j_2 \ldots j_l})$ from (V_n^r, ψ_n^r) to $(\bar{V}^r_n, \bar{\psi}^r_n)$. Note that since $\bar{d}_i = d_i$, we have from (11), $\bar{y}^L_{i_1 i_2 \ldots i_l i_{l+1}} = \bar{d}_{i_{l+1}} \bar{y}^L_{i_1 i_2 \ldots i_l} = d_{i_{l+1}} d_{i_l} \bar{y}^L_{i_1 i_2 \ldots i_{l-1}} = \ldots = d_{i_{l+1}} d_{i_l} \ldots d_{i_1} \bar{y}^L$. In the following formula we sum through all partitions $(I_1, I_2, \ldots, I_p)$ of the set $\{i_1, i_2, \ldots, i_l\}$.

Lemma 3.2. *The functions* $F^K_{i_1 i_2 \ldots i_l}$ *are given by*

$$F^K_{i_1 i_2 \ldots i_l} = \sum_{p=1}^{l} \sum_{(I_1, I_2, \ldots, I_p)} y^{L_1}_{I_1} y^{L_2}_{I_2} \ldots y^{L_p}_{I_p} \frac{\partial^p F^K}{\partial y^{L_1} \partial y^{L_2} \ldots \partial y^{L_p}}. \tag{12}$$

4. Regular velocities

We need a convention on splitting of the sequence $(1, 2, \ldots, n, n+1, \ldots, n+m)$ into two complementary subsequences. Any subsequence, consisting of n integers, is called an *n-subsequence*. Given an n-subsequence $(i) = (i_1, i_2, \ldots, i_n)$, we also have the *complementary subsequence* $(\sigma) = (\sigma_1, \sigma_2, \ldots, \sigma_m)$. By definition $i_1 < i_2 < \ldots < i_n$ and $\sigma_1 < \sigma_2 < \ldots < \sigma_m$.

A velocity $P \in T_n^r Y$ is said to be *regular*, if $P = J_0^r \zeta$, where ζ is an *immersion* at the origin $0 \in \mathbf{R}^n$. This condition is equivalent with the existence of a chart (V, ψ), $\psi = (y^K)$, at $\zeta(0)$ and an n-subsequence

$(i) = (i_1, i_2, \dots, i_n)$ of the sequence $(1, 2, \dots, n, n+1, \dots, n+m)$ such that, with $i \in (i_1, i_2, \dots, i_n)$,

$$\det\left(y^i_j(P)\right) = \det\left(D_j(y^i \circ \zeta)(0)\right) \neq 0. \tag{13}$$

The subset of regular velocities in $T^r_n Y$ is denoted by $\operatorname{imm} T^r_n Y$.

Lemma 4.1. *The set* $\operatorname{imm} T^r_n Y$ *is an open, dense,* L^r_n*-invariant subset of* $T^r_n Y$.

$\operatorname{imm} T^r_n Y$ is called the *manifold of regular n-velocities of order r* over Y.

Our aim now will be to analyze the equivalence $\mathscr{R} \subset \operatorname{imm} T^r_n Y \times \operatorname{imm} T^r_n Y$, defined by the group action (5), "*there exists* $A \in L^r_n$ *such that* $P = Q \circ A$". Fix a chart (V, ψ), $\psi = (y^K)$, on Y, and consider the associated chart (V^r_n, ψ^r_n), $\psi^r_n = (y^K, y^K_{i_1}, y^K_{i_1 i_2}, \dots, y^K_{i_1 i_2 \dots i_r})$, on $\operatorname{imm} T^r_n Y$. We set for every n-subsequence $(i) = (i_1, i_2, \dots, i_n)$ of $(1, 2, \dots, n, n+1, \dots, n+m)$

$$W^{(i)} = \left\{P \in V^r_n \mid \det\left(y^i_j(P)\right) \neq 0\right\}. \tag{14}$$

$W^{(i)}$ is an open, L^r_n-invariant subset of the set V^r_n. Shrinking the coordinates $\psi^r_n = (y^K, y^K_{i_1}, y^K_{i_1 i_2}, \dots, y^K_{i_1 i_2 \dots i_r})$ to $W^{(i)}$, we get a chart, *associated with* (V, ψ) and (i), and denoted by $(W^{(i)}, \chi^{(i)})$. Charts of this type clearly cover V^r_n, and constitute an atlas on $\operatorname{imm} T^r_n Y$. The coordinate transformations between $(W^{(i)}, \chi^{(i)})$ and $(W^{(j)}, \chi^{(j)})$ coincides with the restriction of the identity mapping of V^r_n to $W^{(i)} \cap W^{(j)}$.

We introduce a collection of functions $z^k_i : W^{(i)} \to \mathbf{R}$ by

$$z^k_i y^i_q = \delta^k_q, \tag{15}$$

where $i \in (i)$ and $1 \leq k, q \leq n$. Existence of these functions is guaranteed by the definition of $W^{(i)}$; z^k_i is a rational function of y^i_q.

Lemma 4.2. *Let* $(P, Q) \in \operatorname{imm} T^r_n Y \times \operatorname{imm} T^r_n Y$. *The following conditions are equivalent:*

(a) $(P, Q) \in \mathscr{R}$.

(b) There exist a chart (V, ψ), $\psi = (y^K)$, *and an n-subsequence* (i) *of the sequence* $(1, 2, \dots, n, n+1, \dots, n+m)$, *such that* $P, Q \in W^{(i)}$, *the coordinates* $y^K_{i_1 i_2 \dots i_l}$ *(resp.* $\bar{y}^K_{i_1 i_2 \dots i_l}$, *resp.* a^j_I*) of* P *(resp.* Q, *resp.* A*) satisfy*

$$\bar{y}^K = y^K, \quad \bar{y}^K_{i_1 i_2 \dots i_l} = \sum_{p=1}^{l} y^K_{j_1 j_2 \dots j_p} \sum_{(I_1, I_2, \dots, I_p)} a^{j_1}_{I_1} a^{j_2}_{I_2} \dots a^{j_p}_{I_p}, \tag{16}$$

where $1 \leq i_1, i_2, \ldots, i_l, j_1, j_2, \ldots, j_p \leq n$, $1 \leq l, p \leq r$, *and the recurrence formula*

$$a^q_{i_1 i_2 \ldots i_l} = z^q_i \left(\bar{y}^i_{i_1 i_2 \ldots i_l} - \sum_{p=2}^{l} y^i_{j_1 j_2 \ldots j_p} \sum_{(I_1, I_2, \ldots, I_p)} a^{j_1}_{I_1} a^{j_2}_{I_2} \ldots a^{j_p}_{I_p} \right), \tag{17}$$

where $i \in (i)$, *and* (σ) *is the complementary subsequence of* (i).

We now introduce new charts on $\operatorname{imm} T^r_n Y$, adapted to the group action (5). Let $P \in W^{(i)}$, $P = J^r_0 \zeta$. Any representative ζ defines the $(r-1)$-prolongation $T^{r-1}_n \zeta(t) = J^{r-1}_0 (\zeta \circ \mathrm{tr}_{-t})$. The mapping $\psi^{(i)} \circ \zeta = (y^{i_1}\zeta, y^{i_2}\zeta, \ldots, y^{i_n}\zeta)$, where $(i) = (i_1, i_2, \ldots, i_n)$, is a diffeomorphism at $0 \in \mathbf{R}^n$. We have a well-defined mapping $T^{r-1}_n \zeta \circ (\psi^{(i)} \circ \zeta)^{-1} \circ \psi^{(i)} \circ \tau^{r,0}_n$, and its tangent mapping at P, denoted

$$h^{(i)} : T_{J^r_0 \zeta} \operatorname{imm} T^r_n Y \to T_{J^{r-1}_0 \zeta} \operatorname{imm} T^{r-1}_n Y. \tag{18}$$

Let ξ be a tangent vector to $\operatorname{imm} T^r_n Y$ at P,

$$\xi = \sum_{l=0}^{r} \xi^K_{i_1 i_2 \ldots i_l} \left(\frac{\partial}{\partial y^K_{i_1 i_2 \ldots i_l}} \right)_P . \tag{19}$$

After some computation

$$h^{(i)}(\xi) = \xi^j \Delta_j(P), \tag{20}$$

where

$$\Delta_j = z^q_j d_q. \tag{21}$$

Lemma 4.3. *(a) For every* $i, k \in (i)$, $\Delta_i \Delta_k = \Delta_k \Delta_i$.

(b) Let (i) *and* (j) *be two* n*-subsequences of the sequence* $(1, 2, \ldots, n+m)$. *For any two charts* (V, ψ) *and* $(\bar{V}, \bar{\psi})$, *and the associated charts* $(W^{(i)}, \chi^{(i)})$ *and* $(\overline{W}^{(j)}, \bar{\chi}^{(j)})$, $\bar{\Delta}_j = \bar{z}^s_j y^i_s \Delta_i$.

Theorem 4.1. *Let* (V, ψ), $\psi = (y^K)$, *be a chart on* Y, *let* (i) *be an* n*-subsequence of the sequence* $(1, 2, \ldots, n+m)$, (σ) *the complementary subsequence, and let* $(W^{(i)}, \chi^{(i)})$, $\chi^{(i)} = (y^K, y^K_{i_1}, y^K_{i_1 i_2}, \ldots, y^K_{i_1 i_2 \ldots i_r})$, *be the chart, associated with* (V, ψ) *and* (i).

(a) There exist unique functions $w^\sigma, w^\sigma_{i_1}, w^\sigma_{i_1 i_2}, \ldots, w^\sigma_{i_1 i_2 \ldots i_r}$, *with* $i_1, i_2, \ldots, i_r \in (i)$, $\sigma \in (\sigma)$, *defined on* $W^{(i)}$, *symmetric in the subscripts, such that*

$$y^\sigma = w^\sigma, \quad y^\sigma_{p_1 p_2 \ldots p_k} = \sum_{q=1}^{k} \sum_{(I_1, I_2, \ldots, I_q)} y^{i_1}_{I_1} y^{i_2}_{I_2} \ldots y^{i_q}_{I_q} w^\sigma_{i_1 i_2 \ldots i_q} \tag{22}$$

(summation through partitions $(I_1, I_2, \ldots, I_q)$ of the set $\{p_1, p_2, \ldots, p_k\}$). These functions are L^r_n-invariant, and satisfy the recurrence formulas

$$w^{\sigma}_{i_1 i_2 \ldots i_k i_{k+1}} = \Delta_{i_{k+1}} w^{\sigma}_{i_1 i_2 \ldots i_k}. \tag{23}$$

(b) The pair $(W^{(i)}, \Psi^{(i)})$, where

$$\Psi^{(i)} = (y^i, y^i_{p_1}, y^i_{p_1 p_2}, \ldots, y^i_{p_1 p_2 \ldots p_r}, w^{\sigma}, w^{\sigma}_{i_1}, w^{\sigma}_{i_1 i_2}, \ldots, w^{\sigma}_{i_1 i_2 \ldots i_r}), \tag{24}$$

is a chart on $\operatorname{imm} T^r_n Y$.

(c) The group action (5) is described on $W^{(i)}$ by the equations

$$\begin{gathered} \bar{y}^i = y^i, \quad \bar{w}^{\sigma} = w^{\sigma}, \quad \bar{w}^{\sigma}_{i_1 i_2 \ldots i_k} = w^{\sigma}_{i_1 i_2 \ldots i_k}, \\ \bar{y}^i_{p_1 p_2 \ldots p_k} = \sum_{q=1}^{k} \sum_{(I_1, I_2, \ldots, I_q)} a^{j_1}_{I_1} a^{j_2}_{I_2} \ldots a^{j_q}_{I_q} y^i_{j_1 j_2 \ldots j_q}, \end{gathered} \tag{25}$$

where $i, i_1, i_2, \ldots, i_k \in (i)$, $\sigma \in (\sigma)$, and $1 \leq k \leq r$. Equations of the orbits are

$$w^{\sigma}_{i_1 i_2 \ldots i_k} = c^{\sigma}_{i_1 i_2 \ldots i_k}, \tag{26}$$

where $c^{\sigma}_{i_1 i_2 \ldots i_k} \in \mathbf{R}$.

It is now easy to prove the following result.

Theorem 4.2. *If Y is Hausdorff, then the right action (5) defines the structure of a right principal L^r_n-bundle on* $\operatorname{imm} T^r_n Y$.

If $m = 0$, then $\dim Y = n$, and a regular n-velocity of order r at a point $y \in Y$ is called an *r-frame* at y. In this case we usually write $\operatorname{imm} T^r_n Y = F^r Y$, and call the principal L^r_n-bundle $F^r Y$ the *bundle of r-frames* over Y. If $r = 1$, we get the principal $GL_n(\mathbf{R})$-bundle $FY = F^1 Y$ of *linear* frames.

An L^r_n-orbit, passing through a point $P \in \operatorname{imm} T^r_n Y$, is called an *$n$-contact element of order r*; if $P = J^r_0 \zeta$, we usually denote this contact element by $G^r_0 \zeta$. The orbit manifold $G^r_n Y = \operatorname{imm} T^r_n Y / L^r_n$, i.e., the base of the principal L^r_n-bundle $\operatorname{imm} T^r_n Y$, is the *manifold of n-contact elements of order r*; $G^r_n Y$ is also called the *(n, r)-Grassmann prolongation* of Y.

5. Grassmann prolongations of manifolds

Denote by π^r_n the quotient projection of $\operatorname{imm} T^r_n Y$ onto $G^r_n Y$. From the construction of the Grassmann prolongation, there exists a unique mapping

ρ^r_n of $G^r_n Y$ onto Y such that the following diagram

$$\begin{array}{ccc} \operatorname{imm} T^r_n Y & \xrightarrow{\pi^r_n} & G^r_n Y \\ \downarrow & \swarrow_{\rho^r_n} & \\ Y & & \end{array} \tag{27}$$

commutes. We now describe explicitly the structure of $G^r_n Y$ and ρ^r_n.

Consider a chart (V, ψ), $\psi = (y^K)$, on Y, and the associated chart (V^r_n, ψ^r_n), $\psi^r_n = (y^K, y^K_{i_1}, y^K_{i_1 i_2}, \ldots, y^K_{i_1 i_2 \ldots i_r})$, on $\operatorname{imm} T^r_n Y$. Fix an n-subsequence (i) of the sequence $(1, 2, \ldots, n+m)$, and consider, as in Section 4, Theorem 4.1, the chart $(W^{(i)}, \Psi^{(i)})$, $\Psi^{(i)} = (y^i, y^i_{p_1}, y^i_{p_1 p_2}, \ldots, y^i_{p_1 p_2 \ldots p_r}, w^\sigma, w^\sigma_{i_1}, w^\sigma_{i_1 i_2}, \ldots, w^\sigma_{i_1 i_2 \ldots i_r})$, on $\operatorname{imm} T^r_n Y$. We have

$$w^\sigma_{i_1 i_2 \ldots i_{k-1} i_k} = \Delta_{i_k} \Delta_{i_{k-1}} \ldots \Delta_{i_2} \Delta_{i_1} w^\sigma, \tag{28}$$

where

$$\begin{aligned} \Delta_i = \frac{\partial}{\partial y^i} &+ \sum_{l=0}^{r-1} \sum_{i_1 \le i_2 \le \ldots \le i_l} w^\sigma_{i_1 i_2 \ldots i_l i} \frac{\partial}{\partial w^\sigma_{i_1 i_2 \ldots i_l}} \\ &+ \sum_{l=1}^{r-1} \sum_{i_1 \le i_2 \le \ldots \le i_l} z^s_i y^k_{i_1 i_2 \ldots i_l s} \frac{\partial}{\partial y^k_{i_1 i_2 \ldots i_l}}. \end{aligned} \tag{29}$$

Denoting $W^{(i)}_G = \pi^r_n(W^{(i)})$, $\Psi^{(i)}_G = (y^i, w^\sigma, w^\sigma_{i_1}, w^\sigma_{i_1 i_2}, \ldots, w^\sigma_{i_1 i_2 \ldots i_r})$, we obtain the associated chart on the Grassmann prolongation $G^r_n Y$. Transformation equations between coordinates of these charts are considered in the following lemma.

Lemma 5.1. *Let (V, ψ), $\psi = (y^K)$, and $(\bar V, \bar\psi)$, $\bar\psi = (\bar y^K)$, be two charts on Y such that $V \cap \bar V \neq \emptyset$, with transformation equations*

$$\bar y^j = F^j(y^i, w^\sigma), \quad \bar w^\nu = F^\nu(y^i, w^\sigma). \tag{30}$$

Then

$$\bar w^\nu_{j_1} = \bar z^s_{j_1} y^i_s \left(\frac{\partial F^\nu}{\partial y^i} + w^\sigma_i \frac{\partial F^\nu}{\partial w^\sigma} \right). \tag{31}$$

To obtain transformation formulas for further coordinates, one should apply the recurrence formula (28).

As before, denote $L^r_{n,n-m} = J^r_{(0,0)}(\mathbf{R}^n, \mathbf{R}^{n+m})$, and consider the subset $\operatorname{imm} L^r_{n,n+m}$ of $L^r_{n,n+m}$, formed by jets of immersions at $0 \in \mathbf{R}^n$. Formula (7) defines on $\operatorname{imm} L^r_{n,n+m}$ a right action of the differential group L^r_n; we denote $G^r_{n,n+m} = \operatorname{imm} L^r_{n,n+m}/L^r_n$. $G^r_{n,n+m}$ is a manifold, called the *n-Grassmannian of order r* over $\mathbf{R}^{n+m}$.

$G^r_{n,n+m}$ is endowed with a left action of the differential group L^r_{n+m}. If $G^r_0\zeta$ is the contact element, containing an r-jet $J^r_0\zeta$, and $A \in L^r_{n+m}$, $A = J^r_0\alpha$, then $G^r_0(\alpha\circ\zeta)$ is defined, and depends on A only. We set $A \cdot G^r_0\zeta = G^r_0(\alpha\circ\zeta)$. L^r_{n+m} also acts on the product $F^rY \times G^r_{n,n+m}$; we have the right action

$$\begin{aligned} F^rY \times G^r_{n,n+m} \times L^r_{n+m} \ni (J^r_0\mu, G^r_0\zeta, J^r_0\alpha) \to \\ \to \left(J^r_0(\mu\circ\alpha), G^r_0(\alpha^{-1}\circ\zeta)\right) \in F^rY \times G^r_{n,n+m}. \end{aligned} \tag{32}$$

Let F^rY denote the bundle of r-frames over Y. The principal L^r_{n+m}-bundle structure of F^rY and the structure of the Grassmannian $G^r_{n,n+m}$ allow us to define the mapping $F^rY \times G^r_{n,n+m} \ni (J^r_0\mu, G^r_0\zeta) \to G^r_0(\mu\circ\zeta) \in G^r_nY$; it is immediately seen that this mapping is constant on L^r_{n+m}-orbits, and is smooth, i.e., is a *frame mapping*. Thus, we have the following assertion.

Theorem 5.1. *G^r_nY has the structure of a fiber bundle over Y with fiber $G^r_{n,n+m}$, associated with the principal bundle of frames F^rY.*

6. Example: Scalar differential invariants of order ≤ 3

The action (5) is for $r = 3$ expressed by the equations

$$\begin{aligned} \bar{y}^K &= y^K, \quad \bar{y}^K_{p_1} = a^{k_1}_{p_1} y^K_{k_1}, \quad \bar{y}^K_{p_1p_2} = a^{k_1}_{p_1} a^{k_2}_{p_2} y^K_{k_1k_2} + a^{k_1}_{p_1p_2} y^K_{k_1}, \\ \bar{y}^K_{p_1p_2p_3} &= a^{k_1}_{p_1} a^{k_2}_{p_2} a^{k_3}_{p_3} y^K_{k_1k_2k_3} + \left(a^{k_2}_{p_2p_3} a^{k_1}_{p_1} + a^{k_1}_{p_1p_3} a^{k_2}_{p_2} + a^{k_1}_{p_1p_2} a^{k_2}_{p_3}\right) y^K_{k_1k_2} \\ &\quad + a^{k_1}_{p_1p_2p_3} y^K_{k_1}, \end{aligned} \tag{33}$$

where $K \in (i), (\sigma)$, $1 \leq p_1, p_2, p_3 \leq n$, $1 \leq k_1, k_2, k_3 \leq n$.

Computing the projection $\pi^3_n : \operatorname{imm} T^r_nY \to G^3_nY$ explicitly, we get the following theorem, characterizing all scalar differential invariants of velocities of order ≤ 3 in terms of their basis.

Theorem 6.1. *Every scalar differential invariant of regular velocities of order ≤ 3 depends only on the following functions:*

$$\begin{aligned} w^\sigma_{q_1} &= z^{k_1}_{q_1} y^\sigma_{k_1}, \\ w^\sigma_{q_1q_2} &= z^{k_1}_{q_1} z^{k_2}_{q_2} \left(y^\sigma_{k_1k_2} - z^l_i y^\sigma_l y^i_{k_1k_2}\right), \\ w^\sigma_{q_1q_2q_3} &= z^{k_1}_{q_1} z^{k_2}_{q_2} z^{k_3}_{q_3} \left(y^\sigma_{k_1k_2k_3} - z^l_i y^\sigma_l y^i_{k_1k_2k_3}\right) - z^{k_2}_{q_2} z^{k_3}_{q_3} y^j_{k_2k_3} z^{k_1}_{q_1} z^p_j \\ &\quad \cdot \left(y^\sigma_{k_1p} - z^l_i y^\sigma_l y^i_{k_1p}\right) - z^{k_1}_{q_1} z^{k_3}_{q_3} y^j_{k_1k_3} z^{k_2}_{q_2} z^p_j \left(y^\sigma_{pk_2} - z^l_i y^\sigma_l y^i_{pk_2}\right) \\ &\quad - z^{k_1}_{q_1} z^{k_2}_{q_2} y^j_{k_1k_2} z^{k_3}_{q_3} z^p_j \left(y^\sigma_{pk_3} - z^l_i y^\sigma_l y^i_{pk_3}\right). \end{aligned} \tag{34}$$

Acknowledgments

The authors acknowledge support of grants 201/06/0922 (GACR) and MSM 6198959214 (Czech Ministry of Education).

References

1. C. Ehresmann, Les prolongements d'une variete différentiable I–V, *C. R. Acad. Sc. Paris* **223** (1951) 598–600, 777–779, 1081–1083; **234** (1952) 1028–1030, 1424–1425.
2. D.R. Grigore and D. Krupka, Invariants of velocities and higher order Grassmann bundles, *J. Geom. Phys.* **24** (1998) 244–264.
3. I. Kolář, P. W. Michor and J. Slovák, *Natural Operations in Differential Geometry* (Springer-Verlag, Berlin, 1993).
4. D. Krupka, Natural Lagrangian structures, *In: Semester on Diff. Geom.* (Banach Center, Warsaw, 1979; Banach Center Publications 12, 1984) 185–210.
5. D. Krupka and J. Janyška, *Lectures on Differential Invariants* (J. E. Purkyně University, Faculty of Science, Brno, Czechoslovakia, 1990, pp. 193).
6. D. Krupka and M. Krupka, Jets and contact elements, *In: Proc. Sem. on Diff. Geom.* (MATH. Publications Vol. 2, Silesian Univ. in Opava, Opava, Czech Republic, 2000) 39–85.
7. M. Krupka, Orientability of higher order grassmannians, *Math. Slovaca* **44** (1994) 107–115.
8. P.J. Olver, Symmetry groups and group invariant solutions of partial differential equations, *J. Diff. Geom.* **14** (1979) 497–542.
9. Z. Urban, Grassmann prolongation, thesis, Palacký University, Olomouc, Czech Republic, 2005.

Differential Geometry and its Applications
Proc. Conf., in Honour of Leonhard Euler, Olomouc, August 2007

Like jet prolongation functors of affine bundles

J. Kurek

Institute of Mathematics, Maria Curie-Sklodowska University,
Lublin, Poland
E-mail: kurek@hektor.umcs.lublin.pl

W.M. Mikulski

Institute of Mathematics, Jagiellonian University,
Kraków, Poland
E-mail: mikulski@im.uj.edu.pl

We present a complete description of all fiber product preserving gauge bundle functors F on the category $\mathcal{AB}_m$ of affine bundles with m-dimensional bases and affine bundle maps with local diffeomorphisms as base maps.

Keywords: (fiber product preserving) gauge bundle functors, natural transformations, jets, Weil algebras.

MS classification: 58A05, 58A20.

1. Introduction

It is well-known that product or fiber product preserving (gauge) bundle functors play a very important role in differential geometry. To such bundle functors one can lift some geometric structures as vector fields, forms, connections, e.t.c.

The product preserving bundle functors on the category $\mathcal{M}f$ of manifolds and maps have been classified by means of Weil algebras.[3]

The fiber product preserving bundle functors on the category $\mathcal{FM}_m$ of fibred manifolds with m-dimensional bases and fiber preserving maps with local diffeomorphisms as base maps have been classified in Ref. 4, and studied in Refs. 1,2.

The fiber product preserving gauge bundle functors on the category $\mathcal{VB}_m$ of vector bundles with m-dimensional bases and their vector bundle maps with local diffeomorphisms as base maps have been classified in Ref. 5.

The purpose of the present paper is to describe all fiber product preserving gauge bundle functors on the category $\mathcal{AB}_m$ of affine bundles with m-dimensional bases and affine bundle maps with local diffeomorphisms as base maps.

Let us recall the following definitions (see for ex. Ref. 3).

Let $F : \mathcal{AB}_m \to \mathcal{FM}$ be a covariant functor into the category $\mathcal{FM}$ of fibred manifolds and their fibred maps. Let $B_{\mathcal{AB}_m} : \mathcal{AB}_m \to \mathcal{M}f$ and $B_{\mathcal{FM}} : \mathcal{FM} \to \mathcal{M}f$ be the respective base functors.

A gauge bundle functor on $\mathcal{AB}_m$ is a functor F as above satisfying:

(i) (**Base preservation**) $B_{\mathcal{FM}} \circ F = B_{\mathcal{AB}_m}$. Hence the induced projections form a functor transformation $\pi : F \to B_{\mathcal{AB}_m}$.

(ii) (**Localization**) For every inclusion of an open affine subbundle $i_{E|U} : E|U \to E$, $F(E|U)$ is the restriction $\pi^{-1}(U)$ of $\pi : FE \to B_{\mathcal{AB}_m}(E)$ over U and $Fi_{E|U}$ is the inclusion $\pi^{-1}(U) \to FE$.

(iii) (**Regularity**) F transforms smoothly parametrized systems of $\mathcal{AB}_m$-morphisms into smoothly parametrized systems of $\mathcal{FM}$-morphisms.

A gauge bundle functor $F : \mathcal{AB}_m \to \mathcal{FM}$ is *of finite order* r if from $j^r_x f = j^r_x g$ it follows that $F_x f = F_x g$ for any $\mathcal{AB}_m$-objects $E_1 \to M_1$, $E_2 \to M_2$, any $\mathcal{AB}_m$-maps $f, g : E_1 \to E_2$ and any $x \in M_1$.

Given two gauge bundle functors F_1, F_2 on $\mathcal{AB}_m$, by *a natural transformation* $\mu : F_1 \to F_2$ we shall mean a system of base preserving fibred maps $\mu : F_1E \to F_2E$ for every affine bundle E from $\mathcal{AB}_m$ satisfying $F_2 f \circ \mu = \mu \circ F_1 f$ for every $\mathcal{AB}_m$-morphism $f : E \to G$.

A gauge bundle functor F on $\mathcal{AB}_m$ is *fiber product preserving* if for every fiber product projections $E_1 \xleftarrow{pr_1} E_1 \times_M E_2 \xrightarrow{pr_2} E_2$ in the category $\mathcal{AB}_m$ the mappings $FE_1 \xleftarrow{Fpr_1} F(E_1 \times_M E_2) \xrightarrow{Fpr_2} FE_2$ are fiber product projections in the category $\mathcal{FM}$. In other words $F(E_1 \times_M E_2) = F(E_1) \times_M F(E_2)$ modulo the restriction of (Fpr_1, Fpr_2).

A simple example of fiber product preserving gauge bundle functor on $\mathcal{AB}_m$ is the functor $(\ )^{\to} : \mathcal{AB}_m \to \mathcal{FM}$ sending any $\mathcal{AB}_m$-object $E \to M$ into the corresponding vector bundle $E^{\to} \to M$ and any $\mathcal{AB}_m$-map $f : E_1 \to E_2$ into the corresponding vector bundle map $f^{\to} : E_1^{\to} \to E_2^{\to}$. In fact, $(\ )^{\to} : \mathcal{AB}_m \to \mathcal{VB}_m$.

If we compose $(\ )^{\to} : \mathcal{AB}_m \to \mathcal{VB}_m$ with a fiber product preserving gauge bundle functor $T^{(V,H,t)}$ for so called admissible triple (V, H, t) of some finite order r, see Ref. 5, we obtain fiber product preserving gauge bundle functor $T^{(V,H,t)} \circ (\ )^{\to}$ on $\mathcal{AB}_m$.

The most important example of fiber product preserving gauge bundle functor on $\mathcal{AB}_m$ is the r-jet prolongation functor $J^r : \mathcal{AB}_m \to$

$\mathcal{FM}$, where for an $\mathcal{AB}_m$-object $p : E \to M$ we have $J^rE = \{j^r_x\sigma \mid \sigma$ is a local section of E, $x \in M\}$ and for a $\mathcal{AB}_m$-map $f : E_1 \to E_2$ covering $\underline{f} : M_1 \to M_2$ we have $J^rf : J^rE_1 \to J^rE_2$, $J^rf(j^r_x\sigma) = j^r_{\underline{f}(x)}(f \circ \sigma \circ \underline{f}^{-1})$, $j^r_x\sigma \in J^rE_1$. This functor plays an important role in the theory of higher order connections, Lagrangians, differential equations, e.t.c.

Another example is the so called vertical r-jet prolongation functor $J^r_v : \mathcal{AB}_m \to \mathcal{FM}$, where for a $\mathcal{AB}_m$-object $p : E \to M$ we have $J^r_vE = \{j^r_x\gamma \mid \gamma$ is a local map $M \to E_x$, $x \in M\}$ and for a $\mathcal{AB}_m$-map $f : E_1 \to E_2$ covering $\underline{f} : M_1 \to M_2$ we have $J^r_vf : J^r_vE_1 \to J^r_vE_2$, $J^r_vf(j^r_x\gamma) = j^r_{\underline{f}(x)}(f \circ \gamma \circ \underline{f}^{-1})$, $j^r_x\gamma \in J^r_vE_1$.

Another example is the so called vertical Weil functor $V^A : \mathcal{AB}_m \to \mathcal{FM}$ corresponding to a Weil algebra A, where for a $\mathcal{AB}_m$-object $p : E \to M$ we have $V^AE = \bigcup_{x\in M} T^A(E_x)$ and for a $\mathcal{AB}_m$-map $f : E_1 \to E_2$ we have $V^Af = \bigcup_{x\in M_1} T^A(f_x) : V^AE_1 \to V^AE_2$.

In general, the composition of a fiber product preserving bundle functor $T^{(A,H,t)}$ on $\mathcal{FM}_m$ for some triple (A, H, t), see Ref. 4, with the forgetting functor $\mathcal{AB}_m \to \mathcal{FM}_m$ we obtain fiber product preserving gauge bundle functor $T^{(A,H,t)} : \mathcal{AB}_m \to \mathcal{FM}$.

The fiber product $F_1 \times_{B_{\mathcal{AB}_m}} F_2 : \mathcal{AB}_m \to \mathcal{FM}$ of fiber product preserving gauge bundle functors $F_1, F_2 : \mathcal{AB}_m \to \mathcal{FM}$ is again a fiber product preserving gauge bundle functor. We recall that $(F_1 \times_{B_{\mathcal{AB}_m}} F_2)(E) = F_1E \times_M F_2E$ for any $\mathcal{AB}_m$-object $E \to M$ and $(F_1 \times_{B_{\mathcal{AB}_m}} F_2)(f)(v_1, v_2) = (F_1f(v_1), F_2f(v_2))$ for any $\mathcal{AB}_m$-map $f : E \to G$ and any $(v_1, v_2) \in F_1E \times_M F_2E$.

The composition of some fiber product preserving gauge bundle functors on $\mathcal{AB}_m$ is again a fiber product preserving gauge bundle functor on $\mathcal{AB}_m$. (In Proposition 5.2, it will be proved that every fiber product preserving gauge bundle functor has values in $\mathcal{AB}_m$. So, the composition is possible.)

The first main result in this paper is that all fiber product preserving gauge bundle functors F on $\mathcal{AB}_m$ of finite order r are in bijection with so called admissible systems, i.e. systems $(V, H, t, \mathbf{1})$, where V is a finite dimensional vector space over $\mathbb{R}$, $\mathbf{1} \in V$ is an element, $H : G^r_m \to GL(V)$ is a smooth group homomorphism from $G^r_m = invJ^r_0(\mathbb{R}^m, \mathbb{R}^m)_0$ into $GL(V)$ with $H(\xi)(\mathbf{1}) = \mathbf{1}$ for any $\xi \in G^r_m$, $t : \mathcal{D}^r_m \to gl(V)$ is a G^r_m-equivariant unity preserving associative algebra homomorphism from $\mathcal{D}^r_m = J^r_0(\mathbb{R}^m, \mathbb{R})$ into $gl(V)$.

The second main result is that natural transformations between two fiber product preserving gauge bundle functors on $\mathcal{AB}_m$ of order r are in bijection with the morphisms between corresponding admissible systems.

The third main result is that any fiber product preserving gauge bundle functor on $\mathcal{AB}_m$ is of finite order.

All manifolds are assumed to be finite dimensional. All manifolds and maps are assumed to be smooth, i.e. of class $\mathcal{C}^\infty$.

2. Fiber product preserving gauge bundle functors on $\mathcal{AB}_m$ corresponding to admissible systems

Definition 2.1. An *admissible system of order* r *and dimension* m is a system $(V, H, t, \mathbf{1})$, where V is a finite dimensional vector space over $\mathbb{R}$, $\mathbf{1} \in V$ is an element, $H : G^r_m \to GL(V)$ is a smooth group homomorphism from the Lie group $G^r_m = invJ^r_0(\mathbb{R}^m, \mathbb{R}^m)_0$ of invertible r-jets at $0 \in \mathbb{R}^m$ of diffeomorphisms $\mathbb{R}^m \to \mathbb{R}^m$ preserving 0 into the group $GL(V)$ of linear isomorphisms of V with $H(\xi)(\mathbf{1}) = \mathbf{1}$ for any $\xi \in G^r_m$, and $t : \mathcal{D}^r_m \to gl(V)$ is a G^r_m-equivariant unity preserving algebra homomorphism from the algebra $\mathcal{D}^r_m = J^r_0(\mathbb{R}^m, \mathbb{R})$ of r-jets at $0 \in \mathbb{R}^m$ of maps $\mathbb{R}^m \to \mathbb{R}$ into the associative algebra $gl(V)$ of linear endomorphisms of V.

We recall that G^r_m acts on $\mathcal{D}^r_m$ by $j^r_0\varphi.j^r_0\gamma = j^r_0(\gamma \circ \varphi^{-1})$, $j^r_0\varphi \in G^r_m$, $j^r_0\gamma \in \mathcal{D}^r_m$. This action will be denoted by H^r_m. We also recall that G^r_m acts on $gl(V)$ by $\xi.A = H(\xi) \circ A \circ H(\xi^{-1})$, $\xi \in G^r_m$, $A \in gl(V)$. These actions are by unity preserving algebra isomorphisms.

Let $(V, H, t, \mathbf{1})$ be an admissible system of order r and dimension m. We are going to construct a fiber product preserving gauge bundle functor on $\mathcal{AB}_m$ corresponding to $(V, H, t, \mathbf{1})$.

Example 2.1. For an $\mathcal{AB}_m$-object $p : E \to M$ we put

$$T^{(V,H,t,\mathbf{1})}E = \bigcup_{x \in M} \{\Phi \in Hom_{t_x}(J^r FIBAFF_x(E), \tilde{V}_x M) \mid \Phi(j^r_x 1) = \mathbf{1}_x\}.$$

Here $\tilde{V} : \mathcal{M}f_m \to \mathcal{VB}$ is the vector natural bundle corresponding to the G^r_m space V, i.e. $\tilde{V}M = P^r M[V, H]$ (the associated bundle) for any m-manifold M and $\tilde{V}\varphi = P^r\varphi[id_V] : \tilde{V}M_1 \to \tilde{V}M_2$ for any embedding $\varphi : M_1 \to M_2$ between m-manifolds, $\mathbf{1}_x \in \tilde{V}_x M$ is the canonical element induced by $\mathbf{1}$, $\mathbf{1}_x := \tilde{V}\varphi(\mathbf{1})$ for any local diffeomorphism $\varphi : \mathbb{R}^m \to M$ with $\varphi(0) = x$ (the definition of $\mathbf{1}_x$ is independent of the choice of φ), $Hom_{t_x}(J^r FIBAFF_x(E), \tilde{V}_x M)$ is the space of module homomorphisms over $t_x : J^r_x(M, \mathbb{R}) \to gl(\tilde{V}_x M)$ from the (free) $J^r_x(M, \mathbb{R})$-module $J^r FIBAFF_x(E)$ of r-jets at $x \in M$ of germs at x of fiber affine maps

$E \to \mathbb{R}$ into the $gl(\tilde{V}_x M)$-module $\tilde{V}_x M$, where $t_x : J^r_x(M, \mathbb{R}) \to gl(\tilde{V}_x M)$ is the induced by t unity preserving algebra homomorphism such that $t_x(j^r_x \gamma) = \tilde{V}_0 \varphi \circ t(j^r_0(\gamma \circ \varphi)) \circ (\tilde{V}_0 \varphi)^{-1}$ for any $\gamma : M \to \mathbb{R}$ and any embedding $\varphi : \mathbb{R}^m \to M$ with $\varphi(0) = x$ (t_x is well-defined because of the G^r_m-equivariance of t).

Given an affine bundle trivialization $(x^1 \circ p, ..., x^m \circ p, y^1, ..., y^n) : E|U \to \mathbb{R}^m \times \mathbb{R}^n$ we have an induced fiber bundle trivialization $(\tilde{x}^1, ..., \tilde{x}^m, \tilde{y}^1, ..., \tilde{y}^n) : T^{(V,H,t,\mathbf{1})} E|U \to \mathbb{R}^m \times V^n$ such that $\tilde{x}^i(\Phi) = x^i(x_o) \in \mathbb{R}$ and $\tilde{y}^j(\Phi) = \Phi(j^r_{x_o}(y^i)) \in \tilde{V}_{x_o} M \tilde{=} V$ for any $\Phi \in T^{(V,H,t,\mathbf{1})}_{x_o} E$, $i = 1, ..., m$, $j = 1, ..., n$, where $V \tilde{=} \tilde{V}_{x_o} M$ by $v \tilde{=} \tilde{V}((x^1, ..., x^m)^{-1} \circ \tau_{(x^i(x_o))})(v)$, $v \in V$, $\tau_y : \mathbb{R}^m \to \mathbb{R}^m$ is the translation by $y \in \mathbb{R}^m$. Then $T^{(V,H,t,\mathbf{1})} E$ with obvious projection is a fiber bundle over M.

Every $\mathcal{AB}_m$-map $f : E_1 \to E_2$ covering $\underline{f} : M_1 \to M_2$ induces a fibred map $T^{(V,H,t,\mathbf{1})} E_1 \to T^{(V,H,t,\mathbf{1})} E_2$ covering $\underline{f}$ such that

$$T^{(V,H,t,\mathbf{1})} f(\Phi)(j^r_{\underline{f}(x)} \xi) = \tilde{V} \underline{f} \circ \Phi(j^r_x(\xi \circ f)))$$

for any $\Phi \in T^{(V,H,t,\mathbf{1})}_x E_1$, $x \in M_1$, and any fiber affine map $\xi : E_2 \to \mathbb{R}$.

The correspondence $T^{(V,H,t,\mathbf{1})} : \mathcal{AB}_m \to \mathcal{FM}$ is a fiber product preserving gauge bundle functor of order r.

Definition 2.2. We call $T^{(V,H,t,\mathbf{1})} : \mathcal{AB}_m \to \mathcal{FM}$ *the fiber product preserving gauge bundle functor corresponding to admissible system* $(V, H, t, \mathbf{1})$.

We have the following fact.

Proposition 2.1.

(i) Given an $\mathcal{AB}_m$-object E, $T^{(V,H,t,\mathbf{1})} E$ is an affine bundle with the corresponding vector bundle $T^{(V,H,t,0)} E$.

(ii) Given an $\mathcal{AB}_m$-morphism $f : E \to G$, $T^{(V,H,t,\mathbf{1})} f : T^{(V,H,t,\mathbf{1})} E \to T^{(V,H,t,\mathbf{1})} G$ is an affine bundle map with the corresponding vector bundle map $T^{(V,H,t,0)} f : T^{(V,H,t,0)} E \to T^{(V,H,t,0)} G$.

(iii) The vector bundle $T^{(V,H,t,0)} E$ is canonically isomorphic (by vector bundle isomorphism) with $T^{(V,H,t)}(E^{\to})$ (see Ref. 5 for the definition of $T^{(V,H,t)}$).

Proof. The proof of Parts (i) and (ii) is standard. More precisely, given $\Phi_1, \Phi_2 \in T^{(V,H,t,0)}_z E$ for $z \in M$ and $\alpha \in \mathbb{R}$ we have standardly defined $\Phi_1 + \Phi_2 \in T^{(V,H,t,0)}_z E$ and $\alpha \Phi_1 \in T^{(V,H,t,0)}_z E$. That is why, $T^{(V,H,t,0)} E$ is a vector bundle over M. Similarly, given $\Phi_1 \in T^{(V,H,t,\mathbf{1})}_z E$ and $\Phi_2 \in$

$T_z^{(V,H,t,0)}E$, $z \in M$, we have standardly defined $\Phi_1 + \Phi_2 \in T_z^{(V,H,t,\mathbf{1})}E$. That is why, $T^{(V,H,t,\mathbf{1})}E$ is an affine bundle with the corresponding vector bundle $T^{(V,H,t,0)}E$.

Part (iii) will be clear after Section 10 because $T^{(V,H,t,0)}$ and $T^{(V,H,t)} \circ (\;)^{\rightarrow}$ have isomorphic the corresponding admissible systems. □

3. Admissible systems corresponding to fiber product preserving gauge bundle functors on $\mathcal{AB}_m$

Let $F : \mathcal{AB}_m \to \mathcal{FM}$ be a f.p.p.g.b. functor of order r. We are going to construct an admissible system corresponding to F.

Example 3.1. We put

$$V^F := (F_0(\mathbb{R}^m \times \mathbb{R}), F_0(+), F_0\lambda_a, F_0 0)\ ,$$

where $F_0(\mathbb{R}^m \times \mathbb{R})$ is the fiber of $F(\mathbb{R}^m \times \mathbb{R})$ over $0 \in \mathbb{R}^m$ and $\mathbb{R}^m \times \mathbb{R}$ is the trivial affine bundle over $\mathbb{R}^m$ with fiber $\mathbb{R}$ with the corresponding vector bundle $\mathbb{R}^m \times \mathbb{R}$, and where the fiber sum map $+$, the fiber scalar multiplications λ_a, $a \in \mathbb{R}$, and the zero map 0 of the vector bundle $\mathbb{R}^m \times \mathbb{R}$ are treated as $\mathcal{AB}_m$-morphisms. Then V^F is a finite dimensional vector space over $\mathbb{R}$.

We define $H^F : G^r_m \to GL(V^F)$ by

$$H^F(\xi)(v) := F_0(\varphi \times id_{\mathbb{R}})(v)\ ,\ v \in V^F\ ,\ \xi = j^r_0\varphi \in G^r_m\ .$$

$H^F(\xi)(v)$ is well defined because of F is of order r. By the definition of V^F, $H^F(\xi) \in GL(V^F)$. By the functoriality of F, H^F is a group homomorphism. By the regularity of F, H^F is smooth.

We define $t^F : \mathcal{D}^r_m \to gl(V^F)$ by

$$t^F(\eta)(v) = F_0(\tilde{\gamma})(v)\ ,\ v \in V^F\ \ \eta = j^r_0\gamma \in \mathcal{D}^r_m$$

where $\tilde{\gamma} : \mathbb{R}^m \times \mathbb{R} \to \mathbb{R}^m \times \mathbb{R}$ is an $\mathcal{AB}_m$-map such that $\tilde{\gamma}(x,y) = (x, \gamma(x)y)$, $x \in \mathbb{R}^m$, $y \in \mathbb{R}$. $t^F(\eta)(v)$ is well defined because of F is of order r. By the definition of V^F, $t^F(\eta) \in gl(V^F)$. By the functoriality of F and the definitions of the actions one can standardly verify that t^F is a G^r_m-equivariant unity preserving algebra homomorphism.

We define $\mathbf{1}^F \in V^F$ by

$$\{\mathbf{1}^F\} = \text{the image of } F_0(id_{\mathbb{R}^m}, 1)\ ,$$

where $(id_{\mathbb{R}^m}, 1) : \mathbb{R}^m \to \mathbb{R}^m \times \mathbb{R}$ is the $\mathcal{AB}_m$-morphism.

It is easy to see that $H^F(\xi)(\mathbf{1}^F) = \mathbf{1}^F$ for any $\xi \in G^r_m$.

Then $(V^F, H^F, t^F, \mathbf{1}^F)$ is an admissible system of order r and dimension m.

Definition 3.1. We call $(V^F, H^F, t^F, \mathbf{1}^F)$ *the admissible system corresponding to* F.

4. Admissible systems corresponding to some fiber product preserving gauge bundle functors on $\mathcal{AB}_m$.

In this section we present admissible systems corresponding to the presented in Introduction fiber product preserving gauge bundle functors on $\mathcal{AB}_m$. The results of this section will not be used to prove the main result.

Fact 4.1. The admissible system corresponding to $(\ )^{\to} : \mathcal{AB}_m \to \mathcal{FM}$ is $(\mathbb{R}, Id_{\mathbb{R}}, \kappa_{\mathcal{D}^r_m}, 0)$, where $Id_{\mathbb{R}} : G^r_m \to GL(\mathbb{R})$ is the trivial group homomorphism and $\kappa_{\mathcal{D}^r_m} : \mathcal{D}^r_m \to \mathbb{R} = gl(\mathbb{R})$ is the trivial unity preserving algebra homomorphism.

Fact 4.2. The admissible system corresponding to $T^{(V,H,t)} \circ (\ )^{\to}$ is $(V, H, t, 0)$,where $T^{(A,H,t)} : \mathcal{VB}_m \to \mathcal{FM}$ is described in Ref. 5.

Fact 4.3. The admissible system corresponding to the r-jet prolongation gauge bundle functor $J^r : \mathcal{AB}_m \to \mathcal{FM}$ is $(\mathcal{D}^r_m, H^r_m, t^r_m, j^r_0 1)$, where $H^r_m : G^r_m \to Aut(\mathcal{D}^r_m)$ is defined after Definition 2.1 and $t^r_m : \mathcal{D}^r_m \to gl(\mathcal{D}^r_m)$ is given by $t^r_m(\eta)(\rho) = \eta\rho$,where $\eta, \rho \in \mathcal{D}^r_m$.

Fact 4.4. The admissible system corresponding to the vertical r-jet prolongation gauge bundle functor $J^r_v : \mathcal{AB}_m \to \mathcal{FM}$ is $(\mathcal{D}^r_m, H^r_m, t^r_m \circ \epsilon^r_m, j^r_0 1)$, where $H^r_m : G^r_m \to Aut(\mathcal{D}^r_m)$ is defined after Definition 2.1, $t^r_m : \mathcal{D}^r_m \to gl(\mathcal{D}^r_m)$ is defined above and $\epsilon^r_m : \mathcal{D}^r_m \to \mathbb{R} \subset \mathcal{D}^r_m$ is the algebra homomorphism.

Fact 4.5. The admissible system corresponding to the vertical Weil gauge bundle functor $V^A : \mathcal{AB}_m \to \mathcal{FM}$ corresponding to a Weil algebra A is $(A,\ id_A, \epsilon^A, 1)$, where $id_A : G^r_m \to \{id_A\} \subset GL(A)$ is the trivial group homomorphism and $\epsilon^A : \mathcal{D}^r_m \to gl(A)$ is given by $\epsilon^A(\eta)(a) = \gamma(0)a$, $\eta = j^r_0\gamma \in \mathcal{D}^r_m$, $a \in A$.

Fact 4.6. The admissible triple corresponding to $T^{(A,H,t)} : \mathcal{AB}_m \to \mathcal{FM}$ (obtained from $T^{(A,H,t)} : \mathcal{FM}_m \to \mathcal{FM}$ described in Ref. 4), where A is a Weil algebra of order r. $H : G^r_m \to Aut(A)$ is a smooth group homomorphism and $t : \mathcal{D}^r_m \to A$ is a G^r_m-equivariant unity preserving algebra homomorphism, is $(\tilde{A}, \tilde{H}, \tilde{t}, 1)$, where $\tilde{A}$ is A considered as the vector space, $\tilde{H} : G^r_m \to GL(\tilde{A})$ is $H : G^r_m \to Aut(A) \subset GL(\tilde{A})$, $\tilde{t} : \mathcal{D}^r_m \to gl(\tilde{A})$ is given by $\tilde{t}(\eta)(a) = t(\eta)a$ for $a \in \tilde{A}$ and $\eta \in \mathcal{D}^r_m$, and $1 \in A$ is the unity of the Weil algebra A.

Fact 4.7. The admissible system corresponding to $F_1 \times_{B_{\mathcal{AB}_m}} F_2$ is $(V^{F_1} \oplus V^{F_2}, H^{F_1} \oplus H^{F_2}, t^{F_1} \oplus t^{F_2}, \mathbf{1}^{F_1} \oplus \mathbf{1}^{F_2})$.

An open problem. By Proposition 5.2., any fiber product preserving gauge bundle functor on $\mathcal{AB}_m$ has values in $\mathcal{AB}_m$. So, we can compose fiber product preserving gauge bundle functors on $\mathcal{AB}_m$. Compute the admissible system corresponding to the composition $F_1 \circ F_2$ of two fiber product preserving gauge bundle functors F_1 and F_2 on $\mathcal{AB}_m$ in terms of admissible systems corresponding to F_1 and F_2.

5. Classification of fiber product preserving gauge bundle functors on $\mathcal{AB}_m$ of order r in terms of admissible systems of order r and dimension m

The following classification proposition shows that any fiber product preserving gauge bundle functor on $\mathcal{AB}_m$ of order r is equivalent to some fiber product preserving gauge bundle functor as in Example 2.1.

Proposition 5.1. *Let $F : \mathcal{AB}_m \to \mathcal{FM}$ be a fiber product preserving gauge bundle functor of order r. Let $(V^F, H^F, t^F, \mathbf{1}^F)$ be the admissible system (of order r and dimension m) corresponding to F. Then we have a natural equivalence $\Theta^F : F \tilde{=} T^{(V^F, H^F, t^F, \mathbf{1}^F)}$.*

Proof. Let $p : E \to M$ be an $\mathcal{AB}_m$-object. We construct canonically a diffeomorphism $\Theta^F : FE \to T^{(V^F, H^F, t^F, \mathbf{1}^F)}E$ as follows. Given a point $y \in F_xE$, $x \in M$, we define $\Theta^F(y) : J^rFIBAFF_x(E) \to \tilde{V^F}_xM$ by

$$\Theta^F(y)(\xi) = F_x(f)(y) \in F_x(M \times \mathbb{R}) \tilde{=} \tilde{V^F}_xM, \ \xi = j^r_x f \in J^rFIBAFF_x(E),$$

where a fiber affine map $f : E \to \mathbb{R}$ is (in obvious way) considered as the $\mathcal{AB}_m$-map $f : E \to M \times \mathbb{R}$ covering the identity of M and where the identification $F_x(M \times \mathbb{R}) \tilde{=} \tilde{V^F}_xM$ is given by $F_x(M \times \mathbb{R}) \ni F(\varphi \times id_{\mathbb{R}})(v) \tilde{=} < j^r_0\varphi, v > \in \tilde{V^F}_xM$, $v \in V^F = F_0(\mathbb{R}^m \times \mathbb{R})$, $\varphi : \mathbb{R}^m \to M$ is an embedding with $\varphi(0) = x$. $\Theta^F(y)$ is well defined because F is of order r. Recalling the definition of $(V^F, H^F, t^F, \mathbf{1}^F)$ (see Example 3.1) and using the functoriality of F one can standardly verify that $\Theta^F(y)$ is a module homomorphism over $t^F_x : J^r_x(M,\mathbb{R}) \to gl(\tilde{V^F}_x M)$ with $\Theta^F(y)(j^r_x -) = \mathbf{1}^F_x$, i.e. $\Theta^F(y) \in T_x^{(V^F,H^F,t^F,\mathbf{1}^F)}E$.

It remains to show that $\Theta^F : FE \to T^{(V^F,H^F,t^F,\mathbf{1}^F)}E$ is a diffeomorphism.

Because of $\Theta^F : F \to T^{(V^F,H^F,t^F,\mathbf{1}^F)}$ is natural with respect to $\mathcal{AB}_m$-maps and F and $T^{(V^F,H^F,t^F,\mathbf{1}^F)}$ preserve fiber product and E is locally a (multi) fiber product of $\mathbb{R}^m \times \mathbb{R}$ we may assume that $E = \mathbb{R}^m \times \mathbb{R}$, the trivial affine bundle over $\mathbb{R}^m$ with fiber $\mathbb{R}$. But for $E = \mathbb{R}^m \times \mathbb{R}$ transformation Θ^F is the composition $F(\mathbb{R}^m \times \mathbb{R}) \tilde{=} \mathbb{R}^m \times V^F \tilde{=} T^{(V^F,H^F,t^F,\mathbf{1}^F)}(\mathbb{R}^m \times \mathbb{R})$, where the first identification is given by $F_x(\mathbb{R}^m \times \mathbb{R}) \ni v = (x, F(\tau_{-x} \times id_{\mathbb{R}})(v)) \in \{x\} \times V^F$, $x \in \mathbb{R}^m$, and where the second trivialization is induced (see Example 2.1.) by the obvious trivialization of $\mathbb{R}^m \times \mathbb{R}$. □

From Propositions 2.1 and 5.1 we obtain

Proposition 5.2. *Any fiber product preserving gauge bundle functor F on $\mathcal{AB}_m$ of finite order has values in $\mathcal{AB}_m$.*

6. Classification of admissible systems of order r and dimension m in terms of fiber product preserving gauge bundle functors on $\mathcal{AB}_m$ of order r

The following classification proposition shows that any admissible system of order r and dimension m is isomorphic to some admissible system as in Example 3.1.

Proposition 6.1. *Let $(V,H,t,\mathbf{1})$ be an admissible system of order r and dimension m. Let $F = T^{(V,H,t,\mathbf{1})}$. Then we have an isomorphism $\mathcal{O}^{(V,H,t,\mathbf{1})}$: $(V,H,t,\mathbf{1}) \tilde{=} (V^F,H^F,t^F,\mathbf{1}^F)$ of admissible systems.*

We recall that a morphism $(V_1,H_1,t_1,\mathbf{1}_1) \to (V_2,H_2,t_2,\mathbf{1}_2)$ of admissible systems is a linear map $\mathcal{O} : V_1 \to V_2$ such that $H_2(\xi) \circ \mathcal{O} = \mathcal{O} \circ H_1(\xi)$ for any $\xi \in G^r_m$, $t_2(\eta) \circ \mathcal{O} = \mathcal{O} \circ t_1(\eta)$ for any $\eta \in \mathcal{D}^r_m$ and $\mathcal{O}(\mathbf{1}_1) = \mathbf{1}_2$.

Proof. We have $\mathcal{O}^{(V,H,t,\mathbf{1})} : V \to V^F$ that the composition $V \to V^F = T_0^{(V^F,H^F,t^F,\mathbf{1}^F)}(\mathbb{R}^m \times \mathbb{R}) \tilde{=} \{0\} \times V$ of $\mathcal{O}^{(V,H,t,\mathbf{1})}$ with the isomorphism induced (see Example 2.1) by the usual trivialization of $\mathbb{R}^m \times \mathbb{R}$ is the (almost) identity map. One can show standardly (but long) that $\mathcal{O}^{(V,H,t,\mathbf{1})}$ is a morphism $(V,H,t,\mathbf{1}) \to (V^F,H^F,t^F,\mathbf{1}^F)$ of admissible systems. □

7. Natural transformations of fiber product preserving gauge bundle functors on $\mathcal{AB}_m$ of order r and induced morphisms between corresponding admissible systems

Let $F_1, F_2 : \mathcal{AB}_m \to \mathcal{FM}$ be fiber product preserving gauge bundle functors of order r. Let $(V^{F_1}, H^{F_1}, t^{F_1}, \mathbf{1}^{F_1})$ and $(V^{F_2}, H^{F_2}, t^{F_2}, \mathbf{1}^{F_2})$ be the corresponding admissible systems of order r and dimension m. Let $\mu : F_1 \to F_2$ be a natural transformation.

Example 7.1. Define $\nu^\mu : V^{F_1} \to V^{F_2}$ to be the restriction of $\mu : F_1(\mathbb{R}^m \times \mathbb{R}) \to F_2(\mathbb{R}^m \times \mathbb{R})$ to $V^{F_1} = (F_1)_0(\mathbb{R}^m \times \mathbb{R})$ and $V^{F_2} = (F_2)_0(\mathbb{R}^m \times \mathbb{R})$. Then $\nu^\mu : (V^{F_1}, H^{F_1}, t^{F_1}, \mathbf{1}^{F_1}) \to (V^{F_2}, H^{F_2}, t^{F_2}, \mathbf{1}^{F_2})$ is a morphism of admissible systems. If μ is an isomorphism, then so is ν^μ.

Definition 7.1. We call ν^μ *the morphism corresponding to* μ.

8. Morphisms between admissible systems of order r and dimension m and induced natural transformations between corresponding fiber product preserving gauge bundle functors

Let $(V_1, H_1, t_1, \mathbf{1}_1)$ and $(V_2, H_2, t_2, \mathbf{1}_2)$ be admissible systems of order r and dimension m. Let $\nu : (V_1, H_1, t_1, \mathbf{1}_1) \to (V_2, H_2, t_2, \mathbf{1}_2)$ be a morphism of admissible systems.

Example 8.1. Given an $\mathcal{AB}_m$-object $p : E \to M$ define a base preserving fibred map $\mu^\nu : T^{(V_1,H_1,t_1,\mathbf{1}_1)}E \to T^{(V_2,H_2,t_2,\mathbf{1}_2)}E$ as follows. Let $\Phi \in T_x^{(V_1,H_1,t_1,\mathbf{1}_1)}E$, $x \in M$. Put $\mu^\nu(\Phi) = \tilde{\nu}_x \circ \Phi : J^r FIBAFF_x(E) \to (\tilde{V_2})_x M$, where $\tilde{\nu}_x : (\tilde{V_1})_x M \to (\tilde{V_2})_x M$, $\tilde{\nu}_x(< j^r_x\varphi, v >) =< j^r_x\varphi, \nu(v) >$, $v \in V_1$, $\varphi : \mathbb{R}^m \to M$ is an embedding with $\varphi(0) = x$. We see that $\mu^\nu(\Phi) \in T_x^{(V_2,H_2,t_2,\mathbf{1}_2)}E$ and that $\mu^\nu : T^{(V_1,H_1,t_1,\mathbf{1}_1)} \to T^{(V_2,H_2,t_2,\mathbf{1}_2)}$ is a natural transformation. If ν is an isomorphism, then so is μ^ν.

Definition 8.1. We call μ^ν *the natural transformation corresponding to* ν.

9. Object Classification Theorem

The first main result in this paper is the following theorem.

Theorem 9.1. *The correspondence "$F \to (V^F, H^F, t^F, \mathbf{1}^F)$" induces a bijective correspondence between the equivalence classes of fiber product preserving gauge bundle functors F on $\mathcal{AB}_m$ of order r and the equivalence classes of admissible systems $(V, H, t, \mathbf{1})$ of order r and dimension m.*

The inverse correspondence is induced by the correspondence "$(V, H, t, \mathbf{1}) \to T^{(V,H,t,\mathbf{1})}$".

Proof. The correspondence "$[F] \to [(V^F, H^F, t^F, \mathbf{1}^F)]$" is well-defined. For, if $\mu : F_1 \to F_2$ is an isomorphism, then so is $\nu^\mu : (V^{F_1}, H^{F_1}, t^{F_1}, \mathbf{1}^{F_1}) \to (V^{F_2}, H^{F_2}, t^{F_2}, \mathbf{1}^{F_2})$.

The correspondence "$[(V, H, t, \mathbf{1})] \to [T^{(V,H,t,\mathbf{1})}]$" is well-defined. For, if $\nu : (V_1, H_1, t_1, \mathbf{1}_1) \to (V_2, H_2, t_2, \mathbf{1}_2)$ is an isomorphism, then so is $\mu^\nu : T^{(V_1,H_1,t_1,\mathbf{1}_1)} \to T^{(V_2,H_2,t_2,\mathbf{1}_2)}$.

From Proposition 5.1 it follows that $[F] = [T^{(V^F,H^F,t^F,\mathbf{1}^F)}]$. From Proposition 6.1. it follows that $[(V, H, t, \mathbf{1})] = [(V^F, H^F, t^F, \mathbf{1}^F)]$ if $F = T^{(V,H,t,\mathbf{1})}$. □

10. Morphism Classification Theorem

Let F_1 and F_2 be two fiber product preserving gauge bundle functors on $\mathcal{AB}_m$ of order r. Let $(V^{F_1}, H^{F_1}, t^{F_1}, \mathbf{1}^{F_1})$ and $(V^{F_2}, H^{F_2}, t^{F_2}, \mathbf{1}^{F_2})$ be the corresponding admissible systems of order r and dimension m.

Lemma 10.1. *Let $\nu : (V^{F_1}, H^{F_1}, t^{F_1}, \mathbf{1}^{F_1}) \to (V^{F_2}, H^{F_2}, t^{F_2}, \mathbf{1}^{F_2})$ be a morphism of admissible systems. Let $\mu^{[\nu]} : F_1 \to F_2$ be a natural transformation given by the composition*

$$F_1 \xrightarrow{\Theta^{F_1}} T^{(V^{F_1},H^{F_1},t^{F_1},\mathbf{1}^{F_1})} \xrightarrow{\mu^\nu} T^{(V^{F_2},H^{F_2},t^{F_2},\mathbf{1}^{F_2})} \xrightarrow{(\Theta^{F_2})^{-1}} F_2,$$

where Θ^F is as in Proposition 5.1. and μ^ν is described in Example 8.1. Then $\mu = \mu^{[\nu]}$ is the unique natural transformation $F_1 \to F_2$ such that $\nu^\mu = \nu$, where ν^μ is as in Example 7.1.

Proof. Suppose $\overline{\mu} : F_1 \to F_2$ is another natural transformation such that $\nu^{\overline{\mu}} = \nu$. Then $\overline{\mu}$ coincides with μ on the affine bundle $\mathbb{R}^m \times \mathbb{R}$. Hence $\overline{\mu} = \mu$ because of the same argument as in the proof of Proposition 5.1. □

Now, the following second main result in this paper is clear.

Theorem 10.1. *Let F_1 and F_2 be two fiber product preserving gauge bundle functors on $\mathcal{AB}_m$ of order r. The correspondence "$\mu \to \nu^\mu$" is a bijection between natural transformations $F_1 \to F_2$ and morphisms $(V^{F_1}, H^{F_1}, t^{F_1}, \mathbf{1}^{F_1}) \to (V^{F_2}, H^{F_2}, t^{F_2}, \mathbf{1}^{F_2})$ between corresponding admissible systems. The inverse correspondence is "$\nu \to \mu^{[\nu]}$".*

11. Finite order theorem

Theorem 11.1. *Any fiber product preserving gauge bundle functor $F : \mathcal{AB}_m \to \mathcal{FM}$ is of finite order.*

Proof. Define $A^F : FIBAFF(\mathbb{R}^m \times \mathbb{R}) \to \mathcal{C}^\infty(F(\mathbb{R}^m \times \mathbb{R}), F(\mathbb{R}^m \times \mathbb{R}))$ by $A^F(f) = Ff$, where a fiber affine map $f : \mathbb{R}^m \times \mathbb{R} \to \mathbb{R}$ is considered as a base preserving $\mathcal{AB}_m$-map $\mathbb{R}^m \times \mathbb{R} \to \mathbb{R}^m \times \mathbb{R}$ in obvious way. Clearly, A^F is π-local, where $\pi : F(\mathbb{R}^m \times \mathbb{R}) \to \mathbb{R}^m$ is the projection. The zero map $O \in FIBAFF(\mathbb{R}^m \times \mathbb{R})$ is invariant with respect to the translations of $\mathbb{R}^m$.

Clearly $A^F(tf) = tA^F(f)$ for any f. Here $F(\mathbb{R}^m \times \mathbb{R}) = \mathbb{R}^m \times V^F$ is considered as the trivial vector bundle. Then by the non -linear Peetre theorem[3] and the homogeneous function theorem[3] we see that A^F is linear. Then we can assume that f is fiber linear or f is fibre constant. Moreover, given f in question, $v \in F_0(\mathbb{R}^m \times \mathbb{R})$, a neighbourhood W of $j^\infty_0(0)$ and a neighbourhood U of $0 \in F_0(\mathbb{R}^m \times \mathbb{R})$, there is $t \in \mathbb{R}_+$ such that $j^\infty_0(tf) \in W$ and $F\lambda_t(v) \in U$, where λ_t is the fiber homothety by t. Then standardly by the non-linear (or classical) Peetre theorem we deduce that the operator A^F is of finite order r_1. (More precisely, there exists finite r_1, a neighbourhood W of $j^\infty_0(0)$, a neighbourhood U of $0 \in F_0(\mathbb{R}^m \times \mathbb{R})$ such that for fiber affine $f_1, f_2 : \mathbb{R}^m \times \mathbb{R} \to \mathbb{R}$ with $j^\infty_0(f_1), j^\infty_0(f_2) \in W, v \in U$, from $j^{r_1}_0 f_1 = j^{r_1}_0 f_2$ it follows $Ff_1(v) = Ff_2(v)$. Now, let $f_1, f_2 : \mathbb{R}^m \times \mathbb{R} \to \mathbb{R}$ be fiber linear (resp. fiber constant), $v \in F_0(\mathbb{R}^m \times \mathbb{R})$, $j^{r_1}_0 f_1 = j^{r_1}_0 f_2$. Then there exists $t \in \mathbb{R}_+$ such that $j^\infty_0(tf_1), j^\infty_0(tf_2) \in W, j^{r_1}_0(tf_1) = j^{r_1}_0(tf_2)$ and $F\lambda_t(v) \in U$. Then $F(tf_1)(F\lambda_t(v)) = F(tf_2)(F\lambda_t(v))$. Then $t^2Ff_1(v) = t^2Ff_2(v)$ (resp. $tFf_1(v) = tFf_2(v)$). Then $Ff_1(v) = Ff_2(v)$ as well.)

We have the bundle functor $G^F : \mathcal{M}f_m \to \mathcal{FM}$ such that $G^F M = F(M \times \mathbb{R})$ for any m-manifold M and $G^F\varphi = F(\varphi \times id_\mathbb{R}) : G^F M \to G^F N$ for any embedding $\varphi : M \to N$ between m-manifolds. By the Palais-Terng theorem, see Ref. 3, G^F has finite order r_2.

We prove that F is of order $r = max(r_1, r_2)$. We consider an $\mathcal{AB}_m$-map $f : E_1 \to E_2$ and a point $x \in M_1$. It remains to show that $F_x f$ depends on $j^r_x f$.

Using $\mathcal{AB}_m$-trivialization we can assume that $E_1 = \mathbb{R}^m \times \mathbb{R}^n$, $E_2 = \mathbb{R}^m \times \mathbb{R}^q$ and $x = 0 \in \mathbb{R}^m$. Since F preserves fiber product, we can assume that $q = 1$. Then we can write $f = \tilde{f} \circ (\varphi \times id_{\mathbb{R}^n})$, where $\tilde{f} : \mathbb{R}^m \times \mathbb{R}^n \to \mathbb{R}^m \times \mathbb{R}$ is base preserving $\mathcal{AB}_m$-map and $\varphi : \mathbb{R}^m \to \mathbb{R}^m$ is a 0-preserving embedding. If $\tilde{f}(x, y) = (x, \sum_{i=1}^n (a_i(x)y^i + h(x)))$, we have $F\tilde{f}(v) = \sum_{i=1}^n F(x, a_i(x)y + h(x))(v_i)$ for $v = (v_i) \in F_0(\mathbb{R}^m \times \mathbb{R}^n) = \times^n F_0(\mathbb{R}^m \times \mathbb{R})$, where the sum is the one of the vector space $V^F = F_0(\mathbb{R}^m \times \mathbb{R})$ and where $y : \mathbb{R}^m \times \mathbb{R} \to \mathbb{R}$

is the usual fiber coordinate. Then $F_0 f = F_0 \tilde{f} \circ (\times^n G_0^F \varphi)$ depends on $j_0^{r_1} \tilde{f}$ and $j_0^{r_2} \varphi$. □

References

1. M. Doupovec and I. Kolář, Iteration of fiber product preserving bundle functors, *Monatsh. Math.* **134** (2001) 39–50.
2. I. Kolář, Bundle functors of jet type, *In: Differential Geom. and Appl.* (Proc. of the 7th International Conf. Brno, 1988) 231–237.
3. I. Kolář and P. Michor and J. Slovák, *Natural Operations in Differential Geometry* (Springer-Verlag, 1993).
4. I. Kolář and W.M. Mikulski, On the fiber product preserving bundle functors, *Differential Geometry and Its Appl.* **11** (1999) 105–115.
5. W.M. Mikulski, On the fiber product preserving gauge bundle functors on vector bundles, *Ann. Polon. Math.* **82** (3) (2003) 251–264.

Differential Geometry and its Applications
Proc. Conf., in Honour of Leonhard Euler, Olomouc, August 2007

Two analytical Goodwillie's theorems

Rémi Léandre

Institut de Mathématiques, Université de Bourgogne,
21000, Dijon, France
E-mail: Remi.leandre@u-bourgogne.fr

We define two analytical Goodwillie's theorem, either by using an Hida type Fock space or by using an holomorphic type Fock space.

Keywords: Goodwillie theorem, Fock space.

MS classification: 60H40.

1. Introduction

Cyclic homology is a purely algebraic object. Several works in order to give an analytical meaning to cyclic cohomology were done in the works of Connes,[2] Jaffe-Lesniewski-Osterwalder,[9] Cuntz[3] and Meyer.[15] In these works, cyclic complex is endowed with some topology or some bornologies.

Léandre-Ouerdiane[14] have introduced in this context the interacting Fock space of Accardi-Bozejko and have defined a cyclic complex in Hida sense.

Let us remark that Léandre[11] instead of using the Fock space, has defined Hochschild homology with some auxiliary operators and tensor product of Hilbert Sobolev spaces. Jaffe[8] has done similar works, but with the cyclic homology.

In the algebraic context, people associate to an algebra A the model of $T(A)$, the tensor algebra, given by non-commutative algebraic differential forms of even degree. Cuntz-Quillen consider the X-complex of $T(A)$, and the cyclic complex associated to the tensor algebra $T(A)$.

A particular case of Goodwillie's theorem says that the cohomology of the X-complex of $T(A)$ is equal to the cyclic cohomology associated to $T(A)$. Goodwillie considered only finite sums.

Perrot,[16] by using some bornologies, shows that this theorem remains true in the bornological Calculus.

The goal of this paper is to give an analytical version of Goodwillie's theorem, when A is an algebra of functions on a manifold.

In the first part, we introduce the analogous of some Fock space, and we state a Goodwillie's theorem in Hida sense.

In the second part, we consider the case of a group. We deduce a Goodwillie's theorem, when there are some conditions of analyticity in the derivatives for the elements in A.

We thank D. Arnal for helpful discussions.

2. Goodwillie's theorem and Hida Calculus

Let M be a compact Riemannian manifold. We consider the Hilbert space $H = L^2(M) \otimes C$ and we define the Hilbert Sobolev space H_k of $\phi \in H$ such that $(\Delta + 2)^{k/2}\phi$ belongs to H. H_{k+1} is included in H_k and the intersections of H_k is by the Sobolev imbedding theorem nothing else than the algebra of smooth functions on the manifold M.

We consider $a > 1$ fixed in this part and $S_{k,C}(M)$ the Hilbert space of formal sums $\tilde{\sigma} = \sum \sigma_{2n}$ where σ_{2n} belongs to $H_k \otimes H_k^{\otimes 2n}$ where we take the Hilbert norm

$$\|\tilde{\sigma}\|_{k,C}^2 = \sum \frac{C^n}{(an)!} \|\sigma_{2n}\|_k^2 < \infty \tag{1}$$

If $k > k'$, $\|\tilde{\sigma}\|_{k,C} \geq \|\tilde{\sigma}\|_{k',C}$ and if $C > C'$, $\|\tilde{\sigma}\|_{k,C} \geq \|\tilde{\sigma}\|_{k,C'}$. We denote by $\cap_{k,C} S_{k,C}(M) = S_{\infty-}(M)$. This space is called the space of Hida functionals. We write

$$\sigma_{2n} = \sum_{|J|=2n+1} \lambda_J \phi_{i_0^J} d\phi_{i_1^J} .. d\phi_{i_{|J|}^J} = \sum \lambda_J \tilde{\sigma}_J \tag{2}$$

where ϕ_i constitute an orthonormal basis of the Hilbert space H inherited by diagonalizing the Laplacian Δ_M associated to the eigenvalues C_i of Δ_M. We adjoint the formal unity 1 to H. We write $d(\phi\phi') = (d\phi)\phi' + \phi d\phi'$ and $d^2\phi = 0$. $d1 = 0$ if 1 is the formal unit in $H \oplus C$.

If $\tilde{\sigma}$ and $\tilde{\sigma}_1$ are Hida test functionals, we consider the Fedosov product (See Refs. 4,5,16)

$$\tilde{\sigma} \circ \tilde{\sigma}_1 = \tilde{\sigma}\tilde{\sigma}_1 - d\tilde{\sigma} d\tilde{\sigma}_1 \tag{3}$$

Theorem 2.1. *The Fedosov product is continuous from $S_{\infty-}(M) \times S_{\infty-}(M)$ into $S_{\infty-}(M)$.*

Proof. Let us write:

$$\tilde{\sigma} = \sum \lambda_J \tilde{\sigma}_J \tag{4}$$

$$\tilde{\sigma}_1 = \sum \lambda_J^1 \tilde{\sigma}_J \tag{5}$$

such that

$$\tilde{\sigma}\tilde{\sigma}_1 = \sum \lambda_J \lambda_{J'}^1 \tilde{\sigma}_J \tilde{\sigma}_{J'} \tag{6}$$

We have

$$\|\tilde{\sigma}\tilde{\sigma}_1\|_{k,C} \leq \sum |\lambda_J||\lambda_{J'}^1| \|\tilde{\sigma}_J\tilde{\sigma}_{J'}\|_{k,C} \tag{7}$$

But $(d\phi)\phi' = d(\phi\phi') - \phi d\phi'$. By Sobolev imbedding theorem, we deduce that

$$\|\tilde{\sigma}_J\tilde{\sigma}_{J'}\|_{k,C} \leq C_2^{|J|+|J'|}|J| \|\tilde{\sigma}_{J'}\|_{k_1,C_1} \|\tilde{\sigma}_J\|_{k_1,C_1} \prod_{j\in J}(C_j+2)^{-k'} \prod_{j'\in J'}(C_{j'}+2)^{-k'} \tag{8}$$

for some small C_2, some big k_1, C_1 and k'.

We apply Cauchy-Schwartz inequality in (7) and the bound (8) in order to show that

$$\|\tilde{\sigma}\tilde{\sigma}_1\|_{k,C} \leq K \|\tilde{\sigma}\|_{k_1,C_1} \|\tilde{\sigma}_1\|_{k_1,C_1} \tag{9}$$

where

$$K = \sum_{J,J'} C_2^{2|J|+|J'|} \prod_{j\in J}(C_j+2)^{-2k'} \prod_{j'\in J'}(C_{j'}+2)^{-2k'} \tag{10}$$

But if we fix $|J| = n$, we have:

$$sum \prod_{j\in J}(C_j+2)^{-2k'} = (\sum_j (C_j+2)^{-2k'})^{|J|} = C_3^{|J|} < \infty \tag{11}$$

It remains to choose C_1 big enough in order that $C_2^2 C_3 < 1$ in order to conclude that $K < \infty$.

The proof that $\|d\tilde{\sigma}d\tilde{\sigma}_1\|_{k,C}$ can be estimated by $K\|\tilde{\sigma}\|_{k_1,C_1}\|\tilde{\sigma}_1\|_{k_1,C_1}$ for some big k_1 and some big C_1 is simpler because we we don't have to multiply $d\phi_{i^J_{|J|}}$ by $\phi_{i_0^{J_1}}$ but we have to multiply it by $d\phi_{i_0^{J_1}}$ and the formula is simpler. Therefore the result. □

We follow the indications of Cuntz-Quillen,[4,5] Meyer[15] and Perrot[16] in order to choose as model of the universal tensor algebra associated to $\cap H_k = A$, $S_{\infty-}(M)$ where we apply Ψ:

$$\phi_0 d\phi_1..d\phi_{2n} \to \phi_0 \otimes (\phi_1\phi_2 - \phi_1 \otimes \phi_2)..\otimes(\phi_{2n-1}\phi_{2n} - \phi_{2n-1}\otimes\phi_{2n}) \tag{12}$$

Ψ map the Fedosov product on the traditional tensor product in the tensor algebra. $S_{\infty-}(M)$ is therefore a topological algebra for the Fedosov product $\circ$. There is the same system of orthogonal basis $\tilde{\sigma}_J$ on all $S_{k,C}(M)$.

Let us consider $\Omega^k(S_{k,C}(M))$. There is a basis

$$\tilde{\alpha}_I = \tilde{\sigma}_{J_0^I} d\tilde{\sigma}_{J_1^I} ... d\tilde{\sigma}_{J_{|I|}^I} \tag{13}$$

we denote by $|I| = k$ and by the total weight $\|I\|$ the quantity $\sum |J_i^I| + |I|$.

We consider the Hilbert space $\Omega_{k,C}(S_{k,C}(M))$ as the space of $\tilde{\alpha} = \sum \lambda_I \tilde{\sigma}_I$ such that

$$\|\tilde{\alpha}\|_{k,C} = \sum |\lambda_I|^2 \frac{C^{\|I\|}}{(a\|I\|)!} \prod_{j \in J_i^j \in I} \|\phi_j\|_k^2 < \infty \tag{14}$$

for some fixed $a > 0$.

Definition 2.1. We put:

$$\Omega_{\infty-}(S_{\infty-}(M)) = \cap_{k,C} \Omega_{k,C}(S_{k,C}(M)) \tag{15}$$

Let us recall that the Hochschild boundary $\tilde{b}$ is given (See Refs. 3,15,16) by

$$\tilde{b}(\tilde{\alpha_n} d\tilde{\sigma}) = (-1)^n [\tilde{\alpha}_n, \tilde{\sigma}] \tag{16}$$

Theorem 2.2. *$\tilde{b}$ is continuous on $\Omega_{\infty-}(S_{\infty-}(M))$.*

Proof. Let $\tilde{\alpha} = \sum \lambda_I \tilde{\alpha}_I$. We get:

$$\tilde{b}\tilde{\alpha} = \sum \lambda_I \tilde{b}\tilde{\alpha}_I \tag{17}$$

In $\tilde{b}\tilde{\alpha}_I$, there are $|I| + 1$ terms which appear, whose norms can be bounded by increasing C. The product of some ϕ_i in some $\tilde{\sigma}_{J_r^I}$ by some ϕ_j in some $\tilde{\sigma}_{J_{r'}^I}$ can be estimated as it was done before. We deduce that for some big k_1, some big C_1 and some big k' and some small C_3 that

$$\|\tilde{b}\tilde{\alpha}\|_{k,C} \leq \sum |\lambda_I| K_I \|\tilde{\alpha}_I\|_{k_1,C_1} \tag{18}$$

where

$$K_I = C_3^{\|I\|} \prod_{j \in J_i^I \in I} (C_j + 2)^{-k'} \tag{19}$$

By using the same arguments than in theorem 1, we deduce that $\sum |K_I|^2 < \infty$. We conclude by using the Cauchy-Schwartz inequality. □

Let $\tilde{B}$ be the Connes boundary:

$$\tilde{B}(\tilde{\sigma}_0 d\tilde{\sigma}_1 ... d\tilde{\sigma}_n) = \sum sign\, d\tilde{\sigma}_i ... d\tilde{\sigma}_n d\tilde{\sigma}_0 d\tilde{\sigma}_1 .. d\tilde{\sigma}_{i-1} \tag{20}$$

We get easily:

Theorem 2.3. *$\tilde{B}$ is continuous on $\Omega_{\infty-}(S_{\infty-}(M))$.*

We denote by $H(\Omega_{\infty-}(S_{\infty-}(M))$ the cohomology groups associated to the cyclic complex $\tilde{b}+\tilde{B}$.

We can introduce the X-complex associated to $S_{\infty-}(M)$ following the terminology of Cuntz-Quillen and Perrot.

We consider $\tilde{b}\Omega^2(S_{\infty-}(M))$ and by the same notation we denote its closure for the Frechet topology on $\Omega^1(S_{\infty-}(M))$. We put:

$$X(S_{\infty-}(M)) = S_{\infty-}(M) \oplus \Omega^1(S_{\infty-}(M))/\tilde{b}\Omega^2(S_{\infty-}(M)) \tag{21}$$

where the sum is orthogonal.

We denote by p the projection from $\Omega^1(S_{\infty-}(M))$ into the quotient space $\Omega^1(S_{\infty-}(M))/\tilde{b}\Omega^2(S_{\infty-}(M))$. The X complex is the $Z/2Z$ graded complex

$$p\tilde{d} : S_{\infty-}(M) \to \Omega^1(S_{\infty-}(M))/\tilde{b}\Omega^2(S_{\infty-}(M)) \tag{22}$$

$$\tilde{b} : \Omega^1(S_{\infty-}(M))/\tilde{b}\Omega^2(S_{\infty-}(M)) \to S_{\infty-}(M) \tag{23}$$

We denote by $H(X(S_{\infty-}(M))$ its cohomology group.

We get an extension of Goodwillie's theorem:

Theorem 2.4 (Goodwillie's theorem in Hida sense).
If a is big enough,

$$H(X(S_{\infty-}(M))) = H(\Omega_{\infty-}(S_{\infty-}(M))) \tag{24}$$

Proof. The algebra is the same than the algebra of the proof of Corollary 4.4 of the work of Perrot.[13] Only the analysis is different.

We can define a map, $\tilde{\Phi}$ from $S_{\infty-}(M)$ into $\Omega^2(S_{\infty-}(M))$ satisfying:

$$\tilde{\Phi}(\tilde{\sigma} \circ \tilde{\sigma}') = \tilde{\Phi}(\tilde{\sigma})\tilde{\sigma}' + \tilde{\sigma}\tilde{\Phi}(\tilde{\sigma}') + \tilde{d}\tilde{\sigma}\tilde{d}\tilde{\sigma}' \tag{25}$$

We can write since $d^2 = 0$

$$\tilde{\sigma}_J = \phi_{j_0^J} \circ d\phi_{j_1^J} d\phi_{j_2^J} \circ .. \circ d\phi_{j^J_{|J|-1}} d\phi_{j^J_{|J|}} \tag{26}$$

If

$$\tilde{\sigma} = \phi_0 \circ d\phi_1 d\phi_2 \circ .. \circ d\phi_{2n-1} d\phi_{2n} \tag{27}$$

we get the following formula (See Ref. 16 p 59 for details):

$$\begin{aligned}
\tilde{\Phi}(\tilde{\sigma}) &= \tilde{d}\phi_0 \tilde{d}(d\phi_1 d\phi_2 \circ .. \circ d\phi_{2n-1} d\phi_{2n}) \\
&-\sum \phi_0 \circ d\phi_1 d\phi_2 .. \circ \tilde{d}\phi_{2i-1} \tilde{d}(\phi_{2i} \circ d\phi_{2i+1} d\phi_{2i+2} ... \circ d\phi_{2n-1} d\phi_{2n}) \\
&+\sum \phi_0 \circ d\phi_1 d\phi_2 \circ ... \tilde{d}(\phi_{2i-1}\phi_{2i}) \tilde{d}(d\phi_{2i+1} d\phi_{2i+2} \circ .... \circ d\phi_{2n-1} d\phi_{2n}) \\
&-\sum \phi_0 \circ d\phi_1 d\phi_2 ... \circ \phi_{2i-1} \tilde{d}\phi_{2i} \tilde{d}(d\phi_{2i+1} d\phi_{2i+2} \circ ... \circ d\phi_{2n-1} d\phi_{2n})
\end{aligned} \tag{28}$$

Let us remark that in this formula, there at most $6n$ terms. Moreover, the only product in the algebra $S_{\infty-}(M)$ are in $\tilde{d}(\phi_{2i-1}\phi_{2i})$. In this formula, we can replace Fedosov product by the traditional product because $d^2 = 0$, this in order to do do the estimates, and only to do the estimates, because we are interested by iterating $\tilde{\Phi}$. The proof will arise from the next considerations. □

We extend $\tilde{\Phi}$ to $\Omega^n(S_{\infty-}(M))$ by the following formula (See Ref. 16 formula (2.9) p 56):

$$\tilde{\Phi}(\tilde{\sigma_0}\tilde{d}\tilde{\sigma}_1...\tilde{d}\tilde{\sigma}_n) = \sum_{i=0}^{n}(-1)^{ni}\tilde{\Phi}(\tilde{\sigma_i})\tilde{d}\tilde{\sigma}_{i+1}....\tilde{d}\tilde{\sigma}_{i-1} \tag{29}$$

Lemma 2.1. *$\tilde{\Phi}$ and $(I-\tilde{\Phi})^{-1} = \sum \tilde{\Phi}^l$ are continuous on $\Omega_{\infty-}(S_{\infty-}(M))$.*

Proof. Let us show that $\tilde{\Phi}$ is bounded. Let be $\tilde{\alpha} = \sum \lambda_I \tilde{\alpha}_I$. We have:

$$\|\tilde{\Phi}(\tilde{\alpha})\|_{k,C} \leq \sum |\lambda_I| \|\tilde{\Phi}(\tilde{\alpha}_I)\|_{k,C} \tag{30}$$

Only a product appear of some ϕ_j in some J_i^I and some $\phi_{j'}$ in the same J_i^I. We can estimate its Sobolev norms by the product of some Sobolev norms of each. There are $K\|I\|$ terms which appear in $\tilde{\Phi}(\tilde{\alpha}_I)$ due to (28) and (29). Each term in the sum which appears in $\tilde{\Phi}(\tilde{\alpha}_I)$ has a total weight $\|I\|+1$ or $\|I\|+2$, but the partial weight is $|I|+2$. We deduce that:

$$\|\tilde{\Phi}(\tilde{\alpha}_I)\|_{k,C} \leq CC_2^{\|I\|} \prod_{j\in J_i^I \in I} (C_j+2)^{-k'} \|\tilde{\alpha}_I\|_{k_1,C_1} = K_I \|\tilde{\alpha}_I\|_{k_1,C_1} \tag{31}$$

for some big k', some big C_1, some big k_1 and some small C_2. Under these assumptions, $\sum K_I^2 < \infty$ and we deduce that

$$\|\tilde{\Phi}(\tilde{\alpha})\|_{k,C} \leq K\|\tilde{\alpha}\|_{k_1,C_1} \tag{32}$$

Let us iterate and let us consider $\tilde{\Phi}^l(\tilde{\alpha}_I)$. It is zero if $l > 3\|I\|$. In $\tilde{\Phi}^l(\tilde{\alpha}_I)$, there are at most the product of two ϕ_j belonging to the same $\tilde{\sigma}_{J_i^I}$. The Sobolev norm of each such product can be estimated, by Sobolev imbedding theorem, by the product of some Sobolev norms. If a term which appears in $\tilde{\Phi}^l(\tilde{\alpha_I})$ as a total weight k, when we consider $\tilde{\Phi}^{l+1}(\tilde{\alpha}_I)$, there are deduced from this term bk terms which appears with total weight $k+$ or $k+2$. Therefore, we deduce:

$$\begin{aligned} \|\sum \tilde{\phi}^l \tilde{\alpha}_I\|_{k,C} \leq \sum_{l\leq 3\|I\|} C_2^{\|I\|} \\ \sum_{l_1<l_2<..<l_l} C_2^{l_l} \frac{(b\|I\|)(b(\|I\|+l_1))...(b(\|I\|+l_l))}{((a(\|I\|+l_l))!)^{1/2}} (a\|I\|)!^{1/2} \|\tilde{\alpha}_I\|_{k_1,C_1} \end{aligned} \tag{33}$$

where C_2 is very small while k_1 and C_1 are very big and where l_i differ from l_{i+1} by 1 or 2. But b does not depend on a in (33). We can choose a big enough such that

$$\frac{(b\|I\|)(b(\|I\|+l_1))...(b(\|I\|+l_l))((a\|I\|)!)^{1/2}}{((a(\|I\|+l_l))!)^{1/2}} \leq 1 \tag{34}$$

Therefore, the auxiliary sum is smaller than $C_3^{\|I\|}$ and we get that

$$\|\sum \tilde{\bar{\Phi}}^l(\tilde{\alpha}_J)\|_{k,C} \leq C_4^{\|I\|}\|\tilde{\alpha}_I\|_{k_1,C_1} \tag{35}$$

for some small C_4 and some k_1 and some C_1 enough big. But if C_4 is small enough,

$$\sum C_4^{\|I\|} < \infty \tag{36}$$

which allows to conclude by Cauchy-Schwartz inequality. □

We deduce a connection $\tilde{\nabla}$ from $\Omega^1(S_{\infty-}(M))$ into $\Omega^2(S_{\infty-}(M))$

$$\tilde{\nabla}(\tilde{\sigma}_0\tilde{d}\tilde{\sigma}_1) = \tilde{\sigma}_0\tilde{\Phi}(\tilde{\sigma}_1) \tag{37}$$

which can be extended to $\Omega^n(S_{\infty-}(M))$ by the formula:

$$\tilde{\nabla}(\tilde{\alpha}_{n-1}\tilde{d}\tilde{\sigma}_n) = \tilde{\nabla}(\tilde{\alpha}_{n-1})\tilde{d}\tilde{\sigma}_n \tag{38}$$

if $\tilde{\alpha}_{n-1}$ belongs to $\Omega^{n-1}(S_{\infty-}(M))$ (See Ref. 16 p 56).

Lemma 2.2. $\tilde{\nabla}$ *is continuous from* $\Omega_{\infty-}(S_{\infty-}(M))$ *into* $\Omega_{\infty-}(S_{\infty-}(M))$.

Proof. We get

$$\tilde{\nabla}(\tilde{\alpha}_I) = \tilde{\sigma}_{J_0^I}\tilde{\Phi}(\tilde{\sigma}_{J_1^I})\tilde{d}(\tilde{\sigma}_{J_2^I})...\tilde{d}(\tilde{\sigma}_{J_{|I|}^I}) \tag{39}$$

There are $4|J_1^I|$ terms which appear in $\tilde{\Phi}(\tilde{\sigma}_{J_1^I})$. The total weight of $\tilde{\nabla}(\tilde{\alpha}_I)$ increases by a bounded quantity. By changing of Sobolev norms, we deduce in a simpler way than the proof of Lemma 2.1. that for some small C_4, some big k_1 and some big C_1

$$\|\tilde{\nabla}(\tilde{\alpha}_I)\|_{k,C} \leq C_4^{\|I\|}\|\tilde{\alpha}_I\|_{k_1,C_1} \tag{40}$$

We conclude as in the end of the proof of Lemma 2.1. □

We can finish the Proof of Goodwillie's theorem.

Proof. The map γ from $X(S_{\infty-}(M))$ into the space $\Omega_{\infty-}(S_{\infty-}(M))$ defined by

$$\tilde{\sigma} \to (I-\tilde{\Phi})^{-1}\tilde{\sigma} \tag{41}$$

and by

$$p(\tilde{\sigma}_0 d\tilde{\sigma}_1) \to (I - \tilde{\Phi})^{-1}(\tilde{\sigma}_0 d\tilde{\sigma}_1 + \tilde{b}(\tilde{\sigma}_0 \tilde{\Phi}(\tilde{\sigma}_1)) \tag{42}$$

is continuous and well defined.

The natural projection π from $\Omega_{\infty-}(S_{\infty-}(M))$ into $X(S_{\infty-}(M))$ is continuous.

$Q = \gamma \circ \pi$ is continuous. We put (See Ref. 16 p 58):

$$h = (I - \tilde{\Phi})^{-1}\tilde{\nabla}(I - Q) \tag{43}$$

h is continuous from $\Omega_{\infty-}(S_{\infty-}(M))$ into $\Omega_{\infty-}(S_{\infty-}(M))$. Moreover, see Ref. 16 p 57 Proposition 4.2, iii), $\pi \circ \gamma = I$ on $X(S_{\infty-}(M))$ and

$$\gamma \circ \pi = I + (\tilde{b} + \tilde{B})h + h(\tilde{b} + \tilde{B}) \tag{44}$$

on $\Omega_{\infty-}(S_{\infty-}(M))$.

This shows that γ realizes an isomorphism between $H(X(S_{\infty-}(M)))$ and $H(\Omega_{\infty-}(S_{\infty-}(M)))$. □

3. Holomorphic Goodwillie's theorem

Let us suppose that G is a compact Lie group endowed with its biinvariant metric. Let e_i be an orthonormal basis of its Lie algebra.

If $r = (i_1, ., i_{|r|})$,we denote by $\partial^{(r)} = e_{i_1} ... e_{i_{|r|}}$. We define the Sobolev norm $\|.\|_k$ on $H_k(G)$ by:

$$\|f\|_k^2 = \sum_{|r| \le k} \|\partial^{(r)} f\|_{L^2}^2 \tag{45}$$

We denote by $H_\infty(G) = \cap H_k(G)$.

By Sobolev imbedding theorem, we have the inequality for some convenient C:

$$\|fg\|_k \le C^k \|f\|_{Ck} \|g\|_{Ck} \tag{46}$$

We consider $S_{k,C}^{hol}(G)$ the space of $\tilde{\sigma} = \sum \sigma_{2n}$ where σ_{2n} belongs to $H_\infty(G) \otimes H_\infty(G)^{\otimes 2n}$ such that:

$$\|\tilde{\sigma}\|_{k,C}^2 = \sum_r \sum_n \frac{C^{n+r}}{(an + ar)!} \|\sigma_{2n}\|_{r+kn}^2 < \infty \tag{47}$$

The intersection of $S_{k,C}^{hol}(G)$ is called $S_{\infty-}^{hol}(G)$.

Let us remark that the Fedosov product has no reason to be continuous on $S_{\infty-}^{hol}(G)$.

Let us consider $\Omega_{k,C}^{hol}(G)$ be the space of $\tilde{\alpha} = \sum \tilde{\alpha}_I$ where $I = (J_0^I, .., J_{|I|}^I)$ such that:

$$\|\tilde{\alpha}\|_{k,C}^2 = \sum_{r,I} \frac{C^{r+\|I\|}}{(ar + a\|I\|)!} \|\tilde{\alpha}_I\|_{r+k\|I\|}^2 < \infty \tag{48}$$

Let us remark that by $\tilde{\alpha}_I$ we denote an element of $H_{\infty-}^{J_0^I} \otimes \tilde{d}H_{\infty-}^{J_1^I} \otimes .. \otimes \tilde{d}H_{\infty-}^{J_{|I|}^I}$. Since we consider the Sobolev norms (45), there is no orthogonal basis ϕ_j which is the same for all k of $H_k(G)$. We denote the intersection of the space which are got by this procedure by $\Omega_{\infty-}^{hol}(S_{\infty-}^{hol}(G))$. We get:

Theorem 3.1. *$\tilde{b}$ is continuous on $\Omega_{\infty-}^{hol}(S_{\infty-}^{hol}(G))$.*

Proof. Let us consider an orthonormal basis of $\Omega_{k,C}^{hol}(S_{k,C}^{hol}(G))$ for k big. It can be written as in the previous part as $\tilde{\alpha}_I = \tilde{\sigma}_{J_0^I} \tilde{d}\tilde{\sigma}_{J_1^I} ... \tilde{d}\tilde{\sigma}_{J_{|I}^I}$ where $\tilde{\sigma}_J = \phi_{i_0^J} d\phi_{i_1^J} ... d\phi_{i_{|J|}^J}$.

But

$$\|\varphi_i \phi_j\|_{\frac{k}{C}} \leq C^k \|\phi_i\|_k \|\phi_j\|_k \tag{49}$$

for C big independent of k by (46).

For this k, with the ϕ_i associated to the Hilbert structure $H_k(G)$, we can use the notation of the previous part and show the following statement: let us consider a sum $\tilde{\alpha} = \sum \tilde{\lambda}_I \tilde{\alpha}_I$ such that

$$\tilde{b}\tilde{\alpha} = \sum \tilde{\lambda}_I \tilde{b}\tilde{\alpha}_I \tag{50}$$

Then

$$\|\tilde{b}\tilde{\alpha}_I\|_{k/C_1, C_2} \leq C_4^{\|I\|} \|\tilde{\alpha}_I\|_{k,C} \tag{51}$$

if C_2 is small enough tending to ∞ when $C \to \infty$ and C" small enough. We deduce that:

$$\|\tilde{b}\tilde{\alpha}\|_{k/C_1, C_2} \leq (\sum C_3^{\|I\|})^{1/2} \|\tilde{\alpha}\|_{k,C} \tag{52}$$

where C_3 can be chosen small enough that the series in the right hand side of (52) converges. Moreover, when k and C tend to ∞, k/C_1 and C_2 tend to ∞. □

Theorem 3.2. *The Connes boundary $\tilde{B}$ is continuous over $\Omega_{\infty-}^{hol}(S_{\infty-}^{hol}(G))$.*

The proof is straightforward.

Lemma 3.1. *$\tilde{\Phi}$ and $(I-\tilde{\Phi})^{-1} = \sum \tilde{\Phi}^l$ are continuous on $\Omega^{hol}_{\infty-}(S^{hol}_{\infty-}(G))$ if a in (48) is big enough.*

Proof. Let $\tilde{\alpha} = \sum \lambda_I \tilde{\alpha}_I$ where $\tilde{\alpha}_I$ is an orthogonal basis of $\Omega^{hol}_{k,C}(S^{hol}_{k,c}(G))$ for k big enough. Each $\tilde{\Phi}\tilde{\alpha}_I$ can be decomposed into a sum of at most $A\|I\|$ elementary terms, whose $k/C_1 + r$ Sobolev norms can be estimated by $C^{\|I\|+r}$ the Sobolev norm of $\tilde{\alpha}_I$ for $C_1 r + k$. If we consider the norm $\|.\|_{k/C_1,C_2}$ of each term, it can be estimated by

$$\sum_r \frac{C_2^{\|I\|+r}}{((ar + a\|I\|)!)^{1/2}} \|\tilde{\alpha}_I\|_{r+k\|I\|} \tag{53}$$

which is smaller by Cauchy-Schwartz inequality to $\|\tilde{\alpha}_I\|_{k,C} C_4^{\|I\|}$ for some small C_4 if we choose C big enough, which will tends to infinity later.

We deduce as before that:

$$\|\tilde{\Phi}\tilde{\alpha}\|_{k/C_1, C/C_1} \leq C\|\tilde{\alpha}\|_{k,C} \tag{54}$$

We will do as in Lemma 6 in order to show the continuity of $(I-\tilde{\Phi})^{-1}$.

$$(I-\tilde{\Phi})^{-1}\tilde{\alpha}_I = \sum_{l=0}^{K\|I\|} \tilde{\Phi}^l \tilde{\alpha}_I \tag{55}$$

There is a sum of elementary terms on $l \leq 3\|I\|$ and on $l_1 < l_2 < .. < l_l$ where $l_j = l_{j-1}+1$ or $l_j = l_{j-1}+2$ of terms of total weight $\|I\| + l_l$. Therefore, corresponding to a sequence $l_1, .., l_l$, there are at most $b^l \|I\|(\|I\| = l_1)...(\|I\| + l_l)$ terms. We deduce that we have the following bound:

$$\begin{aligned} \|(I-\tilde{\Phi})^{-1}\tilde{\alpha}_I\|_{k/C_1,C_1} \leq \sum_{r,l\leq 3\|I\|} C^{\|I\|+r} \\ \sum_{l_1<l_2<..l_l} b^l C^{l_l} \frac{\|I\|(\|I\|+l_1)(\|I\|+l_l)}{((ar+a\|I\|+al_l)!)^{1/2}} \|\tilde{\alpha}_I\|_{C_1 r + k\|I\|} \end{aligned} \tag{56}$$

If a is big enough,

$$\frac{b^l \|I\|(\|I\|+l_1)..(\|I\|+l_l)}{((ar+a\|I\|+al_l)!)^{1/2}} \leq C((ar+a\|I\|)!)^{-1/2} \tag{57}$$

We deduce that there exist a C_1 big enough such that for C_2 small enough

$$\|(I-\tilde{\Phi})^{-1}\tilde{\alpha}_I\|_{k/C_1,C/C_1} \leq C_2^{\|I\|} \|\tilde{\alpha}_I\|_{k,C} \tag{58}$$

The result arises by Cauchy-Schwartz inequality since $\sum C_2^{2\|I\|}$ is finite if C_2 is small enough. □

Lemma 3.2. *$\tilde{\nabla}$ is continuous on $\Omega^{hol}_{\infty-}(S^{hol}_{\infty-}(G))$.*

Proof. We use (40) for an orthonormal basis $\tilde{\alpha}_I$ of $\Omega^{hol}_{k,C}(S^{hol}_{k,C}(G))$. We deduce that

$$\|\tilde{\nabla}\tilde{\alpha}_I\|_{k/C_1,C/C_1} \leq C_2^{\|I\|}\|\tilde{\alpha}_I\|_{k,C} \tag{59}$$

for C_1 big enough and C_2 small enough. We conclude by using Cauchy-Schwartz inequality. □

We deduce as in Ref. 16:

Theorem 3.3 (Holomorphic Goodwillie's theorem).
If a is big enough:

$$H(X(S^{hol}_{\infty-}(G)) = H(\Omega^{hol}_{\infty-}(S^{hol}_{\infty-}(G)) \tag{60}$$

References

1. L. Accardi and M. Bozejko, Intracting Fock spaces and gaussianization of probability measures, *Inf. Dim. Ana. Quant. Probab. Rel. Top.* **1** (1998) 663–670.
2. A. Connes, Entire cyclic cohomology of Banach algebras and characters of θ-summable Fredhom modules, *K-theory* **1** (1988) 519–541.
3. J. Cuntz, Cyclic Theory, Bivariant K-Theory and the Bivariant Chern-Connes Character, *In: Cyclic homology in non-commutative geometry* (Springer, 2001) 2–69.
4. J. Cuntz and D. Quillen, Algebra extensions and nonsingularity, *J.A.M.S.* **8** (1995) 251–289.
5. J. Cuntz and D. Quillen, Cyclic homology and nonsingularity, *J.A.M.S.* **8** (1995) 373–442.
6. E. Getzler and A.K. Szenes, On the Chern character of a theta-summable Fredholm module, *J.F.A.* **84** (1989) 343–357.
7. T.G. Goodwillie, Cyclic homology, derivations and the free loop space, *Topology* **24** (1985) 187–215.
8. A. Jaffe, Quantum harmonic analysis and geometric invariants, *Adv. Maths.* **143** (1999) 1–110.
9. A. Jaffe, A. Lesniewski and K. Osterwalder, Quantum K-theory I. The Chern character, *Com. Math. Phys.* **118** (1988) 1–14.
10. J.D.S. Jones and R. Léandre, L^p Chen forms on loop spaces, *In: Stochastic analysis* ((M. Barlow and N. Bingham, Eds.) Cambridge. Univ. Press, 1991) 104–162.
11. R. Léandre, Cohomologie de Bismut-Nualart-Pardoux et cohomologie de Hochschild entiere, *In: Séminaire de probabilités XXX in honour of P.A. Meyer and J. Neveu* ((J. Azéma, M. Emery, M. Yor, Eds.) L.N.M. 1626, Springer, 1996) 68–100.

12. R. Léandre, Wiener Analysis and cyclic cohomology, *In: Stochastic analysis and mathematical physics* ((R. Rebolledo and J.C. Zambrini, Eds.) World Scientific, 2004) 115–129.
13. R. Léandre, Connes-Hida Calculus in the Index theory, *In: XIVth Int. Cong. Math. Phys* ((J.C. Zambrini, Ed.) World Scientific, 2005) 493–498.
14. R. Léandre and H. Ouerdiane, Connes-Hida Calculus and Bismut-Quillen superconnections, *In: Stochastic Analysis: Classical and Quantum* ((T. Hida, Ed.) World Scientific, 2005) 72–86.
15. R. Meyer, Analytic cyclic cohomology, PhD thesis, University of Muenster 1999.
16. D. Perrot, A bivariant Chern character for families of spectral triples, *Com. Math. Phys.* **231** (2002) 45–95.

Differential Geometry and its Applications
Proc. Conf., in Honour of Leonhard Euler, Olomouc, August 2007

De Rham like cohomology of geometric structures

M.A. Malakhaltsev

Kazan State University,
Kremlevskaya 18, Kazan, 420008, Russia
E-mail: mikhail.malakhaltsev@ksu.ru

In this paper we give a brief review of differential complexes associated with geometric structures on manifolds, like the complex structure, the foliation structure, etc. These complexes can be constructed using the Spencer P-complex of the Lie derivative of the section corresponding to an integrable G-structure.[18] Also we give an example of differential complex associated to a non regular integrable G-structure.

Keywords: de Rham complex, Dolbeaux cohomology, foliated cohomology, Spencer complex, regular Poisson structure, manifold over algebra.

MS classification: 14F05, 14F10, 14F40.

1. Introduction

Let M be a smooth manifold, $\Omega^q(M)$ be the set of differential q-forms on M, and $d : \Omega^q(M) \to \Omega^{q+1}(M)$ be the exterior differential. The *de Rham complex* of M is defined as follows:

$$0 \to \Omega^0(M) \xrightarrow{d} \Omega^1(M) \xrightarrow{d} \dots \xrightarrow{d} \Omega^{n-1}(M) \xrightarrow{d} \Omega^n(M) \xrightarrow{d} 0 \to \dots$$

The cohomology of this complex in dimension k:

$$H^k(M) = H^k_{DR}(M) = \frac{\operatorname{Ker} d : \Omega^k(M) \to \Omega^{k+1}(M)}{\operatorname{im} d : \Omega^{k-1}(M) \to \Omega^k(M)}$$

is called the *kth de Rham cohomology group* of M. Note that the wedge product induces the structure of a graded algebra on $H_{DR}(M) = \oplus_{k=1}^n H^k_{DR}(M)$.

The de Rham cohomology has nice properties, in particular, the correspondence $M \to H_{DR}(M)$ is a cofunctor from the category of smooth manifolds to the category of vector spaces (for a detailed exposition of the de Rham cohomology theory and its applications we refer the reader to Ref. 1). Another important property of the de Rham cohomology is that the de Rham complex is defined in terms of the differential structure, in part, d is a differential operator, however the de Rham cohomology is a topological invariant. This follows from the fact that, at the sheaf level, the de Rham complex gives a fine resolution for the sheaf $\mathbb{R}_M$ of locally constant functions on a manifold M:

$$0 \to \mathbb{R}_M \xrightarrow{i} \Omega^0_M \xrightarrow{d} \Omega^1_M \xrightarrow{d} \dots \xrightarrow{d} \Omega^{n-1}_M \xrightarrow{d} \Omega^n_M \xrightarrow{d} 0 \to \dots,$$

where Ω^k_M is the sheaf of differential k-forms on M. Hence follows that the de Rham cohomology $H_DR(M)$ of a manifold M is isomorphic to the cohomology $H(M; \mathbb{R}_M)$ of M with coefficients in the sheaf $\mathbb{R}_M$.

In the present paper we describe how to construct a complex of sheaves associated to an integrable G-structure which provides a fine resolution to the sheaf of infinitesimal automorphism of this structure (see Ref. 18 and also Ref. 20). In this way one can obtain the well-known cohomology of classical differential geometric structures: the Dolbeaux cohomology for the complex structure, the foliated cohomology for the foliation structure, etc. Also, we construct de Rham like complexes associated to closed differential forms of non-constant rank (these ones are examples of non-regular integrable G-structures).[27–29]

2. Examples of de Rham like complexes for geometrical structures

2.1. *De Rham like cohomology for complex manifold*

Let (M, J) be an n-dimensional complex manifold. Each tangent space T_pM, $p \in M$, is a vector bundle over $\mathbb{C}$, therefore we have the decomposition: $T^*M \otimes \mathbb{C} \cong T^{1,0}M \oplus T^{0,1}M$, where $T^{1,0}M$ ($T^{0,1}M$) consists of $\mathbb{C}$-linear ($\mathbb{C}$-antilinear) $\mathbb{C}$-valued covectors. If z^i are local complex coordinates on M, then dz^i ($d\,\overline{z}^i$) are sections of $T^{1,0}M$ ($T^{0,1}M$).

Denote by $\Omega_{\mathbb{C}}(M)$ the algebra of $\mathbb{C}$-valued differential forms. An $\omega \in \Omega_{\mathbb{C}}(M)$ is called *holomorphic* if ω and $d\omega$ are $\mathbb{C}$-linear. Locally, a holomorphic form is written as $\omega = \omega_{i_1 \dots i_p}(z^k) dz^{i_1} \wedge \dots \wedge dz^{i_p}$, where $\omega_{i_1 \dots i_p}(z^k)$ are holomorphic functions. Denote by $\Omega_h(M)$ the algebra of holomorphic forms on (M, J). It is clear that, if ω is a holomorphic form, $d\omega$ is also a holomorphic form, therefore we get the *de Rham complex of holomorphic forms* (Ω_h, d), which is a subcomplex of $(\Omega_{\mathbb{C}}(M), d)$.

The space of $\mathbb{C}$-valued differential forms decomposes as follows:

$$\Omega^k_{\mathbb{C}}(M) = \Omega^k(M) \otimes \mathbb{C} = \oplus_{p+q=k} \Omega^{p,q}(M)$$

where

$$\alpha \in \Omega^{p,q}(M) \iff \alpha = \alpha_{i_1 \ldots i_p j_1 \ldots j_q} dz^{i_1} \wedge \cdots \wedge dz^{i_p} \wedge d\overline{z}^{j_1} \wedge \cdots \wedge d\overline{z}^{j_q} \quad (1)$$

For the differential df of a function $f : M \to \mathbb{C}$ we have $df = \partial f + \overline{\partial} f$, where $\partial f = \frac{\partial}{\partial z^k} f dz^k$, $\overline{\partial} f = \frac{\partial}{\partial \overline{z}^k} f d\overline{z}^k$. This gives rise to the differential operators $\partial : \Omega^{p,q}(M) \to \Omega^{p+1,q}(M)$, $\overline{\partial} : \Omega^{p,q}(M) \to \Omega^{p,q+1}(M)$ such that $d = \partial + \overline{\partial}$, $\partial^2 = 0$, $\overline{\partial}^2 = 0$, and $\partial\overline{\partial} = \overline{\partial}\partial$. The complex $(\Omega^{p,*}(M), \overline{\partial})$ is called the *Dolbeaux complex* of (M, J), and $H^{p,q}(M)$ is called the *Dolbeaux cohomology* of (M, J). It is clear that all these complexes can be defined at the sheaf level, and we have the following

Theorem 2.1. *Let (M, J) be a complex manifold. Denote by Ω^p_h the sheaf of holomorphic p-forms on M, and by $\Omega^{p,q}$ the sheaf of forms of type (p, q) on M. The sheaf sequence*

$$0 \to \Omega^p_h \xrightarrow{i} \Omega^{p,0} \xrightarrow{\overline{\partial}} \Omega^{p,1} \ldots \xrightarrow{\overline{\partial}} \Omega^{p,q} \xrightarrow{\overline{\partial}} \Omega^{p,q+1} \xrightarrow{\overline{\partial}} \ldots.$$

is a fine resolution for the sheaf of holomorphic p-forms

In terms of the Dolbeaux cohomology many results concerning complex manifolds, in part algebraic manifolds, are formulated (see, e. g. Refs. 2–4). Note also that the structure of bicomplex on $\Omega_{\mathbb{C}}(M)$ (see (1)) gives rise to the *Frölicher spectral sequence* converging to the de Rham cohomology of M.

2.2. *De Rham like complexes for foliations*

A foliation $\mathcal{F}$ of codimension m on an $(m + r)$-dimensional manifold M is given by a totally integrable distribution Δ on M. The integral submanifolds of Δ are called the *leaves of foliation* $\mathcal{F}$. The distribution Δ is called the tangent bundle of the foliation $\mathcal{F}$ and is denoted by $T\mathcal{F}$. For any point $p \in M$, one can find an *adapted coordinate system* (x^a, x^α), $a = \overline{1, m}$, $\alpha = \overline{m + 1, m + r}$, in a neighborhood U of p such that Δ restricted to U is given by $dx^a = 0$. Each connected component of intersection of any integral submanifold of Δ with U is given with respect to this coordinate system by equations $x^a = \text{const}$.

A manifold endowed by a foliation is called a *foliated manifold*. For foliated manifolds one can define de Rham like cohomology of two types: *basic cohomology* and *foliated cohomology* (see Refs. 5,7,8).

2.2.1. *Basic cohomology of foliation*

Let $(M, \mathcal{F})$ be a foliated manifold. A form $\omega \in \Omega^k(M)$ is called *basic* if, with respect to the adapted coordinate system (x^a, x^α), $\omega = \omega_{a_1 \dots a_k}(x^a) dx^{a_1} \wedge \dots \wedge dx^{a_k}$. Let us denote by $\Omega_b(M, \mathcal{F})$ the algebra of basic forms. It is clear that, for any $\omega \in \Omega^k_b(M, \mathcal{F})$, we have $d\omega \in \Omega^{k+1}_b(M, \mathcal{F})$. Thus we obtain the subcomplex $(\Omega_b(M, \mathcal{F}), d_b)$ of the de Rham complex $(\Omega(M), d)$. Cohomology $H_b(M, \mathcal{F})$ of $(\Omega_b(M, \mathcal{F}), d_b)$ is called *basic cohomology of the foliated manifold* $(M, \mathcal{F})$.

Note that $H_b(M, \mathcal{F})$ can be infinite-dimensional even if M is compact. If a foliation $\mathcal{F}$ is simple (the leaves of $\mathcal{F}$ are fibers of a submersion $\pi : M \to B$), then $H_b(M, \mathcal{F}) \cong H_{DR}(B)$.

2.2.2. *Foliated cohomology*

For a foliated manifold $(M, \mathcal{F})$, where $\dim M = n + r$, $\operatorname{codim} \mathcal{F} = m$, we take an r-dimensional distribution N complementary to $T\mathcal{F}$: $TM = T\mathcal{F} \oplus N$. Then, for any $p \in M$, we can take a local coframe (dx^a, η^α), $a = \overline{1, m}$, $\alpha = \overline{m+1, m+r}$, on a neighborhood U of p such that N is given by equations $\eta^\alpha = 0$. We say that a form $\theta \in \Omega^{p+q}(M)$ has type (p, q) if locally ω is written as follows:

$$\theta = \theta_{a_1 \dots a_p \alpha_1 \dots \alpha_q} dx^{a_1} \wedge \dots dx^{a_p} \wedge \eta^{\alpha_1} \wedge \dots \eta^{\alpha_q}.$$

(it is evident that the type is defined correctly and does not depend on the choice of adapted coordinate system and the forms η^α). Let us denote by $\Omega^{p,q}(M, \mathcal{F})$ the space of forms of type (p, q).

Evidently, $\Omega^k(M) \cong \oplus_{p+q=k} \Omega^{p,q}(M)$, and the exterior differential $d = d_{1,0} + d_{0,1}$, where $d_{1,0} : \Omega^{p,q}(M, \mathcal{F}) \to \Omega^{p+1,q}(M, \mathcal{F})$, $d_{0,1} : \Omega^{p,q}(M, \mathcal{F}) \to \Omega^{p,q+1}(M, \mathcal{F})$. Then $d^2_{1,0} = 0$, $d^2_{0,1} = 0$, $d_{1,0} d_{0,1} = d_{0,1} d_{1,0}$. Therefore, for any $p \geq 0$, we obtain the complex $(\Omega^{p,*}(M), \partial_{0,1})$, and its cohomology is called *foliated cohomology* or *leafwise cohomology*. Note that the *foliated cohomology does not depend on a choice of* N. We set

$$H^{p,q}(M, \mathcal{F}) = H^q(\Omega^{p,*}(M, \mathcal{F}), \partial_{0,1}).$$

Theorem 2.2. *Let $(M, \mathcal{F})$ be a foliated manifold. Let Ω_b be the sheaf of basic forms on M, and $\Omega^{p,q}$ be the sheaf of forms of type (p, q). Then, for all $p \geq 0$, the sheaf sequence*

$$0 \to \Omega^p_b \xrightarrow{i} \Omega^{p,0} \xrightarrow{\partial} \Omega^{p,1} \xrightarrow{\partial} \dots \xrightarrow{\partial} \Omega^{p,k} \xrightarrow{\partial} \Omega^{p,k+1} \xrightarrow{\partial} \dots$$

is a fine resolution for the sheaf Ω_b. Hence $H^q(M; \Omega^p_b) \cong H^{p,q}(M, \mathcal{F})$.

The foliated cohomology is used in various constructions of characteristic classes of foliations and obstructions to existence of basic geometric objects (see, e. g. Refs. 5,6). Note also that the bicomplex $(\Omega^{*,*}(M), d_{1,0}, d_{0,1})$ generates a spectral sequence converging to the de Rham cohomology of M.[3,8]

2.3. *Cohomology of manifolds over algebras*

Let $\mathbb{A}$ be a finite-dimensional commutative algebra with unit over $\mathbb{R}$. As $\mathbb{A}^n$ is a vector space isomorphic to $\mathbb{R}^{nm}$, where $m = \dim_{\mathbb{R}} \mathbb{A}$, we have the natural topology on $\mathbb{A}^n$. Let $U \subset \mathbb{A}^n$, $V \subset \mathbb{A}^k$ be open subsets. We say that $F : U \to V$ is $\mathbb{A}$-differentiable if and only if $dF_a : T_a\mathbb{A}^n \cong \mathbb{A}^n \to T_{F(a)}\mathbb{A}^k \cong \mathbb{A}^k$ is $\mathbb{A}$-linear.

Let us denote by Γ the pseudogroup of all local $\mathbb{A}$-diffeomorphisms of $\mathbb{A}^n$. A maximal Γ-atlas on a topological space M is called a *structure of manifold over* $\mathbb{A}$ *on* M, and M is called *a manifold over* $\mathbb{A}$. The structure of manifold over algebra is a generalization of the complex structure. Also this structure determines a family of foliations associated with the ideals of the algebra (for the theory of manifolds over algebras we refer the reader to Refs. 9–11).

Recall that $\mathbb{A}$ is called *local* if and only if $\mathbb{A} \cong \mathbb{R} \oplus \overset{\circ}{\mathbb{A}}$, where $\overset{\circ}{\mathbb{A}}$ is the radical of $\mathbb{A}$. For manifolds over local algebras, V. V. Shurygin constructed de Rham like cohomology which provides analogs both for the Dolbeaux cohomology and the foliated cohomology.[11]

Let us consider the complex $(\Omega(M) \otimes \mathbb{A}, d)$ of $\mathbb{A}$-valued differentiable forms, where d is the exterior differential extended to $\Omega(M)\otimes\mathbb{A}$ by linearity. We say that $\omega \in \Omega^k(M) \otimes \mathbb{A}$ is $\mathbb{A}$-differentiable if, for any $p \in M$, $\omega(p) : T_pM \times \ldots \times T_pM \to \mathbb{A}$ is $\mathbb{A}$-linear (note that T_pM is a finite-dimensional $\mathbb{A}$-module) and $d\omega$ is also $\mathbb{A}$-linear. Let us denote by $\Omega_{\mathbb{A}-diff}(M)$ the algebra of $\mathbb{A}$-differentiable forms, and it is clear that, for any $\omega \in \Omega_{\mathbb{A}-diff}(M)$, we have $d\omega \in \Omega_{\mathbb{A}-diff}(M)$. Therefore $(\Omega_{\mathbb{A}-diff}(M), d)$ is a subcomplex of $(\Omega(M) \otimes \mathbb{A}, d)$.

Theorem 2.3 (Poincare Lemma for $\mathbb{A}$-differentiable forms[10,11]). *Let U be an open coordinate parallelepiped in $\mathbb{A}^n$, and let φ be an $\mathbb{A}$-differentiable k-form on U such that $d\varphi = 0$; then there exists an $\mathbb{A}$-smooth $(k-1)$-form ψ on U such that $\varphi = d\psi$.*

For manifolds over algebra $\mathbb{A}$, V.V. Shurygin constructed complexes of differential forms determined by ideals of $\mathbb{A}$ and found properties of their cohomology. Also, he constructed a bicomplex of differential forms, which, on one hand generalizes the Dolbeaux bicomplex of complex manifold, and on the other hand, the Vaisman bicomplex of foliated manifold;[10,11] see also Ref. 12, where these complexes are constructed in a more general situation). For the properties of the de Rham complex of manifold over algebras see also Refs. 14,15.

As an example, consider the algebra of dual numbers $\mathbb{R}(\varepsilon) = \{a + b\varepsilon \mid \varepsilon^2 = 0\}$. Let M be a manifold over $\mathbb{R}(\varepsilon)$ (a simple example of manifold over $\mathbb{R}(\varepsilon)$ is a two-dimensional torus; also the total space of a tangent bundle can be endowed by a structure of manifold over $\mathbb{R}(\varepsilon)$). Then, on M we have an atlas with coordinates $z^k = x^k + \varepsilon x^{n+k}$ and the coordinate change $z^{k'} = z^{k'}(z^k)$ is $\mathbb{R}(\varepsilon)$-differentiable, from this follows that $x^{k'} = x^{k'}(x^i)$ and $x^{n+k'} = \frac{\partial x^{k'}}{\partial x^s} x^{n+s} + g^{k'}(x^i)$. The tangent bundle $T\mathcal{F}$ of the *canonical foliation* $\mathcal{F}$ of the structure of manifold over $\mathbb{R}(\varepsilon)$ is given locally by the equations $dx^k = 0$.

Now let N be a distribution on M such that $N \oplus T\mathcal{F} = TM$. Then $\Omega^k(M) \otimes \mathbb{R}(\varepsilon) = \oplus_{p+q=k} \Omega^{p,q}(M)$, where $\omega \in \Omega^{p,q}(M)$ are written locally as follows:

$$\omega = \omega_{i_1 \dots i_p n+j_1 \dots n+j_q} dz^{i_1} \wedge \dots \wedge dz^{i_p} \wedge \theta^{j_1} \wedge \dots \wedge \theta^{j_q}.$$

Here $dz^k = dx^k + \varepsilon dx^{n+k}$, θ^j is the basis of $Ann(N) \subset T^*M$, and $\omega_{i_1 \dots i_p n+j_1 \dots n+j_q}$ are $\mathbb{R}(\varepsilon)$-valued functions. We have the differential $\overline{\partial} : \Omega^{p,q}(M) \to \Omega^{p,q+1}(M)$ determined by $\overline{\partial} f = (\partial_{n+s} - \varepsilon \partial_s) f \, \theta^{n+s}$, and thus obtain the complex $(\Omega^{p,*}, \overline{\partial})$.

It is clear that the above complexes can be constructed at the sheaf level. From the results in Ref. 11 it follows that the following sequence of sheaves on M:

$$0 \to \Omega^p_h \xrightarrow{i} \Omega^{p,0} \xrightarrow{\overline{\partial}} \Omega^{p,1} \dots \xrightarrow{\overline{\partial}} \Omega^{p,q} \xrightarrow{\overline{\partial}} \Omega^{p,q+1} \xrightarrow{\overline{\partial}} \dots$$

is a fine resolution of the sheaf $\Omega_{\mathbb{R}(\varepsilon)-diff}$ of $\mathbb{R}(\varepsilon)$-differentiable forms on M.

3. Complex associated to integrable G-structure

3.1. *Integrable G-structures*

Let M be a manifold, $L(M)$ be the linear frame bundle of M. Consider a Lie subgroup $G \subset \mathrm{GL}(n)$. A G-subbundle $P \subset L(M)$ is called a *G-structure on M*. A G-structure P is said to be *integrable* if for any $p \in M$ there exists a chart (U, u^i) such that $\{\frac{\partial}{\partial u^i}\} \in P|_U$. This chart is called *adapted* to the integrable G-structure P. Recall also that the set of G-structures on M is in one-to-one correspondence with the set of all sections $s : M \to L(M)/G$.[16]

3.1.1. *Examples of integrable G-structures*

Complex structure. An almost complex structure on a $2n$-dimensional manifold M is a G-structure $P \subset L(M)$ with $G = \mathrm{GL}(n, \mathbb{C}) \subset \mathrm{GL}(2n, \mathbb{R})$. The corresponding section s is an affinor field such that $s^2 = -I$. A complex structure on M is an integrable almost complex structure.

Foliation structure. A distribution Δ of codimension m on an $(m+r)$-dimensional manifold M determines a G-structure $P \subset L(M)$ with

$$G = \{\begin{pmatrix} A & 0 \\ B & C \end{pmatrix}\} \subset \mathrm{GL}(m+r),$$

where $A \in \mathrm{GL}(r)$, $C \in \mathrm{GL}(m)$, B is an $m \times r$-matrix. Then the section s is just the distribution Δ

The G-structure P is integrable if and only if the distribution Δ is totally integrable, i. e. determines a foliation.

3.1.2. *The Lie derivative of a section $s : M \to L(M)/G$*

Let us denote by $X(M)$ the Lie algebra of vector fields on M. Then, for an $X \in X(M)$, the flow φ_t of X determines the flow $\overline{\phi}_t$ on $E_G(M) = L(M)/G$: if $\{e_i\}$ is a frame of T_pM, then $\overline{\phi}_t(\{e_i\}) = \{d\phi(e_i)\}$ is a frame of $T_{\phi(p)}M$. The vector field $\overline{X} = d\overline{\phi}_t/dt \in X(E_G(M))$ is called the *complete lift* of X.

Denote by VE_G the vertical subbundle of $TE_G(M)$. Let $s : M \to E_G(M)$ be a section. Then, for any $p \in M$, we have an isomorphism $\Pi_p \colon (VE_G)_{s(p)} \to (s^*(VE_G))_p$. The Lie derivative of s with respect to X is defined as follows:

$$(\mathcal{L}_X s)_p = \Pi_p(\overline{X}(s(p)) - ds_p(X(p))),$$

(we refer the reader to Ref. 19 for the general definition of the Lie derivative of a section of natural bundle, see also Ref. 18).

3.2. *Spencer P-complex for the Lie derivative*

From the definition it follows that the Lie derivative of a section $s : M \to E_G(M)$ is a first order differential operator from the tangent bundle TM to the bundle $s^*(VE_G(M)$, call it $\mathcal{L}(s)$. In Ref. 18, J.F.Pommaret considered the Spencer P-complex for a differential operator and, in particular, for the Lie derivative.

Given an integrable G-structure $P \subset L(M)$ on an n-dimensional manifold M, we denote by E_g the subbundle of the affinor bundle $T^1_1 M$ consisting of affinors A whose coordinates A^i_j with respect to the charts adapted to P constitute a matrix $||A^i_j||$ lying in the Lie algebra $g \subset \mathfrak{gl}(n)$ of $G \subset \mathrm{GL}(n)$.

Theorem 3.1 (Ref. 20). *The Spencer P-complex for $\mathcal{L}(s)$ is isomorphic to the complex $(C^q(P), \delta)$, where*

$$C^q(P) = \frac{\Omega^q(M) \otimes TM}{\mathrm{Alt}(\Omega^{q-1}(M) \otimes E_{\mathfrak{g}})},$$

and the differential $\delta\colon C^q(P) \to C^{q+1}(P)$ is induced by the differential operator $D = \mathrm{Alt} \circ \nabla$, where Alt *is the alternation and ∇ is the covariant derivative of torsion-free connection adapted to P (i. e. $(D\omega)_{i_1 \dots i_{q+1}}^{\;j} = \nabla_{[i_1} \omega_{i_2 \dots i_{q+1}]}^{\;j}$ with respect to local coordinates adapted to P).*

The restriction $P|_U$ of an integrable G-structure $P \subset L(M)$ to any open subset $U \subset M$ is an integrable G-structure on U. Therefore, we can construct sheaves $\mathcal{C}^q(P)$ on M by setting $U \to C^q(P|_U)$. Denote by X the sheaf of vector fields on M, and by X_s the sheaf of infinitesimal automorphisms of s, i. e. the sheaf of vector fields V such that $\mathcal{L}_V s = 0$. From the following examples we can see that, for the classical integrable G-structures, the complex of sheaves $0 \to X_s \to X \to \mathcal{C}_0(P) \xrightarrow{\delta} \mathcal{C}_1(P) \xrightarrow{\delta} \cdots$ gives a fine resolution for the sheaf of infinitesimal automorphisms of s; in particular this fact can be used in finding the space of essential infinitesimal deformations of s (see also Ref. 18).

3.3. *Examples of P-complex for classical G-structures*

3.3.1. *Foliation structure*

Let P be an integrable G-structure determining a foliation structure on a smooth manifold M, and let Δ be the corresponding integrable distribution (see 3.1.1). Then, $E_g = \{A \in T^1_1(M) \mid A(\Delta) \subset \Delta\}$ and $C^p = \Omega^p(\Delta) \otimes N$, where $N = TM/\Delta$. We have, with respect to the coordinate system adapted to the foliation structure, $(\delta\omega)_{\alpha_1 \dots \alpha_{q+1}}^{\;i} = \partial_{[\alpha_1} \omega_{\alpha_2 \dots \alpha_{q+1}]}^{\;j}$. Thus, the complex $(C^q(P), \delta)$ is isomorphic to the complex $(\Omega^{0,*}(M, \mathcal{F}) \otimes X_b(N), \partial_{0,1})$, where $X_b(N)$ is the space of basic sections of the bundle N (see 2.2.2).

3.3.2. *Complex structure, manifolds over algebras*

For the complex structure, the complex (C^q, δ) is the Dolbeaut complex tenzoried by the sheaf of holomorphic vector fields.[18] And, for the structure of manifold over algebra, a similar result (with the complex constructed by V. V. Shurygin in Ref. 1_ instead of the Dolbeaux complex; see 2.3) was proved by T. I. Gaisin in Ref. 21.

3.3.3. *Symplectic structure*

Let θ be a symplectic form given on a $2n$-dimensional manifold M, and $P \subset L(M)$ the corresponding integrable $\mathrm{Sp}(n)$-structure on M. Denote by $S^l(M)$ the space of symmetric tensors of type $(l, 0)$, and by $\mathfrak{d}$ the algebraic Spencer operator: $\mathfrak{d} : \Omega^k(M) \otimes S^l(M) \to \Omega^{k+1}(M) \otimes S^{l-1}(M)$, $\mathfrak{d}(t_{i_1 \dots i_k i_{k+1} \dots i_{k+l}}) = t_{[i_1 \dots i_k i_{k+1}] i_{k+2} \dots i_{k+l}}$. Then $E_{\mathfrak{g}} = \{A \mid \theta(AX, Y) + \theta(X, AY) = 0\}$ and

$$C^q(P) \cong \frac{\Omega^q(M) \otimes T^*(M)}{\mathfrak{d}(\Omega^{q-1}(M) \otimes S^2(M))} \cong \Omega^{q+1}(M)$$

(we use the isomorphism $T^1_1(M) \cong T^2_0(M)$ determined by θ). The differential δ is the exterior differential. Thus *the P-complex for symplectic structure is the de Rham compex.*

3.3.4. *Regular Poisson structure*

Let M be a smooth manifold. A Poisson structure on M is a skew-symmetric tensor field $\Psi \in \Omega_2(M)$ such that the bracket $\{f, g\} = \Psi(df, dg)$ on the vector space $F(M)$ of functions on M satisfies the Jacobi identity: $\{\{f, g\}, h\} + \{\{h, f\}, g\} + \{\{g, h\}, f\} = 0$. A *Poisson manifold* is (M, Ψ), where Ψ is a Poisson structure on M (for the theory of Poisson manifolds and associated cohomology we refer the reader to Refs. 22–25).

A Poisson structure is said to be *regular* if the rank of Ψ is constant. In this case, the distribution $\Delta = \widetilde{\Psi}(T^*M)$, where $\widetilde{\Psi} : T^*M \to TM$, $\widetilde{\Psi}(\alpha)^k = \Psi^{ks}\alpha_s$, is integrable and determines a foliation $\mathcal{F}$ on M. Denote by $\Omega_{\mathcal{F}}(M)$ the complex $(\Omega^{0,*}(M, \mathcal{F}), \partial_{0,1})$ (see 2.2.2), and set $\overset{+}{\Omega}{}^k_{\mathcal{F}}(M) = \Omega^{k+1}_{\mathcal{F}}(M)$ for $k \geq 0$, and $\overset{+}{\Omega}{}^k_{\mathcal{F}}(M) = 0$ for $k < 0$.

Theorem 3.2 (Ref. 26). *Let $N = TM/T\mathcal{F}$ be the normal bundle of the foliation $\mathcal{F}$, and $X_b(N)$ be the space of its basic sections. The Spencer P-complex $(C^q(P), \delta)$ for a regular Poisson structure is included in the exact* *sequence*

$$0 \to \overset{+}{\Omega}_{\mathcal{F}}(M) \to C(P) \to \Omega_{\mathcal{F}}(M) \otimes X_b(N) \to 0,$$

which generates the long exact cohomology sequence:

$$\ldots \to H^k(\overset{+}{\Omega}_{\mathcal{F}}(M) \to H^k C(M) \to H^k(\Omega_{\mathcal{F}}) \otimes N \xrightarrow{\partial} H^{k+1}\overset{+}{\Omega}_{\mathcal{F}}(M) \to \ldots$$

4. Example of de Rham complex for an non-regular integrable G-structure[29]

4.1. *Complex associated to a morphism $\xi : E_\xi \to M$*

Let $\xi : E_\xi \to M$ be a vector bundle, $A : \xi \to \Lambda^q M$ be a vector bundle morphism, and $\mathcal{A} : \Gamma_\xi \to \Omega_M$ be the corresponding sheaf morphism.

Consider an open $U \subset M$ such that $\xi|_U$ and Λ_U are trivial. Then, on U we have q-forms ω_a which span $\mathcal{A}(U)$. We set

$$\mathcal{F}^k(U) = \{\phi^a \wedge \omega_a + \psi^a \wedge d\omega_a \mid \phi^a \in \Omega^k(U), \psi^a \in \Omega^{k-1}(U)\}.$$

Thus, *to any morphism $A : \xi \to \Lambda^q M$ we associate the sheaf of differential complexes $(\mathcal{F}^*, d)$ which is a subsheaf of the sheaf of de Rham complexes* (Ω_M, d). We have the exact sequence of sheaves

$$0 \to \mathcal{F} \to \Omega_M \to \mathcal{G} = \Omega_M/\mathcal{F} \to 0.$$

Assume now that A is surjective on an open everywhere dense set $M \setminus \Sigma$, where $\iota : \Sigma \hookrightarrow M$ is an embedded submanifold. If $U \cap \Sigma = \emptyset$, then $\mathcal{F}^k(U) = \Omega^k(U)$, and $\mathcal{G}(U) = 0$. Hence $\mathcal{G}$ is supported on Σ.

4.2. *Complex associated with a differential form*

Consider $\omega \in \Omega^{q+1}(M)$, $d\omega = 0$. Denote by X_h the sheaf of vector fields V such that $L_V\omega = 0$ (the sheaf of infinitesimal automorphisms of ω).

Theorem 4.1 (Ref. 29). *$D(V) = L_V\omega$ is included to the differential complex of sheaves associated to the vector bundle morphism $I : TM \to \Lambda^q$, $I(V) = i_V\omega$:*

$$0 \to X_h \to X_M \xrightarrow{D} \Omega_M^{q+1} \xrightarrow{d} \Omega_M^{q+2} \xrightarrow{d} \ldots,$$

where all the sheaves, except for X_h, are fine.

4.2.1. *Case q=1*

Let us consider $\eta \in \Omega^1(M)$ such that $d\eta = 0$. Let $\Sigma = \{p \in M \mid \eta(p) = 0\}$ be an embedded submanifold of codimension k. Denote by $\Omega^q_{\Sigma,0}$ the sheaf over M consisting of q-forms θ such that $\iota^*\theta = 0$. Then d maps $\Omega^q_{\Sigma,0}$ to $\Omega^{q+1}_{\Sigma,0}$.

Theorem 4.2 (Ref. 29). *Assume that for each $p \in \Sigma$ there exists a coordinate system (u^a, u^α), $a = \overline{1,k}$, $\alpha = \overline{k+1,n}$ such that Σ is given by the equations $u^a = 0$ and*

$$\eta = A_{ab} u^a du^b,$$

where A_{ab} is a constant symmetric matrix of rank k. Then the sequence of sheaves and their morphisms

$$0 \to X_\Gamma \xrightarrow{i} X \xrightarrow{\mathcal{D}} \Omega^1_{\Sigma,0} \xrightarrow{d} \Omega^2_{\Sigma,0} \xrightarrow{d} \ldots \Omega^n_{\Sigma,0}$$

is a fine resolution of the sheaf X_Γ of infinitesimal automorphisms of η.

4.2.2. *Case q=2; Martinet singularities*

Let us consider $\omega \in \Omega^2(M)$ such that $d\omega = 0$. We set $E = \mathrm{Ker}\omega$. Assume that for any $p \in M$ a coordinate system (U, u^i) exists such that

$$\omega = u^1 du^1 \wedge du^2 + du^3 \wedge du^4 \cdots + du^{2n-1} \wedge du^{2n}.$$

Theorem 4.3 (Ref. 29). *a) Let $(\mathcal{F}^*, d)$ be the complex of sheaves associated to ω. Then $\mathcal{F}^1$ is the subsheaf of Ω^1_M consisting of forms which vanish on the subbundle $E \subset TM|_\Sigma$, and $\mathcal{F}^k = \Omega^k_M$ for $k \geq 2$.*

b) For $\mathcal{D} : X_M \to \Omega^2_M$, $D(V) = L_V\omega$, the sequence of sheaves

$$0 \to X_\Gamma \xrightarrow{i} X_M \xrightarrow{D} \Omega^2 \xrightarrow{d} \Omega^3 \xrightarrow{d} \ldots$$

is a fine resolution for the sheaf X_Γ.

Now consider another type of Martinet singularities. Assume that, for any $p \in M$ a coordinate system (U, u^i) exists such that

$$\begin{aligned}\omega =& du^1 \wedge du^2 + u^3 du^1 \wedge du^4 \\ &+ u^3 du^2 \wedge du^3 + u^4 du^2 \wedge du^4 + (u^1 - (u^3)^2) du^3 \wedge du^4,\end{aligned}$$

Theorem 4.4 (Ref. 29). *Let $(\mathcal{F}^*, d)$ be the complex of sheaves associated ω Then $\mathcal{F}^1$ is the subsheaf of Ω^1_M consisting of forms vanishing on the subbundle $E \subset TM|_\Sigma$, and $\mathcal{F}^k = \Omega^k_M$ for $k \geq 2$.*

Note that in this case the sequence of sheaves

$$0 \to X_\Gamma \xrightarrow{i} X_M \xrightarrow{D} \Omega^2 \xrightarrow{d} \Omega^3 \xrightarrow{d} \dots$$

is not exact.

The properties of cohomology of the complexes described in the theorems of this section and the relation of this cohomology to infinitesimal deformations of the symplectic structures with Martinet singularities was studied in Refs. 27,28.

References

1. R. Bott and L.W. Tu, *Differential forms in algebraic topology* (Graduate Texts in Mathematics, 82, New York - Heidelberg - Berlin: Springer-Verlag, XIV, pp 331, 2001).
2. Ph. Griffiths and J. Harris, *Principles of algebraic geometry* (2nd ed., Wiley Classics Library, New York, NY: John Wiley & Sons Ltd. xii, pp 813, 1994).
3. I. Vaisman, *Cohomology and differential forms* (Translation editor: Samuel I. Goldberg. Pure and Applied Mathematics, 21. Marcel Dekker, Inc., New York, pp. 284, 1973).
4. R.O. Wells, *Differential analysis on complex manifolds* (Prentice-Hall, Inc. Englewood Cliffs, N.J., 1973).
5. P. Molino, *Riemannian foliations* (Birkhäuser, 1988).
6. F.W. Kamber and Ph. Tondeur, Characteristic classes of foliated bundles, *Lect.Notes Math.* V.494 (1975).
7. F.W. Kamber and Ph. Tondeur, De Rham-Hodge theory for Riemannian foliations, *Math. Ann.* **277** (1987) 415–431.
8. A. El Kacimi-Alaoui and A. Tihami, Cohomologie bigraduée de certains feuilletages, *Bull. Soc. Math. Belg., Sér. B* **38** (1986) 144–156.
9. A.P. Shirokov, V.V. Shurygin and V.V. Vishnevskij, *Spaces over algebras.* (Prostranstva nad algebrami, in Russian, Kazan: Izdatel'stvo Kazanskogo Universiteta, pp. 263, 1985).
10. V.V. Shurygin, Smooth manifolds over local algebras and Weil bundles, *J. Math. Sci., New York* **108** (2) (2002) 249–294; translation from *Itogi Nauki Tekh., Ser. Sovrem Mat. Prilozh., Temat. Obz.* **73** VINITI, Moscow (2002) 162–236 .
11. V.V. Shurygin, On the cohomology of manifolds over local algebras, *Russ. Math.* **40** (9) (1996) 67–81; translation from *Izv. Vyssh. Uchebn. Zaved., Mat.* **1996** (9) (412) (1996) 71–85.
12. V.V. Shurygin and L.B. Smolyakova, An analog of the Vaisman-Molino cohomology for manifolds modelled on some types of modules over Weil algebras and its application, *Lobachevskii J. Math.* **9** (2001) 55–75.
13. V.V. Shurygin, The Atiyah-Molino classes of a smooth manifold over a local algebra A as obstructions for prolongation of transverse connections to A-smooth connections, *Tr. Geom. Sem.* **23** Kazan Univ., Kazan, (1997) 199–210.

14. M.A. Mikenberg, Manifolds over the algebra of dual numbers whose canonical foliation has an everywhere dense leaf, *Russ. Math.* **47** (3) (2003) 28–30; translation from *Izv. Vyssh. Uchebn. Zaved., Mat.* **2003** (3) (2003) 31–33.
15. T.I. Gaisin, To the question about the maximum principle for manifolds over local algebras, *Sib. Mat. Zh.* **46** (2005) 79–89.
16. Sh. Kobayashi and K Nomizu, *Foundations of differential geometry. I* (New York-London: Interscience Publishers, a division of John Wiley & Sons. XI, 1963).
17. Sh. Kobayashi and K. Nomizu, *Foundations of differential geometry. II* (New York-London-Sydney Interscience Publishers a division of John Wiley and Sons, 1969).
18. J.F. Pommaret, Systems of partial differential equations and Lie pseudogroups, *Math. and Appl.* **14** (1978).
19. I. Kolář, P.W. Michor and J. Slovák, *Natural operators in Differential Geometry* (Springer-Verlag, 1993).
20. M.A. Malakhaltsev, The Lie derivative and cohomology of G-structures, *Lobachevskii J. Math.* **3** (1999) 215–220.
21. T.I. Gaisin, *The Spencer complex for manifolds over algebras, In: Proceedings of Geometrical Seminar* (Issue 23, 1997).
22. P. Xu, Poisson cohomology of regular Poisson manifolds, *Ann. Inst. Fourier Grenoble* **42** (1992) 967–988.
23. I. Vaisman, Lectures on the Geometry of Poisson Manifolds, *In: Progress in Math.* (118, Birkhäuser, Basel, 1994).
24. N. Nakanishi, Poisson cohomology of plane quadratic Poisson structures, *Publ. RIMS, Kyoto Univ.* **33** (1997) 73–89.
25. C. Roger and P. Vanhaecke, Poisson cohomology of the affine plane, *J. Algebra* **251** (2002) 448–460.
26. M.A. Malakhaltsev, Vaisman complex and hamiltonian vector fields *In: Aktualnye problemy matematiki i metodiki ee prepodavaniya* (Penza State University, 2001) 55-58.
27. M.A. Malakhaltsev, Infinitesimal deformations of a symplectic structure with singularities, *Russ. Math.* **47** (11) (2003) 38–46; translation from *Izv. Vyssh. Uchebn. Zaved., Mat.* **2003** (11) (2003) 42–50.
28. M.A. Malakhaltsev, Sheaf of local Hamiltonians of symplectic manifolds with Martinet singularities, *Russ. Math.* (11) (2004) 45–52.
29. M.A. Malakhaltsev, Differential complex associated to closed differential forms of nonconstant rank, *Lobachevskii J. Math.* **23** 183–192.

Differential Geometry and its Applications
Proc. Conf., in Honour of Leonhard Euler, Olomouc, August 2007

Submanifolds in pseudo-Euclidean spaces and Dubrovin–Frobenius structures

O.I. Mokhov

Department of Geometry and Topology, Faculty of Mechanics and Mathematics, M.V.Lomonosov Moscow State University, Moscow, 119991, Russia
E-mail: mokhov@mi.ras.ru

We introduce a natural class of *potential* submanifolds in pseudo-Euclidean spaces and prove that each N-dimensional Frobenius manifold can locally be represented as an N-dimensional potential submanifold in a $2N$-dimensional pseudo-Euclidean space. We show that all potential submanifolds bear natural Dubrovin–Frobenius structures on their tangent spaces; in particular, Weingarten operators of each N-dimensional potential submanifold give N-parameter integrable deformations of structural constants of Frobenius algebras. We prove that the associativity equations of two-dimensional topological quantum field theories are very natural reductions of the fundamental nonlinear equations of the theory of submanifolds in pseudo-Euclidean spaces and define locally the class of potential submanifolds. The problem of explicit realization of an arbitrary concrete Frobenius manifold as a potential submanifold in a pseudo-Euclidean space is reduced to solving a linear system of second-order partial differential equations. For concrete Frobenius manifolds, this realization problem can be solved explicitly in elementary and special functions. Moreover, we consider a nonlinear system, which is a natural generalization of the associativity equations, namely, the system describing all flat torsionless submanifolds in pseudo-Euclidean spaces and prove that this system is integrable by the inverse scattering method.

Keywords: Frobenius manifold, Frobenius algebra, deformation of algebra, topological quantum field theory, Dubrovin–Frobenius structure, associativity equations, potential submanifold in pseudo-Euclidean space, integrable system.

MS classification: 53D45, 53A07, 53B25, 53B30, 57R56, 35Q58, 37K15, 37K25, 81T40, 81T45.

1. Introduction

We prove that the associativity equations of two-dimensional topological quantum field theories (the Witten–Dijkgraaf–Verlinde–Verlinde equations[1]) for a function (a *potential*) $\Phi = \Phi(u^1, \ldots, u^N)$, namely,

$$\sum_{k,l=1}^{N} \frac{\partial^3 \Phi}{\partial u^i \partial u^j \partial u^k} \eta^{kl} \frac{\partial^3 \Phi}{\partial u^l \partial u^m \partial u^n} = \sum_{k,l=1}^{N} \frac{\partial^3 \Phi}{\partial u^i \partial u^m \partial u^k} \eta^{kl} \frac{\partial^3 \Phi}{\partial u^l \partial u^j \partial u^n}, \tag{1}$$

where η^{ij} is an arbitrary constant nondegenerate symmetric matrix, $\eta^{ij} = \eta^{ji}$, $\eta^{ij} = \text{const}$, $\det(\eta^{ij}) \neq 0$, are very natural reductions of the fundamental nonlinear equations of the theory of submanifolds in pseudo-Euclidean spaces (namely, the Gauss equations, the Codazzi equations and the Ricci equations) and give a very natural class of *potential* submanifolds. All potential submanifolds in pseudo-Euclidean spaces bear natural structures of special Frobenius algebras on their tangent spaces. These Dubrovin–Frobenius structures are generated by the corresponding flat first fundamental form and the set of the second fundamental forms of the submanifolds (in fact, the structural constants are given by the set of the Weingarten operators of the submanifolds).

We recall that each solution $\Phi(u^1, \ldots, u^N)$ of the associativity equations (1) gives N-parameter deformations of special Frobenius algebras (some special commutative associative algebras equipped with nondegenerate invariant symmetric bilinear forms). Indeed, consider algebras $A(u)$ in an N-dimensional vector space with the basis $e_1, \ldots, e_N$ and the multiplication (see Ref. 1)

$$e_i \circ e_j = c^k_{ij}(u) e_k, \qquad c^k_{ij}(u) = \eta^{ks} \frac{\partial^3 \Phi}{\partial u^s \partial u^i \partial u^j}. \tag{2}$$

For all values of the parameters $u = (u^1, \ldots, u^N)$ the algebras $A(u)$ are commutative, $e_i \circ e_j = e_j \circ e_i$, and the associativity condition

$$(e_i \circ e_j) \circ e_k = e_i \circ (e_j \circ e_k) \tag{3}$$

in the algebras $A(u)$ is equivalent to the associativity equations (1). The matrix η_{ij} inverse to the matrix η^{ij}, $\eta^{is}\eta_{sj} = \delta^i_j$, defines a nondegenerate invariant symmetric bilinear form on the algebras $A(u)$,

$$\langle e_i, e_j \rangle = \eta_{ij}, \qquad \langle e_i \circ e_j, e_k \rangle = \langle e_i, e_j \circ e_k \rangle. \tag{4}$$

Recall that locally the tangent space at every point of any Frobenius manifold (see Ref. 1) possesses the structure of Frobenius algebra (2)–(4), which is determined by a solution of the associativity equations (1) and smoothly depends on the point. We prove that each N-dimensional Frobenius manifold can locally be represented as an N-dimensional potential submanifold in a $2N$-dimensional pseudo-Euclidean space. The problem of explicit realization of an arbitrary concrete N-dimensional Frobenius manifold as an N-dimensional potential submanifold in a $2N$-dimensional pseudo-Euclidean space is reduced to solving a linear system of second-order partial differential equations. For concrete Frobenius manifolds, this realization problem can be solved explicitly in elementary and special functions.

Moreover, we consider a nonlinear system, which is a natural generalization of the associativity equations (1), namely, the system describing all flat torsionless submanifolds in pseudo-Euclidean spaces, and prove that this system is integrable by the inverse scattering method.

2. Frobenius algebras and Frobenius manifolds

Recall the notion of Frobenius algebra over a field $\mathbb{K}$ (in this paper we consider Frobenius algebras only over $\mathbb{R}$ or $\mathbb{C}$) and the notion of Frobenius manifold.

Definition 2.1. A bilinear form $f : \mathcal{A} \times \mathcal{A} \to \mathbb{K}$ in an algebra $\mathcal{A}$ is called invariant (or associative) if $f(a \circ b, c) = f(a, b \circ c)$ for all $a, b, c \in \mathcal{A}$.

Definition 2.2. A finite dimensional algebra $\mathcal{A}$ over a field $\mathbb{K}$ is called Frobenius if it is equipped with a nondegenerate invariant bilinear form.

Consider an N-dimensional pseudo-Riemannian manifold M with a metric g and a structure of a Frobenius algebra on each tangent space T_uM at any point $u \in M$ smoothly depending on the point, $(T_uM, \circ, g)$, $T_uM \times T_uM \xrightarrow{\circ} T_uM$, such that the metric g is the corresponding nondegenerate invariant symmetric bilinear form on each tangent space T_uM,

$$g(X \circ Y, Z) = g(X, Y \circ Z),$$

where X, Y and Z are arbitrary vector fields on M.

This class of pseudo-Riemannian manifolds equipped with Frobenius structures could be naturally called Frobenius, but in this paper we shall consider well-known and generally accepted Dubrovin's definition of Frobenius manifolds,[1] which is motivated by two-dimensional topological quantum field theories and quantum cohomology and imposes very severe additional constraints on Frobenius structures of Frobenius manifolds.

Definition 2.3 (Dubrovin, Ref. 1). An N-dimensional pseudo-Riemannian manifold M with a metric g and a structure of a Frobenius algebra on each tangent space T_uM at any point $u \in M$ smoothly depending on the point, $T_uM \times T_uM \xrightarrow{\circ} T_uM$, is called *Frobenius* if

(i) the metric g is a nondegenerate invariant symmetric bilinear form on each tangent space $T_u M$,

$$g(X \circ Y, Z) = g(X, Y \circ Z), \tag{5}$$

(ii) the Frobenius algebra is commutative,

$$X \circ Y = Y \circ X \tag{6}$$

for all vector fields X and Y,

(iii) the Frobenius algebra is associative,

$$(X \circ Y) \circ Z = X \circ (Y \circ Z) \tag{7}$$

for all vector fields X, Y and Z,

(iv) the metric g is flat,

(v) $A(X, Y, Z) = g(X \circ Y, Z)$ is a symmetric tensor on M such that the tensor $(\nabla_W A)(X, Y, Z)$ is symmetric with respect to all vector fields X, Y, Z, W, where ∇ is the covariant differentiation generated by the Levi-Civita connection of the metric g,

(vi) the Frobenius algebra possesses a unit, and the unit vector field U ($X \circ U = U \circ X = X$ for each vector field X) is covariantly constant,

$$\nabla U = 0, \tag{8}$$

where ∇ is the covariant differentiation generated by the Levi-Civita connection of the metric g,

(vii) the manifold M is equipped with a vector field E (*Euler vector field*) such that

$$\nabla \nabla E = 0, \tag{9}$$

$$\mathcal{L}_E (X \circ Y) - (\mathcal{L}_E X) \circ Y - X \circ (\mathcal{L}_E Y) = X \circ Y, \tag{10}$$

$$\mathcal{L}_E\, g(X, Y) - g(\mathcal{L}_E X, Y) - g(X, \mathcal{L}_E Y) = K\, g(X, Y), \tag{11}$$

$$\mathcal{L}_E U = -U, \tag{12}$$

where K is an arbitrary fixed constant, $\mathcal{L}_E$ is the Lie derivative along the Euler vector field, and ∇ is the covariant differentiation generated by the Levi-Civita connection of the metric g.

A beautiful theory of these very special Frobenius structures and Frobenius manifolds and many important examples were constructed by Dubrovin.[1] No doubt that these special Frobenius structures should be called *Dubrovin–Frobenius structures*. A lot of very important examples of Frobenius manifolds arises in the theory of Gromov–Witten invariants, the quantum cohomology, the singularity theory, the enumerative geometry, the topological field theories and the modern differential geometry, mathematical and theoretical physics.

In this paper we describe a natural special class of submanifolds in pseudo-Euclidean spaces possessing Dubrovin–Frobenius structures satisfying the conditions (i)–(v). Moreover, we show that each manifold satisfying the conditions (i)–(v) can be realized as a submanifold of this special type in a pseudo-Euclidean space. For any concrete Dubrovin–Frobenius structure satisfying the conditions (i)–(v) and for any given Frobenius manifold, the corresponding realization problem is reduced to solving a system of linear second-order partial differential equations.

Consider an arbitrary manifold satisfying the conditions (i)–(v). Let $u = (u^1, \dots, u^N)$ be arbitrary flat coordinates of the flat metric g. In flat local coordinates, the metric $g(u)$ is a constant nondegenerate symmetric matrix η_{ij}, $\eta_{ij} = \eta_{ji}$, $\det(\eta_{ij}) \neq 0$, $\eta_{ij} =$ const, $g(X, Y) = \eta_{ij} X^i Y^j$.

In these flat local coordinates, for structural functions $c^i_{jk}(u)$ of Frobenius structure on the manifold and for the symmetric tensor $A_{ijk}(u)$, we have $X \circ Y = W$, $W^i(u) = c^i_{jk}(u) X^j(u) Y^k(u)$, $A(X, Y, Z) = A_{ijk}(u) X^i(u) Y^j(u) Z^k(u) = g(X \circ Y, Z) = g(W, Z) = \eta_{ij} W^i(u) Z^j(u) = \eta_{ij} c^i_{kl}(u) X^k(u) Y^l(u) Z^j(u)$,

$$A_{ijk}(u) = \eta_{sk} c^s_{ij}(u). \tag{13}$$

Since $(\nabla_l A_{ijk})(u)$ is a symmetric tensor according to (v), in the flat local coordinates we also have

$$\frac{\partial A_{ijk}}{\partial u^l} = \frac{\partial A_{ijl}}{\partial u^k}. \tag{14}$$

Hence there locally exist a function (a *potential*) $\Phi(u)$ such that

$$A_{ijk}(u) = \frac{\partial^3 \Phi}{\partial u^i \partial u^j \partial u^k}. \tag{15}$$

From (13) for the structural functions $c^i_{jk}(u)$ we have

$$c^i_{jk}(u) = \eta^{is} A_{sjk}(u) = \eta^{is} \frac{\partial^3 \Phi}{\partial u^s \partial u^j \partial u^k}, \tag{16}$$

where the matrix η^{ij} is inverse to the matrix η_{ij}, $\eta^{is} \eta_{sj} = \delta^i_j$.

For any values of parameters $u = (u^1, \dots, u^N)$, the structural functions (16) give a commutative Frobenius algebra equipped with a symmetric invariant nondegenerate bilinear form for arbitrary constant nondegenerate symmetric matrix η_{ij} and arbitrary function $\Phi(u)$, but, generally speaking, this algebra is not associative. All conditions (i)–(v) except the associativity condition (iii) are obviously satisfied for all these N-parameter deformations of nonassociative Frobenius algebras.

The associativity condition (iii) is equivalent to a nontrivial overdetermined system (1) of nonlinear partial differential equations for the potential $\Phi(u)$, which are well known as the associativity equations of two-dimensional topological quantum field theories (the Witten–Dijkgraaf–Verlinde–Verlinde equations[1]). The associativity equations (1) are consistent, integrable by the inverse scattering method, and possess a rich set of nontrivial solutions (see Ref. 1).

It is obvious that each solution $\Phi(u^1, \dots, u^N)$ of the associativity equations (1) gives N-parameter deformations of commutative associative Frobenius algebras (2) equipped with nondegenerate invariant symmetric bilinear forms (4). These Frobenius structures satisfy to all the conditions (i)–(v).

3. Flat submanifolds with zero torsion in pseudo-Euclidean spaces

Let us consider totally nonisotropic smooth N-dimensional flat submanifolds with zero torsion in an arbitrary $(N + L)$-dimensional pseudo-Euclidean space. Locally, each of the submanifolds is given by a smooth vector function $r(u^1, \dots, u^N)$ of N independent variables $(u^1, \dots, u^N)$, $r(u^1, \dots, u^N) = (z^1(u^1, \dots, u^N), \dots, z^{N+L}(u^1, \dots, u^N))$, where $(z^1, \dots, z^{N+L})$ are pseudo-Euclidean coordinates, $\operatorname{rank}(\partial z^i/\partial u^j) = N$ (here $1 \le i \le N + L$, $1 \le j \le N$). Then $\partial r/\partial u^i = r_{u^i}$, $1 \le i \le N$, are tangent vectors at an arbitrary point $u = (u^1, \dots, u^N)$. Let $\mathbf{N}_u$ be the normal space of the submanifold at an arbitrary point $u = (u^1, \dots, u^N)$. In the normal spaces $\mathbf{N}_u$, we use the bases n_α, $1 \le \alpha \le L$, with arbitrary admissible constant Gram matrices $\mu_{\alpha\beta}$, $(n_\alpha, n_\beta) = \mu_{\alpha\beta}$, $\mu_{\alpha\beta} = \text{const}$, $\mu_{\alpha\beta} = \mu_{\beta\alpha}$, $\det(\mu_{\alpha\beta}) \ne 0$, $(n_\alpha, r_{u^i}) = 0$, $1 \le \alpha, \beta \le L$, $1 \le i \le N$. Then $\mathbf{I} = ds^2 = g_{ij}(u)du^i du^j$, $g_{ij}(u) = (r_{u^i}, r_{u^j})$, is the first fundamental form, and $\mathbf{II}_\alpha = \omega_{\alpha,ij}(u)du^i du^j$, $\omega_{\alpha,ij}(u) = (n_\alpha, r_{u^i u^j})$, $1 \le \alpha \le L$, are the second fundamental forms of the submanifold.

For N-dimensional flat torsionless submanifolds in an $(N + L)$-dimensional pseudo-Euclidean space, we have the following system of fundamental equations for the first fundamental form $g_{ij}(u)$ and the second fundamental forms $\omega_{\alpha,ij}(u)$: the Gauss equations

$$\sum_{\alpha=1}^{L}\sum_{\beta=1}^{L}\mu^{\alpha\beta}(\omega_{\alpha,ik}(u)\omega_{\beta,jl}(u)-\omega_{\alpha,il}(u)\omega_{\beta,jk}(u))=0, \tag{17}$$

where $\mu^{\alpha\beta}$ is inverse to the matrix $\mu_{\alpha\beta}$, $\mu^{\alpha\gamma}\mu_{\gamma\beta}=\delta^{\alpha}_{\beta}$, the Codazzi equations

$$\nabla_k(\omega_{\alpha,ij}(u))=\nabla_j(\omega_{\alpha,ik}(u)), \tag{18}$$

and the Ricci equations

$$\sum_{i=1}^{N}\sum_{j=1}^{N}g^{ij}(u)\,(\omega_{\alpha,ik}(u)\omega_{\beta,jl}(u)-\omega_{\alpha,il}(u)\omega_{\beta,jk}(u))=0, \tag{19}$$

where $g^{ij}(u)$ is the contravariant metric inverse to the first fundamental form g_{ij}, $g^{is}g_{sj}=\delta^i_j$.

Now we can consider that $u=(u^1,\ldots,u^N)$ are certain flat coordinates of the metric $g_{ij}(u)$. In flat coordinates, the metric is a constant nondegenerate symmetric matrix η_{ij}, $\eta_{ij}=\eta_{ji}$, $\eta_{ij}=\text{const}$, $\det(\eta_{ij})\neq 0$, and the Codazzi equations (18) have the form

$$\frac{\partial\omega_{\alpha,ij}}{\partial u^k}=\frac{\partial\omega_{\alpha,ik}}{\partial u^j}. \tag{20}$$

Therefore, there locally exist some functions $\psi_\alpha(u)$, $1\le\alpha\le L$, such that

$$\omega_{\alpha,ij}(u)=\frac{\partial^2\psi_\alpha}{\partial u^i\partial u^j}. \tag{21}$$

Lemma 3.1. *All the second fundamental forms of each flat torsionless submanifold in a pseudo-Euclidean space are Hessians in any flat coordinates in any simply connected domain on the submanifold.*

It follows from Lemma 3.1 that in any flat coordinates, the Gauss equations (17) have the form

$$\sum_{\alpha=1}^{L}\sum_{\beta=1}^{L}\mu^{\alpha\beta}\left(\frac{\partial^2\psi_\alpha}{\partial u^i\partial u^k}\frac{\partial^2\psi_\beta}{\partial u^j\partial u^l}-\frac{\partial^2\psi_\alpha}{\partial u^i\partial u^l}\frac{\partial^2\psi_\beta}{\partial u^j\partial u^k}\right)=0, \tag{22}$$

and the Ricci equations (19) have the form

$$\sum_{i=1}^{N}\sum_{j=1}^{N}\eta^{ij}\left(\frac{\partial^2\psi_\alpha}{\partial u^i\partial u^k}\frac{\partial^2\psi_\beta}{\partial u^j\partial u^l}-\frac{\partial^2\psi_\alpha}{\partial u^i\partial u^l}\frac{\partial^2\psi_\beta}{\partial u^j\partial u^k}\right)=0, \tag{23}$$

where η^{ij} is inverse to the matrix η_{ij}, $\eta^{is}\eta_{sj}=\delta^i_j$.

Theorem 3.1. *The class of N-dimensional flat torsionless submanifolds in $(N+L)$-dimensional pseudo-Euclidean spaces is described (in flat coordinates) by the system of nonlinear equations (22), (23) for functions $\psi_\alpha(u)$, $1 \le \alpha \le L$. Here, η^{ij} and $\mu^{\alpha\beta}$ are arbitrary constant nondegenerate symmetric matrices, $\eta^{ij} = \eta^{ji}$, $\eta^{ij} = \mathrm{const}$, $det(\eta^{ij}) \neq 0$, $\mu^{\alpha\beta} = \mathrm{const}$, $\mu^{\alpha\beta} = \mu^{\beta\alpha}$, $det(\mu^{\alpha\beta}) \neq 0$; the signature of the ambient $(N+L)$-dimensional pseudo-Euclidean space is the sum of the signatures of the metrics η^{ij} and $\mu^{\alpha\beta}$; $\mathbf{I} = ds^2 = \eta_{ij} du^i du^j$ is the first fundamental form, where η_{ij} is inverse to the matrix η^{ij}, $\eta^{is}\eta_{sj} = \delta^i_j$; $\mathbf{II}_\alpha = (\partial^2 \psi_\alpha / (\partial u^i \partial u^j)) du^i du^j$, $1 \le \alpha \le L$, are the second fundamental forms given by the Hessians of the functions $\psi_\alpha(u)$, $1 \le \alpha \le L$.*

According to the Bonnet theorem, any solution $\psi_\alpha(u)$, $1 \le \alpha \le L$, of the nonlinear system (22), (23) determines a unique (up to motions) N-dimensional flat torsionless submanifold of the corresponding $(N+L)$-dimensional pseudo-Euclidean space with the first fundamental form $\eta_{ij} du^i du^j$ and the second fundamental forms $\omega_\alpha(u) = (\partial^2 \psi_\alpha / (\partial u^i \partial u^j)) du^i du^j$, $1 \le \alpha \le L$, given by the Hessians of the functions $\psi_\alpha(u)$, $1 \le \alpha \le L$. It is obvious that we can always add arbitrary terms linear in the coordinates $(u^1, \ldots, u^N)$ to any solution of the system (22), (23), but the set of the second fundamental forms and the corresponding submanifold will be the same. Moreover, any two sets of the second fundamental forms of the form $\omega_{\alpha,ij}(u) = \partial^2 \psi_\alpha / (\partial u^i \partial u^j)$, $1 \le \alpha \le L$, coincide if and only if the corresponding functions $\psi_\alpha(u)$, $1 \le \alpha \le L$, coincide up to terms linear in the coordinates; hence we must not distinguish solutions of the nonlinear system (22), (23) up to terms linear in the coordinates $(u^1, \ldots, u^N)$.

Consider the following linear problem with parameters for vector functions $\partial a(u)/\partial u^i$, $1 \le i \le N$, and $b_\alpha(u)$, $1 \le \alpha \le L$:

$$\frac{\partial^2 a}{\partial u^i \partial u^j} = \lambda\, \mu^{\alpha\beta} \omega_{\alpha,ij}(u) b_\beta(u), \qquad \frac{\partial b_\alpha}{\partial u^i} = \rho\, \eta^{kj} \omega_{\alpha,ij}(u) \frac{\partial a}{\partial u^k}, \tag{24}$$

where η^{ij}, $1 \le i,j \le N$, and $\mu^{\alpha\beta}$, $1 \le \alpha,\beta \le L$, are arbitrary constant nondegenerate symmetric matrices, $\eta^{ij} = \eta^{ji}$, $\eta^{ij} = \mathrm{const}$, $\det(\eta^{ij}) \neq 0$, $\mu^{\alpha\beta} = \mathrm{const}$, $\mu^{\alpha\beta} = \mu^{\beta\alpha}$, $\det(\mu^{\alpha\beta}) \neq 0$; λ and ρ are arbitrary constants (parameters). Of course, only one of the parameters is essential (but it is really essential). It is obvious that the coefficients $\omega_{\alpha,ij}(u)$, $1 \le \alpha \le L$, here must be symmetric matrix functions, $\omega_{\alpha,ij}(u) = \omega_{\alpha,ji}(u)$.

The consistency conditions for the linear system (24) are equivalent to the nonlinear system (22), (23) describing the class of N-dimensional flat torsionless submanifolds in $(N + L)$-dimensional pseudo-Euclidean spaces.

Theorem 3.2. *The nonlinear system (22), (23) is integrable by the inverse scattering method.*

4. Reduction to the associativity equations and potential submanifolds in pseudo-Euclidean spaces

We now also find some natural and very important integrable reductions of the nonlinear system (22), (23). We show that the class of flat torsionless submanifolds in pseudo-Euclidean spaces is quite rich, and we describe a nontrivial and very important family of submanifolds of this class. This family is generated by the associativity equations of two-dimensional topological quantum field theories (the WDVV equations). First of all, we note that although the Gauss equations (22) and the Ricci equations (23) for flat torsionless submanifolds in pseudo-Euclidean spaces are essentially different, they are fantastically similar. The case of a natural reduction under which the Gauss equations (22) and the Ricci equations (23) merely coincide is of particular interest. Such a reduction readily leads to the associativity equations of two-dimensional topological quantum field theories.

Theorem 4.1. *If $L = N$, $\mu^{ij} = c\eta^{ij}$, $1 \leq i,j \leq N$, c is an arbitrary nonzero constant, and $\psi_\alpha(u) = \partial\Phi/\partial u^\alpha$, $1 \leq \alpha \leq N$, where $\Phi = \Phi(u^1, \ldots, u^N)$, then the Gauss equations (22) coincide with the Ricci equations (23), and each of them coincides with the associativity equations (1) of two-dimensional topological quantum field theories (the WDVV equations) for the potential $\Phi(u)$.*

Definition 4.1. A flat torsionless N-dimensional submanifold in a $2N$-dimensional pseudo-Euclidean space with a flat first fundamental form $g_{ij}(u)du^i du^j$ is called *potential* if there locally exist a certain function $\Phi(u)$ in a neighborhood on the submanifold such that the second fundamental forms of this submanifold locally in this neighborhood have the form

$$(\omega_i)_{jk}(u)du^j du^k = (\nabla_i\nabla_j\nabla_k\Phi(u))\, du^j du^k, \quad 1 \leq i \leq N, \tag{25}$$

where ∇_i is the covariant differentiation defined by the Levi-Civita connection generated by the flat metric $g_{ij}(u)$.

Theorem 4.2. *The associativity equations of two-dimensional topological quantum field theories describe a special class of N-dimensional flat submanifolds without torsion in $2N$-dimensional pseudo-Euclidean spaces, namely, exactly the class of potential submanifolds.*

According to the Bonnet theorem, any solution $\Phi(u)$ of the associativity equations (1) (with an arbitrary fixed constant metric η_{ij}) determines a unique (up to motions) N-dimensional potential flat torsionless submanifold of the corresponding $2N$-dimensional pseudo-Euclidean space with the first fundamental form $\eta_{ij}du^i du^j$ and the second fundamental forms $\omega_n(u) = (\partial^3\Phi/(\partial u^n \partial u^i \partial u^j))du^i du^j$ given by the third derivatives of the potential $\Phi(u)$. Here, we do not distinguish solutions of the associativity equations (1) up to terms quadratic in the coordinates u.

Theorem 4.3. *On each potential submanifold in a pseudo-Euclidean space, there is a structure of a Frobenius algebra given (in flat coordinates) by the flat first fundamental form* η_{ij} *and the Weingarten operators* $(A_s)^i_j(u) = -\eta^{ik}(\omega_s)_{kj}(u)$*:*

$$\langle e_i, e_j \rangle = \eta_{ij}, \qquad e_i \circ e_j = c^k_{ij}(u) e_k, \qquad e_i = \frac{\partial}{\partial u^i}, \tag{26}$$

$$c^k_{ij}(u^1, \dots, u^N) = -(A_i)^k_j(u) = \eta^{ks}(\omega_i)_{sj}(u^1, \dots, u^N). \tag{27}$$

In arbitrary local coordinates, this Frobenius structure has the form

$$\langle e_i, e_j \rangle = g_{ij}, \qquad e_i \circ e_j = c^k_{ij}(u) e_k, \qquad e_i = \frac{\partial}{\partial u^i}, \tag{28}$$

$$c^k_{ij}(u^1, \dots, u^N) = -(A_i)^k_j(u) = g^{ks}(u^1, \dots, u^N)(\omega_i)_{sj}(u^1, \dots, u^N), \tag{29}$$

where $g^{ij}(u)$ *is the contravariant metric inverse to the first fundamental form* $g_{ij}(u)$*,* $g^{is}(u) g_{sj}(u) = \delta^i_j$*, and* $(\omega_k)_{ij}(u) du^i du^j$*,* $1 \le k \le N$*, are the second fundamental forms.*

Theorem 4.4. *Each N-dimensional Frobenius manifold can be locally represented as a potential flat torsionless N-dimensional submanifold in a $2N$-dimensional pseudo-Euclidean space. This submanifold is uniquely determined up to motions in the corresponding ambient $2N$-dimensional pseudo-Euclidean space.*

5. Realization of Frobenius manifolds as submanifolds in pseudo-Euclidean spaces

It is important to note that we have at least two essentially different possibilities for signature of the corresponding ambient $2N$-dimensional pseudo-Euclidean space, namely, we can always consider the ambient $2N$-dimensional pseudo-Euclidean space of zero signature, and we can also consider the ambient $2N$-dimensional pseudo-Euclidean space whose signature is equal to doubled signature of the metric η_{ij}. Thus, if the metric η_{ij} of a Frobenius manifold has a nonzero signature, then according to our construction we have two essentially different possibilities for realization of the Frobenius manifold as a potential flat torsionless submanifold.

Theorem 5.1. *For an arbitrary Frobenius manifold, which is locally given by a solution $\Phi(u^1, \dots, u^N)$ of the associativity equations (1), the corresponding potential flat torsionless submanifold in a $2N$-dimensional pseudo-Euclidean space that realizes this Frobenius manifold is given by the $2N$-component vector function $r(u^1, \dots, u^N)$ satisfying the following compatible linear system of second-order partial differential equations:*

$$\frac{\partial^2 r}{\partial u^i \partial u^j} = c\eta^{kl} \frac{\partial^3 \Phi}{\partial u^i \partial u^j \partial u^k} \frac{\partial n}{\partial u^l}, \tag{30}$$

$$\frac{\partial^2 n}{\partial u^i \partial u^j} = -\eta^{kl} \frac{\partial^3 \Phi}{\partial u^i \partial u^j \partial u^k} \frac{\partial r}{\partial u^l}, \tag{31}$$

where $n(u^1, \dots, u^N)$ is a $2N$-component vector function, c is an arbitrary nonzero constant (a deformation parameter preserving the corresponding Frobenius structure). In particular, two essentially different cases $c = 1$ and $c = -1$ correspond to ambient $2N$-dimensional pseudo-Euclidean spaces of different signatures (if the metric η_{ij} has a nonzero signature). The consistency of the linear system (30), (31) is equivalent to the associativity equations (1).

Acknowledgments

This research was supported by the Max-Planck-Institut für Mathematik (Bonn, Germany), the Russian Foundation for Basic Research (Grant No. 08-01-00054) and the Program for Supporting Leading Scientific Schools (Grant No. NSh-1824.2008.1).

References

1. B. Dubrovin, Geometry of 2D topological field theories, *In: Lecture Notes in Math.* (Vol. 1620, Springer-Verlag, Berlin, 1996) 120–348; arXiv:hep-th/9407018.
2. O.I. Mokhov, Non-local Hamiltonian operators of hydrodynamic type with flat metrics, and the associativity equations, *Russian Mathematical Surveys* **59** (1) (2004) 191–192.
3. O.I. Mokhov, Nonlocal Hamiltonian operators of hydrodynamic type with flat metrics, integrable hierarchies, and the associativity equations, *Functional Analysis and its Applications* **40** (1) (2006) 11–23; arXiv:math.DG/0406292.

4. O.I. Mokhov, Theory of submanifolds, the associativity equations in 2D topological quantum field theories, and Frobenius manifolds, *In: Proceedings of the Workshop "Nonlinear Physics. Theory and Experiment. IV,"* (Gallipoli, Lecce, Italy, 22 June – 1 July, 2006, published in Theoretical and Mathematical Physics, Vol. 152, No. 2, 2007) 1183–1190; Preprint MPIM2006-152, Max-Planck-Institut für Mathematik, Bonn, Germany, 2006; arXiv:math.DG/0610933.

Differential Geometry and its Applications
Proc. Conf., in Honour of Leonhard Euler, Olomouc, August 2007

The Cartan form, 20 years on

D.J. Saunders

Palacký University,
Olomouc, Czech Republic
E-mail: david@symplectic.demon.co.uk

The period around twenty years ago saw significant activity in the specification of "Cartan forms" for variational field theory. As well as the positive results, there were some negative ones and also some open questions. We review these results and compare them with recent developments in the homogeneous version of the theory. An important tool in the latter theory is a bicomplex comprising spaces of vector-valued forms, and an associated global homotopy operator.

Keywords: Calculus of variations, Cartan form, Lepage equivalent.

MS classification: 58E19, 53A20.

1. The Cartan form

Twenty years ago, the author published a paper on the Cartan form in higher-order field theories.[19] A number of papers on this topic had appeared during the previous decade, and more were to follow; these presented both positive and negative results, although there was one significant open question. The purpose of the present paper is to describe an alternative perspective on this problem, and to suggest that this perspective might be able both to throw some light on the known results and to suggest a resolution of the open question.

The original 'Poincaré-Cartan form' of course dates back, in one form or another, nearly a century rather than a mere two decades. This construction arose when considering variational problems in classical mechanics. Given a Lagrangian L and a variational problem

$$\delta \int L(t, q^\alpha, \dot{q}^\alpha) dt = 0 \,,$$

the Poincaré-Cartan form is the differential 1-form

$$L\, dt + \frac{\partial L}{\partial \dot{q}^\alpha}(dq^\alpha - \dot{q}^\alpha\, dt)$$

and, in modern parlance, is defined on the first jet bundle $J^1\pi$ of a fibration. The problem is to extend this construction to higher-order field theories, and to see how one might approach the problem we shall reformulate it in terms of *Lepage equivalents* (and, also, use a slightly different convention for coordinates).

Consider a fibred manifold $\pi : E \to \mathbb{R}$, with fibred coordinates (x, y^α). In this approach to mechanics, a Lagrangian is a 1-form $\lambda = L\,dx$ defined on $J^1\pi$; a similar construction applies also in higher-order mechanics, where the Lagrangian is defined on $J^k\pi$. The Cartan form Θ_λ then has several good properties:

- Θ_λ is the unique Lepage equivalent of λ (in other words, $\Theta_\lambda - \lambda$ is a contact form, and if $Y \in \mathcal{X}(J^1\pi)$ is vertical over E then $i_Y d\Theta_\lambda$ is also a contact form);
- if λ is regular (so that L has a non-degenerate Hessian) then extremals of Θ_λ are prolongations of extremals of λ; and
- if λ is regular then the Euler-Lagrange equations can be written in normal form (as the coefficients of a vector field).

When trying to extend this construction to the case of m independent variables, we have an immediate problem regarding uniqueness. If λ is a Lagrangian and Θ_λ is any Lepage equivalent then so is $\Theta_\lambda + \Phi$ where Φ is any 2-contact m-form. On the other hand, there is certainly a local construction available: if $\lambda = L\,d^m x$ is defined on $J^k\pi$ with coordinates (x^i, y^α_I) then

$$\Theta_\lambda = L\,d^m x + \left(\sum_{|I|=0}^{k-1} \sum_{|J|=0}^{k-|I|-1} (-1)^{|J|} \frac{(I+J+1_i)!|I|!|J|!}{|I+J+1_i|!I!J!} \times \right. \\ \left. \frac{d^{|J|}}{dx^J} \frac{\partial L}{\partial y^\alpha_{I+J+1_i}} (dy^\alpha_I - y^\alpha_{I+1_k} dx^k) \wedge d^{m-1}x_i \right)$$

is a local Lepage equivalent of λ. So we now have two important questions to answer.

- Does a global Lepage equivalent exist?
- Are there any global Lepage equivalents which are *geometric objects*? (A geometric object is, informally, one whose local coordinate representation retains the same form under an allowable change of coordinates. A rigorous definition uses the ideas of Nijenhuis,[16] and a good

exposition of these ideas may be found in the monograph by Kolář, Michor and Slovák.[13])

There is a positive answer to the first question. Various approaches to global existence have been investigated, using, for instance, partitions of unity,[15] connections[7,12] and reduction of order.[19] On the other hand, there is a mixed answer to the second question. If the order of the Lagrangian satisfies $k \geq 3$ then there is no Lepage equivalent of the Lagrangian which at the same time is a geometric object;[10,11] but in the first and second-order cases several geometric objects have been found. In coordinates, two of the geometric Lepage equivalents for a first-order Lagrangian are

- $L\omega + p^i_\alpha \theta^\alpha \wedge \omega_i$ (1)
- $\dfrac{1}{L^{m-1}} \bigwedge_{i=1}^{m} \left(L\, dx^i + p^i_\alpha \theta^\alpha\right) \qquad \text{where } p^i_\alpha = \dfrac{\partial L}{\partial y^\alpha_i}$ (2)

and, similarly, two such equivalents for a second-order Lagrangian are

- $L\omega + (p^i_\alpha \theta^\alpha + p^{ij}_\alpha \theta^\alpha_j) \wedge \omega_i$ (3)
- $\dfrac{1}{L^{m-1}} \bigwedge_{i=1}^{m} \left(L\, dx^i + p^i_\alpha \theta^\alpha + p^{ij}_\alpha \theta^\alpha_j\right)$ (4)

$$\text{where } p^{ij}_\alpha = \frac{1}{\#(ij)} \frac{\partial L}{\partial y^\alpha_{ij}}, \qquad p^i_\alpha = \frac{\partial L}{\partial y^\alpha_i} - \frac{dp^{ij}}{dx^j},$$

where in these formulæ we have written

$$\theta^\alpha = dy^\alpha - y^\alpha_k dx^k, \quad \theta^\alpha_j = dy^\alpha_j - y^\alpha_{jk} dx^k, \quad \omega = d^m x, \quad \omega_i = i_{\partial/\partial x^i}\, \omega.$$

There is, however, a third geometric Lepage equivalent available for a first-order Lagrangian:[3,14]

$$\Theta_\lambda = \sum_{r=0}^{\min\{m,n\}} \frac{1}{(r!)^2} \frac{\partial^r L}{\partial y^{\alpha_1}_{j_1} \cdots \partial y^{\alpha_r}_{j_r}} \theta^{\alpha_1} \wedge \cdots \wedge \theta^{\alpha_r} \wedge \omega_{i_1 \cdots i_r} \tag{5}$$

where $\omega_{i_1 \cdots i_r} = i_{\partial/\partial x^{i_r}}\, \omega_{i_1 \cdots i_{r-1}}$; this particular Lepage equivalent has the property that

$$d\Theta_\lambda = 0 \Leftrightarrow \lambda \text{ is null.}$$

The question of whether a geometric object with a similar property may be found for second-order Lagrangians is still open; later in this paper we suggest that the answer to this question is negative.

There is, though, a rather different approach to Cartan forms; this might be called the *lifting approach*. A Cartan form might be defined, not on a jet manifold $J^{2k-1}\pi$, but on a bundle $C \to J^{2k-1}\pi$, so that there are 'undetermined coefficients'.[6] For instance, consider the bundle $\bigwedge^m T^* J^{2k-1}\pi \to J^{2k-1}\pi$ and the restriction of the tautological m-form Θ to a certain submanifold $C_\lambda \subset \bigwedge^m T^* J^{2k-1}\pi$;[9] pull-backs of this form are Lepage equivalents of the Lagrangian.

In the rest of this paper we consider an alternative version of the lifting approach. Here, the Lagrangian itself will be lifted to a new bundle. The lift will take specific account of the problem that higher-order Cartan forms are not invariant under a change of independent variables: " ... indeed it is the non-linear changes of independent variable that destroy the invariance of the semi-invariant Cartan form".[17]

2. Homogeneous variational systems

Homogeneous variational systems are those known in the literature as 'parametric systems' because the extremals are specified in parametric form: see, for example, Rund[18] or Giaquinta and Hildebrandt.[8] The basic examples of such systems arise in Finsler geometry, of which Riemannian geometry is a special case.

In the general case, multiple integral problems (of order k) are defined on the bundle of k-th order m-frames in E, $\mathcal{F}^k_{(m)}E$, which is the bundle of k-jets (at the origin) of non-singular maps $\mathbb{R}^m \to E$; this is also known as the bundle of non-degenerate k-th order velocities. Given local coordinates (u^a) on E, the corresponding coordinates on $\mathcal{F}^k_{(m)}E$ are (u^a, u^a_I) where $I \in \mathbb{N}^m$ is a multi-index satisfying $0 \leq |I| \leq k$. We also consider the manifold $J^k_+(E, m)$ of *oriented k-th order m-dimensional contact elements*; this comes from taking the quotient by diffeomorphisms of $\mathbb{R}^m$ preserving the origin and the orientation. The projection $\rho : \mathcal{F}^k_{(m)}E \to J^k_+(E, m)$ defines a principal bundle whose structure group is the group of k-jets of such diffeomorphisms. For any fibration $\pi : E \to M$ the jet bundle $J^k\pi$ is an open submanifold of $J^k_+(E, m)$.

There are two important sets of intrinsic objects defined on the frame bundle $\mathcal{F}^k_{(m)}E$. These are the *total derivatives* d_i and the *vertical endomorphisms* S^j $(1 \leq i, j \leq m)$. These objects may be defined intrinsically,[4] and in coordinates they are given by

$$d_i = \sum_{|I|=0}^{k-1} u^a_{I+1_i} \frac{\partial}{\partial u^a_I}, \qquad S^j = \sum_{|J|=0}^{k-1} (J(j) + 1) du^b_J \otimes \frac{\partial}{\partial u^b_{J+1_j}};$$

note that the total derivatives are vector fields along the projection $\mathcal{F}^k_{(m)}E \to \mathcal{F}^{k-1}_{(m)}E$, whereas the vertical endomorphisms are tensor fields on

$\mathcal{F}^k_{(m)}E$ itself. Note also that, whereas the indices a, b refer to a local chart, the indices i, j and the multi-indices I, J refer ultimately to the canonical coordinates on $\mathbb{R}^m$ and so are not affected by coordinate changes on E. In this way we separate out the effect of changes in the independent variables for our variational problems, as these changes appear only when we take the projection to $J^k_+(E, m)$.

We observe that $d_i d_j = d_j d_i$ and that $S^i S^j = S^j S^i$, so that we may use a multi-index notation for composites of both types of object. With this notation we define the vector fields $\Delta_i^J = S^J(d_i)$, where in this case the d_i are vector fields along $\mathcal{F}^{k+1}_{(m)} \to \mathcal{F}^k_{(m)}$; in principle, therefore, Δ_i^J would be vector fields along the same projection, but the properties of S^J imply that they are in fact well-defined as vector fields on the manifold $\mathcal{F}^k_{(m)}E$. They are, indeed, the fundamental vector fields of the principal bundle ρ.

We are now in a position to describe a general homogeneous variational problem. The Lagrangian $L : \mathcal{F}^k_{(m)}E \to \mathbb{R}$ is a function, as in the case of Finsler geometry, and it is homogeneous if $\Delta_j^i L = \delta_j^i L$ and $\Delta_j^I L = 0$ for $|I| > 1$. The corresponding variational problem is

$$\delta \int_C ((j^k\sigma)^* L)\, d^m t = 0\,, \qquad C \subset \mathbb{R}^m \text{ compact}\,;$$

homogeneity implies that if σ is an extremal then so is $\sigma \circ \phi$ for any orientation-preserving diffeomorphisms $\phi : \mathbb{R}^m \to \mathbb{R}^m$ satisfying $\phi(0) = 0$. Starting with an m-form $\lambda \in \Omega^m J^k(E, m)$ horizontal over $J^{k-1}(E, m)$, the function

$$L = \langle (d_1, \ldots, d_m), \rho^* \lambda \rangle$$

is then a homogeneous Lagrangian function on $\mathcal{F}^k_{(m)}E$ with the same extremals as λ. A Lagrangian form on the jet bundle $J^k\pi$ satisfies these conditions (on an open submanifold of $J^{k-1}(E, m)$ rather than on the whole manifold) and so may be used to define a homogeneous Lagrangian function on an open submanifold of $\mathcal{F}^k_{(m)}E$.

Given a homogeneous Lagrangian, we now construct its *Hilbert forms.*[4] These are 1-forms ϑ^i $(1 \leq i \leq m)$ on $\mathcal{F}^{2k-1}_{(m)}E$ defined by the formula

$$\vartheta^i = \sum_{|I|=0}^{k-1} \frac{(-1)^{|I|}}{(|I| + 1)\, I!} d_I S^{I+1_i} dL\,, \tag{6}$$

and are so named because they generalise the Hilbert form of Finsler geometry.[2] We use the Hilbert forms in a version of the First Variation Formula to construct the Euler-Lagrange form ε for the variational problem,

$$\varepsilon = dL - d_i \vartheta^i \, ; \tag{7}$$

the coordinate expression for this form is

$$\varepsilon = \sum_{|I|=0}^{k} (-1)^{|I|} d_I \left(\frac{\partial L}{\partial u^a_I} \right) du^a \, .$$

We may see from this First Variation Formula that, given a compact submanifold $C \subset \mathbb{R}^m$, an immersion $\sigma : C \to E$ and a vector field X on $\sigma(C)$, we have

$$\begin{aligned} \int_C (j^k\sigma)^*(i_{X^k} dL) d^m t &= \int_C (j^{2k}\sigma)^*(i_{X^{2k}} \varepsilon) d^m t \\ &\quad + \int_C (j^{2k}\sigma)^*(i_{X^{2k}} d_i \vartheta^i) d^m t \end{aligned}$$

where X^k, X^{2k} denote the prolongations of X to $\mathcal{F}^k_{(m)}E$ and $\mathcal{F}^{2k}_{(m)}E$ respectively; but

$$\begin{aligned} \int_C (j^{2k}\sigma)^*(i_{X^{2k}} d_i \vartheta^i) d^m t &= \int_C (j^{2k}\sigma)^*(d_i i_{X^{2k-1}} \vartheta^i) d^m t \\ &= \int_C \frac{\partial}{\partial t^i} (j^{2k-1}\sigma)^*(i_{X^{2k-1}} \vartheta^i) d^m t \\ &= \int_{\partial C} (j^{2k-1}\sigma)^*(i_{X^{2k-1}} \vartheta^i) d^{m-1} t_i \\ &= 0 \end{aligned}$$

and then a standard argument shows that the Euler-Lagrange form ε vanishes along extremals of the variational problem.

Of course the Hilbert forms themselves provide us with a 'vector' of 1-forms on $\mathcal{F}^{2k-1}_{(m)}E$. To relate our construction to those on jet bundles, we should like instead to find an m-form, which we can integrate, and which has the same extremals as the Lagrangian. The m-form

$$\widetilde{\Theta} = \frac{1}{L^{m-1}} \bigwedge\nolimits_{i=1}^{m} \vartheta^i \tag{8}$$

satisfies this condition,[4] and is by construction a geometric object. It is significant that the condition holds regardless of the order of the Lagrangian, in contrast to the case with jet bundles: the reason is that, by considering the problem on a frame bundle, we have effectively chosen a fixed parametrisation for the extremals.

To see the effect of a reparametrisation, we can now consider whether the m-form (8) we have constructed is projectable to the bundle of contact elements. When $m = 1$ this is always the case. For the multiple-integral case when $m \geq 2$ the form is projectable when the order satisfies $k \leq 2$, and then the restriction to the jet bundle of any fibration $\pi : E \to M$ gives the decomposable Lepage equivalent (2) or (4),[4] providing some insight into the known results described earlier.

3. The fundamental form

We now address the open question of the existence of a geometrical Lepage equivalent for a second-order Lagrangian analogous to the form (5) associated with a first-order Lagrangian. We do this by rewriting the homogeneous formulation in terms of certain vector-valued forms: namely, forms on a frame bundle $\mathcal{F}^k_{(m)}E$ taking values in the vector space $\mathbb{R}^{m*}$ and its exterior powers.

Put

$$\Omega^{r,s}_k = \left(\Omega^r \mathcal{F}^k_{(m)} E\right) \otimes \left(\textstyle\bigwedge^s \mathbb{R}^{m*}\right) ;$$

if the standard basis for $\mathbb{R}^{m*}$ is denoted by (dt^i) then a typical element $\Phi \in \Omega^{r,s}_k$ may be written (globally) as

$$\Phi = \phi_{i_1 \cdots i_s} \otimes dt^{i_1} \wedge \ldots \wedge dt^{i_s}$$

where the scalar forms $\phi_{i_1 \cdots i_s}$ are skew-symmetric in their indices. Two natural coboundary operators on these vector-valued forms are

$$d : \Omega^{r,s}_k \to \Omega^{r+1,s}_k , \qquad d_T : \Omega^{r,s}_k \to \Omega^{r,s+1}_{k+1}$$

where d is the usual de Rham differential acting on the scalar components, and the *total derivative operator* d_T is defined by its action on decomposable forms:

$$\begin{aligned} d(\phi \otimes \omega) &= d\phi \otimes \omega , \\ d_T(\phi \otimes \omega) &= d_i\phi \otimes dt^i \wedge \omega . \end{aligned}$$

It is immediate that the properties $dd_T = d_T d$, $d_T^2 = 0$ hold, and we may construct a bicomplex which is in some ways similar to the variational bicomplex of the jet bundle theory.[1] The figure below shows an important part of this bicomplex; in the first column, where the elements are vector-valued 0-forms (functions), the overline denotes a quotient by the constant functions. (Of course each value of k gives a different bicomplex.)

$$
\begin{array}{ccccccccc}
& & \downarrow & & \downarrow & & \downarrow & & \downarrow \\
0 & \longrightarrow & \overline{\Omega}_k^{0,m-2} & \xrightarrow{\ d\ } & \Omega_k^{1,m-2} & \xrightarrow{\ d\ } & \Omega_k^{2,m-2} & \xrightarrow{\ d\ } & \Omega_k^{3,m-2} & \longrightarrow \\
& & \Big\downarrow{\scriptstyle d_T} & & \Big\downarrow{\scriptstyle d_T} & & \Big\downarrow{\scriptstyle d_T} & & \Big\downarrow{\scriptstyle d_T} \\
0 & \longrightarrow & \overline{\Omega}_{k+1}^{0,m-1} & \xrightarrow{\ d\ } & \Omega_{k+1}^{1,m-1} & \xrightarrow{\ d\ } & \Omega_{k+1}^{2,m-1} & \xrightarrow{\ d\ } & \Omega_{k+1}^{3,m-1} & \longrightarrow \\
& & \Big\downarrow{\scriptstyle d_T} & & \Big\downarrow{\scriptstyle d_T} & & \Big\downarrow{\scriptstyle d_T} & & \Big\downarrow{\scriptstyle d_T} \\
0 & \longrightarrow & \overline{\Omega}_{k+2}^{0,m} & \xrightarrow{\ d\ } & \Omega_{k+2}^{1,m} & \xrightarrow{\ d\ } & \Omega_{k+2}^{2,m} & \xrightarrow{\ d\ } & \Omega_{k+2}^{3,m} & \longrightarrow
\end{array}
$$

$$
\begin{array}{cccc}
\overline{\Omega}^{0,m}\Big/{d_T \overline{\Omega}^{0,m-1}} & \Omega^{1,m}\Big/{d_T \Omega^{1,m-1}} & \Omega^{2,m}\Big/{d_T \Omega^{2,m-1}} & \Omega^{3,m}\Big/{d_T \Omega^{3,m-1}} \\
\downarrow & \downarrow & \downarrow & \downarrow \\
0 & 0 & 0 & 0
\end{array}
$$

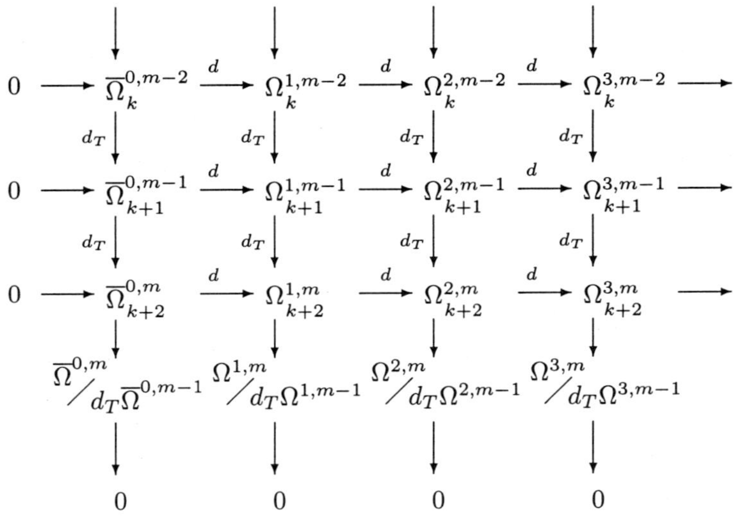

In any given bicomplex the map $d_T : \Omega_k^{r,s} \to \Omega_{k+1}^{r,s+1}$ is not exact, even locally. But it is globally exact modulo pull-backs (for $r \geq 1$),[20] and indeed there is a canonical homotopy operator $P : \Omega_k^{r,s} \to \Omega_{(r+1)k-1}^{r,s-1}$ given by

$$
P(\Phi) = P_{(s)}^j (\phi_{i_1 \cdots i_s}) \otimes \left\{ \frac{\partial}{\partial t^j} \ \lrcorner \ \left(dt^{i_1} \wedge \ldots \wedge dt^{i_s} \right) \right\}
$$

where

$$
P_{(s)}^j = \sum_{|J|=0}^{rk-1} \frac{(-1)^{|J|}(m-s)!|J|!}{r^{|J|+1}(m-s+|J|+1)!J!} d_J S^{J+1_j} .
$$

We now use this homotopy operator to construct the *fundamental form* of a homogeneous Lagrangian. If the Lagrangian has order k then we consider in more detail the bottom left-hand corner of the bicomplex, where now we show the homotopy operator and the pull-back map explicitly.

$$
\begin{array}{ccccc}
& & & & \Omega_{2k-1}^{1,m-1} \\
& & & {\scriptstyle P}\!\!\nearrow & \Big\downarrow{\scriptstyle d_T} \\
\overline{\Omega}_k^{0,m} & \xrightarrow{\ d\ } & \Omega_k^{1,m} & \longrightarrow & \Omega_{2k}^{1,m} \\
& & & & \downarrow \\
& & & & \Omega^{1,m}\Big/{d_T \Omega^{1,m-1}}
\end{array}
$$

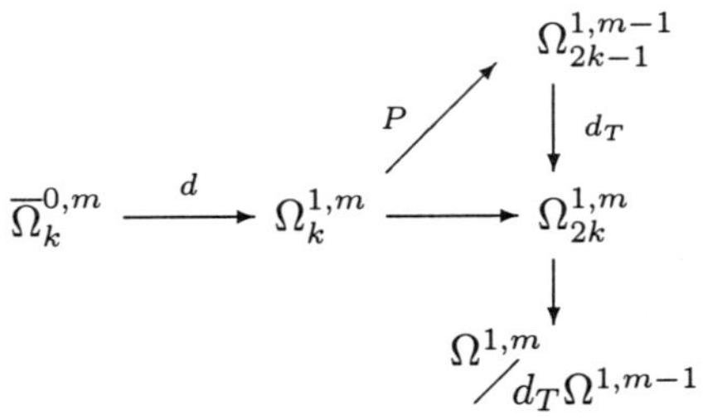

We regard the Lagrangian as a vector-valued function and the Euler-Lagrange form as a vector-valued 1-form, and the Hilbert forms as the scalar forms of a vector-valued 1-form:

$$\begin{aligned}\Lambda &= [L\, d^m t] \in \overline{\Omega}_k^{0,m} \\ \mathcal{E} &= \varepsilon\, d^m t \in \Omega_{2k}^{1,m} \\ \Theta_1 &= \vartheta^i \otimes d^{m-1} t_i \in \Omega_{2k-1}^{1,m-1}\,.\end{aligned}$$

It is clear that the definition of the individual Hilbert forms (6) may be rewritten in terms of the homotopy operator as $\Theta_1 = Pd\Lambda$, and that the first variation formula (7) may be rewritten (modulo a pull-back on $d\Lambda$) as $\mathcal{E} = d\Lambda - d_T\Theta_1$.

We now generalise the definition of Θ_1 by putting $\Theta_r = (Pd)^r\Lambda \in \Omega^{r,m-r}$ for $1 \leq r \leq m$; we call $\Theta_m \in \Omega^{m,0}$ the *fundamental form* of Λ. Our conjecture is that

$$\Theta_r = \frac{1}{m-r} i_T \Theta_{r+1} \quad \text{for} \quad 1 \leq r \leq m\,.$$

It is straightforward to show that if the conjecture holds then $d\Theta_m = 0$ precisely when Λ is null. The following results are known:

- the conjecture holds when $m = 1$, for any order k;
- the conjecture holds when $k = 1$, for any m;
- for any order k, if $m \geq 2$ the conjecture holds for $r = 1$ and $r = 2$.

For the third of these results,[22] the computations for the case $r = 1$ are straightforward, whereas those for $r = 2$ are quite complicated. The computations when $r \geq 3$ are more complicated again and we have not completed them, although there do not appear to be any intrinsic obstructions to a successful proof.

We are particularly interested in the multiple-integral case when $m \geq 2$. For a first-order Lagrangian, the fundamental form Θ_m satisfies the condition that $d\Theta_m = 0$ precisely when Λ is null, and it is also projectable to $J^1_+(E, m)$; the restriction to the jet bundle $J^1\pi$ of any fibration $\pi : E \to M$ gives the Lepage equivalent (5).[5] When the order k is greater than one we have definite information about the fundamental form only in the case $m = 2$, and now Θ_2 again satisfies the condition that $d\Theta_2 = 0$ precisely when Λ is null; but it is significant that the order of Θ_2 is greater than $2k - 1$. An example using a second-order Lagrangian[21] shows that, in general, Θ_2 (which in this case is defined on a fourth-order frame bundle) is not invariant under the action of the jet group, and so does not project to the bundle of contact elements. As we might expect that a generalisation of (5) to second-order Lagrangians would have been obtained by just such a projection, it seems reasonable to suspect that such a generalisation does not exist.

Acknowledgments

This work was supported by the Czech Science Foundation (grant no. 201/06/0922 for Global Analysis and its Applications)

References

1. I.M. Anderson, *The variational bicomplex* (Utah State University Technical Report, 1989); http://www.math.usu.edu/~fg_mp/.
2. D. Bao, S.-S. Chern and Z. Shen, *An introduction to Riemann-Finsler geometry* (Springer, Berlin, 2000).
3. D.E. Betounes, Extensions of the classical Cartan form, *Phys. Rev. D* **29** (1984) 599–606.
4. M. Crampin and D.J. Saunders, The Hilbert-Carathéodory and Poincaré-Cartan forms for higher-order multiple-integral variational problems, *Houston J. Math.* **30** (3) (2004) 657–689.
5. M. Crampin and D.J. Saunders, On null Lagrangians, *Diff. Geom. Appl.* **22** (2) (2005) 131–146.
6. P. Dedecker, private communication (1990).
7. M. Ferraris, Fibred connections and global Poincaré-Cartan forms in higher-order calculus of variations, *In: Differential Geometry and its Applications* (Proc. Conf. 1983, Vol 2, (D. Krupka, Ed.) J. E. Purknyě University, Brno, 1984) 61–91.
8. M. Giaquinta and S. Hildebrandt, *Calculus of variations* (Springer, Berlin, 1996).
9. M. Gotay, A multisymplectic framework for classical field theory and the calculus of variations I: Covariant Hamiltonian formalism, *In: Mechanics, Analysis and Geometry: 200 years after Lagrange* ((M. Francaviglia, Ed.) Elsevier, New York, 1991) 203–235.
10. M. Horák and I. Kolář, On the higher-order Poincar-Cartan form, *Czech. Math. J.* **33** (1983) 467–475.
11. I. Kolář, Natural Operators Related With The Variational Calculus, *In: Differential Geometry and its Applications* (Proc. Conf. Opava, 1992, Silesian University, Opava, 1993, (O. Kowalski and D. Krupka, Eds.)) 461–472.
12. I. Kolář, A geometrical version of the higher-order Hamilton formalism in fibred manifolds, *J. Geom. Phys* **1 (2)**, 127–137 (1984).
13. I. Kolář, P. W. Michor and J. Slovák, *Natural Operations in Differential Geometry* (Springer, Berlin, 1993).
14. D. Krupka, A map associated to the Lepagean forms in the calculus of variations, *Czech. Math. J.* **27** (1977) 114–118.
15. D. Krupka, Lepagean forms in higher order variational theory, *In: Modern*

developments in analytical mechanics I (Accad. delle Scienze, Torino, 1983) 197-238.
16. A. Nijenhuis, Natural bundles and their general properties, *In: Differential Geometry in Honor of K. Yano* (Kinokuniya, Tokyo, 1972) 317–334.
17. P.J. Olver, Equivalence and the Cartan form, *Acta Appl. Math.* **31** (1993) 99–136.
18. H. Rund, *The Hamilton-Jacobi theory in the calculus of variations* (Van Nostrand, London, 1966).
19. D.J. Saunders, An alternative approach to the Cartan form in Lagrangian field theories, *J. Phys. A* **20** (1987) 339–349.
20. D.J. Saunders, Homogeneous variational complexes and bicomplexes; arXiv:math.DG/0512383.
21. D.J. Saunders, The fundamental form of a second-order homogeneous Lagrangian in two variables; arXiv:math.DG/0512384.
22. D.J. Saunders and M. Crampin, The fundamental form of a homogeneous Lagrangian in two independent variables; arXiv:0709.3092.

Differential Geometry and its Applications
Proc. Conf., in Honour of Leonhard Euler, Olomouc, August 2007

Calculus along the tangent bundle projection and projective metrizability

József Szilasi

Institute of Mathematics, University of Debrecen, Debrecen, Hungary
E-mail: szilasi@math.klte.hu

Abstract. We sketch an economical framework and a simple index-free calculus for direction-dependent objects, based almost exclusively on Berwald derivative. To illustrate the efficiency of these tools, we characterize projectively Finslerian sprays, and derive Hamel's PDEs in an analytic version of Hilbert's fourth problem.

Keywords: Direction dependence, Berwald derivative, sprays, Finsler metrizability, Hilbert's fourth problem.

MS classification: 58C20, 53C60, 53C22, 58G30.

1. Introduction

A part of calculus of growing importance for current research can be classified as 'calculus of direction dependent objects'. It seems to me that the most convenient and economical setting for their study is the pull-back of the tangent bundle of a base manifold over the natural projection $\mathring{\tau}$ of the slit tangent bundle. Then our direction dependent objects can be interpreted as tensors along $\mathring{\tau}$. A comprehensive calculus for this kind of objects was elaborated by E. Martínez, J. Cariñena and W. Sarlet in the early 1990's;[15,16] see also Chapter 2/E of Ref. 21. I think, for most purposes a simplified version of this calculus, with the Berwald derivative in the focus, is already sufficient. Berwald's derivative is a covariant derivative operator in our pull-back bundle, built up from a canonical vertical part and a horizontal part depending on an Ehresmann connection. In the most important applications, e.g., in the presence of a Finsler function, an Ehresmann connection can 'naturally' be constructed from the structure.

In this article I give a bold outline of a calculus based mainly on a Berwald derivative, with special emphasis on the fundamental geometric

objects which may be constructed in terms of the Berwald derivative. To present some non trivial applications, I discuss the following
Metrizability problem: *Under what necessary and sufficient conditions has a spray common geodesics (as unparametrized curves) with a Finsler function?*

It will turn out that this question leads naturally to an analytic version of Hilbert's fourth problem.

2. Basic constructions and conventions

Notations. Our base manifold M will be an n-dimensional ($n \geq 2$), connected, paracompact smooth manifold. $C^\infty(M)$ stands for the ring of real-valued smooth functions on M.

T_pM is the tangent space to M at the point $p \in M$, TM is the $2n$-dimensional tangent manifold of M, and $\mathring{T}M$ is the open submanifold of the non-zero tangent vectors to M. τ and $\mathring{\tau}$ denote the natural projections of TM and $\mathring{T}M$ onto M, resp. The *tangent bundle* of M is the triplet (TM, τ, M), denoted also simply by τ or, less consequently, by TM. Similarly, $(\mathring{T}M, \mathring{\tau}, M)$ is the *slit tangent bundle* of M. The shorthand for the tangent bundle of TM and $\mathring{T}M$ will be TTM and $T\mathring{T}M$, respectively. The *vertical lift of a function* $f \in C^\infty(M)$ is $f^{\mathsf{v}} := f \circ \tau \in C^\infty(TM)$; the *complete lift* $f^{\mathsf{c}} \in C^\infty(TM)$ of f is defined by $f^{\mathsf{c}}(v) := v(f)$, $v \in TM$.

If $\varphi\colon M \to N$ is a smooth map, then φ_* will denote the smooth map of TM into TN induced by φ, the *tangent map* (or *derivative*) of φ.

Vector fields. A *rough vector field* on M is a map $X\colon M \to TM$ which satisfies $\tau \circ X = 1_M$. If, in addition, X is smooth, then it is a *vector field* on M. The $C^\infty(M)$-module of vector fields on M is denoted by $\mathfrak{X}(M)$. If $X \in \mathfrak{X}(M)$, i_X and $\mathcal{L}_X$ denote the *substitution operator* induced by X and the *Lie derivative* with respect to X, respectively. The operator of the *exterior derivative* will be denoted by d on every *manifold.*
To make the text typographically more transparent, capitals $X, Y, Z, \dots$ will stand for vector fields on the base manifold, while vector fields on TM will usually (but not exclusively) be denoted by Greek letters $\xi, \eta, \zeta, \dots$.

Given a vector field X on M, there is a unique vector field X^{c} on TM such that $X^{\mathsf{c}} f^{\mathsf{c}} = (Xf)^{\mathsf{c}}$ for all $f \in C^\infty(M)$. X^{c} is called the *complete lift of* X.

By a *rough second-order vector field* over M we mean a rough vector field S on TM such that $\tau_* \circ S$ is the identity on TM. S is said to be a *semispray* or a *second-order vector field* over M if it is smooth on $\mathring{T}M$ or on TM, respectively.

Basic setup. The majority of our objects will live on the pull-back of the tangent bundle of M over its projection τ or over $\mathring{\tau}$. These are vector bundles over TM and $\mathring{T}M$ whose total manifolds are the fibre products $TM \times_M TM$ and $\mathring{T}M \times_M TM$, and which will be denoted by π and $\mathring{\pi}$, respectively. For the $C^\infty(TM)$-module of sections of π we use the notation $\Gamma(\pi)$. Any section in $\Gamma(\pi)$ is of the form

$$\widetilde{X}: v \in TM \longmapsto \widetilde{X}(v) = (v, \underline{X}(v)) \in TM \times_M TM,$$

where $\underline{X}: TM \to TM$ is a smooth map such that $\tau \circ \underline{X} = \tau$. In particular, we have the *canonical section* δ given by $\delta(v) := (v, v)$, $v \in TM$.

An important class of sections of π can be obtained from vector fields on M: if $X \in \mathfrak{X}(M)$, then the map

$$\widehat{X}: v \in TM \longmapsto \widehat{X}(v) := (v, X(\tau(v))) \in TM \times_M TM$$

is a section of π, called the *lift of* X *into* $\Gamma(\pi)$ or a *basic section* of π. *If* $(X^i)_{i=1}^n$ *is a local basis for* $\mathfrak{X}(M)$, *then* $(\widehat{X}^i)_{i=1}^n$ *is a local basis for* $\Gamma(\pi)$; this simple observation proves to be useful in calculations.

The module $\Gamma(\pi)$ may naturally be identified with the $C^\infty(TM)$-module

$$\mathfrak{X}(\tau) := \{\underline{X}: TM \to TM \mid \underline{X} \text{ is smooth, and } \tau \circ \underline{X} = \tau\}$$

of *vector fields along* τ by the map $\widetilde{X} \in \Gamma(\pi) \longmapsto \underline{X} \in \mathfrak{X}(\tau)$. We shall use this harmless identification tacitly, whenever it is convenient. Then the identity map 1_{TM} corresponds to the canonical section, while the maps $X \circ \tau$ ($X \in \mathfrak{X}(M)$) correspond to basic sections.

We may describe the $C^\infty(\mathring{T}M)$-module $\Gamma(\mathring{\pi}) \cong \mathfrak{X}(\mathring{\tau})$ analogously. The tensor algebras of the modules $\Gamma(\pi)$ and $\Gamma(\mathring{\pi})$ will be denoted by $\mathcal{T}(\pi)$ and $\mathcal{T}(\mathring{\pi})$, respectively. It is natural again to interpret the elements of these tensor algebras as 'tensors along τ or $\mathring{\tau}$', and throughout the paper, as a rule, the term 'tensor' will be used in this sense. We note finally that $\Gamma(\pi)$ will be considered as a subalgebra of $\Gamma(\mathring{\pi})$.

Coordinates. In coordinate descriptions we shall specify a chart $(U, (u^i)_{i=1}^n)$ on M and use the *induced chart* $(\tau^{-1}(U), (x^i, y^i)_{i=1}^n)$ on TM, where $x^i := (u^i)^{\mathsf{v}}$, $y^i := (u^i)^{\mathsf{c}}$. These charts lead to the local bases

$$\left(\frac{\partial}{\partial u^i}\right)_{i=1}^n, \qquad \left(\frac{\partial}{\partial x^i}, \frac{\partial}{\partial y^i}\right)_{i=1}^n \qquad \text{and} \qquad \left(\widehat{\frac{\partial}{\partial u^i}}\right)_{i=1}^n \tag{1}$$

of $\mathfrak{X}(M)$, $\mathfrak{X}(TM)$ and $\mathfrak{X}(\overset{\circ}{\tau})$, respectively. Note that $\frac{\partial}{\partial x^i} = \left(\frac{\partial}{\partial u^i}\right)^{\mathsf{c}}$.

Einstein's convention on repeated indices will be used. If $X = X^i \frac{\partial}{\partial u^i}$ is a vector field on U, then its lift into $\mathfrak{X}(\overset{\circ}{\tau})$ is locally given by

$$\widehat{X} = (X^i \circ \tau)\widehat{\frac{\partial}{\partial u^i}}.$$

The coordinate expression of the canonical section is

$$\delta \restriction \tau^{-1}(U) = y^i \widehat{\frac{\partial}{\partial u^i}}.$$

If S is a semispray over M, then in local coordinates

$$S \restriction \tau^{-1}(U) = y^i \frac{\partial}{\partial x^i} - 2G^i \frac{\partial}{\partial y^i} \tag{2}$$

for some smooth functions G^i on $\overset{\circ}{\tau}^{-1}(U)$. These functions will be called the *semispray coefficients* of S with respect to the chosen chart. We note that the factor -2 in the coordinate expression of S proves to be convenient, and is more or less traditional.

3. Some vertical calculus

Canonical objects. We recall that the *fundamental exact sequence* over TM is the sequence of vector bundle maps

$$0 \longrightarrow TM \times_M TM \xrightarrow{\mathbf{i}} TTM \xrightarrow{\mathbf{j}} TM \times_M TM \longrightarrow 0, \tag{3}$$

where $\mathbf{j}(w) := (v, \tau_*(w))$, if $w \in T_vTM$, while $\mathbf{i}$ identifies the fibre $\{v\} \times T_{\tau(v)}M$ with the tangent space $T_vT_{\tau(v)}M$ for all $v \in TM$.

The bundle maps $\mathbf{i}$ and $\mathbf{j}$ induce $C^\infty(TM)$-homomorphisms between the modules of sections, denoted by the same symbols. Thus we also have the exact sequence

$$0 \longrightarrow \mathfrak{X}(\tau) \xrightarrow{\mathbf{i}} \mathfrak{X}(TM) \xrightarrow{\mathbf{j}} \mathfrak{X}(\tau) \longrightarrow 0 \tag{4}$$

of module homomorphisms. $\mathbf{i}$ and $\mathbf{j}$ act on the local bases in (1) by

$$\mathbf{i}\left(\widehat{\frac{\partial}{\partial u^i}}\right) = \frac{\partial}{\partial y^i}, \quad \mathbf{j}\left(\frac{\partial}{\partial x^i}\right) = \widehat{\frac{\partial}{\partial u^i}}, \quad \mathbf{j}\left(\frac{\partial}{\partial y^i}\right) = 0; \quad i \in \{1, \ldots, n\}. \tag{5}$$

$\mathfrak{X}^{\mathsf{v}}(TM) := \mathbf{i}\mathfrak{X}(\tau)$ is the module of *vertical vector fields* on TM, $X^{\mathsf{v}} := \mathbf{i}(\widehat{X})$ is the *vertical lift* of the vector field X on M. We have a canonical vertical vector field on TM, the *Liouville vector field* $C := \mathbf{i}\delta$. Its bracket with any vertical lift X^{v} is given by

$$[C, X^{\mathsf{v}}] = -X^{\mathsf{v}}. \tag{6}$$

From the local form of δ we obtain by (5) that $C \upharpoonright \tau^{-1}(U) = y^i \frac{\partial}{\partial y^i}$.

The type $\binom{1}{1}$ tensor field $\mathbf{J} := \mathbf{i} \circ \mathbf{j}$ will be called the *vertical endomorphism* of TTM. (Other terms, e.g., 'the canonical almost tangent structure on TM' are also frequently used.) Evidently $\mathbf{J}^2 = \mathbf{J} \circ \mathbf{J} = 0$. In terms of vertical and complete lifts

$$\mathbf{J}X^{\mathsf{v}} = 0, \qquad \mathbf{J}X^{\mathsf{c}} = X^{\mathsf{v}}; \qquad X \in \mathfrak{X}(M). \tag{7}$$

Thus, in particular, $\mathbf{J}\left(\frac{\partial}{\partial y^i}\right) = \mathbf{J}\left(\frac{\partial}{\partial u^i}\right)^{\mathsf{v}} = 0$, $\mathbf{J}\left(\frac{\partial}{\partial x^i}\right) = \mathbf{J}\left(\frac{\partial}{\partial u^i}\right)^{\mathsf{c}} = \frac{\partial}{\partial y^i}$ $(i \in \{1, \dots, n\})$.

We shall use the *vertical differentiation* $d_{\mathbf{J}}$ which associates the semibasic 1-form $d_{\mathbf{J}}f := df \circ \mathbf{J}$ to a smooth function f on TM. In induced coordinates,

$$d_{\mathbf{J}}f \upharpoonright \tau^{-1}(U) = \frac{\partial f}{\partial y^i} dx^i.$$

For a detailed treatment of vertical differentiation, see Ref. 8.

We define the *Frölicher-Nijenhuis bracket* $[\mathbf{J}, \xi]$ of $\mathbf{J}$ and a vector field ξ on TM by $[\mathbf{J}, \xi] := -\mathcal{L}_\xi \mathbf{J}$. Then for any vector field η on TM,

$$[\mathbf{J}, \xi]\eta = [\mathbf{J}\eta, \xi] - \mathbf{J}[\eta, \xi].$$

Observe that if S is a semispray over M, then we have

$$\mathbf{J}S = C \quad \textit{or, equivalently,} \quad \mathbf{j}S = \delta. \tag{8}$$

Conversely, condition $\tau_* \circ S = 1_{TM}$ in the definition of a semispray may be replaced by (8).

If S is a semispray over M, then

$$\mathbf{J}[\mathbf{J}\xi, S] = \mathbf{J}\xi, \quad \textit{for all} \quad \xi \in \mathfrak{X}(TM) \tag{9}$$

(*Grifone identity*). For a simple proof we refer to Ref. 22.

Vertical derivatives. Let $\widetilde{X}$ be a section in $\mathfrak{X}(\tau)$. We define an operator $\nabla^{\mathsf{v}}_{\widetilde{X}}$ specifying its action on functions in $C^\infty(TM)$ and sections in $\mathfrak{X}(\tau)$ as follows:

$$\nabla^{\mathsf{v}}_{\widetilde{X}} f := (\mathbf{i}\widetilde{X})f = (df \circ \mathbf{i})(\widetilde{X}), \qquad f \in C^\infty(TM); \tag{10}$$

$$\nabla^{\mathsf{v}}_{\widetilde{X}} \widetilde{Y} := \mathbf{j}[\mathbf{i}\widetilde{X}, \eta]; \qquad \widetilde{Y} = \mathbf{j}(\eta) \in \mathfrak{X}(\tau), \quad \eta \in \mathfrak{X}(TM). \tag{11}$$

Then $\nabla^{\mathsf{v}}_{\widetilde{X}}\widetilde{Y}$ is well-defined: it is easy to check its independence of the choice of $\eta \in \mathfrak{X}(TM)$ satisfying $\mathbf{j}(\eta) = \widetilde{Y}$. The map ∇^{v} which assigns to each pair $(\widetilde{X},\widetilde{Y})$ of sections the section $\nabla^{\mathsf{v}}_{\widetilde{X}}\widetilde{Y}$ obeys the same formal rules as a covariant derivative operator; the crucial derivation rule takes the form

$$\nabla^{\mathsf{v}}_{\widetilde{X}} f\widetilde{Y} = ((\mathbf{i}\widetilde{X})f)\widetilde{Y} + f\nabla^{\mathsf{v}}_{\widetilde{X}}\widetilde{Y} = (\nabla^{\mathsf{v}}_{\widetilde{X}} f)\widetilde{Y} + f\nabla^{\mathsf{v}}_{\widetilde{X}}\widetilde{Y}, \quad f \in C^\infty(TM).$$

The *(canonical) v-covariant derivative* $\nabla^{\mathsf{v}}_{\widetilde{X}}$ so obtained can be extended to operate on an arbitrary tensor, the idea is to make sure that Leibniz's rule holds. Thus, for example, if $\widetilde{A} \in \mathcal{T}^1_1(\pi)$, then for any 1-form $\widetilde{\alpha}$ and vector field $\widetilde{Y}$ along τ,

$$\nabla^{\mathsf{v}}_{\widetilde{X}}\widetilde{A}(\widetilde{\alpha},\widetilde{Y}) := (\mathbf{i}\widetilde{X})\widetilde{A}(\widetilde{\alpha},\widetilde{Y}) - \widetilde{A}(\nabla^{\mathsf{v}}_{\widetilde{X}}\widetilde{\alpha},\widetilde{Y}) - \widetilde{A}(\widetilde{\alpha},\nabla^{\mathsf{v}}_{\widetilde{X}}\widetilde{Y}),$$

where the 1-form $\nabla^{\mathsf{v}}_{\widetilde{X}}\widetilde{\alpha}$ is given by $(\nabla^{\mathsf{v}}_{\widetilde{X}}\widetilde{\alpha})(\widetilde{Z}) = \mathbf{i}\widetilde{X}(\widetilde{\alpha}(\widetilde{Z})) - \widetilde{\alpha}(\nabla^{\mathsf{v}}_{\widetilde{X}}\widetilde{Z})$, $\widetilde{Z} \in \mathfrak{X}(\tau)$.

Finally, we define the *(canonical) v-covariant differential* of a type $\binom{k}{l}$ tensor $\widetilde{A}$ as the type $\binom{k}{l+1}$ tensor $\nabla^{\mathsf{v}}\widetilde{A}$, which 'collects all the v-covariant derivatives of $\widetilde{A}$'. For example, if $k = l = 1$, then

$$\nabla^{\mathsf{v}}\widetilde{A}(\widetilde{\alpha},\widetilde{X},\widetilde{Y}) := (\nabla^{\mathsf{v}}_{\widetilde{X}}\widetilde{A})(\widetilde{\alpha},\widetilde{Y}).$$

The v-covariant differential of the canonical section is the identity operator. Indeed, using Grifone's identity (9), for any section $\widetilde{X} \in \mathfrak{X}(\tau)$ we get

$$\mathbf{i}(\nabla^{\mathsf{v}}\delta)(\widetilde{X}) = \mathbf{i}\nabla^{\mathsf{v}}_{\widetilde{X}}\delta \overset{(11)}{=} \mathbf{J}[\mathbf{i}\widetilde{X}, S] = \mathbf{i}\widetilde{X},$$

whence our claim.

For any vector field X on M we have $\nabla^{\mathsf{v}}\widehat{X} = 0$. In fact, if $Y \in \mathfrak{X}(M)$, then $\nabla^{\mathsf{v}}\widehat{X}(\widehat{Y}) = \nabla^{\mathsf{v}}_{\widehat{Y}}\widehat{X} = \mathbf{j}[Y^{\mathsf{v}}, X^{\mathsf{c}}] = 0$, since $[Y^{\mathsf{v}}, X^{\mathsf{c}}]$ is vertical. But any section of π can locally be combined from basic sections, so it follows that $\nabla^{\mathsf{v}}\widehat{X}(\widetilde{Y}) = 0$ for all $\widetilde{Y} \in \mathfrak{X}(\tau)$.

Notice that the vertical differentiation $d_{\mathbf{J}}$ and the v-covariant differential ∇^{v} are related on $C^\infty(TM)$ by $d_{\mathbf{J}} f = \nabla^{\mathsf{v}} f \circ \mathbf{j}$.

To conclude our brief overview on the 'vertical part' of the tensor calculus in $\Gamma(\pi)$, we define the *v-exterior differential* $d^{\mathsf{v}}\widetilde{\alpha}$ of a k-form $\widetilde{\alpha}$ along τ by

$$d^{\mathsf{v}}\widetilde{\alpha} := (k+1)\mathrm{Alt}\nabla^{\mathsf{v}}\widetilde{\alpha}, \tag{12}$$

where Alt is the operator of alternation. Then d^{v} has the expected property $d^{\mathsf{v}} \circ d^{\mathsf{v}} = 0$. For details, we refer to Ref. 21.

4. Horizontal extension of the vertical calculus

Ehresmann connections. In order to differentiate our tensors not only in vertical directions, we need some further structure in general. This further structure will be an Ehresmann connection in what follows. Since the term has various meanings in differential geometry, we have to fix our usage precisely. By an *Ehresmann connection* on TM we mean a map

$$\mathcal{H}: TM \times_M TM \to TTM$$

satisfying the following four conditions:

Ehr 1. *$\mathcal{H}$ is fibre-preserving and fibrewise linear.*
Ehr 2. $\mathbf{j} \circ \mathcal{H} = 1_{TM \times_M TM}$.
Ehr 3. *$\mathcal{H}$ is smooth over $\mathring{T}M \times_M TM$.*
Ehr 4. *If $\sigma\colon p \in M \longmapsto \sigma(p) := 0_p \in T_pM$ is the zero section of τ, then $\mathcal{H}(\sigma(p), v) := (\sigma_*)_p(v)$ for all $p \in M$, $v \in T_pM$.*

Thus, roughly speaking, *an Ehresmann connection is a right splitting of the fundamental exact sequence (3).* If we specify an Ehresmann connection $\mathcal{H}$ on TM, there is a unique left splitting of (3) 'complementary' to $\mathcal{H}$, i.e., a fibre-preserving fibrewise linear map $\mathcal{V}\colon TTM \to TM \times_M TM$, smooth on $T\mathring{T}M$, such that

$$\mathcal{V} \circ \mathbf{i} = 1_{TM \times_M TM} \qquad \text{and} \qquad \operatorname{Ker}(\mathcal{V}) = \operatorname{Im}(\mathcal{H}).$$

$\mathcal{V}$ is called the *vertical map* associated to $\mathcal{H}$. The maps

$$\mathbf{h} := \mathcal{H} \circ \mathbf{j}, \quad \mathbf{v} := \mathbf{i} \circ \mathcal{V} \quad \text{and} \quad \mathbf{F} := \mathcal{H} \circ \mathcal{V} - \mathbf{i} \circ \mathbf{j}$$

are bundle endomorphisms of TTM, called the *horizontal projector*, the *vertical projector* and the *almost complex structure* associated to $\mathcal{H}$, respectively. $\mathbf{h}$ and $\mathbf{v}$ are indeed complementary projection operators ($\mathbf{h}^2 = \mathbf{h}$, $\mathbf{v}^2 = \mathbf{v}$, $\mathbf{h} \circ \mathbf{v} = \mathbf{v} \circ \mathbf{h} = 0$), while $\mathbf{F}^2 = -1_{TTM}$.

An Ehresmann connection, as well as its associated maps, induce module homomorphisms at the level of sections, denoted by the same letters. $\mathfrak{X}^{\mathsf{h}}(\mathring{T}M) := \mathcal{H}(\mathfrak{X}(\mathring{\tau}))$ is the module of *horizontal vector fields* on $\mathring{T}M$. The *horizontal lift* of a vector field X on M to TM is $X^{\mathsf{h}} := \mathcal{H}(\widehat{X}) = \mathbf{h}X^{\mathsf{c}}$. In particular, the horizontal lifts of the coordinate vector fields $\frac{\partial}{\partial u^j}$ can be represented in the form

$$\left(\frac{\partial}{\partial u^j}\right)^{\mathsf{h}} = \frac{\partial}{\partial x^j} - N^i_j \frac{\partial}{\partial y^i}, \qquad j \in \{1, \dots, n\}. \tag{13}$$

The functions N^i_j, defined on $\tau^{-1}(U)$ and smooth on $\mathring{\tau}^{-1}(U)$, are called the *Christoffel symbols* of $\mathscr{H}$ with respect to the chosen chart; the minus sign in (13) is quite traditional.

The Berwald derivative. Let an Ehresmann connection $\mathscr{H}$ be given on TM, and let $\widetilde{X}$ be a section of $\mathring{\pi}$. Following the scheme of construction of $\nabla^{\mathsf{v}}_{\widetilde{X}}$, we define a differential operator $\nabla^{\mathsf{h}}_{\widetilde{X}}$, prescribing its action

$$\text{on } functions \text{ by } \nabla^{\mathsf{h}}_{\widetilde{X}} f := (\mathscr{H}\widetilde{X})f = (df \circ \mathscr{H})(\widetilde{X}), \ f \in C^\infty(\mathring{T}M); \qquad (14)$$

$$\text{on } sections \text{ by } \nabla^{\mathsf{h}}_{\widetilde{X}} \widetilde{Y} := \mathscr{V}[\mathscr{H}\widetilde{X}, \mathbf{i}\widetilde{Y}], \ \widetilde{Y} \in \mathfrak{X}(\mathring{\tau}),$$

and extending it to the whole tensor algebra $\mathcal{T}(\mathring{\pi})$ in such a way that $\nabla^{\mathsf{h}}_{\widetilde{X}}$ satisfies the derivation property. Finally, we define the ∇^{h}-*differential* of a tensor formally in the same way as in the vertical case. The operator ∇^{h} so obtained will be called the *h-Berwald derivative* determined by the Ehresmann connection $\mathscr{H}$.

Putting together the canonical v-covariant derivative and the h-Berwald derivative, we get an all-important covariant derivative operator in $\mathring{\pi}$, called the **Berwald derivative** associated to (or determined by) $\mathscr{H}$. Explicitly, for any vector field ξ on $\mathring{T}M$ and section $\widetilde{Y}$ in $\mathfrak{X}(\mathring{\tau})$,

$$\nabla_\xi \widetilde{Y} := \nabla^{\mathsf{v}}_{\mathscr{V}\xi}\widetilde{Y} + \nabla^{\mathsf{h}}_{\mathbf{j}\xi}\widetilde{Y} = \mathbf{j}[\mathbf{v}\xi, \mathscr{H}\widetilde{Y}] + \mathscr{V}[\mathbf{h}\xi, \mathbf{i}\widetilde{Y}]. \qquad (15)$$

Then, in particular,

$$\nabla_{\mathbf{i}\widetilde{X}}\widetilde{Y} = \nabla^{\mathsf{v}}_{\widetilde{X}}\widetilde{Y}, \quad \nabla_{\mathscr{H}\widetilde{X}}\widetilde{Y} = \nabla^{\mathsf{h}}_{\widetilde{X}}\widetilde{Y}; \quad \widetilde{X}, \widetilde{Y} \in \mathfrak{X}(\mathring{\tau}); \qquad (16)$$

$$\nabla_{X^{\mathsf{v}}}\widehat{Y} = 0, \quad \mathbf{i}\nabla_{X^{\mathsf{h}}}\widehat{Y} = [X^{\mathsf{h}}, Y^{\mathsf{v}}]; \quad X, Y \in \mathfrak{X}(M). \qquad (17)$$

From the last relation and (13) we obtain

$$\mathbf{i}\nabla_{\left(\frac{\partial}{\partial u^j}\right)^{\mathsf{h}}} \widehat{\frac{\partial}{\partial u^k}} = \frac{\partial N^i_j}{\partial y^k}\frac{\partial}{\partial y^i} \quad (j, k \in \{1, \ldots, n\});$$

the functions $N^i_{jk} := \frac{\partial N^i_j}{\partial y^k}$ are the *Christoffel symbols* of the Berwald derivative determined by $\mathscr{H}$.

Geometric data. By the *tension* of an Ehresmann connection we mean the h-Berwald differential of the canonical section, i.e., the type $\binom{1}{1}$ tensor $\mathbf{t} := \nabla^{\mathsf{h}}\delta$. An Ehresmann connection is said to be *homogeneous* if its tension vanishes. For any vector field X on M,

$$\mathbf{it}(\widehat{X}) = \mathbf{i}\nabla_{X^{\mathsf{h}}}\delta \overset{(14)}{=} \mathbf{i}\mathscr{V}[X^{\mathsf{h}}, C] = \mathbf{v}[X^{\mathsf{h}}, C] = [X^{\mathsf{h}}, C],$$

therefore for each $j \in \{1, \dots, n\}$,

$$\mathbf{it}\left(\widehat{\frac{\partial}{\partial u^j}}\right) = \left[\left(\frac{\partial}{\partial u^j}\right)^{\mathsf{h}}, C\right] \overset{(13)}{=} \left[\frac{\partial}{\partial x^j} - N^i_j \frac{\partial}{\partial y^i}, C\right] = (CN^i_j - N^i_j)\frac{\partial}{\partial y^i}.$$

It follows that *an Ehresmann connection is homogeneous, if and only if, its Christoffel symbols are positive-homogeneous of degree 1.*

The canonical surjection $\mathbf{j}$ in (3) may be interpreted as a $\mathring{\pi}$-valued 1-form on $\mathring{T}M$, so we can form its covariant exterior derivative $d^{\nabla}\mathbf{j}$ with respect to the Berwald derivative associated to an Ehresmann connection $\mathscr{H}$. By the *torsion of* $\mathscr{H}$ we mean the type $\binom{1}{2}$ tensor $\mathbf{T}$ defined by

$$\mathbf{T}(\widetilde{X}, \widetilde{Y}) := d^{\nabla}\mathbf{j}(\mathscr{H}\widetilde{X}, \mathscr{H}\widetilde{Y}); \quad \widetilde{X}, \widetilde{Y} \in \mathfrak{X}(\mathring{\tau}).$$

Evaluating at basic sections $\widehat{X}$, $\widehat{Y}$, we get the more suggestive formula

$$\mathbf{i}\mathbf{T}(\widehat{X}, \widehat{Y}) = [X^{\mathsf{h}}, Y^{\mathsf{v}}] - [Y^{\mathsf{h}}, X^{\mathsf{v}}] - [X, Y]^{\mathsf{v}}.$$

h-exterior and v-Lie derivative. Having specified an Ehresmann connection $\mathscr{H}$, we may define the operator d^{h} of the *h-exterior derivative* with respect to $\mathscr{H}$, prescribing its action

$$\begin{aligned} &\text{on } \textit{functions} \text{ by} && d^{\mathsf{h}} f(\widetilde{X}) = \nabla^{\mathsf{h}}_{\widetilde{X}} f; \quad f \in C^\infty(\mathring{T}M), \widetilde{X} \in \mathfrak{X}(\mathring{\tau}); \\ &\text{on } \textit{basic 1-forms} \text{ by} && d^{\mathsf{h}}(\alpha \circ \tau) := (d\alpha) \circ \tau, \quad \alpha \in \mathfrak{X}^*(M). \end{aligned} \tag{18}$$

Then d^{h} is a graded derivation of the Grassmann algebra $\Omega(\mathring{\tau})$ of differential forms along $\mathring{\tau}$. It may be shown that *if the torsion of the Ehresmann connection vanishes, then for any k-form $\widetilde{\alpha}$ along $\mathring{\tau}$ we have*

$$d^{\mathsf{h}}\widetilde{\alpha} = (k+1)\mathrm{Alt}\nabla^{\mathsf{h}}\widetilde{\alpha} \tag{19}$$

(cf. (12)), *and*

$$d^{\mathsf{h}} \circ d^{\mathsf{v}} + d^{\mathsf{v}} \circ d^{\mathsf{h}} = 0. \tag{20}$$

For a proof, see Ref. 21.

To complete our calculus summary, we define a kind of *Lie derivative* $\mathcal{L}^{\mathsf{v}}_{\xi}$ in $\mathfrak{T}(\mathring{\pi})$ $(\xi \in \mathfrak{X}(\mathring{T}M))$ by

$$\begin{aligned} \mathcal{L}^{\mathsf{v}}_{\xi} f &:= \xi(f), \quad \text{if } f \in C^\infty(\mathring{T}M); \\ \mathcal{L}^{\mathsf{v}}_{\xi}\widetilde{Y} &:= \mathscr{V}[\xi, \mathbf{i}\widetilde{Y}], \quad \text{if } \widetilde{Y} \in \mathfrak{X}(\mathring{\tau}) \end{aligned} \tag{21}$$

where $\mathscr{V}$ is the vertical map associated to $\mathscr{H}$.

Then, in particular, for any vector fields X, Y on M,

$$\mathbf{i}\mathcal{L}^{\mathsf{v}}_{X^{\mathsf{c}}}\widehat{Y} := \mathbf{i}\mathscr{V}[X^{\mathsf{c}}, Y^{\mathsf{v}}] = \mathbf{v}[X^{\mathsf{c}}, Y^{\mathsf{v}}] = [X^{\mathsf{c}}, Y^{\mathsf{v}}] = \mathcal{L}_{X^{\mathsf{c}}}Y^{\mathsf{v}},$$

so our generalization of the classical Lie derivative is reasonable. This is also confirmed by the fact that the operators $\mathcal{L}^{\mathsf{v}}$, d^{v} and the substitution operator $i_{\widetilde{X}}$ ($\widetilde{X} \in \mathfrak{X}(\mathring{\tau})$) are related by

$$\mathcal{L}^{\mathsf{v}}_{\mathbf{i}\widetilde{X}} = i_{\widetilde{X}} \circ d^{\mathsf{v}} + d^{\mathsf{v}} \circ i_{\widetilde{X}}, \tag{22}$$

which is a just a mutation of H. Cartan's magic formula.

Homogeneous tensors. Let k be an integer. A covariant or a $\mathring{\pi}$-valued covariant tensor $\widetilde{\alpha}$ along $\mathring{\tau}$ is said to be *homogeneous of degree* k, briefly k*-homogeneous*, if $\nabla_C\widetilde{\alpha} = k\alpha$.

Lemma 4.1. *If* $\widetilde{\alpha}$ *is a 0-homogeneous 2-form along* $\mathring{\tau}$, *then we have*

$$\widetilde{\alpha} = \frac{1}{2}(i_\delta d^{\mathsf{v}}\widetilde{\alpha} + d^{\mathsf{v}} i_\delta \widetilde{\alpha}).$$

Proof. By 'Cartan's formula' (22),

$$i_\delta d^{\mathsf{v}}\widetilde{\alpha} + d^{\mathsf{v}} i_\delta\widetilde{\alpha} = \mathcal{L}^{\mathsf{v}}_{\mathbf{i}\delta}\widetilde{\alpha} = \mathcal{L}^{\mathsf{v}}_{C}\widetilde{\alpha}.$$

For any vector fields X, Y on M,

$$\begin{aligned}\mathcal{L}^{\mathsf{v}}_C\widetilde{\alpha}(\widehat{X},\widehat{Y}) &= C\widetilde{\alpha}(\widehat{X},\widehat{Y}) - \widetilde{\alpha}(\mathcal{L}^{\mathsf{v}}_C\widehat{X},\widehat{Y}) - \widetilde{\alpha}(\widehat{X},\mathcal{L}^{\mathsf{v}}_C\widehat{Y}) \overset{(21)}{=} \\ &= C\widetilde{\alpha}(\widehat{X},\widehat{Y}) - \widetilde{\alpha}(\mathscr{V}[C,X^{\mathsf{v}}],\widehat{Y}) - \widetilde{\alpha}(\widehat{X},\mathscr{V}[C,Y^{\mathsf{v}}]) \overset{(6)}{=} \\ &= C\widetilde{\alpha}(\widehat{X},\widehat{Y}) + 2\widetilde{\alpha}(\widehat{X},\widehat{Y}) = \\ &= (\nabla_C\widetilde{\alpha})(\widehat{X},\widehat{Y}) + 2\widetilde{\alpha}(\widehat{X},\widehat{Y}) = 2\widetilde{\alpha}(\widehat{X},\widehat{Y}).\end{aligned}$$

This proves our assertion. □

5. Curvatures of an Ehresmann connection

Throughout this section we assume that an Ehresmann connection $\mathscr{H}$ *is specified on* TM. ∇ *is the Berwald derivative determined by* $\mathscr{H}$, d^{∇} *is the corresponding covariant exterior derivative. We denote by* R^{∇} *the curvature of* ∇.

Affine curvatures. The vertical map $\mathscr{V}$ associated to $\mathscr{H}$ may also be interpreted as a $\mathring{\pi}$-valued 1-form on $\mathring{T}M$, so it makes sense to speak of its covariant exterior derivative. By the *fundamental affine curvature of* $\mathscr{H}$ we mean the type $\binom{1}{2}$ tensor field $\mathbf{R}$ along $\mathring{\tau}$ defined by

$$\mathbf{R}(\widetilde{X},\widetilde{Y}) := d^{\nabla}\mathscr{V}(\mathscr{H}\widetilde{X},\mathscr{H}\widetilde{Y}); \qquad \widetilde{X},\widetilde{Y} \in \mathfrak{X}(\mathring{\tau}). \tag{23}$$

Then for vector fields X, Y on M we get

$$\mathbf{iR}(\widehat{X},\widehat{Y}) = -\mathbf{v}[X^{\mathsf{h}}, Y^{\mathsf{h}}], \tag{24}$$

so $\mathbf{R}$ is just the obstruction tensor to integrability of the horizontal distribution $\operatorname{Im}(\mathscr{H}) \subset T\mathring{T}M$.

We define the *affine curvature tensor* $\mathbf{H}$ of the Ehresmann connection by

$$\mathbf{H}(\widetilde{X},\widetilde{Y})\widetilde{Z} := R^{\nabla}(\mathscr{H}\widetilde{X},\mathscr{H}\widetilde{Y})\widetilde{Z}. \tag{25}$$

It is related to the fundamental affine curvature by

$$\mathbf{H}(\widetilde{X},\widetilde{Y})\widetilde{Z} = (\nabla^{\mathsf{v}}_{\widetilde{Z}}\mathbf{R})(\widetilde{X},\widetilde{Y}). \tag{26}$$

We remark that the terms 'fundamental affine curvature' and 'affine curvature' are borrowed from Berwald's paper,[5] whose terminology we try to adopt as far as possible.

The Berwald curvature. An immediate calculation shows that for any sections $\widetilde{X},\widetilde{Y},\widetilde{Z}$ along $\mathring{\tau}$, $R^{\nabla}(\mathbf{i}\widetilde{X},\mathbf{i}\widetilde{Y})\widetilde{Z} = 0$. However, the type $\binom{1}{3}$ tensor $\mathbf{B}$ along $\mathring{\tau}$ defined by

$$\mathbf{B}(\widetilde{X},\widetilde{Y})\widetilde{Z} := R^{\nabla}(\mathbf{i}\widetilde{X},\mathscr{H}\widetilde{Y})\widetilde{Z}, \tag{27}$$

and called the *Berwald curvature* of the Ehresmann connection, comprises important new information. Notice first that for any vector fields X, Y, Z on M,

$$\mathbf{iB}(\widehat{X},\widehat{Y})\widehat{Z} = [[X^{\mathsf{v}}, Y^{\mathsf{h}}], Z^{\mathsf{v}}]. \tag{28}$$

From this relation, using the Jacobi identity and the fact that the Lie brackets of vertically lifted vector fields vanish, we infer: *the Berwald curvature is symmetric in its first and third variable. If,* in addition, *the Ehresmann connection is homogeneous or torsion-free, then the Berwald curvature is totally symmetric.*[21]

To clarify the meaning of the Berwald curvature, we recall that any covariant derivative operator D on M gives rise to an Ehresmann connection $\mathscr{H}_D$ on TM by the rule

$$(v,w) \in TM \times_M TM \longrightarrow \mathscr{H}_D(v,w) := Y_*(w) - \mathbf{i}(v, D_wY) \in TTM,$$

where $Y \in \mathfrak{X}(M)$ is any vector field such that $Y(\tau(v)) = v$. Then $\mathscr{H}_D$ *is homogeneous and has vanishing Berwald curvature.* The torsions T^D of D and $\mathbf{T}$ of $\mathscr{H}_D$ are related by

$$(T^D(X,Y))^{\mathsf{v}} = \mathbf{i}\mathbf{T}(\widehat{X},\widehat{Y}); \qquad X,Y \in \mathfrak{X}(M), \tag{29}$$

while the relation between the curvature R^D of D and the affine curvature $\mathbf{H}$ of $\mathscr{H}_D$ is given by

$$(R^D(X,Y)Z)^{\mathsf{v}} = \mathbf{i}\mathbf{H}(\widehat{X},\widehat{Y})\widehat{Z}; \qquad X,Y,Z \in \mathfrak{X}(M). \tag{30}$$

Conversely, if $\mathscr{H}$ is a *homogeneous* Ehresmann connection *of class* C^1 on $TM \times_M TM$, then there exists a (necessarily unique) covariant derivative D on M such that for any vector fields X, Y on M we have

$$(D_X Y)^{\mathsf{v}} = \mathbf{i}\nabla^{\mathsf{h}}_{\widehat{X}}\widehat{Y} = [X^{\mathsf{h}}, Y^{\mathsf{v}}]. \tag{31}$$

The Ehresmann connection arising from D is just the given connection $\mathscr{H}$, therefore *the Berwald curvature of* $\mathscr{H}$ *vanishes.*

6. Semisprays and Ehresmann connections

A semispray S over M is said to be a *spray* if it is of class C^1 on TM and positive-homogeneous of degree 2 (briefly, 2^+-homogeneous) in the sense that $[C,S] = S$. By an *affine spray* we mean a spray which is of class C^2 (and hence is smooth) on its whole domain TM. Equivalently, an affine spray is a 2^+-homogeneous second-order vector field.

There is an important relation between affine sprays and covariant derivatives, formulated clearly and explicitly by Ambrose, Palais and Singer[1] first. In our terms, *if* $S \in \mathfrak{X}(TM)$ *is an affine spray, then there exists a covariant derivative* D *on* M *such that* $S = \mathscr{H}_D \circ \delta$. Conversely, if D is a covariant derivative on M and $\mathscr{H}_D$ is the Ehresmann connection determined by D, then $S := \mathscr{H}_D \circ \delta$ is an affine spray. This construction was generalized to a great extent by M. Crampin and J. Grifone in the early 1970s, independently (see Ref. 6 and Ref. 9).

Theorem 6.1 (M. Crampin and J. Grifone). *If* S *is a semispray over* M*, then there exists a unique Ehresmann connection* $\mathscr{H}_S$ *on* TM *such that the horizontal lift of a vector field on* M *with respect to* $\mathscr{H}_S$ *is given by*

$$X^{\mathsf{h}} := \mathscr{H}_S(\widehat{X}) = \frac{1}{2}(X^{\mathsf{c}} + [X^{\mathsf{v}}, S]) \qquad \text{(Crampin's formula)}, \tag{32}$$

or, equivalently, the horizontal projector associated to $\mathscr{H}_S$ *is*

$$\mathbf{h}_S = \frac{1}{2}(1_{\mathfrak{X}(TM)} + [\mathbf{J}, S]) \qquad \text{(Grifone's formula)}. \tag{33}$$

We list some basic properties of $\mathscr{H}_S$.

CrGr 1. *The torsion of $\mathscr{H}_S$ vanishes.*

This can be checked by a direct calculation.

CrGr 2. *If $\mathscr{H}$ is an Ehresmann connection with vanishing torsion, then there is a semispray S such that $\mathscr{H} = \mathscr{H}_S$.*

This result of M. Crampin[7] is far from being trivial, see also E. K. Ayassou's *Thèse de doctorat.*[2]

CrGr 3. *$\mathscr{H}_S$ is homogeneous, if and only if, there is a 2^+-homogeneous semispray $\bar{S}$ and a vector field X on M such that $S = \bar{S} + X^{\mathsf{v}}$.*

For a proof we refer to Ref. 21.

CrGr 4. *The semispray $\mathscr{H}_S \circ \delta$ coincides with S, if and only if, $[C, S] = S$.*

Applying, e.g., Grifone's formula (33) and Grifone's identity (9), the proof of this claim is routine.

By the (affine and Berwald) *curvatures of a semispray S* we shall mean the (corresponding) curvatures of $\mathscr{H}_S$. When there is no risk of confusion, instead of $\mathscr{H}_S$ we shall simply write $\mathscr{H}$.

The *affine deviation tensor* (L. Berwald[5]) or the *Jacobi endomorphism* (W. Sarlet et al.[15]) of a semispray S is the type $\binom{1}{1}$ tensor $\mathbf{K}$ defined by

$$\mathbf{K}(\widetilde{X}) := \mathscr{V}[S, \mathscr{H}\widetilde{X}]; \qquad \widetilde{X} \in \mathfrak{X}(\mathring{\tau}), \quad \mathscr{H} := \mathscr{H}_S. \tag{34}$$

The Jacobi endomorphism and the fundamental affine curvature $\mathbf{R}$ of S are related by

$$\mathbf{R}(\widetilde{X}, \widetilde{Y}) = \frac{1}{3}(\nabla^{\mathsf{v}}\mathbf{K}(\widetilde{Y}, \widetilde{X}) - \nabla^{\mathsf{v}}\mathbf{K}(\widetilde{X}, \widetilde{Y})); \quad \widetilde{X}, \widetilde{Y} \in \mathfrak{X}(\mathring{\tau}), \tag{35}$$

for a proof see Ref. 21.

If S is a *spray*, then the tensors $\mathbf{K}$, $\mathbf{R}$ and $\mathbf{H}$ carry the same information: each of them can be expressed from another one. Adopting Z. Shen's terminology,[20] we say that a spray is *R-flat*, if one (hence every) of its affine curvatures vanishes. By a *flat spray* we mean an affine R-flat spray. Property **CrGr 3** and our remarks concerning the Berwald curvature at the end of the preceding section imply immediately the following

Characterization of affine sprays. *A spray is affine, if and only if, its Berwald curvature vanishes.*

The next important observation goes back to the 1920s, i.e., to the first golden age of the 'geometry of paths'.

Characterization of flat sprays. *A spray S is flat, if and only if, there is a coordinate system $(u^i)_{i=1}^n$ for M at any point $p \in M$ such that the*

coordinate expression of S in the induced coordinate system $(x^i, y^i)_{i=1}^n$ is $S = y^i \frac{\partial}{\partial x^i}$.

Then (x^i, y^i) will be called a *rectilinear coordinate system* on TM.

We sketch here the main steps of the proof, cf. Ref. 20: 8.1.6. The sufficiency of the condition is obvious. To prove the converse, suppose that S is flat. Then $\mathbf{B} = 0$ and $\mathbf{H} = 0$. Let the coordinate expression of S in an induced coordinate system $(x_0^i, y_0^i)_{i=1}^n$ $(x_0^i := u_0^i \circ \tau)$ be

$$S = y_0^i \frac{\partial}{\partial x_0^i} - 2\overset{\circ}{G}{}^i \frac{\partial}{\partial y_0^i}.$$

Condition $\mathbf{B} = 0$ implies the existence of a covariant derivative D on M satisfying (31). If the Christoffel symbols of D with respect to $(u_0^i)_{i=1}^n$ are the functions $\overset{\circ}{\Gamma}{}^i_{jk}$, then

$$\overset{\circ}{\Gamma}{}^i_{jk} \circ \tau = \overset{\circ}{G}{}^i_{jk} := \frac{\partial^2 \overset{\circ}{G}{}^i}{\partial y_0^j \partial y_0^k}; \qquad i, j, k \in \{1, \ldots, n\}. \tag{36}$$

Our assumption $\mathbf{H} = 0$ implies by (30) that $R^D = 0$. From this we infer that there is a chart $(U, (u^i)_{i=1}^n)$ on M such that in the u^i*-coordinates the Christoffel symbols of D vanish.* If the spray coefficients of S in the induced coordinate system $(x^i, y^i)_{i=1}^n$ are the functions G^i, then $0 = (\Gamma^k_{ij} \circ \tau) y^k \overset{(36)}{=} G^i_{jk} y^k = G^i_j$ and $0 = G^i_j y^j = 2G^i$ by the 2^+-homogeneity of the G^is, therefore $S \restriction \overset{\circ}{\tau}{}^{-1}(U) = y^i \frac{\partial}{\partial x^i}$.

We recall that two sprays S and $\bar{S}$ over M are said to be *projectively related* if there is a function P, smooth on $\overset{\circ}{T}M$, of class C^1 on TM such that $\bar{S} = S - 2PC$. Then the *projective factor* P is necessarily positive-homogeneous of degree 1. We shall find useful the following simple

Observation. *A spray S is projectively related to a given spray $\bar{S}$, if and only if, S is contained by the $C^\infty(\overset{\circ}{T}M)$-submodule of $\mathfrak{X}(\overset{\circ}{T}M)$ generated by $\bar{S}$ and C.*

Indeed, the necessity of the condition is obvious. Conversely, if $S = f\bar{S} + hC$ $(f, h \in C^\infty(\overset{\circ}{T}M))$, then using relation (8) we get $C = fC$. Hence f is the constant function with the value 1, and $S = \bar{S} + hC$, as we claimed.

7. Basic facts on Finsler functions

By a *Finsler function* we mean a function $F\colon TM \to \mathbb{R}$ satisfying the following conditions:

Fins 1. *F is smooth on $\mathring{T}M$.*
Fins 2. *$CF = F$, i.e., F is positive-homogeneous of degree 1.*
Fins 3. *The metric tensor $g := \frac{1}{2}\nabla^{\mathsf{v}}\nabla^{\mathsf{v}}F^2$ is (fibrewise) non-degenerate.*

A pair (M, F) consisting of a manifold and a Finsler function on its tangent manifold is called a *Finsler manifold.* A Finsler manifold (M, F) is *positive definite* if the condition

Fins 4. *$F(v) > 0$ for all $v \in \mathring{T}M$ and $F(0) = 0$*

is also satisfied. It may be shown (see Ref. 14) that the nomenclature is correct: **Fins 4** implies that the metric tensor is positive definite. It is immediately verified that a positive definite Finsler manifold (M, F) reduces to a Riemannian manifold (M, γ) in the sense that $g = \gamma \circ \tau$, if and only if, $\nabla^{\mathsf{v}}\nabla^{\mathsf{v}}\nabla^{\mathsf{v}}F^2 = 0$.

If F is a Finsler function, then $\theta := d_{\mathbf{J}}F$ is a 1-form, $\omega := dd_{\mathbf{J}}F^2$ is a 2-form on $\mathring{T}M$, called the *Hilbert 1-form* and the *fundamental 2-form* of (M, F), respectively. **Fins 3** implies that ω is non-degenerate, and the converse is also true. The 1-forms θ and $d_{\mathbf{J}}F^2$, and the 2-form ω have the following properties:

$$\mathcal{L}_C\theta = 0, \quad \mathcal{L}_C d_{\mathbf{J}}F^2 = d_{\mathbf{J}}F^2, \quad \mathcal{L}_C\omega = \omega; \tag{37}$$

$$i_C\omega = d_{\mathbf{J}}F^2; \tag{38}$$

$$\omega(\mathbf{J}\xi, \eta) + \omega(\xi, \mathbf{J}\eta) = 0 \quad (\xi, \eta \in \mathfrak{X}(\mathring{T}M)), \ \textit{i.e., the vertical endomorphism is skew-symmetric with respect to the fundamental 2-form.} \tag{39}$$

The next result is the miracle of Finsler geometry.

Proposition 7.1 (the fundamental lemma of Finsler geometry) and definition. *If (M, F) is a Finsler manifold, then there exists a unique spray S such that $i_S\omega = -dF^2$. The Ehresmann connection $\mathcal{H}$ determined by S according to Theorem 6.1 is homogeneous and* conservative *in the sense that $d^{\mathsf{h}}F = 0$. $\mathcal{H}$ is said to be the* canonical connection *of (M, F); this is the only torsion-free, homogeneous, conservative Ehresmann connection on TM.*

The first intrinsic formulation and index-free proof of the fundamental lemma is due to J. Grifone.[9] For our formulation and recent proofs we refer to Refs. 21,23.

Lemma 7.1. *If S is spray over M, then it is related to the canonical spray $\bar{S}$ of a Finsler manifold $(M, \bar{F})$ over $\mathring{T}M$ by*

$$\bar{S} = S - \frac{S\bar{F}}{\bar{F}}C - \bar{F}(i_S dd_{\mathbf{J}}\bar{F})^{\sharp} \tag{40}$$

where the sharp operator is taken with respect to the fundamental 2-form of $(M, \bar{F})$.

A proof of this useful observation can be found in Ref. 24.

8. Projectively Finslerian sprays

We say that a spray is *Finsler metrizable in a broad sense*, or *projectively Finslerian* if there is a Finsler function whose canonical spray is *projectively related* to the given spray. From relation (40) it follows at once that a spray S is projectively related to the canonical spray of a Finsler manifold $(M, \bar{F})$, if and only if, $i_S dd_{\mathbf{J}}\bar{F} = 0$. In this section we give a more illuminating derivation of this important observation, and deal with several equivalent characterizations of Finsler metrizability in a broad sense.

We need some preparatory results.

Lemma 8.1. *A function $\bar{F}\colon TM \to \mathbb{R}$ satisfying* ***Fins 1*** *and* ***Fins 2*** *is a Finsler function, if and only if, the 2-form $dd_{\mathbf{J}}\bar{F}$ on $\mathring{T}M$ is of rank $2n-2$.*

The proof is quite immediate.

Lemma 8.2. *If $\bar{F}$ is a Finsler function, then the nullspace of the 2-form $dd_{\mathbf{J}}\bar{F}$ is generated by the canonical spray of $\bar{F}$ and the Liouville vector field.*

By the preceding Lemma it is enough to check that the canonical spray and the Liouville vector field indeed belong to the nullspace, and this is a routine verification.

Lemma 8.3. *Let $\bar{F}\colon TM \to \mathbb{R}$ be a function satisfying* ***Fins 1*** *and* ***Fins 2****. Let S be a spray over M. If $\nabla = (\nabla^{\mathsf{h}}, \nabla^{\mathsf{v}})$ is the Berwald derivative determined by S, then relations*

$$i_S dd_{\mathbf{J}}\bar{F} = 0 \qquad \text{and} \qquad \nabla_S \nabla^{\mathsf{v}}\bar{F} = \nabla^{\mathsf{h}}\bar{F}$$

are both equivalent to the 'Euler–Lagrange equation'

$$S(X^{\mathsf{v}}\bar{F}) - X^{\mathsf{c}}\bar{F} = 0, \qquad X \in \mathfrak{X}(M). \tag{41}$$

The proof is just a calculation again: it is enough to evaluate the 1-form $i_S dd_J\bar{F}$ on $\mathring{T}M$ at a complete lift and the Finsler 1-form $\nabla_S\nabla^{\mathsf{v}}\bar{F} - \nabla^{\mathsf{h}}\bar{F}$ at a basic vector field.

Theorem 8.1. *Let S be a spray over M, and let $\nabla = (\nabla^{\mathsf{h}}, \nabla^{\mathsf{v}})$ be the Berwald derivative determined by S. If $\bar{F}$ is a Finsler function, then the following four statements are equivalent:*

Proj. *S and the canonical spray of $(M, \bar{F})$ are projectively related.*
Rap 1. $i_S dd_J\bar{F} = 0$.
Rap 2. $\nabla_S\nabla^{\mathsf{v}}\bar{F} = \nabla^{\mathsf{h}}\bar{F}$.
Rap 3. $d^{\mathsf{h}}d^{\mathsf{v}}\bar{F} = 0$.

If in an induced coordinate system $(x^i, y^i)_{i=1}^n$ on TM the spray coefficients of S are the functions G^i, and $G^i_j := \frac{\partial G^i}{\partial y^j}$, then the common local expression of **Rap 1** *and* **Rap 2** *is*

$$y^i \frac{\partial^2 \bar{F}}{\partial x^i \partial y^j} - 2G^i \frac{\partial^2 \bar{F}}{\partial y^i \partial y^j} = \frac{\partial \bar{F}}{\partial x^j} \qquad (j \in \{1, \dots, n\}), \tag{42}$$

while the local expression of **Rap 3** *is*

$$\frac{\partial^2 \bar{F}}{\partial x^j \partial y^k} - G^i_j \frac{\partial^2 \bar{F}}{\partial y^i \partial y^k} = \frac{\partial^2 \bar{F}}{\partial x^k \partial y^j} - G^i_k \frac{\partial^2 \bar{F}}{\partial y^i \partial y^j} \qquad (j, k \in \{1, \dots, n\}) \tag{43}$$

Proof. Let $\bar{S}$ be the canonical spray of $(M, \bar{F})$. By the Observation in the end of section 6, S and $\bar{S}$ are projectively related, if and only if, S is in the submodule generated by $\bar{S}$ and C. This submodule is just the nullspace of $dd_J\bar{F}$ by Lemma 8.2, so we obtain the equivalence of **Proj** and **Rap 1**. Since for any vector field X on M we have

$$d^{\mathsf{h}}d^{\mathsf{v}}\bar{F}(\delta, \widehat{X}) = S(X^{\mathsf{v}}\bar{F}) - X^{\mathsf{c}}\bar{F},$$

applying Lemma 8.3 we conclude that **Rap 3** implies **Rap 1**. To show the converse, observe that the 2-form $d^{\mathsf{h}}d^{\mathsf{v}}\bar{F}$ is 0^+-homogeneous, i.e., $\nabla_C d^{\mathsf{h}}d^{\mathsf{v}}\bar{F} = 0$, and hence it may be represented in the form

$$d^{\mathsf{h}}d^{\mathsf{v}}\bar{F} = \frac{1}{2}(i_\delta d^{\mathsf{v}}d^{\mathsf{h}}d^{\mathsf{v}}\bar{F} + d^{\mathsf{v}}i_\delta d^{\mathsf{h}}d^{\mathsf{v}}\bar{F}) \tag{44}$$

by Lemma 4.1. Using (20) and relation $d^{\mathsf{v}} \circ d^{\mathsf{v}} = 0$, it follows that the first term in the right-hand side of (44) vanishes. As to the second term, taking into account Lemma 8.3 again, we get for any vector field X on M

$$i_\delta d^{\mathsf{h}}d^{\mathsf{v}}\bar{F}(\widehat{X}) = d^{\mathsf{h}}d^{\mathsf{v}}\bar{F}(\delta, \widehat{X}) = S(X^{\mathsf{v}}\bar{F}) - X^{\mathsf{c}}\bar{F} = i_S dd_J\bar{F}(\widehat{X}),$$

hence assuming **Rap 1**, we conclude that $d^{\mathrm{h}} d^{\mathrm{v}} \bar{F} = 0$.
Our statements concerning the coordinate expressions can immediately be checked. □

Relations (42) and (43), as criteria for Finsler metrizability of a spray in a broad sense are due to A. Rapcsák.[18] The index-free form **Rap 1** and an equivalent of **Rap 3** were first published in Ref. 24. Later I realized that the equivalence of **Proj** and **Rap 1** has already been discovered by J. Klein and A. Voutier.[13] There is nothing new under the sun...

Now I am in a position to make some remarks about the following

Analytic version of Hilbert's 4th problem: *find the Finsler functions whose canonical spray is projectively related to a given* **flat spray**.

This reformulation of Hilbert's problem 4 differs very much from Hilbert's original formulation,[12] in which he immediately connected the problem to his actual research on the axiomatic foundations of geometry. However, the first thorough approach to the problem was analytic in G. Hamel's thesis.[10] Hamel's work was supervised by Hilbert, immediately after his famous lecture in Paris in 1900. This indicates that the reformulation of the problem in differential geometric terms alien to classical geometry was found relevant by Hilbert himself. The next corollary of Theorem 8.1 makes it clear that our formulation is just an index-free expression of Hamel's analytic interpretation of Hilbert's 4th problem.

Corollary 8.1. *A flat spray is (locally) projectively related to the canonical spray of a Finsler function $\bar{F}$, if and only if, one of the following relations holds in any rectilinear coordinate system:*

$$\textbf{Ham 1.} \qquad y^i \frac{\partial^2 \bar{F}}{\partial x^i \partial y^j} = \frac{\partial \bar{F}}{\partial x^j}, \qquad j \in \{1, \dots, n\}.$$

$$\textbf{Ham 2.} \qquad \frac{\partial^2 \bar{F}}{\partial x^i \partial y^k} = \frac{\partial^2 \bar{F}}{\partial x^k \partial y^j}, \qquad j, k \in \{1, \dots, n\}.$$

In Hamel's paper[11] equation **Ham 2** was derived and solved in 3 dimensions. In the 2-dimensional case an extremely elegant solution of **Ham 2** was given by A. V. Pogorelov;[17] see also Álvarez Paiva's delightful account.[3]

The next observation makes it possible to transform the problem into the quest of a suitable 0^+-homogeneous function instead of a 1^+-homogeneous Finsler function.

Proposition 8.1. *A flat spray S is (locally) projectively Finslerian, if and only if, there is a 0^+-homogeneous function f on TM, smooth on $\mathring{T}M$, such that $\nabla^{\mathsf{v}}\nabla^{\mathsf{v}}(Sf)^2$ is a non-degenerate 2-form along $\mathring{\tau}$, and in any rectilinear coordinate system $(x^i, y^i)_{i=1}^n$ on TM we have*

$$\frac{\partial f}{\partial y^j \partial x^k} y^k = 0, \qquad j \in \{1, \dots, n\}. \tag{45}$$

A sketchy solution of (45) can be found in Rapcsák's paper,[19] but his solution is probably incomplete. On the other side, an interesting new strategy has been shown to solve Hilbert's fourth problem by Álvarez Paiva in Ref. 4. Notice that the Liouville vector field and the given spray S generate an integrable distribution, and we can (at least locally) take the quotient by its leaves. The result is a manifold of dimension $2n-2$, the so-called *path space*, whose points represent the unparametrized geodesics of S. It follows from Lemma 8.2 that $dd_{\mathbf{J}}\bar{F}$ defines a symplectic 2-form on the path space. The idea of Álvarez Paiva is to specify the properties of the symplectic form on the path space. It would be illuminating to establish an exact relation between the solutions of (45) and the general form of Álvarez Paiva's 'admissible symplectic forms'.

Acknowledgments

Supported by National Science Research Foundation OTKA No.NK68040.

I am grateful to Mike Crampin for the stimulating conversations during the Conference in Olomouc, and for his '*Some thoughts on Hilbert's 4th problem*' shared with me after the Conference.

References

1. W. Ambrose, R.S. Palais and I.M. Singer, Sprays, *An. Acad. Bras. Ciênc.* **32** (1960) 163–178.
2. E. K. Ayassou, Cohomologies définies pour une 1-forme vectorielle plate, Thèse de Doctorat, Université de Grenoble, 1985.
3. J.C. Álvarez Paiva, Hilbert's fourth problem in two dimensions, *In: Mass Selecta: teaching and learning advanced undergraduate mathematics* (Amer. Math. Soc., Providence, RI, 2003) 165–183.
4. J.C. Álvarez Paiva, Symplectic geometry and Hilbert's fourth problem, *J. Diff. Geom.* **69** (2005) 353–378.
5. L. Berwald, Ueber Finslersche und Cartansche Geometrie IV, *Ann. Math.* **48** (1947) 755–781.

6. M. Crampin, On horizontal distributions on the tangent bundle of a differentiable manifold, *J. London Math. Soc. (2)* **3** (1971) 178–182.
7. M. Crampin, Generalized Bianchi identities for horizontal distributions, *Math. Proc. Camb. Phil. Soc.* **94** (1983) 125–132.
8. M. de León and P.R. Rodrigues, *Methods of differential geometry in analytical mechanics* (North-Holland, Amsterdam, 1989).
9. J. Grifone, Structure presque tangente et connexions, I, *Ann. Inst. Fourier, Grenoble* **22** (1) (1972) 287–334.
10. G. Hamel, Über die Geometrien, in denen die Geraden die Kürzesten sind, Dissertation, Göttingen, 1901, pp. 90.
11. G. Hamel, Über die Geometrien, in denen die Geraden die Kürzesten sind, *Math. Ann.* **57** (1903) 231–264.
12. D. Hilbert, Mathematical Problems, *Bull. Am. Math. Soc.* **37** (4) (2000) 407–436.
13. J. Klein et A. Voutier, Formes extérieures génératrices de sprays, *Ann. Inst. Fourier, Grenoble* **18** (1) (1968) 241–260.
14. R.L. Lovas, A note on Finsler-Minkowski norms, *Houston J. Math.* **33** (2007) 701–707.
15. E. Martínez, J.F. Cariñena and W. Sarlet, Derivations of differential forms along the tangent bundle projection, *Diff. Geometry and its Applications* **2** (1992) 17–43.
16. E. Martínez, J.F. Cariñena and W. Sarlet, *Derivations of differential forms along the tangent bundle projection, II, Ibid.* **3** (1993) 1–29.
17. A.V. Pogorelov, *Hilbert's fourth problem* (Scripta Series in Mathematics, Winston and Sons, New York, 1979).
18. A. Rapcsák, Über die bahntreuen Abbildungen metrischer Räume, *Publ. Math. Debrecen* **8** (1961) 285–290.
19. A. Rapcsák, Die Bestimmung der Grundfunktionen projektiv-ebener metrischer Räume, *Publ. Math. Debrecen* **9** (1962) 164–167.
20. Z. Shen, *Differential Geometry of Spray and Finsler Spaces* (Kluwer Academic Publishers, Dordrecht, 2001).
21. J. Szilasi, A Setting for Spray and Finsler Geometry, *In: Handbook of Finsler Geometry* (Kluwer Academic Publishers, Dordrecht, 2003) 1183–1426.
22. J. Szilasi and Á. Győry, A generalization of a theorem of H. Weyl, *Rep. Math. Phys.* **53** (2004) 261–273.
23. J. Szilasi and R.L. Lovas, Some aspects of differential theories, *In: Handbook of Global Analysis* (Elsevier, 2007) 1071–1116.
24. J. Szilasi and Sz. Vattamány, On the Finsler-metrizabilities of spray manifolds, *Periodica Mathematica Hungarica* **44** (2002) 81–100.

Differential Geometry and its Applications
Proc. Conf., in Honour of Leonhard Euler, Olomouc, August 2007

Distance functions of Finsler spaces and distance spaces

Lajos Tamássy

Department of Mathematics, University of Debrecen,
4010 Debrecen, P.O. Box 12, Hungary
E-mail: tamassy@math.klte.hu

In a Finsler space $F^n = (M, \mathcal{F})$ with base manifold M and Finsler metric (fundamental function) $\mathcal{F}$ the basic notion is the arc length of curves $[a, b] \to c(t) \subset M$ given by $\int_a^b \mathcal{F}(c, \dot{c})dt$. The distance $\varrho^F(p, q)$ of two points $p, q \in M$ is a derived notion defined as the infimum of the arc length of the curves from p to q.

Contrarily to this, in a distance space (M, ϱ) the distance $\varrho : M \times M \to R^+$, $(p, q) \mapsto \varrho(p, q)$ is the basic notion. According to the theorem of H. Busemann and W. Mayer not only $\mathcal{F}$ determines a distance function ϱ^F, but this ϱ^F also uniquely determines the Finsler metric $\mathcal{F}$ (and thus the Finsler space F^n). Thus the relation $F \leftrightarrow \varrho^F$ is $1 : 1$.

It is easy to see that not every ϱ determines a Finsler metric $\mathcal{F}$. Those, which do this (and have certain further properties of ϱ^F) are denoted by ϱ^*. But even if a ϱ^* determines an $\mathcal{F}$, it may happen (Theorem 3.1) that $\varrho^* \mapsto \mathcal{F} \mapsto \varrho^F \neq \varrho^*$, that is the relation between ϱ^* and $\mathcal{F}$ is not $1 : 1$. Thus $\{\varrho\} \supset \{\varrho^*\} \supset \{\varrho^F\}$. Here both signs mean proper containment. Then we find necessary and sufficient conditions (Theorems 4.2 and 4.3) in order that a ϱ^* be a ϱ^F.

A more detailed text and complete proofs on these can be seen in Ref. 1.

Keywords: Finsler metric from distance function.

MS classification: 53C60, 51K05, 51K10.

1. Introduction

A Finsler space $F^n = (M, \mathcal{F})$ is a couple consisting of an n-dimensional manifold M, which here will be supposed to be connected, and a fundamental function (Finsler metric) (see Ref. 2, Chap. 1)

$$\mathcal{F} : TM \to R^+, \quad (p, y) \mapsto \mathcal{F}(p, y), \qquad p \in M, \quad y \in T_pM,$$

which must satisfy the following properties:

(F i) regularity: $\mathcal{F} \in C^\infty$ if $y \neq 0$, and $\mathcal{F} \in C^\circ$ for $y = 0$

(F ii) positive homogeneity: $\mathcal{F}(p, \lambda y) = \lambda \mathcal{F}(p, y)$, $\lambda \in R^+$
(F iii) strong convexity: $\mathcal{F}(p, y_1 + y_2) < \mathcal{F}(p, y_1) + \mathcal{F}(p, y_2)$,

$$y_2 \neq \lambda y_1, \qquad \lambda \in R.$$

A property stronger than (F ii) is

(F iv) absolute homogeneity: $\mathcal{F}(p, \lambda y) = |\lambda| \mathcal{F}(p, y)$, $\lambda \in R$.

$\|y\| = \mathcal{F}(p, y)$ is the Finsler norm of $y \in T_pM$.
The arc length of the curve $[a, b] \to c(t) \subset M$ is defined by

$$s = \int_a^b \mathcal{F}(c, \dot{c}) dt.$$

This is clearly a generalization of the arc length of $c(t)$ in a Riemann space $V^n = (M, g)$:

$$s = \int_a^b \langle c, \dot{c} \rangle_g dt.$$

(F ii) is equivalent to the invariance of the arc length s with respect to orientation preserving parameter transformations of the curve $c(t)$. (F iv) is equivalent to the invariance of s with respect to any parameter transformation of $c(t)$. (F iv) has also the consequence $\mathcal{F}(p, y) = \mathcal{F}(p, -y)$, $\forall\, y$. This means the symmetry of the indicatrix $I(p) \subset T_pM$ defined by

$$I(p_0) := \{y \in T_{p_0}M \mid \mathcal{F}(p_0, y) = 1\}.$$

$I(p)$ plays the role of the unit sphere S^{n-1} of the Euclidean space E^n.

A Minkowski space $\mathcal{M}^n$ is a special Finsler space $(R^n, \mathcal{F})$ over the coordinate space R^n, with the property that on R^n there exists a coordinate system (x), called adapted coordinate system, in which $\mathcal{F}$ is independent of the point $x : \mathcal{F} = \mathcal{F}(y)$. Thus F^n makes each of its tangent spaces T_pM into a Minkowski space. If also absolute homogeneity (F iv) is supposed, then these tangent spaces become finite dimensional regular Banach spaces.

In a Finsler space F^n the distance $\varrho^F(p, q)$ from the point $p \in M$ to $q \in M$ is defined by

$$\varrho^F(p, q) := \inf_{\Gamma(p,q)} \int_a^b \mathcal{F}(c, \dot{c}) dt, \tag{1}$$

where $\Gamma(p, q)$ means the collection of the curves $c(t)$ with $c(a) = p$, $c(b) = q$.

On the other hand, a distance space (M, ϱ) is given by M, and the distance function[3]

$$\varrho : M \times M \to R^+, \qquad p, y \mapsto \varrho(p, q),$$

which may have the following properties:

(R i) $\varrho(p,q) \leq 0$, $\varrho(p,q) = 0 \Leftrightarrow p = q$, positive definiteness
(R ii) $\varrho(p,q) = \varrho(q,p)$, symmetry
(R iii) $\varrho(p,q) + \varrho(q,r) \geq \varrho(p,r)$, $p,q,r \in M$, triangle inequality.

A distance space with properties (R i, ii, iii) is called metric; with (R i) and (R ii) semi-metric; and with (R i) and (R iii) quasi-metric. Z. Shen calls this last class metric (Ref. 4, p. 72). Quasi-metric spaces often occur in the investigations of metrizability and other questions on topological spaces (see Refs. 5–7).

2. Properties of the distance function ϱ^F

It is easy to see that a ϱ^F defined by (1) is positive definite, that is, it satisfies (R i).

Nevertheless, without the absolute homogeneity (F iv) of $\mathcal{F}$ (R ii) may not be satisfied. Consider namely a Minkowski space $\mathcal{M}^n = (R^n, \mathcal{F})$ in an adapted coordinate system (x). In this coordinate system $\mathcal{F}(x,y)$ is independent of $x \equiv p$, and thus

$$\varrho^F(p,q) = \inf_{\Gamma(p,q)} \int_a^b \mathcal{F}(\dot{c})dt, \qquad c(a) = p, \quad c(b) = q.$$

Furthermore the geodesics of $\mathcal{M}^n$ in this coordinate system are the straight lines, for which $\dot{c}(t) \equiv \dot{x}(t) = \text{const.} = y_0$. Thus

$$\varrho^F(p,q) = \int_a^b \mathcal{F}(y_0)dt = |b-a|\mathcal{F}(y_0), \tag{a}$$

while $\varrho^F(q,p)$ is obtained by the integral of $\mathcal{F}(-y_0)$ from $q = c(b)$ to $p = c(a)$:

$$\varrho^F(q,p) = \left| \int_b^a \mathcal{F}(-y_0)dt \right| = \left| -\int_a^b \mathcal{F}(-y_0)dt \right| = |b-a|\mathcal{F}(-y_0). \tag{b}$$

(a) and (b) are equal on every line segment $\overline{pq}$ only if $\mathcal{F}(y_0) = \mathcal{F}(-y_0)$ holds. Thus the absolute homogeneity (F iv) is necessary to the symmetry (R ii) of ϱ^F. – However, absolute homogeneity of $\mathcal{F}$ is also sufficient. Namely $\Gamma(p,q) = \{c(t)\}$ and $\Gamma(q,p) = \{\bar{c}(\tau)\}$ consist of the same curves, only with opposite orientation: $\bar{c}(\tau) = c(t(\tau))$ $t = -\tau + a + b$. Thus in this case $\mathcal{F}(\bar{c}(\tau), \frac{d\bar{c}}{d\tau}) = \mathcal{F}(c(t), -\frac{dc}{dt}) = \mathcal{F}(c(t), \frac{dc}{dt})$, and hence $\varrho^F(p,q) = \varrho^F(q,p)$.

Finally (R iii) is always satisfied by ϱ^F. Indeed for any given $\varepsilon > 0$ there exist curves $c_1(t)$, $t \in [0,1]$, $c_1(0) = p$, $c_1(1) = q$, and $c_2(t)$, $t \in [1,2]$, $c_2(1) = q$, $c_2(2) = r$, such that

$$\int_0^1 \mathcal{F}(c_1, \dot{c}_1)dt < \varrho^F(p,q) + \frac{\varepsilon}{2}$$

and

$$\int_1^2 \mathcal{F}(c_2, \dot{c}_2)dt < \varrho^F(q,r) + \frac{\varepsilon}{2}.$$

Thus for $c_3(t) = c_1(t) \cup c_2(t), \quad t \in [0,2]$

$$\int_0^2 \mathcal{F}(c_3, \dot{c}_3)dt < \varrho^F(p,q) + \varrho^F(q,r) + \varepsilon, \qquad \forall \varepsilon > 0.$$

Since

$$\varrho^F(p,r) \overset{(1)}{=} \inf_{\Gamma(p,r)} \int_a^b \mathcal{F}(c, \dot{c})dt, \quad c(a) = p, \ c(b) = r$$

and $c_3 \in \Gamma(p,r)$, we obtain

$$\varrho^F(p,r) \leq \varrho^F(p,q) + \varrho^F(q,r).$$

So (M, ϱ^F) is metric, provided $\mathcal{F}$ is absolutely homogeneous, and it is quasi-metric if $\mathcal{F}$ is positively homogeneous only. In the sequel (M, ϱ) is supposed to be a quasi-metric space. Metric spaces (M, ϱ) are included as special cases.

We collect some further properties of ϱ^F. Let $B_{p_0}(r)$ be a geodesic ball of F^n centered at p_0 and of radius r. $\varrho^F(p_0, q)$, $q \in B_{p_0}(r)$ is the length of the geodesic arc $\overline{p_0, q}$, which is smaller than r. Thus, if $r < \varepsilon$, then $|\varrho^F(p_0,q) - \varrho^F(p_0,p_0)| < \varepsilon$, and hence $\varrho^F(p_0,q)$ is continuous at p_0.

Moreover, $\varrho^F(p_0,q) \in C^\infty$ at any $q \in B_{p_0}(r)$, $q \neq p_0$. Indeed, if we use in $B_{p_0}(r)$ a geodesic polar coordinate system $(v^1, \ldots, v^{n-1}, s)$, where $v^1, \ldots, v^{n-1}$ are directional coordinates of the geodesic ray emanating from p_0 and s is the arc length of the geodesic from p_0 to q, then $s = \varrho^F(p_0,q)$ is simply the value of the n-th coordinate, which is of C^∞.

Let $q(t) \subset B_{p_0}(r)$, $0 \leq t \leq a$, be a geodesic of F^n with $q(0) = p_0$, $\dot{q}(0) = y_0 \neq 0$. Thus

$$\lim_{t\to 0+}\left[\frac{d}{dt}\varrho^F(p_0, q(t))\right] = \lim_{t\to 0+}\left[\frac{d}{dt}\int_0^t \mathcal{F}(q(\tau), \dot{q}(\tau))d\tau\right] = \mathcal{F}(p_0, y_0) > 0, \quad (2)$$

where $\frac{d}{dt}\varrho^F(p_0, q(t)) = \frac{d}{dt}\Big|_{q(t),\dot{q}(t)} \varrho^F(p_0, q)$ is the directional derivative of ϱ^F at $q(t)$ in the direction $\dot{q}(t)$. Since directional derivatives depend on the point and the direction only, in (2) $q(t)$ can be replaced by any other curve $c(t)$ emanating from p_0, and having the same (one sided) tangent y_0 and $-y_0$ resp. at p_0. Thus

$$\lim_{t\to 0^+}\left[\frac{d}{dt}\varrho^F(p_0,c(t))\right]=\mathcal{F}(p_0,y_0). \tag{3}$$

This is the content of the Bussemann-Mayer theorem (Ref. 8, p. 186. In a more comfortable form in Ref. 2, p. 153 or Ref. 4, p. 72).

Thus we obtain:

(R iv) (a) $\varrho^F(p_0,q)\in C^\infty$ at $q=p_0$

(b) $\varrho^F(p_0,q)\in C^\infty$ in an open domain around, but without p_0

(c) there exists $\lim_{t\to 0}\frac{d}{dt}\Big|_{c(t),\dot c(t)}\varrho^F(p_0,q)$ for any curve $c(t)$, $0\le t\le b$ emanating from $p_0=c(0)$. The value of this limit is positive and of class C^∞ at p_0, $\dot c(0)\neq 0$, and of class C° if $\dot c(0)=0$.

It follows from the property of the directional derivative that

$$\text{(R v)}\qquad \lim_{t\to 0}\frac{d}{dt}\Big|_{c(t),\dot c(t)}\varrho^F(p_0,q)=\frac{1}{\lambda}\lim_{t\to 0}\frac{d}{dt}\Big|_{\bar c(t),\dot{\bar c}(t)}\varrho^F(p_0,q),\qquad \lambda\in R^+$$

if $\bar c(0)=c(0)$ and $\dot{\bar c}(0)=\lambda\dot c(0)$.

Let $c_1(t)$, $c_2(t)$, $c_3(t)$, $0\le t\le b$ be curves emanating from p_0 with non-null and non-parallel tangents $\dot c_1(0)=y_1$, $\dot c_2(0)=y_2$, $\dot c_3(0)=y_1+y_2$. By (4), (F iii) and (R iv,c) we have

$$\text{(R vi)}\qquad \begin{aligned}&\lim_{t\to 0}\frac{d}{dt}\Big|_{c_1(t),\dot c_1(t)}\varrho^F(p_0,q)+\lim_{t\to 0}\frac{d}{dt}\Big|_{c_2(t),\dot c_2(t)}\varrho^F(p_0,q)>\\ &\quad >\lim_{t\to 0}\frac{d}{dt}\Big|_{c_3(t),\dot c_3(t)}\varrho^F(p_0,q).\end{aligned}$$

(R iv–vi) hold also for $\varrho^F(q,p_0)$.

We can summarize these statements in

Proposition 2.1. *The distance function ϱ^F defined by (1) possesses the properties* (Ri, iii – vi). (Rii) *is added iff $\mathcal{F}$ is absolutely homogeneous.*

Distance functions with properties (R i, iii–vi) ≡ (R *) will be denoted by ϱ^*. Clearly not every ϱ satisfies (R *). Thus $\{\varrho^*\}$ is a proper part of $\{\varrho\}$ (over the same manifold).

3. Finsler spaces determined by quasi-metric spaces

Given a Finsler metric $\mathcal{F}$, formula (1) yields a function $\varrho^F(p,q)$. This ϱ^F satisfies (R i, iii), thus it is a quasi-metric distance function, and (M,ϱ^F) is a quasi-metric distance space:

$$\mathcal{F} \mapsto \varrho^F \quad \text{or} \quad F^n = (M,\mathcal{F}) \mapsto (M,\varrho^F).$$

But according to (3) (Busemann-Mayer theorem) from ϱ^F we can retrieve that Finsler metric $\mathcal{F}$, from which the ϱ^F was derived:

$$\varrho^F \mapsto \mathcal{F} \quad \text{or} \quad (M,\varrho^F) \mapsto (M,\mathcal{F}) = F^n.$$

Thus the relation between $\mathcal{F}$ and ϱ^F is $1:1$

$$\mathcal{F} \rightleftarrows \varrho^F \quad \text{or} \quad F^n = (M,\mathcal{F}) \rightleftarrows (M,\varrho^F).$$

Now consider a distance function ϱ^*. It satisfies (R *), and by (R iv,c) it determines an $\overline{\mathcal{F}}(p,y)$

$$\lim_{t\to 0\downarrow}\left[\frac{d}{dt}\varrho^*(p,c(t))\right] = \bar{\mathcal{F}}(p,y), \quad y = \lim_{t\to 0+}\dot{c}(t). \tag{4}$$

By (R iv,c) this $\bar{\mathcal{F}}$ is not negative, it is of class C^∞ for $y \neq 0$, and of class C° if $y=0$. Thus $\bar{\mathcal{F}}$ satisfies (F i). By (R v) it satisfies (F ii) too. Finally by (R vi) $\bar{\mathcal{F}}$ fulfills (F iii). Thus we obtain

Proposition 3.1. *By* (4) *every ϱ^* determines a Finsler metric $\bar{\mathcal{F}}$ or every quasi-metric distance space (M,ϱ^*) determines a Finsler space $(M,\bar{\mathcal{F}})$. $\bar{\mathcal{F}}$ is absolute homogeneous iff ϱ^* satisfies* (Rii).

Every ϱ^F satisfies (R *). Thus every ϱ^F is a ϱ^*. But the converse is not true. We know that

$$\varrho^F \mapsto \mathcal{F} \mapsto \varrho^F. \tag{5}$$

We show that there are ϱ^* for which

$$\varrho^* \mapsto \mathcal{F} \mapsto \varrho^F \neq \varrho^*, \tag{6}$$

and thus (5) is not satisfied by this ϱ^*. Such a ϱ^* is no ϱ^F. We remark that if in the case of $\varrho^* = \varrho^F$ we want to preserve (5), then for $\varrho^* \mapsto \mathcal{F}$ (4) is the only possibility.

Theorem 3.1. *$\{\varrho^F\}$ is a proper part of $\{\varrho^*\}$, and $\{\varrho^*\}$ is a proper part of $\{\varrho\}$: $\{\varrho^F\} \subset \{\varrho^*\} \subset \{\varrho\}$.*

$\{\varrho^*\} \subset \{\varrho\}$ is clear, for not every ϱ must satisfy (R iv–vi), e.g. $\varrho(p_0, q)$ does not need to be differentiable at $q \neq p_0$.

We show our theorem by giving examples, where $\varrho^* \neq \varrho^F$. In the first example M is 1-dimensional: $M = R^1$ $(n = 1)$. Then this will be extended to an example, where $M = R^n$.

Example 3.1. (R^1, ϱ). We define a function $\varrho(x_0, x)$ on $R \equiv R^1$, where x means canonical coordinates. Let $\varrho(0, x)$, $x \in [0, \infty)$ be a strictly increasing C^∞ function with strictly decreasing first derivative, satisfying $\varrho(0, 0) = 0$, and

$$\lim_{x \to 0+} \frac{d}{dx} \varrho(0, x) = 1 \tag{7}$$

(e.g. $\varrho(0, x) = \log(x + 1)$). We define ϱ for $\bar{x} < 0$ by

$$\varrho(0, \bar{x}) = \varrho(0, |\bar{x}|), \tag{8}$$

and for $x_0 \neq 0$ by

$$\varrho(x_0, x) = \varrho(0, x - x_0). \tag{9}$$

These mean that $\varrho(x_1, x_2)$ is positive definite (see (R i)), and the functions $\varrho(x_0, x)$ for different x_0 are parallel translates of each others.

Also (R ii) is satisfied. Namely by (8)

$$\varrho(0, b - a) = \varrho(0, a - b). \tag{10}$$

By (9) we can add the same value to both arguments of ϱ. Thus the left side of (10) equals $\varrho(a, b)$, and the right side equals $\varrho(b, a)$, what gives (R ii).

Our $\varrho(x_1, x_2)$ satisfies also (R iii). The proof is a little longer, therefore it is omitted. (Proofs omitted in this article can be found in Ref. 1.)

Thus (R, ϱ) is a metric space.

$c(t) \equiv x(t) = x_0 + at \subset R^1$, $0 \leq t$ is a curve on R. Now

$$\lim_{t \to 0} \frac{d}{dt}\Big|_{c(t), \dot{c}(t)} \varrho(x_0, x) = \lim_{t \to 0+} \frac{d}{dt} \varrho(x_0, x_0 + at) \overset{(3)}{=} \mathcal{F}(x_0, a),$$

and it is easy to see that also (R iv–vi) are satisfied ((R vi) with the sign of equality).

Thus, by (3) and Proposition 3.1, our ϱ defines a Finsler space $\bar{F^1} = (R^1, \bar{\mathcal{F}})$ ((F iii) with the sign of equality). By (3), (7), (8) and (9) $\bar{\mathcal{F}}(x_0, a) = \bar{\mathcal{F}}(x_0, -a)$. Thus $\bar{\mathcal{F}}$ is absolute homogeneous. Because of (3), (7) and (9) $\bar{\mathcal{F}}(x_0, a)$ is independent of x_0. Therefore $\bar{F^1}$ is a Minkowski space with symmetric indicatrix, and, because of $n = 1$ it is a Euclidean space E^1.

Hence $\varrho^{\bar{F}}(x_1,x_2) = |x_1 - x_2|$. We can suppose that $x_1 < x_2$. Nevertheless by the integral mean theorem

$$\varrho(x_1,x_2) = \int_{x_1}^{x_2} \varrho'(x_1,x)dx = |x_1 - x_2|\varrho'(x_1,x)|_{x_0}, \quad x_0 \in (x_1,x_2).$$

By (7) and (9) and the strict decrease of $\varrho'(x_1,x)$ on $x > x_1$ we obtain

$$\varrho'(x_1,x)|_{x_0} < 1 = \lim_{x \to x_0^+} \varrho'(x_0,x).$$

Thus, for $x_1 \neq x_2$, $\varrho(x_1,x_2) < |x_1 - x_2| = \varrho^{\bar{F}}(x_1,x_2)$, i.e. $\varrho \overset{(3)}{\longmapsto} \bar{\mathcal{F}} \overset{(1)}{\longmapsto} \varrho^{\bar{F}} \neq \varrho$.

Example 3.2. (R^n, ϱ) The previous result can be extended to $M = R^n$. Let us endow R^n with a Euclidean metric, and let (x) be a Cartesian coordinate system on it. We define $z = \varrho(0,x)$, $x \in R^n$ over each ray $x^i = r^i t$, $0 \leq t$ emanating from the origin 0, as in the previous paragraph $(n = 1)$. Then $z = \varrho(0,x)$ is given as a surface of revolution around the z axis in $R^{n+1}(x;z)$. We define $\varrho(x_0,x)$, $x_0 \neq 0$ by (8).

These $\varrho(x_0,x)$ satisfy (R i–vi). The fulfillment of (R i) is trivial, and (R ii) is immediate from the definition. (R iv–vi) easily follow from the previous considerations. The proof of the triangle inequality is omitted again.

Because of (9) the graphs θ of the distance functions $z = \varrho(x_0,x)$ constructed for different x_0 are parallel translates of each other. Therefore the Finsler metric $\bar{\mathcal{F}}(x,y)$ defined by our ϱ according to (3) is independent of x. This means that $\bar{F^n} = (R^n, \bar{\mathcal{F}})$ is a Minkowski space $\mathcal{M}^n$, and (x) is an adapted coordinate system for it. Let a be a unit vector of $T_{x_0}R^n$. Then $(ta, \varrho(x_0,ta)) \subset (T_{x_0}R^n) \times R^1(z)$, $0 \leq t$ is a curve on θ. Its (one sided) tangent at x_0 $(t = 0)$ is a tangent of θ too. For different a these tangents form a cone centered at x_0 and tangent to θ. Since θ is a surface of revolution, the cone is a rotation cone around the z axis with its cape at x_0, and thus its section with the hyperplane $z = 1$ is a sphere. The orthogonal projection of this sphere on $T_{x_0}R^n$ is the indicatrix $\mathcal{I}(x_0) = \{\bar{\mathcal{F}}(y) = 1 \mid y \in T_{x_0}R^n\}$, and it is again a sphere. Since different θ-s are parallel translates of each other, the different indicatrices are congruent spheres, and thus $\bar{F^n}$ is a Euclidean space E^n. Here again $\varrho \neq \varrho^F$.

Example 3.3. (a sketch) Similar examples can be constructed on a manifold M different from R^n, provided that M admits a locally Minkowski structure. This is possible iff M admits an open cover $M = \cup_\alpha U_\alpha$, and on each U_α there exists a coordinate system $\underset{\alpha}{(x)}$, such that the transitions

$\underset{\alpha}{(x)} \leftrightarrow \underset{\beta}{(x)}$ on $U_\alpha \cap U_\beta$ are linear (Ref. 9, section 2). The torus has this property, but the sphere does not (Ref. 10, p. 250; Ref. 2, p. 14).

4. Conditions for $\varrho = \varrho^F$

Examples of the previous section show that (R *) does not assure that (M, ϱ^*) and $(M, \bar{\mathcal{F}})$ with $\bar{\mathcal{F}}$ given by (3) yield the same distance, i.e. that

$$\varrho^*(p,q) = \inf_{\Gamma(p,q)} \int_a^b \bar{\mathcal{F}}(c(t), \dot{c}(t))dt \quad \text{or} \quad \varrho^* \overset{(8)}{\longmapsto} \bar{\mathcal{F}} \overset{(14)}{\longmapsto} \varrho^{\bar{F}} = \varrho^*. \qquad (11)$$

So we look for conditions assuring (11). Our line of thought is the following: In a Finsler space F^n we consider a minimizing geodesic $g(t)$ with arc length parameter t. Then the Finsler distance $\varrho^F(g(t_0), g(t))$ is the arc length $t - t_0$ of the geodesic. We will be looking for essential properties of geodesics of $(M, \bar{\mathcal{F}})$, which can be expressed by terms of an (M, ϱ^*). Such a property is the parallelism property, and a curve $p(t)$ having this property will be called parallelism curve. With the aid of these curves $p(t)$ we will be able to express conditions for $\varrho^* = \varrho^F$.

This means the following: In a Finsler space we have distance functions $\varrho^F(g(\tau), q)$ (we write τ if we consider a fixed point of g, and t if we consider the curve $g(t)$) measuring the Finsler distance from a fixed point $g(\tau)$ of g to an arbitrary point $q \in M$. For every fixed τ we have the graph of $z = \varrho^F(g(\tau), q)$. It is called distance (hyper-) surface and denoted by $\theta^F_{g(\tau)}$. It lies in $U \times Z \subset R^{n+1}$, $U \subset M^n$. For different τ the $\theta^F_{\varrho(\tau)}$ form a family of surfaces. By a lifting of $g(t)$ to the distance surfaces $\theta^F_{g(\tau)}$ we obtain in R^{n+1} a 1-parameter family of curves $\zeta_\tau(t)$ over g. The tangents of the curves of the family $\zeta_\tau(t)$ over any fixed point $g(\tau_1)$ turn out to be parallel. These tangents lie in the planes tangent to the corresponding distance surfaces $\theta^F_{g(\tau)}$ over a point $g(\tau_1)$ (see Figure 1).

This is called the "parallelism property". This is expressed in

Theorem 4.1. *If $g(t)$, $0 \le t \le T$ is a minimizing geodesic of a Finsler space $F^n = (M, \mathcal{F})$, then the lifts of $\dot{g}(\tau_1)$, $\tau_1 \in (0,T)$ to the tangent planes of the distance surfaces $\theta^F_{g(\tau)}$, $0 \le \tau \le \tau_1$ are parallel.*

We consider a quasi-metric space (M, ϱ^*) and an arbitrary curve $p(t) \subset M$. Along $p(t)$ we again have distance functions $\varrho^*(p(\tau), q)$, distance surfaces $\theta^\varrho_{p(\tau)}$, and a similarly constructed 1-parameter family of curves $\xi_\tau(t)$ over $p(t)$. Nevertheless, in general, the tangents of the curves of this family $\xi_\tau(t)$ over a point $p(t_0)$ are not parallel, the parallelism property is not satisfied. If it is still satisfied, then $p(t)$ is called a "*parallelism curve*". Parallelism curves give us the appropriate notion and tool to find conditions for $\varrho^* = \varrho^F$, this is what we are looking for.

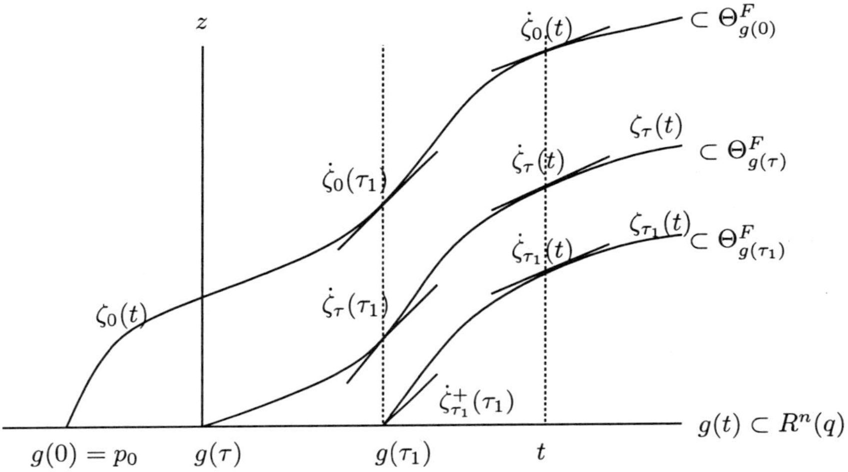

Figure 1

According to Theorem 4.1 in a Finsler space $F^n = (M, \mathcal{F})$ the parallelism property is satisfied along any minimizing geodesic.

We state some relations between quasi-metric distance spaces, their induced Finsler metric, and parallelism curves.

Proposition 4.1. *For any curve $c(t)$, $t \in [0, T)$ of a quasi-metric distance space (M, ϱ^*), and for the Finsler metric $\bar{\mathcal{F}}$ determined by ϱ^* according to (8) we obtain*

a) $\varrho^*(c(0), c(T)) \leq \int_0^T \bar{\mathcal{F}}(c(t), \dot c(t))dt,$

b) *If $c(t)$ is a parallelism curve, then*

$$\varrho^*(c(\tau), c(t)) = \int_\tau^t \bar{\mathcal{F}}(c(u), \dot c(u))du, \ \forall \tau, t, \ 0 \leq \tau < t < T. \tag{12}$$

c) *If along $c(t)$ (12) holds for $\forall\, \tau,\ t,\ 0 \leq \tau < t < T$, then $c(t)$ is a parallelism curve $p(t)$.*

From these we obtain the

Corollary 4.1. *In a Finsler space parallelism curves and minimizing geodesics coincide.*

Now we give conditions (Theorems 4.2 and 4.3) which assure that ϱ^* is the distance function of a Finsler space. By (3) a distance space (M, ϱ^*) always induces a Finsler metric $\mathcal{F}$ (we omit the $^-$ from $\bar{\mathcal{F}}$, for here we do not need to distinguish it from another $\mathcal{F}$), and the Finsler space $F^n = (M, \mathcal{F})$ determines a distance function ϱ^F.

Theorem 4.2. *If the distance function ϱ^* of a quasi-metric space (M, ϱ^*) with properties* (R$*$) *satisfies the parallelism property along any minimizing geodesic $g(t)$ of the Finsler space $F^n = (M, \mathcal{F})$, where $\mathcal{F}$ is obtained from ϱ^* by* (3)*, then ϱ^* is the distance function ϱ^F obtained from $\mathcal{F}$ by* (1)*. Also conversely, if $\varrho^* = \varrho^F$, then the parallelism property is satisfied along any minimizing geodesic.*

In Theorem 4.2 we required the parallelism property on curves determined by an F^n, and not on certain curves determined by the distance space (M, ϱ^*), so we applied an outer space too. Thus the characterization of $\varrho^* = \varrho^F$ was extrinsic. Now we replace the parallelism property (the essential condition of Theorem 4.2) by another condition, which is expressed directly in terms of (M, ϱ^*). This yields an intrinsic characterization of $\varrho^* = \varrho^F$.

Our tool for this will be the osculation curve. Let us consider in a Euclidean space E^n two points a, b. Let $S_a^E(t)$ be a sphere around a with radius $t < \varrho^E(a,b)$, and $S_b^E(\tau)$ a sphere around b with radius τ, such that they osculate each other from outside at a point $\sigma(t)$ (in an E^n $\tau = \varrho^E(a,b) - t$). If t runs from o to $\varrho^E(a,b)$, $\sigma(t)$ yields a curve. This curve is a segment of a straight line in E^n, and $\sigma(t)$ is a short geodesic if E^n is replaced by a Riemannian or Finslerian space, and the spheres of the E^n are replaced by geodesic spheres. This notion transplanted to distance spaces (M, ϱ^*) gives the *osculation curve*, which will help us to express the intrinsic condition for $\varrho^* = \varrho^F$.

Because of the triangle inequality (R iii)

$$d = \varrho^*(a,b) \leq \varrho^*(a, \sigma(t)) + \varrho^*(\sigma(t), b) = t + \tau. \tag{13}$$

Equality holds in special cases only.

Osculation curves of (quasi-) metric spaces are generalizations of the straight line and of the minimizing geodesic. If in (13) $d = t + \tau$, $\forall 0 < t < \varrho^*(a,b)$, then the line is called straight (Ref. 11, p. 58) or a Hilbert curve (Ref. 8, p. 176).

Finally we give an intrinsic criterion for the ϱ^* of a distance space (M, ϱ^*) to be the distance function ϱ^F of a Finsler space F^n.

Theorem 4.3. *(M, ϱ^*) is a distance space determined by a Finsler space $F^n = (M, \mathcal{F})$ (i.e. $\varrho^* = \varrho^F$) if and only if between any pair of points $a, b \in M$ there exists an osculation curve $\sigma(t; a, b)$ of (M, ϱ^*), along which the parallelism property is satisfied.*

Theorem 4.3 could be formulated also in such form that $\varrho^* = \varrho^F$ iff between any pair of points $a, b \in M$ there exists a parallelism curve of (M, ϱ^*). Our original formulation gives a little more than this. Namely an osculation curve $\sigma(t)$ is constructible, while a parallelism curve $p(t)$ is not. The proof of Theorem 4.3 is applicable in this last case too.

References

1. L. Tamássy, Relations between metric spaces and Finsler spaces, *Diff. Geom. Appl.*, to appear in 2008.
2. D. Bao, S.S. Chern and Z. Shen, *An Introduction to Riemann-Finsler Geometry* (Springer, New York, 2000).
3. L.M. Blumenthal, *Theory and Application of Distance Geometry* (Clarendon Press, Oxford, 1953).
4. Z. Shen, *Lecture notes on Finsler geometry* (Indiana Univ., 1998).
5. H.P.A. Künzi, Nonsymmetric distances and their associated topologies: About the origins of basic ideas, *In: Handbook of the History of General Topology* ((C.E. Aull and L. Lowen, Eds.) vol. 3, Kluwer, Dordrecht, 2001) 853–968.
6. T.G. Raghavan and I.L. Reilly, Metrizability of quasi-metric spaces, *J. London Math. Soc.* **15** (1977) 169–172.
7. S. Romaguera and M. Schellenkens, Quasi-metric properties of complexity spaces, *Topology. Appl.* **98** (1999) 311–323.
8. H. Busemann and W. Mayer, On the foundation of calculus of variation, *Trans. AMS.* **49** (1948) 173–198.
9. L. Tamássy, Point Finsler spaces with metrical linear connections, *Publ. Math. Debrecen* **56** (2000) 643–655.
10. D. Bao and S.S. Chern, A note on the Gauss-Bonnet theorem for Finsler spaces, *Ann. Math.* **143** (1996) 233–252.
11. W. Benz, Reelle Abstandräume und hyperbolische Geometrie, *Results. Math.* **34** (1998) 56–68.

PART 4

Geometric methods in physics

Differential Geometry and its Applications
Proc. Conf., in Honour of Leonhard Euler, Olomouc, August 2007

Differential forms relating twistors to Dirac fields

I.M. Benn

School of Mathematics and Physical Sciences, University of Newcastle, Newcastle NSW 2308, Australia
E-mail: Ian.Benn@newcastle.edu.au

J.M. Kress

School of Mathematics and Statistics, University of New South Wales, Sydney NSW 2052, Australia
E-mail: j.kress@unsw.edu.au

We look at first-order operators taking solutions to the twistor equation to solutions to the Dirac equation. This leads us to an interesting family of conformally-covariant equations for differential forms of arbitrary degree. This family of equations includes the conformally-covariant Laplace equation and the generalised (to middle-forms) Maxwell equation. First-order symmetry operators for this family give us second-order ones for the homogeneous members. We outline how this works in the conformally-flat case.

Keywords: Twistor equation, conformal Killing-Yano equation, symmetry operator, Dirac equation.

MS classification: 53Z05, 81Q05, 83C50.

1. Introduction

In a four-dimensional Lorentzian signatured space conformal-Killing-Yano tensors (or CKY tensors for short) are geometrically related to shear-free congruences, aligned with the conformal tensor. CKY tensors are also known as Killing forms, twistor forms or, in the four-dimensional case, Penrose-Floyd tensors.

In this four-dimensional case CKY tensors have many interesting consequences, amongst them the following. They underpin the construction of Debye and Hertz potentials for the Maxwell equations. They give rise to first-order symmetry operators for the Dirac equation, and second-order symmetry operators for the scalar wave equation and Maxwell's equations. It is natural to ask which of these relations generalise to arbitrary dimensions and signatures.

One way of addressing this question is via Penrose's helicity changing construction. In a conformally-flat four-dimensional space one can use a twistor to 'raise the helicity' of a scalar field to produce a solution to the massless Dirac equation. One can alternatively regard this construction as using a scalar field to produce a first-order operator that takes solutions to the twistor equation to solutions to the Dirac equation. This suggests a fruitful generalisation: what is the most general such operator?

We shall show how, in a conformally flat space of arbitrary dimension and signature, one can express the most general first-order operator taking twistors to massless Dirac solutions. Such operators involve forms of various degrees satisfying an interesting conformally-covariant equation. This equation includes the scalar wave equation and Maxwell-like equations. Since we know all the first-order symmetry operators for the Dirac equation, we can construct symmetry operators for this equation. This gives a generalised Hertz-Debye scheme.

2. Helicity Changing in Conformally-flat space

The twistor equation for a spinor u is

$$\nabla_X u - \frac{1}{n} X^\flat D u = 0 \tag{1}$$

where D is the Dirac operator

$$D = e^a \nabla_{X_a} .$$

Here ∇ denotes the spin connection, and the basis 1-forms $\{e^a\}$ act on a spinor with the Clifford action. The twistor equation is covariant under the following conformal scalings:

$$\hat{g} = e^{2\lambda} g \quad \hat{u} = e^{\frac{\lambda}{2}} u$$

It has the (rather severe) integrability conditions

$$\nabla_{X_a} D u = -\frac{n}{2(n-2)} P_a u + \frac{n\mathcal{R}}{4(n-1)(n-2)} e_a u \tag{2}$$

$$C_{ab} u = 0 . \tag{3}$$

P_a are Ricci 1-forms, C_{ab} the conformal 2-forms and $\mathcal{R}$ the curvature scalar.

Suppose that u satisfies the twistor equation (1), and ψ satisfies the (massless) Dirac equation:

$$\begin{aligned} \nabla_X u - \frac{1}{n} X^\flat D u &= 0 \\ D\psi &= 0 . \end{aligned} \tag{4}$$

Then if

$$f = (u, \psi)\,,$$

where (,) denotes the Spin-invariant inner product on spinors, in flat space we have

$$\triangle f = 0\,.$$

The twistor u lowers the helicity of ψ to produce the scalar f.[5] Since everything is conformally covariant, all this generalises to the conformally flat case.

Suppose now that we have a function f that satisfies Laplace's Equation, and a spinor u that satisfies the twistor equation:

$$\triangle f = 0$$
$$\nabla_X u - \frac{1}{n} X^\flat D u = 0\,.$$

Then if

$$\psi = df\,u + \frac{n-2}{n} f D u$$

we have, in flat space,

$$D\psi = 0\,.$$

This time the twistor u raises the helicity of f by 1/2.[5] Again conformal covariance ensures that all goes through in the conformally-flat case.

3. Dirac Symmetry Operators from Helicity Changing

Suppose that $D\psi = 0$. Then if u and v satisfy the twistor equation we can use u to lower the helicity of ψ, then v to raise it back up:[2]

$$\psi \longrightarrow f = (u, \psi) \longrightarrow df\,v + \frac{n-2}{n} f D v \equiv \psi'$$

So we have a first-order symmetry operator for the massless Dirac equation, taking one solution ψ to another ψ'. This operator can be written in terms of the (inhomogeneous in general) form K constructed from u and v,

$$K = v \otimes \bar{u}\,.$$

Floyd obtained the equation satisfied by this K in the 4-D Lorentzian case.[4] Floyd's equation had previously been given by Tachibana[7] generalising Yano's Killing tensor equation.[8]

4. Conformal Killing-Yano Tensors

The conformal-Killing-Yano equation for a p-form K can be written as[2]

$$\nabla_X K = \frac{1}{p+1} X \lrcorner dK - \frac{1}{n+1-p} X^\flat \wedge d^* K \,.$$

This equation is conformally covariant and when $p = 1$ this is just the conformal Killing equation. When we additionally have $d^* K = 0$ then this becomes the Killing-Yano equation.

The CKY equation is invariant under (Hodge) duality. So in four dimensions we only have (other than conformal-Killing vectors) CKY 2-forms, the best known example being in the Kerr Spacetime.

Whereas in general pairs of twistors do give CKY tensors,[6] not all CKY tensors are thus given. In fact, while the integrability conditions for the CKY equation are quite restrictive, they are not as restrictive as those required for twistors.

5. CKY Tensors and Dirac Symmetry Operators

The helicity changing route gave us a first-order Dirac symmetry operator that could be written in terms of a CKY tensor. In fact any CKY tensor gives a first-order symmetry operator of the massless Dirac equation, and all such symmetry operators are thus given:[1,3]

$$\mathcal{L}(K) = X^a \lrcorner K \nabla_{X_a} + \frac{p}{2(p+1)} dK - \frac{n-p}{2(n+1-p)} d^* K$$

(We remind that the forms act on the spinor by the Clifford action.)

For the case in which K is a 1-form the above expression reduces to the Lie derivative (with the appropriate conformal weight).

We will need these symmetry operators later, but there is also a lesson that we hope proves useful: the twistor construction was useful in guiding us to the form of the symmetry operator, but we still have the symmetry operator when we don't have the twistor solutions.

6. Another View of Helicity Raising

In flat space we can take the scalar f and spinor u satisfying

$$\triangle f = 0$$

$$\nabla_X u - \frac{1}{n} X^\flat D u = 0$$

to construct the spinor ψ,

$$\psi = df u + \frac{n-2}{n} f D u$$

which then satisfies

$$D\psi = 0\,.$$

We can think of the twistor u as raising the helicity-0 f to the helicity-1/2 ψ. Alternatively we can think of the function f as providing a first-order operator mapping solutions to the twistor equation to solutions to the (massless) Dirac equation. This suggests the question: "What is the most general first-order operator mapping solutions to the twistor equation to solutions to the (massless) Dirac equation?"

We may parameterize any first-order operator L on solutions to the twistor equation as

$$Lu = \omega D u + \Omega u\,.$$

Lu satisfies the massless Dirac equation when

$$(d - d^*)\omega + \frac{1}{n}(n - 2\Pi)\eta\Omega = 0$$
$$(d - d^*)\Omega + \frac{n}{n-2} P_a \wedge X^a \lrcorner \eta\omega - \frac{n(n+2\Pi-2)}{4(n-1)(n-2)}\mathcal{R}\eta\omega = 0$$

In general ω and Ω are inhomogeneous (consisting of forms of mixed degrees). The operators Π and η act on a form according to its degree, $\Pi\omega = \sum p\omega_p$ and $\eta\omega_p = (-1)^p\omega_p$. P_a are Ricci 1-forms and $\mathcal{R}$ the curvature scalar.

In (conformally) flat space it is necessary to satisfy these equations; but not in general.

We are guaranteed to have conformal covariance of these equations. (We simply use the separate, but different, conformal covariance of Dirac and twistor equations.)

In even dimensions we have a 0'th order operator with $\omega = 0$ and Ω a 'middle-form' satisfying $d\Omega = d^*\Omega = 0$. This is a 'helicity-lowering' operator.

The first equation expresses Ω in terms of derivatives of ω (except for the middle-form part). The second equation then gives a second-order equation for ω:

$$(n-2p-2)dd^*\omega + (n-2p+2)d^*d\omega$$
$$+(n-2p-2)(n-2p+2)\left(\frac{n+2p-2}{4(n-1)(n-2)}\mathcal{R}\omega - \frac{1}{n-2}P_a \wedge X^a \rfloor \omega\right) = 0\,.$$

Solutions to this are homogeneous in the degree of ω. We will (for want of a better name) refer to such solutions as 'potential forms'.

The equation for potential forms is (Hodge) duality invariant, and has the following conformal covariance:

$$\hat{g} = e^{2\lambda} g\,, \qquad \hat{\omega}_p = e^{(p+1-n/2)\lambda}\omega_p$$

The equation is also self-adjoint.

This family of equations (labelled by the degree p) includes some important special cases. For $p = 0$ we have the conformally-covariant Laplace equation. For $p = n/2 - 1$ (in even dimensions) ω is a potential for the middle-form-Maxwell equations. (For $p = n/2 + 1$ ω is a co-potential.)

7. Symmetry Operators for Potential Forms

Suppose that we somehow found a first-order symmetry operator L for the potential equation. We do not require that such an operator preserve the degree of the forms. Rather we envisage that we have an operator of degree q taking a p-form solution to a $p+q$-form one. Now the potential equation is homogeneous, and so the homogeneous parts of L would also be symmetry operators. Since the potential equation is self-adjoint, if L is a symmetry operator then so is L^*. If L is homogeneous of degree q then L^* is of degree $-q$. Therefore LL^* (and L^*L) is a second-order symmetry operator that is homogeneous of degree 0 (it preserves the degree of the forms it acts on).

Since the conformally-covariant Laplace equation and 'generalised' Maxwell equations (middle-form-Maxwell) are special cases of the potential equation, any such L would give a second-order symmetry operator for these equations.

In a conformally flat space we have such an L. The potential equation was arrived at by requiring that we had an operator $L(\omega)$ mapping twistors to Dirac solutions. If K is a CKY tensor then we have the Dirac symmetry operator $\mathcal{L}(K)$. Then $\mathcal{L}(K)L(\omega)$ maps twistors to Dirac solutions.

The integrability conditions for the twistor equation mean that all second derivatives can be expressed in terms of curvature. So we have another first-order operator. In conformally-flat space all such operators correspond to a potential form. i.e

$$\mathcal{L}(K)L(\omega) = L(\omega')$$

when acting on twistors, for some potential form ω'. So we must have some (first-order) symmetry operator $L(K)$,

$$L(K)\omega = \omega'\,.$$

8. Example: Hertz-Debye Potentials

We consider a specific example of how this scheme works (in a conformally-flat even-dimensional space).

Supose that f satisfies Laplace's equation, and that K is a CKY 'middle-form'. Then the above scheme gives us a co-potential ω' for the middle-form-Maxwell equations:

$$\begin{aligned} \omega' &= \frac{n}{2(n+2)} f dK + \frac{n}{2(n-2)} df \wedge K \\ dd^*\omega' &= 0\,. \end{aligned}$$

The harmonic function f is a 'Debye potential'[2] for the (generalised) Maxwell field $F = d^*\omega'$.

Now suppose that we start with a middle-form-Maxwell field F. We can use a middle-form CKY K to form a harmonic f,

$$f = K \cdot F$$

(where the $\cdot$ denotes the metric on middle-forms). So we have second-order symmetry operators on middle-form-Maxwell fields by going 'down then up': and on harmonic functions by going 'up then down'.

In this example we go 'down to the bottom' then 'up to the middle'. In sufficiently high dimensions we could go down some way, then back up. Thus any of the solutions to the 'potential equation' could act as generalised Hertz potentials.

9. Conclusions

By looking at the helicity changing scheme in a new way we were lead to the family of potential equations. These equations are interesting independent of their ancestory. They are a conformally-covariant family that includes the scalar Laplace equation and generalised (middle-form) Maxwell equation.

Any first-order symmetry operators for the family of potentials will give us homogeneous second-order symmetry operators, and hence symmetry operators for the scalar Laplace and generalised Maxwell equations. We have seen how in conformally-flat space we have such operators, and gave a specific example which was a generalised Hertz-Debye scheme.

In general there are potentially a range of such second-order operators. What we did not do was to express these operators directly in terms of the CKY tensors used in their construction. The important question we did not answer is: "Do (any of) these second-order operators give us symmetries without conformal flatness?" Clearly we should address this question.

References

1. I.M. Benn and Ph. Charlton, Dirac symmetry operators from conformal Killing-Yano tensors, *Class. Quantum Gravity* **14** (1997) 1037–1042.
2. I.M. Benn, Ph.R. Charlton and J.M. Kress, Debye Potentials for Maxwell and Dirac Fields from a Generalisation of the Killing-Yano Equation, *J. Math. Phys.* **38** (1997) 4504–4527; and references therein.
3. I.M. Benn and J.M. Kress, First-Order Dirac Symmetry Operators, *Class. Quantum Grav.* **21** (2004) 427–431.
4. N. Kamran and R.G. McLenaghan, Symmetry operators for the neutrino and Dirac fields on curved spacetime, *Physical Review D* **30** (1984) 357–362.
5. R. Penrose and W. Rindler, *Spinors and Space-time* (volume 2, Cambridge University Press, Cambridge, 1986).
6. U. Semmelmann, Conformal Killing forms on Riemannian manifolds, *Math. Z.* **245** (2003) 503–527.
7. S. Tachibana, On conformal Killing tensor in a Riemannian space, *Tôhoku Mathematical Journal* **21** (1969) 56–64.
8. K. Yano, Some remarks on tensor fields and curvature, *Annals of Mathematics* **55** (1952) 328–347.

Differential Geometry and its Applications
Proc. Conf., in Honour of Leonhard Euler, Olomouc, August 2007

The gravitational field of the Robertson-Walker spacetime

Włodzimierz Borgiel

Faculty of Mathematics and Information Science,
Warsaw University of Technology,
Plac Politechniki 1, 00-661 Warsaw, Poland
E-mail: borgie@mini.pw.edu.pl

The purpose of this paper is to study a gravitational field of the Robertson-Walker spacetime. The type of the gravitational field is determined by the characteristic of λ-tensor.

Keywords: Gravitational field, Robertson-Walker spacetime.

MS classification: 58A05, 58D17, 83C15.

1. Introduction

A spacetime (M, g) which is locally isometric to a 4-dimensional infinitesimally isotropic Lorentzian manifold is called a Robertson-Walker spacetime. The metric given by Equation

$$g = -dt^2 + \boldsymbol{a}^2(t)\left(\frac{1}{1-\varepsilon r^2}dr^2 + r^2(d\theta^2 + \sin^2(\theta)d\phi^2)\right), \qquad (1)$$

where $\varepsilon \in \{-1, 0, 1\}$ and $\boldsymbol{a}\colon \mathbb{R} \to \mathbb{R}^+$ is a differentiable function, is called the Robertson-Walker metric (see Ref. 1).

The function $\boldsymbol{a}$ is totally independent of any geometrical consideration. It can be specified only within a theory of gravity. Actually, general relativity is of little use in most cosmological questions. The basic equation of general relativity relates the geometrical tensor G_{ab} to the energy-momentum tensor T_{ab}:

$$G_{ab} + \Lambda g_{ab} = kT_{ab},$$

where Λ and k are constants (see Ref. 2). There exists a coordinates system, called the comoving coordinates, in which the matter is at rest and the tensor T_{ab} is diagonal with $T_{00} = \rho$ and $T_{11} = T_{22} = T_{33} = p$, ρ being the density and p the pressure.

From the above rules, we can easily derive the equation for $\boldsymbol{a}$. Let us have a comoving spherical coordinates system (t, r, θ, ϕ) with the observer at the origin of the spatial coordinates $(t_0, r = 0, \theta = 0, \phi = 0)$. Let us assume the observed source is emitting an electromagnetic wave traveling along the r direction, with $\theta = 0$ and $\phi = 0$. As this trajectory is a null geodesic, we have

$$dt^2 - \boldsymbol{a}^2(t)\frac{dr^2}{1-\varepsilon r^2} = 0 \tag{2}$$

so the variables can be separated and the integration over r is analytical. The solution of Equation (2) one can find in the book, Ref. 3.

2. Forms of the Function $\boldsymbol{a}$

From Equation (1) the metric tensor g in comoving coordinates $q = (t, r, \theta, \phi)$ has the form

$$g_{ab}(q) = \begin{pmatrix} -1 & & & 0 \\ & \dfrac{\boldsymbol{a}^2(t)}{1-\varepsilon r^2} & & \\ & & r^2\boldsymbol{a}^2(t) & \\ 0 & & & r^2\boldsymbol{a}^2(t)\sin^2(\theta) \end{pmatrix}, \tag{3}$$

where $a, b = 0, 1, 2, 3$, and it has the signature $(1, 3)$ if and only if $1-\varepsilon r^2 > 0$ and $\sin(\theta) \neq 0$. Hence, it follows that if $\varepsilon = -1$ or $\varepsilon = 0$ then $r \in \mathbb{R}^+$, and if $\varepsilon = 1$ then $r \in (0, 1)$ as well as $\theta \in (0, \pi)$. Since the metric tensor (3) is a diagonal this is a simple task to calculate the Christoffel symbols using the formula

$$\Gamma^b_{ac} = \frac{1}{2}g^{bd}(\partial_c g_{ad} + \partial_a g_{dc} - \partial_d g_{ca}).$$

Here we denote the derivative with respect to t with $(\cdot)'$ and we note that the values of the Christoffel symbols are

$$\Gamma^1_{10} = \Gamma^1_{01} = \frac{\boldsymbol{a}'(t)}{\boldsymbol{a}(t)},$$

$$\Gamma^2_{20} = \Gamma^2_{02} = \frac{\boldsymbol{a}'(t)}{\boldsymbol{a}(t)},$$

$$\Gamma^3_{30} = \Gamma^3_{03} = \frac{\boldsymbol{a}'(t)}{\boldsymbol{a}(t)},$$

$$\Gamma^0_{11} = \frac{\boldsymbol{a}(t)\boldsymbol{a}'(t)}{1-\varepsilon r^2},$$

$$\Gamma^1_{11} = \frac{\varepsilon r}{1-\varepsilon r^2},$$

$$\Gamma^2_{21} = \Gamma^2_{12} = \frac{1}{r},$$
$$\Gamma^3_{31} = \Gamma^3_{13} = \frac{1}{r},$$
$$\Gamma^0_{22} = r^2 \boldsymbol{a}(t)\boldsymbol{a}'(t),$$
$$\Gamma^1_{22} = -r(1 - \varepsilon r^2),$$
$$\Gamma^3_{32} = \Gamma^3_{23} = \frac{\cos(\theta)}{\sin(\theta)},$$
$$\Gamma^0_{33} = r^2 \boldsymbol{a}(t)\boldsymbol{a}'(t)\sin^2(\theta),$$
$$\Gamma^1_{33} = -r(1 - \varepsilon r^2)\sin^2(\theta),$$
$$\Gamma^2_{33} = -\sin(\theta)\cos(\theta),$$

where it is understood that all other Christoffel symbols vanish.

Now, let us recall the well-known formulas

$$R_{abcd} = \frac{1}{2}(\partial^2_{bc} g_{ad} + \partial^2_{ad} g_{bc} - \partial^2_{bd} g_{ac} - \partial^2_{ac} g_{bd}) + g_{kl}(\Gamma^k_{bc}\Gamma^l_{ad} - \Gamma^k_{bd}\Gamma^l_{ac}),$$
$$R_{abcd} = -R_{bacd} = -R_{abdc},$$
$$R_{abcd} = R_{cdab}.$$

Thus the components, different from zero, of the Riemann tensor R_{abcd} are

$$R_{1010}(q) = -\frac{\boldsymbol{a}(t)\boldsymbol{a}''(t)}{1 - \varepsilon r^2},$$
$$R_{2020}(q) = -r^2 \boldsymbol{a}(t)\boldsymbol{a}''(t),$$
$$R_{3030}(q) = -r^2 \boldsymbol{a}(t)\boldsymbol{a}''(t)\sin^2(\theta),$$
$$R_{1212}(q) = \frac{r^2 \boldsymbol{a}^2(t)}{1 - \varepsilon r^2}(\varepsilon + (\boldsymbol{a}'(t))^2),$$
$$R_{3131}(q) = \frac{r^2 \boldsymbol{a}^2(t)}{1 - \varepsilon r^2}(\varepsilon + (\boldsymbol{a}'(t))^2)\sin^2(\theta),$$
$$R_{2323}(q) = r^4 \boldsymbol{a}^2(t)(\varepsilon + (\boldsymbol{a}'(t))^2)\sin^2(\theta).$$

From the formula $R_{ab} = g^{kl} R_{kalb}$ we yield the Ricci tensor R_{ab}:

$$R_{00}(q) = \frac{-3\boldsymbol{a}''(t)}{\boldsymbol{a}(t)},$$
$$R_{11}(q) = \frac{1}{1 - \varepsilon r^2}(\boldsymbol{a}(t)\boldsymbol{a}''(t) + 2(\varepsilon + (\boldsymbol{a}'(t))^2)),$$
$$R_{22}(q) = r^2(\boldsymbol{a}(t)\boldsymbol{a}''(t) + 2(\varepsilon + (\boldsymbol{a}'(t))^2)),$$
$$R_{33}(q) = r^2(\boldsymbol{a}(t)\boldsymbol{a}''(t) + 2(\varepsilon + (\boldsymbol{a}'(t))^2))\sin^2(\theta),$$
$$R_{ab}(q) = 0 \text{ for } a \neq b$$

and the Ricci scalar is

$$R(t) = \frac{6}{\boldsymbol{a}^2(t)}(\boldsymbol{a}(t)\boldsymbol{a}''(t) + \varepsilon + (\boldsymbol{a}'(t))^2). \tag{4}$$

Then components of the Einstein tensor G_{ab} have the form

$$\begin{aligned}
G_{00}(q) &= \frac{3(\varepsilon + (\boldsymbol{a}'(t))^2)}{\boldsymbol{a}^2(t)},\\
G_{11}(q) &= -\frac{1}{1-\varepsilon r^2}(2\boldsymbol{a}(t)\boldsymbol{a}''(t) + \varepsilon + (\boldsymbol{a}'(t))^2),\\
G_{22}(q) &= -r^2(2\boldsymbol{a}(t)\boldsymbol{a}''(t) + \varepsilon + (\boldsymbol{a}'(t))^2),\\
G_{33}(q) &= -r^2(2\boldsymbol{a}(t)\boldsymbol{a}''(t) + \varepsilon + (\boldsymbol{a}'(t))^2)\sin^2(\theta),\\
G_{ab}(q) &= 0 \text{ for } a \neq b.
\end{aligned}$$

Thus

$$\begin{aligned}
G_{ab}(q) &= R_{ab}(q) - \frac{1}{2}R(t)g_{ab}(q)\\
&= 2\begin{pmatrix} \frac{\varepsilon + (\boldsymbol{a}'(t))^2}{\boldsymbol{a}^2(t)} & & & 0\\ & -\frac{\boldsymbol{a}(t)\boldsymbol{a}''(t)}{1-\varepsilon r^2} & & \\ & & -r^2\boldsymbol{a}(t)\boldsymbol{a}''(t) & \\ 0 & & & -r^2\boldsymbol{a}(t)\boldsymbol{a}''(t)\sin^2(\theta)\end{pmatrix}\\
&\quad - \frac{\varepsilon + (\boldsymbol{a}'(t))^2}{\boldsymbol{a}^2(t)}\begin{pmatrix} -1 & & & 0\\ & \frac{\boldsymbol{a}^2(t)}{1-\varepsilon r^2} & & \\ & & r^2\boldsymbol{a}^2(t) & \\ 0 & & & r^2\boldsymbol{a}^2(t)\sin^2(\theta)\end{pmatrix}\\
&= 2T_{ab}(q) - \Lambda(t)g_{ab}(q),
\end{aligned}$$

where

$$T_{ab}(q) = \begin{pmatrix} \frac{\varepsilon + (\boldsymbol{a}'(t))^2}{\boldsymbol{a}^2(t)} & & & 0\\ & -\frac{\boldsymbol{a}(t)\boldsymbol{a}''(t)}{1-\varepsilon r^2} & & \\ & & -r^2\boldsymbol{a}(t)\boldsymbol{a}''(t) & \\ 0 & & & -r^2\boldsymbol{a}(t)\boldsymbol{a}''(t)\sin^2(\theta)\end{pmatrix}, \tag{5}$$

$$\Lambda(t) = \frac{\varepsilon + (\boldsymbol{a}'(t))^2}{\boldsymbol{a}^2(t)}. \tag{6}$$

So we have

$$G_{ab}(q) + \Lambda(t)g_{ab}(q) = 2T_{ab}(q). \tag{7}$$

The problem of finding a coordinates system in which the tensor T_{ab} is diagonal relies on the finding of the canonical form of the λ–tensor $T_{ab} - \lambda g_{ab}$. The canonical forms of symmetric covariant λ-tensors have been given in the paper, Ref. 4.

In our case the determinant of the λ-matrix $T_{ab}(q) - \lambda g_{ab}(q)$, where $g_{ab}(q)$ and $T_{ab}(q)$ are given by formulas (3) and (5), respectively, is equal to zero if and only if

$$\lambda_0(t) = -\frac{1}{\boldsymbol{a}^2(t)}(\varepsilon + (\boldsymbol{a}'(t))^2)$$

or

$$\lambda_1(t) = \lambda_2(t) = \lambda_3(t) = -\frac{\boldsymbol{a}''(t)}{\boldsymbol{a}(t)},$$

so we have

$$g_{a'b'} = \begin{pmatrix} -1 & & & 0 \\ & 1 & & \\ & & 1 & \\ 0 & & & 1 \end{pmatrix},$$

$$T_{a'b'} = \begin{pmatrix} \frac{\varepsilon + (\boldsymbol{a}'(t))^2}{\boldsymbol{a}^2(t)} & & & 0 \\ & -\frac{\boldsymbol{a}''(t)}{\boldsymbol{a}(t)} & & \\ & & -\frac{\boldsymbol{a}''(t)}{\boldsymbol{a}(t)} & \\ 0 & & & -\frac{\boldsymbol{a}''(t)}{\boldsymbol{a}(t)} \end{pmatrix}.$$

Hence we have

$$\rho(t) = \frac{\varepsilon + (\boldsymbol{a}'(t))^2}{\boldsymbol{a}^2(t)} \tag{8}$$

and

$$p(t) = -\frac{\boldsymbol{a}''(t)}{\boldsymbol{a}(t)}. \tag{9}$$

Then the tensor T_{ab} given by formula (5) has the form

$$T_{ab}(q) = (\rho(t) + p(t))U_a \otimes U_b + p(t)g_{ab}(q), \tag{10}$$

where U_a is a vector such that $g^{ab}(q)U_aU_b = -1$.

Let now

$$G^{ab} = g^{ak}g^{bl}G_{kl}$$

and

$$T^{ab} = g^{ak} g^{bl} T_{kl},$$

where G_{kl} is an Einstein's tensor definite earlier and T_{kl} is given by the formula (5). In order to you may say that the equation (7) is an Einstein's equation it should obey the conservation equation and this means that the covariant divergence of each side vanishes identically, that is,

$$\nabla_a (G^{ab}(q) + \Lambda(t) g^{ab}(q)) = 0$$

and

$$\nabla_a T^{ab}(q) = 0,$$

where the symbol ∇_a stands for the (Christoffel) covariant derivative.[2]

We note that

$$\begin{aligned}
\nabla_a G^{a0}(q) &= 0, \\
\nabla_a G^{a1}(q) &= 0, \\
\nabla_a G^{a2}(q) &= 0, \\
\nabla_a G^{a3}(q) &= 0, \\
\nabla_a (\Lambda(t) g^{a0}(q)) &= -\frac{2}{\boldsymbol{a}^3(t)} \cdot \boldsymbol{a}'(t)(\boldsymbol{a}(t)\boldsymbol{a}''(t) - \varepsilon - (\boldsymbol{a}'(t))^2), \\
\nabla_a (\Lambda(t) g^{a1}(q)) &= 0, \\
\nabla_a (\Lambda(t) g^{a2}(q)) &= 0, \\
\nabla_a (\Lambda(t) g^{a3}(q)) &= 0
\end{aligned}$$

and

$$\begin{aligned}
\nabla_a T^{a0}(q) &= -\frac{1}{\boldsymbol{a}^3(t)} \cdot \boldsymbol{a}'(t)(\boldsymbol{a}(t)\boldsymbol{a}''(t) - \varepsilon - (\boldsymbol{a}'(t))^2), \\
\nabla_a T^{a1}(q) &= 0, \\
\nabla_a T^{a2}(q) &= 0, \\
\nabla_a T^{a3}(q) &= 0.
\end{aligned}$$

Therefore the covariant divergence of the left side of the equation (7) is equal to the covariant divergence of the right side and it is equal to zero if and only if

$$\boldsymbol{a}'(t)(\boldsymbol{a}(t)\boldsymbol{a}''(t) - (\boldsymbol{a}'(t))^2 - \varepsilon) = 0. \tag{11}$$

Hence we have

$$\boldsymbol{a}'(t) = 0 \quad \text{or} \quad \boldsymbol{a}(t)\boldsymbol{a}''(t) - (\boldsymbol{a}'(t))^2 = \varepsilon.$$

From the first equation it follows that

$$(C_\varepsilon) \qquad \boldsymbol{a}(t) = \text{constant}$$

and the second equation we obtain a differential equation

$$\boldsymbol{a}''(t) = \frac{\varepsilon + (\boldsymbol{a}'(t))^2}{\boldsymbol{a}(t)}. \tag{12}$$

Thus we have

$$(Q_0) \qquad \boldsymbol{a}''(t) = \frac{(\boldsymbol{a}'(t))^2}{\boldsymbol{a}(t)} \quad \text{for } \varepsilon = 0,$$

$$(Q_1) \qquad \boldsymbol{a}''(t) = \frac{1+(\boldsymbol{a}'(t))^2}{\boldsymbol{a}(t)} \quad \text{for } \varepsilon = 1,$$

$$(Q_{-1}) \qquad \boldsymbol{a}''(t) = \frac{-1+(\boldsymbol{a}'(t))^2}{\boldsymbol{a}(t)} \quad \text{for } \varepsilon = -1.$$

The general solutions of the equations (Q_0), (Q_1) and (Q_{-1}) are, respectively

$$(S_0) \qquad \boldsymbol{a}(t) = e^{\alpha_1 t + \alpha_2},$$

where $\alpha_1 > 0$ and α_2 are constants,

$$(S_1) \qquad \boldsymbol{a}(t) = \frac{1}{\beta_1}\cosh(\beta_1 t + \beta_2),$$

where $\beta_1 > 0$ and β_2 are constants,

$$(S_{-1,1}) \qquad \boldsymbol{a}(t) = \frac{1}{\gamma_1}\sinh(\gamma_1 t + \gamma_2) \quad \text{if } \boldsymbol{a}'(t) - 1 \neq 0,$$

where $\gamma_1 > 0$ and γ_2 are constants,

$$(S_{-1,2}) \qquad \boldsymbol{a}(t) = \gamma_3 t + \gamma_4 \quad \text{if } \boldsymbol{a}'(t) - 1 = 0,$$

where γ_3 and γ_4 are constants. Hence, it follows that $\gamma_3 = 1$.

Let us notice that solutions $(S_{-1,1})$ and $(S_{-1,2})$ are not positive, so they do not come up to our assumption that the function $\boldsymbol{a}$ should be positive for $t \in \mathbb{R}$.

The condition $\rho \geq 0$ guarantees that if the function $\boldsymbol{a}$ has one of three the following forms

- $\boldsymbol{a}(t) = \text{constant} > 0$ if $\varepsilon \in \{0, 1\}$,
- $\boldsymbol{a}(t) = e^{\alpha_1 t + \alpha_2}$ if $\varepsilon = 0$,
- $\boldsymbol{a}(t) = \frac{1}{\beta_1}\cosh(\beta_1 t + \beta_2)$ if $\varepsilon = 1$

then we have

Proposition 2.1. *The Robertson-Walker solution of an Einstein's equation is a Lorentzian nonsingular spacetime and satisfies*

$$R_{ab} - \frac{1}{2}Rg_{ab} + \Lambda g_{ab} = 2T_{ab}.$$

It corresponds therefore to a solution with the cosmological constant Λ.

We note that constants R and Λ have the following values given in the table below

Case	Function $\boldsymbol{a}$	ε	R	Λ
C_0	$\boldsymbol{a}(t) = c > 0$ (constant)	0	0	0
S_0	$\boldsymbol{a}(t) = e^{\alpha_1 t + \alpha_2}$	0	$12\alpha_1^2$	α_1^2
C_1	$\boldsymbol{a}(t) = c > 0$ (constant)	1	$\frac{6}{c^2}$	$\frac{1}{c^2}$
S_1	$\boldsymbol{a}(t) = \frac{1}{\beta_1}\cosh(\beta_1 t + \beta_2)$	1	$12\beta_1^2$	β_1^2

Let us notice that formula (10) as well as Cases S_0 and S_1 show that the energy-momentum tensor of the nonsingular universe takes the same form as for a perfect fluid. From formula (10) and the Case C_1 it follows that the energy-momentum tensor of the nonsingular universe has the form as for a dust. In the Case C_0 the nonsingular universe is a vacuum.

Let now $\theta = 0$ or $\theta = \pi$. Since $d\theta = 0$ if θ=constant so the metric given by Equation (1) to takes the form

$$\widetilde{g} = -dt^2 + \frac{\boldsymbol{a}^2(t)}{1 - \varepsilon r^2}dr^2.$$

The metric tensor $\widetilde{g}$ is comoving coordinates $\widetilde{q} = (t, r)$ has the form

$$\widetilde{g}_{ab}(\widetilde{q}) = \begin{pmatrix} -1 & 0 \\ 0 & \dfrac{\boldsymbol{a}^2(t)}{1 - \varepsilon r^2} \end{pmatrix}, \tag{13}$$

where a, b = 0, 1. Thus the hypersurface $\widetilde{H}_0$ or $\widetilde{H}_\pi$ degenerates to a 2-dimensional surface. The Riemann tensor is easily calculated for the metric (13):

$$\widetilde{R}_{1010}(\widetilde{q}) = -\frac{\boldsymbol{a}(t)\boldsymbol{a}''(t)}{1 - \varepsilon r^2},$$

$$\widetilde{R}_{abdc}(\widetilde{q}) = 0 \quad \text{for remaining indices } a,\ b,\ c, \text{ and } d.$$

It is also easy to check that the Gauss curvature of the surface $\widetilde{H}_0$ or $\widetilde{H}_\pi$ is

$$K(\widetilde{q}) = \frac{\widetilde{R}_{1010}}{\det(\widetilde{g}_{ab}(\widetilde{q}))} = \frac{\boldsymbol{a}''(t)}{\boldsymbol{a}(t)}.$$

We still consider a case when $r > 0$, $\theta \in (0, \pi)$ and $\varphi = 0$. Since $d\varphi = 0$ if $\varphi =$ constant so the metric given by Equation (1) takes the form

$$\overline{g} = -dt^2 + \boldsymbol{a}^2(t)\left(\frac{1}{1-\varepsilon r^2}dr^2 + r^2 d\theta^2\right).$$

The metric tensor $\overline{g}$ in comoving coordinates $\overline{q} = (t, r, \theta)$ in the hypersurface $\overline{H}_0$ has the form

$$\overline{g}_{ab}(\overline{q}) = \begin{pmatrix} -1 & & 0 \\ & \dfrac{\boldsymbol{a}^2(t)}{1-\varepsilon r^2} & \\ 0 & & r^2\boldsymbol{a}^2(t) \end{pmatrix}, \tag{14}$$

where a, $b = 0$, 1, 2. Then the Riemann tensor calculated for the metric tensor (14) is

$$\begin{aligned}
\overline{R}_{1010}(\overline{q}) &= -\frac{\boldsymbol{a}(t)\boldsymbol{a}''(t)}{1-\varepsilon r^2}, \\
\overline{R}_{2020}(\overline{q}) &= -r^2\boldsymbol{a}(t)\boldsymbol{a}''(t), \\
\overline{R}_{1212}(\overline{q}) &= \frac{r^2\boldsymbol{a}^2(t)}{1-\varepsilon r^2}(\varepsilon + (\boldsymbol{a}'(t))^2), \\
\overline{R}_{abcd}(\overline{q}) &= 0 \text{ for remaining indices } a,\ b,\ c \text{ and } d.
\end{aligned}$$

A curvature of a 3-dimensional space is so determined in each point $\overline{q}$ by three quantities:

$$K_\theta(\overline{q}) = \frac{\overline{R}_{1010}(\overline{q})}{\det\begin{pmatrix} -1 & 0 \\ 0 & \frac{\boldsymbol{a}^2(t)}{1-\varepsilon r^2}\end{pmatrix}} = \frac{\boldsymbol{a}''(t)}{\boldsymbol{a}(t)},$$

$$K_r(\overline{q}) = \frac{\overline{R}_{2020}(\overline{q})}{\det\begin{pmatrix} -1 & 0 \\ 0 & r^2\boldsymbol{a}^2(t)\end{pmatrix}} = \frac{\boldsymbol{a}''(t)}{\boldsymbol{a}(t)},$$

$$K_t(\overline{q}) = \frac{\overline{R}_{1212}(\overline{q})}{\det\begin{pmatrix} \frac{\boldsymbol{a}^2(t)}{1-\varepsilon r^2} & 0 \\ 0 & r^2\boldsymbol{a}^2(t)\end{pmatrix}} = \frac{\varepsilon + (\boldsymbol{a}'(t))^2}{\boldsymbol{a}^2(t)}.$$

If we assume that $K_\theta(\overline{q}) = K_r(\overline{q}) = K_t(\overline{q})$ then we yield

$$\boldsymbol{a}''(t) = \frac{\varepsilon + (\boldsymbol{a}'(t))^2}{\boldsymbol{a}(t)},$$

that is, Equation (12) whose solutions are (S_0) and (S_1). We still note that $K(\widetilde{q}) = K_\theta(\overline{q})$. Therefore, $\widetilde{H}_0$ or $\widetilde{H}_\pi$ is a surface of the constant curvature $K = \alpha_1^2$ for the solution (S_0), $K = \beta_1^2$ for the solution (S_1). $\overline{H}_0$ is a hypersurface of the constant curvature $K = \alpha_1^2$, $K = \beta_1^2$, respectively to the solutions (S_0) and (S_1).

3. Types of the Gravitational Field

One of the most immediate examples of the utility of canonical forms of a pair of second order symmetric covariant tensors[4] is the classification of the pair: the Riemann tensor R_{abcd} and the tensor $g_{abcd} = g_{ac}g_{bd} - g_{ad}g_{bc}$, where g_{ab} is the metric tensor (of signature (1,3)). In order to give the types of the gravitational field of the Robertson-Walker nonsingular spacetime we will be to use this classification.

Let us take the components of the Riemann tensor R_{abcd} given in Section 2. In a nonsingular case is convenient to introduce a 6-dimensional formalism in the pseudo-Eucliden space $\mathbb{R}^6$ (the Klein space). The rule changing to the 6-dimensional formalism is the following (see Ref. 5)

$$\begin{array}{rcccccc} ab: & 23 & 31 & 12 & 10 & 20 & 30 \\ A: & 1 & 2 & 3 & 4 & 5 & 6. \end{array}$$

Now, we introduce the metric tensor defined by

$$g_{ac}g_{bd} - g_{ad}g_{bc} = g_{abcd} \longrightarrow g_{AB},$$

where g_{ab} are components a metric tensor at an arbitrary point of the Robertson-Walker nonsingular spacetime and collective indices are the skew-symmetric pairs $ab \to A$, $cd \to B$. The tensor g_{AB} $(A, B = 1, \ldots, 6)$ of signature $(3, 3)$ is symmetric and nonsingular.

Thus components of tensors g_{AB} and R_{AB} are, respectively

$$\begin{aligned} g_{11}(q) &= r^4 \boldsymbol{a}^4(t) \sin^2(\theta), \\ g_{22}(q) &= \frac{r^2 \boldsymbol{a}^4(t)}{1 - \varepsilon r^2} \sin^2(\theta), \\ g_{33}(q) &= \frac{r^2 \boldsymbol{a}^4(t)}{1 - \varepsilon r^2}, \\ g_{44}(q) &= -\frac{\boldsymbol{a}^2(t)}{1 - \varepsilon r^2}, \\ g_{55}(q) &= -r^2 \boldsymbol{a}^2(t), \\ g_{66}(q) &= -r^2 \boldsymbol{a}^2(t) \sin^2(\theta) \end{aligned}$$

and g_{AB} vanishing for $A \neq B$,

$$\begin{aligned}
R_{11}(q) &= r^4 \boldsymbol{a}^2(t)(\varepsilon + (\boldsymbol{a}'(t))^2)\sin^2(\theta),\\
R_{22}(q) &= \frac{r^2\boldsymbol{a}^2(t)}{1-\varepsilon r^2}(\varepsilon + (\boldsymbol{a}'(t))^2)\sin^2(\theta),\\
R_{33}(q) &= \frac{r^2\boldsymbol{a}^2(t)}{1-\varepsilon r^2}(\varepsilon + (\boldsymbol{a}'(t))^2),\\
R_{44}(q) &= -\frac{\boldsymbol{a}(t)\boldsymbol{a}''(t)}{1-\varepsilon r^2},\\
R_{55}(q) &= -r^2\boldsymbol{a}(t)\boldsymbol{a}''(t),\\
R_{66}(q) &= -r^2\boldsymbol{a}(t)\boldsymbol{a}''(t)\sin^2(\theta)
\end{aligned}$$

and R_{AB} vanishing for $A \neq B$.

Therefore, it only remains to find a canonical form of the λ-tensor $R_{AB} - \lambda g_{AB}$. The canonical forms of symmetric λ-tensors have been given in the paper, Ref. 6.

In our case the determinant of the λ-matrix $R_{AB}(q) - \lambda g_{AB}(q)$ is equal to zero if and only if

$$\lambda_1(t) = \lambda_2(t) = \lambda_3(t) = \frac{1}{\boldsymbol{a}^2(t)}(\varepsilon + (\boldsymbol{a}'(t))^2)$$

or

$$\lambda_4(t) = \lambda_5(t) = \lambda_6(t) = \frac{\boldsymbol{a}''(t)}{\boldsymbol{a}(t)},$$

so we have

$$g_{A'B'} = \begin{pmatrix} 1 & & & & & 0\\ & 1 & & & & \\ & & 1 & & & \\ & & & -1 & & \\ & & & & -1 & \\ 0 & & & & & -1 \end{pmatrix}$$

$$R_{A'B'} = \begin{pmatrix} \frac{\varepsilon+(\boldsymbol{a}'(t))^2}{\boldsymbol{a}^2(t)} & & & & & 0\\ & \frac{\varepsilon+(\boldsymbol{a}'(t))^2}{\boldsymbol{a}^2(t)} & & & & \\ & & \frac{\varepsilon+(\boldsymbol{a}'(t))^2}{\boldsymbol{a}^2(t)} & & & \\ & & & -\frac{\boldsymbol{a}''(t)}{\boldsymbol{a}(t)} & & \\ & & & & -\frac{\boldsymbol{a}''(t)}{\boldsymbol{a}(t)} & \\ 0 & & & & & -\frac{\boldsymbol{a}''(t)}{\boldsymbol{a}(t)} \end{pmatrix}$$

The gravitational field determined by the characteristic of the λ-tensor $R_{AB} - \lambda g_{AB}$ in each case from C_0 to S_1 (see Table in Section 2) is

Case C_0

$$g_{A'B'} = \begin{pmatrix} 1 & & & & & 0 \\ & 1 & & & & \\ & & 1 & & & \\ & & & -1 & & \\ & & & & -1 & \\ 0 & & & & & -1 \end{pmatrix}$$

$$R_{A'B'} = \begin{pmatrix} 0 & & & & & 0 \\ & 0 & & & & \\ & & 0 & & & \\ & & & 0 & & \\ & & & & 0 & \\ 0 & & & & & 0 \end{pmatrix}$$

of the type $G_7[(111111)]$,

Case S_0

$$g_{A'B'} = \begin{pmatrix} 1 & & & & & 0 \\ & 1 & & & & \\ & & 1 & & & \\ & & & -1 & & \\ & & & & -1 & \\ 0 & & & & & -1 \end{pmatrix}$$

$$R_{A'B'} = \begin{pmatrix} \alpha_1^2 & & & & & 0 \\ & \alpha_1^2 & & & & \\ & & \alpha_1^2 & & & \\ & & & -\alpha_1^2 & & \\ & & & & -\alpha_1^2 & \\ 0 & & & & & -\alpha_1^2 \end{pmatrix}$$

of the type $G_7[(111111)]$,

Case C_1

$$g_{A'B'} = \begin{pmatrix} 1 & & & & & 0 \\ & 1 & & & & \\ & & 1 & & & \\ & & & -1 & & \\ & & & & -1 & \\ 0 & & & & & -1 \end{pmatrix}$$

$$R_{A'B'} = \begin{pmatrix} c^{-2} & & & & & 0 \\ & c^{-2} & & & & \\ & & c^{-2} & & & \\ & & & 0 & & \\ & & & & 0 & \\ 0 & & & & & 0 \end{pmatrix}$$

of the type $G_1[(111)(111)]$.

Case S_1

$$g_{A'B'} = \begin{pmatrix} 1 & & & & & 0 \\ & 1 & & & & \\ & & 1 & & & \\ & & & -1 & & \\ & & & & -1 & \\ 0 & & & & & -1 \end{pmatrix},$$

$$R_{A'B'} = \begin{pmatrix} \beta_1^2 & & & & & 0 \\ & \beta_1^2 & & & & \\ & & \beta_1^2 & & & \\ & & & -\beta_1^2 & & \\ & & & & -\beta_1^2 & \\ 0 & & & & & -\beta_1^2 \end{pmatrix}$$

of the type $G_7[(111111)]$ in the Segre symbols (see Ref. 6).

4. Conclusion

We have studied the gravitational field of the Robertson-Walker spacetime and we have given in a nonsingular case analytical forms of the function $\boldsymbol{a}$. None of authors of papers, Refs. 1,3,7 has given list of analytical forms of the function $\boldsymbol{a}$.

In this paper we have gone after Landau's statement that the function $\boldsymbol{a}$ is determined by an Einstein's equation (see Ref. 8). It has led us to the differential equation (11). A list of analytical forms of the function $\boldsymbol{a}$ consists of solutions to this equation.

We have demonstrated for Cases S_0 and S_1 that the Robertson–Walker spacetime is completely isotropic, i.e.,

$$\begin{aligned} R_{abcd} &= K(g_{ac}g_{bd} - g_{ad}g_{bc}), \\ \widetilde{R}_{abcd} &= K(\widetilde{g}_{ac}\widetilde{g}_{bd} - \widetilde{g}_{ad}\widetilde{g}_{bc}), \\ \overline{R}_{abcd} &= K(\overline{g}_{ac}\overline{g}_{bd} - \overline{g}_{ad}\overline{g}_{bc}), \end{aligned}$$

where $K = \alpha_1^2$ for the solution (S_0) and $K = \beta_1^2$ for the solution (S_1). Constants α_1 and β_1 are integral constants of the equation (11), so they are dependent on given initial data. Further, we have expressed components T_{ab} of the energy-momentum tensor at q with respect to an orthonormal basis E^0, E^1, E^2, E^3 in the canonical form $T_{a'b'}$. In Cases S_0 and S_1 we have yielded that $\rho = -p_i$ $(i = 1, 2, 3)$. The eigenvalue ρ represents the energy-density and the eigenvalues p_i $(i = 1, 2, 3)$ represent the principal pressures in the three spacelike directions E^i.

Components R_{AB} of the Riemann tensor at q represented in the pseudo-Eucliden space $\mathbb{R}^6$ by a 6-dimensional formalism we have also expressed with respect to an orthonormal basis $e^1, e^2, e^3, e^4, e^5, e^6$ in the canonical form $R_{A'B'}$. In Cases S_0 and S_1 we have yielded that the eigenvalues in the directions e^i $(i = 1, \ldots, 6)$ are equal to each other. Since they are constants and with respect to that the Robertson–Walker spacetime is completely isotropic we may say that the gravitational field of the Robertson–Walker spacetime is constant.

References

1. M. Kriele, *Spacetime: foundations of general relativity and differential geometry* (Springer-Verlag, Berlin-Heidelberg, 1999).
2. B.F. Schutz, *A First Course in General Relativity* (Cambridge University Press, 1985).
3. S. Weinberg, *Gravitational and Cosmology: Principles and Applications of the General Theory of Relativity* (John Wiley and Sons, Inc. New York, London, Sydney, Toronto, 1971).
4. W. Borgiel, Canonical forms of a pair of second order symmetric covariant tensors in General Relativity, *Acta Cosmologica* **21** (1) (1995) 23–45.
5. F.A.E. Pirani, Invariant formulation of gravitational radiation theory, *Phys. Rev.* **105** (1957) 1089–1099.
6. W. Borgiel, Classification of Gravitational Fields, *Acta Cosmologica* **22** (1) (1996) 47–71.
7. J. Rich, *Fundamentals of Cosmology* (Springer-Verlag, Berlin-Heidelberg, 2001).
8. L.D. Landau and E.M. Lifszyc, *Field Theory* (in Polish, Polish Scientific Publishers, Warsaw, 1980).

Differential Geometry and its Applications
Proc. Conf., in Honour of Leonhard Euler, Olomouc, August 2007

Lie algebra pairing and the Lagrangian and Hamiltonian equations in gauge-invariant problems

Marco Castrillón López

*Departamento de Geometría y Topología, Facultad de Matemáticas,
Universidad Complutense de Madrid,
28040-Madrid, Spain
E-mail: mcastri@mat.ucm.es*

Jaime Muñoz Masqué

*Instituto de Física Aplicada, CSIC,
C/ Serrano 144, 28006-Madrid, Spain
E-mail: jaime@iec.csic.es*

Let $P \to M$ be a principal G-bundle over a pseudo-Riemannian manifold (M, g). If G is semisimple, the Euler-Lagrange and the Hamilton-Cartan equations of the Yang-Mills Lagrangian defined by g are proved to remain unchanged if the Cartan-Killing metric is replaced by any other non-degenerate, adjoint-invariant bilinear form on the Lie algebra $\mathfrak{g}$. Moreover, it is shown that the Hamilton-Cartan equations for any gauge invariant problem are essentially the same as their Euler-Lagrange equations if some simple regularity assumption is made. Actually, the set of solutions to Hamilton equations trivially fibers over the set of extremals of the variational problem.

Keywords: Adjoint-invariant pairing, gauge invariance, Hamilton-Cartan equations, jet bundles, principal connection, Yang-Mills fields.

MS classification: 70S15, 53C07, 58A20, 58E15, 58E30, 70S05, 70S10.

1. Introduction and preliminaries

1.1. *Notations*

Let (M, g) be an n-dimensional pseudo-Riemannian manifold with volume form $\mathbf{v}_g$. The isomorphism $T_xM \to T_x^*M$, $X \mapsto X^\flat$ induced by g defines a metric $g^{(r)}$ on $\wedge^r T^*M$ for every $r \geq 0$. Moreover, the Hodge star operator is extended to vector-valued forms as follows. If $V \to M$ is a vector bundle, then $\star\colon \wedge^\bullet T^*M \otimes V \to \wedge^\bullet T^*M \otimes V$ is defined by $\star(\omega_r \otimes v) = (\star\omega_r) \otimes v$, $\forall \omega_r \in \wedge^\bullet T_x^*M$, $\forall v \in V_x$.

A symmetric bilinear form $\langle \cdot, \cdot \rangle$ in $\mathfrak{g}$ is said to be invariant under the adjoint representation if $\langle \mathrm{ad}_g B, \mathrm{ad}_g C\rangle = \langle B, C\rangle$, $\forall g \in G$, $\forall B, C \in \mathfrak{g}$. The infinitesimal version of this equation is

$$\langle [A,B], C\rangle + \langle B, [A,C]\rangle = 0, \quad \forall A, B, C \in \mathfrak{g}. \tag{1}$$

Both equations are equivalent if the group G is connected.

Let $\pi\colon P \to M$ be a principal G-bundle and let $\pi_{\mathrm{ad}\, P}\colon \mathrm{ad}\, P \to M$ be the adjoint bundle; i.e., the bundle associated with P under the adjoint representation of G on its Lie algebra $\mathfrak{g}$. For all $B \in \mathfrak{g}$, $u \in P$, we set $(u,B)_G = (u,B) \bmod G$. Every adjoint-invariant symmetric bilinear form $\langle\cdot,\cdot\rangle\colon \mathfrak{g}\times\mathfrak{g} \to \mathbb{R}$ induces a fibred metric $\langle\!\langle\cdot,\cdot\rangle\!\rangle : \mathrm{ad}\, P \oplus \mathrm{ad}\, P \to \mathbb{R}$ by the formula,

$$\langle\!\langle (u,B)_G, (u,C)_G\rangle\!\rangle = \langle B, C\rangle, \quad \forall u \in P,\ \forall B, C \in \mathfrak{g}. \tag{2}$$

Furthermore, g and $\langle\!\langle\cdot,\cdot\rangle\!\rangle$ induce a fibred metric on the bundle of $\mathrm{ad}\, P$-valued differential r-forms on M given by

$$((\alpha_r \otimes a, \beta_r \otimes b)) = g^{(r)}(\alpha_r, \beta_r)\, \langle\!\langle a, b\rangle\!\rangle\,,$$

$\forall \alpha, \beta \in \wedge^r T_x^* M$, $\forall a, b \in (\mathrm{ad}\, P)_x$, and (2) gives rise to an exterior product (cf. Ref. 1),

$$\begin{aligned} &\dot\wedge\colon\ (\wedge^\bullet T^*M \otimes \mathrm{ad}\, P) \oplus (\wedge^\bullet T^*M \otimes \mathrm{ad}\, P) \to \wedge^\bullet T^*M, \\ &(\alpha_q \otimes a\ )\,\dot\wedge\,(\beta_r \otimes b) = (\alpha_q \wedge \beta_r)\, \langle\!\langle a, b\rangle\!\rangle\,. \end{aligned}$$

2. Lagrangians of the Yang-Mills type

Let $p\colon C \to M$ be the bundle of connections of P; see Refs. 1,2 for the basic definitions and properties of such a bundle. With the aid of the metric g and an adjoint invariant bilinear form $\langle\cdot,\cdot\rangle$ on $\mathfrak{g}$, a quadratic Lagrangian density $L\mathbf{v}_g\colon J^1C \to \wedge^n T^*M$ is defined by

$$\begin{aligned} (L\mathbf{v}_g)\left(j_x^1\sigma_\Gamma\right) &= \left(\left(\Omega^\Gamma(x), \Omega^\Gamma(x)\right)\right) \mathbf{v}_g(x) \\ &= \Omega^\Gamma(x)\dot\wedge \star\, \Omega^\Gamma(x), \end{aligned} \tag{3}$$

where σ_Γ is the section of p determined the the principal connection Γ defined around $x \in M$, whose curvature form is denoted by Ω^Γ. We recall that the mapping $\Omega\colon J^1C \to \wedge^2 T^*M \otimes \mathrm{ad}\, P$, $j_x^1\sigma_\Gamma \mapsto \Omega^\Gamma(x)$, is a submersion called the curvature mapping (cf. Ref. 1). When $\langle\cdot,\cdot\rangle = \langle\cdot,\cdot\rangle_{\mathrm{CK}}$ is the Cartan-Killing pairing, the Lagrangian (3) is the standard Yang-Mills Lagrangian. In addition, if G is semisimple, then the Euler-Lagrange equations are the well-known Yang-Mills equations. Theorem 3.1 below states that these equations can also be obtained when $\langle\cdot,\cdot\rangle_{\mathrm{CK}}$ is replaced by an arbitrary adjoint-invariant non-degenerate symmetric bilinear form on $\mathfrak{g}$.

The interest of this result is motivated from the fact that there are different invariant symmetric bilinear forms on $\mathfrak{g}$. Indeed, we have

Proposition 2.1. *If $\mathfrak{g}$ is a semisimple Lie algebra and $\mathfrak{g} = \mathfrak{g}_1 \oplus \ldots \oplus \mathfrak{g}_k$ is its decomposition into simple Lie algebras, then the adjoint-invariant symmetric bilinear forms on $\mathfrak{g}$ are as follows:*

$$F\left((A_1, \ldots, A_k), (A'_1, \ldots, A'_k)\right) = \sum_{i=1}^{k} \langle A_i, A'_i \rangle_i, \quad A_i, A'_i \in \mathfrak{g}_i,$$

$\langle \cdot, \cdot \rangle_i$ being any adjoint-invariant symmetric bilinear form on $\mathfrak{g}_i$.

Moreover, it is known (cf. Ref. 10) that a simple real Lie algebra $\mathfrak{g}$ is either a real form of a simple complex Lie algebra $\mathfrak{g}^{\mathbb{C}} = \mathfrak{g} \otimes_{\mathbb{R}} \mathbb{C}$ or the underlying simple real Lie algebra of a given simple complex Lie algebra $\bar{\mathfrak{g}}$. By using the theory of real semisimple Lie algebra (e.g., see Ref. 10), the following can be proved (cf. Ref. 4):

Proposition 2.2. *If $\mathfrak{g}$ is the real form of a simple complex Lie algebra, then any adjoint-invariant bilinear form f on $\mathfrak{g}$ is a scalar multiple of the Cartan-Killing metric on $\mathfrak{g}$.*

If $\mathfrak{g}$ is the underlying real Lie algebra of a simple complex Lie algebra $\bar{\mathfrak{g}}$, then any adjoint invariant bilinear form $\langle \cdot, \cdot \rangle$ in $\mathfrak{g}$ decomposes as

$$\langle \cdot, \cdot \rangle = \lambda \operatorname{Re}(\langle \cdot, \cdot \rangle_{\mathrm{CK}}^{\bar{\mathfrak{g}}}) + \mu \operatorname{Im}(\langle \cdot, \cdot \rangle_{\mathrm{CK}}^{\bar{\mathfrak{g}}}), \quad \lambda, \mu \in \mathbb{R},$$

$\langle \cdot, \cdot \rangle_{\mathrm{CK}}^{\bar{\mathfrak{g}}}$ being the complex Cartan-Killing metric on $\bar{\mathfrak{g}}$.

Remark 2.1. It is straightforward to see that $\langle \cdot, \cdot \rangle_{\mathrm{CK}}^{\mathfrak{g}} = 2 \operatorname{Re}(\langle \cdot, \cdot \rangle_{\mathrm{CK}}^{\bar{\mathfrak{g}}})$. On the other hand, $\operatorname{Im}(\langle \cdot, \cdot \rangle_{\mathrm{CK}}^{\bar{\mathfrak{g}}})$ is not generally a multiple of $\operatorname{Re}(\langle \cdot, \cdot \rangle_{\mathrm{CK}}^{\bar{\mathfrak{g}}})$ as it can be readily check for $\mathfrak{g} = \mathfrak{sl}(2, \mathbb{C})$.

3. Independence of E-L equations

For the proof of the following result we refer the reader to Ref. 3.

Theorem 3.1. *Let $\pi \colon P \to M$ be principal G-bundle on a pseudo-Riemannian compact oriented connected manifold (M, g) and let $p \colon C \to M$ be its bundle of connections. Let $L\mathbf{v}_g$ be the Lagrangian density defined in the formula (3), where $((\cdot, \cdot))$ is the fibred metric induced on $\wedge^\bullet T^*M \otimes \operatorname{ad} P$ by g and an arbitrary non-degenerate symmetric bilinear form on $\mathfrak{g}$, which is invariant under the adjoint representation. Then, the Euler-Lagrange equations for $L\mathbf{v}_g$ are, $\nabla^\Gamma \star \Omega^\Gamma = 0$ and, therefore, they are independent of $\langle \cdot, \cdot \rangle$.*

3.1. *Examples*

- $G = SU(m)$. In this case, the Lie algebra is the real form of the complex simple Lie algebra $\mathfrak{sl}(m, \mathbb{C})$. Hence, any adjoint invariant nondegenerate bilinear form is a scalar multiple of the Cartan-Killing metric and the Yang-Mills Lagrangian is the only Lagrangian density of the type (3).
- $G = Spin(4)$. In this case, $\mathfrak{spin}(4) = \mathfrak{su}(2) \oplus \mathfrak{su}(2)$ and therefore any adjoint invariant bilinear form can be written as $\lambda\langle\cdot,\cdot\rangle_1 + \mu\langle\cdot,\cdot\rangle_2$, $\langle\cdot,\cdot\rangle_i$ being the Cartan-Killing metric on the i-th summand for $i = 1, 2$. There are essentially two different cases: $\lambda = \mu$, for which the Yang-Mills Lagrangian is recovered; and $\lambda = -\mu$, for which the Lagrangian differs from YM Lagrangian, though it provides the same variational equations.
- $G = U(1) \times SU(2)$. This case is relevant in the formulation of the electroweak theory of particles (e.g., see Ref. 11). This group is not semisimple and the set of adjoint invariant bilinear forms are to be computed by hand. Every adjoint invariant quadratic form F is proved to be as follows:

$$F(B, C) = \lambda B^2 + \mu\langle C, C\rangle_{\mathrm{CK}}, \quad B \in \mathfrak{u}(1), C \in \mathfrak{su}(2),$$

  all of them giving rise to the same Euler-Lagrange equations.

4. Independence of H-C equations

4.1. *Preliminaries on H-C equations*

Let $p\colon E \to M$ be any fibred manifold, $\dim M = n$, $\dim E = m + n$. Assume the base manifold M is connected and oriented by a volume form $\mathbf{v}$. The solutions to the Hamilton-Cartan equations for a density on p, say $\Lambda = L\mathbf{v}$, $L \in C^\infty(J^1E)$, are the sections $\bar{s}\colon M \to J^1E$ of the canonical projection $p_1\colon J^1E \to M$ such that,

$$\bar{s}^* (i_X d\Theta_\Lambda) = 0, \quad \forall X \in \mathfrak{X}^v(J^1E), \tag{4}$$

where

$$\Theta_\Lambda = (-1)^{i-1}\frac{\partial L}{\partial y_i^\alpha}\theta^\alpha \wedge \mathbf{v}_i + L\mathbf{v}$$

is the Poincaré-Cartan form defined by Λ (cf. Refs. 5,9), $\mathfrak{X}^v(J^1E)$ denotes the Lie algebra of p_1-vertical vector fields, $\theta^\alpha = dy^\alpha - y_i^\alpha dx^i$ are the contact forms on the J^1E for a fiber coordinate system (x^i, y^α) in E such that

$$\mathbf{v} = dx^1 \wedge \ldots \wedge dx^n,$$
$$\mathbf{v}_i = dx^1 \wedge \ldots \wedge \widehat{dx^i} \wedge \ldots \wedge dx^n.$$

If $\bar{s} = j^1 s$ is a holonomic section, then $\bar{s}$ is a solution to the Hamilton-Cartan equations if and only if s is a solution to the Euler-Lagrange equations. If Λ is regular, that is

$$\det \left(\frac{\partial^2 L}{\partial y_i^\alpha \partial y_j^\beta} \right) \neq 0,$$

then any solution to the Hamilton-Cartan equations is of the form $\bar{s} = j^1 s$, with s a solution to the Euler-Lagrange equations.

For a general discussion about the notion of regularity, including different approaches by using other Lepagean equivalents to a Lagrangian density, we refer the reader to Ref. 7 and Ref. 8.

4.2. *H-C equations for YM Lagrangians*

In the case $E = C \to M$, the Yang-Mills Lagrangian (3) defined by any adjoint invariant non-degenerate bilinear form in $\mathfrak{g}$ is not regular. Hence, the Hamilton-Cartan equations are not equivalent to the Euler-Lagrange equations, though we now show their independence with respect to the pairing in $\mathfrak{g}$. Let $p_{10}\colon J^1C \to C$ be the standard projection.

We have (cf. Ref. 3)

Theorem 4.1. *Let $p\colon C \to M$ be the bundle of connections of a principal G-bundle $\pi\colon P \to M$. The Hamilton-Cartan equations of the Yang-Mills Lagrangian (3), induced by any non-degenerate adjoint invariant bilinear form on $\mathfrak{g}$, for a section $\bar{s}\colon M \to J^1C$ of $p_1\colon J^1C \to M$ are,*

(a) The standard Yang-Mills equation $\nabla^\Gamma \star \Omega^\Gamma = 0$ for the connection Γ defined by the section $s = p_{10} \circ \bar{s}$ of p.
(b) The difference $j^1 s - \bar{s}$ must be symmetric.

Remark 4.1. As $C \to M$ is an affine fiber bundle modelled on

$$T^*M \otimes \operatorname{ad} P \to M,$$

the jet bundle $J^1C \to M$ is also affine, modelled on

$$T^*M \otimes T^*M \otimes \operatorname{ad} P.$$

The difference of two sections of this bundle can thus be understood as a 2-tensor on M taking values in $\operatorname{ad} P$. The condition of the item 2 in Theorem 4.1 thus means that this 2-tensor is symmetric.

5. H-C equations for gauge invariant problems

A gauge transformation $\Phi \colon P \to P$, that is, a vertical G-equivariant diffeomorphism of P, naturally induces a diffeomorphism $\Phi_C \colon C \to C$ by the condition $\Phi_C \circ s_\Gamma = s_{\Phi(\Gamma)}$, where $\Phi(\Gamma)$ is the image of the connection Γ induced by the section s_Γ of the bundle of connections by Φ. A Lagrangian $L \colon J^1C \to \mathbb{R}$ is said to be gauge invariant if $L \circ \Phi_C = L$ for any gauge transformations Φ. A celebrated result by Utiyama (see for example Ref. 1) gives the structure the gauge invariant Lagrangian: a Lagrangian $L \colon J^1C \to \mathbb{R}$ is gauge invariant if and only if $L = \bar{L} \circ \Omega$ for a function $\bar{L} \colon \wedge^2 T^*M \otimes \operatorname{ad} P \to \mathbb{R}$ invariant under the adjoint action on $\operatorname{ad} P$.

A gauge invariant Lagrangian $L \colon J^1C \to \mathbb{R}$ is said to be "weakly regular" if

$$\det \left(\frac{\partial^2 \bar{L}}{\partial R^\alpha_{ij} \partial R^\beta_{kl}} \right) \neq 0, \tag{5}$$

where (x^i, R^α_{ij}) is the natural fibred coordinate system on $\wedge^2 T^*M \otimes \operatorname{ad} P$ induced from a coordinate system (U, x^i) on M such that $\pi^{-1}(U) \cong U \times G$, and a basis (B_α) for $\mathfrak{g}$; i.e.,

$$\eta_2 = \sum_{i<j} R^\alpha_{ij}(\eta_2) \left(dx^i\right)_x \wedge \left(dx^j\right)_x \otimes \left(\tilde{B}_\alpha\right)_x,$$
$$\forall \eta_2 \in \wedge^2 T^*_x M \otimes \operatorname{ad} P,\ x \in U,$$

where $\tilde{B}_\alpha$ is the infinitesimal generator of the one-parameter group of gauge transformations $\tau_t \colon U \times G \to U \times G$, $\tau_t(x,g) = (x, \exp(tB_\alpha) \cdot g)$.

For example, for every non-degenerate adjoint-invariant symmetric bilinear form on $\mathfrak{g}$ the Yang-Mills Lagrangian (3) is weakly regular.

A natural generalization of Theorem 4.1 for every weakly regular Lagrangian is as follows:

Theorem 5.1. *Let $\bar{\mathcal{S}}_\Lambda$ (resp. $\mathcal{S}_\Lambda$) denote the set of solutions to H-C (resp. E-L) equations of a weakly regular gauge-invariant Lagrangian L on the bundle of connections $p \colon C \to M$ of a principal bundle $\pi \colon P \to M$. If $\bar{s} \colon M \to J^1C$ belongs to $\bar{\mathcal{S}}_\Lambda$, then the section $s = p_{10} \circ \bar{s}$ belongs to $\mathcal{S}_\Lambda$. Hence a natural projection exists $\varrho \colon \bar{\mathcal{S}}_\Lambda \to \mathcal{S}_\Lambda$, $\varrho(\bar{s}) = p_{10} \circ \bar{s}$, which is an* *affine bundle modelled as follows:*

$$\varrho^{-1}(s) = \left\{ j^1 s + t : t \in \Gamma(S^2T^*M \otimes \operatorname{ad} P) \right\}, \quad \forall s \in \mathcal{S}_\Lambda.$$

The basic steps of the proof are the following:

(a) As $L = \bar{L} \circ \Omega$, we obtain

$$\frac{\partial \bar{L}}{\partial A^\alpha_{i,i}} = 0, \qquad \frac{\partial L}{\partial A^\alpha_{i,k}} = \frac{\partial \bar{L}}{\partial R^\alpha_{ik}} \circ \Omega, \tag{6}$$

where $R^\alpha_{ik} = -R^\alpha_{ki}$ for $i > k$. From the H-C equations for $X = \partial/\partial A^\alpha_{i,k}$ we deduce the following local expressions:

$$(s^\beta_{h,j} - \bar{s}^\beta_{h,j}) \left(\frac{\partial^2 \bar{L}}{\partial R^\alpha_{ik} \partial R^\beta_{hj}} \circ (\Omega \circ \bar{s}) \right) = 0.$$

If the weak regularity condition (5) holds, then the previous equation yields

$$(s^\beta_{h,j} - \bar{s}^\beta_{h,j}) - (s^\beta_{j,h} - \bar{s}^\beta_{j,h}) = 0, \qquad \forall \beta, h, j. \tag{7}$$

If we write $\bar{s} = j^1 s + t$, for a 2-tensor t, the condition above means that t is symmetric.

(b) Taking the equations (6) and (7) into account, from the H-C equations for $X = \partial/\partial A^\alpha_i$ we deduce

$$-\frac{\partial}{\partial x^j} \left(\frac{\partial L}{\partial A^\alpha_{i,j}} \circ \bar{s} \right) + \frac{\partial L}{\partial A^\alpha_i} \circ \bar{s} = 0, \tag{8}$$

which is precisely the E-L equation, but evaluated at $\bar{s}$ instead of $j^1 s$.

(c) Finally, the following formula is readily checked:

$$\frac{\partial L}{\partial A^\alpha_i} = 2 \left(c^\beta_{\alpha\gamma} A^\gamma_j \frac{\partial \hat{L}}{\partial R^\beta_{ij}} \right) \circ \Omega. \tag{9}$$

Moreover, from (7) we deduce

$$\begin{aligned} \Omega \circ \bar{s} &= (\bar{s}^\alpha_{i,j} - \bar{s}^\alpha_{j,i} - c^\alpha_{\beta\gamma} s^\beta_i s^\gamma_j) dx^i \wedge dx^j \otimes \tilde{B}_\beta \\ &= (s^\alpha_{i,j} - s^\alpha_{j,i} - c^\alpha_{\beta\gamma} s^\beta_i s^\gamma_j) dx^i \wedge dx^j \otimes \tilde{B}_\beta \\ &= \Omega \circ j^1 s. \end{aligned} \tag{10}$$

Therefore, from the formulas (6) and (9) we conclude that the equation (8) coincides with the E-L equation for s.

Acknowledgments

Supported by Ministerio de Educación y Ciencia of Spain under grants MTM2005–00173 and MTM2007–60017.

References

1. D. Bleecker, *Gauge theory and variational principles* (Global Analysis Pure and Applied Series A, 1, Addison-Wesley Publishing Co., Reading, MA, 1981).
2. M. Castrillón López and J. Muñoz Masqué, The geometry of the bundle of connections, *Math. Z.* **236** (2001) 797–811.
3. M. Castrillón López and J. Muñoz Masqué, Independence of Yang-Mills Equations with Respect to the Invariant Pairing in the Lie Algebra, *Int. J. Theor. Phys.* **46** (2007) (4) 1020–1026.
4. M. Castrillón López and J. Muñoz Masqué, Gauge-Invariant Characterization of Yang-Mills-Higgs Equations, *Ann. Henri Poincaré* **8** (2007) 203–217.
5. H. Goldschmidt and S. Sternberg, The Hamilton-Cartan formalism in the calculus of variations, *Ann. Inst. Fourier (Grenoble)* **23** (1973) 203–267.
6. S. Kobayashi and K. Nomizu, *Foundations of Differential Geometry* (John Wiley & Sons, Inc. (Interscience Division), New York, Volume I, 1963).
7. O. Krupková, Hamiltonian field theory, *J. Geom. Phys.* **43** (2-3) (2002) 93–132.
8. O. Krupková and D. Smetanová, On regularization of variational problems in first-order field theory, *In: Proceedings of the 20th Winter School "Geometry and Physics"* (Srní, 2000, Rend. Circ. Mat. Palermo (2) Suppl. No. 66, 2001) 133–140.
9. J. Muñoz Masqué and L.M. Pozo Coronado, Parameter Invariance in Field Theory and the Hamiltonian Formalism, *Fortschr. Phys.* **48** (2000) 361–405.
10. A.L. Onishchik, *Lectures on real semisimple Lie algebras and their representations* (ESI Lectures in Mathematics and Physics, European Mathematical Society, Zürich, 2004).
11. A. Zee, *Quantum Field Theory in a nutshell* (Princeton University Press, New Jersey, 2003).

Differential Geometry and its Applications
Proc. Conf., in Honour of Leonhard Euler, Olomouc, August 2007

Non-holonomic constraint forces in field theory in physics

L. Czudková and J. Musilová
Institute of Theoretical Physics and Astrophysics, Masaryk University, Brno, Czech Republic
E-mail: czudkova@physics.muni.cz, janam@physics.muni.cz
http://www.physics.muni.cz/

A problem of non-holonomic constraint forces in field theory is discussed, using the geometrical theory of non-holonomic constrained systems as proposed by Krupková.[8] Geometrically, fibered manifolds with n-dimensional bases ($n > 1$ for field theory and $n = 1$ for mechanics) and their prolongations are there chosen as underlying structures. From the physical point of view, this theory is based on constraint forces of a special type, called *Chetaev-type forces*, in analogy with well-known Chetaev forces. The main purpose of the paper is to discuss possibilities of the use of the theory for realistic physical situations, in our case in continuum mechanics, or, more precisely, in elasticity theory. By means of an elementary but in physical experiments frequently occurring example we show some limitations of the choice of physical problems suitable for the application of the mentioned geometrical theory. They lie primarily in the fact that another types of constraint forces can be taken into account. Then we sketch a more general approach to a problem of non-holonomic constraint forces.

Keywords: Fibered manifold, field theory, non-holonomic constraints, constraint submanifold, canonical distribution, constraint forces, Chetaev forces, reduced equations, deformed equations.

MS classification: 35G20, 58A10, 58J90.

1. Introduction

Equations describing the time-space evolution of systems in physics and engineering (*equations of motion*) have often the form of differential equations, ordinary in mechanics and partial in field theories, subjected to various types of constraints, holonomic or non-holonomic. In physics, these constraints are always interpreted by means of some additional forces, called *constraint* ones. Using such a description of constraints, equations of constrained motion can be easily obtained by modification of unconstrained equations, the corresponding constraint variational principle can be formulated and the constraint variationality of equations can be discussed (for mechanics see e.g. Refs. 5,9,11). In mechanics, theories of constrained systems are well elaborated for both holonomic and non-holonomic case. We refer here to the geometrical theory of constrained mechanical systems on fibered manifolds with one-dimensional bases proposed in 1990's by Krupková.[6,7] This theory involves *two equivalent approaches* to a problem of non-holonomic constrained systems (for details see Ref. 7). In the first of them purely geometrical methods are used, leading to a system of differential equations for unknown trajectories of the constrained motion, called *reduced equations.* The second approach is "more physical". It is based on the concept of a special type of constraint forces, called the *Chetaev-type constraint forces.* They are introduced in the analogy with well-known Chetaev forces (cf. Ref. 2) and they represent their generalisation for non-holonomic cases. These forces, containing *Lagrange multipliers*, are introduced into original (unconstrained) equations of motion, giving rise to the system of *deformed equations* for unknown trajectories and Lagrange multipliers. There were studied various aspects of the geometrical theory of non-holonomic systems in mechanics, especially:

- concept of constraint variationality (e.g. Refs. 4,9),
- constraint variational principle (e.g. Ref. 12),
- application to concrete examples and discussions of results from both geometrical and physical point of view (e.g. Refs. 3,4).

Having on mind the topic considered bellow, let us recall one of the results of the last item: While the Chetaev force is in physics *only possible type of constraint force* for systems subjected to *holonomic constraints* (e.g. a body on an inclined plane, a spherical pendulum, etc.), the situation for non-holonomic constraints is more general — there exist constraint forces that *are not* of Chetaev-type.

Recently, the geometrical theory was extended also to fibered manifolds with generally n-dimensional bases, which for $n > 1$ corresponds to field theories in physics.[8] As in mechanics, there are two equivalent approaches to a problem of constrained systems: "geometrical" one, leading to reduced equations without Lagrange multipliers, and "physical" one, giving deformed equations with Chetaev-type constraint forces and Lagrange multipliers.

The main purpose of our paper is to discuss possibilities of using the geometrical theory to describe the constrained motion of non-holonomic systems in realistic physical situations. More precisely, we choose for our discussion a typical example included in the theory of elasticity — the elastic deformation of an isotropic body. Since the general scheme of reduced equations is somewhat complicated, we sketch here (in Sec. 2) only main ideas of the corresponding approach and basic steps of the procedure leading to mentioned equations. However, for analysis of our example we prefer the equivalent but better arranged deformed equations. This moreover permits to illustrate basic properties of Chetaev-type constraint forces. We disclose some limitations on the choice of physical situations appropriate for treating them within the geometrical theory. These limitations are related to parameters determining underlying geometrical structure on the one hand, and to the possible existence of another types of constraint forces on the other. Finally, we sketch an idea of a more general approach to the choice of constraint forces.

2. Geometrical background — a brief outline

Notation:
The standard notation of geometrical objects is used. Let $\pi : Y \longrightarrow X$ be a fibered manifold with n-dimensional base X ($n > 1$ for field theory) and total space Y of dimension $m + n$, $\pi_r : J^rY \longrightarrow X$, $0 \leq r \leq 2$, $J^0Y = Y$, $\pi_0 = \pi$, its r-jet prolongations and $\pi_{r,s} : J^rY \longrightarrow J^sY$, $0 \leq s < r \leq 2$, canonical projections. Denote by (V, ψ), $\psi = (x^i, y^\sigma)$, $1 \leq i \leq n$, $1 \leq \sigma \leq m$, a local fibered chart on Y and (V_s, ψ_s) for $1 \leq s \leq 2$, $\psi_1 = (x^i, y^\sigma, y^\sigma_j)$, $\psi_2 = (x^i, y^\sigma, y^\sigma_j, y^\sigma_{jk})$, $1 \leq j \leq k \leq n$, the associated fibered charts on J^sY. A section $\delta : I \longrightarrow J^rY$ of π_r, where $I \subset X$ is an open set, is called *holonomic* if there is a section γ of π such that $\delta = J^r\gamma$.

A vector field ξ on J^rY is called π_r-*projectable* if there is a vector field ξ_0 on X such that $T\pi_r\,\xi = \xi_0 \circ \pi_r$. It is called π_r-*vertical* if $T\pi_r\,\xi = 0$. A form η on J^rY is called π_r-*horizontal* if $i_\xi\eta = 0$ for every π_r-vertical vector field on J^rY. A form η is called *contact* if $J^r\gamma^*\eta = 0$, $r > 0$, for every section γ of π. Analogously, a $\pi_{r,s}$-projectable vector field, $\pi_{r,s}$-vertical vector field and $\pi_{r,s}$-horizontal form are defined. We use the basis $(dx^i, \omega^\sigma, dy^\sigma_j)$ of 1-forms on J^1Y and the basis $(dx^i, \omega^\sigma, \omega^\sigma_j, dy^\sigma_{jk})$ of 1-forms on J^2Y adapted to the contact structure, where $\omega^\sigma = dy^\sigma - y^\sigma_j\,dx^j$ and $\omega^\sigma_j = dy^\sigma_j - y^\sigma_{jk}\,dx^k$ are contact 1-forms. Recall that for every k-form η on J^rY we have the unique decomposition $\pi^*_{r+1,r}\eta = \sum_{q=0}^{k} p_q\eta$, where $p_0\eta = h\eta$ is the horizontal component of η and $p_s\eta$, $1 \leq s \leq k$, is the s-contact component of η.

A *distribution* $\mathcal{D}$ on J^rY is a mapping assigning to every point $x \in J^rY$ a vector subspace $\mathcal{D}(x) \subset T_xJ^rY$. It is generated either by (local) vector fields ξ_ι on J^rY, $\iota \in \mathcal{I}$, or, equivalently, by (local) 1-forms η_κ on J^rY, $\kappa \in \mathcal{K}$, $i_{\xi_\iota}\eta_\kappa = 0$ for every $\iota \in \mathcal{I}$, $\kappa \in \mathcal{K}$, where $\mathcal{I}$, $\mathcal{K}$ are sets of indices. These forms generate the annihilator $\mathcal{D}^0$ of $\mathcal{D}$. A section δ of π_r is called an *integral section* of a distribution $\mathcal{D}$ if $\delta^*\eta = 0$ for every $\eta \in \mathcal{D}^0$.

Unconstrained partial differential equations:
Consider a system of m partial differential equations (PDE) affine in accelerations

$$E_\sigma \equiv A_\sigma + B^{ij}_{\sigma\nu}\, y^\nu_{ij} = 0\,, \qquad 1 \le \sigma, \nu \le m\,, \tag{1}$$

where A_σ and $B^{ij}_{\sigma\nu}$ are functions on J^1Y. (Such system is typical for equations of motion in physics.) In the geometrical theory, the system (1) is described by the *dynamical form* (1-contact $\pi_{2,0}$-horizontal $(n+1)$-form on J^2Y)

$$E = E_\sigma \omega^\sigma \wedge \omega_0\,, \qquad\qquad \omega_0 = dx^1 \wedge \cdots \wedge dx^n\,,$$

or equivalently, by the *Lepage class* $[\alpha]$ of $(n+1)$-forms on J^1Y such that $p_1\alpha' = E$ for all $\alpha' \in [\alpha]$. It is easy to check that $\pi^*_{2,1}\alpha' = E + F$, where F is at least 2-contact form on J^1Y.

A section γ of π is called a *path* of dynamical form E if

$$E \circ J^2\gamma = 0\,, \quad \text{i.e.} \quad A_\sigma + B^{ij}_{\sigma\nu}\, y^\nu_{ij} = 0 \quad \text{along} \quad J^2\gamma\,.$$

Proposition 2.1. *A section γ of π is a path of dynamical form iff for every $\alpha' \in [\alpha]$ it holds $J^1\gamma^* i_\xi \alpha' = 0$ for every π_1-vertical vector field ξ on J^1Y.*

Partial differential equations with non-holonomic constraints:
Consider again the system of m partial differential equations (1) together with *equations of non-holonomic constraints*

$$f^\alpha\left(x^i, y^\sigma, y^\sigma_j\right) = 0\,, \quad 1 \le \alpha \le \kappa\,, \quad 1 \le \kappa \le mn - 1\,, \tag{2}$$

satisfying the rank conditions

- rank $\left(\frac{\partial f^\alpha}{\partial y^\sigma_j}\right) = \kappa$, where α and (σ, j) are row and column indices, respectively,
- rank $\left(\frac{\partial f^\alpha}{\partial y^\sigma_j}\right) = k$ for some k such that $1 \le k \le m-1$, where (α, j) and σ are row and column indices, respectively.

The first condition means that equations of constraints (2) define a *constraint submanifold* $Q \subset J^1Y$ of codimension κ fibered over Y. This submanifold is called *regular constraint of corank* (κ, k). The second condition is important for the possibility to introduce the so-called *canonical distribution* on Q defined by the following theorem (for details see Ref. 8):

Theorem 2.1. *Let Q be a regular constraint on J^1Y of corank (κ, k) and let $\iota : Q \longrightarrow J^1Y$ be the canonical embedding. Let $\varphi^a = (M^a_\sigma \circ \iota)\, \iota^* \omega^\sigma$, $1 \leq a \leq k$, where the $(k \times m)$-matrix $M = (M^a_\sigma)$, $M^a_\sigma = \frac{1}{n} c^a_{\alpha j} \frac{\partial f^\alpha}{\partial y^\sigma_j}$, has for some set of functions $c^a_{\alpha j}$ defined on an open set $U \subset V_1$ maximal rank k. Then $\mathcal{C}$=annih$\{\varphi^a, 1 \leq a \leq k\}$ is a distribution of corank k on Q.*

The meaning of this distribution is the following: Let $\gamma : I \longrightarrow Y$, where $I \subset X$ is an open set, be a section of π. Then γ is a *constrained path* (i.e. $f^\alpha \circ J^1\gamma = 0$, $1 \leq \alpha \leq \kappa$), iff $J^1\gamma$ is an integral section of $\mathcal{C}$. Moreover, 1-forms annihilating $\mathcal{C}$ generate the *constraint ideal* factorizing the Lepage class of the dynamical form as follows: Let $[\alpha]$ be the Lepage class of the dynamical form E and $\iota : Q \longrightarrow J^1Y$ be the canonical embedding of Q into J^1Y. Consider the equivalence class $[[\alpha]]$ such that

$$\bar{\alpha} \in [[\alpha]] \quad \text{if} \quad \bar{\alpha} = \iota^* \alpha' + \bar{F} + \phi\,,$$

where α' is a $(n+1)$-form belonging to the Lepage class $[\alpha]$, $\bar{F}$ is at least 2-contact form and ϕ is a form belonging to the constraint ideal. The class $[[\alpha]]$ is called *constrained system.* Then, the *reduced equations* are introduced as follows:

$$J^1\gamma^* i_\xi \bar{\alpha} = 0 \qquad \text{for every } \pi_1\text{−vertical vector field } \xi \text{ belonging to } \mathcal{C}\,,$$

$$f^\alpha \circ J^1\gamma = 0\,, \quad 1 \leq \alpha \leq \kappa\,.$$

This represents a system of $(m-k)$ second-order PDE and κ first-order PDE for unknown components of γ. In adapted fibered coordinates, the reduced equations are of the form summarized in the following proposition (for details and proofs see Ref. 8):

Proposition 2.2. *Denote by (x^i, y^σ, z^J), $1 \leq i \leq n$, $1 \leq \sigma \leq m$, $1 \leq J \leq$ $\leq nm - \kappa$, $1 \leq \alpha \leq \kappa$, adapted fibered coordinates on Q and $\bar{A}_\sigma = A_\sigma \circ \iota$, $\bar{B}^{ji}_{\sigma\nu} = B^{ji}_{\sigma\nu} \circ \iota$. Then the reduced equations are of the form*

$$(\mathcal{A}_p + \mathcal{B}^i_{pJ} z^J_i) \circ J^2\gamma = 0\,, \qquad f^\alpha \circ J^1\gamma = 0\,, \quad 1 \leq p \leq m-k\,,$$

where

$$\mathcal{A}_p = \bar{A}_p + \bar{A}_{m-k+a} \bar{G}^a_p + \left(\bar{B}^{ji}_{p\nu} + \bar{B}^{ji}_{m-k+a\,\nu} \bar{G}^a_p\right) \frac{d'_c g^\nu_j}{dx^i}\,,$$

$$\mathcal{B}^i_{pJ} = \left(\bar{B}^{ji}_{p\nu} + \bar{B}^{ji}_{m-k+a\,\nu} \bar{G}^a_p\right) \frac{\partial g^\nu_j}{\partial z^J}\,.$$

Here $g^\sigma_j = y^\sigma_j \circ \iota$ and $\bar{G}^a_p = G^a_p \circ \iota$. Functions G^a_p are at each point of J^1Y determined by a fundamental system $\Xi_p = (\Xi^\sigma_p)$ of solutions of the algebraic system of equations $M^a_\sigma \Xi^\sigma = 0$ for m unknowns Ξ^σ, where for the choice $\Xi^{k+q}_p = \delta^a_p$, $1 \le p, q \le m - k$, we put $G^a_p = \Xi^a_p$, $1 \le a \le k$. Further, $\frac{d'_c}{dx^i} = \frac{\partial_c}{\partial x^i} + g^p_i \frac{\partial_c}{\partial y^p}$ is the *i-th cut constraint derivative.*

The above mentioned geometrical approach is equivalent to the "physical one" leading to *deformed equations* based on the concept of Chetaev-type constraint forces, introduced by their components $\Phi_\sigma = \Lambda_{\alpha l} \frac{\partial f^\alpha}{\partial y^\sigma_l}$ (cf. Ref. 2), where $\Lambda_{\alpha l} = \Lambda_{\alpha l}(x^i, y^\sigma, y^\sigma_j)$ stand for *Lagrange multipliers*:

$$A_\sigma + B^{ij}_{\sigma\nu}\, y^\nu_{ij} = \Lambda_{\alpha l}\, \frac{\partial f^\alpha}{\partial y^\sigma_l} \qquad \text{along} \quad J^2\gamma\,,$$

$$f^\alpha \circ J^1\gamma = 0\,, \qquad 1 \le \alpha \le \kappa\,.$$

Proposition 2.3. *A section γ of π is a solution of deformed equations iff for every $\alpha'_\Phi \in [\alpha_\Phi] = [\alpha - \Phi]$ it holds $J^1\gamma^* i_\xi \alpha'_\Phi = 0$ for every π_1-vertical vector field ξ on J^1Y and $f^\alpha \circ J^1\gamma = 0$.*

Deformed equations represent the system of m second-order PDE and κ first-order PDE for the components of section γ and for Lagrange multipliers $\Lambda_{\alpha l}$. Contrary to mechanics (cf. Ref. 7), general form of deformed equations is in field theory much simpler then the general form of corresponding reduced equations. Moreover, deformed equations enable us to determine directly components of Chetaev-type constraint force. Thus, we will use them for our next calculations.

3. Example

In this section we apply the results of the geometrical theory to an example typical for theory of elasticity. Let us take into account well-known system of PDE describing a small elastic deformation of an isotropic body (Refs. 1,10)

$$\frac{\partial \sigma_{ji}}{\partial x_j} + G_i = \varrho\, \frac{\partial^2 u_i}{\partial t^2}\,, \qquad 1 \le i, j \le 3\,, \tag{3}$$

$$\sigma_{ij} = \lambda \delta_{ij}\,(e_{11} + e_{22} + e_{33}) + 2\mu e_{ij}\,, \tag{4}$$

where the notation standard in physics is used, i.e. x_i are cartesian coordinates, t is time, $\sigma_{ij} = \sigma_{ij}(x_1, x_2, x_3, t)$ are components of the stress tensor that is symmetrical, $G_i = G_i(x_1, x_2, x_3, t)$ are components of the net force per unit volume, $\varrho = \varrho(x_1, x_2, x_3, t)$ is the local density of the body,

$u_i = u_i(x_1, x_2, x_3, t)$ are components of the displacement, λ and μ are Lamé coefficients and $e_{ij} = \frac{1}{2}\left(\frac{\partial u_i}{\partial x_j} + \frac{\partial u_j}{\partial x_i}\right)$ are components of the (symmetrical) strain tensor. We will study equations (3) and (4) under two different constraint conditions:

(A) Consider the system of constraints

$$\sigma_{ij} = const_{\,ij}$$

that refer to so-called static method of loading of material (creep). As one can see, there are two possible approaches to the system of equations (3) and (4).

Approach 1:
Equations (3) and (4) give the system of 9 equations for 9 unknowns $u_1, u_2, u_3, \sigma_{11}, \sigma_{22}, \sigma_{33}, \sigma_{23}, \sigma_{13}, \sigma_{12}$. Then the underlying fibered manifold is $\pi : Y \longrightarrow X$, where $X = \mathbf{R}^4$, $Y = \mathbf{R}^4 \times \mathbf{R}^9$ with the global charts

$$(X, \varphi)\,, \quad \varphi = (x^1, x^2, x^3, x^4) = (x_1, x_2, x_3, t)\,,$$

$$\begin{aligned}(Y, \psi)\,, \quad \psi &= (x^1, x^2, x^3, x^4, y^1, y^2, y^3, y^4, y^5, y^6, y^7, y^8, y^9) = \\ &= (x_1, x_2, x_3, t, u_1, u_2, u_3, \sigma_{11}, \sigma_{22}, \sigma_{33}, \sigma_{23}, \sigma_{13}, \sigma_{12})\,,\end{aligned}$$

$n = \dim X = 4$, $m = \dim Y - n = 9$. Constraints $\sigma_{ij} = const_{\,ij}$, i.e $y^\tau = const^\tau$, $4 \leq \tau \leq 9$, are holonomic. Corresponding non-holonomic constraints are of the form (see Ref. 8)

$$y_l^\tau = \frac{\partial \sigma_{ij}}{\partial x^l} = 0\,, \qquad 4 \leq \tau \leq 9\,, \qquad 1 \leq l \leq 4\,.$$

For this system of constraints it holds $\kappa = 24$, $k = 6$ and both rank conditions (see Sec. 2) are fulfilled. Computing components of the constraint force, we come to deformed equations

$$\frac{\partial \sigma_{ji}}{\partial x_j} + G_i = \varrho \frac{\partial^2 u_i}{\partial t^2}\,, \qquad \sigma_{ij} = \lambda \delta_{ij}\,(e_{11} + e_{22} + e_{33}) + 2\mu e_{ij} + \tilde{\Lambda}_{ij}\,, \qquad y_l^\tau = 0\,,$$
$$1 \leq i, j \leq 3\,, \quad 1 \leq l \leq 4\,, \quad 4 \leq \tau \leq 9\,,$$

where every $\tilde{\Lambda}_{ij}$ stands for a definitely given linear combination of Lagrange multipliers $\Lambda_{\alpha l}$, $1 \leq \alpha \leq 24$. In adapted fibered coordinates $(x^l, y^\sigma, y_l^\rho)$,

$1 \leq l \leq 4$, $1 \leq \sigma \leq 9$, $1 \leq \rho \leq 3$, these equations have the following form:

$$y_1^4 + y_2^9 + y_3^8 + G_1 - \varrho y_{44}^1 = 0\,, \tag{5}$$
$$y_1^9 + y_2^5 + y_3^7 + G_2 - \varrho y_{44}^2 = 0\,, \tag{6}$$
$$y_1^8 + y_2^7 + y_3^6 + G_3 - \varrho y_{44}^3 = 0\,, \tag{7}$$
$$y^4 - \lambda\left(y_1^1 + y_2^2 + y_3^3\right) - 2\mu y_1^1 = \tilde{\Lambda}_{11}\,, \tag{8}$$
$$y^5 - \lambda\left(y_1^1 + y_2^2 + y_3^3\right) - 2\mu y_2^2 = \tilde{\Lambda}_{22}\,, \tag{9}$$
$$y^6 - \lambda\left(y_1^1 + y_2^2 + y_3^3\right) - 2\mu y_3^3 = \tilde{\Lambda}_{33}\,, \tag{10}$$
$$y^7 - \mu\left(y_3^2 + y_2^3\right) = \tilde{\Lambda}_{23}\,, \tag{11}$$
$$y^8 - \mu\left(y_3^1 + y_1^3\right) = \tilde{\Lambda}_{13}\,, \tag{12}$$
$$y^9 - \mu\left(y_2^1 + y_1^2\right) = \tilde{\Lambda}_{12}\,, \tag{13}$$
$$y_1^4 = \cdots = y_4^4 = y_1^5 = \cdots = y_4^5 = \cdots = y_1^9 = \cdots = y_4^9 = 0\,. \tag{14}$$

Note that $B_{11}^{44} = B_{22}^{44} = B_{33}^{44} = -\varrho$ and $B_{\sigma\nu}^{ij} = 0$ for other cases. A_σ are the remaining parts of left-hand sides of Eqs. (5-13). Let us emphasize that Eqs. (5-14) form the system of 33 PDE for 9 unknowns components $y^\sigma\gamma$ of the constrained path γ and for 96 primary Lagrange multipliers $\Lambda_{\alpha l}$.

As for reduced equations, matrices M and G can be easily obtained because of the simplicity of constraint equations. M is the (6×9)-matrix, $M_\sigma^a = \delta_\sigma^{a+3}$, and G is zero (6×3)-matrix. Thus, the system of reduced equations is given by Eqs. (5,6,7,14).

Approach 2:
Substituting Eq. (4) into Eq. (3) we obtain 3 equations for 3 unknowns u_1, u_2, u_3. This system will be studied on the fibered manifold $\pi : Y \longrightarrow X$, where $X = \mathbf{R}^4$, $Y = \mathbf{R}^4 \times \mathbf{R}^3$ with the global charts

$$(X, \varphi)\,, \quad \varphi = \left(x^1, x^2, x^3, x^4\right) = \left(x_1, x_2, x_3, t\right)\,,$$

$$(Y, \psi)\,, \quad \psi = \left(x^1, x^2, x^3, x^4, y^1, y^2, y^3\right) = \left(x_1, x_2, x_3, t, u_1, u_2, u_3\right)\,,$$

$n = \dim X = 4$, $m = \dim Y - n = 3$. Constraints $\sigma_{ij} = const_{\,ij}$, i.e.

$$\lambda\delta_{ij}\left(\frac{\partial u_1}{\partial x_1} + \frac{\partial u_2}{\partial x_2} + \frac{\partial u_3}{\partial x_3}\right) + \mu\left(\frac{\partial u_i}{\partial x_j} + \frac{\partial u_j}{\partial x_i}\right) - const_{\,ij} = 0\,,$$

i.e.

$$\begin{aligned}
&\lambda\left(y_1^1 + y_2^2 + y_3^3\right) + 2\mu y_1^1 - const_{\,11} = 0\,, && \mu\left(y_3^2 + y_2^3\right) - const_{\,23} = 0\,,\\
&\lambda\left(y_1^1 + y_2^2 + y_3^3\right) + 2\mu y_2^2 - const_{\,22} = 0\,, && \mu\left(y_3^1 + y_1^3\right) - const_{\,13} = 0\,,\\
&\lambda\left(y_1^1 + y_2^2 + y_3^3\right) + 2\mu y_3^3 - const_{\,33} = 0\,, && \mu\left(y_2^1 + y_1^2\right) - const_{\,12} = 0\,,
\end{aligned}$$

are non-holonomic with $\kappa = 6$ and $k = 3$. Thus, the second rank condition is not satisfied ($k = 3$, $m - 1 = 2$) and the geometrical theory in its non-trivial form ($k \leq m - 1$) is not applicable. This situation is similar to another realistic case, the stress tensor without torsion ($\sigma_{ij} = 0$ for $i \neq j$).

(B) Consider a more general constraint

$$\sigma_{12} = const_{12}\,.$$

Analogously to above procedure we obtain the following results:

Approach 1:
Equations of non-holonomic constraints are

$$y_l^9 = \frac{\partial \sigma_{12}}{\partial x^l} = 0\,, \qquad 1 \leq l \leq 4\,.$$

It holds $\kappa = 4$, $k = 1$, i.e. both rank conditions are satisfied. Deformed equations are

$$y_1^4 + y_2^9 + y_3^8 + G_1 - \varrho y_{44}^1 = 0\,, \tag{15}$$
$$y_1^9 + y_2^5 + y_3^7 + G_2 - \varrho y_{44}^2 = 0\,, \tag{16}$$
$$y_1^8 + y_2^7 + y_3^6 + G_3 - \varrho y_{44}^3 = 0\,, \tag{17}$$
$$y^4 - \lambda\left(y_1^1 + y_2^2 + y_3^3\right) - 2\mu y_1^1 = 0\,, \tag{18}$$
$$y^5 - \lambda\left(y_1^1 + y_2^2 + y_3^3\right) - 2\mu y_2^2 = 0\,, \tag{19}$$
$$y^6 - \lambda\left(y_1^1 + y_2^2 + y_3^3\right) - 2\mu y_3^3 = 0\,, \tag{20}$$
$$y^7 - \mu\left(y_3^2 + y_2^3\right) = 0\,, \tag{21}$$
$$y^8 - \mu\left(y_3^1 + y_1^3\right) = 0\,, \tag{22}$$
$$y^9 - \mu\left(y_2^1 + y_1^2\right) = \tilde{\Lambda}\,, \tag{23}$$
$$y_1^9 = y_2^9 = y_3^9 = y_4^9 = 0\,, \tag{24}$$

where $\tilde{\Lambda}$ stands for a definitely given linear combination of Lagrange multipliers. As for reduced equations, they are given by Eqs. (15-22) together with Eq. (24).

Approach 2:
We have only one equation of constraint

$$\mu\left(\frac{\partial u_1}{\partial x_2} + \frac{\partial u_2}{\partial x_1}\right) - const_{12} = 0\,, \quad \text{i.e.} \quad \mu\left(y_2^1 + y_1^2\right) - const_{12} = 0\,,$$

$\kappa = 1$, $k = 2$, so the rank conditions are fulfilled. Deformed equations are

$$\lambda\left(y^1_{11}+y^2_{12}+y^3_{13}\right)+\mu\left(2y^1_{11}+y^1_{22}+y^2_{12}+y^1_{33}+y^3_{13}\right)+G_1-\varrho y^1_{44} = \tilde{\Lambda}_1\,, \quad (25)$$

$$\lambda\left(y^1_{12}+y^2_{22}+y^3_{23}\right)+\mu\left(2y^2_{22}+y^1_{12}+y^2_{11}+y^2_{33}+y^3_{23}\right)+G_2-\varrho y^2_{44} = \tilde{\Lambda}_2\,, \quad (26)$$

$$\lambda\left(y^1_{13}+y^2_{23}+y^3_{33}\right)+\mu\left(2y^3_{33}+y^1_{13}+y^3_{11}+y^2_{23}+y^3_{22}\right)+G_3-\varrho y^3_{44} = 0\,, \quad (27)$$

$$\mu\left(y^1_2+y^2_1\right)-const_{12} = 0\,, \quad (28)$$

where $\tilde{\Lambda}_1$ and $\tilde{\Lambda}_2$ stand for definitely given linear combinations of Lagrange multipliers. As for reduced equations, they are given by Eq. (27) together with Eq. (28).

Conclusions:
We have discussed the system of PDE given by Eqs. (3,4) representing a situation typical for one of field theories in physics (theory of elasticity) and frequently studied not only theoretically, but also by standard experiments. We presented two different starting formulations of the problem to discuss the possibility to use the geometrical field theory of non-holonomic systems as the mathematical background. The theory is, in principle, convenient for formulation of practical problems. On the other hand, there can be some physical situations not appropriate for treating them by this theory. The main reason of such a result lies in the possible existence of more general constraint forces than the Chetaev-type ones.

4. Constraint forces — an idea of a more general approach

Having all the time on mind the geometrical meaning of solutions of a constrained system of PDE, i.e. the fact that they are just sections γ of π such that $J^1\gamma$ are integral sections of the canonical distribution, we can generalise the concept of constraint forces to a larger family by the following way: We choose $(m-k)$ components of constraint forces arbitrarily, and the remaining k components are then given by initial unconstrained PDE using the constraint equations. (Note that such constraint forces are only local in general, the same choice of $(m-k)$ "free" components (e.g. zeros) in a different chart represents the different but from the physical point of view again admissible constraint force.) A detailed analysis of such a more general approach is not the goal of our paper.

Acknowledgments

The work is supported by the grant 201/06/0922 of the Czech Grant Agency and by the grant MSM 0021622409 of the Ministry of Education, Youth and Sports of the Czech Republic.

References

1. M. Brdička, *Continuum mechanics* (NČSAV, Praha, 1959, in Czech).
2. N.G. Chetaev, On the Gauss principle, *Izv. Kazan. Fiz.-Mat. Obsc.* 3 (1932-33) 323-326, in Russian.
3. L. Czudková and J. Musilová, Variational non-holonomic systems in physics, *In: Global Analysis and Applied Mathematics* (International Workshop on Global Analysis, Ankara, Turkey, April 15-17, 2004, (K. Tas, D. Krupka, O. Krupková and D. Baleanu, Eds.) AIP Conference Proceedings, vol. 729, Melville, New York, 2004) 131–140.
4. L. Czudková, J. Janová and J. Musilová, Non-holonomic mechanical systems and variational principle, *In: Differential Geometry and its Applications* (Proceedings of 9th International Conference, August 30-September 3, 2004, Prague, Czech Republic, (J. Bureš, O. Kowalski, D. Krupka and J. Slovák, Eds.) Matfyzpress, Prague, 2005) 571–579.
5. X. Grácia, J. Marín-Solano and M.-C. Muñoz-Lecanda, Some geometric aspects of variational calculus in constrained systems, *Rep. Math. Phys.* **51** (2003) 127–148.
6. O. Krupková, *The Geometry of Ordinary Variational Equations* (Lecture Notes in Mathematics 1678, Springer, Berlin, 1997).
7. O. Krupková, Mechanical systems with nonholonomic constraints, *J. Math. Phys.* **38** (1997) 5098–5126.
8. O. Krupková, Partial differential equations with differential constraints, *J. Differential Equations* **220** (2006) 354–359.
9. O. Krupková and J. Musilová, Non-holonomic variational systems, *Rep. Math. Phys.* **55** (2005) 211–220.
10. L.D. Landau and E.M. Lifshitz, *Course of Theoretical Physics, Volume 7: Theory of Elasticity* (3rd edition, Butterworth-Heinemann, Oxford, 1986).
11. P. Morando and S. Vignolo, A geometric approach to constrained mechanical systems, symmetries and inverse problems, *J. Phys. A.: Math. Gen.* **31** (1998) 8233–8245.
12. M. Swaczyna, On the non-holonomic variational principle, *In: Global Analysis and Applied Mathematics* (International Workshop on Global Analysis, Ankara, Turkey, April 15-17, 2004, (K. Tas, D. Krupka, O. Krupková and D. Baleanu, Eds.) AIP Conference Proceedings, vol. 729, Melville, New York, 2004) 297–306.

Differential Geometry and its Applications
Proc. Conf., in Honour of Leonhard Euler, Olomouc, August 2007

On a generalization of the Dirac bracket in the De Donder-Weyl Hamiltonian formalism

I. Kanatchikov

Institute of Theoretical Physics, TU Berlin,
D-10623 Berlin, Germany
E-mail: ivar@itp.physik.tu-berlin.de

The elements of the constrained dynamics algorithm in the De Donder-Weyl (DW) Hamiltonian theory for degenerate Lagrangian theories are discussed. A generalization of the Dirac bracket to the DW Hamiltonian theory with second class constraints (defined in the text) is presented.

Keywords: De Donder-Weyl Hamiltonian formalism, Poisson-Gerstenhaber brackets, degenerate Lagrangians, constrained dynamics, Dirac brackets.

MS classification: 70S05, 81T99.

1. Introduction

In Refs. 1–4 a generalization of the Poisson bracket to the De Donder-Weyl (DW) Hamiltonian formulation[5,6] in field theory (and the multiparametric calculus of variations) has been proposed. It leads to a Gerstenhaber algebra of brackets defined on specific horizontal differential forms (called Hamiltonian forms). The construction assumes that the corresponding Lagrangian theory is regular (in the sense of DW Hamiltonian formulation):

$$\det \left|\left| \frac{\partial^2 L}{\partial \phi^\mu_a \partial \phi^\nu_a} \right|\right| \neq 0. \tag{1}$$

However, many field theories of physical interests, such as spinor fields, are not regular in the above sense. In this paper we discuss a corresponding extension of the formalism to the degenerate case and generalize the above mentioned Poisson-Gerstenhaber bracket similarly to the construction of the Dirac bracket in the constrained systems with second-class constraints.[7]

Note that the regularity condition in the DW Hamiltonian formalism is different from the one in the standard Hamiltonian formalism: $\det \left|\left|\partial^2 L/\partial\dot\phi_a \partial\dot\phi_a\right|\right| \neq 0$, which involves only the time derivatives of fields

$\dot{\phi}_a$. As a consequence, the theories which are regular from the point of view of the DW formalism can be irregular from the point of view of the standard Hamiltonian formalism and *vice versa.*

This work is motivated by the project of manifestly space-time symmetric *precanonical* quantization of field theory[8–11] based on the DW Hamiltonian formalism, which requires a well understood formalism for degenerate theories on the classical level.

2. The polysymplectic structure and the Poisson-Gerstenhaber brackets

In this section we briefly recall the construction and properties of the Poisson-Gerstenhaber brackets in DW Hamiltonian formalism which are used in the following section. For further generalizations and a detailed geometrical treatment of the related issues I refer to the existing literature.[12–17] Here I closely follow my previous work.[2–4,10]

The DW Hamiltonian formalism[5,6] is a space-time symmetric generalization of the Hamiltonian formalism from mechanics to field theory. Given the first order Lagrangian density $L = L(y^a, y^a_\mu, x^\nu)$ we introduce the variables (polymomenta)

$$p^\mu_a := \partial L / \partial y^a_\mu$$

and the DW Hamiltonian function

$$H := y^a_\mu p^\mu_a - L = H(y^a, p^\mu_a, x^\mu).$$

Then the Euler-Lagrange equations take the form

$$\partial_\mu y^a(x) = \partial H / \partial p^\mu_a, \quad \partial_\mu p^\mu_a(x) = -\partial H / \partial y^a, \tag{2}$$

which is equivalent to the Euler-Lagrange field equations if L is regular in the sense of Eq. (1).

Geometrically, classical fields $y^a = y^a(x)$ are sections in the *covariant configuration bundle* $Y \to X$ over an oriented n-dimensional space-time manifold X with the volume form ω. The local coordinates in Y are (y^a, x^μ). If $\bigwedge^p_q(Y)$ denotes the space of p-forms on Y which are annihilated by $(q+1)$ arbitrary vertical vectors of Y, then $\bigwedge^n_1(Y) \to Y$ generalizes the cotangent bundle and serves as a model of a *multisymplectic phase space*[6] which possesses the canonical n-form structure

$$\Theta_{MS} = p^\mu_a dy^a \wedge \omega_\mu + p\,\omega \tag{3}$$

called multisymplectic, where $\omega_\mu := \partial_\mu \rfloor \omega$ are the basis of $\bigwedge^{n-1} T^*X$. A section $p = -H(y^a, p^\mu_a, x^\nu)$ yields the multidimensional *Hamiltonian Poincaré-Cartan form* Θ_{PC}.

In order to introduce the Poisson brackets which reflect the dynamical structure of DW Hamiltonian equations (2) we need a structure which is independent of p or a choice of H. We define the *extended polymomentum phase space* as the quotient bundle Z: $\bigwedge^n_1(Y)/\bigwedge^n_0(Y) \to Y$. The local coordinates on Z are $(y^a, p^\nu_a, x^\nu) =: (z^v, x^\mu) = z^M$. We introduce a canonical structure on Z as an equivalence class of forms $\Theta := [p^\mu_a dy^a \wedge \omega_\mu \mod \bigwedge^n_0(Y)]$. The *polysymplectic structure* on Z is defined as an equivalence class of forms Ω given by

$$\Omega := [d\Theta \mod \textstyle\bigwedge^{n+1}_1(Y)] = [-dy^a \wedge dp^\mu_a \wedge \omega_\mu \mod \bigwedge^{n+1}_1(Y)]. \tag{4}$$

The equivalence classes are introduced as an alternative to the explicit introduction of a non-canonical connection on the multisymplectic phase space in order to define the polysymplectic structure as a "vertical part" of the multisymplectic form $d\Theta_{MS}$. The fundamental constructions, such as the Poisson bracket below, are designed to be independent of the choice of representatives in the equivalence classes and the choice of a connection.

A multivector field of degree p, $\overset{p}{X} \in \bigwedge^p TZ$, is called *vertical* if $\overset{p}{X} \rfloor F = 0$ for any form $F \in \bigwedge^*_0(Z)$ The polysymplectic form establishes a map of horizontal p–forms $\overset{p}{F} \in \bigwedge^p_0(Z)$, $p = 0, 1, ..., (n-1)$, to vertical multivector fields of degree $(n-p)$, $\overset{n-p}{X}_F$, called *Hamiltonian*:

$$\overset{n-p}{X}_F \rfloor \Omega = d\overset{p}{F}. \tag{5}$$

The forms for which the map (5) exists are also called *Hamiltonian*. More precisely, horizontal forms are mapped to the *equivalence classes* of Hamiltonian multivector fields modulo the *characteristic* multivector fields $\overset{p}{X}_0$: $\overset{p}{X}_0 \rfloor \Omega = 0$, $p = 2, \ldots n$. It is important to note that the space of Hamiltonian forms is not stable with respect to the exterior product. The natural product operation of Hamiltonian forms is the *co-exterior* product

$$\overset{p}{F} \bullet \overset{q}{F} := *^{-1}(*\overset{p}{F} \wedge *\overset{q}{F}) \tag{6}$$

which is graded commutative and associative.

The Poisson bracket of Hamiltonian forms is given by the formula

$$\{\![\overset{p}{F}_1, \overset{q}{F}_2]\!\} = (-1)^{(n-p)} \overset{n-p}{X}_1 \rfloor d\overset{q}{F}_2 = (-1)^{(n-p)} \overset{n-p}{X}_1 \rfloor \overset{n-q}{X}_2 \rfloor \Omega. \tag{7}$$

In fact, it is induced by the Schouten-Nijenhuis bracket $[\![\ ,\ ]\!]$ of the corresponding Hamiltonian multivector fields:

$$-d\{\![\overset{p}{F},\overset{q}{F}]\!\} := [\![\overset{n-p}{X},\overset{n-q}{X}]\!] \rfloor \Omega. \tag{8}$$

The algebraic properties of the bracket are summarized in the following

Theorem 2.1. *The space of Hamiltonian forms with the operations* $\{\![\ ,\ ]\!\}$ *and* $\bullet$ *is a* ***(Poisson-)Gerstenhaber algebra****, i.e.*

$$\begin{aligned}
\{\![\overset{p}{F},\overset{q}{F}]\!\} &= -(-1)^{g_1 g_2}\{\![\overset{q}{F},\overset{p}{F}]\!\},\\
(-1)^{g_1 g_3}\{\![\overset{p}{F},\{\![\overset{q}{F},\overset{r}{F}]\!\}]\!\} &+ (-1)^{g_1 g_2}\{\![\overset{q}{F},\{\![\overset{r}{F},\overset{p}{F}]\!\}]\!\}\\
&+ (-1)^{g_2 g_3}\{\![\overset{r}{F},\{\![\overset{p}{F},\overset{q}{F}]\!\}]\!\} = 0,\\
\{\![\overset{p}{F},\overset{q}{F}\bullet\overset{r}{F}]\!\} &= \{\![\overset{p}{F},\overset{q}{F}]\!\}\bullet\overset{r}{F} + (-1)^{g_1(g_2+1)}\overset{q}{F}\bullet\{\![\overset{p}{F},\overset{r}{F}]\!\},
\end{aligned} \tag{9}$$

where $g_1 = n-p-1$, $g_2 = n-q-1$, $g_3 = n-r-1$.

The graded Lie algebra properties are induces by the graded Lie properties of the Schouten-Nijenhuis bracket. The graded Leibniz property is a consequence of the Frölicher-Nijenhuis classification of graded derivations adapted to the co-exterior algebra of forms.

Let us mention a few applications of the Poisson-Gerstenhaber (PG) brackets. They enable us to identify the pairs of canonically conjugate variables which may become a starting point of quantization:

$$\{\![p_a^\mu \omega_\mu, y^b]\!\} = \delta_a^b, \quad \{\![p_a^\mu \omega_\mu, y^b \omega_\nu]\!\} = \delta_a^b \omega_\nu, \quad \{\![p_a^\mu, y^b \omega_\nu]\!\} = \delta_a^b \delta_\nu^\mu. \tag{10}$$

In fact, geometric prequantization of PG brackets has been discussed in our previous paper.[10] The brackets can be used also in order to write the DW Hamiltonian equations in the bracket form: for any Hamiltonian $(n-1)$–form $F := F^\mu \omega_\mu$ the equations of motion take the form

$$\mathbf{d}\bullet F = -\sigma(-1)^n \{\![H, F]\!\} + d^h \bullet F, \tag{11}$$

where $\mathbf{d}\bullet$ denotes the "total co-exterior differential"

$$\mathbf{d}\bullet\overset{p}{F} := \frac{1}{(n-p)!}\partial_M F^{\mu_1 \dots \mu_{n-p}} \partial_\mu z^M dx^\mu \bullet \partial_{\mu_1 \dots \mu_{n-p}} \rfloor\ \omega, \tag{12}$$

d^h is the "horizontal co-exterior differential":

$$d^h \bullet \overset{p}{F} := \frac{1}{(n-p)!}\partial_\mu F^{\mu_1 \dots \mu_{n-p}} dx^\mu \bullet \partial_{\mu_1 \dots \mu_{n-p}} \rfloor\ \omega, \tag{13}$$

and $\sigma = \pm 1$ for the Euclidean/Minkowskian signature of the base manifold X. Consequently, the conservation of the quantity represented by $(n-1)$-form F is equivalent to the condition $\{\![H, F]\!\} = 0$.

3. The constrained dynamics and the Dirac bracket of forms

The violation of the regularity condition (1) implies the noninvertibility of the space-time gradients of fields ϕ^a_ν as functions of the field variables ϕ^a and polymomenta p^ν_a. In this case the polymomenta are not all independent, and there exist relations

$$C_A(\phi^a, p^\nu_a) = 0 \tag{14}$$

that follow from the definition of the polymomenta and the form of the Lagrangian. The conditions (14), as usual, will be called **primary constraints** to emphasize that the field equations are not used to obtain these relations and that they imply no restriction on the field variables and their space-time gradients.

The canonical DW Hamiltonian function is now not unique. It can be replaced by the effective DW Hamiltonian function which is weakly equal to H:

$$\tilde{H} := H + u_A C_A \approx H. \tag{15}$$

In the present approach we restrict our attention to the constraints which can be organized into Hamiltonian $(n-1)$–forms

$$C_m = C^\mu_m \omega_\mu. \tag{16}$$

In this case the effective Hamiltonian can be written as

$$\tilde{H} := H + u_m \bullet C_m, \tag{17}$$

where the Lagrange multipliers are organized into one-forms u_m.

Definition 3.1. *A dynamical variable represented by a semi-basic Hamiltonian form F of degree $|F| = f$ is said to be* ***first class*** *if its PG bracket with every constraints $(n-1)$-form C_m weakly vanishes,*

$$\{\![F, C_m]\!\} \approx 0. \tag{18}$$

Definition 3.2. *A semi-basic Hamiltonian form F is said to be* ***second class*** *if there is at least one constraint such that its PG bracket with F does not vanish weakly.*

Proposition 3.1. *The first-class property is* ***preserved*** *under the PG bracket operation, i.e. the PG bracket of two first-class Hamiltonian forms is first class.*

Proof. If F and G are first class, then

$$\{\![F, C_m]\!\} = f_m^n \bullet C_n, \quad \{\![G, C_m]\!\} = g_m^n \bullet C_n \tag{19}$$

for some $(f+1)$-forms f_m^n and $(g+1)$-form g_m^n, where $|G| = g$, $|F| = f$. By the graded Jacobi identity

$$-(-1)^{d_G}\{\![C_m, \{\![F, G]\!\}]\!\} = (-1)^{d_F}\{\![F, \{\![G, C_m]\!\}]\!\} + (-1)^{d_F d_G}\{\![G, \{\![C_m, F]\!\}]\!\},$$

where $d_G := n - g - 1$, $d_F := n - f - 1$. Using Eq. (19), the graded Leibniz rule, and the graded antisymmetry of the bracket in the right hand side, we obtain

$$(-1)^{d_G}\{\![F, g_m^n \bullet C_n]\!\} - (-1)^{d_F d_G + d_F}\{\![G, f_m^n \bullet C_n]\!\} \approx 0.$$

Thus,

$$\{\![C_m, \{\![F, G]\!\}]\!\} \approx 0$$

for any first class Hamiltonian forms F and G. □

Next, we consider the necessary condition for the conservation of the constraints $C_m forms : \mathbf{d} \bullet C_m = 0$ (c.f. Eq. (11)). They give rise to the necessary consistency conditions

$$\{\![\tilde{H}, C_m]\!\} = \{\![H, C_m]\!\} + u^n \bullet \{\![C_n, C_m]\!\} \approx 0, \tag{20}$$

where u^n are one-form coefficients. Eq. (20) can either impose a restriction on the u's or it may reduce to a new relation independent of the u's. In the latter case, i.e. if the new relation between p's and ϕ's is independent of the primary constraints, it will be called a **secondary constraint**, in accordance with the conventional terminology introduced by Dirac and Bergmann.

While the primary constraints are merely consequences of the definition of the polymomentum variables, the secondary constraints make use of the equations of motion (the DW Hamiltonian field equations) as well. If there is a secondary constraint written as an $(n-1)$-form $B_m(\phi, p) := B_m^\mu \omega_\mu$ appearing, a new consistency condition in the form (20)

$$\{\![H, B_m]\!\} + u^n \bullet \{\![B_n, C_m]\!\} \approx 0 \tag{21}$$

must be imposed. Next, we must again check whether Eq. (21) implies new secondary constraints or whether it only restructures the u's, and so on. It is similar to the standard procedure originally proposed by Dirac.

The **second-class constraints** are present whenever the $(n-1)$–form valued matrix

$$C_{mn} := \{\![C_m, C_n]\!\} \tag{22}$$

does not vanish on the constraints surface. As usual, when computing C_{mn} we must not use the constraints equations until after calculating the Poisson-Gerstenhaber bracket. For simplicity, we assume that the constraints are irreducible and the rank of C_{mn} is constant on the constraints surface. Since C_{mn} is a nonsingular matrix whose components are $(n-1)$-forms, its "inverse" C^{-1}_{mn} exists such that

$$C^{-1}_{mn} \wedge C_{nk} = \delta_{mk}\ \omega. \tag{23}$$

The components of C^{-1}_{mn} are one-forms. The key observation is that

Proposition 3.2. *For any Hamiltonian form F we can construct a new Hamiltonian form $F' \approx F$,*

$$F' := F + \sigma \{\![F, C_m]\!\} \bullet (C^{-1}_{mn} \wedge C_n), \tag{24}$$

which has vanishing brackets with all second class constraints.

Here, σ is the signature of the metric whose appearance will be clarified later. The parentheses in Eq. (24) are important because the algebraic system of exterior forms equipped with two products $\wedge$ and $\bullet$ is **not associative**.

Proof. Using the (graded) Leibniz rule with respect to the $\bullet$-product, which is fulfilled by the Poisson-Gerstenhaber bracket of Hamiltonian forms, we obtain

$$\begin{aligned} \{\![F', C_k]\!\} &= \{\![F, C_k]\!\} + \sigma \{\![\{\![F, C_m]\!\} \bullet (C^{-1}_{mn} \wedge C_n), C_k]\!\} \\ &= \{\![F, C_k]\!\} + \sigma \{\![F, C_m]\!\} \bullet \{\![C^{-1}_{mn} \wedge C_n, C_k]\!\} \\ &+ \sigma \{\![\{\![F, C_m]\!\}, C_k]\!\} \bullet (C^{-1}_{mn} \wedge C_n). \end{aligned} \tag{25}$$

As the last term weakly vanishes,

$$\{\![F', C_k]\!\} \approx \{\![F, C_k]\!\} + \sigma \{\![F, C_m]\!\} \bullet \{\![C^{-1}_{mn} \wedge C_n, C_k]\!\}. \tag{26}$$

Let us consider the bracket

$$\{\![C_k, C^{-1}_{mn} \wedge C_n]\!\} = -\pounds_{X_{C_k}} (C^{-1}_{mn} \wedge C_n), \tag{27}$$

where $\pounds_{X_{C_k}}$ is the Lie derivative with respect to the vector field X_{C_k} associated with the Hamiltonian form C_k via the correspondence given by the polysymplectic form Ω

$$X_{C_k} \lrcorner\ \Omega = dC_k$$

(c.f. Eq. (7)). As the Lie derivative is a derivation on the $\wedge$-algebra,

$$\{\![C_k, C^{-1}_{mn} \wedge C_n]\!\} = -\pounds_{X_{C_k}}(C^{-1}_{mn}) \wedge C_n + C^{-1}_{mn} \wedge \pounds_{X_{C_k}}(C_n), \qquad (28)$$

we obtain

$$\{\![C_k, C^{-1}_{mn} \wedge C_n]\!\} \approx -C^{-1}_{mn} \wedge \{\![C_k, C_n]\!\} = C^{-1}_{mn} \wedge C_{nk} = \delta_{mk}\, \omega. \qquad (29)$$

Further, using $*\omega = \sigma$, so that for any form F: $F \bullet \omega = \sigma F$, we obtain

$$\{\![F', C_k]\!\} \approx \{\![F, C_k]\!\} - \sigma\{\![F, C_m]\!\} \bullet \delta_{mk}\omega = \{\![F, C_k]\!\} - \sigma^2\{\![F, C_k]\!\} = 0. \quad (30)$$

Hence,

$$\{\![F', C_k]\!\} \approx 0.$$

□

Now, following a well known way of introducing the Dirac bracket, we may try to define it as $\{\![F', G']\!\}$. Though $F' \approx F$ and $G' \approx G$, the bracket $\{\![F', G']\!\}$ is not weakly equal to $\{\![F, G]\!\}$ and the desired properties

$$\{\![F', G']\!\} \approx \{\![F', G]\!\} \approx \{\![F, G']\!\} \qquad (31)$$

are satisfied if F and G have the degree $(n-1)$. In this case we obtain the following formula for the analogue of the Dirac bracket

$$\{\![F, G]\!\}^D := \{\![F, G]\!\} + \sigma\{\![F, C_m]\!\} \bullet (C^{-1}_{mn} \wedge \{\![C_n, G]\!\}). \qquad (32)$$

The properties of the bracket (32) are described by the following

Proposition 3.3. *The Dirac bracket of any Hamiltonian (n-1)- form with a second class constraint vanishes.*

Proof. The statement follows from Eq. (31) and Proposition 3.2. □

Proposition 3.4. *The Dirac bracket fulfills the Lie algebra properties*

$$\begin{aligned} &\{\![F, G]\!\}^D \approx -\{\![G, F]\!\}^D, \\ &\{\![\{\![F, G]\!\}^D, K]\!\}^D + \{\![\{\![G, K]\!\}^D, F]\!\}^D + \{\![\{\![K, F]\!\}^D, G]\!\}^D \approx 0. \end{aligned} \qquad (33)$$

Proof. These weak identities follow from the properties of the Poisson-Gerstenhaber bracket in Theorem 2.1 and Eq. (31). □

The space of Hamiltonian $(n-1)$ forms on which the bracket has been defined is not closed with respect the product operations $\bullet$ and $\wedge$. We, therefore, can not discuss the Leibniz property without extending the space of forms. A minimal extension would involve forms of degree 0 or n. Let us extend the definition of the bracket in Eq. (32) to the case when one of the arguments is a 0-form k (the bracket of two 0-forms vanishes identically):

$$\{\![F, k]\!\}^D = \{\![F, k]\!\} + \sigma\{\![F, C_m]\!\} \bullet (C^{-1}_{mn} \wedge \{\![C_n, k]\!\}). \tag{34}$$

Then the following analogues of the Leibniz property follow:

$$\{\![F, kG]\!\}^D = \{\![F, k]\!\}^D G + k\{\![F, G]\!\}^D, \tag{35}$$

$$\{\![F, kl]\!\}^D = \{\![F, k]\!\}^D l + k\{\![F, l]\!\}^D. \tag{36}$$

The Leibniz properties for the $\bullet$−multiplication with n-forms are similar because $F \bullet k\omega = \sigma k F$.

4. Conclusion

We presented a generalization of the Dirac constrained dynamics algorithm and a generalization of the Dirac bracket formula to the De Donder-Weyl Hamiltonian formalism of degenerate Lagrangian theories. The Dirac bracket is defined for dynamical variables given by Hamiltonian $(n-1)$−forms if the constraints can also be organized into Hamiltonian $(n-1)$−forms. A possible generalization to forms of arbitrary degree and non-Hamiltonian forms remains an open issue. We also left beyond the scope of our discussion a possible geometrical interpretation of the construction in terms of the restriction of the polysymplectic structure to the subspace of constraints.

Let us recall that in the geometric calculus of variations the issues related to the degeneracy of the Lagrangian can be treated with the help of Lepagean equivalents of the Cartan form[19] which effectively modify the definitions of polymomentum variables and the corresponding regularity conditions. It would be interesting to understand if our formal Dirac-like treatment of constraints in the DW theory could be understood geometrically using the theory of Lepagean equivalents. Yet another promising approach to the geometrical understanding of the formalism could be based on the extension of the Cartan form to the constrained variational problems as presented in Ref. 20.

Note in conclusion, that the approach developed here can be applied to the Dirac spinor field[21] which is a singular theory with second-class constraints from the point of view of the DW Hamiltonian formulation.

Acknowledgments

I thank Prof. Hellwig for his warm hospitality at TU Berlin. A discussion with Prof. Krupková is gratefully acknowledged.

References

1. I.V. Kanatchikov, On the canonical ctructure of the De Donder-Weyl covariant Hamiltonian formulation of field theory I. Graded Poisson brackets and the equations of motion; arXiv:hep-th/9312162.
2. I.V. Kanatchikov, Canonical structure of classical field theory in the polymomentum phase space, *Rep. Math. Phys.* **41** (1998) 49–90; arXiv:hep-th/9709229.
3. I.V. Kanatchikov, On field theoretic generalizations of a Poisson algebra, *Rep. Math. Phys.* **40** (1997) 225–34; arXiv:hep-th/9710069.
4. I.V. Kanatchikov, Novel algebraic structures from the polysymplectic form in field theory; arXiv:hep-th/9612255.
5. Th. De Donder, *Théorie Invariantive du Calcul des Variations* (Gauthier-Villars, Paris, 1935).
   H. Weyl, Geodesic fields in the calculus of variations, *Ann. Math. (2)* **36** (1935) 607–29.
   H. Rund, *The Hamilton-Jacobi Theory in the Calculus of Variations* (D. van Nostrand, Toronto, 1966).
6. M.J. Gotay, J. Isenberg and J. Marsden, Momentum maps and classical relativistic fields, Part I: Covariant field theory, arXiv:physics/9801019; Part II: Canonical Analysis of Field Theories, arXiv:math-ph/0411032.
7. M. Henneaux and C. Teitelboim, *Quantization of Gauge Systems* (Princeton Univ. Press, Princeton, NJ 1992).
8. I.V. Kanatchikov, De Donder-Weyl theory and a hypercomplex extension of quantum mechanics to field theory, *Rep. Math. Phys.* **43** (1999) 157–70; arXiv:hep-th/9810165;
   I.V. Kanatchikov, On quantization of field theories in polymomentum variables; arXiv:hep-th/9811016.
9. I.V. Kanatchikov, Precanonical quantum gravity: quantization without the space-time decomposition, *Int. J. Theor. Phys.* **40** (2001) 1121–49, arXiv:gr-qc/0012074.
10. I.V. Kanatchikov, Geometric (pre)quantization in the polysymplectic approach to field theory; arXiv:hep-th/0112263.
11. I.V. Kanatchikov, Precanonical quantization of Yang-Mills fields and the functional Schrödinger representation, *Rep. Math. Phys.* **53** (2004) 181–193, arXiv:hep-th/0301001.
12. N. Roman-Roy, A. M. Rey, M. Salgado and S. Vilarino, On the k-Symplectic, k-Cosymplectic and Multisymplectic Formalisms of Classical Field Theories; arXiv:0705.4364 [math-ph].
13. M. Francaviglia, M. Palese and E. Winterroth, A new geometric proposal for the Hamiltonian description of classical field theories; arXiv:math-ph/0311018.
14. M. Forger, C. Paufler and H. Römer, Hamiltonian multivector fields and poisson forms in multisymplectic field theory, *J. Math. Phys.* **46** (2005) 112901; arXiv:math-ph/0407057.
15. M. de León, M. McLean, L. K. Norris, A. Rey-Roca and M. Salgado, Geometric structures in field theory; arXiv:math-ph/0208036.

A. Echeverria-Enriquez, M. de León, M. C. Munoz-Lecanda and N. Roman-Roy, Hamiltonian systems in multisymplectic field theories; arXiv:math-ph/0506003.
16. O. Krupková, Hamiltonian field theory, *J. Geom. Phys.* **43** (2002) 93–132.
17. G. Giachetta, L. Mangiarotti and G. Sardanashvily, *New Lagrangian and Hamiltonian Methods in Field Theory* (World Scientific, Singapore 1997).
18. F. Hélein and J. Kouneiher, Covariant Hamiltonian formalism for the calculus of variations with several variables; arXiv:math-ph/0211046.
19. O. Krupková and D. Smetanová, Legendre transformation for regularizable Lagrangians in field theory, *Lett. Math. Phys.* **58** (2001) 189–204;
O. Krupková and D. Smetanová, On regularization of variational problems in first-order field theory, *In: Proceedings of the 20th Winter School "Geometry and Physics"* (Srní, 2000, Rend. Circ. Mat. Palermo (2) Suppl. No. 66, 2001), 133–140.
D. Smetanová, Hamiltonian systems in dimension 4, to appear *In: Proc. 10th Int. Conf. Diff. Geom. and Its Applications* (Olomouc, Czech Republic, 2007).
20. A. Garcia, P.L. García and C. Rodrigo, Cartan forms for first order constrained variational problems, *J. Geom. Phys.* **56** (2006) 571–610.
21. I. Kanatchikov, in preparation.

Differential Geometry and its Applications
Proc. Conf., in Honour of Leonhard Euler, Olomouc, August 2007

Constrained Lepage forms

Olga Krupková

Department of Algebra and Geometry, Faculty of Science, Palacký University, Tomkova 40, 779 00 Olomouc, Czech Republic
and
Department of Mathematics, La Trobe University, Bundoora, Victoria 3086, Australia
E-mail: krupkova@inf.upol.cz

Jana Volná

Department of Mathematics, Tomáš Baťa University, Nad Stráněmi 4511, 760 05 Zlín, Czech Republic
E-mail: volna@fai.utb.cz

Petr Volný

Department of Mathematics and Descriptive Geometry, VŠB - Technical University of Ostrava, 17. listopadu 15, Ostrava Poruba, Czech Republic
E-mail: Petr.Volny@vsb.cz

Lepage forms, introduced by D. Krupka in 1973, are known to be fundamental objects in the global calculus of variations on fibred manifolds. We extend this concept to the case of non-holonomic constraints. We introduce constrained Lepage form as a differential form defined on a constraint submanifold in the first jet bundle over a fibred manifold, and obtain the constrained Euler–Lagrange form, globally representing variational equations with non-holonomic constraints.

Keywords: Fibred manifold, non-holonomic constraint, canonical distribution, Lepage form, Lagrangian, Euler–Lagrange form.

MS classification: 35A30, 58A10, 70S05

1. Introduction

Lepage forms, introduced by Krupka,[3] represent a concept unifying and generalising Cartan form, Poincaré–Cartan form and Carthéodory form, used in the classical calculus of variations. On fibred manifolds, Lepage forms play a fundamental role in global and intrinsic setting for the first variation, Euler–Lagrange equations, Hamilton and Hamilton–Jacobi theory and invariant variational problems (see e.g. Refs. 2,5,6,10). The aim of this paper is to extend the concept of Lepage form to the case of non-holonomic constraints. We introduce constrained Lepage form as a differential form defined on a constraint submanifold in the first jet bundle over a fibred manifold, and obtain the constrained Euler–Lagrange form, globally representing variational equations with non-holonomic constraints. Our setting concerns first and second order variational problems in non-holonomic field theory, and includes, as a special case constrained Cartan forms and Euler–Lagrange equations in non-holonomic mechanics,[7,8,13,14] and constrained Poincaré–Cartan forms and constrained Euler–Lagrange equations, associated with first order Lagrangians in field theory[9,11,12] (cf. also Refs. 1,15).

2. Non-holonomic constraints

Throughout the paper, manifolds and mappings are smooth, and summation over repeated indices is used. We consider a fibred manifold $\pi : Y \to X$, $\dim X = n$, $\dim Y = m + n$, and its jet prolongations $\pi_r : J^rY \to X$, $r = 1, 2$. Canonical projections are denoted by $\pi_{r,s} : J^rY \to J^sY$, $r > s \geq 0$. Next we denote by (x^i, y^σ) fibred coordinates on Y, and by $(x^i, y^\sigma, y^\sigma_j)$ and $(x^i, y^\sigma, y^\sigma_j, y^\sigma_{jk})$, $j \leq k$, the associated coordinates on J^1Y and J^2Y, respectively. A q-form η on J^rY is called *contact* if $J^r\gamma^*\eta = 0$ for every section γ of π. η is called $\pi_{r,s}$*-horizontal* if $i_\xi\eta = 0$ for every $\pi_{r,s}$-vertical vector field ξ on J^rY. A π_r*-horizontal* form η is also called 0*-contact.* A contact q-form η is called k*-contact*, $1 \leq k \leq q$, if for every vertical vector field ξ the form $i_\xi\eta$ is $(k-1)$-contact.[4] If lifted to J^2Y, every q-form η on J^1Y can be uniquely expressed as a sum of k-contact forms η_k, $0 \leq k \leq q$. If we put $h\eta = \eta_0$ and $p_k\eta = \eta_k$, $1 \leq k \leq q$, we have $\pi^*_{2,1}\eta = h\eta + p_1\eta + \cdots + p_{q-1}\eta + p_q\eta$.[4] We call $h\eta$ and $p_k\eta$ the *horizontal* and k*-contact part* of η, respectively. For more details on horizontal and contact forms we refer to Refs. 4,5.

A *nonholonomic constraint* in J^1Y is defined to be a submanifold $\iota : Q \to J^1Y$, fibred over Y. We assume $\operatorname{codim} Q = \kappa$, where $1 \leq \kappa \leq mn - 1$. If (x^i, y^σ) are fibred coordinates on Y, Q is locally expressed by a system of first order partial differential equations

$$f^\alpha(x^i, y^\sigma, y^\sigma_j) = 0, \quad 1 \leq \alpha \leq \kappa, \tag{1}$$

such that $\operatorname{rank}(\partial f^\alpha / \partial y^\sigma_j) = \kappa$, where α label rows and σ, j columns. We put

$$y^\sigma_j \circ \iota = g^\sigma_j, \tag{2}$$

and consider on the submanifold Q adapted fibred coordinates (x^i, y^σ, z^J), where $z^J = g^\sigma_j$ for appropriate indices σ and j; we denote the set of admissible indices for the z^J's by $\mathcal{J}$ and write $J = (\sigma, j) \in \mathcal{J}$.

Lifting the constraint Q to J^2Y we get a submanifold $\hat{Q}$ locally defined by equations $f^\alpha = 0$, $df^\alpha/dx^i = 0$. Continuing the procedure, higher lifts of Q in J^rY are obtained. The manifold $\hat{Q}$ is fibred over Q, Y and X, we denote $\overline{\pi}_2 : \hat{Q} \to X$, $\overline{\pi}_{2,0} : \hat{Q} \to Y$, $\overline{\pi}_{2,1} : \hat{Q} \to Q$, and $\overline{\pi}_1 : Q \to X$, $\overline{\pi}_{1,0} : Q \to Y$ restrictions of the canonical fibred projections of the corresponding "unconstrained" fibred manifolds. On $\hat{Q}$ we consider local adapted fibred coordinates $(x^i, y^\sigma, z^J, z^J_i)$, and a basis of contact 1-forms

$$\bar{\omega}^\sigma = dy^\sigma - g^\sigma_j dx^j, \quad \hat{\omega}^J = dz^J - z^J_j dx^j, \tag{3}$$

on the next lift we have additional coordinates z^J_{ip}, $i \leq p$, and contact 1-forms

$$\hat{\omega}^J_i = dz^J_i - z^J_{ip} dx^p. \tag{4}$$

A form η on Q is called *horizontal* (or *0-contact*) if $i_\xi \eta = 0$ for every $\overline{\pi}_1$-vertical vector field ξ on Q. Let $k \geq 1$. A contact form η on Q is called *k-contact*, if for every $\overline{\pi}_1$-vertical vector field ξ on Q, the form $i_\xi \eta$ is $(k-1)$-contact. Analogous definitions apply to forms on $\hat{Q}$. Is easy to check that for differential forms on Q we get a similar *theorem on decomposition into contact components* as in the "unconstrained case". For the sake of simplicity of notations, we denote the corresponding operators of horizontalisation and k-contactisation by the same symbols, h, and p_k. We also put

$$d^c_i = \frac{\partial}{\partial x^i} + g^\sigma_i \frac{\partial}{\partial y^\sigma} + z^J_i \frac{\partial}{\partial z^J}. \tag{5}$$

3. Constrained Lepage forms

A differential n-form ρ on J^1Y is called *Lepage form* if $p_1 d\rho$ is $\pi_{2,0}$-horizontal; the horizontal n-form $\lambda = h\rho$ is then called a *Lagrangian*, and ρ itself is said to be its *Lepage equivalent.*[3]

Note that the Lagrangian associated with a first order Lepage form is generally of order 2. If ρ is $\pi_{1,0}$-horizontal (the fibred coordinate expression does not contain the dy^σ_j's) then the Lagrangian is defined on J^1Y.

Theorem 3.1 (Krupka[3,4]). *Let ρ be an n-form on J^1Y. The following conditions are equivalent:*

(1) *ρ is a Lepage form*

(2) *In every fibred chart,*

$$\pi^*_{2,1}\,\rho = \Theta_\lambda + d\nu + \mu, \tag{6}$$

where $\lambda = h\rho$,

$$\Theta_\lambda = L\,\omega_0 + \Big(\frac{\partial L}{\partial y^\sigma_i} - d_p\,\frac{\partial L}{\partial y^\sigma_{pi}}\Big)\,\omega^\sigma \wedge \omega_i + \frac{\partial L}{\partial y^\sigma_{ji}}\,\omega^\sigma_j \wedge \omega_i, \tag{7}$$

ν *is a contact* $(n-1)$*-form, and* μ *is an at least* 2*-contact form.*

(3) *The* 1*-contact part* $p_1 d\rho$ *of* $d\rho$ *depends only on* $h\rho = \lambda = L\omega_0$ *and reads*

$$p_1 d\rho = E_\lambda = \Big(\frac{\partial L}{\partial y^\sigma} - d_i\,\frac{\partial L}{\partial y^\sigma_i} + d_i d_j\,\frac{\partial L}{\partial y^\sigma_{ij}}\Big)\omega^\sigma \wedge \omega_0. \tag{8}$$

The $(n+1)$-form E_λ is called *the Euler–Lagrange form of* λ,[3] and its fibred-chart components are called *Euler–Lagrange expressions.*

Definition 3.1. Let $Q \subset J^1Y$ be a non-holonomic constraint. We call a differential n-form ρ on Q *constrained Lepage form* if $p_1 d\rho$ is $\overline{\pi}_{2,0}$-horizontal.

From the definition we get the structure of constrained Lepage forms:

Theorem 3.2. *Let* ρ *be an* n*-form on* Q*.* ρ *is a constrained Lepage form if and only if in every adapted fibred chart,*

$$\begin{aligned}\overline{\pi}^*_{2,1}\rho = L\omega_0 + &\sum_{J=(\sigma,i)\in\mathcal{J}} \Big(\frac{\partial L}{\partial z^J} - d^c_j\,\frac{\partial L}{\partial z^J_j} - \sum_{(\nu,p)\notin\mathcal{J}} L^p_\nu \frac{\partial g^\nu_p}{\partial z^J}\Big)\bar\omega^\sigma \wedge \omega_i \\ &+ \sum_{(\nu,p)\notin\mathcal{J}} L^p_\nu \bar\omega^\nu \wedge \omega_p + \frac{\partial L}{\partial z^J_j}\hat\omega^J \wedge \omega_j + \mu,\end{aligned} \tag{9}$$

where μ *is an arbitrary at least* 2*-contact* n*-form on* Q*.*

Proof. Let ρ be an n-form on Q. Then

$$\overline{\pi}^*_{2,1}\rho = L\omega_0 + L^i_\sigma\bar\omega^\sigma \wedge \omega_i + f^i_J\hat\omega^J \wedge \omega_i + \cdots, \tag{10}$$

where ... stand for the at least 2-contact part of ρ. Computing the 1-contact part of $d\rho$ we obtain:

$$\begin{aligned}p_1 d\rho &= dL \wedge \omega_0 + (hdL^i_\sigma \wedge \bar\omega^\sigma + L^i_\sigma d\bar\omega^\sigma + hdf^i_J \wedge \hat\omega^J + f^i_J d\hat\omega^J) \wedge \omega_i \\ &= \Big(\frac{\partial L}{\partial y^\sigma} - d^c_i L^i_\sigma - L^i_\nu \frac{\partial g^\nu_i}{\partial y^\sigma}\Big)\bar\omega^\sigma \wedge \omega_0 \\ &\quad + \Big(\frac{\partial L}{\partial z^J} - d^c_i f^i_J - L^i_\nu \frac{\partial g^\nu_i}{\partial z^J}\Big)\hat\omega^J \wedge \omega_0 + \Big(\frac{\partial L}{\partial z^J_i} - f^i_J\Big)\hat\omega^J_i \wedge \omega_0.\end{aligned} \tag{11}$$

Since $g^\sigma_j = z^J$ for $J = (\sigma, j) \in \mathcal{J}$, conditions for ρ be a constrained Lepage form read as follows:

$$f^i_J = \frac{\partial L}{\partial z^J_i}, \quad L_J = \frac{\partial L}{\partial z^J} - d^c_i \frac{\partial L}{\partial z^J_i} - \sum_{(\nu,p)\notin\mathcal{J}} L^p_\nu \frac{\partial g^\nu_p}{\partial z^J}. \tag{12}$$

Finally, we obtain formula (9) as desired. □

Definition 3.2. Given a constrained Lepage form ρ on $Q \subset J^1Y$, we call the at most 1-contact part $\bar{\Theta}$ of ρ *constrained Poincaré-Cartan form*, and the $(n+1)$-form $\bar{E} = p_1 d\rho$ *constrained Euler–Lagrange form.*

Both the constrained Poincaré–Cartan and the constrained Euler–Lagrange form depend not only upon the horizontal part $\lambda = h\rho$ of ρ, but also upon the embedding ι of Q into J^1Y, and the components L^p_ν of ρ where $(\nu, p) \notin \mathcal{J}$. Explicitly,

$$\begin{aligned}\bar{\Theta} = L\omega_0 &+ \sum_{J=(\sigma,i)\in\mathcal{J}} \Big(\frac{\partial L}{\partial z^J} - d^c_j \frac{\partial L}{\partial z^J_j} - \sum_{(\nu,p)\notin\mathcal{J}} L^p_\nu \frac{\partial g^\nu_p}{\partial z^J}\Big)\bar{\omega}^\sigma \wedge \omega_i \\ &+ \sum_{(\nu,p)\notin\mathcal{J}} L^p_\nu \bar{\omega}^\nu \wedge \omega_p + \frac{\partial L}{\partial z^J_j}\hat{\omega}^J \wedge \omega_j,\end{aligned} \tag{13}$$

and

$$\begin{aligned}\bar{E} = p_1 d\rho &= \Big(\frac{\partial L}{\partial y^\sigma} - d^c_i L^i_\sigma - L^i_\nu \frac{\partial g^\nu_i}{\partial y^\sigma}\Big)\bar{\omega}^\sigma \wedge \omega_0 \\ &= \Big(\frac{\partial L}{\partial y^\sigma} - \sum_{i,(\sigma,i)=J\in\mathcal{J}} d^c_i \Big(\frac{\partial L}{\partial z^J} - d^c_j \frac{\partial L}{\partial z^J_j}\Big) \\ &\quad - \sum_{(\nu,p)\notin\mathcal{J}} L^p_\nu \Big(\frac{\partial g^\nu_p}{\partial y^\sigma} - \sum_{i,(\sigma,i)=J\in\mathcal{J}} d^c_i \frac{\partial g^\nu_p}{\partial z^J}\Big) \\ &\quad + \sum_{(\nu,p)\notin\mathcal{J}} d^c_i L^p_\nu \Big(\frac{\partial g^\nu_p}{\partial z^{(\sigma,i)}} - \delta^\nu_\sigma \delta^i_p\Big)\Big)\bar{\omega}^\sigma \wedge \omega_0.\end{aligned} \tag{14}$$

Corollary 3.1. *Assume that ρ is horizontal with respect to the projection onto Y. Then all the contact components $p_k\rho$, $0 \le k \le n$, are defined on Q, i.e. $\lambda = h\rho$ and $\bar{\Theta}$ are first order forms, and the constrained Euler–Lagrange form $\bar{E}$ is of order 2 (defined on $\hat{Q}$). The corresponding coordinate formulas* *read as follows:*

$$\bar{\Theta} = L\omega_0 + \Big(\frac{\partial L}{\partial z^J} - \sum_{(\nu,p)\notin\mathcal{J}} L^p_\nu \frac{\partial g^\nu_p}{\partial z^J}\Big)\bar{\omega}^\sigma \wedge \omega_i + \sum_{(\nu,p)\notin\mathcal{J}} L^p_\nu \bar{\omega}^\nu \wedge \omega_p,$$

$$\bar{E} = \Big(\frac{\partial L}{\partial y^\sigma} - d^c_i \frac{\partial L}{\partial z^J} - \sum_{(\nu,p)\notin\mathcal{J}} L^p_\nu \Big(\frac{\partial g^\nu_p}{\partial y^\sigma} - d^c_i \frac{\partial g^\nu_p}{\partial z^J}\Big) \tag{15}$$

$$+ \sum_{(\nu,p)\notin\mathcal{J}} d^c_i L^p_\nu \Big(\frac{\partial g^\nu_p}{\partial z^J} - \delta^\nu_\sigma \delta^i_p\Big)\Big)\bar{\omega}^\sigma \wedge \omega_0,$$

where $J = (\sigma, i) \in \mathcal{J}$, *and summation over* σ, i *applies.*

Theorem 3.3. *Let* $\iota : Q \to J^1Y$ *be a non-holonomic constraint,* $\hat{\iota} : \hat{Q} \to J^2Y$ *the lift of* Q. *If* ρ *is a Lepage form on* J^1Y *then* $\iota^*\rho$ *is a constrained Lepage form on* Q, $\hat{\iota}^*\Theta_\lambda$ *is a constrained Poincaré–Cartan form, and* $\hat{\iota}^*E_\lambda$ *is a constrained Euler–Lagrange form.*

Proof. Theorem follows immediately form the formula $\iota^* d\rho = d\iota^*\rho$ and the corresponding definitions. □

Acknowledgments

Research supported by grants GACR 201/06/0922 of the Czech Science Foundation, and MSM 6198959214 of the Czech Ministry of Education, Youth and Sports.

References

1. E. Binz, M. de León, D.M. de Diego and D. Socolescu, Nonholonomic constraints in classical field theories, *Rep. Math. Phys.* **49** (2002) 151–166.
2. H. Goldschmidt and S. Sternberg, The Hamilton-Cartan formalism in the calculus of variations, *Ann. Inst. Fourier* **23** (1973) 203–267.
3. D. Krupka, Some geometric aspects of variational problems in fibered manifolds, *Folia Fac. Sci. Nat. Univ. Purk. Brunensis, Physica* **14**, Brno, Czechoslovakia (1973) 65pp; arXiv:math-ph/0110005.
4. D. Krupka, Lepagean forms in higher order variational theory, *In: Modern Developments in Analytical Mechanics I: Geometrical Dynamics* (Proc. IUTAM-ISIMM Symp., Torino, 1982, (S. Benenti, M. Francaviglia and A. Lichnerowicz, Eds.) Accad. delle Scienze di Torino, Torino, 1983) 197–238.
5. D. Krupka, Global variational theory in fibred spaces, *In: Handbook of Global Analysis* (Elsevier, 2008) 755–839.
6. O. Krupková, *The Geometry of Ordinary Variational Equations* (Lecture Notes in Mathematics 1678, Springer, Berlin, 1997).

7. O. Krupková, Mechanical systems with nonholonomic constraints, *J. Math. Phys.* **38** (1997) 5098–5126.
8. O. Krupková, Recent results in the geometry of constrained systems, *Rep. Math. Phys.* **49** (2002) 269–278.
9. O. Krupková, Partial differential equations with differential constraints, *J. Differential Equations* **220** (2006) 354–395.
10. O. Krupková, Lepage forms in the calculus of variations, *In: Variations, Geometry and Physics* (Nova Science Publishers, USA, 2008) 33–65.
11. O. Krupková and P. Volný, Euler-Lagrange and Hamilton equations for nonholonomic systems in field theory, *J. Phys. A: Math. Gen.* **38** (2005) 8715–8745.
12. O. Krupková and P. Volný, Differential equations with constraints in jet bundles: Lagrangian and Hamiltonian approach, *Lobachevskii J. Math.* **23** (2006) 95–150.
13. W. Sarlet, A direct geometrical construction of the dynamics of nonholonomic Lagrangian systems, *Extracta Mathematicae* **11** (1996) 202–212.
14. M. Swaczyna, On the nonholonomic variational principle, *In: Global Analysis and Applied Mathematics* (Proc. International Workshop on Global Analysis, Ankara, Turkey, 2004, AIP Conference Proceedings, Melville, New York, 2004) 297–306.
15. J. Vankerschaver, F. Cantrijn, M. de León, D. Martín de Diego, Geometric aspects of nonholonomic field theories, *Rep. Math. Phys.* **56** (3) (2005) 387–411.

Differential Geometry and its Applications
Proc. Conf., in Honour of Leonhard Euler, Olomouc, August 2007

Variationality of geodesic circles in two dimensions

R.Ya. Matsyuk

Institute for Applied Problems in Mechanics and Mathematics,
15 Dudayev St., L'viv, 79005, Ukraine
E-mail: matsyuk@lms.lviv.ua

This note treats the notion of Lagrange derivative for the third order mechanics in the context of covariant Riemannian geometry. The variational differential equation for geodesic circles in two dimensions is obtained. The influence of the curvature tensor on the Lagrange derivative leads to the emergence of the notion of quasiclassical spin in the pseudo-Riemannian case.

Keywords: Ostrohrads'kyj mechanics, inverse variational problem, concircular geometry, classical spin.

MS classification: 53A40, 70H50, 49N45, 83C10.

1. Introduction

This is a note on the variational formulation for the differential equations of geodesic circles in two-dimensional Riemannian space, although the results apply straightforward to the pseudo-Riemannian case. The geodesic curves $x^i(t)$ obey with respect to the natural parameter s the third order differential equation

$$\frac{D^3x^i}{ds^3} + g_{lj}\frac{D^2x^l}{ds^2}\frac{D^2x^j}{ds^2}\frac{Dx^i}{ds} = 0\,,$$

and they are exactly characterized by the property that the (signed) Frenet curvature k keeps constant along them. In view of the Proposition 2.1 below we could have immediately stated that the variational functional $\int k ds$ provides an answer to the problem, all the more that in two dimensions $\sqrt{k^2}$ depends linearly on the second derivatives of the coordinates along the curve thus producing exactly *the third order* variational (called Euler-Poisson) equation.

However, we wish to investigate, to what extent this answer in predefined by the limiting case of the Euclidean space — the local model of the Riemannian one. With this idea in mind we start by recalling one solution of the *invariant inverse variational problem* in two-dimensional Euclidean space for a third order variational equation possessing the first integral k. Before proceeding further, it is necessary to agree about some notations and to recall some basic calculus on the second order Ehresmann velocity space $T^2M \overset{\text{def}}{=} J^2_0(\mathbb{R}, M)$ of jets from $\mathbb{R}$ to our manifold M starting at $0 \in \mathbb{R}$ (as possible source of references we can recommend, for example, Refs. 2–4)

2. Calculus on the higher order velocities space

Let $u^i, \dot{u}^i$ denote the standard fiber coordinates in T^2M. In case of an affine space M, we use the vector notations $\boldsymbol{u}, \dot{\boldsymbol{u}}$ for that tuple. In future we shall profoundly also use another tuple of coordinates, namely, $\boldsymbol{u}, \boldsymbol{u}'$, where

$$u'^i = \dot{u}^i + \Gamma^i_{lj} u^l u^j \tag{1}$$

stands for the covariant derivative of $\boldsymbol{u}$. Let us recall some operators acting in the algebra of differential forms, defined on the velocity spaces of the sequential orders:

- The total derivative:

$$d_T f = u^i \frac{\partial f}{\partial x^i} + \dot{u}^i \frac{\partial f}{\partial u^i} + \ddot{u}^i \frac{\partial f}{\partial \dot{u}^i} \cdots$$

  This is a derivation of degree zero and of the type d, *i.e.* who commutes with the exterior differential: $dd_T = d_T d$.

- For each $k = 1, 2, 3$, let $u^i_{(k)} = \overbrace{\dot{\cdots}}^{k}{u}^i$, $u^i_{(0)} = u^i$, $x^i_{(k)} = u^i_{(k-1)}$, and $x^i_{(0)} = x^i$. For each $r = 0, 1, 2, 3, \ldots$, we recall the following derivations of degree zero and of the type i, *i.e.* who produce zeros while acting on the ring of functions:

$$\begin{gathered} \iota_r(f) = 0\,, \\ \iota_r(dx^i_{(k)}) = \frac{k!}{(k-r)!} dx^i_{(k-r)}, \quad \text{and} \quad \iota_r(dx^i_{(k)}) = 0, \quad \text{if} \quad r > k. \end{gathered} \tag{2}$$

- The Lagrange derivative δ:

$$\delta = (\iota_0 - d_T \iota_1 + \frac{1}{2!} d_T^2 \iota_2 - \frac{1}{3!} d_T^3 \iota_3 + \ldots)\, d\,, \tag{3}$$

  which satisfies $\delta^2 = 0$.

Let some system of the third order ordinary differential equations $\mathcal{E}_i(x^j, u^j, \dot{u}^j, \ddot{u}^j)$ be put in the shape of a covariant object ϵ:

$$\epsilon = \mathcal{E}_i(x^j, u^j, \dot{u}^j, \ddot{u}^j)dx^i\,.$$

The variationality criterion reads: If $\delta\epsilon = 0$, then the system $\mathcal{E}_i$ is variational, *i.e.* locally there exists some function L, such that $\epsilon = \delta L$.

The right action of the prolonged group $GL_{(2)}(\mathbb{R}) \stackrel{\text{def}}{=} \overset{\circ}{J}{}^2_0(\mathbb{R},\mathbb{R})_0$ of parameter transformations (invertible transformations of the independent variable t) on T^2M gives rise to the so-called fundamental fields on T^2M:

$$\zeta_1 = u^i\frac{\partial}{\partial u^i} + 2\dot{u}^i\frac{\partial}{\partial \dot{u}^i}, \quad \zeta_2 = u^i\frac{\partial}{\partial \dot{u}^i}\,.$$

A function f defined on T^2M does not depend on the change of independent variable t (so–called parameter–independence) if and only if

$$\zeta_1 f = 0, \quad \zeta_2 f = 0\,. \tag{4}$$

On the other hand, a function L defined on T^2M constitutes a parameter–independent variational problem with the functional $\int L(x^j, u^j, \dot{u}^j)dt$ if and only if the following Zermelo conditions are satisfied:

$$\zeta_1 L = L, \quad \zeta_2 L = 0\,. \tag{5}$$

Let us introduce the generalized momenta:

$$p_i^{(1)} = \frac{\partial L}{\partial \dot{u}^i}, \quad p_i = \frac{\partial L}{\partial u^i} - d_T p_i^{(1)}\,.$$

These satisfy the relation:

$$p^{(1)}{}_i du^i + p_i dx^i = \iota_1 dL - \frac{1}{2}d_T\iota_2 dL\,. \tag{6}$$

The Euler–Poisson equation is given by $\delta L = 0$, or, equivalently, by

$$\dot{p}_i dx^i = \frac{\partial L}{\partial x^i}dx^i\,.$$

The Hamilton function is given by:

$$H = p_i^{(1)}\dot{u}^i + p_i u^i - L\,.$$

Lemma 2.1.

$$H = \zeta_1 L - d_T\zeta_2 L - L\,.$$

Proposition 2.1. *If a function L_{II} is parameter–independent and a function L_{I} constitutes a parameter-independent variational problem, then L_{II} is constant along the extremals of $L = L_{\text{II}} + L_{\text{I}}$.*

Proof. By Lemma 2.1 and in course of the properties (4) and (5) we calculate $H_{L_{\mathrm{II}}+L_{\mathrm{I}}} = \zeta_1(L_{\mathrm{II}} + L_{\mathrm{I}}) - d_T\zeta_2(L_{\mathrm{II}} + L_{\mathrm{I}}) - L = -L_{\mathrm{II}}$. But as far as the Hamilton function is constant of motion, so is the L_{II}. □

3. The Lagrange derivative in Riemannian space

In Riemannian space with symmetric connection the covariant differential of a vector field $\boldsymbol{\xi}$ is a vector field valued semibasic differential form $D\boldsymbol{\xi}$ calculated according to the formula

$$(D\boldsymbol{\xi})^i = d\xi^i + \Gamma^i_{lj}\xi^j dx^l. \tag{7}$$

The fundamental application of the curvature tensor, from which this note profits, provides the commutator of the subsequent derivations, the one that substitutes the known Schwarz lemma:

$$(D\boldsymbol{u})'^i = (D(\boldsymbol{u}'))^i + R_{ljq}{}^i u^j u^q dx^l\,. \tag{8}$$

We also recall that, on the other hand, the first order derivations commute:

$$(dx)' = D\boldsymbol{u}\,. \tag{9}$$

Given some local coordinate expression of a function,

$$L(x^i, u^i, \dot{u}^i)\,,$$

we wish to introduce generalized momenta π_i and $\pi^{(1)}{}_i$, calculated with respect to the alternative set of coordinates in T^2M, namely, x^i,u^i, u'^i, where the transition functions are presented by (1).

Definition 3.1. Let

$$\pi^{(1)}{}_i = \frac{\partial L}{\partial u'^i}\,, \qquad \pi_i = \frac{\partial L}{\partial u^i} - \pi^{(1)\prime}{}_i\,.$$

Proposition 3.1. *In Riemannian space the generalized momenta satisfy the relation, analogous to (6):*

$$\boldsymbol{\pi}^{(1)} D\boldsymbol{u} + \boldsymbol{\pi} dx = \iota_1 dL - \frac{1}{2}(\iota_2 dL)'$$

Proof. First let us calculate by the reason of formulas (2) and (1):

$$\iota_1 d\boldsymbol{u} = dx, \quad \iota_1 d\boldsymbol{u}' = 2D\boldsymbol{u}, \quad \iota_2 d\boldsymbol{u}' = 2dx\,.$$

For the differential of Lagrange function,

$$dL = \frac{\partial L}{\partial x} + \frac{\partial L}{\partial \boldsymbol{u}} d\boldsymbol{u} + \frac{\partial L}{\partial \boldsymbol{u}'} d\boldsymbol{u}'\,, \tag{10}$$

we then check:

$$\iota_2 dL = 2\,\frac{\partial L}{\partial \boldsymbol{u}'}dx\,.$$

Now calculate:

$$\begin{aligned}\iota_1 dL &= \frac{\partial L}{\partial \boldsymbol{u}}dx + 2\,\frac{\partial L}{\partial \boldsymbol{u}'}D\boldsymbol{u}\\ &= \frac{\partial L}{\partial \boldsymbol{u}}dx + 2\left(\frac{\partial L}{\partial \boldsymbol{u}'}dx\right)' - 2\left(\frac{\partial L}{\partial \boldsymbol{u}'}\right)' dx\\ &= \frac{\partial L}{\partial \boldsymbol{u}}dx + (\iota_2 dL)' - 2\left(\frac{\partial L}{\partial \boldsymbol{u}'}\right)' dx\,,\end{aligned}$$

from where and from the Definition 3.1 it follows immediately that

$$\begin{aligned}\boldsymbol{\pi} dx &= \iota_1 dL - (\iota_2 dL)' + \left(\frac{\partial L}{\partial \boldsymbol{u}'}\right)' dx\\ &= \iota_1 dL - \frac{1}{2}(\iota_2 dL)' - \frac{1}{2}(\iota_2 dL)' + (\boldsymbol{\pi}^{(1)})'dx\\ &= \iota_1 dL - \frac{1}{2}(\iota_2 dL)' - (\boldsymbol{\pi}^{(1)}dx)' + (\boldsymbol{\pi}^{(1)})'dx\\ &= \iota_1 dL - \frac{1}{2}(\iota_2 dL)' - \boldsymbol{\pi}^{(1)}D\boldsymbol{u}\end{aligned}$$

by virtue of (9). □

Proposition 3.2. *In Riemannian space the Euler–Poisson equation for a second order Lagrange function reads:*

$$\boldsymbol{\pi}'dx + \pi^{(1)}{}_i R_{ljq}{}^i u^j u^q dx^l = \frac{\partial L}{\partial x^l}dx^l - \frac{\partial L}{\partial u^i}\Gamma^i_{lj}u^j dx^l - \frac{\partial L}{\partial u'^i}\Gamma^i_{lj}u'^j dx^l \quad (11)$$

Proof. From (3) and from Proposition 3.1 we obtain

$$\delta L = \iota_0 dL - d_T(\boldsymbol{\pi} dx + \boldsymbol{\pi}^{(1)}D\boldsymbol{u}).$$

While the expression in the parenthesis constitutes a geometrical invariant, it is possible to replace d_T by the covariant derivative, after what by direct calculation we obtain in virtue of (9) and of (8):

$$\begin{aligned}\delta L &= \iota_0 dL - (\boldsymbol{\pi} dx + \boldsymbol{\pi}^{(1)}D\boldsymbol{u})' = \iota_0 dL - \boldsymbol{\pi}'dx - (\boldsymbol{\pi} + \boldsymbol{\pi}^{(1)\prime})D\boldsymbol{u} - \boldsymbol{\pi}^{(1)}(D\boldsymbol{u})'\\ &= \iota_0 dL - \boldsymbol{\pi}'dx - \frac{\partial L}{\partial \boldsymbol{u}}D\boldsymbol{u} - \pi^{(1)}{}_i\left(D(\boldsymbol{u}')^i + R_{ljq}{}^i u^j u^q dx^l\right),\end{aligned}$$

and the proof ends by substituting (10) into $i_0 dL \equiv dL$ here and by applying (7). □

4. The two-dimensional variational concircular geometry

As promised, we first cite one result, concerning the invariant inverse variational problem in two dimensional Euclidean space.[5]

Proposition 4.1. *Let some system of third order differential equations*

$$\mathcal{E}_i(x^j, u^j, \dot{u}^j, \ddot{u}^j) = 0 \tag{12}$$

satisfy the conditions:

(i) $\delta\, \mathcal{E}_i dx^i \; = 0$
(ii) The system (12) possesses Euclidean symmetry
(iii) The Euclidean geodesics $\dot{\boldsymbol{u}} = \boldsymbol{0}$ *enter in the set of solutions of (12)*
(iv) $d_T k = 0$ *along the solutions of (12)*

Then

$$\mathcal{E}_i = \frac{e_{ij}\ddot{u}^j}{\|\boldsymbol{u}\|^3} - 3\,\frac{(\dot{\boldsymbol{u}}\cdot\boldsymbol{u})}{\|\boldsymbol{u}\|^5} e_{ij}\dot{u}^j + m\,\frac{\|\boldsymbol{u}\|^2\dot{u}_i - (\dot{\boldsymbol{u}}\cdot\boldsymbol{u})u_i}{\|\boldsymbol{u}\|^3}\,.$$

This system may be obtained from the Lagrange function

$$L = \frac{e_{ij}u^i\dot{u}^j}{\|\boldsymbol{u}\|^3} - m\,\|\boldsymbol{u}\|\,. \tag{13}$$

The first addend in (13) is sometimes called *the signed Frene curvature* in $\mathbb{E}^2$. This, along with the observation that in two dimensional Riemannian space the Frenet curvature

$$k = \frac{\|\boldsymbol{u}\wedge\boldsymbol{u}'\|}{\|\boldsymbol{u}\|^3} = \pm\frac{*\,(\boldsymbol{u}\wedge\boldsymbol{u}')}{\|\boldsymbol{u}\|^3} \tag{14}$$

depends linearly on $\boldsymbol{u}'$ and thus produces at most third order Euler–Poisson equation, suggests the next assertion, based on Proposition 2.1:

Proposition 4.2. *The variational functional* $\int (k - m\,\|\boldsymbol{u}\|)\,dt$ *produces geodesic circles in two dimensional Riemannian space.*

It remains to calculate the Euler–Poisson expression for the Lagrange function (14). In the process of calculations it is convenient to profit from the exceptional properties of vector operations in two dimensions. Namely, the following two relations for arbitrary vectors hold:

$$(\boldsymbol{a}\wedge\boldsymbol{b})\cdot(\boldsymbol{v}\wedge\boldsymbol{w}) = \pm\,\|\boldsymbol{a}\wedge\boldsymbol{b}\|\,\|\boldsymbol{v}\wedge\boldsymbol{w}\|\,,$$

and

$$\|\boldsymbol{a}\wedge\boldsymbol{b}\|(\boldsymbol{b}\cdot\boldsymbol{c}) + \|\boldsymbol{b}\wedge\boldsymbol{c}\|(\boldsymbol{a}\cdot\boldsymbol{b}) = \|\boldsymbol{a}\wedge\boldsymbol{c}\|(\boldsymbol{b}\cdot\boldsymbol{b})\,.$$

The above simplifications bring much release to otherwise very laborious calculations.

We start with the momentum $\pi^{(1)}$:

$$\pm\pi^{(1)}dx = -\frac{(dx\wedge \boldsymbol{u})\cdot(\boldsymbol{u}\wedge\boldsymbol{u}')}{\|\boldsymbol{u}\|^3\|\boldsymbol{u}\wedge\boldsymbol{u}'\|} = -\frac{\|dx\wedge\boldsymbol{u}\|}{\|\boldsymbol{u}\|^3};$$

$$\pm\pi^{(1)\prime}dx = -\frac{\|dx\wedge\boldsymbol{u}'\|}{\|\boldsymbol{u}\|^3} + 3\,\frac{\|dx\wedge\boldsymbol{u}\|}{\|\boldsymbol{u}\|^5}(\boldsymbol{u}\cdot\boldsymbol{u}').$$

Based on Definition 3.1 we now calculate π:

$$\pm\pi dx = 2\,\frac{\|dx\wedge\boldsymbol{u}'\|}{\|\boldsymbol{u}\|^3} - 3\,\frac{\|dx\wedge\boldsymbol{u}\|(\boldsymbol{u}\cdot\boldsymbol{u}')}{\|\boldsymbol{u}\|^5} - 3\,\frac{(dx\cdot\boldsymbol{u})\|\boldsymbol{u}\wedge\boldsymbol{u}'\|}{\|\boldsymbol{u}\|^5} = -\frac{\|dx\wedge\boldsymbol{u}'\|}{\|\boldsymbol{u}\|^3}$$

In terms of the Hodge star operator the derivative of the momentum π may be put in the form

$$\pi' = \frac{*\boldsymbol{u}''}{\|\boldsymbol{u}\|^3} - 3\,\frac{*\boldsymbol{u}'}{\|\boldsymbol{u}\|^5}(\boldsymbol{u}\cdot\boldsymbol{u}'),$$

which agrees with the flat Euclidean case.

For the Lagrange function (14) it is easy to verify that

$$\frac{\partial k}{\partial x^l}dx^l - \frac{\partial k}{\partial u^i}\Gamma^i_{lj}u^j dx^l - \frac{\partial k}{\partial u'^i}\Gamma^i_{lj}u'^j dx^l = 0.$$

The proof consists in direct calculations and founds on the skew-symmetric property of the Christoffel symbols in Riemannian geometry:

$$g_{jl}\Gamma^l_{qi} + g_{il}\Gamma^l_{qj} = \frac{\partial g_{ij}}{\partial x^q}.$$

Going back to the Euler–Poisson equation (11) it is now facile to obtain the variational equation for the full Lagrange function $L = k - m\|\boldsymbol{u}\|$:

$$-\frac{*\boldsymbol{u}''}{\|\boldsymbol{u}\|^3} + 3\,\frac{*\boldsymbol{u}'}{\|\boldsymbol{u}\|^5}(\boldsymbol{u}\cdot\boldsymbol{u}') + m\,\frac{\|\boldsymbol{u}\|^2\boldsymbol{u}' - (\boldsymbol{u}'\cdot\boldsymbol{u})\boldsymbol{u}}{\|\boldsymbol{u}\|^3} = \pi^{(1)}{}_i R_{ljq}{}^i u^j u^q. \quad (15)$$

The term on the right in pseudo-Riemannian case physically may be interpreted as a spin force[7] if, following Ref. 6, we formally introduce spin tensor as $S = (\boldsymbol{u}\wedge\boldsymbol{u}')$.

In fact, one checks that in terms of the tensor S the right hand side of equation (15) may be rewritten as $\dfrac{R_{ljqi}u^j S^{qi}}{\|\boldsymbol{u}\|\|\boldsymbol{u}\wedge\boldsymbol{u}'\|}$.

Acknowledgments

This work was supported by the Grant GAČR 201/06/0922 of Czech Science Foundation.

References

1. K. Yano, Concircular geometry I. Concircular transformations, *Proc. Impt. Academy. Tokyo* **16** (6) (1940) 195–200.
2. L.A. Ibort and C. López-Lacasta, On the existence of local and global Lagrangians for ordinary differential equations, *J. Phys. A* **23** (21) (1990) 4779–4792.
3. M. de Léon and P.R. Rodrigues, *Generalized Classical Mechanics and Field Theory* (Elsevier, Amsterdam, 1985).
4. W.M. Tulczyjew, Sur la différentielle deLagrange, *C. R. Acad. Sci. Paris. Série A* **280** (19) (1975) 1295–1298.
5. R.Ya. Matsyuk, Existence of a Lagrangian for a nonautonomous system of ordinary differential equations, *Mat. Metody Fiz.-Mekh. Polya* **20** (1984) 16–19, in Russian.
6. S.G. Leĭko, Extremals of rotation functionals of curves in a pseudo-Riemannian space, and trajectories of spinning particles in gravitational fields, *Russian Acad. Sci. Dokl. Math.* **46** (1) (1993) 84–87.
7. M. Mathisson, Neue Mechanik materieller Systeme, *Acta phys. polon.* **6** (1937) 163–200.

Differential Geometry and its Applications
Proc. Conf., in Honour of Leonhard Euler, Olomouc, August 2007

Noether identities in Einstein–Dirac theory and the Lie derivative of spinor fields

M. Palese and E. Winterroth

Department of Mathematics, University of Torino, Torino, I-10123, Italy
E-mail: marcella.palese@unito.it, ekkehart.winterroth@unito.it

We characterize the Lie derivative of spinor fields from a variational point of view by resorting to the theory of the Lie derivative of sections of gauge-natural bundles. Noether identities from the gauge-natural invariance of the first variational derivative of the Einstein(–Cartan)–Dirac Lagrangian provide restrictions on the Lie derivative of fields.

Keywords: Jet, gauge-natural bundle, Noether identity, spinor field.

MS classification: 58A20, 58A32, 58E30, 58E40.

1. Introduction

It is nowadays widely recognized the prominent *rôle* played by the gauge-natural functorial approach to the geometric description of (classical) field theories.[1–4] Physical fields are assumed to be geometrically represented by sections of fiber bundles functorially associated with some jets prolongation of the relevant principal bundle by means of left actions of Lie groups on manifolds, usually tensor spaces. Such an approach enables to functorially define the Lie derivative of physical fields with respect to gauge-natural lifts of (prolongations of) infinitesimal principal automorphisms of the underlying principal bundle.[5]

This concept generalizes that of the natural lift of an infinitesimal automorphism of the basis manifold to the bundle of higher order frames.[5] The structure group of the total space is generalized to be the semidirect product of a differential group on a Lie group $\boldsymbol{G}$ — the gauge group — which is not in general a subgroup of a differential group. In particular, in the Einstein(-Cartan)-Dirac theory, the coupling between gravitational and fermionic fields requires the use of the concept of a *spin-tetrad*, which turns out to be a gauge-natural object (see *e.g.* Refs. 3,6–9 and references therein).

However, since this construction involves the enlargement of the class of morphisms of the category, such an approach yields an indeterminacy in the concept of conserved quantities. In fact the vertical components of an infinitesimal principal automorphism are completely independent from the components of its projection on the tangent bundle to the basis manifold. This implies that there is *a priori* no natural way of relating infinitesimal gauge transformations with infinitesimal transformations of the basis manifold, *e.g.* of space-time (see *e.g.* in particular Ref. 8). It is generally believed that such an indeterminacy is somehow unavoidable, and that *ad hoc* restrictions[10,11] on the allowed automorphisms of the gauge-natural bundle must be performed, in order to coherently and uniquely define a geometric concept of the Lie derivative of sections of gauge-natural bundles. For a quite exhaustive review see *e.g.* Refs. 6–8,12.

For reasons which will be clear later, due to its invariance with respect to contact structure induced by jets, we consider the geometric framework *finite order variational sequences*[13] as the most suitable for the definition of the class of infinitesimal automorphisms of the gauge-natural bundle with respect to which the Lie derivative of fields can be defined unambiguously. In particular, in Refs. 15,17–20 a variational sequence on gauge-natural bundles was considered, pointing out some important properties of the Lie derivative of sections of bundles[1,5] and relative consequences on the content of Noether Theorems.[21]

We stress the very important fact — although underestimated — that in the category of gauge-natural bundles it is possible to relate the Lie derivative of sections of bundles with *the vertical part* — *not* the vertical component — of jet prolongations of gauge-natural lifts of infinitesimal principal automorphisms (they *coincide* up to a sign). The concept of the vertical part of a projectable vector field, together with other important related decompositions induced by the contact structure on jet bundles will be defined in the next Section.

In Ref. 18 for the first time the fact has been pointed out that — when taken as variation vector fields, in order to derive covariantly and canonically conserved Noether currents — *such vertical parts are constrained by generalized Jacobi equations.* A restriction on the Lie derivative of fields is then immediately derived by the simple request of the covariance of conserved quantities generated by gauge-natural symmetries. We notice that necessary and sufficient conditions (Bergmann-Bianchi identities[22]) for the existence of canonical covariant conserved currents and associated superpotentials can be suitably interpreted as Noether identities.[23] By representing the Noether Theorems[21] in terms of the generalized gauge-natural Jacobi morphism, the Lie derivative of spinor fields is then accordingly characterized as the above mentioned indeterminacy disappears along the kernel of the generalized gauge-natural Jacobi morphism.

2. Variational sequences on jets of gauge-natural bundles

Consider a fibered manifold $\pi : \boldsymbol{Y} \to \boldsymbol{X}$, with $\dim \boldsymbol{X} = n$ and $\dim \boldsymbol{Y} = n+m$. For $s \geq q \geq 0$ integers we are concerned with the s–jet space $J_s\boldsymbol{Y}$ of s–jet prolongations of (local) sections of π; in particular, we set $J_0\boldsymbol{Y} \equiv \boldsymbol{Y}$. We recall the natural fiberings $\pi^s_q : J_s\boldsymbol{Y} \to J_q\boldsymbol{Y}$, $s \geq q$, $\pi^s : J_s\boldsymbol{Y} \to \boldsymbol{X}$, and, among these, the *affine* fiberings π^s_{s-1}.[24]

By adopting a multiindex notation, the charts induced on $J_s\boldsymbol{Y}$ are denoted by $(x^\sigma, y^i_{\boldsymbol{\alpha}})$, with $0 \leq |\boldsymbol{\alpha}| \leq s$; in particular, we set $y^i_{\boldsymbol{0}} \equiv y^i$. The local vector fields and forms of $J_s\boldsymbol{Y}$ induced by the above coordinates are denoted by $(\partial^{\boldsymbol{\alpha}}_i)$ and $(d^i_{\boldsymbol{\alpha}})$, respectively.

Let $\mathcal{C}^*_{s-1}[\boldsymbol{Y}] \simeq J_s\boldsymbol{Y} \underset{J_{s-1}\boldsymbol{Y}}{\times} V^*J_{s-1}\boldsymbol{Y}$. For $s \geq 1$, we have a natural splitting:

$$J_s\boldsymbol{Y} \underset{J_{s-1}\boldsymbol{Y}}{\times} T^*J_{s-1}\boldsymbol{Y} = \left(J_s\boldsymbol{Y} \underset{J_{s-1}\boldsymbol{Y}}{\times} T^*\boldsymbol{X}\right) \oplus \mathcal{C}^*_{s-1}[\boldsymbol{Y}]\,. \tag{1}$$

Given a vector field $\Xi : J_s\boldsymbol{Y} \to TJ_s\boldsymbol{Y}$, the splitting (1) yields $\Xi \circ \pi^{s+1}_s = \Xi_H + \Xi_V$ where, if $\Xi = \Xi^\gamma \partial_\gamma + \Xi^i_{\boldsymbol{\alpha}} \partial^{\boldsymbol{\alpha}}_i$, then we have $\Xi_H = \Xi^\gamma D_\gamma$ and the invariant expression $\Xi_V = (\Xi^i_{\boldsymbol{\alpha}} - y^i_{\boldsymbol{\alpha}+\gamma}\Xi^\gamma)\partial^{\boldsymbol{\alpha}}_i$. We shall call Ξ_H and Ξ_V the horizontal and *the vertical part of* Ξ, respectively. The splitting (1) induces also a decomposition of the exterior differential on $\boldsymbol{Y}$, $(\pi^s_{s-1})^* \circ d = d_H + d_V$, where d_H and d_V, the *horizontal* and *vertical differential*.[14,24,25]

Let $\boldsymbol{P} \to \boldsymbol{X}$ be a principal bundle with structure group $\boldsymbol{G}$. Let $r \leq k$ be integers and $\boldsymbol{W}^{(r,k)}\boldsymbol{P}$ be the gauge-natural prolongation of $\boldsymbol{P}$ with structure group $\boldsymbol{W}^{(r,k)}_n\boldsymbol{G}$.[1,5] Let $\boldsymbol{F}$ be any manifold and $\zeta : \boldsymbol{W}^{(r,k)}_n\boldsymbol{G} \times \boldsymbol{F} \to \boldsymbol{F}$ be a left action of $\boldsymbol{W}^{(r,k)}_n\boldsymbol{G}$ on $\boldsymbol{F}$. There is a naturally defined right action of $\boldsymbol{W}^{(r,k)}_n\boldsymbol{G}$ on $\boldsymbol{W}^{(r,k)}\boldsymbol{P} \times \boldsymbol{F}$ so that we can associate in a standard way to $\boldsymbol{W}^{(r,k)}\boldsymbol{P}$ the *gauge-natural bundle* of order (r,k) $\boldsymbol{Y}_\zeta \doteq \boldsymbol{W}^{(r,k)}\boldsymbol{P} \times_\zeta \boldsymbol{F}$.[1,5]

Let $\mathcal{A}^{(r,k)} \doteq T\boldsymbol{W}^{(r,k)}\boldsymbol{P}/\boldsymbol{W}^{(r,k)}\boldsymbol{G}$ $(r \leq k)$ be the *vector* bundle over $\boldsymbol{X}$ of right invariant infinitesimal automorphisms of $\boldsymbol{W}^{(r,k)}\boldsymbol{P}$. The *gauge-natural lift* $\mathfrak{G}$ functorially associates with any right-invariant local automorphism (Φ, ϕ) of the principal bundle $W^{(r,k)}\boldsymbol{P}$ a unique local automorphism (Φ_ζ, ϕ)

of the associated bundle $\boldsymbol{Y}_\zeta$. An infinitesimal version can be defined:

$$\mathfrak{G} : \boldsymbol{Y}_\zeta \underset{\boldsymbol{X}}{\times} \mathcal{A}^{(r,k)} \to T\boldsymbol{Y}_\zeta : (\boldsymbol{y}, \bar{\Xi}) \mapsto \hat{\Xi}(\boldsymbol{y}) , \tag{2}$$

where, for any $\boldsymbol{y} \in \boldsymbol{Y}_\zeta$, one sets: $\hat{\Xi}(\boldsymbol{y}) = \frac{d}{dt}[(\Phi_{\zeta\, t})(\boldsymbol{y})]_{t=0}$, and $\Phi_{\zeta\, t}$ denotes the (local) flow corresponding to the gauge-natural lift of Φ_t. This mapping fulfills important linearity properties.[4]

Although the jet prolongation of up to a given order of a gauge-natural bundle is again a gauge-natural bundle associated with some gauge natural prolongation of the underlying principal bundle,[5] in general — as it can be easily seen also from the corresponding local (invariant) expression — the generalized vector field $j_s\Xi_V \doteq (j_s\Xi)_V$ *is not the gauge-natural lift of some infinitesimal principal automorphism.*

Following Ref. 5 we give the definition of the Lie derivative of a section of a gauge-natural bundle. Notice that this object is uniquely and functorially defined by the right invariant vector field $\bar{\Xi}$.

Definition 2.1. Let γ be a (local) section of $\boldsymbol{Y}_\zeta$, $\bar{\Xi} \in \mathcal{A}^{(r,k)}$ and $\hat{\Xi}$ its gauge-natural lift. We define the *generalized Lie derivative* of γ along the vector field $\hat{\Xi}$ to be the (local) section $\pounds_{\bar{\Xi}}\gamma : \boldsymbol{X} \to V\boldsymbol{Y}_\zeta$, given by $\pounds_{\bar{\Xi}}\gamma = T\gamma \circ \xi - \hat{\Xi} \circ \gamma$.

Remark 2.1. The Lie derivative of sections is an homomorphism of Lie algebras; furthermore for any local section γ of $\boldsymbol{Y}_\zeta$, the mapping $\bar{\Xi} \mapsto \pounds_{\bar{\Xi}}\gamma$ is a linear differential operator. As a consequence, for any gauge-natural lift $\hat{\Xi}$, the fundamental relation hold true:

$$\hat{\Xi}_V = -\pounds_{\bar{\Xi}} . \quad \square \tag{3}$$

2.1. *Variational derivatives and Noether identities*

Let us now introduce variational sequences on gauge-natural bundles. For $s \geq 0$ (*resp.* $0 \leq q \leq s$) we consider the sheaves Λ^p_s of p–forms on $J_s\boldsymbol{Y}_\zeta$ (*resp.* $\mathcal{H}^p_{(s,q)}$ and $\mathcal{H}^p_s$ of *horizontal forms* with respect to the projections π^s_q and π^s_0). Furthermore, for $0 \leq q < s$, let $\mathcal{C}^p_{(s,q)} \subset \mathcal{H}^p_{(s,q)}$ and $\mathcal{C}^p{}_s \subset \mathcal{C}^p_{(s+1,s)}$ be the sheaves of *contact forms*, *i.e.* horizontal forms valued into $\mathcal{C}^*_s[\boldsymbol{Y}_\zeta]$. The fibered splitting (1) yields the *sheaf splitting* $\mathcal{H}^p_{(s+1,s)} = \bigoplus_{t=0}^p \mathcal{C}^{p-t}_{(s+1,s)} \wedge\mathcal{H}^t_{s+1}$. Let the surjective map h be the restriction to Λ^p_s of the projection of the above splitting onto the non–trivial summand with the highest value of t. Set $\Theta^*_s := \mathrm{Ker}h + d\mathrm{Ker}h$.[13,25]

The following *s–th order variational sequence* associated with the fibered manifold $\boldsymbol{Y}_\zeta \to \boldsymbol{X}$, where the integer I depends on the dimension of the fibers of $\boldsymbol{Y}_\zeta$, is an exact resolution of the constant sheaf $I\!R_{\boldsymbol{Y}_\zeta}$ over $\boldsymbol{Y}_\zeta$:[13]

$$0 \longrightarrow I\!R_{\boldsymbol{Y}_\zeta} \longrightarrow \Lambda^0_s \xrightarrow{\mathcal{E}_0} \Lambda^1_s/\Theta^1_s \xrightarrow{\mathcal{E}_1} \dots \xrightarrow{\mathcal{E}_{I-1}} \Lambda^I_s/\Theta^I_s \xrightarrow{\mathcal{E}_I} \Lambda^{I+1}_s \xrightarrow{d} 0\,.$$

For our purposes we refer to the representation of a truncated variational sequence due to Vitolo[25] where $\Lambda^p_s/\Theta^p_s \equiv \mathcal{C}^{p-n}_s \wedge \mathcal{H}^{n,h}{}_{s+1}/h(d\mathrm{Ker}h)$ with $0 \leq p \leq n+2$. Further developments can be found *e.g.* in Refs. 14–20,25,26.

Let now $\eta \in \mathcal{C}^1_s \wedge \mathcal{C}^1_{(s,0)} \wedge \mathcal{H}^{n,h}{}_{s+1}$; then there is a unique morphism[25,27] $K_\eta \in \mathcal{C}^1_{(2s,s)} \otimes \mathcal{C}^1_{(2s,0)} \wedge \mathcal{H}^{n,h}{}_{2s+1}$ such that, for all $\Xi : \boldsymbol{Y}_\zeta \to V\boldsymbol{Y}_\zeta$, $C^1_1(j_{2s}\Xi \otimes K_\eta) = E_{j_s\Xi\rfloor\eta}$, where E is the the generalized Euler–Lagrange form;[13] C^1_1 stands for tensor contraction on the first factor and $\rfloor$ denotes inner product.

Let $\mathcal{L}_{j_s\Xi}$ be the *variational Lie derivative*.[14] The First and the Second Noether Theorem[21] then read as follows (compare with Ref. 28).

Theorem 2.1. *Let $[\alpha] = h(\alpha) \in \mathcal{V}^n_s$. Then we have locally*

$$\mathcal{L}_{j_s\Xi}(h(\alpha)) = \Xi_V \rfloor \mathcal{E}_n(h(\alpha)) + d_H(j_{2s}\Xi_V \rfloor p_{d_V h(\alpha)} + \xi \rfloor h(\alpha))\,,$$

where $p_{d_V h(\alpha)}$ is the generalized momentum associated with $h(\alpha)$.[14,25]

Theorem 2.2. *Let $\alpha \in \Lambda^{n+1}_s$. Then we have globally*

$$\mathcal{L}_{j_s\Xi}[\alpha] = \mathcal{E}_n(j_{s+1}\Xi_V \rfloor h(\alpha)) + C^1_1(j_s\Xi_V \otimes K_{hd\alpha})\,.$$

Definition 2.2. We say λ to be a *gauge-natural invariant Lagrangian* if the gauge-natural lift $(\hat{\Xi}, \xi)$ of *any* vector field $\Xi \in \mathcal{A}^{(r,k)}$ is a symmetry for λ, *i.e.* if $\mathcal{L}_{j_{s+1}\Xi}\,\lambda = 0$. In this case the projectable vector field $\hat{\Xi}$ is called a *gauge-natural symmetry* of λ.

Let λ be a Lagrangian and let $\hat{\Xi}_V$ be considered a variation vector field. Let us set $\chi(\lambda, \hat{\Xi}_V) \doteq C^1_1(\hat{\Xi}_V \otimes K_{hd\mathcal{L}_{j_{2s}\hat{\Xi}_V}\lambda}) \equiv E_{j_s\hat{\Xi}\rfloor hd\mathcal{L}_{j_{2s+1}\hat{\Xi}_V}\lambda}$. Because of linearity properties of $K_{hd\mathcal{L}_{j_{2s}\hat{\Xi}_V}\lambda}$,[27] by using a global decomposition formula due to Kolář,[29] we can decompose the morphism defined above as $\chi(\lambda, \hat{\Xi}_V) = E_{\chi(\lambda,\hat{\Xi}_V)} + F_{\chi(\lambda,\hat{\Xi}_V)}$, where $F_{\chi(\lambda,\hat{\Xi}_V)}$ is a *local* horizontal differential which can be globalized by means of the fixing of a connection; however we will not fix any connection *a priori* in the present paper. Such a decomposition is a kind of integration by parts, which provides us with a globally defined gauge-natural morphism playing a relevant *rôle.*[18]

Definition 2.3. We call the morphism $\mathcal{J}(\lambda, \hat{\Xi}_V) \doteq E_{\chi(\lambda,\hat{\Xi}_V)}$ the *gauge-natural generalized Jacobi morphism* associated with the Lagrangian λ and the variation vector field $\hat{\Xi}_V$.

Such a morphism has also been represented on finite order variational sequence modulo horizontal differentials[15] and thereby proved to be self-adjoint along solutions of the Euler–Lagrange equations, a result already well known for first order field theories.[30] The same property has been also proved in finite order variational sequences on gauge-natural bundles[20] *without quotienting out horizontal differentials.*

Because of linearity properties of the Lie derivative of sections of gauge-natural bundles, we can consider the form $\omega(\lambda, \hat{\Xi}_V) \doteq -\mathcal{L}_{\Xi} \rfloor \mathcal{E}_n(\lambda)$ as a new Lagrangian defined on an extended space $J_{2s}(\mathcal{A}^{(r,k)} \times_{\boldsymbol{Y}} \boldsymbol{Y})$. This Lagrangian plays a very important *rôle* in the study of conserved quantities. In fact, it is for example remarkable that when $\omega(\lambda, \hat{\Xi}_V)$ is an horizontal differential (*i.e.* a null Lagrangian) from the First Noether Theorem 2.1 we get a conservation law which holds true along any section of the gauge natural bundle (not only along solutions of the Euler–Lagrange equations).

It is also remarkable that the new Lagrangian ω, in principle, *is not gauge-natural invariant.* In fact from the gauge-natural invariance of λ we only infer that, for any $\hat{\Xi}$, $\mathcal{L}_{j_{s+1}\hat{\Xi}}[\mathcal{L}_{j_{s+1}\hat{\Xi}_V}\lambda] = \mathcal{L}_{j_{s+1}[\hat{\Xi},\hat{\Xi}_V]}\lambda + \mathcal{L}_{j_{s+1}\hat{\Xi}_V}\mathcal{L}_{j_{s+1}\hat{\Xi}}\lambda = \mathcal{L}_{j_{s+1}[\hat{\Xi}_H,\hat{\Xi}_V]}\lambda$ and *a priori* neither $[\hat{\Xi}_H, \hat{\Xi}_V] = 0$ *nor* it is the gauge-natural lift of some infinitesimal principal automorphism. Nevertheless it is still possible to derive some invariance properties of the new Lagrangian $\omega(\lambda, \hat{\Xi}_V)$ restricted along the kernel of the gauge-natural generalized gauge-natural Jacobi morphism as well as corresponding Noether conservation laws and Noether identities.[21]

Let then $\delta^2_{\mathfrak{G}}\lambda \doteq \mathcal{L}_{j_{s+1}\hat{\Xi}_V}\mathcal{L}_{j_{s+1}\hat{\Xi}_V}\lambda$ be the *gauge-natural second variation* of λ taken with respect to vertical parts of gauge-natural lifts of infinitesimal principal automorphisms. First we generalize a classical result.[30]

Proposition 2.1. *Let $\hat{\Xi}_V \in \mathfrak{K}$, where $\mathfrak{K}$ is the kernel of the gauge-natural Jacobi morphism. We have:*

$$\mathcal{L}_{j_{s+1}\hat{\Xi}_H}[\mathcal{L}_{j_{s+1}\hat{\Xi}_V}\lambda] \equiv 0\,.$$

Proof.

$$\mathcal{L}_{j_{s+1}\hat{\Xi}}[\mathcal{L}_{j_{s+1}\hat{\Xi}_V}\lambda] = \mathcal{L}_{j_{s+1}\hat{\Xi}_H}[\mathcal{L}_{j_{s+1}\hat{\Xi}_V}\lambda]$$
$$+ \mathcal{L}_{j_{s+1}\hat{\Xi}_V}[\mathcal{L}_{j_{s+1}\hat{\Xi}_V}\lambda] = \mathcal{L}_{j_{s+1}\hat{\Xi}_H}[\mathcal{L}_{j_{s+1}\hat{\Xi}_V}\lambda] + \delta^2_{\mathfrak{G}}\lambda\,.$$

Theorem 2.2 implies $\delta^2_{\mathfrak{G}}\lambda = \mathcal{J}(\lambda, \hat{\Xi}_V)$, thus we get the assertion. □

Theorem 2.3. *The existence of canonical covariant conserved currents and superpotentials associated with a gauge-natural invariant Lagrangian is* *equivalent to the existence of Noether identities associated with the invariance properties of the first variational derivative of the Lagrangian λ taken with respect to vertical parts of gauge-natural lifts lying in $\mathfrak{K}$.*

Proof. From the above Proposition we see that $\mathcal{L}_{j_{s+1}\bar{\Xi}}[\mathcal{L}_{j_{s+1}\hat{\Xi}_V}\lambda] \equiv 0$. This last condition means that $\bar{\Xi}$ is a gauge-natural symmetry of the new Lagrangian $\mathcal{L}_{j_{s+1}\hat{\Xi}_V}\lambda = \omega(\lambda, \mathfrak{K})$. This invariance implies also the existence of Noether identities from the gauge-natural invariance of the Euler–Lagrange morphism $\mathcal{E}_n(\hat{\Xi}_V \rfloor \mathcal{E}_n(\lambda))$ and ultimately from the corresponding invariance properties of the first variational derivative of the Lagrangian λ. It is easy to verify that Bergmann-Bianchi identities for the existence of canonical covariant conserved currents and superpotentials associated with the invariance of λ coincide with the condition $\hat{\Xi}_V \rfloor \mathcal{E}_n(\hat{\Xi}_V \rfloor \mathcal{E}_n(\lambda)) = 0$ (see *e.g.* Refs. 18,20 and Refs. 15,30) equivalent to $\mathcal{L}_{j_{s+1}\hat{\Xi}_V}\mathcal{E}_n(\hat{\Xi}_V \rfloor \mathcal{E}_n(\lambda)) = 0$. □

3. Einstein(–Cartan)–Dirac theory

Let $\boldsymbol{X}$ be a n-dimensional manifold admitting Lorentzian structures ($SO(1,3)^e$-reductions) and let Λ be the epimorphism wich exhibits $SPIN(1,3)^e$ as the twofold covering of $SO(1,3)^e$.[2,3]

We recall that a free spin structure on $\boldsymbol{X}$ is a $SPIN(1,3)^e$-principal bundle $\pi : \Sigma \to \boldsymbol{X}$ and a bundle map inducing a spin-frame on Σ given by $\tilde{\Lambda} : \Sigma \to L(\boldsymbol{X})$ defining a metric g *via* the reduced subbundle $SO(\boldsymbol{X}, g) = \tilde{\Lambda}(\Sigma)$ of $L(\boldsymbol{X})$.[3,7,8]

Now, let ρ be the left action of the group $W^{(1,0)}SPIN(1,3)^e$ on the manifold $GL(4, I\!R)$ given by $\rho : ((A, S), \theta) \mapsto \Lambda(S) \circ \theta \circ A^{-1}$ and consider the associated bundle (a gauge-natural bundle of order $(1,0)$)

$$\Sigma_\rho \doteq W^{(1,0)}\Sigma \times_\rho GL(4, I\!R) \; :$$

the *bundle of spin-tetrads* θ.[9]

The induced metric is $g_{\mu\nu} = \theta^a_\mu \theta^b_\nu \eta_{ab}$, where θ^a_μ are local components of a spin tetrad θ and η_{ab} the Minkowski metric.

Let $\mathfrak{so}(1,3) \simeq \mathfrak{spin}(1,3)$ be the Lie algebra of $SO(1,3)$. One can consider the left action of $W^{(1,1)}SPIN(1,3)^e$ on the vector space $(I\!R^4)^* \otimes \mathfrak{so}(1,3)$. The associated bundle is a gauge-natural bundle of order $(1,1)$:

$$\Sigma_l \doteq W^{(1,1)}\Sigma \times_l ((I\!R^4)^* \otimes \mathfrak{so}(1,3)) \simeq J_1(\Sigma/Z_2)/SO(1,3)^e \,,$$

the *bundle of spin-connections* ω.

If $\hat{\gamma}$ is the linear representation of $SPIN(1,3)^e$ on the vector space $\mathbb{C}^4$ induced by the choice of matrices γ we get a $(0,0)$-gauge-natural bundle

$$\Sigma_{\hat{\gamma}} \doteq \Sigma \times_{\hat{\gamma}} \mathbb{C}^4 \,,$$

the bundle of spinors. A spinor connection $\tilde{\omega} = (id \otimes (T_e\Lambda)^{-1})(\omega)$ - where $T_e\Lambda$ defines the isomorphism of Lie algebras $\mathfrak{spin}(1,3) \simeq \mathfrak{so}(1,3)$ - is locally given by $\tilde{\omega}_\mu = -\frac{1}{4}\omega_\mu^{ab}\gamma_{ab}$.

In the following the Einstein–Cartan Lagrangian will be the base preserving morphism $\lambda_{EC} : \Sigma_\rho \underset{\boldsymbol{X}}{\times} J_1\Sigma_l \to \Lambda^4 T^*\boldsymbol{X}$, locally given by $\lambda_{EC}(\theta, J^1\omega) \doteq -\frac{1}{2k}\Omega_{ab} \wedge \epsilon^{ab}$, where Ω_{ab} is the curvature of ω, $k = \frac{8\pi G}{c^4}$, $\epsilon^{ab} = e_a\rfloor(e_b\rfloor\epsilon)$ and ϵ is the standard volum form on $\boldsymbol{X}$. Locally $\epsilon = det||\theta||d_0 \wedge \ldots d_3$, $e_a \doteq e_a^\mu\partial_\mu$, where $||e_a^\mu||$ is the inverse of $||\theta||$.

The Dirac Lagrangian is the base preserving morphism $\lambda_D : \Sigma_\rho \underset{\boldsymbol{X}}{\times} \Sigma_l \underset{\boldsymbol{X}}{\times} J_1\Sigma_{\hat{\gamma}} \to \Lambda^4 T^*\boldsymbol{X}$, locally given by $\lambda_D(\theta, \omega, J^1\psi) \doteq (\frac{i\alpha}{2}(\bar{\psi}\gamma^a\nabla_a\psi - \nabla_a\bar{\psi}\gamma^a\psi) - m\bar{\psi}\psi)\epsilon$, whit $\alpha = hc$, ∇ the covariant derivative with respect to a connection on Σ_ρ, and $\bar{\psi}$ the adjoint with respect to the standard $SPIN(1,3)^e$-invariant product on $\mathbb{C}^4$. Under the assumption of a minimal coupling the total Lagrangian of a gravitational field interacting with spinor matter is $\lambda = \lambda_{EC} + \lambda_D$.

Let now $\bar{\Xi}$ be a $SPIN(1,3)^e$-invariant vector field on Σ. The lagrangian λ is invariant with respect to any lift $\hat{\bar{\Xi}}$ of $\bar{\Xi}$ to the total space of the theory. By the First Noether Theorem the following conserved Noether current has been found in Ref. 31:

$$\epsilon(\lambda, \bar{\Xi}) = \xi^b(-\frac{1}{k}G_b^a + T_b^a)\epsilon_a + \bar{\Xi}_v{}^a{}_b(\frac{1}{2k}\mathfrak{D}\epsilon_{ab} - u_{ab}^c\epsilon_c) + d_H(\frac{1}{2k}\bar{\Xi}_v{}^a{}_b\epsilon_{ab})\,,$$

where $\pounds_{\bar{\Xi}}\omega_b^a = \xi\rfloor\Omega_b^a + \mathfrak{D}\bar{\Xi}_v{}^a{}_b$, $\mathfrak{D}$ is the covariant exterior derivative with respect to the connection ω, and $\bar{\Xi}_v{}^a{}_b = \bar{\Xi}_b^a - \omega_{b\mu}^a\xi^\mu$ is the vertical part of $\bar{\Xi}$ with respect to ω. The corresponding superpotential is $\nu(\lambda, \bar{\Xi}) \doteq -\frac{1}{2k}\bar{\Xi}_v^{ab}\epsilon_{ab}$.

3.1. *Lie derivative of spinor fields*

The natural splitting induced by the contact structure provides us with the vertical part $\hat{\bar{\Xi}}_V^{\boldsymbol{A}} = \bar{\Xi}^{\boldsymbol{A}} - u_\mu^{\boldsymbol{A}}\xi^\mu$, where $(x^\mu, u^{\boldsymbol{A}})$ are local coordinates on the total gauge-natural bundle and $\boldsymbol{A}$ is a multiindex. On the other hand , since the vertical part with respect to the spinor-connection $\tilde{\omega}$ is given by $\hat{\bar{\Xi}}_v^{\boldsymbol{A}} = \hat{\bar{\Xi}}^{\boldsymbol{A}} - \tilde{\omega}_\mu^{\boldsymbol{A}}\xi^\mu$, we get the simple invariant relation:

$$\hat{\bar{\Xi}}_V^{\boldsymbol{A}} = \hat{\bar{\Xi}}_v^{\boldsymbol{A}} + \tilde{\omega}_\mu^{\boldsymbol{A}}\xi^\mu - u_\mu^{\boldsymbol{A}}\xi^\mu\,. \tag{4}$$

In the above Section we saw that, without the fixing of a connection *a priori*, the existence of *canonical* global conserved quantities in field theory is related with gauge-natural invariant properties of $-\pounds_{\bar{\Xi}}\rfloor\mathcal{E}_n(\lambda)$ and corresponding Noether identities:

$$
\begin{aligned}
&(-1)^{|\boldsymbol{\sigma}|} D_{\boldsymbol{\sigma}} \,(D_{\boldsymbol{\mu}}(-\pounds_{\bar{\Xi}}\psi)^{ab}(\partial_{cd}(\partial^{\boldsymbol{\mu}}_{ab}(-\frac{1}{2k}\Omega_{ab}\wedge\epsilon^{ab} \\
&+(\frac{i\alpha}{2}(\bar{\psi}\gamma^a\nabla_a\psi-\nabla_a\bar{\psi}\gamma^a\psi)-m\bar{\psi}\psi)\epsilon)) \\
&-\sum_{|\boldsymbol{\alpha}|=0}^{s-|\boldsymbol{\mu}|}(-1)^{|\boldsymbol{\mu}+\boldsymbol{\alpha}|}\frac{(\boldsymbol{\mu}+\boldsymbol{\alpha})!}{\boldsymbol{\mu}!\boldsymbol{\alpha}!}D_{\boldsymbol{\alpha}}\partial^{\boldsymbol{\alpha}}_{cd}(\partial^{\boldsymbol{\mu}}_{ab}(-\frac{1}{2k}\Omega_{ab}\wedge\epsilon^{ab} \\
&+(\frac{i\alpha}{2}(\bar{\psi}\gamma^a\nabla_a\psi-\nabla_a\bar{\psi}\gamma^a\psi)-m\bar{\psi}\psi)\epsilon)))) = 0\,.,
\end{aligned}
\tag{5}
$$

with $0 \leq |\boldsymbol{\sigma}| \leq 1$, $0 \leq |\boldsymbol{\mu}| \leq 1$ and the fibered local coordinates in the total bundle we denote by $(x^\mu, \theta^a_\mu, \omega^{ab}, \omega^{ab}_\mu, \psi)$. Such identities imply, after some manipulations, that $\bar{\Xi}^{ab}_v = -\tilde{\nabla}^{[a}\xi^{b]}$ (the so-called Kosmann lift[10]), where $\tilde{\nabla}$ is the covariant derivative with respect to the standard transposed connection on Σ_ρ. On the other hand, the Lie derivative of spinor fields can be expressed in terms of $\bar{\Xi}^{ab}_v$ as follows:

$$
\begin{aligned}
\pounds_{\bar{\Xi}}\psi &= \xi^a e_a\psi + \frac{1}{4}\hat{\Xi}_{ab}\gamma^a\gamma^b\psi = \xi^a\nabla_a\psi - \frac{1}{4}\nabla_{[a}\xi_{b]}\gamma^a\gamma^b\psi \\
&= \xi^\mu\partial_\mu\psi + \frac{1}{4}\hat{\Xi}_{h\,ab}\gamma^a\gamma^b\psi - \frac{1}{4}\nabla_{[a}\xi_{b]}\gamma^a\gamma^b\psi\,,
\end{aligned}
$$

where $\hat{\Xi}_h$ is the horizontal part of $\hat{\Xi}$ *with respect to the spinor-connection.* We see that, *because of relation* (3), once the expression of $\bar{\Xi}^{ab}_v$ derived by Eq. (5) has been substituted in Eq. (4), we obtain a condition involving the spinor-connection $\tilde{\omega}$.

This result has been pointed out in Ref. 23. It agrees with an analogous one obtained in Ref. 12 within a different approach to conservation laws for the Einstein–Cartan theory of gravitation.

In Refs. 6,7 a geometric interpretation of the Kosmann lift as a reductive lift has been recovered for the definition of a $SO(1,3)^e$-reductive Lie derivative of spinor fields. We justify the "naturality" of the Kosmann lift from a variational point of view: it charaterizes the only gauge-natural lift which ensures the naturality condition $\mathcal{L}_{j_{s+1}\hat{\Xi}_H}[\mathcal{L}_{j_{s+1}\hat{\Xi}_V}\lambda] \equiv 0$ holds true. Along such a lift not only the initial Lagrangian λ is by assumption invariant, but also its first variational derivative, *taken with respect to vertical parts of gauge-natural lifts lying in the kernel of the generalized Jacobi morphism,* $\omega(\lambda, \mathfrak{K})$ is it, thus implying that either $[\hat{\Xi}_H, \hat{\Xi}_V] = 0$ or it is a gauge-natural lift. The Hamiltonian content of such a naturality condition and implications on conserved quantities have been investigated in Refs. 19,23,32. The two approaches are strictly related: in a forthcoming paper[33] we study how the kernel of the Jacobi morphism induces a split structure which is also reductive.

Acknowledgments

Research supported by MIUR–PRIN (2005) and University of Torino. Thanks are due to Prof. I. Kolář for interesting discussions and for the kind invitation to take part to the DGA2007 Conference.

References

1. D.J. Eck, Gauge-natural bundles and generalized gauge theories, *Mem. Amer. Math. Soc.* **247** (1981) 1–48.
2. L. Fatibene, M. Ferraris, M. Francaviglia and M. Godina, Gauge formalism for general relativity and fermionic matter, *Gen. Rel. Grav.* **30** (9) (1998) 1371–89.
3. L. Fatibene and M. Francaviglia, *Natural and gauge natural formalism for classical field theories. A geometric perspective including spinors and gauge theories* (Kluwer Academic Publishers, Dordrecht, 2003).
4. L. Fatibene, M. Francaviglia and M. Palese, Conservation laws and variational sequences in gauge-natural theories, *Math. Proc. Camb. Phil. Soc.* **130** (2001) 555–569.
5. I. Kolář, P.W. Michor and J. Slovák, *Natural Operations in Differential Geometry* (Springer–Verlag, N.Y., 1993).
6. M. Godina and P. Matteucci, Reductive G-structures and Lie derivatives, *J. Geom. Phys.* **47** (1) (2003) 66–86.
7. M. Godina and P. Matteucci, The Lie derivative of spinor fields: theory and applications, *Int. J. Geom. Methods Mod. Phys.* **2** (2) (2005) 159–188.
8. P. Matteucci, Einstein-Dirac theory on gauge-natural bundles, *Rep. Math. Phys.* **52** (1) (2003) 115–139.
9. S. Weinberg, *Gravitation and cosmology: principles and applications of the general theory of relativity* (Wiley, New York, 1972).
10. Y. Kosmann, Dérivée de Lie de spineurs, *C. R. Acad. Sci. Paris Sér. A-B* **262** (1966) A289–A292; – Dérivée de Lie de spineurs. Applications, *C. R. Acad. Sci. Paris Sér. A-B* **262** (1966) A394–A397.
11. A. Lichnerowicz, Spineurs harmoniques, *C.R.Acad.Sci.Paris* **257** (1963) 7–9.
12. M. Ferraris, M. Francaviglia and M. Raiteri, Conserved Quantities from the Equations of Motion (with applications to natural and gauge natural theories of gravitation) *Class. Quant. Grav.* **20** (2003) 4043–4066.
13. D. Krupka, Variational Sequences on Finite Order Jet Spaces, *In: Proc. Diff. Geom. and its Appl.* (Brno, 1989, (J. Janyška and D. Krupka, Eds.) World Scientific Singapore, 1990) 236–254.

14. M. Francaviglia, M. Palese and R. Vitolo, Symmetries in Finite Order Variational Sequences, *Czech. Math. J.* **52** (127) (2002) 197–213.
15. M. Francaviglia, M. Palese and R. Vitolo, The Hessian and Jacobi Morphisms for Higher Order Calculus of Variations, *Diff. Geom. Appl.* **22** (1) (2005) 105–120.
16. M. Krbek, J. Musilová, Representation of the variational sequence by differential forms, *Rep. Math. Phys.* **51** (2003) 251–258.
17. M. Palese and E. Winterroth, Covariant gauge-natural conservation laws, *Rep. Math. Phys.* **54** (3) (2004) 349–364.
18. M. Palese and E. Winterroth, Global Generalized Bianchi Identities for Invariant Variational Problems on Gauge-natural Bundles, *Arch. Math. (Brno)* **41** (3) (2005) 289–310.
19. M. Palese and E. Winterroth, Gauge-natural field theories and Noether theorems: canonical covariant conserved currents, *Rend. Circ. Mat. Palermo (2) Suppl.* **79** (2006) 161–174.
20. M. Palese and E. Winterroth, The relation between the Jacobi morphism and the Hessian in gauge-natural field theories, *Theoret. Math. Phys.* **152** (2) (2007) 1191–1200.
21. E. Noether, Invariante Variationsprobleme, *Nachr. Ges. Wiss. Gött., Math. Phys. Kl.* **II** (1918) 235–257.
22. J.L. Anderson and P.G. Bergmann, Constraints in Covariant Field Theories, *Phys. Rev.* **83** (5) (1951) 1018–1025.
23. E.H.K. Winterroth, Variational derivatives of gauge-natural invariant Lagrangians and conservation laws, PhD thesis University of Torino, 2007.
24. D.J. Saunders, *The Geometry of Jet Bundles* (Cambridge Univ. Press, Cambridge, 1989).
25. R. Vitolo, Finite Order Lagrangian Bicomplexes, *Math. Proc. Camb. Phil. Soc.* **125** (1) (1999) 321–333.
26. J.F. Pommaret, Spencer Sequence and Variational Sequence, *Acta Appl. Math.* **41** (1995) 285–296.
27. I. Kolář and R. Vitolo, On the Helmholtz operator for Euler morphisms, *Math. Proc. Cambridge Phil. Soc.* **135** (2) (2003) 277–290.
28. A. Trautman, Noether equations and conservation laws, *Comm. Math. Phys.* **6** (1967) 248–261.
29. I. Kolář, A Geometrical Version of the Higher Order Hamilton Formalism in Fibred Manifolds, *J. Geom. Phys.* **1** (2) (1984) 127–137.
30. H. Goldschmidt and S. Sternberg, The Hamilton–Cartan Formalism in the Calculus of Variations, *Ann. Inst. Fourier, Grenoble* **23** (1) (1973) 203–267.
31. M. Godina, P. Matteucci and J.A. Vickers, Metric-affine gravity and the Nester-Witten 2-form, *J. Geom. Phys.* **39** (4) (2001) 265–275.
32. M. Francaviglia, M. Palese and E. Winterroth, Second variational derivative of gauge-natural invariant Lagrangians and conservation laws, *In: Proc. IX Int. Conf. Diff. Geom. Appl.* (Prague, 2004, (J. Bureš *et al.* Eds.) Charles University, 2005) 591–604.
33. M. Palese and E. Winterroth, Lagrangian reductive structures on gauge-natural bundles, preprint (2007).

Differential Geometry and its Applications
Proc. Conf., in Honour of Leonhard Euler, Olomouc, August 2007

Hilbert–Yang–Mills functional: Examples

A. Paták
Institute of Theoretical Physics and Astrophysics, Masaryk University,
Kotlářská 2, 61137 Brno, Czech Republic
E-mail: patak@physics.muni.cz
www.muni.cz

The geometric structure of gauge natural theories is investigated. We study especially the Einstein–Yang–Mills theory. We discuss the first variation formula by means of the principal Lepage equivalent of the Hilbert–Yang–Mills Lagrangian. We establish basic invariant properties and the first variation formula for induced variations of this Lagrangian. Noether currents split naturally to several terms, one of which is the exterior derivative of the Komar–Yang–Mills superpotential. We give some examples of Komar–Yang–Mills superpotentials corresponding to solutions of the Einstein equations.

Keywords: Variational Principle, Lagrangian, Noether's Invariance, Jet Prolongation, Gauge Natural Structure.

MS classification: 58E15, 58E30, 70S10, 70S15, 83C22.

1. Introduction

The goal of this paper is to study geometric structure of variational principles of the Einstein–Yang–Mills theory. This theory is an example of the so called gauge natural field theory and it describes gravity in interaction with the Yang–Mills field. The variational principle is based on the sum of the Hilbert Lagrangian and the Yang–Mills Lagrangian. Especially, we analyze invariance of this total Lagrangian and its consequences for conservations laws.

The paper is organized as follows. In Sec. 2 we give a survey of the general variational theory[3,9,10] and we focus our attention on the concepts needed in the Einstein–Yang–Mills theory. In Sec. 3 we present the gauge natural structure of the Einstein–Yang–Mills theory. In Sec. 4, we give some examples and we analyze the Komar–Yang–Mills superpotential for some solutions of the Einstein–Yang–Mills equations; it includes commentary on the conserved quantities (mass, electric charge, angular momentum).

2. Variational Theory on Fibered Manifolds

Recall our standard notation. In this paper we suppose that we have a fibered manifold $\pi : Y \to X$, and write $n = \dim X$, $n + m = \dim Y$. J^rY is the *r-jet prolongation of* Y, and $\pi^{r,s} : J^rY \to J^sY$, $\pi^r : J^rY \to X$ are the *canonical jet projections.* The *r-jet prolongation* of a section γ is defined to be the mapping $x \to J^r\gamma(x) = J^r_x\gamma$. For any set $W \subset Y$ we denote $W^r = (\pi^{r,0})^{-1}(W)$. Any *fibered chart* (V, ψ), $\psi = (x^i, y^\sigma)$, on Y induces the *associated charts* on X and J^rY, denoted by (U, φ), $\varphi = (x^i)$, and (V^r, ψ^r), $\psi^r = (x^i, y^\sigma, y^\sigma_{j_1}, y^\sigma_{j_1j_2}, \ldots, y^\sigma_{j_1j_2\ldots j_r})$, respectively; here $1 \le i \le n$, $1 \le \sigma \le m$, and $V^r = (\pi^{r,0})^{-1}(V)$, $U = \pi^r(V)$. We denote $\omega_0 = dx^1 \wedge dx^2 \wedge \cdots \wedge dx^n$, and its contractions $\omega_i = i_{\partial/\partial x^i}\omega_0$, $\omega_{ij} = i_{\partial/\partial x^j}\omega_i$. We define the *formal derivative operator* with respect to x^i by $d_i = \partial/\partial x^i + y^\sigma_i(\partial/\partial y^\sigma) + y^\sigma_{i_1 i}(\partial/\partial y^\sigma_{i_1}) + \cdots + y^\sigma_{i_1i_2\ldots i_r i}(\partial/\partial y^\sigma_{i_1i_2\ldots i_r})$.

For any open set $W \subset Y$, let $\Omega^r_0 W$ be the ring of functions on W^r. The $\Omega^r_0 W$-module of differential q-forms on W^r is denoted by $\Omega^r_q W$, and the exterior algebra of forms on W^r is denoted by $\Omega^r W$. The module of $\pi^{r,0}$-horizontal (π^r-horizontal) q-forms is denoted by $\Omega^r_{q,Y}W$ ($\Omega^r_{q,X}W$, respectively). The fibered structure of Y induces a morphism of exterior algebras $h : \Omega^r W \to \Omega^{r+1} W$, called the *horizontalization.* In a fibered chart (V, ψ), $\psi = (x^i, y^\sigma)$, h is defined by $hf = f \circ \pi^{r+1,r}$, $hdx^i = dx^i$, $hdy^\sigma_{j_1j_2\ldots j_p} = y^\sigma_{j_1j_2\ldots j_p k}dx^k$, where f is a real function on W^r, and $0 \le p \le r$.

We say that a form $\eta \in \Omega^r_k W$ is contact, if $h\eta = 0$. For any fibered chart (V, ψ), $\psi = (x^i, y^\sigma)$, the 1-forms $\eta^\sigma_{j_1j_2\ldots j_p} = dy^\sigma_{j_1j_2\ldots j_p} - y^\sigma_{j_1j_2\ldots j_p k}dx^k$, where $1 \le p \le r - 1$, are examples of contact 1-forms, defined on V^r. Note that these forms define a basis of 1-form on V^r, $(dx^i, \eta^\sigma_{j_1j_2\ldots j_p}, dy^\sigma_{j_1j_2\ldots j_r})$.

A *Lagrangian* for Y is a π^r-horizontal n-form λ on the r-jet prolongation J^rY of Y. The number r is called the *order* of λ. In a fibered chart (V, ψ), $\psi = (x^i, y^\sigma)$, on Y, and the associated chart on J^rY, (V^r, ψ^r), $\psi^r = (x^i, y^\sigma, y^\sigma_{j_1}, y^\sigma_{j_1j_2}, \ldots, y^\sigma_{j_1j_2\ldots j_r})$, a Lagrangian of order r has an expression $\lambda = \mathcal{L}\omega_0$, where $\mathcal{L} : V^r \to R$ is the *component* of λ with respect to (V, ψ) (the *Lagrange function* associated with (V, ψ)). The *Euler–Lagrange form* of λ is defined to be an $(n+1)$-form E_λ on $J^{2r}Y$, defined by

$$E_\lambda = E_\sigma(\mathcal{L})\eta^\sigma \wedge \omega_0, \; E_\sigma(\mathcal{L}) = \sum_{l=0}^{r} (-1)^l d_{p_1} d_{p_2} \ldots d_{p_l} \frac{\partial \mathcal{L}}{\partial y^\sigma_{p_1p_2\ldots p_l}},$$

$E_\sigma(\mathcal{L})$ are the *Euler–Lagrange expressions.* A Lagrangian can equivalently be defined as a morphism of fibered manifolds from J^rY to $\wedge^n T^*X$ over the identity morphism of X.

In this paper we consider a variational functional defined by a second order Lagrangian λ. Lepage equivalents of second order Lagrangians are very well known; we use the *principal Lepage equivalent* Θ_λ of λ introduced in Ref. 9. For each Lagrangian $\lambda \in \Omega^2_{n,X}Y$ there exists a principal Lepage equivalent $\Theta_\lambda \in \Omega^3_n Y$ such that in any fibered chart $(V,\psi), \psi = (x^i, y^\sigma)$,

$$\Theta_\lambda = \mathcal{L}\omega_0 + \left(\frac{\partial \mathcal{L}}{\partial y^\sigma_i} - d_p \frac{\partial \mathcal{L}}{\partial y^\sigma_{pi}}\right)\eta^\sigma \wedge \omega_i + \frac{\partial \mathcal{L}}{\partial y^\sigma_{ji}}\eta^\sigma_j \wedge \omega_i. \tag{1}$$

It can be shown that the principal Lepage equivalent of the Hilbert Lagrangian is of first order[12] and so the principal Lepage equivalent of the total Hilbert–Yang–Mills Lagrangian is of first order too (see Theorem 3.1). For a Lepage equivalent $\Theta_\lambda \in \Omega^1_n Y$ of a Lagrangian $\lambda \in \Omega^2_{n,X}Y$ the Lie derivative $\partial_{J^2\xi}\lambda$ of λ with respect to the second jet prolongation $J^2\xi$ of a π-projectable vector field ξ on Y can be expressed by the *(infinitesimal) first variation formula* $\partial_{J^2\xi}\lambda = hi_{J^1\xi}d\Theta_\lambda + hdi_{J^1\xi}\Theta_\lambda$.

Theorem 2.1. *Let ξ be a vector field on J^1Y, which has a chart expression $\xi = \xi^i(\partial/\partial x^i) + \xi^\sigma(\partial/\partial y^\sigma) + \xi^\sigma_j(\partial/\partial y^\sigma_j)$, and let Θ_λ be the principal Lepage equivalent* (1). *Then the Euler–Lagrange term has a chart expression*

$$hi_\xi d\Theta_\lambda = \left(\frac{\partial \mathcal{L}}{\partial y^\sigma} - d_i\left(\frac{\partial \mathcal{L}}{\partial y^\sigma_i} - d_p\frac{\partial \mathcal{L}}{\partial y^\sigma_{pi}}\right)\right)(\xi^\sigma - y^\sigma_j\xi^j)\omega_0$$

and the boundary term has a chart expression

$$hdi_\xi\Theta_\lambda = d_i\left(\mathcal{L}\xi^i + \left(\frac{\partial \mathcal{L}}{\partial y^\sigma_i} - d_p\frac{\partial \mathcal{L}}{\partial y^\sigma_{pi}}\right)(\xi^\sigma - y^\sigma_k\xi^k) + \frac{\partial \mathcal{L}}{\partial y^\sigma_{ij}}(\xi^\sigma_j - y^\sigma_{jk}\xi^k)\right)\omega_0.$$

Recall that an *automorphism* of Y is a diffeomorphism $f : W \to Y$, where $W \subset Y$ is an open set, such that there exists a diffeomorphism $f_0 : \pi(W) \to X$ such that $\pi \circ f = f_0 \circ \pi$. An automorphism $f : W \to Y$ of the fibered manifold Y is said to be an *invariance transformation* of a form $\rho \in \Omega^s_p W$, if $J^s f^*\rho = \rho$ holds. Let ξ be a projectable vector field on Y. We say that ξ is the *generator* of invariance transformations of ρ, if $\partial_{J^s\xi}\rho = 0$. The following consequence of the first variation formula is known as the *Noether's theorem* and the term $i_{J^1\xi}\Theta_\lambda$ in the conservation law is called the *current* associated with ξ. Let $\Theta_\lambda \in \Omega^1_n W$ be a Lepage equivalent of a Lagrangian $\lambda \in \Omega^2_{n,X}W$ and let a section $\gamma \in \Gamma_{\Omega,W}(Y)$ be an extremal ($E_\lambda \circ J^2\gamma = 0$). For any generator ξ of invariance transformations of λ a (differential) conservation law $dj^1\gamma^* i_{J^1\xi}\Theta_\lambda = 0$ holds.

3. Gauge Natural Structure of the Einstein–Yang–Mills Theory

Many physical theories can be described as a gauge natural field theory, i.e. they have the following *gauge natural structure*:

(a) a *structure bundle* P which is a principal bundle over an n-dimensional manifold X with a Lie group G,
(b) a *configuration bundle* C which is a gauge natural bundle[7,11] of order (s,r), i.e. C is associated to the (s,r)-*th principal prolongation* $W^{s,r}P$ of P, [a]
(c) a Lagrangian λ of order r for C which is *gauge natural*, i.e. λ is a r-th order gauge natural operator[7] from C to $\wedge^n T^*B$, where B is the *base functor.*

For the details and the omitted proofs in the rest of this section see Ref. 14. The gauge natural structure of the *Einstein–Yang–Mills theory* is made of the following items:

(a) a structure bundle (P, p, X, G), where the n-dimensional manifold X is interpreted as a spacetime,
(b) the configuration bundle $C = C_g \times_X C_e$ which is a gauge natural bundle of order 1, where C_g is the bundle of Lorentzian metrics [b] over X, C_e is the bundle of principal connections corresponding to the Yang–Mills part,
(c) the *Hilbert–Yang–Mills Lagrangian* $\lambda = \lambda_H + \lambda_{YM}$ on J^2C, where in any fibered coordinates x^i, g_{ij}, Γ_i^P on C we have [c]
$\lambda_H = \mathcal{L}_H\omega_0 = R\sqrt{g}\omega_0$, $\lambda_{YM} = \mathcal{L}_{YM}\omega_0 = -\frac{1}{4}R_{ij}^P g^{ik} g^{jl} R_{kl}^Q h_{PQ}\sqrt{g}\omega_0$.

We define the Christoffel symbols by $\Gamma^i_{jk} = (1/2)g^{is}(g_{sj,k}+g_{sk,j}-g_{jk,s})$. The chart expressions of the curvature tensor, the Ricci tensor and the scalar curvature are given by $R^i_{jkl} = d_k\Gamma^i_{jl} - d_l\Gamma^i_{jk} + \Gamma^i_{sk}\Gamma^s_{jl} - \Gamma^i_{sl}\Gamma^s_{jk}$, $R_{jl} =$

---

[a] In what follows we write C for the corresponding functor too.

[b] $C_g = F^1X \times_{l_1} \mathrm{LMet}\,\mathbb{R}^n$ is an associated bundle to the first order frame bundle F^1X of a manifold X, $\mathrm{LMet}\,\mathbb{R}^n$ is the set of bilinear, symmetric, non-degenerate forms with the Lorentzian signature $(1, n-1)$ and $l_1 : L^1_n \times \mathrm{LMet}\,\mathbb{R}^n \to \mathrm{LMet}\,\mathbb{R}^n$, $l_1(a,g) = g \circ (a^{-1} \times a^{-1})$ is a left action of the first differential group $L^1_n \cong GL(\mathbb{R}^n)$.

[c] In these equations R is the scalar curvature, $g = |\det g_{ij}|$, $R^P_{ij} = \Gamma^P_{j,i} - \Gamma^P_{i,j} + c^P_{QR}\Gamma^Q_i\Gamma^R_j$ is the curvature (field strength) of the principal connection (Yang–Mills field) Γ^P_i and h denotes an *Ad-invariant form* on the Lie algebra $\mathfrak{g}$, i.e. invariant with respect to the adjoint representation Ad of G in the sense that for all $g \in G$, $X, Y \in \mathfrak{g}$ the relation $h(\mathrm{Ad}_g(X), \mathrm{Ad}_g(Y)) = h(X,Y)$ holds h_{PQ} are the components of an *Ad*-invariant form on the Lie algebra $\mathfrak{g}$.

R^m_{jml}, $R = g^{jl}R_{jl}$. We denote the contact forms for the gravitational part by $\eta_{ij} = dg_{ij} - g_{ij,l}dx^l$, $\eta_{ij\,k} = dg_{ij,k} - g_{ij,kl}dx^l$ and for the Yang–Mills part by $\eta_i^P = d\Gamma_i^P - \Gamma_{i,j}^P dx^j$ (similarly for the vector field ξ). Furthermore, we denote by $G_{st} = R_{st} - (1/2)g_{st}R$ the components of the Einstein tensor, by $T^{ab} = (1/2)(R^{Pal}R_P{}^b{}_l - (1/4)R^P_{ij}R^{ij}_P g^{ab})$ the components corresponding to the stress-energy tensor and by $\nabla_j R^{ji}_P = d_j R^{ji}_P + R^{ji}_R c^R_{PS}\Gamma^S_j - R^{ij}_P\Gamma^k_{kj}$ the components of the covariant derivative with respect to the C-prolongation[7] of the principal connection with respect to the Levi-Civita connection.

Theorem 3.1. *The principal Lepage equivalent Θ_λ of the Hilbert–Yang–Mills Lagrangian has a chart expression*

$$\begin{aligned}\Theta_\lambda &= \sqrt{g}(R - \frac{1}{4}R^P_{ij}R^{ij}_P)\omega_0 + \sqrt{g}R^{ij}_P\eta^P_i \wedge \omega_j \\ &+ \frac{1}{2}\sqrt{g}(g^{nm}g^{ia}g^{bj} - 2g^{in}g^{ma}g^{bj} + g^{ma}g^{nb}g^{ij})g_{mn,i}\eta_{ab} \wedge \omega_j \\ &+ \sqrt{g}(g^{ad}g^{jb} - g^{jd}g^{ab})\eta_{ab,d} \wedge \omega_j.\end{aligned}$$

The Euler–Lagrange term $hi_\xi d\Theta_\lambda$ has a chart expression

$$hi_\xi d\Theta_\lambda = \sqrt{g}(-G^{ab} + T^{ab})(\xi_{ab} - g_{ab,m}\xi^m)\omega_0 + \sqrt{g}\nabla_j R^{ji}_P(\xi^P_i - \Gamma^P_{i,m}\xi^m)\omega_0.$$

The boundary term $hdi_\xi\Theta_\lambda$ has a chart expression

$$\begin{aligned}hdi_\xi\Theta_\lambda = d_j\Big(\sqrt{g}\Big(R - \frac{1}{4}R^P_{ki}R^{ki}_P\Big)\xi^j + \sqrt{g}(g^{nb}g^{ij} - g^{in}g^{bj})\Gamma^a_{ni}(\xi_{ab} - g_{ab,m} \\ \xi^m) + \sqrt{g}R^{ij}_P(\xi^P_i - \Gamma^P_{i,m}\xi^m) + \sqrt{g}(g^{ad}g^{jb} - g^{jd}g^{ba})(\xi_{ab,d} - g_{ab,dm}\xi^m)\Big)\omega_0.\end{aligned}$$

Therefore we get from Theorem 3.1 the *Einstein–Yang–Mills equations* (cf. Refs. 1,16, it differs from Ref. 6)

$$G^{ab} = T^{ab}, \quad \nabla_j R^{ji}_P = 0. \tag{2}$$

Now we will discuss the first variation formula for the so called induced variations for a gauge natural Lagrangian λ and then specially for the Hilbert–Yang–Mills Lagrangian. By the *induced variation* we mean the variation induced by the lifted vector field $C\xi = (\frac{d}{dt}C\mathrm{Fl}^\xi_t)_0$ on the configuration bundle C determined by an *(infinitesimal) generator of automorphisms* ξ on a principal bundle P which is a vector field such that $\mathrm{Fl}^\xi_t \in \mathrm{Aut}\,(P)$. We have the local expression $\xi(x,a) = \xi^i(x)(\partial/\partial x^i) + \xi^P(x)R_{e_P}(a)$ of a generator of automorphisms ξ, where R_{e_P} denotes the right invariant vector field on G corresponding to e_P. It can be seen that $C\xi$ is the generator of invariance transformations of λ.

We denote by $\mathcal{E}^{ab} = \sqrt{g}(-G^{ab} + T^{ab})$, $\mathcal{F}^i_P = \sqrt{g}\nabla_j R^{ji}_P$ the Euler–Lagrange expressions of the Hilbert–Yang–Mills Lagrangian, and by $\xi^J_V = \xi^J - \Gamma^J_i \xi^i$ the components of the vertical part ξ_V of the generator of automorphisms ξ with respect to the principal connection. We define the *Komar–Yang–Mills superpotential* by $\nu_\xi = \frac{1}{2}\sqrt{g}(\nabla^{[i}\xi^{j]} - R^{ij}_J \xi^J_V)\omega_{ij}$, where ∇ denotes the covariant derivative with respect to the Levi-Civita connection and $[\,]$ denotes antisymmetrization without a factor 1/2. The first term in the Komar–Yang–Mills superpotential is the so called *Komar potential.*[8] The following basic theorem clarifies the structure of the currents associated with vector fields on the underlying principal bundle P.

Theorem 3.2. *For the principal Lepage equivalent of the Hilbert–Yang–Mills Lagrangian Θ_λ the Euler–Lagrange term has a chart expression*

$$hi_{J^1C\xi} d\Theta_\lambda = -\mathcal{E}^{ab}(g_{ac} d_b \xi^c + g_{bc} d_a \xi^c + g_{ab,c}\xi^c)\omega_0 + \mathcal{F}^i_P (c^P_{RQ}\xi^R \Gamma^Q_i + d_i \xi^P - \Gamma^P_j d_i \xi^j - \Gamma^P_{i,m}\xi^m)\omega_0.$$

The current has a chart expression

$$i_{J^1C\xi}\Theta_\lambda = (2\mathcal{E}^{ib} g_{bj}\xi^j - \mathcal{F}^i_J \xi^J_V)\omega_i + hd\nu_\xi + \sqrt{g}\xi^i[(g^{jc}g^{ab} - g^{bc}g^{ja})(\eta_{ab,c} - \Gamma^d_{ab}\eta_{dc}) + R^{jk}_P \eta^P_k] \wedge \omega_{ij}.$$

Since we could add the contact part of $d\nu_\xi$ to the last term, we see that the current splits into three terms, the first term vanishes along solutions of the Euler–Lagrange equations, the second term is exact, the third term is contact.

4. Examples

First we apply the result of the previous section to the *Levi–Civita–Bertotti–Robinson* solution[5,13,15] of the Einstein equations. We take as the structure bundle $(X \times U(1), \mathrm{pr}_1, X, U(1))$ over the Levi–Civita-Bertotti–Robinson spacetime (X, g), we suppose there exist coordinates (t, r, θ, φ) on X such that the metric g is given by

$$g = \frac{e^2}{r^2}[-dt^2 + dr^2 + r^2(d\theta^2 + \sin^2\theta\, d\varphi^2)].$$

This g together with the $U(1)$-connection $\Gamma = -2(e/r)(dt + dr)$ is a solution of the Einstein–Maxwell equations. [d] We denote this solution by γ_{LBR}.

[d] It means a solution of Eqs. (2) corresponding to the Hilbert–Yang–Mills Lagrangian λ with the only component of the Ad-invariant form h on $\mathfrak{u}(1)$ equal to 1. We will not write the base vector e_1 in $\mathfrak{u}(1)$ for the $U(1)$-connections explicitly.

We write a generator ξ of automorphisms of the structure bundle in the form

$$\xi = \xi^1 \frac{\partial}{\partial t} + \xi^2 \frac{\partial}{\partial r} + \xi^3 \frac{\partial}{\partial \theta} + \xi^4 \frac{\partial}{\partial \varphi} + \zeta R_{e_1}, \tag{3}$$

where R_{e_1} denotes the right invariant vector field on $U(1)$ corresponding to the base vector e_1 in $\mathfrak{u}(1)$. Then we get the following coordinate expression for the pull-back $J^1\gamma^*_{LBR}\nu_\xi$ of the Komar–Yang–Mills superpotential ν_ξ along the solution γ_{LBR}:

$$\begin{aligned}
J^1\gamma^*_{LBR}\nu_\xi = &\left(-2\zeta e - e^2 \left(\frac{2}{r}(\xi^1 + 2\,\xi^2) + \frac{\partial \xi^1}{\partial r} + \frac{\partial \xi^2}{\partial t} \right) \right) \sin\theta \, d\theta \wedge d\varphi \\
&+ \left(\frac{\partial \xi^3}{\partial t} + \frac{1}{r^2} \frac{\partial \xi^1}{\partial \theta} \right) e^2 \sin\theta \, dr \wedge d\varphi - \left(\sin\theta \frac{\partial \xi^4}{\partial t} + \frac{1}{r^2 \sin\theta} \frac{\partial \xi^1}{\partial \varphi} \right) e^2 \, dr \wedge d\theta \\
&+ \left(\frac{\partial \xi^3}{\partial r} - \frac{1}{r^2} \frac{\partial \xi^2}{\partial \theta} \right) e^2 \sin\theta \, dt \wedge d\varphi + \left(\frac{1}{r^2 \sin\theta} \frac{\partial \xi^2}{\partial \varphi} - \sin\theta \frac{\partial \xi^4}{\partial r} \right) e^2 \, dt \wedge d\theta \\
&+ \left(2\cos\theta \, \xi^4 - \frac{1}{\sin\theta} \frac{\partial \xi^3}{\partial \varphi} + \sin\theta \frac{\partial \xi^4}{\partial \theta} \right) \frac{e^2}{r^2} \, dt \wedge dr.
\end{aligned}$$

If we choose ξ as ζR_{e_1}, then we get $J^1\gamma^*_{LBR}\nu_{\zeta R_{e_1}} = -2\zeta e \sin\theta \, d\theta \wedge d\varphi$. Taking ζ as an appropriate constant and integrating on spatial spheres we obtain the electric charge e.

Our second example is the *Kerr–Newman* solution[15] of the Einstein equations. We take as the structure bundle $(X \times U(1), \mathrm{pr}_1, X, U(1))$ over the Kerr–Newman spacetime (X, g), we suppose there exist coordinates (t, r, θ, φ) on X such that the metric g is given by

$$\begin{aligned}
g = &-\left(1 - \frac{2mr - e^2}{u} \right) dt^2 + \frac{u}{s} \, dr^2 + u \, d\theta^2 + \sin^2\theta \Big(r^2 + a^2 \\
&+ \frac{a^2 \sin^2\theta}{u} (2mr - e^2) \Big) \, d\varphi^2 - \frac{a \sin^2\theta}{u} (2mr - e^2)(dt \otimes d\varphi + d\varphi \otimes dt), \\
s = &\; r^2 + a^2 + e^2 - 2mr, \; u = r^2 + a^2 \cos^2\theta.
\end{aligned}$$

This g together with the connection $\Gamma = -2e(r/u)(dt - a\sin^2\theta \, d\theta)$ is a solution of the Einstein–Maxwell equations. We denote this solution by γ_{KN}.

The Komar–Yang–Mills superpotential ν_ξ along γ_{KN} for the general generator ξ of automorphisms of the structure bundle (3) is too long. We will consider only a special choices of the generator. If we choose ξ as

$\xi^1(\partial/\partial t)$, where ξ^1 is constant, then we have

$$\begin{aligned} J^1\gamma^*_{KN}\nu_{\xi^1(\partial/\partial t)} &= 2\xi^1 \frac{\sin\theta}{u^3}[A(a^2+r^2)\,d\theta\wedge d\varphi - B\sin\theta\cos\theta a^2\,dr\wedge d\varphi \\ &\quad - Aa\,dt\wedge d\theta - Ba\cos\theta\,dt\wedge dr], \\ A &= -mr^4 + ma^4\cos^4\theta - r^3e^2 + 3\,re^2a^2\cos^2\theta, \\ B &= (2\,mr - e^2)a^2\cos^2\theta + 2\,mr^3 + 3\,e^2r^2. \end{aligned}$$

Thus for an appropriate choice of the constant ξ^1 we obtain, after integrating on spatial spheres at spatial infinity, the mass m.

If we choose ξ as $\xi^4(\partial/\partial\varphi)$, where ξ^4 is constant, then we have

$$\begin{aligned} J^1\gamma^*_{KN}\nu_{\xi^4(\partial/\partial\varphi)} &= 2\xi^4\frac{\sin\theta}{u^3}(-Ca\sin^2\theta\,d\theta\wedge d\varphi - Ba^3\sin^3\theta\cos\theta\,dr\wedge d\varphi \\ &\quad + D\,dt\wedge d\theta - E\mathrm{cotg}\theta\,dt\wedge dr), \\ C &= a^4(-mr^2 + a^2m + re^2)\cos^4\theta + a^2r(-4\,mr^3 + 5\,e^2r^2 + 3\,a^2e^2)\cos^2\theta \\ &\quad - r^3(3\,mr^3 + a^2mr + a^2e^2), \\ D &= a^6(m-r)\cos^6\theta + a^4(2\,re^2 + 2\,mr^2 - a^2m - 3\,r^3)\cos^4\theta - 3\,a^2r(r^4 \\ &\quad - mr^3 + e^2r^2 + a^2e^2)\cos^2\theta + r^3(-r^4 + 2\,mr^3 - e^2r^2 + a^2mr + a^2e^2), \\ E &= a^6\cos^6\theta + a^4(e^2 + 3\,r^2 - 2\,mr)\cos^4\theta - a^2(3\,e^2r^2 + 2\,mr^3 + a^2e^2 \\ &\quad - 2\,a^2mr - 3\,r^4)\cos^2\theta + r^2(r^4 + 2\,a^2mr + 3\,a^2e^2). \end{aligned}$$

Thus for an appropriate choice of the constant ξ^4 we obtain, after integrating on spatial spheres at spatial infinity, the angular momentum ma.

If we choose ξ as ζR_{e_1}, where ζ is constant, then we get

$$\begin{aligned} J^1\gamma^*_{KN}\nu_{\zeta R_{e_1}} &= 2\zeta e\frac{\sin\theta}{u^2}[(-r^2 + a^2\cos^2\theta)(a^2+r^2)\,d\theta\wedge d\varphi \\ &\quad - a^2r\sin\theta\cos\theta\,dr\wedge d\varphi + a(-r^2 + a^2\cos^2\theta)\,dt\wedge d\theta - ra\mathrm{cotg}\theta\,dt\wedge dr]. \end{aligned}$$

Taking ζ as an appropriate constant and integrating on spatial spheres at spatial infinity we obtain the electric charge e.

As a third example we take one of the simplest non-Abelian black hole solution of the Einstein–Yang–Mills equations – one of the so called *colored black holes.*[2,4,16] We take as the structure bundle $(X\times SU(2), \mathrm{pr}_1, X, SU(2))$ over the Reissner–Nordström-like spacetime (X, g), we suppose there exist coordinates (t, r, θ, φ) on X such that the metric g is given by

$$\begin{aligned} g &= -\left(1 - 2\,\frac{m}{r} + \frac{e^2+q^2}{r^2}\right)dt^2 + \left(1 - 2\,\frac{m}{r} + \frac{e^2+q^2}{r^2}\right)^{-1} dr^2 \\ &\quad + r^2\left(d\theta^2 + \sin^2\theta\,d\varphi^2\right). \end{aligned}$$

Let e_P for $1 \le P \le 3$ be a basis of the Lie algebra $\mathfrak{su}(2)$ given by $e_P = -\frac{i}{2}\sigma_P$ with σ_P being the Pauli matrices. Then for the structure constants we have $c^R_{PQ} = \varepsilon_{PQR}$. This g together with the $SU(2)$-connection $\Gamma = (-2(e/r)(dt+dr) + 2\,q(1-\cos\theta)\,d\varphi)\,e_3$ is a solution of the Einstein–Yang–Mills equations. [e] We denote this solution by γ_{CBH}.

We write a generator ξ of automorphisms of the structure bundle in the form $\xi = \xi^1(\partial/\partial t) + \xi^2(\partial/\partial r) + \xi^3(\partial/\partial\theta) + \xi^4(\partial/\partial\varphi) + \zeta^P R_{e_P}$, where R_{e_P} denote the right invariant vector fields on $SU(2)$ corresponding to the base vectors e_P in $\mathfrak{su}(2)$. Then we get the following coordinate expression for the pull-back $J^1\gamma^*_{CBH}\nu_\xi$ of the Komar–Yang–Mills superpotential ν_ξ along the solution γ_{CBH}:

$$
\begin{aligned}
J^1\gamma^*_{CBH}\nu_\xi = {}& \Bigg(\frac{2}{r}[\xi^1(q^2-e^2-mr) - 2\xi^2e^2] + 4\xi^4 eq(1-\cos\theta) - 2\zeta^3 e \\
& - s\frac{\partial\xi^1}{\partial r} - \frac{r^4}{s}\frac{\partial\xi^2}{\partial t}\Bigg)\sin\theta\,d\theta\wedge d\varphi + \left(\frac{r^4}{s}\frac{\partial\xi^3}{\partial t} + \frac{\partial\xi^1}{\partial\theta}\right)\sin\theta\,dr\wedge d\varphi \\
& - \left(\frac{r^4\sin\theta}{s}\frac{\partial\xi^4}{\partial t} + \frac{1}{\sin\theta}\frac{\partial\xi^1}{\partial\varphi}\right)dr\wedge d\theta + \left(\frac{2}{r}s\xi^3 + s\frac{\partial\xi^3}{\partial r} - \frac{\partial\xi^2}{\partial\theta}\right)\sin\theta\,dt\wedge d\varphi \\
& - \left(\frac{2s}{r}\sin\theta\xi^4 + s\sin\theta\frac{\partial\xi^4}{\partial r} - \frac{1}{\sin\theta}\frac{\partial\xi^2}{\partial\varphi}\right)dt\wedge d\theta + \Bigg(-\frac{4eq}{r^3}(\xi^1+\xi^2) \\
& + 2\left(\cos\theta - \frac{2q^2}{r^2}(1+\cos\theta)\right)\xi^4 - 2\frac{q}{r^2}\zeta - \frac{1}{\sin\theta}\frac{\partial\xi^3}{\partial\varphi} + \sin\theta\frac{\partial\xi^4}{\partial\theta}\Bigg)dt\wedge dr, \\
s = {}& r^2 - 2\,mr + e^2 + q^2.
\end{aligned}
$$

It is easy to see, similarly as before, that the mass m corresponds to $\partial/\partial t$ and the electric charge e to R_{e_3}. Moreover, if we choose ξ as $\xi^4(\partial/\partial\phi)$, where ξ^4 is constant, then after integrating on spatial spheres we see that e times the magnetic charge q must be a constant.

The current evaluated along an extremal can be computed directly or as the exterior derivative of the Komar–Yang–Mills superpotential evaluated along a solution. This is straightforward but the result for an arbitrary generator of automorphisms of the structure bundle is quite long.

---

[e] It means a solution of Eqs. (2) corresponding to the Hilbert–Yang–Mills Lagrangian λ with the components of the Ad-invariant form h on $\mathfrak{su}(2)$ given by $h_{PQ} = \delta_{PQ}$ (h is up to a factor the Killing form of $\mathfrak{su}(2)$).

Acknowledgments

The author thanks D. Krupka for stimulating discussions and M. Lenc, J. Musilová for their support.

References

1. D. Bleecker, *Gauge Theory and Variational Principles* (Global analysis, pure and applied, no. 1, Addison-Wesley, Reading, MA, 1981).
2. F.A. Bais and R.J. Russel, Magnetic-monopole solution of non-Abelian gauge theory in curved spacetime, *Phys. Rev. D* **11** (1975) 2692–2695.
3. J. Brajerčík and D. Krupka, Variational principles for locally variational forms, *J. Math. Phys.* **46** 052903 (2005).
4. Y.M. Cho and P.G.O. Freund, Gravitating 't Hooft monopoles, *Phys. Rev. D* **12** (1975) 1588–1589.
5. P. Dolan, A Singularity Free Solution of the Maxwell-Einstein Equations, *Commun. Math. Phys.* **9** (1968) 161–168.
6. L. Fatibene and M. Francaviglia, *Natural and Gauge Natural Formalism for Classical Field Theories* (Kluwer Academic Publishers, Dordrecht, 2003).
7. I. Kolář, P.W. Michor and J. Slovák, *Natural Operations in Differential Geometry* (Springer-Verlag, Berlin, 1993).
8. A. Komar A., Covariant Conservation Laws in General Relativity, *Phys. Rev.* **113**, (1959) 934–936.
9. D. Krupka, Some Geometric Aspects of Variational Problems in Fibred Manifolds, *Folia Fac. Sci. Nat. Univ. Purk. Brunensis, Physica* **14**, Brno (1973); arXiv:math-ph/0110005.
10. D. Krupka, The Geometry of Lagrange Structures, Preprint Series in Global Analysis GA 7/1997, Silesian University, Opava, 1997.
11. D. Krupka and J. Janyška, *Lectures on Differential Invariants* (UJEP, Brno, 1990).
12. D. Krupka and M. Lenc, The Hilbert variational principle, Preprint 3/2002 GACR 201/00/0724, Masaryk University, Brno, 2002.
13. D. Lovelock, A Spherically Symmetric Solution of the Maxwell-Einstein Equations, *Commun. Math. Phys.* **5** (1967) 257–261.
14. A. Paták and D. Krupka, Geometric Structure of the Hilbert–Yang–Mills Functional, Preprint Series in Global Analysis and Applications, 2, Palacky University, Olomouc, 2007.
15. H. Stephani, D. Kramer, M. MacCallum, C. Hoenselaers and E. Herlt, *Exact Solutions to Einstein's Field Equations* (Cambridge monographs on mathematical physics, Cambridge University Press, Cambridge, 2003).
16. M.S. Volkov, D.V. Gal'tsov, Gravitating Non-Abelian Solitons and Black Holes with Yang–Mills Fields, *Phys. Rept.* **319** (1999) 1–83, arXiv:hep-th/9810070v2.

Differential Geometry and its Applications
Proc. Conf., in Honour of Leonhard Euler, Olomouc, August 2007

Hamiltonian formalism on Lie algebroids and its applications

Liviu Popescu
Dept. of Applied Mathematics in Economy, University of Craiova,
13, Al. I. Cuza st., Craiova, Romania
E-mail: liviupopescu@central.ucv.ro

In this paper we use the Hamiltonian formalism on Lie algebroids in order to solve a problem of control affine system with holonomic distribution.

Keywords: Lie algebroids, Hamilton equation, distributional system.

MS classification: 17B66, 70H25, 49J15.

1. Introduction

It is well known that the cotangent bundle plays a very important role in symplectic geometry and its applications, since this carries a canonical symplectic structure induced by the Liouville form. The notion of prolongation of Lie algebroid over the vector bundle projection of the dual bundle generalize the concept of cotangent bundle. A. Weinstein[8] gives a generalized theory of Lagrangian on Lie algebroids and obtains the equations of motion. The Hamilton equations were later obtained by E. Martinez[6] using the symplectic formalism and the notion of prolongation of Lie algebra over a mapping introduced by P.J. Higgins and K. Mackenzie.[1] The Hamiltonian formalism on Lie algebroids has studied by M. de Leon, H. Marrero and E. Martinez.[4]

The purpose of the present paper is to study an optimal control affine system using the geometrical structures on the prolongation of Lie algebroid and to show that the framework of Lie algebroid is better than cotangent bundle in order to solve some distributional systems with holonomic distribution (see D. Hrimiuc, L. Popescu[2] for nonholonomic case). In the section 2 the known results on Lie algebroid and its prolongation are presented. In sections 3 we investigate an optimal control problem with the solution provided by Pontryagin Maximum Principle and use the Hamilton equations at the level of a Lie algebroid.

2. Preliminaries on Lie algebroids

Let M be a differentiable, n-dimensional manifold. A Lie algebroid[5] over the manifold M is the triple $(E, [\cdot,\cdot], \sigma)$ where $\pi : E \to M$ is a vector bundle of rank m over M, whose $C^\infty(M)$-module of sections $\Gamma(E)$ is equipped with a Lie algebra structure $[\cdot,\cdot]$ and $\sigma : E \to TM$ is a vector bundle homomorphism (called *the anchor*) which induces a Lie algebra homomorphism from $\Gamma(E)$ to $\chi(M)$, satisfying the rules

$$[s_1, fs_2] = f[s_1, s_2] + (\sigma(s_1)f)s_2, \quad f \in C^\infty(M),\ s_1, s_2 \in \Gamma(E),$$
$$[\sigma(s_1), \sigma(s_2)] = \sigma[s_1, s_2], \quad [s_1,[s_2,s_3]] + [s_2,[s_3,s_1]] + [s_3,[s_1,s_2]] = 0.$$

The image of the anchor map $\sigma(E) \subseteq TM$ defines an integrable smooth distribution on M. The manifold M is foliated by integral leaves of $\sigma(E)$, which are called the leaves of Lie algebroid. A curve $u : [t_0, t_1] \to E$ is called admissible if $\sigma(u(t)) = \dot{c}(t)$, where $c(t) = \pi(u(t))$ is the base curve. It follows that $u(t)$ is admissible if and only if the base curve $c(t)$ lies on a leaf of the Lie algebroid, and that two points can be joint by an admissible curve if and only if are situated in the same leaf. If $\omega \in \Gamma(\bigwedge^k E^*)$ then the exterior derivative $d\omega \in \Gamma(\bigwedge^{k+1} E^*)$ is given by the formula

$$d\omega(s_1, ..., s_{k+1}) = \sum_{i=1}^{k+1} (-1)^{i+1} \sigma(s_i)\omega(s_1, ..., \hat{s}_i, ..., s_{k+1}) +$$
$$+ \sum_{1 \le i < j \le k+1} (-1)^{i+j} \omega([s_i, s_j], s_1, ..., \hat{s}_i, ..., \hat{s}_j, ...s_{k+1}).$$

Also, for $\xi \in \Gamma(E)$ on can define the *Lie derivative* with respect to ξ by $\mathcal{L}_\xi = i_\xi \circ d + d \circ i_\xi$, where i_ξ is the contraction with ξ. Let us consider the local coordinates (x^i) on an open $U \subset M$ and a local basis s_α of $\Gamma(E)$. The functions $\sigma^i_\alpha(x)$, $L^\gamma_{\alpha\beta}(x)$ on M given by

$$\sigma(s_\alpha) = \sigma^i_\alpha \frac{\partial}{\partial x^i}, \quad [s_\alpha, s_\beta] = L^\gamma_{\alpha\beta} s_\gamma, \quad i = \overline{1,n}, \quad \alpha, \beta, \gamma = \overline{1,m},$$

satisfy the so called *structure equations* on Lie algebroid

$$\sigma^j_\alpha \frac{\partial \sigma^i_\beta}{\partial x^j} - \sigma^j_\beta \frac{\partial \sigma^i_\alpha}{\partial x^j} = \sigma^i_\gamma L^\gamma_{\alpha\beta}, \quad \sum_{(\alpha,\beta,\gamma)} \left(\sigma^i_\alpha \frac{\partial L^\delta_{\beta\gamma}}{\partial x^i} + L^\delta_{\alpha\eta} L^\eta_{\beta\gamma} \right) = 0.$$

Locally, if $f \in C^\infty(M)$ then $df = \frac{\partial f}{\partial x^i} \sigma^i_\alpha s^\alpha$ and if $\theta \in \Gamma(E^*)$, $\theta = \theta_\alpha s^\alpha$ then

$$d\theta = (\sigma^i_\alpha \frac{\partial \theta_\beta}{\partial x^i} - \frac{1}{2} \theta_\gamma L^\gamma_{\alpha\beta}) s^\alpha \wedge s^\beta,$$

where $\{s^\alpha\}$ is the dual basis of $\{s_\alpha\}$.

2.1. *The prolongation of a Lie algebroid over the vector bundle projection of the dual bundle*

Let $\tau : E^* \to M$ be the dual of $\pi : E \to M$ and $(E, [\cdot,\cdot], \sigma)$ a Lie algebroid structure over M. One can construct the prolongation of a Lie algebroid over the vector bundle projection of a dual bundle (see Refs. 1,4,6) given by:
(i) The vector bundle is $(\mathcal{T}E^*, \tau_1, E^*)$ with $\mathcal{T}E^* = \cup_{u^* \in E^*} \mathcal{T}_{u^*}E^*$

$$\mathcal{T}_{u^*}E^* = \{(u_x, v_{u^*}) \in E_x \times T_{u^*}E^* | \sigma(u_x) = T_{u^*}\tau(v_{u^*}), \tau(u^*) = x \in M\}$$

and the projection $\tau_1 : \mathcal{T}E^* \to E^*$, $\tau_1(u_x, v_{u^*}) = u^*$.
(ii) The Lie algebra structure $[\cdot,\cdot]$ on $\Gamma(\mathcal{T}E^*)$ is defined in the following way: if $\rho_1, \rho_2 \in \Gamma(\mathcal{T}E^*)$ are such that $\rho_i(u^*) = (X_i(\tau(u^*)), U_i(u^*))$ where $X_i \in \Gamma(E)$, $U_i \in \chi(E^*)$ and $\sigma(X_i(\tau(u^*)) = T_{u^*}\tau(U_i(u^*))$, $i = 1, 2$, then

$$[\rho_1, \rho_2](u^*) = ([X_1, X_2](\tau(u^*)), [U_1, U_2](u^*))$$

(iii) The anchor is the projection $\sigma^1 : \mathcal{T}E^* \to TE^*$, $\sigma^1(u, v) = v$.
Notice that if $\mathcal{T}\tau : \mathcal{T}E^* \to E$, $\mathcal{T}\tau(u, v) = u$ then $(V\mathcal{T}E^*, \tau_{1|V\mathcal{T}E^*}, E^*)$ with $V\mathcal{T}E^* = Ker\mathcal{T}\tau$ is a subbundle of $(\mathcal{T}E^*, \tau_1, E^*)$, called the *vertical subbundle*. If (x^i, μ_α) are local coordinates on E^* at u^* and $\{s_\alpha\}$ is a local basis of $\Gamma(E)$ then a local basis of $\Gamma(\mathcal{T}E^*)$ is $\{\mathcal{X}_\alpha, \mathcal{P}^\alpha\}$ where

$$\mathcal{X}_\alpha(u^*) = \left(s_\alpha(\tau(u^*)), \sigma^i_\alpha \frac{\partial}{\partial x^i}|_{u^*}\right), \quad \mathcal{P}^\alpha(u^*) = \left(0, \frac{\partial}{\partial \mu_\alpha}|_{u^*}\right).$$

The structure functions on $\mathcal{T}E^*$ are given by following formulas

$$\sigma^1(\mathcal{X}_\alpha) = \sigma^i_\alpha \frac{\partial}{\partial x^i}, \quad \sigma^1(\mathcal{P}^\alpha) = \frac{\partial}{\partial \mu_\alpha}, \tag{1}$$

$$[\mathcal{X}_\alpha, \mathcal{X}_\beta] = L^\gamma_{\alpha\beta}\mathcal{X}_\gamma, \quad [\mathcal{X}_\alpha, \mathcal{P}^\alpha] = 0, \quad [\mathcal{P}^\alpha, \mathcal{P}^\beta] = 0, \tag{2}$$

where $\{\mathcal{X}^\alpha, \mathcal{P}_\alpha\}$ is the dual basis of $\{\mathcal{X}_\alpha, \mathcal{P}^\alpha\}$. The *canonical symplectic section* of Lie algebroid $\mathcal{T}E^*$ is given by $\omega_E = -d\theta$, where $\theta = \mu_\alpha \mathcal{X}^\alpha$ is the *Liouville form*. In the local coordinates we have

$$\omega_E = \mathcal{X}^\alpha \wedge \mathcal{P}_\alpha + \frac{1}{2}\mu_\alpha L^\alpha_{\beta\gamma}\mathcal{X}^\beta \wedge \mathcal{X}^\gamma.$$

Let $\mathcal{H} : E^* \to \mathbb{R}$ be a regular Hamiltonian. Then, since ω_E is a symplectic section on $(\mathcal{T}E^*, [\cdot,\cdot], \sigma^1)$ and $d\mathcal{H} \in \Gamma(\mathcal{T}E^*)$, there exists a unique section $\xi_\mathcal{H} \in \Gamma(\mathcal{T}E^*)$ such that

$$i_{\xi_\mathcal{H}}\omega_E = d\mathcal{H},$$

called the Hamilton section. With respect to the local basis $\{\mathcal{X}_\alpha, \mathcal{P}^\alpha\}$, the local expression of $\xi_\mathcal{H}$ is

$$\xi_\mathcal{H} = \frac{\partial \mathcal{H}}{\partial \mu_\alpha}\mathcal{X}_\alpha - (\sigma^i_\alpha \frac{\partial \mathcal{H}}{\partial x^i} + \mu_\gamma L^\gamma_{\alpha\beta}\frac{\partial \mathcal{H}}{\partial \mu_\beta})\mathcal{P}^\alpha$$

Thus, the vector field $\sigma^1(\xi_\mathcal{H})$ on E^* is given by

$$\sigma^1(\xi_\mathcal{H}) = \sigma^i_\alpha \frac{\partial \mathcal{H}}{\partial \mu_\alpha}\frac{\partial}{\partial x^i} - (\sigma^i_\alpha \frac{\partial \mathcal{H}}{\partial x^i} + \mu_\gamma L^\gamma_{\alpha\beta}\frac{\partial \mathcal{H}}{\partial \mu_\beta})\frac{\partial}{\partial \mu_\alpha}$$

and consequently, the integral curves of $\xi_\mathcal{H}$ (i.e. the integral curves of the vector field $\sigma^1(\xi_\mathcal{H})$) satisfy the Hamilton equations[4]

$$\frac{dx^i}{dt} = \sigma^i_\alpha \frac{\partial \mathcal{H}}{\partial \mu_\alpha}, \quad \frac{d\mu_\alpha}{dt} = -\sigma^i_\alpha \frac{\partial \mathcal{H}}{\partial x^i} - \mu_\gamma L^\gamma_{\alpha\beta}\frac{\partial \mathcal{H}}{\partial \mu_\beta} \tag{3}$$

3. Applications of Lie algebroids to distributional systems

We consider the distributional system (driftless control-affine system)

$$\dot{x} = \sum_{i=1}^m u_i X_i(x),$$

where $x \in M$, $X_1, X_2, ..., X_m$ are smooth vector fields on M and the control $u = (u_1, u_2, .., u_m)$ takes values in an open subset Ω of $\mathbb{R}^m$. The vectors fields $X_i, i = \overline{1,m}$ generate a holonomic distribution $D \subset TM$ such that the rank of D is assumed to be constant. Let x_0 and x_1 be two points of M. An optimal control problem consists in finding the trajectories of our distributional system which connects x_0 and x_1 and minimizing the cost

$$\min_{u(.)} \int_I \mathcal{F}(u(t))dt$$

where $\mathcal{F}$ is a Minkowski norm (positive homogeneous) on D.

In order to apply the theory of Lie algebroids we consider $E = D$ and inclusion as anchor $\sigma : E \to TM$. We can associate to any positive homogeneous cost $\mathcal{F}$ on Lie algebroid E a cost F on $Im\sigma \subset TM$ defined by

$$F(v) = \{\mathcal{F}(u)\,|u \in E_x, \quad \sigma(u) = v\},$$

for each $v \in (Im\sigma)_x \subset T_xM$, $x \in M$. A piecewise smooth curve $c : I \subset \mathbb{R} \to M$ is called horizontal if the tangent vectors are in D, i.e. $\dot{c}(t) \in D_{c(t)} \subset TM$ for almost every $t \in I$. Let $u : I \to E$ be an admissible curve projecting by

π onto the horizontal curve $c : I \to M$. The length of the horizontal curve c is defined by

$$length(c) = \int_I \mathcal{F}(u(t))dt = \int_I F(\dot{c}(t))dt,$$

and the distance is given by $d(x_0, x_1) = \inf length(c)$, where the infimum is taken over all horizontal curves connecting x_0 and x_1. The distance is infinite if there is no admissible curve that connects these two points. It results by Frobenius theorem that the distribution D is integrable and it determines a foliation on M and two points can be joint if and only if are situated in the same leaf. The energy of a horizontal curve is

$$E(c) = \frac{1}{2}\int_I F^2(\dot{c}(t))dt$$

and it can easily be proved that if a curve is parameterized to a constant speed, then it minimize the length integral if and only if it minimize the energy integral. For 2-homogeneous Lagrangians $L = \frac{1}{2}F^2$ and $\mathcal{L} = \frac{1}{2}\mathcal{F}^2$ we have $\mathcal{L} = L \circ \sigma$ and L on TM is a Lagrangian with constraints.

Theorem 3.1. *The relation between the Hamiltonian H on T^*M and the Hamiltonian $\mathcal{H}$ on E^* is given by*

$$H(x,p) = \mathcal{H}(\mu), \quad \mu = \sigma^\star(p), \quad p \in T_x^*M, \quad \mu \in E_x^*. \tag{4}$$

Proof. The Fenchel-Legendre dual of Lagrangian L is the Hamiltonian H

$$\begin{aligned} H(x,p) &= \sup_v \{\langle p, v\rangle - L(v)\} = \sup_v \{\langle p, v\rangle - \mathcal{L}(u); \sigma(u) = v\} \\ &= \sup_u \{\langle p, \sigma(u)\rangle - \mathcal{L}(u)\} = \sup_u \{\langle \sigma^\star(p), u\rangle - \mathcal{L}(u)\} = \mathcal{H}(\sigma^\star(p)), \end{aligned}$$

and we get $H(x,p) = \mathcal{H}(\mu)$, $\mu = \sigma^\star(p)$, $p \in T_x^*M$, $\mu \in E_x^*$, or locally $\mu_\alpha = \sigma_\alpha^i p_i$, where the Hamiltonian $H(x,p)$ is degenerate on T^*M □

We consider the distributional system with positive homogeneous cost

$$\dot{X} = u_1X_1 + u_2X_2, \; X = (x,y,z)^t \in \mathbb{R}^3, \; X_1 = (1,0,0)^t, \; X_2 = (x,1,z)^t$$

$$\min_{u(.)} \int_0^T \mathcal{F}(u(t))dt. \quad \mathcal{F} = \|u\| + \langle b, u\rangle, \qquad b = (\varepsilon, 0)^t, \qquad u = (u_1, u_2)^t,$$

$0 \leq \varepsilon < 1.$

and are looking for the trajectories starting from the point $(1,0,1)^t$ and parameterized by arclength. The distribution $D = \langle X_1, X_2\rangle$ is holonomic,

$X_1 = \frac{\partial}{\partial x}$, $X_2 = x\frac{\partial}{\partial x} + \frac{\partial}{\partial y} + z\frac{\partial}{\partial z}$ and $[X_1, X_2] = X_1$. In the case of Lie algebroid we consider $E = \langle X_1, X_2 \rangle$ and the anchor has local components

$$\sigma^i_\alpha = \begin{pmatrix} 1 & 0 & 0 \\ x & 1 & z \end{pmatrix}$$

We get the Lagrangian $\mathcal{L} = \frac{1}{2}\left(\sqrt{(u_1)^2 + (u_2)^2} + \varepsilon u_1\right)^2$ and it results the Hamiltonian $\mathcal{H}$ on E^* given by (see Ref. 3)

$$\mathcal{H}(\mu) = \frac{1}{2}\left(\sqrt{\frac{(\mu_1)^2}{(1-\varepsilon^2)^2} + \frac{(\mu_2)^2}{1-\varepsilon^2}} - \frac{\varepsilon\mu_1}{1-\varepsilon^2}\right)^2$$

Remark 3.1. Using relation (4) we calculate the Hamiltonian H on T^*M

$$H(x,p) = \frac{1}{2}\left(\sqrt{\frac{(p_1)^2}{(1-\varepsilon^2)^2} + \frac{(p_1 x + p_2 + p_3 z)^2}{1-\varepsilon^2}} - \frac{\varepsilon p_1}{1-\varepsilon^2}\right)^2$$

Unfortunately, the Hamilton's equations on T^*M, $\dot{x}^i = \frac{\partial H}{\partial p_i}$, $\dot{p}_i = -\frac{\partial H}{\partial x^i}$, lead to a complicated system of implicit partial differential equations.

We will use a different approach, considering the framework of Lie algebroid. From the relation $[X_\alpha, X_\beta] = L^\gamma_{\alpha\beta} X_\gamma$ we obtain the non-zero components $L^1_{12} = 1$, $L^1_{21} = -1$ and will use the Hamilton's equations on Lie algebroid (3), provided by Pontryagin Maximum Principle[7]

$$\dot{x} = \frac{\partial \mathcal{H}}{\partial \mu_1} + x\frac{\partial \mathcal{H}}{\partial \mu_2}, \quad \dot{y} = \frac{\partial \mathcal{H}}{\partial \mu_2}, \quad \dot{z} = z\frac{\partial \mathcal{H}}{\partial \mu_2}, \quad \dot{\mu}_1 = -\mu_1\frac{\partial \mathcal{H}}{\partial \mu_2}, \quad \dot{\mu}_2 = -\mu_1\frac{\partial \mathcal{H}}{\partial \mu_1}$$

where

$$\frac{\partial \mathcal{H}}{\partial \mu_1} = \frac{(1+\varepsilon^2)\,\mu_1}{(1-\varepsilon^2)^2} - \frac{\varepsilon\sqrt{\frac{(\mu_1)^2}{(1-\varepsilon^2)^2} + \frac{(\mu_2)^2}{1-\varepsilon^2}}}{1-\varepsilon^2} - \frac{\varepsilon\mu_1^2}{(1-\varepsilon^2)^3\sqrt{\frac{(\mu_1)^2}{(1-\varepsilon^2)^2} + \frac{(\mu_2)^2}{1-\varepsilon^2}}}$$

$$\frac{\partial \mathcal{H}}{\partial \mu_2} = \frac{\mu_2}{1-\varepsilon^2} - \frac{\varepsilon\mu_1\mu_2}{(1-\varepsilon^2)^2\sqrt{\frac{(\mu_1)^2}{(1-\varepsilon^2)^2} + \frac{(\mu_2)^2}{1-\varepsilon^2}}}$$

The form of the Hamiltonian $\mathcal{H}$ leads to the following change of variables

$$\mu_1(t) = (1-\varepsilon^2)\frac{r(t)}{\cosh\theta(t)}, \quad \mu_2(t) = \sqrt{1-\varepsilon^2}\,r(t)\tanh\theta(t),$$

In this conditions we get $\sqrt{\frac{(\mu_1)^2}{(1-\varepsilon^2)^2} + \frac{(\mu_2)^2}{1-\varepsilon^2}} = |r|$, and from $\dot{\mu}_1 = -\mu_1\frac{\partial \mathcal{H}}{\partial \mu_2}$ we have

$$\sqrt{1-\varepsilon^2}\left(\frac{\dot{r}}{r} - \dot{\theta}\tanh\theta\right) = r(-\tanh\theta + \varepsilon\sec h\theta\tanh\theta) \qquad (5)$$

and follows $\dot{\mu}_2 = -\mu_1 \frac{\partial \mathcal{H}}{\partial \mu_1}$ we get

$$\sqrt{1-\varepsilon^2}\left(\frac{\dot{r}}{r}\tanh\theta+\dot{\theta}\sec h^2\theta\right)=r\sec h\left((1+\varepsilon)^2\sec h\theta-\varepsilon\sec h^2\theta-\varepsilon\right) \quad (6)$$

Now, reducing $\dot{\theta}$ and $\frac{\dot{r}}{r}$ from the relations (5) and (6), we obtain

$$\sqrt{1-\varepsilon^2}\dot{r} = r^2\varepsilon\sec h\theta\tanh\theta(\varepsilon\sec h\theta - 1).$$
$$\sqrt{1-\varepsilon^2}\dot{\theta} = r(\varepsilon\sec h\theta - 1)^2. \quad (7)$$

The last two relations lead to

$$\frac{\dot{r}}{\dot{\theta}} = \frac{r\varepsilon\sec h\theta\tanh\theta}{\varepsilon\sec h\theta - 1}$$

or

$$\frac{1}{r}dr = \frac{\varepsilon\sec h\theta\tanh\theta}{\varepsilon\sec h\theta - 1}d\theta$$

with the solution $\ln|r| = -\ln(\varepsilon\sec h\theta - 1) - \ln c$ and therefore

$$|r| = \frac{1}{c\,(\varepsilon\sec h\theta - 1)}$$

But the geodesics are parameterized by arclength, that corresponds to fix the level 1/2 of the Hamiltonian and we have

$$\mathcal{H} = \frac{r^2}{2}(1-\varepsilon\sec h\theta)^2 = \frac{1}{2c^2}$$

so $c = \pm 1$ and therefore

$$r = \pm\frac{1}{\varepsilon\sec h\theta - 1}$$

From the relation (7) we have

$$\frac{d\theta}{dt} = \frac{\sqrt{1-\varepsilon^2}}{1-\varepsilon\sec h\theta}$$

and we obtain

$$t = \sqrt{1-\varepsilon^2}\int\frac{1}{1-\varepsilon\sec h\theta}d\theta$$

Relation $\dot{\mu}_1 = -\mu_1 \dot{y}$ leads to

$$y(\theta) = \ln\frac{c_1(1-\varepsilon\sec h\theta)}{(1-\varepsilon^2)\sec h\theta}, \quad c_1 \in R$$

Since we are looking for the geodesics starting from the point $(1,0,1)^t$ we have

$$\ln\frac{c_1}{1+\varepsilon}=0\Rightarrow c_1=1+\varepsilon$$

and therefore

$$y(\theta)=\ln\frac{1-\varepsilon\sec h\theta}{(1-\varepsilon)\sec h\theta}=\ln\frac{\cosh\theta-\varepsilon}{1-\varepsilon}$$

The relation

$$\frac{\dot z}{z}=-\frac{\dot\mu_1}{\mu_1}$$

leads to

$$z(\theta)=\frac{c_2(1-\varepsilon\sec h\theta)}{(1-\varepsilon^2)\sec h\theta}$$

and from $z(0)=1$ we get $c_2=1+\varepsilon$ and obtain

$$z(\theta)=\frac{\cosh\theta-\varepsilon}{1-\varepsilon}.$$

Also we have

$$\dot\mu_2=\mu_1(\dot x-x\frac{\partial\mathcal{H}}{\partial\mu_2})=\mu_1\dot x+x\dot\mu_1$$

and it results $\mu_2=\mu_1 x+c_3$, so

$$x(\theta)=\frac{\sinh\theta}{\sqrt{1-\varepsilon^2}}\pm\frac{c_3(1-\varepsilon\sec h\theta)}{(1-\varepsilon^2)\sec h\theta}$$

From $x(0)=1$ we obtain $c_3=1+\varepsilon$ and it results

$$x(\theta)=\frac{\sinh\theta}{\sqrt{1-\varepsilon^2}}+\frac{\cosh\theta-\varepsilon}{1-\varepsilon}$$

Remark 3.2. If $\varepsilon=0$ we obtain the case of distributional systems with quadratic cost with the solution

$$x(t)=\sinh t+\cosh t,\quad y(t)=\ln\cosh t,\quad z(t)=\cosh t.$$

Acknowledgments

This paper is partially supported by CNCSIS grant 179/26C/2007 and research funds of Dragoş Hrimiuc. Also, many thanks to organizers for reduced fee.

References

1. P.J. Higgins and K. Mackenzie, Algebraic constructions in the category of Lie algebroids, *Journal of Algebra* **129** (1990) 194–230.
2. D. Hrimiuc and L. Popescu, Geodesics of sub-Finslerian geometry, *In: Differential Geometry and its Applications* (Proc. Conf. Prague, August 30-September 3, 2004, Charles University, Prague, 2005) 59–67.
3. D. Hrimiuc and H. Shimada, On the L-duality between Lagrange and Hamilton Manifold, *Nonlinear World* **3** (1996) 613–641.
4. M. de Leon, J. C. Marrero and E. Martinez, Lagrangian submanifolds and dynamics on Lie algebroids, *J. Phys. A: Math. Gen.* **38** (2005) 241–308.
5. K. Mackenzie, *Lie groupoids and Lie algebroids in differential geometry* (London Mathematical Society Lecture Note Series, Cambridge, no.124, 1987).
6. E. Martinez, Geometric formulation of mechanics on Lie algebroids. *In: Proc. of the VIII Workshop on Geometry and Physics* (Medina del Campo, 1999, vol. 2 of Publ. R. Soc. Mat. Esp., 2001) 209–222.
7. E. Martinez, Reduction of optimal control systems, *Rep. Math. Phys.* **53** (1) (2004) 79–90.
8. A. Weinstein, Lagrangian mechanics and groupoids, *Fields Inst. Comm.* **7** (1996) 206–231.

Differential Geometry and its Applications
Proc. Conf., in Honour of Leonhard Euler, Olomouc, August 2007

Variational theory of balance systems

Serge Preston

Department of Mathematics and Statistics, Portland State University, Portland, OR, 97207-0751, USA
E-mail: serge@pdx.edu
www.mth.pdx.edu

In this talk we are presenting the systems of balance equations of Continuum Thermodynamics (balance systems) in a variational form using Poincare-Cartan formalism. A constitutive relation $\mathcal{C}$ of a balance system $\mathcal{B}_{\mathcal{C}}$ is realized as a mapping between a k-jet bundle $J^k(\pi)$ and the extended dual bundle similar to the Legendre mapping of the Lagrangian Field Theory. Invariant (variational) form of the balance system $\mathcal{B}_{\mathcal{C}}$ is studied, the class of admissible variations is defined. A case of semi-Lagrangian balance systems is characterized in geometrical terms. Action of automorphisms of the bundle π on the constitutive mappings $\mathcal{C}$ is studied and it is shown that the symmetry group $Sym(\mathcal{C})$ of $\mathcal{C}$ acts on the space of solutions of balance system $\mathcal{B}_{\mathcal{C}}$. Suitable version of Noether Theorem for an action of a symmetry group is presented together with the special form for semi-Lagrangian balance systems and examples of energy momentum and gauge symmetries balance laws.

Keywords: k-jet bundle, Balance law, Poincare-Cartan form, Noether Theorem.

MS classification: 58E30, 53B50.

1. Notations and preliminaries

Throughout this paper $\pi : Y \to X$ is a (*configurational*) fibred bundle with a n-dim base X and a total space Y, $dim(Y) = n + m$. Base manifold X is endowed with a (pseudo)-Riemannian metric G. Volume form of metric G will be denoted by η. As an basic example, we consider the case where $X = T \times B$ is the product of time axis T and an open material manifold (or a domain in the physical space) B.

We will be using fiber charts (W, x^μ, y^i) in the bundle π, where $(\pi(W), x^\mu)$ is a chart in X and y^i are coordinates along the fibers. Corresponding frame will be denoted by $(\partial_{x^\mu}, \partial_{y^i})$, coframe - by (dx^μ, dy^i). For a mutliindex $\sigma = \{\sigma_1, \ldots, \sigma_n\}, \sigma_i \in \mathbb{N}$ denote by ∂^σ the differential operator in $C^\infty(X)$ $\partial^\sigma f = \partial^{\sigma_1}_{x^1} \cdot \ldots \partial^{\sigma_n}_{x^n} f$. Introduce the contracted forms

$\eta_\mu = i_{\partial_{x^\mu}}\eta,\ \eta_{\mu\nu} = i_{\partial_{x^\nu}} i_{\partial_{x^\mu}}\eta$.

Sections $s : V \to Y,\ V \subset X$ of the bundle π represent the collection of (classical) fields y^i defined in the domain $V \subset X$. Usually these fields are components of some tensor or tensor densities fields.

1.1. *The k-jet bundles $J^k(\pi)$*

Given a fiber bundle $\pi : Y \to X$ denote by $J^k_x(\pi),\ 1 \leqq k \leqq \infty$ the k-jet bundle of section of bundle π.[4,5] Denote by $\pi^k : J^k(\pi) \to X$ the projection to the base manifold X, by $\pi_{kr} : J^k(\pi) \to J^r(\pi)$, where $k > r$ the projection between two jet bundles.

To every fiber chart (V, x^μ, y^i) in Y there corresponds the fiber chart $(x^\mu, y^i, z^i_\sigma, |\sigma| = \sum_i \sigma_i \leqq k)$ in $\pi^{-1}_{k0}(V) \subset J^k(\pi)$. This chart is defined by the condition $z^i_\sigma(j^1_x s) = \partial^\sigma s^i(x)$.

We will be using standard notations $T(M)$ for the tangent bundle of a manifold M, $V(M) \subset T(M)$ for the vertical subbundle of $T(M)$, $\mathcal{X}(M)$ - Lie algebra of vector fields in M, $\Lambda^r(M)$ for the bundle of exterior r-forms on the manifold M, $(\Lambda^* M = \oplus \Lambda^r(M), d)$ for the differential algebra of exterior forms on the manifold M.[4]

For $k \geqq 1$ the k-jet bundle $J^k(\pi) \to X$ is endowed with the Cartan distribution Ca defined by the basic contact forms

$$\omega^i = dy^i - z^i_\mu dx^\mu, \ldots \omega^i_\sigma = dz^i_\sigma - z^i_{\sigma+1_\nu} dx^\nu,\ |\sigma| < k,\ dz^i_\sigma,\ |\sigma| = k.$$

In the case where $k = \infty$ the basic contact forms dz^i_σ are absent. We refer to Refs. 6,7 for the notion of horizontal and contact forms and only remind the basic result of D.Krupka on the contact decomposition of a form (see Ref. 6) $\nu \in \Lambda^q(J^k(\pi))$ such that $\nu = \pi^*_{k(k-1)}\nu_1, \nu_1 \in \Lambda^q(J^{k-1}(\pi))$. The form ν has the unique *contact* decomposition:

$$\nu = \nu_0 + \nu_1 + \ldots + \nu_q,$$

where $\nu_i, 0 \leqq i \leqq q$ is a i-contact form on $J^k(\pi)$.

In particular, for a function $f \in \pi^*_{\infty k} C^\infty(J^k(\pi))$,

$$df = (d_\mu f) dx^\mu + \sum_{(i,\sigma)} f_{,z^i_\sigma} \omega^i_\sigma, \tag{1}$$

where $d_\mu f = \partial_{x^\mu} f + \sum_{\sigma||\sigma|\geqq 0} z^i_{\sigma+1_\mu} \partial_{z^i_\sigma} f$ is the total derivative of the function f by x^μ.

Contact decomposition can be used to present the differential $d\nu$ of a q-form $\nu \in \Lambda^*(J^\infty)(\pi))$ as the sum of horizontal and vertical forms: $d\nu = d_h\nu + d_v\nu$. Operators d_h, d_v are called respectively horizontal and vertical differentials.[1,5] We recall also that $d_h^2 = 0$.

1.2. *Lagrangian Poincare-Cartan formalism, k=1*

Here we remind the basic notions of Lagrangian (multisymplectic) Field Theory based on the use of Poincare-Cartan form, see Refs. 2,8. For a Lagrangian n-form $L\eta$, $L \in C^{\infty}(J^1(\pi))$ the Poincaré-Cartan n-form are defined as follows:

$$\Theta_L = L\eta + S^*_\eta(dL) = (L - z^i_\mu \frac{\partial L}{\partial z^i_\mu})\eta + \frac{\partial L}{\partial z^i_\mu} dy^i \wedge \eta_\mu, \tag{2}$$

where S^*_η is the adjoint operator of the vertical endomorphism S_η. An *extremal* of L is a section s of the bundle π such that for any vector field $\hat{\xi}$ on the manifold $J^1(\pi)$,

$$(j^1(s))^*(i_{\hat{\xi}} d\Theta_L) = 0, \tag{3}$$

where $j^1(s)$ is the first jet prolongation of s. A section s is an extremal of L if and only if it satisfies to the conventional system of Euler-Lagrange Equations (see Refs. 1,6).

1.3. *Bundles $\Lambda^r_p Y$ and canonical forms*

Introduce the bundle $\Lambda^r_p(Y)$ of the exterior forms on Y which are annulated if p of its arguments are vertical: $\omega^r \in \Lambda^r_p(Y) \Leftrightarrow i_{\xi_1} \dots i_{\xi_p} \omega^r = 0$, $\xi_i \in V(Y)$, see Ref. 8. We will be using these bundles for $r = n, n+1$ and $p = 1, 2$. Elements of the space $\Lambda^n_1 Y$ are n-forms locally expressed as $p(x,y)\eta$. Elements of the space $\Lambda^n_2 Y$ have, in a fiber coordinates (x^μ, y^i) the form

$$p(x,y)\eta + p^\mu_i dy^i \wedge \eta_\mu.$$

This introduces coordinates (x^μ, y^i, p) on the manifold $\Lambda^n_1 Y$ and (x^μ, y^i, p, p^μ_i) on the manifold $\Lambda^n_2 Y$.

For $r = n+1$ the forms $dy^i \wedge \eta$ form the basis of $\Lambda^{n+1}_2(Y)$ while the bundle $\Lambda^{n+1}_1(Y)$ is *zero bundle*.

Introduce the notation $\Lambda^{n+(n+1)}_p(Y) = \Lambda^n_p(Y) \oplus \Lambda^{n+1}_p(Y), p = 1, 2$, for the direct sum of the bundles on the right side.

It is clear that $\Lambda^k_1(Y) \subset \Lambda^k_2(Y)$ and, therefore $\Lambda^{n+(n+1)}_1(Y) \subset \Lambda^{n+(n+1)}_2(Y)$. This allows to define the factor-bundle

$$\Lambda^{n+(n+1)}_{2/1}(Y) = \Lambda^{n+(n+1)}_2(Y)/\Lambda^{n+(n+1)}_1(Y),$$

with the projection $q : \Lambda^{n+(n+1)}_2(Y) \to \Lambda^{n+(n+1)}_{2/1}(Y)$.

On the bundles $\Lambda_2^{n+(n+1)}(Y)$ there are defined the canonical forms[8] with the coordinate expression

$$\Theta_2^{n+(n+1)} = \Theta_2^n + \Theta_2^{n+1} = p\eta + p_i^\mu dy^i \wedge \eta_\mu + p_i dy^i \wedge \eta. \tag{4}$$

On the factor-bundle $\Lambda_{2/1}^{n+(n+1)}(Y)$ the form $\widetilde{\Theta}_{2/1}^{n+(n+1)} = p_i^\mu dy^i \wedge \eta_\mu + p_i dy^i \wedge \eta$ is defined *mod* η.

1.4. *"Iglesias" differential*

Below we will be using the differential $\tilde{d}$ which is a special case of operators introduced by D. Iglesias and used in Ref. 3. This differential is defined on couples of exterior forms $\alpha^k + \beta^{k+1}$:

$$\begin{cases} \tilde{d} : \Omega^{k+(k+1)}(X) = \Omega^k(X) \oplus \Omega^{k+1}(X) \to \Omega^{(k+1)+(k+2)}(X) \\ \qquad\qquad\qquad\qquad = \Omega^{k+1}(X) \oplus \Omega^{k+2}(X) : \\ \tilde{d}(\alpha^k + \beta^{k+1}) = ((-d\alpha + \beta) + d\beta). \end{cases} \tag{5}$$

It is easy to check that $\tilde{d} \circ \tilde{d} = 0$.

1.5. *Systems of balance equations*

Let the base manifold X be the material or physical space-time with (local) coordinates $x^1 = t,\ x^A, A = 2, \ldots, n$. Typical system of balance equations for the fields y^i is determined by a choice of flux fields F_i^μ (including the densities for $\mu = 1$ and source fields Π_i as functions on a k-jet space $J^k(\pi),\ k \geqq 0$. A choice of the functions $F_i^\mu, \Pi_i \in C^\infty(J^k(\pi))$ is codified in physics as the choice of the **constitutive relation** of a given material.[9,10] After such a choice has been done, the *closed* system of balance equations

$$(F_i^\mu \circ j^k s)_{;\mu} = \Pi_i \circ j^k s \tag{6}$$

can be solved for a section $s : V \to Y$ if one add to the balance system the appropriate boundary (including initial) conditions.

2. Constitutive relations (CR) and their Poincare-Cartan forms

Here we introduce the constitutive and covering constitutive relations of order k as generalized Legendre transformation from the k-jet bundle

$J^k(\pi)$ to the extended multisymplectic bundles of $n+(n+1)$-forms on the manifold Y, see the commutative diagram

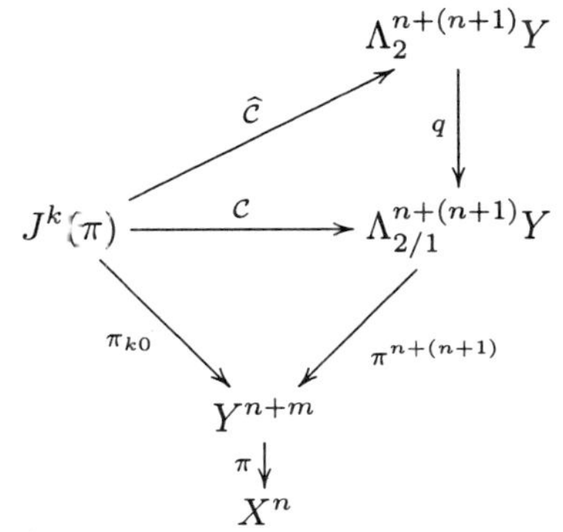

$$
\begin{array}{c}
\Lambda_2^{n+(n+1)}Y \\
\hat{\mathcal{C}} \nearrow \qquad \downarrow q \\
J^k(\pi) \xrightarrow{\;\mathcal{C}\;} \Lambda_{2/1}^{n+(n+1)}Y \\
\pi_{k0} \searrow \qquad \swarrow \pi^{n+(n+1)} \\
Y^{n+m} \\
\pi \downarrow \\
X^n
\end{array}
\qquad (7)
$$

Definition 2.1.

(a) **A constitutive relation (CR)** C **of order** k is a morphism of bundles $C : J^k(\pi) \to \Lambda_{2/1}^{n+(n+1)}Y$ over Y:

$$
C(x^\mu, y^i, z_\mu^i, \dots z_{\mu_1 \dots \mu_k}^i) = (x^\mu, y^i; F_i^\mu; \Pi_i), \; F_i^\mu, \Pi_i \in C^\infty(J^k(\pi)).
$$

The Poincare-Cartan form of the CR $\mathcal{C}$ is the form defined *mod* η

$$
\Theta_C = \mathcal{C}^*(\Theta_2^n - \Theta_2^{n+1} \; mod \; \eta) = F_i^\mu dy^i \wedge \eta_\mu + \Pi_i dy^i \wedge r \; mod \; \eta.
$$

(b) **A covering constitutive relation (CCR)** $\hat{C}$ **of order** k is morphism of bundles $\hat{C} : J^k(\pi) \to \Lambda_2^{n+(n+1)}Y$ over Y:

$$
\hat{C}(x^\mu, y^i, z_\mu^i, \dots, z_{\mu_1 \dots \mu_k}^i) = (x^\mu, y^i; p; F_i^\mu; \Pi_i), \; p, F_i^\mu, \Pi_i \in C^\infty(J^k(\pi)).
$$

The Poincare-Cartan form of the CCR $\hat{C}$ is the form

$$
\Theta_{\hat{C}} = \hat{C}^*(\Theta_2^n + \Theta_2^{n+1}) = p\eta + F_i^\mu dy^i \wedge \eta_\mu + \Pi_i dy^i \wedge \eta.
$$

Any covering constitutive relation defines the corresponding constitutive relation by the projection - $C = q \circ \hat{C}$. On the other hand, there are many ways to lift a constitutive relation to the CCR. In between all possible CCR corresponding to a given CR there is one privileged (as we will see below)

Definition 2.2. The **lifted CCR** $\tilde{C}$ **of a constitutive relation** C (of order k) is defined by

$$
\Theta_{\tilde{C}} = F_i^\mu \omega^i \wedge \eta_\mu - \Pi_i \omega^i \wedge \eta = -(\sum_{(i,\mu)} z_\nu^j F_j^\nu)\eta + F_i^\mu dy^i \wedge \eta_\mu + \Pi_i dy^i \wedge \eta. \quad (3)
$$

Remark 2.1. Let $k = 1$. For a 1-form $\chi = F_i^\mu dz_\mu^i + \Pi_i dy^i \in \Lambda^1(J^1(\pi))$, one has

$$\Theta_{\tilde{C}} = S_\eta^*(F_i^\mu dz_\mu^i + \Pi_i dy^i).$$

Next we present three examples of different types of constitutive relations or covering constitutive relations and corresponding Poincare-Cartan forms. In all these examples $k = 1$.

Example 2.1. A **Lagrange constitutive relation** defined by a function $L \in C^\infty(J^1(\pi))$ is:

$$C_L(x^\mu, y^i, z_\mu^i) = (p_i^\mu = F_i^\mu = \frac{\partial L}{\partial z_\mu^i}; \Pi_i = \frac{\partial L}{\partial y^i}).$$

Example 2.2. A **semi-Lagrangian CCR** is defined by a functions $L, Q_i, i = 1, \ldots, m \in C^\infty(J^1(\pi))$:

$$\hat{C}_{L,Q_i}(x^\mu, y^i, z_\mu^i) = (p = L - z_\mu^i L_{,z_{,\mu}^i}, p_i^\mu = \frac{\partial L}{\partial z_\mu^i}; \Pi_i = Q_i(x^\mu, y^i, z_\mu^i)).$$

Example 2.3. *L* + *D*-**system**. Let L, D ("dissipative potential") $\in C^\infty(J^1(\pi))$. Let The **CR** $\mathcal{C}_{L,D}$ is defined by its Poincare-Cartan form

$$\Theta_{L,D} = L_{z_\mu^i} dy^i \wedge \eta_\mu + (D_{z_{,1}^i} - L_{,y^i}) dy^i \wedge \eta.$$

Remark that in the continuum thermodynamics where dissipative potentials are used, $x^1 = t$ is the time.

3. Variational form of balance systems

In this section we study the invariant variational form of balance system and the separation of this invariant form into the m separate balance laws by independent variations. In difference to the Lagrangian case in general one have to put a condition on the variations that can be used for separating equations. For simplicity we will consider that a CR $\mathcal{C}$ is defined on the infinite partial jet bundle $J^\infty(\pi)$ but is $\pi_{\infty k}$-projectable for some $k < \infty$ (i.e constitutive relation depends on the derivatives of fields y^i up to the order k.

We start with an arbitrary covering constitutive relation $\widehat{C}$ and change the sign of source term, i.e. we consider $\Theta_{\widehat{C}_-} = p\eta + F_i^\mu dy^i \wedge \eta_\mu - \Pi_i dy^i \wedge \eta$.

For a section $s \in \Gamma_V(\pi)$, $V \subset X$ we request the fulfilment of the equation (Invariant Balance System)

$$j^k(s)^*(i_\xi \tilde{d}\Theta_{\hat{C}_-}) = 0, \; \xi \in \mathcal{X}(J^k(\pi)) \qquad (9)$$

for a large enough family of variations ξ - sufficient for separation of individual balance laws (see Def. 3.1 below).

Thus, we take the $n+(n+1)$-form $\Theta_{\hat{C}_-}$ as in Def.2.1(2) and apply first the Iglesias differential $\tilde{d}$ and then i_ξ for a vector field $\xi = \xi^\nu \partial_\nu + \xi^j \partial_{y^j} + \sum_{|\sigma|>1} \xi^i_\sigma \partial_{z^i_\sigma}$.

We will denote by $kCon$ the k-contact forms that appears in calculations. We introduce the notation $d\eta_\mu = \lambda_{G,\mu}\eta$. It will be convenient to include variables y^i into the family of variables z^i_σ for $|\sigma| = 0$ taking $z^i = y^i$.

We will also use the contact splitting (1) of the lift of differential of a function p to the $J^\infty(\pi)$ and similar contact decomposition for the flux components F^ν_j. We have, $\tilde{d}\Theta_{\hat{C}_-} = d(p\eta + F^\mu_i dy^i \wedge \eta_\mu) + \Pi_i dy^i \wedge \eta = dp \wedge \eta + dF^\mu_i \wedge dy^i \wedge \eta_\mu - F^\mu_i dy^i \wedge \lambda_{G,\mu}\eta + \Pi_i dy^i \wedge \eta =$

To continue the calculation we replace in the last two terms the dy^i by ω^i and using $dy^j \wedge \eta_\nu = \omega^j \wedge \eta_\nu + z^j_\sigma dx^\sigma \wedge \eta_\nu = \omega^j \wedge \eta_\nu + z^j_\nu \eta$ we get

$$
\begin{aligned}
&= [(d_\mu p)dx^\mu + \sum_{(i,\sigma)} p_{,z^i_\sigma}\omega^i_\sigma] \wedge \eta + [(d_\mu F^\nu_j)dx^\mu \\
&+ \sum_{(i,\sigma)} (F^\nu_j)_{,z^i_\sigma}\omega^i_\sigma] \wedge dy^j \wedge \eta_\nu - \lambda_{G,\mu}F^\mu_i\omega^i \wedge \eta + \Pi_i\omega^i \wedge \eta \\
&= (d_\mu F^\nu_j)dx^\mu \wedge dy^i \wedge \eta_\nu - \lambda_{G,\mu}F^\mu_i\omega^i \wedge \eta + \Pi_i\omega^i \wedge \eta \\
&+ [\sum_{(i,\sigma)} p_{,z^i_\sigma}\omega^i_\sigma] \wedge \eta + [\sum_{(i,\sigma)} (F^\nu_j)_{,z^i_\sigma}\omega^i_\sigma] \wedge (\omega^j \wedge \eta_\nu + z^j_\nu \eta) \\
&= (-d_\nu F^\nu_j)\omega^j \wedge \eta - \lambda_{G,\mu}F^\mu_i\omega^i \wedge \eta + \Pi_i\omega^i \wedge \eta \\
&+ \sum_{(i,\sigma)} [p_{,z^i_\sigma} + z^j_\nu (F^\nu_j)_{,z^i_\sigma}]\omega^i_\sigma \wedge \eta + 2con \\
&= [-d_\mu F^\mu_i) - \lambda_{G,\mu}F^\mu_i + \Pi_i]\omega^i \wedge \eta + F^\nu_j\omega^j_\nu \wedge \eta \\
&+ [\sum_{(i,\sigma)} (p + z^j_\nu F^\nu_j)_{,z^i_\sigma}\omega^i_\sigma \wedge \eta + 2Con.
\end{aligned} \tag{10}
$$

Here we have used the equality $(d_\mu F^\nu_j)dx^\mu \wedge dy^i \wedge \eta_\nu = -(d_\nu F^\nu_j)dy^i \wedge \eta = -(d_\nu F^\nu_j)\omega^i \wedge \eta$. Last formula proves the first statement of the next

Theorem 3.1. *Let $\hat{C}$ be a CCR defined in a domain of the partial k-jet bundle $J^k(\pi)$. Then,*

(a) There is the following contact decomposition

$$\tilde{d}\Theta_{\hat{C}_-} = [-d_\mu F_i^\mu) - \lambda_{G,\mu} F_i^\mu + \Pi_i]\omega^i \wedge \eta + F_j^\nu \omega_\nu^j \wedge \eta + \left[\sum_{(i,\sigma)} (p + \sum_{(j,\nu)} z_\nu^j F_j^\nu)_{,z_\sigma^i} \omega_\sigma^i\right] \wedge \eta + 2Con. \tag{11}$$

Term in the brackets is the ***vertical differential*** $d_v(p + z_\nu^j F_j^\nu)$ *of the function* $p + z_\nu^j F_j^\nu$*, see Refs. 1,5.*

(b) Let $\xi \in \mathcal{X}(J^k(\pi))$ *be any vector field. Then,*

$$i_\xi \tilde{d}\Theta_{\hat{C}_-} = -\omega_{\hat{C}}^1(\xi)\eta - \omega_{\hat{C}}^{k+1}(\xi^{k+1})\eta + Con, \tag{12}$$

for a prolongation ξ^{k+1} *to the vector field* ξ *to* $J^{k+1}(\pi)$*, where*

$$\begin{cases} \omega_{\hat{C}}^1(\xi) = \omega^i(\xi)[d_\mu F_i^\mu + \lambda_{G,\mu} F_i^\mu - \Pi_i], \\ \omega_{\hat{C}}^{k+1}(\xi^{k+1}) = \sum_{(j,\nu)} F_j^\nu \omega_\nu^j(\xi^{k+1}) \\ \qquad + \left[\sum_{(i,\sigma)} (p + \sum_{(j,\nu)} z_\nu^j F_j^\nu)_{,z_\sigma^i} \omega_\sigma^i(\xi^{k+1})\right]. \end{cases} \tag{13}$$

(c) For the lifted CCR $\tilde{C}$ *of a constitutive relation* C *and an arbitrary vector field* $\xi \in \mathcal{X}(J^k(\pi))$*,*

$$i_\xi \tilde{d}\Theta_{\hat{C}_-} = -\omega_{\hat{C}}^1(\xi)\eta - \sum_{(j,\nu)} F_j^\nu \omega_\nu^j(\xi^{k+1})\eta + Con \tag{14}$$

Proof. Previous discussion proves the validity of the first decomposition. Second and third statements follow from the first one. Covariance of obtained decomposition follows from the tensorial behavior of both components of decompositions with regard of such coordinate changes.[11] □

Consider now the case where $\omega_{\hat{C}}^{k+1}$ is identically zero and no restriction to the variations of Poincare-Cartan form appears.

Proposition 3.1. *Let* $\hat{C}$ *be a CCR* ***of order 1****. Then,*

$$\omega_{\hat{C}}^2 = \sum_{(i,\mu)} \left[F_i^\mu - \partial_{z_\mu^i}(p + \sum_{(j,\nu)} z_\nu^j F_j^\nu)\right] \omega_\mu^i = 0 \Leftrightarrow F_i^\mu = \partial_{z_\mu^i} L, L \in C^\infty(J^1(\pi)),$$

i.e. CR C is ***(locally) semi-Lagrangian****. In this case,* $p = L - \sum_{(i,\mu)} z_\mu^i \partial_{z_\mu^i} L + l(x,y)$ *with an arbitrary* $l(x,y) \in C^\infty(Y)$*.*

Proof. Rewrite the first equality as $\partial_{z^i_\sigma} p = -z^i_\mu \partial_{z^i_\sigma} F^\mu_i$. To see that the implication $\Rightarrow$ holds we notice that provided left statement is true, the right sides of these equalities satisfy to the mixed derivative test

$$\partial_{z^j_\lambda}(-z^k_\mu \partial_{z^i_\sigma} F^\mu_k) = \partial_{z^i_\sigma}(-z^k_\mu \partial_{z^j_\lambda} F^\mu_k),$$

or $\partial_{z^j_\lambda} F^\sigma_i = \partial_{z^i_\sigma} F^\lambda_j$. From this equality valid for all couples of indices $(j,\lambda),(i,\sigma)$ the second statement follows. To prove the opposite - reverse the arguments. $\square$

Theorem 3.2. *Let $C_{L,Q}$ be a semi-Lagrangian CR* ***of order 1***: $F^\mu_i = L_{,z^i_\mu}$. *Let $\hat{C}_{L,Q}$ be a CCR covering C with $p = L - z^i_\mu \partial_{z^i_\mu} L$. Then the following statements are equivalent*

(a) For a section $s \in \Gamma(\pi)$ and for all $\xi \in \mathcal{X}(J^1(\pi))$

$$j^1(s)^* i_\xi \tilde{d}\Theta_{\hat{C}_{L,Q}} = 0.$$

(b) Section s is the solution of the system of balance equations

$$= \sum_\mu (L_{,z^i_\mu} \circ j^1(s))_{;\mu} - L_{,y^i} \circ j^1(s) = Q_i(j^1(s)).$$

In general case we have to introduce the class of *admissible variations.*

Definition 3.1. A vector field $\xi \in \mathcal{X}(U)$, $U \subset J^k(\pi)$ is called C-admissible ($\xi \in \mathcal{X}_C(U)$) if

(a) In the case $k = 1$, for some (=any) prolongation of ξ to the vector field $\xi^2 \in \mathcal{X}(J^2(\pi_{21}^{-1}(U)), \omega^2_C(\xi^2) = \sum_{(i,\mu)}(\xi^i_\mu - \xi^\sigma z^i_{\mu\sigma})F^\mu_i = 0$.
(b) For $k > 1$, $\omega^2_C(\xi) = \sum_{(i,\mu)}(\xi^i_\mu - \xi^\sigma z^i_{\mu\sigma})F^\mu_i = 0$.

As the next result shows, locally any CR is separable.

Lemma 3.1. *Let $W \subset Y$ be the domain of a fiber chart (x^μ, y^i). Any CR C is* ***separable*** *in $W^k = \pi_{k0}^{-1}(W)$, more specifically, $\forall y \in W$, $z \in W^k, \pi_{k0}(z) = y$, the mapping $\mathcal{X}_C(W^k) \ni \xi \to \{\omega^i_z(\xi)\} \in R^m$ is the epimorphism.*

Proof. Consider vertical vector fields that are constant along the fiber $U \cap W$. $\square$

As a result, in general case we have the following result

Theorem 3.3. *For any CR C and a domain $U \subset J^k(\pi)$ the following statements for a section $s \in \Gamma(\pi)(U)$, $U \subset X$ are equivalent:*

(a) $\mathcal{C}$ is separable in U and for all $\xi \in \Gamma(U), \mathcal{X}_{\mathcal{C}}(J^k(\pi))$,

$$j^k(s)^*(i_\xi \tilde{d}\Theta_{\tilde{C}_-}) = 0.$$

(b) Section $s \in \Gamma(U,\pi)$ is the solution of the system

$$(F_i^\mu \circ j^k(s))_{,x^\mu} + F_i^\mu \circ j^k(s)\lambda_{G,\mu} = \Pi_i(j^k(s)),\ i = 1,\ldots,m. \qquad (\bigstar)$$

Remark 3.1. Using the following Lepage form (see Ref. 7) on the (k+1)-jet bundle $J^{k+1}(\pi)$ equivalent to the covering Poincare-Cartan form $\Theta_{\hat{C}_-}$

$$\widehat{\Theta}_{\hat{C}_-} = p\eta + F_i^\mu dy^i \wedge \eta_\mu - \Pi_i dy^i \wedge \eta - F_i^\mu \omega_\mu^i \wedge \eta. \qquad (15)$$

one can remove the condition of admissibility of a variations ξ.

4. Action of automorphisms and symmetry groups of CR

Let $\hat{C}$ be a CCR with the domain $J^k(\pi)$, let $\phi \in Aut(\pi)$. Define the action of ϕ on the CCR $\hat{C}$ by

$$\hat{C}^\phi = \phi^{k*} \circ \hat{C} \circ \phi^{1\ -1}, \qquad (16)$$

where ϕ^k is the flow lift of ϕ to $J^k(\pi)$, ϕ^{1*} - its canonical flow lift to $\Lambda_2^{n+(n+1)}Y$. Similarly one defines the action of ϕ on a CR C, using the projection $\widetilde{\phi}^{1*}$ of ϕ^{1*} to $\Lambda_{2/1}^{n+(n+1)}Y$ instead of ϕ^{1*}, see Refs. 4,11. It is easy to see that in such a case $\Theta_{\hat{C}^\phi} = \phi^{1\ -1*}\Theta_{\hat{C}}$.

Definition 4.1.

(a) An automorphism $\phi \in Aut(\pi)$ is called a **geometrical symmetry transformation** of a CR $\mathcal{C}$ (resp. of a CCR $\widehat{\mathcal{C}}$) if $\mathcal{C}^\phi = \mathcal{C}$ (respectively, $\widehat{\mathcal{C}}^\phi = \widehat{\mathcal{C}}$).

(b) Let $\xi \in \mathcal{X}(\pi)$ be a π-projectable vector field. ξ is called a **geometrical infinitesimal symmetry** of the CR $\mathcal{C}$ (resp. of a CCR $\widehat{\mathcal{C}}$) if $L_{(\xi^k,\tilde{\xi}^{1*})}\mathcal{C} = 0$, (respectively, $L_{(\xi^k,\xi^{1*})}\mathcal{C} = 0$). In such a case for the local phase flow ϕ_t of ξ one has

$$\Theta_{C^{\phi_t}} = \phi_t^{k\ -1\ *}\Theta_C = \Theta_C \Leftrightarrow \mathcal{L}_{\xi^1}\Theta_C = 0,$$

(respectively, $\mathcal{L}_{\xi^k}\Theta_{\hat{C}} = 0$).

In the next statement we collect the basic properties of action of automorphisms $\phi \in Aut_p(\pi)$, and of geometrical symmetries of constitutive relations, on the balance systems $\mathcal{B}_{\mathcal{C}}$. Proof of this result is given in Ref. 12.

Theorem 4.1.

(a) *Mapping* $\xi \to \phi_* \xi$ *maps the space of* $\mathcal{C}$*-admissible vector fields* $\mathcal{X}_{\mathcal{C}}(W)$ *at a domain* W *onto the space of* $\mathcal{C}^{\phi}$*-admissible vector fields* $\mathcal{X}_{\mathcal{C}^{\phi}}(\phi(W))$ *in the domain* $\phi(W)$ *for any open subset* $W \subset J^k(\pi)$. *This mapping defines the isomorphism of sheaves of admissible vector fields* $\mathcal{X}_{\mathcal{C}} \rightleftarrows \mathcal{X}_{\mathcal{C}^{\phi}}$.

(b) *Mapping* $s \to s^{\phi} = \phi \circ s \circ \bar{\phi}^{-1}$, *on the sections of the configurational bundle* $\pi : Y \to X$ *maps the sheaf of solutions* $Sol(\mathcal{C})$ *onto the sheaf of solutions of the balance system* $\mathcal{B}_{\mathcal{C}^{\phi}}$:

$$Sol(C) \rightleftarrows Sol(C^{\phi}).$$

(c) *Let* $\phi \in Aut(\pi)$ *be a symmetry of the CR* $\mathcal{C}$. *Then the mapping* $s \to s^{\phi} =$ *maps the sheaf* $Sol(\mathcal{C})$ *of solutions of the balance system* ★ *into itself.*

5. Noether Theorem

Let $G \subset Sym(C) \subset Aut_p(\pi)$ be a Lie group of the geometrical symmetries of a CR $\mathcal{C}$. Denote by $\mathfrak{g}$ - the Lie algebra of the group G, $\mathfrak{g}^*$ - its dual space, $\xi \to \xi_Y \in \mathcal{X}(Y)$ - action mapping on the manifold Y and ξ^k_Y be its flow lift to the k-jet bundle $J^k(\pi)$.

Condition that the group G is the symmetry group of the CR $\mathcal{C}$ (and its lifted CCR $\tilde{C}$) has the infinitesimal form $\mathcal{L}_{\xi^k_Y}(\Theta^n_{\tilde{C}} + \Theta^{n+1}_{\tilde{C}}) = 0$. This condition splits into two corresponding to the order of the forms:

$$\begin{cases} \mathcal{L}_{\xi^k_Y}\Theta^n_{\tilde{C}} = (i_{\xi^k_Y}d + di_{\xi^k_Y})\Theta^n_{\tilde{C}} = 0, \\ \mathcal{L}_{\xi^k_Y}\Theta^{n+1}_{\tilde{C}} = (i_{\xi^k_Y}d + di_{\xi^k_Y})\Theta^{n+1}_{\tilde{C}} = 0. \end{cases} \tag{17}$$

For the first condition we have, using Theorem 3.3; $di_{\xi^k_Y}\Theta^n_{\tilde{C}} = -i_{\xi^k_Y}d\Theta^n_{\tilde{C}} = i_{\xi^k_Y}(\Pi_i\omega^i \wedge \eta) - \omega^1_C(\xi_Y) - \sum_{(j,\nu)} F^{\nu}_j \omega^j_{\nu}(\xi^{k+1}_Y)\eta + Con$. Let now $s : X \to Y$ be a solution of the balance system $\mathcal{B}_{\mathcal{C}}$. Taking pullback of the last equality with respect to the $j^{k+1}s$ we see that the second and last terms in the right side vanishes and we get

$$\begin{aligned} j^k(s)^* di_{\xi^k_Y}\Theta^n_{\tilde{C}} &= dj^k(s)^* i_{\xi^k_Y}\Theta^n_{\tilde{C}} \\ &= (\Pi_i\xi^i_Y) \circ j^k(s)\eta - \left[\sum_{(j,\nu)} F^{\nu}_j \omega^j_{\nu}(\xi^{k+1}_Y)\right] \circ j^{k+1}(s)\eta. \end{aligned} \tag{18}$$

Recall that $J^{\tilde{C}}(z)(\xi^k_Y) = i_{\xi^k_Y}\Theta^n_{\tilde{C}}(z)$ is the conventional **multimomentum mapping**, see Ref 2. Obtained formula delivers the proof of the first statement of the next Theorem. Second statement follows from the first.

Theorem 5.1 (Noether Theorem (general)). *Let $\mathcal{C}$ be a constitutive relation defined on a k-jet bundle $J^k(\pi)$ and $\tilde{\mathcal{C}}$ - its lifted CCR. Let a Lie group $G \subset Sym(\tilde{C}) \subset Aut(\pi)$ be a* geometrical symmetry group of the flux part $\Theta^n_{\tilde{C}}$ *of CR $\tilde{\mathcal{C}}$.*

(a) Then for all $\xi \in \mathfrak{g}$ and for all solutions $s \in Sol(X, \mathcal{B}_C)$ of the balance system ★,

$$d[J^{\tilde{C}}(j^k(s)(x)(\xi^k_Y)] = (\Pi_i \xi^i_Y) \circ j^k(s) + \left[\sum_{(j,\nu)} F^\nu_j \omega^j_\nu(\xi^{k+1}_Y)\right] \circ j^{k+1}(s). \quad (19)$$

(b) If $\mathfrak{g}_Y \subset \mathcal{X}_C(Y)$ on Y, last term in the right side drops out.

In the special case of semi-Lagrangian constitutive relation the formulation of Noether Theorem simplifies.

Theorem 5.2 (Noether Theorem for $C_{L,\Pi}, k = 1$). *Let $\hat{C}_{L,\Pi}$ be a semi-Lagrangian CCR with*

$$\Theta_{\hat{C}} = (L - z^i_\mu L_{z^i_\mu})\eta + \sum_{(i,\mu)} L_{z^i_\mu} dy^i \wedge \eta_\mu + (\Pi_i - L_{,y^i}) dy^i \wedge \eta.$$

Then, if $\xi \in \mathcal{X}(Y)$ is an infinitesimal variational symmetry of the CR $\hat{C}$, then for all $s \in Sol(\mathcal{C})$ of the system ★

$$d[(j^1(s))^* J^{\hat{C}}(z)(\xi^1)] = (\omega^i(\xi^1)\Pi_i) \circ j^1(s))^* \eta.$$

Consider now the case where G is the symmetry group of both components of Poincare-Cartan form, i.e. where the second of conditions (17) is fulfilled as well.

Proposition 5.1. *Let G is the symmetry Lie group of of a regular constitutive relation $\tilde{C}$. Then*

(a) For all sections s, $di_{\xi^k_Y}\Theta^{n+1}_{\tilde{C}} = Cont \Rightarrow dj^{k+1}(s)^* i_{\xi^k_Y}\Theta^{n+1}_{\tilde{C}} = 0$,
(b) Form $Q = \pi^*_{(k+1)k} i_{\xi^k_Y}\Theta^{n+1}_{\tilde{C}}$ defines the class of cohomology $[Q]$ in the complex $(\bigwedge^*(J^\infty(\pi)), d_h)$,
(c) If the class $[Q]$ is zero, there exists a $\mathfrak{g}^*$-valued (n+1)-form $\Phi_C(\xi^k_Y)$ ($\mathfrak{g}$- potential of the source Θ^{n+1}_C*) (linear by ξ) such that $Q = d\Phi_C(\xi^k_Y) + Con$. In this case locally (and in a topologically trivial domain, globally)*

$$j^k(s)^* i_{\xi^k_Y}\Theta^{n+1}_{\tilde{C}} = dj^2(s)^*\Phi_C(\xi^k_Y),$$

(d) *In the last case in the conditions of Noether Theorem for all solutions* $s \in Sol(\mathcal{B}_C)$ *and all* $\xi \in \mathfrak{g}$ *the following* conservation law *holds*

$$d[J^{\tilde{C}}(j^k(s)(x))(\xi_Y^k) - j^{k+1}(s)^*\Phi_C(\xi_Y^k)] = 0. \tag{20}$$

Proof. Let both conditions (13.1) are fulfilled i.e. G is symmetry group of *flux **and** source terms* of a CCR $\tilde{C}$. Then,

$$\begin{aligned} d(\Pi_i\omega^i \wedge \eta) &= (d_\mu \Pi_i dx^u + Con) \wedge \omega^i \wedge \eta + \Pi_i d\omega^i \wedge \eta + \Pi_i \omega^i \wedge d\eta \\ &= Con \wedge \eta^i \wedge \eta + \Pi_i(dz^i_\mu \wedge dx^\mu) \wedge \eta = 2Con. \end{aligned} \tag{21}$$

Acting by $i_{\xi_Y^k}$ we see that second condition in () takes the form $di_{\xi_Y^k}\Theta_C^{n+1} = Con$ i.e. the form in the left side is the contact form.

Using the decomposition $d = d_h + d_v$ of the differential as the sum of *vertical and horizontal differentials* (see) we see that the (n+1)-form $Q = \pi^*_{(k+1)k} i_{\xi_Y^k} \Theta_{\tilde{C}}^{n+1} = i_{\xi_Y^{k+1}} \pi^*_{(k+1)k} \Theta_{\tilde{C}}^{n+1}$ has *zero horizontal differential*: $d_h Q = 0$. Thus, it defines the **class of cohomology** $[Q]$ in the complex $(\bigwedge^*(J_p^\infty(\pi)), d_h)$. If the class $[Q]$ is zero, then $Q = d_h\Phi + Con$ is the sum of horizontal differential of a form Φ and a contact form and, therefore is also the sum of a differential $d\Phi$ and the contact form $Q = d\Phi + Con$.

Applying now the pullback by $j^{k+1}(s)$ (or $j^k(s)$ where appropriate) we finish the proof. □

Example 5.1 (Energy-Momentum Balance Law). Let ν be a connection in the bundle $\pi : Y \to X$. Let $\xi_\mu = \partial_{x^\mu} + \Gamma^i_\mu \partial_{y^i}$ be the horizontal lift of ∂_{x^μ} in Y. Let $\xi^k_\mu = \partial_{x^\mu} + \Gamma^i_\mu \partial_{y^i} + d_\nu \Gamma^i_\mu \partial_{z^i_\nu} + \ldots$ is the flow lift of ξ_μ to $J^k(\pi)$. Consider a constitutive law $\mathcal{C}$ and the corresponding balance system $\mathcal{B}_\mathcal{C}$ and **assume that** $\xi_\mu \in \mathcal{X}_\mathcal{C}(\mathcal{Y})$. Assume that the CR C is ν-homogeneous in the sense that it is invariant with respect to the local flows of vector fields ξ_Y^k, see Sec.4.

The Noether balance law for ξ_μ, $\mu = 1, \ldots, n$ (the energy-momentum balance law) has the form

$$[(F_i^\nu(\Gamma^i_\mu + z^i_\mu) \circ j^k(s) - \delta^\nu_\mu F_i^\sigma s^i_{,\sigma}]_{,x^\nu} = [\Pi_i(\Gamma^i_\mu - z^i_\mu)] \circ j^k(s). \tag{22}$$

Thus, the energy-momentum tensor for the CR $\mathcal{C}$ has the form $T^\nu_u = F_i^\nu(\Gamma^i_\mu + z^i_\mu) - \delta^\nu_\mu F_i^\sigma z^i_\sigma$.

Example 5.2 (Pure gauge symmetry transformation). Let $\xi = \xi^i \partial_{y^i}$ be a vertical (pure gauge) symmetry transformation of a constitutive relation $\mathcal{C}$. Then, the Noether balance law corresponding to the vector field ξ has the form

$$[(\xi^i F_i^\mu)(j^1(s))]_{;x^\mu} = (\xi^i \Pi_i)(j^1(s)). \tag{23}$$

6. Conclusion

In this work we've presented a variational theory of system of balance equations, realization of the constitutive relations of such system as a generalized Legendre transformation, invariant form of the balance system and the Noether Theorem associating with the geometrical symmetries of the constitutive relation the corresponding balance law. For the more detailed exposition of these and further results, we refer to the works, Refs.11,12.

References

1. G. Giachetta, L. Mangiarotti and G. Sardanashvily, *New Lagrangian and Hamiltonian Methods in Field Theory* (World Scientific, Singapure, 1997).
2. M.J. Gotay, J. Isenberg and J.E. Marsden, Momentum maps and classical relativistic fields, Part I: Covariant Field Theory, Preprint, 1998; arXiv:physics9801019.
3. D. Iglesias-Ponte and A. Wade, Ĉontact Manifolds and generalized complex structures, 5 May 2004; arXive:math.DG/0404519.
4. I. Kolar, P. Michor and J. Slovak, *Natural Operations in Differential Geometry* (Springer-Verlag, Berlin, 1996).
5. I. Krasil'shchick and A. Vinogradov, Eds., *Symmetries and Conservative Laws for Differential Equations of Mathematical Physics* (AMS,Providence, 1999).
6. D. Krupka, Some Geometrical Aspects of Variational Problems in fibred Manifolds, *Folia Fac. Sci. Nat. Univ., Purk., Brunensis, Physica* **14**, Brno (1973), Electr transcr. 2001.
7. D. Krupka, Lepagean Forms in Higher Order Variaional Theory, *In: Modern Developments in analytic Mechanics, I Geometrical dynamics* ((S. Benenti, M.Francaviglia and A. Lichnerovich, Eds.) Proc. IUTAM-ISIMM Symp., Torino, It, 1982, Acc. delle Scienze di Torino, Torino, 1983).
8. M. de Leon, J. Marrero and D. Martin de Diego, *A new geometric setting for classical field theories* (Banach Center Publ., Warszawa, 2002).
9. I. Muller, *Thermodynamics* (Pitman Adv. Publ. co., 1985).
10. W. Muschik, C. Papenfuss and H. Ehrentraut, A sketch of continuum thermodynamics, *J. Non-Newtonian Fluid Mechanics* **96** (2001) 255–290.
11. S. Preston, Multisymplectic Theory of Balance Systems and the Entropy Principle, 2006; arXiv:math-ph/0611079v1.
12. S. Preston, Geometrical Theory of Balance Systems and the Entropy Principle, *In: Proceedings of GCM7* (Lancaster, UK, Journal of Physics: Conference Series, 62, 2007) 102–154.

Differential Geometry and its Applications
Proc. Conf., in Honour of Leonhard Euler, Olomouc, August 2007

Constraint Lepage one forms in higher order mechanics with nonholonomic constraints on the dynamical space

M. Swaczyna

Department of Mathematics, Faculty of Science, University of Ostrava, 30.dubna 22, 701 03, Ostrava, Czech Republic
E-mail: Martin.Swaczyna@osu.cz

A geometric setting for higher order mechanical systems subjected to nonholonomic constraints of an arbitrary order has been proposed in Ref. 1. In this paper we study the simplest case, when nonholonomic constraints are of order $s-1$, $s \geq 2$, i.e. are given on the dynamical space $J^{s-1}Y$, where the unconstrained dynamics proceeds. Analogously as in the case of first order constraints we introduce a constraint Lepage form, which provides the invariant decomposition of the Lie derivative into a "constraint" Euler–Lagrange term and a boundary term. Coordinate expression of the constraint Lepage one form associated with a nonholonomic constraint structure of order $s-1$ is presented, corresponding constraint Euler–Lagrange equations for "constraint extremals" are obtained.

Keywords Lagrangian system of order $s-1$, system of k nonholonomic constraints of order $s-1$, canonical distribution, constraint ideal, nonholonomic constraint structure, constraint Lepage one form, constraint first variational formula.

MS classification: 70H03, 70H50, 70F25, 70H30.

1. Introduction

In last decades the theory of nonholonomic constraints of the first order has been intensively studied by means of modern methods of differential geometry, global analysis and calculus of variations by many authors, e.g. Refs. 2–8.

On the contrary, there are only a few works generalizing the subject to higher order. In particular, some results on first order mechanical systems with second order, and higher order constraints were obtained by classical methods in Refs. 9–11. A geometric setting for r-th order Lagrangian system with constraints of order $2r-1$ on tangent bundles has been proposed in Ref. 12. A geometric theory for higher order mechanical systems subjected to nonholonomic constraints of an arbitrary order was developed in Ref. 1, where mechanical systems are considered in general, without restricting to Lagrangian systems, or regular systems. Consequently, one can obtain geometric characterization of first order mechanical systems subject to higher order constraints, higher order Lagrangian systems subject to constraints of an arbitrary order, etc., as particular cases of the general theory.

In this paper we study the simplest case, when nonholonomic constraints are given on the dynamical space $J^{s-1}Y$, where the unconstrained dynamics of Lagrangian systems proceeds, i.e. constraint conditions depend on time, positions, velocities, accelerations and higher derivatives up to order $s-1$. The basis for the present approach are papers,[1,6] where a geometric setting for the first order case was developed, and which is suitable for generalization to an arbitrary order.

The aim of this contribution is to formulate a *variational principle* for nonholonomic mechanical systems of order $s-1$ subjected to constraints of order $s-1$. For this purpose we define *constraint Lepage one form*, we present the corresponding *constraint first variational formula* and finally we obtain *constraint Euler–Lagrange equations* for "constraint extremals."

The paper is based on higher order generalization of the variational principle for nonholonomic mechanical systems of the first order (subjected to constraints of the first order) formulated in Ref. 14.

2. Higher order Lagrangian systems on fibered manifolds

Throughout the paper we will consider higher order mechanics on jet prolongations of fibered manifolds over one dimensional bases. Let $\pi\colon Y \to X$ be a fibered manifold with $\dim X = 1$, $\dim Y = m+1$, its jet prolongations $\pi_r\colon J^rY \to X$ and $\pi_{r+1}\colon J^{r+1}Y \to X$ and the jet projections $\pi_{r,0} : J^rY \to Y$ and $\pi_{r+1,r} : J^{r+1}Y \to J^rY$.

Local fibered coordinates on Y are denoted by (t, q^σ), where $1 \leq \sigma \leq m$. We also use the notation $Y = J^0Y$ and $q^\sigma = q^\sigma_0$. The associated coordinates on J^rY are denoted by $(t, q^\sigma_0, q^\sigma_1, q^\sigma_2, \dots, q^\sigma_r)$. In calculations we use either a canonical basis of one forms on J^rY, $(dt,\, dq^\sigma_0,\, dq^\sigma_1,\, dq^\sigma_2, \dots, dq^\sigma_r)$, or a basis adapted to the contact structure, $(dt,\, \omega^\sigma_0,\, \omega^\sigma_1, \dots, \omega^\sigma_{r-1}, dq^\sigma_r)$, where

$$\omega^\sigma_N = dq^\sigma_N - q^\sigma_{N+1}\, dt, \quad 1 \leq \sigma \leq m, \quad 0 \leq N \leq r-1, \tag{1}$$

are canonical contact 1-forms. Instead ω^σ_0 we simply write ω^σ.

If $f(t, q^\sigma, q^\sigma_1, q^\sigma_2, \dots, q^\sigma_r)$ is a function defined on J^rY, by the total derivative of the function f we mean the expression

$$\frac{df}{dt} = \frac{\partial f}{\partial t} + \frac{\partial f}{\partial q^\sigma} q_1^\sigma + \frac{\partial f}{\partial q_1^\sigma} q_2^\sigma + \cdots + \frac{\partial f}{\partial q_r^\sigma} q_{r+1}^\sigma . \tag{2}$$

A (local) section δ of π_r is called *holonomic* if $\delta = J^r\gamma$ for a section γ of π.

On fibered manifolds and their jet prolongations there arise specific geometric objects as vector fields, differential forms, distributions etc., which are adapted to the fibered structure and its jet prolongations. In this paper we will not to recall well known concepts of π_r-vertical, resp. $\pi_{r,0}$- vertical vector fields, horizontal and contact forms and its properties used in the calculus of variations on fibered manifolds. For this purpose we refer to Ref. 13 or Ref. 1.

A 2-form E on J^sY is called a *dynamical form* if it is 1-contact and $\pi_{s,0}$-horizontal. In fibered coordinates a dynamical form E is represented by

$$E = E_\sigma(t, q^\nu, q_1^\nu, \ldots, q_s^\nu)\, \omega^\sigma \wedge dt, \quad 1 \le \sigma, \nu \le m, \tag{3}$$

where components E_σ are *affine in the derivatives of order* s, i.e. are of the form

$$E_\sigma = A_\sigma(t, q^\nu, q_1^\nu, \ldots, q_{s-1}^\nu) + B_{\sigma\rho}(t, q^\nu, q_1^\nu, \ldots, q_{s-1}^\nu)\, q_s^\rho . \tag{4}$$

A section γ of π is called a *path* of E if $E \circ J^s\gamma = 0$. In fibered coordinates this equation represents a system of m ordinary differential equations of order s for the components $\gamma^\nu(t)$, $1 \le \nu \le m$, of a section γ.

Let E is a dynamical form on J^sY with components affine in the derivatives of order s. A 2-form α defined on an open set W in $J^{s-1}Y$ is called Lepage 2-form associated to the dynamical form E if $p_1\alpha = E$. Two Lepage 2-forms α, α' with the same domain of definition W will be called *equivalent* if $p_1\alpha = p_1\alpha'$. Immediately one gets for E given by (3) and (4), the following *equivalence class* of associated Lepage 2-forms on $J^{s-1}Y$

$$[\alpha] = (A_\sigma\, dt + B_{\sigma\rho}\, dq_{s-1}^\rho) \wedge \omega^\sigma + F, \tag{5}$$

where F is an arbitrary 2-contact 2-form on $J^{s-1}Y$. This equivalence class of order $s-1$ associated with a dynamical form E is then called a *mechanical system of order* $s-1$ on a fibered manifold π. The manifold $J^{s-1}Y$ is called the *dynamical space* for $[\alpha]$.

Consider now the particular case when a mechanical system is defined by a dynamical form E which is the Euler-Lagrange form of a Lagrangian. If λ is a Lagrangian of order r (i.e. defined on an open set in J^rY), we denote by θ_λ its *Lepage equivalent* or *Cartan form.* In fibered coordinates where $\lambda = L(t, q^\sigma, q_1^\sigma, q_2^\sigma, \ldots, q_r^\sigma)dt$ we have

$$\theta_\lambda = L(t, q^\sigma, q_1^\sigma, q_2^\sigma, \ldots, q_r^\sigma)dt + \sum_{N=0}^{r-1} \sum_{A=0}^{r-1-N} (-1)^A \frac{d^A}{dt^A} \left(\frac{\partial L}{\partial q_{N+1+A}^\sigma} \right) \omega_N^\sigma. \quad (6)$$

Recall that Lepage equivalent of a Lagrangian of order r is a 1-form on $J^{s-1}Y$, which is in general of order $2r-1$, (i.e. $s-1 \leq 2r-1$), and which is horizontal with respect to the projection $\pi_{s-1,r-1}$.

If θ_λ is Lepage equivalent of Lagrangian λ, then by relation $E_\lambda = p_1 d\theta_\lambda$ is defined *Euler-Lagrange form of the Lagrangian* λ. In fibered coordinates $E_\lambda = E_\sigma(L)\omega^\sigma \wedge dt$, $1 \leq \sigma \leq m$, where functions $E_\sigma(L)$

$$E_\sigma(L) = \sum_{A=0}^{r} (-1)^A \frac{d^A}{dt^A} \left(\frac{\partial L}{\partial q_A^\sigma} \right), \quad 1 \leq \sigma \leq m, \quad (7)$$

are called *Euler-Lagrange expressions.*

The equivalence class of Lepage 2-forms (of order $s-1$) associated to the Euler-Lagrange form E_λ

$$[\alpha]_\lambda = d\theta_\lambda + F, \quad (8)$$

where F is an arbitrary 2-contact 2-forms on $J^{s-1}Y$ is called a *Lagrangian system of order* $s-1$. The corresponding Euler–Lagrange equations $E_\lambda \circ J^s\gamma = 0$ can be written in an intrinsic form as follows

$$J^{s-1}\gamma^* i_\xi d\theta_\lambda = 0, \quad (9)$$

where ξ is a π_{s-1}-vertical vector field on $J^{s-1}Y$, or quite equivalently in the form

$$J^{s-1}\gamma^* i_\xi \alpha = 0, \quad (10)$$

where α is any representative of the class $[\alpha]_\lambda$.

3. Nonholonomic constraint structure on dynamical space

Let $s \geq 2$, and $1 \leq k \leq m-1$. Recall that the fibered manifold $J^{s-1}Y$ is the dynamical space for unconstrained mechanical (Lagrangian) systems of order $s-1$. By a *constraint submanifold* in $J^{s-1}Y$ we mean a fibered submanifold $\pi_{s-1,s-2}|_Q : Q \to J^{s-2}Y$ of the fibered manifold $\pi_{s-1,s-2} : J^{s-1}Y \to J^{s-2}Y$. We denote by ι the canonical embedding of Q into $J^{s-1}Y$, and suppose $\operatorname{codim} Q = k$ (see Ref. 1). Locally, Q can be given by equations

$$f^i(t, q^\sigma, q_1^\sigma, \ldots, q_{s-1}^\sigma) = 0, \quad 1 \leq i \leq k, \quad (11)$$

where

$$\operatorname{rank}\left(\frac{\partial f^i}{\partial q^\sigma_{s-1}}\right) = k, \tag{12}$$

or, equivalently in an explicit form

$$q^{m-k+i}_{s-1} = g^i(t, q^\sigma, q^\sigma_1, \dots, q^\sigma_{s-2}, q^1_{s-1}, \dots, q^{m-k}_{s-1}), \quad 1 \leq i \leq k, \tag{13}$$

called a *system of k nonholonomic constraints of order* $s-1$.

Remark 3.1. Quite analogously one can consider constraints of an arbitrary order r with $r \geq 0$, possible different from $s-1$, however in this paper we consider nonholonomic constraints ($r > 0$), which are given on the dynamical space $J^{s-1}Y$, where the unconstrained dynamics proceeds.

Let $1 \leq p \leq s-1$. A nonholonomic constraint in $J^{s-1}Y$ given by the equations (11) will be called *semiholonomic of degree p* if

$$f^i = \frac{d^p u^i}{dt^p} \tag{14}$$

for some functions u^i. $1 \leq i \leq k$. In fact, any semiholonomic constraint represents a nonholonomic constraint of any lower order.

Remark 3.2. We will deal with nonholonomic constraints, which are really of order $s-1$, it means that considered constraints are not semiholonomic constraints.

The presence of a constraint submanifold in $J^{s-1}Y$ gives rise to a concept of a *constrained section* as a local section $\bar{\delta}$ of the fibered manifold π_{s-1} such that for every $x \in \operatorname{dom}\bar{\delta} : \bar{\delta}(x) \in Q$, and a *Q-admissible* section as a section $\bar{\gamma}$ of the fibered manifold π such that $J^{s-1}\bar{\gamma}(x) \in Q$ for every $x \in \operatorname{dom}\bar{\gamma}$. The set of all Q-admissible sections $\bar{\gamma}$ of π will be denoted by $\bar{\Gamma}^Q(\pi)$.

The submanifold Q is naturally endowed with a distribution, called the *canonical distribution*,[1] or *Chetaev bundle of order* $s-1$, and denoted by C. It is generated by the following vector fields

$$\begin{aligned} \zeta_{(0)} &= \frac{\partial}{\partial t} + \sum_{\sigma=1}^{m} q^\sigma_1 \frac{\partial}{\partial q^\sigma}, & \zeta_{(\sigma p)} &= \frac{\partial}{\partial q^\sigma_p}, \quad 1 \leq p \leq s-2, \\ \zeta_{(l)} &= \frac{\partial}{\partial q^l} + \sum_{i=1}^{k} \frac{\partial g^i}{\partial q^l_{s-1}} \frac{\partial}{\partial q^{m-k+i}}, & \tilde{\zeta}_{(l)} &= \frac{\partial}{\partial q^l_{s-1}} \quad 1 \leq l \leq m-k. \end{aligned} \tag{15}$$

For the annihilator of C we have $C^0 = \mathrm{span}\{\varphi^i, 1 \le i \le k\}$, where in fibered coordinates

$$\varphi^i = -\sum_{l=1}^{m-k} \frac{\partial g^i}{\partial q^l_{s-1}}\, \omega^l + \omega^{m-k+i}, \quad 1 \le i \le k, \tag{16}$$

for the case, when order $s-1 \ge 2$, and

$$\varphi^i = -\sum_{l=1}^{m-k} \frac{\partial g^i}{\partial q^l_1}\, \omega^l + \iota^*\omega^{m-k+i} = -\sum_{l=1}^{m-k} \frac{\partial g^i}{\partial q^l_1}\, \omega^l + dq^{m-k+i} - g^i dt \tag{17}$$

in the case of first order constraints (i.e. $s-1=1$). Forms φ^i are called *canonical constraint* 1*-forms.* The ideal in the exterior algebra of forms on Q generated by the annihilator of C is called the *constraint ideal*, and denoted by $I(C^0)$, or simply I; its elements are called *constraint forms.* The pair (Q, C) is then called a *(nonholonomic) constraint structure of order* $s-1$ on the fibered manifold π (see Ref. 1).

Remark 3.3. Notice that the holonomic integral sections of the canonical distribution C coincide with the holonomic constrained sections of π_{s-1}, i.e. have the form $\bar{\delta} = J^{s-1}\bar{\gamma}$, where $\bar{\gamma}$ is a Q-admissible section of π.

Let $\tilde{Q}$ be the *lift* of Q to J^sY, i.e. the manifold of all points $J^s_x\gamma \in J^sY$ such that $J^{s-1}_x\gamma \in Q$. If Q is given by (13) then equations of the submanifold $\tilde{Q}$ in J^sY are

$$q^{m-k+i}_{s-1} = g^i(t, q^\sigma, q^\sigma_1, \ldots, q^\sigma_{s-2}, q^1_{s-1}, \ldots, q^{m-k}_{s-1}), \quad 1 \le i \le m-k$$
$$q^{m-k+i}_s = \frac{\partial g^i}{\partial t} + \frac{\partial g^i}{\partial q^\sigma} q^\sigma_1 + \cdots + \frac{\partial g^i}{\partial q^l_{s-2}} q^l_{s-1} + \frac{\partial g^i}{\partial q^{m-k+j}_{s-2}} g^j + \frac{\partial g^i}{\partial q^l_{s-1}} q^l_s.$$

We denote by $\rho : \tilde{Q} \to Q$ the corresponding jet projection (i.e. $\rho = \pi_{s,s-1}|_{\tilde{Q}}$). The distribution $\tilde{C}$ on $\tilde{Q}$, such that for every $y \in \tilde{Q}$

$$T_y\rho\,(\tilde{C}(y)) = C(\rho(y)) \tag{18}$$

is called the *lift of the canonical distribution* C. Since $\mathrm{rank}\,\tilde{C} = (s+1)\cdot m + 1 - 3k$, and $\dim\tilde{Q} = (s+1)\cdot m + 1 - 2k$ one has $\mathrm{corank}\,\tilde{C} = \mathrm{corank}\,C = k$. The annihilator of $\tilde{C}$ is locally spanned by 1-forms $\tilde{\varphi}^i = \rho^*\varphi^i$, $1 \le i \le k$. But since canonical constraint 1-forms are $\pi_{s-1,0}$-horizontal, $\rho^*\varphi^i = \varphi^i$, and thus $\tilde{C}^0 = C^0 = \mathrm{span}\,\{\varphi^i, 1 \le i \le k\}$. We denote by $\tilde{I}$ the ideal on $\tilde{Q}$, generated by $\tilde{C}^0$.

4. Constraint horizontal and contact forms on the constraint manifold

Constraint horizontal and contact forms on constraint manifolds defined in Ref. 15 for the case of first order constraints, can be naturally generalized to an arbitrary order. Let us recall some basic facts on the constraint calculus.

Let Q be a constraint submanifold in $J^{s-1}Y$, $\tilde{Q}$ its lift to J^sY. We denote by $\Lambda^q(Q)$, resp. $\Lambda^q(\tilde{Q})$ the module of differential q-forms on Q, resp. $\tilde{Q}$. Canonical basis of 1-forms on Q is $(dt, dq_0^\sigma, dq_1^\sigma, \dots, dq_{s-2}^\sigma, dq_{s-1}^l)$. In calculations we frequently use a basis adapted to the contact structure, $(dt, \iota^*\omega_N^\sigma, dq_{s-1}^l)$, where $1 \leq \sigma \leq m$, $0 \leq N \leq s-2$, and $1 \leq i \leq k$, and where ι is the canonical embedding of Q into $J^{s-1}Y$. Notice, that contact 1-forms $\iota^*\omega_N^\sigma$, on Q can be assorted into three kinds

$$
\begin{aligned}
\iota^*\omega_N^\sigma &= \omega_N^\sigma = dq_N^\sigma - q_{N+1}^\sigma\, dt, \quad 1 \leq \sigma \leq m, \quad 0 \leq N \leq s-3, \\
\iota^*\omega_{s-2}^l &= \omega_{s-2}^l = dq_{s-2}^l - q_{s-1}^l\, dt, \quad 1 \leq l \leq m-k, \\
\iota^*\omega_{s-2}^{m-k+i} &= dq_{s-2}^{m-k+i} - g^i\, dt, \quad 1 \leq i \leq k.
\end{aligned}
$$

The concepts of vertical vector fields, horizontal and contact forms directly transfer to forms on Q. Mappings h and p for forms on Q are defined in a similar way as in the unconstrained case, making use of the projection $\rho : \tilde{Q} \to Q$. For a form η on Q, $h\eta$ and $p\eta$ are defined on $\tilde{Q}$. If I is the constraint ideal, we denote by $\Lambda^q(I)$ the submodule of constraint q-forms. Similar notations are used if $\tilde{I}$, the constraint ideal on $\tilde{Q}$ is considered.

Constraint ideal I enables the construction of equivalence classes $[\eta]_{\Lambda^q(I)}$, resp. $[\eta]_{\Lambda^q(\tilde{I})}$ of q-forms modulo constraint q-forms. The corresponding module operations, as well as the wedge product of classes and contraction of a class by a vector field from the canonical distribution C are defined as usual. Note that the contraction of forms restricted to vector fields belonging to the canonical distribution transfers to a well-defined operation of classes modulo constraint q-forms, for $\Xi \in C$ we put

$$
i_\Xi[\eta]_{\Lambda^q(I)} = [i_\Xi\eta]_{\Lambda^{q-1}(I)}. \tag{19}
$$

A 1-form η on Q is called *constraint horizontal* if $i_\xi\eta = 0$ for every π_{s-1}-vertical vector field $\xi \in C$. Constraint horizontal 1-forms take the form $\eta = \eta_0 + \varphi$, where η_0 is a horizontal form and φ is a constraint 1-form. If h and p are the standard horizontalization and contactization mappings for forms on the submanifold Q, we can introduce *constraint horizontalization* mapping $\bar{h}$ and *constraint contactization* mapping $\bar{p}$ that will be defined on equivalence classes of forms modulo constraint forms. In particular, for 1-forms

$$\bar{h}[\eta]_{\Lambda^1(I)} = [h\eta]_{\Lambda^1(\tilde{I})} = h\eta + \varphi, \tag{20}$$

$$\bar{p}[\eta]_{\Lambda^1(I)} = [p\eta]_{\Lambda^1(\tilde{I})} = p\eta + \varphi, \tag{21}$$

where φ is an arbitrary constraint 1-form defined on $\tilde{Q}$. Recall that every equivalence class of 1-forms on Q admits a unique decomposition

$$\rho^*[\eta]_{\Lambda^1(I)} = \bar{h}[\eta]_{\Lambda^1(I)} + \bar{p}[\eta]_{\Lambda^1(I)}. \tag{22}$$

In particular, for $\eta = df$, where f is an arbitrary function on Q, the formula (22) gives a unique decomposition into a constraint-horizontal differential $\bar{h}[df]_{\Lambda^1(I)}$ and constraint-contact differential $\bar{p}[df]_{\Lambda^1(I)}$. It holds

$$\bar{h}[df]_{\Lambda^1(I)} = \frac{d_C\, f}{dt}\, dt + \varphi, \tag{23}$$

$$\begin{aligned} \bar{p}[df]_{\Lambda^1(I)} = {} & \sum_{l=1}^{m-k} \frac{\partial_C\, f}{\partial q^l}\, \omega^l + \sum_{N=1}^{s-3} \sum_{\sigma=1}^{m} \frac{\partial f}{\partial q^\sigma_N}\, \omega^\sigma_N + \sum_{l=1}^{m-k} \frac{\partial f}{\partial q^l_{s-2}}\, \omega^l_{s-2} \\ & + \sum_{i=1}^{k} \frac{\partial f}{\partial q^{m-k+i}_{s-2}}\, (\iota^* \omega^{m-k+i}_{s-2}) + \sum_{l=1}^{m-k} \frac{\partial f}{\partial q^l_{s-1}}\, \omega^l_{s-1} + \varphi, \end{aligned} \tag{24}$$

where φ runs over constraint 1-forms on $\tilde{Q}$, and the operators

$$\begin{aligned} \frac{d_C\, f}{dt} = {} & \frac{\partial f}{\partial t} + \sum_{N=0}^{s-3} \sum_{\sigma=1}^{m} \frac{\partial f}{\partial q^\sigma_N}\, q^\sigma_{N+1} \\ & + \sum_{l=1}^{m-k} \frac{\partial f}{\partial q^l_{s-2}}\, q^l_{s-1} + \sum_{j=1}^{k} \frac{\partial f}{\partial q^{m-k+j}_{s-2}}\, g^j + \frac{\partial f}{\partial q^l_{s-1}}\, q^l_s, \\ \frac{\partial_C\, f}{\partial q^l} = {} & \frac{\partial f}{\partial q^l} + \sum_{i=1}^{k} \frac{\partial f}{\partial q^{m-k+i}}\, \frac{\partial g^i}{\partial q^l_{s-1}}, \end{aligned} \tag{25}$$

are called *constraint total derivative* and *constraint partial derivative* respectively.

If p_1 and p_2 are standard 1-contactization and 2-contactization mappings defined for 2-forms on Q, we can introduce *constraint* 1*-contactization* mapping $\bar{p}_1$ and *constraint* 2*-contactization* mapping $\bar{p}_2$ that will be defined on equivalence classes of 2-forms modulo constraint 2-forms, as follows

$$\bar{p}_1[\eta]_{\Lambda^2(I)} = [p_1\eta]_{\Lambda^2(\tilde{I})} = p_1\eta + \varphi, \tag{26}$$

$$\bar{p}_2[\eta]_{\Lambda^2(I)} = [p_2\eta]_{\Lambda^2(\tilde{I})} = p_2\eta + \varphi, \tag{27}$$

where φ runs over constraint 2-forms on $\tilde{Q}$. Every class of 2-forms on Q admits a decomposition

$$\rho^*[\eta]_{\Lambda^2(I)} = \bar{p}_1[\eta]_{\Lambda^2(I)} + \bar{p}_2[\eta]_{\Lambda^2(I)}. \tag{28}$$

Consequently, every class of the exterior derivative of a 1-form η is decomposable in the form :

$$[d(\rho^*\eta)]_{\Lambda^2(\tilde{I})} = [\rho^*(d\eta)]_{\Lambda^2(\tilde{I})} = \bar{p}_1[d\eta]_{\Lambda^2(I)} + \bar{p}_2[d\eta]_{\Lambda^2(I)}. \tag{29}$$

5. Constraint Lepage one form associated with the constraint structure on dynamical space $J^{s-1}Y$

This section is based on higher order generalization of the variational principle for nonholonomic mechanical systems of the first order (subject to constraints of the first order) formulated in our paper.[14]

Consider a 1-form η on the constraint submanifold $Q \subset J^{s-1}Y$. Let Ω be a piece of X, i.e. a compact submanifold of X with boundary $\partial\Omega$ (without loss of generality we can assume $\Omega = [a,b] \subset \mathbb{R}$). Denote by $\bar{\Gamma}^Q_\Omega(\pi)$ the set of all Q-admissible sections $\bar{\gamma}$ of π such that $\mathrm{dom}\,\bar{\gamma} \supset \Omega$. The function

$$\bar{\Gamma}^Q_\Omega(\pi) \ni \bar{\gamma} \longmapsto \int_\Omega J^{s-1}\bar{\gamma}^*\eta \in \mathbb{R}. \tag{30}$$

is called the *action function of the* 1*-form* η *over* Ω.

Remark 5.1. Notice that the action function does not change if one takes instead of η the 1-form $\eta+\varphi$, where φ is an arbitrary constraint 1-form, since $J^{s-1}\bar{\gamma}$ are integral sections of the canonical distribution C. This enables one to define the *action function of the class* $[\eta]_{\Lambda^1(I)}$

$$\bar{\Gamma}^Q_\Omega(\pi) \ni \bar{\gamma} \longmapsto \int_\Omega J^{s-1}\bar{\gamma}^*[\eta]_{\Lambda^1(I)} \in \mathbb{R}. \tag{31}$$

Let Ξ be a π_{s-1}-projectable vector field on Q belonging to the canonical distribution C, and ξ its π_{s-1}-projection. Denote ϕ_u (resp. ϕ_{0u}) the local one-parameter group of Ξ (resp. ξ). Let $\bar{\gamma}$ be a Q-admissible section of π. We define the *constraint deformation* of the section $\bar{\gamma}$ induced by Ξ as the one-parameter family of sections $\{\bar{\delta}_u\} = \{\phi_u \circ J^{s-1}\bar{\gamma} \circ \phi_{0u}^{-1}\}$. Notice that deformed sections $\bar{\delta}_u$ remain in the set of constrained sections, but need not be the $(s-1)$-th jet prolongations of Q-admissible sections of π.

Consider the real valued function :

$$(-\varepsilon, \varepsilon) \ni u \longmapsto \int_{\phi_{0u}(\Omega)} \bar{\delta}_u^*\eta \in \mathbb{R}, \tag{32}$$

where ε is a suitable number. This function is differentiable. Differentiating it with respect to u at $u = 0$ we obtain

$$\left(\frac{d}{du}\int_{\phi_{0u}(\Omega)} \bar{\delta}_u^* \eta\right)_{u=0} = \int_\Omega J^{s-1}\bar{\gamma}^* \partial_\Xi \eta. \tag{33}$$

The arising action function of the 1-form $\partial_\Xi \eta$ over Ω will be called *the constraint first variation of the action function* (30), induced by Ξ. In view of Remark 2 the action function of $\partial_\Xi \eta$ does not change if one takes instead of $\partial_\Xi \eta$ a form $\partial_\Xi \eta + \varphi$, where φ is any constraint 1-form. This enables us better to consider the function (33) defined on the class $[\partial_\Xi \eta]_{\Lambda^1(I)}$.

Theorem 5.1. *Let η be a 1-form on the constraint manifold $Q \subset J^{s-1}Y$. The following conditions are equivalent:*

1. $[d\eta]_{\Lambda^2(I)}$ is decomposable in the form :

$$\rho^*[d\eta]_{\Lambda^2(I)} = [E]_{\Lambda^2(\tilde{I})} + [F]_{\Lambda^2(\tilde{I})}, \tag{34}$$

where $[E]_{\Lambda^2(\tilde{I})} = \bar{p}_1[d\eta]_{\Lambda^2(I)}$ is horizontal with respect to the projection $\pi_{s,0}\,|_{\tilde{Q}}\colon \tilde{Q} \to Y$.

2. The constraint 1*-contact part $\bar{p}_1[d\eta]_{\Lambda^2(I)}$ is the equivalence class of a dynamical form on $\tilde{Q}$.*

3. Let Ξ be a π_{s-1}-vertical vector field on Q belonging to the canonical distribution C given by

$$\Xi = \sum_{l=1}^{m-k} a^l \zeta_{(l)} + \sum_{p=1}^{s-2}\sum_{\sigma=1}^{m} b^{\sigma p}\zeta_{(\sigma p)} + \sum_{l=1}^{m-k} c^l \tilde{\zeta}_{(l)},$$

where $\zeta_{(l)}, \zeta_{(\sigma p)}, \tilde{\zeta}_{(l)}$ are π_{s-1}-vertical generators of C. The coordinate representation of the class $\bar{h}\, i_\Xi [d\eta]_{\Lambda^2(I)}$ depends only on components a^l of the vector field Ξ.

4. In every fibered chart,

$$\eta = \bar{L}dt + \sum_{l=1}^{m-k} \eta_l^0\,\omega^l + \sum_{l=1}^{m-k}\sum_{j=1}^{k} \eta_j^0 \frac{\partial g^j}{\partial q_{s-1}^l}\,\omega^l + \sum_{l=1}^{m-k}\sum_{N=1}^{s-2} \eta_l^N\,\omega_N^l$$
$$+ \sum_{j=1}^{k}\sum_{N=1}^{s-3} \eta_j^N\,\omega_N^{m-k+j} + \sum_{j=1}^{k} \bar{M}_j\,(\iota^*\omega_{s-2}^{m-k+j}) + \sum_{j=1}^{k} \eta_j^0\,\varphi^j, \tag{35}$$

where $\bar{L}$ and $\bar{M}_i$ are (local) functions on Q, φ^i are canonical constraint 1*-forms, and components $\eta_l^0, \eta_j^0, \eta_l^N, \eta_j^N$ are determined by the following for-*

mulas

$$\begin{aligned}
\eta_l^0 &= \sum_{A=0}^{s-2}(-1)^A\left(\frac{d_C}{dt}\right)^A\left(\frac{\partial\bar{L}}{\partial q_{A+1}^l}-\bar{M}_i\frac{\partial g^i}{\partial q_{A+1}^l}\right),\\
\eta_j^0 &= \sum_{A=0}^{s-3}(-1)^A\left(\frac{d_C}{dt}\right)^A\left(\frac{\partial\bar{L}}{\partial q_{A+1}^{m-k+j}}-\bar{M}_i\frac{\partial g^i}{\partial q_{A+1}^{m-k+j}}\right)\\
&\quad+(-1)^{(s-2)}\frac{d_C^{(s-2)}}{dt^{(s-2)}}(\bar{M}_j),\\
\eta_l^N &= \sum_{A=0}^{s-2-N}(-1)^A\left(\frac{d_C}{dt}\right)^A\left(\frac{\partial\bar{L}}{\partial q_{A+N+1}^l}-\bar{M}_i\frac{\partial g^i}{\partial q_{A+N+1}^l}\right),\\
\eta_j^N &= \sum_{A=0}^{s-3-N}(-1)^A\left(\frac{d_C}{dt}\right)^A\left(\frac{\partial\bar{L}}{\partial q_{A+N+1}^{m-k+j}}-\bar{M}_i\frac{\partial g^i}{\partial q_{A+N+1}^{m-k+j}}\right)\\
&\quad+(-1)^{(s-2-N)}\frac{d_C^{(s-2-N)}}{dt^{(s-2-N)}}(\bar{M}_j).
\end{aligned} \tag{36}$$

We call a form η satisfying any of the equivalent conditions above a *constraint Lepage one form* associated with the constraint structure (Q, C) of order $s-1$. Notice that this form is in general defined on the $(s-2)$-th jet lift $\tilde{Q}^{(s-2)} \subset J^{2(s-1)-1}Y$ of the constraint submanifold Q, and it is horizontal with respect to the projection $\pi_{2(s-1)-1,s-2} : \tilde{Q}^{(s-2)} \to J^{s-2}Y$.

The (non-invariant) constraint horizontal part of η, i.e. the local form

$$\lambda_C = \bar{L}dt + \sum_{j=1}^{k}\eta_j^0\varphi^j, \tag{37}$$

is then said to be a *constraint Lagrangian* associated with the constraint structure (Q, C) of order $s-1$. Notice that a constraint Lagrangian is determined by $(k+1)$ functions $\bar{L}$, $\bar{M}_i$.

Remark 5.2. In particular, in the case of the first order constraint, i.e. $s=2$, in the coordinate expression of the constraint Lepage one form (35) do not arise components η_j^0, and η_l^N, $1 \le N \le s-2$, and $\eta_j^N, 1 \le N \le s-3$, since corresponding formulas (36) for $s<3$ are not defined. Only for components η_l^0 we obtain

$$\eta_l^0 = \frac{\partial\bar{L}}{\partial q_1^l} - \bar{M}_i\frac{\partial g^i}{\partial q_1^l}.$$

Since in the case of the first order constraint one can express contact 1-forms

$\iota^* \omega_0^{m-k+i}$ by means of canonical constraint 1-forms φ^i (17), we have

$$\iota^* \omega^{m-k+i} = dq^{m-k+i} - g^i \, dt = \varphi^i + \sum_{l=1}^{m-k} \frac{\partial g^i}{\partial q_1^l} \, \omega^l ,$$

and we obtain final coordinate expression of the constraint Lepage one form associated with the first order constraint structure in the form

$$\eta = \bar{L} dt + \frac{\partial \bar{L}}{\partial q_1^l} \, \omega^l + \bar{M}_i \, \varphi^i ,$$

presented in Ref. 14.

6. Constraint Euler-Lagrange equations associated with the constraint structure on dynamical space

The meaning of the constraint Lepage one form is similar as in the unconstrained case; they provide the invariant decomposition of the Lie derivative into a "constraint" Euler–Lagrange term and a boundary term. Namely, if η is a constraint Lepage form, then for every π_{s-1}-vertical vector field on Q belonging to the canonical distribution C,

$$\bar{h}[\partial_\Xi \eta]_{\Lambda^1(I)} = \bar{h} \left(i_\Xi [d\eta]_{\Lambda^2(I)} \right) + \bar{h} \, d \left(i_\Xi [\eta]_{\Lambda^1(I)} \right) , \tag{38}$$

and for every Q-admissible section $\bar{\gamma}$ of π

$$\begin{aligned} \int_a^b J^{s-1}\bar{\gamma}^* [\partial_\Xi \eta]_{\Lambda^1(I)} &= \int_a^b J^{s-1}\bar{\gamma}^* \left(i_\Xi [d\eta]_{\Lambda^2(I)} \right) \\ &+ \; J^{s-1}\bar{\gamma}^* \left(i_\Xi [\eta]_{\Lambda^1(I)} \right) (b) \; - \; J^{s-1}\bar{\gamma}^* \left(i_\Xi [\eta]_{\Lambda^1(I)} \right) (a). \end{aligned} \tag{39}$$

The formulas above are an infinitesimal and integral form of the *constraint first variation formula*, respectively. Indeed, computing the first term on the right-hand side of (38) we obtain

$$\bar{p}_1 [d\eta]_{\Lambda^2(I)} = [E]_{\Lambda^2(\bar{I})} , \tag{40}$$

where

$$E = E_l(\bar{L}, \bar{M}_i) \, \omega^l \wedge dt, \quad 1 \leq l \leq m-k, \tag{41}$$

take the form of the constraint Euler–Lagrange expressions, as follows :

$$\begin{aligned}
E_l = {} & \frac{\partial_C \bar{L}}{\partial q^l} - \sum_{i=1}^{k} \bar{M}_i \frac{\partial_C g^i}{\partial q^l} \\
& + \sum_{A=0}^{s-2} (-1)^{A+1} \left(\frac{d_C}{dt} \right)^{A+1} \left(\frac{\partial \bar{L}}{\partial q^l_{A+1}} - \sum_{i=1}^{k} \bar{M}_i \frac{\partial g^i}{\partial q^l_{A+1}} \right) \\
& + \sum_{j=1}^{k} \sum_{A=0}^{s-3} (-1)^{A+1} \left(\frac{d_C}{dt} \right)^{A+1} \left(\frac{\partial \bar{L}}{\partial q^{m-k+j}_{A+1}} - \sum_{i=1}^{k} \bar{M}_i \frac{\partial g^i}{\partial q^{m-k+j}_{A+1}} \right) \frac{\partial g^j}{\partial q^l_{s-1}} \\
& + \sum_{i=1}^{k} (-1)^{s-1} \frac{d_C^{s-1}(\bar{M}_i)}{dt^{s-1}} \frac{\partial g^i}{\partial q^l_{s-1}},
\end{aligned} \tag{42}$$

where operators d_C/dt and $\partial_C/\partial q^l$ are defined by (25).

Remark 6.1. Evidently, in the case of the first order constraint (i.e. $s = 2$), the constraint Euler–Lagrange expressions reduce to the form

$$E_l = \frac{\partial_C \bar{L}}{\partial q^l} - \sum_{i=1}^{k} \bar{M}_i \frac{\partial_C g^i}{\partial q^l_1} - \frac{d_C}{dt} \left(\frac{\partial \bar{L}}{\partial q^l_1} \right) + \sum_{i=1}^{k} \bar{M}_i \frac{d_C}{dt} \left(\frac{\partial g^i}{\partial q^l_1} \right).$$

In the analogy with the unconstrained variational calculus we can now define the concept of a *constraint extremal of* η *on* Ω as a Q-admissible section such that the constraint first variation over Ω is zero for every deformation vanishing over the boundary of Ω, and the concept of a *constraint extremal of* η as a Q-admissible section $\bar{\gamma}$ of π the restriction of which to any piece $\Omega \subset X$ is a constraint extremal of η on Ω.

Theorem 6.1. *Let* η *be a constraint Lepage one form associated with the constraint structure* (Q, C) *of order* $s - 1$. *Let* $\bar{\gamma}$ *be a* Q*-admissible section of* π. *The following conditions are equivalent:*

1. $\bar{\gamma}$ *is a constraint extremal of* η.

2. For every vector field Ξ *on* Q *belonging to the canonical distribution* C

$$J^{s-1}\bar{\gamma}^* i_\Xi \alpha = 0 \tag{43}$$

for every element $\alpha \in [d\eta]_{\Lambda^2(I)}$.

3. The constraint Euler-Lagrange expressions vanish along $J^{2s-2}\bar{\gamma}$, *i.e.*

$$E_l \circ J^{2s-2}\bar{\gamma} = 0, \quad 1 \leq l \leq m - k, \tag{44}$$

where E_l *are given by (42).*

Acknowledgments

Research supported by the grant GAČR 201/06/0922 of the Czech Grant Agency. The author is gratefull to Prof. Olga Krupková for her valuable stimulation and fruitful discussions in course of the present paper.

References

1. O. Krupková, Higher order mechanical systems with constraints, *J. Math. Phys.* **41** (8) (2000) 5304–5324.
2. G. Giachetta, Jet methods in nonholonomic mechanics, *J. Math. Phys.* **33** (1992) 1652–1665.
3. J.F. Cariñena and M.F. Rañada, Lagrangian systems with constraints: a geometric approach to the method of Lagrange multipliers, *J. Phys. A: Math. Gen.* **26** (1993) 1335–1351.
4. W. Sarlet, F. Cantrijn, and D.J. Saunders, A geometrical framework for the study of non-holonomic Lagrangian systems, *J. Phys. A: Math. Gen.* **28** (1995) 3253–3268.
5. W. Sarlet, A direct geometrical construction of the dynamics of non-holonomic Lagrangian systems, *Extracta Mathematicae* **11** (1996) 202–212.
6. O. Krupková, Mechanical systems with nonholonomic constraints, *J. Math. Phys.* **38** (1997) 5098–5126.
7. M. de León, J.C. Marrero and D.M. de Diego, Non-holonomic Lagrangian systems in jet manifolds, *J. Phys. A: Math. Gen.* **30** (1997) 1167–1190.
8. D.J. Saunders, The geometry of nonholonomic Lagrangian systems, *In: Proceedings of the Conference on Differential Geometry and Applications* (Brno, August 1998, Masaryk University, Brno, 1999) 575–579.
9. V. Valcovici, Une extension des liasions non holonomes et des principes variationelles, *Ber. Ver. Schs. Akad. d. Wiss. Leipzig* **102** (1958) 1–39.
10. V.I. Kirgetov, On possible displacements of mechanical systems with second order linear differential constraints, *Prikladnaya mat. i mech.* **23** (1959) 666–671, in Russian.
11. D. Shan, Equations of motion for system subject to second order nonholonomic constraints, *Prikladnaya mat. i mech.* **2** (1973) 349–354, in Russian.
12. M. de León and P.R. Rodrigues, Higher order mechanical systems with constraints, *Int. J. Theor. Phys.* **32** (1992) 1303–1313.
13. O. Krupková, *The Geometry of Ordinary Variational Equations* (Lecture Notes in Mathematics No.1678, Springer, Berlin, 1997).
14. M. Swaczyna, On the nonholonomic variational principle, *In: Global Analysis and Applied Mathematics* (Proc. of the International Workshop on Global Analysis, Ankara, 2004, AIP Conference Proceedings, Vol. 729, Melville, New York, 2004) 297–306.
15. O. Krupková and M. Swaczyna, Horizontal and contact forms on constraint manifolds, *Supplemento ai Rendiconti del Circolo Matematico di Palermo*, Serie II (Suppl. **75**) (2005) 259–267.

Differential Geometry and its Applications
Proc. Conf., in Honour of Leonhard Euler, Olomouc, August 2007

A connection-theoretic approach to the analysis of symmetries of nonholonomic field theories

J. Vankerschaver

Dept. of Mathematical Physics and Astronomy, Ghent University,
Krijgslaan 281, B-9000 Ghent, Belgium
E-mail: Joris.Vankerschaver@UGent.be
Present address: Control and Dynamical Systems, California Institute of Technology,
1200 E California Blvd, MC 107-81, Pasadena CA 91125

For mechanical systems or classical field theories with nonholonomic constraints, the presence of symmetries does not immediately imply the existence of associated conservation laws, *i.e.* Noether's theorem is no longer valid and is replaced instead by an equation describing the evolution of the conserved quantity under the flow of the nonholonomic system. In this paper, we use the De Donder-Weyl formalism for nonholonomic field theories to derive this nonholonomic momentum equation.

Keywords: Nonholonomic constraints, classical field theories, symmetry.

MS classification: 70S05, 37K05, 53D20, 70F25.

1. Introduction

Symmetries play a fundamental role in classical mechanics and field theory. This is epitomized in Noether's first theorem: every continuous symmetry gives rise to a conserved quantity. In the case where nonholonomic constraints are present, however, Noether's theorem is no longer valid, but the analysis of symmetries is still a very powerful tool, an observation which goes back at least to the work of Vierkandt.[11]

For nonholonomic mechanical systems with symmetry, instead of Noether's theorem, there exists an equation describing the evolution of the associated "conserved" quantities under the nonholonomic flow. This equation is know as the *nonholonomic momentum equation* and was derived by Bloch *et al.*[2] for mechanical systems (see also the work of Cantrijn *et al.*[3]).

In this paper, we derive a nonholonomic momentum equation for classical field theories with nonholonomic constraints (theorem 4.1). The main tool in our derivation is the nonholonomic De Donder-Weyl equation, which we recall in section 2. In section 3 we introduce nonholonomic symmetries as sections of a certain bundle of Lie algebras. Section 4 is devoted to the proof of the nonholonomic momentum equation.

This paper is meant as a companion paper to Ref. 9, where proofs of some of the technical lemmata can be found, as well as a number of applications.

2. Lagrangian first-order field theories

2.1. *First-order jet bundles*

Throughout this paper, we will represent fields as sections of a given fibre bundle $\pi : Y \to X$, whose base space is an $(n+1)$-dimensional oriented manifold with volume form η. A typical coordinate system on X is denoted as (x^μ), $\mu = 0, \ldots, n$, and is assumed to be adapted to the volume form in the sense that η can locally be written as

$$\eta = \mathrm{d}^{n+1}x := \mathrm{d}x^0 \wedge \cdots \wedge \mathrm{d}x^n.$$

In addition, a typical coordinate system on Y will be assumed to be adapted to the projection π, meaning that, if $(x^\mu; y^a)$ are coordinates on Y $(a = 1, \ldots, m)$, then π is locally given by $\pi(x^\mu, y^a) = (x^\mu)$.

The *first jet bundle* $J^1\pi$ is the manifold of equivalence classes of local sections of π, where two sections are said to be equivalent if their first-order Taylor expansions agree at a point. Elements of $J^1\pi$ are denoted as $j^1_x\phi$; there exists a projection $\pi_{1,0} : J^1\pi \to Y$ defined by $\pi_{1,0}(j^1_x\phi) = \phi(x)$. Furthermore, we define the projection $\pi_1 : J^1\pi \to X$ as the composition $\pi \circ \pi_{1,0}$.

Recall that a *connection* on π_1 is a vector-valued one-form $\mathbf{h}$ on $J^1\pi$ such that $\mathbf{h} \lrcorner\, \sigma = \sigma$ for every semi-basic one-form σ. Here, $\mathbf{h} \lrcorner\, \sigma$ is the contraction of $\mathbf{h}$ with σ; see section A.1 for its definition. In coordinates, a connection can be represented as follows:

$$\mathbf{h} = \mathrm{d}x^\mu \otimes \left(\frac{\partial}{\partial x^\mu} + \Gamma^a_\mu \frac{\partial}{\partial y^a} + \Gamma^a_{\mu\nu} \frac{\partial}{\partial y^a_\nu} \right), \tag{1}$$

For more information about jet bundles and connections, see Ref. 7.

The Cartan form

For the purpose of this paper, a *Lagrangian* is a function L on $J^1\pi$. Without going into any details, we mention here the existence of a distinguished $(n+1)$-form on $J^1\pi$, called the *Cartan form*, which is a central object in the geometric formulation of the field equations. If L is a Lagrangian, then the associated Cartan form has the following coordinate expression:

$$\Theta_L = \frac{\partial L}{\partial y^a_\mu}(\mathrm{d}y^a - y^a_\nu \mathrm{d}x^\nu) \wedge \mathrm{d}^n x_\mu + L\mathrm{d}^{n+1}x.$$

We define the Poincaré-Cartan form Ω_L as $-\mathrm{d}\Theta_L$.

2.2. *Nonholonomic constraints*

A nonholonomic field theory is determined by the specification of three elements (see also Refs. 5,6,8,9):

(a) a (regular) first-order Lagrangian $L : J^1\pi \to \mathbb{R}$;
(b) a constraint submanifold $\mathcal{C} \hookrightarrow J^1\pi$, such that the restriction of the projection $(\pi_{1,0})_{|\mathcal{C}}$ defines a subbundle of $J^1\pi$;
(c) a bundle of reaction forces F, where the elements Φ of F are $(n+1)$-forms defined along $\mathcal{C}$ which satisfy the following requirements:
   (a) Φ is n-horizontal, *i.e.* Φ vanishes when contracted with any two π_1-vertical vector fields;
   (b) Φ is 1-contact, *i.e.* $(j^1\phi)^*\Phi = 0$ for any section ϕ of π.

For the sake of simplicity, we assume that $\mathcal{C}$ is defined by the vanishing of k functionally independent functions φ^α on $J^1\pi$, and that F is globally generated by l generators Φ^κ of the following form:

$$\Phi^\kappa = A^{\kappa\mu}_a(\mathrm{d}y^a - y^a_\nu \mathrm{d}x^\nu) \wedge \mathrm{d}^n x_\mu \quad (\kappa = 1, \ldots, l).$$

In practice, the dimension l of F will be equal to the codimension k of $\mathcal{C}$. There seems to be no a priori reason for supposing that $k = l$. In most cases, however, F will be determined by $\mathcal{C}$ through application of the Chetaev principle (see Refs. 8,9) but this is not necessary for the present treatment.

2.3. *The nonholonomic De Donder-Weyl equation*

The dynamics of a nonholonomic field theory is specified by the *nonholonomic Euler-Lagrange equations*, which are given below in coordinate form:

$$\left[\frac{\partial L}{\partial y^a} - \frac{\mathrm{d}}{\mathrm{d}x^\mu}\frac{\partial L}{\partial y^a_\mu}\right](j^2\phi) = \lambda_{\alpha\kappa}A^{\alpha\kappa}_a(j^1\phi) \quad \text{and} \quad \varphi^\alpha(j^1\phi) = 0. \qquad (2)$$

Here, $\lambda_{\alpha\kappa}$ are unknown Lagrange multipliers, to be determined from the constraints. Note that the terms on the right-hand side represent the reaction forces. A derivation of these equations can be found in Ref. 1.

The dynamics of a field theory can also be approached through the so-called De Donder-Weyl equations, a set of algebraic equations specifying a connection $\mathbf{h}$ on π_1 with the following property: if ψ is an integral section of $\mathbf{h}$, then ψ is holonomic, *i.e.* there exists a section ϕ of π such that $\psi = j^1\phi$, and in addition ϕ satisfies the Euler-Lagrange equations. In Ref. 8, the De Donder-Weyl equations were extended to the case of nonholonomic field theories and take the following form:

$$i_{\mathbf{h}}\Omega_L - n\Omega_L \in \mathcal{I}(F) \quad \text{and} \quad \operatorname{Im}\mathbf{h} \subset T\mathcal{C},$$

where $\mathcal{I}(F)$ is the ideal generated by F. These equations again specify a connection on π_1, whose integral sections are now the solutions of (2).

Remark 2.1. Throughout this paper, all Lagrangians are assumed to be *regular*, in the sense that the associated Hessian is nonsingular. If this is not the case, the correspondence between integral sections of the De Donder-Weyl equation and solutions of the Euler-Lagrange equations is not so straightforward.

3. Nonholonomic symmetries

Let G be a Lie group acting on π by bundle automorphisms, and assume that G leaves invariant L, $\mathcal{C}$ and F, *i.e.* there exist smooth actions $\overline{\Phi}$: $G \times Y \to Y$ and $\underline{\Phi} : G \times X \to X$ such that $\pi(\overline{\Phi}(g, y)) = \underline{\Phi}(g, \pi(y))$ for all $g \in G$ and $y \in Y$. The Lie group G then also acts on $J^1\pi$ by prolonged bundle automorphisms.

We consider first the bundle $\mathfrak{g}^F$ over Y, defined as follows: $\mathfrak{g}^F(y)$ is the linear subspace of $\mathfrak{g}$ consisting of all $\xi \in \mathfrak{g}$ such that

$$j^1\xi_Y(\gamma) \lrcorner F = 0 \quad \text{for all } \gamma \in \mathcal{C} \cap \pi_{1,0}^{-1}(y), \tag{3}$$

where ξ_Y is the infinitesimal generator of the action corresponding to ξ, that is,

$$\xi_Y(y) = \left.\frac{\mathrm{d}}{\mathrm{d}t}\right|_{t=0} \overline{\Phi}_{\exp(t\xi)}(y).$$

Next, we assume that the disjoint union of all $\mathfrak{g}^F(y)$, for all $y \in Y$ can be given the structure of a vector bundle over Y, which we denote by $\mathfrak{g}^F$. Note that to any section $\bar{\xi}$ of $\mathfrak{g}^F$, one can associate a vector field $\tilde{\xi}_Y$ on Y according to the following prescription:

$$\tilde{\xi}_Y(y) = \left[\bar{\xi}(y)\right]_Y(y). \tag{4}$$

Definition 3.1. A *nonholonomic symmetry* is a section $\bar{\xi}$ of $\mathfrak{g}^F$ such that the associated vector field $\tilde{\xi}_Y$ is π-projectable; *i.e.* there exists a vector field $\tilde{\xi}_X$ on X such that $T\pi \circ \tilde{\xi}_Y = \tilde{\xi}_X \circ \pi$.

The following lemma, taken from Ref. 9, will be useful in the proof of the nonholonomic momentum equation.

Lemma 3.1. *Let $\bar{\xi}$ be a section of $\mathfrak{g}^F$. For $y \in Y$, put $\xi := \bar{\xi}(y)$ and consider any $\gamma \in \pi_{1,0}^{-1}(y) \cap \mathcal{C}$. Then there exists a $\pi_{1,0}$-vertical vector $v_\gamma \in T_\gamma J^1\pi$ such that*

$$j^1\tilde{\xi}_Y(\gamma) = j^1\xi_Y(\gamma) + v_\gamma.$$

4. The nonholonomic momentum equation

Theorem 4.1. *Let $\mathbf{h}$ be a solution of the nonholonomic De Donder-Weyl equation and consider an integral section $j^1\phi$ of $\mathbf{h}$. Then for any nonholonomic symmetry $\bar{\xi}$ the associated component of the momentum map $J^{\text{n.h.}}_{\bar{\xi}}$ satisfies the following* nonholonomic momentum equation*:*

$$(j^1\phi)^*(\mathrm{d}J^{\text{n.h.}}_{\bar{\xi}}) = (j^1\phi)^*\left(\mathscr{L}_{j^1\tilde{\xi}_Y}(L\eta)\right).$$

Proof. Note first of all that if $j^1\phi$ is an integral section of $\mathbf{h}$, then the following holds:

$$(j^1\phi)^*(\mathrm{d}J^{\text{n.h.}}_{\bar{\xi}}) = (j^1\phi)^*(\mathrm{d}_{\mathbf{h}}J^{\text{n.h.}}_{\bar{\xi}}).$$

Using lemma A.1, the $\mathbf{h}$-derivative of $J^{\text{n.h.}}$ on the right-hand side can be expanded as follows.

$$\begin{aligned}
\mathrm{d}_{\mathbf{h}}J^{\text{n.h.}}_{\bar{\xi}} &= (i_{\mathbf{h}}\mathrm{d} - \mathrm{d}i_{\mathbf{h}})i_{j^1\tilde{\xi}_Y}\Theta_L \\
&= i_{\mathbf{h}}\mathscr{L}_{j^1\tilde{\xi}_Y}\Theta_L - i_{\mathbf{h}}i_{j^1\tilde{\xi}_Y}\mathrm{d}\Theta_L - \mathrm{d}i_{\mathbf{h}}i_{j^1\tilde{\xi}_Y}\Theta_L \\
&= i_{\mathbf{h}}\mathscr{L}_{j^1\tilde{\xi}_Y}\Theta_L + i_{j^1\tilde{\xi}_Y}i_{\mathbf{h}}\Omega_L + i_{\mathbf{h}(j^1\tilde{\xi}_Y)}\mathrm{d}\Theta_L - \mathrm{d}i_{\mathbf{h}}i_{j^1\tilde{\xi}_Y}\Theta_L.
\end{aligned}$$

Each of these terms will now be treated separately. Bear in mind that we will eventually pull back $\mathrm{d}_{\mathbf{h}}J^{\text{n.h.}}_{\bar{\xi}}$ to X using a prolonged section $j^1\phi$: this will get rid of all contact forms in the above expression, and for that purpose we introduce the following equivalence relation: we say that two forms α and β on $J^1\pi$ are equivalent (denoted by $\alpha \simeq \beta$) if they agree up to a contact form, *i.e.* $\alpha \simeq \beta$ iff $\alpha = \beta + \theta$, where θ is contact. Note that this is equivalent to saying that $(j^1\phi)^*\alpha = (j^1\phi)^*\beta$ for all sections ϕ of π.

Term 1: $i_{\mathbf{h}}\mathscr{L}_{j^1\tilde{\xi}_Y}\Theta_L$. Using lemma A.1, we have

$$i_{\mathbf{h}}\mathscr{L}_{j^1\tilde{\xi}_Y}\Theta_L = \mathscr{L}_{j^1\tilde{\xi}_Y} i_{\mathbf{h}}\Theta_L - i_{[j^1\tilde{\xi}_Y,\mathbf{h}]}\Theta_L.$$

To the first term, lemma A.2 can be applied. The second term vanishes, since $[j^1\tilde{\xi}_Y, \mathbf{h}]$ takes values in $V\pi_{1,0}$ (proposition 4.1) and Θ_L is semi-basic. In conclusion, term 1 is equal to

$$i_{\mathbf{h}}\mathscr{L}_{j^1\tilde{\xi}_Y}\Theta_L = n\mathscr{L}_{j^1\tilde{\xi}_Y}\Theta_L + \mathscr{L}_{j^1\tilde{\xi}_Y}(L\eta).$$

Term 2: $i_{j^1\tilde{\xi}_Y} i_{\mathbf{h}}\Omega_L$. The nonholonomic De Donder-Weyl equations give us

$$i_{j^1\tilde{\xi}_Y} i_{\mathbf{h}}\Omega_L = n\, i_{j^1\tilde{\xi}_Y}\Omega_L + i_{j^1\tilde{\xi}_Y}\zeta \simeq n\, i_{j^1\tilde{\xi}_Y}\Omega_L,$$

since ζ belongs to $\mathcal{I}(F)$, and hence can be written as $\zeta = \sum_\alpha \lambda^\alpha \wedge \Phi_\alpha$, for one-forms λ^α. The contraction of $j^1\tilde{\xi}_Y$ with ζ then becomes

$$i_{j^1\tilde{\xi}_Y}\zeta = \sum_\alpha (i_{j^1\tilde{\xi}_Y}\lambda^\alpha)\Phi_\alpha - \sum_\alpha \lambda^\alpha \wedge (i_{j^1\tilde{\xi}_Y}\Phi_\alpha).$$

The first term is contact, and the second term vanishes since $j^1\tilde{\xi}_Y$ is admissible.

Term 3: $i_{\mathbf{h}(j^1\tilde{\xi}_Y)}\mathrm{d}\Theta_L$. In the case of vertical symmetries, as treated in Ref. 10, this term is automatically zero. In the general case, it is zero up to a contact form; the proof is rather technical. More precisely, we will show the following. If $j^1\phi$ is an integral section of the nonholonomic connection $\mathbf{h}$, then

$$(j^1\phi)^*(i_{\mathbf{h}(j^1\tilde{\xi}_Y)}\mathrm{d}\Theta_L) = 0.$$

Consider any such integral section $j^1\phi$, and fix a point x in X. Put $y = \phi(x)$ and $\gamma = j^1_x\phi$ and define ξ as $\tilde{\xi}(y)$. Then lemma 3.1 allows us to conclude that

$$\mathbf{h}(j^1\tilde{\xi}(\gamma)) \lrcorner\, \mathrm{d}\Theta_L(\gamma) = \mathbf{h}(j^1\xi_Y(\gamma)) \lrcorner\, \mathrm{d}\Theta_L(\gamma).$$

As $j^1\xi_Y$ is projectable (onto ξ_X), the last term is equal to $T_x j^1\phi(\xi_X(x)) \lrcorner\, \mathrm{d}\Theta_L(\gamma)$.

For the pullback of term 3 under $j^1\phi$ we therefore have

$$\left[(j^1\phi)^*(\mathbf{h}(j^1\tilde{\xi}_Y) \lrcorner\, \mathrm{d}\Theta_L)\right]_x = \xi_X(x) \lrcorner\, \left[(j^1\phi)^*\mathrm{d}\Theta_L\right]_x, \tag{5}$$

but the right-hand side is zero since $\mathrm{d}\Theta_L$ is an $(n+2)$-form pulled back to an $(n+1)$-dimensional space.

Term 4: $\mathrm{d}i_{\mathbf{h}}i_{j^1\tilde{\xi}_Y}\Theta_L$. Using lemma A.1, it follows that this term can be rewritten as

$$\mathrm{d}i_{\mathbf{h}}i_{j^1\tilde{\xi}_Y}\Theta_L = \mathrm{d}i_{j^1\tilde{\xi}_Y}i_{\mathbf{h}}\Theta_L - \mathrm{d}i_{\mathbf{h}(j^1\tilde{\xi}_Y)}\Theta_L.$$

When considering the pullback of these terms under $j^1\phi$, a similar reasoning as the one leading to (5) shows that

$$\mathrm{d}(j^1\phi)^*(i_{\mathbf{h}(j^1\tilde{\xi}_Y)}\Theta_L) = \mathrm{d}(\xi_X \lrcorner\, (j^1\phi)^*\Theta_L),$$

and this is in turn equal to $\mathrm{d}(\xi_X \lrcorner\, (j^1\phi)^*(L\eta))$, since $\Theta_L \simeq L\eta$. Using again lemma A.2, we conclude that

$$\begin{aligned}(j^1\phi)^*(\mathrm{d}i_{\mathbf{h}}i_{j^1\tilde{\xi}_Y}\Theta_L) &= n(j^1\phi)^*(\mathrm{d}i_{j^1\tilde{\xi}_Y}\Theta_L) + (j^1\phi)^*(\mathrm{d}i_{j^1\tilde{\xi}_Y}(L\eta) \\ &\quad + \mathrm{d}(\xi_X \lrcorner\, (j^1\phi)^*(L\eta)) = n(j^1\phi)^*(\mathrm{d}i_{j^1\tilde{\xi}_Y}\Theta_L).\end{aligned}$$

Conclusion. Taking the pullback of $\mathrm{d}_{\mathbf{h}}J_{\xi}^{\text{n.h.}}$ under an integral section $j^1\phi$ of $\mathbf{h}$, the foregoing allows us to write

$$(j^1\phi)^*(\mathrm{d}_{\mathbf{h}}J_{\xi}^{\text{n.h.}}) = (j^1\phi)^*(\mathscr{L}_{j^1\tilde{\xi}_Y}(L\eta)).$$

This is the desired form of the nonholonomic momentum equation. □

Acknowledgments

I would like to thank D. Saunders and F. Cantrijn for their suggestions and comments.

I am a Postdoctoral Fellow from the Fund for Scientific Research – Flanders (FWO-Vlaanderen), and a Fulbright Research Scholar at the California Institute of Technology. Additional financial support from the Fonds Professor Wuytack is gratefully acknowledged.

Appendix A. Technical lemmata

The proof of the nonholonomic momentum equation uses a number of rather technical lemmata, which have been collected in this appendix.

Appendix A.1. *The Fröhlicher-Nijenhuis bracket*

The following lemma is a special case of lemma 8.6 in Ref. 4; a direct proof can be found in Ref. 10. Recall that the contraction $i_{\mathbf{h}}\alpha$ of a 1-1 tensor $\mathbf{h}$ with a k-form α is a k-form defined as follows:

$$(i_{\mathbf{h}}\alpha)(v_1, \ldots, v_k) = \sum_{i=1}^{k}(-1)^{i+1}\alpha(\mathbf{h}(v_i), v_1, \ldots, \hat{v}_i, \ldots, v_k).$$

This k-form will sometimes be denoted as $\mathbf{h} \lrcorner \alpha$.

Lemma A.1. *Let X be a vector field on M and $\mathbf{h}$ a vector-valued one-form. Then, for any k-form α on M, the following holds:*

(a) $i_X i_{\mathbf{h}} \alpha = i_{\mathbf{h}} i_X \alpha + i_{\mathbf{h}(X)} \alpha$;
(b) $i_{\mathbf{h}} \mathscr{L}_X \alpha = \mathscr{L}_X i_{\mathbf{h}} \alpha - i_{[X,\mathbf{h}]} \alpha$.

Appendix A.2. *Semi-holonomic connections*

Recall that a connection $\mathbf{h}$ on π_1 is said to be *semi-holonomic* if $i_{\mathbf{h}}\theta = 0$ for all contact forms θ on $J^1\pi$. In coordinates, if $\mathbf{h}$ is locally represented as in (1), then $\mathbf{h}$ is semi-holonomic if $\Gamma^a_\mu = y^a_\mu$.

Lemma A.2. *For each semi-holonomic connection Υ with horizontal projector $\mathbf{h}$, the following holds:*

$$i_{\mathbf{h}} \Theta_L = n\Theta_L + L\eta.$$

Proof. This is Lemma 1 in Ref. 10. □

Proposition 4.1. *Let X be a projectable vector field, and $\mathbf{h}$ a semi-holonomic connection on π_1. Then the Fröhlicher-Nijenhuis bracket $[j^1X, \mathbf{h}]$ is a vector-valued one-form taking values in $V\pi_{1,0}$.*

Proof. Let us write X in coordinates as

$$X = X^\mu(x)\frac{\partial}{\partial x^\mu} + X^a(x,y)\frac{\partial}{\partial y^a},$$

and $\mathbf{h}$ as

$$\mathbf{h} = \mathrm{d}x^\mu \otimes \left(\frac{\partial}{\partial x^\mu} + \Gamma^a_\mu \frac{\partial}{\partial y^a} + \Gamma^a_{\mu\nu}\frac{\partial}{\partial y^a_\nu}\right).$$

Note that Γ^a_μ is equal to y^a_μ since $\mathbf{h}$ is semi-holonomic.

The bracket $[j^1X, \mathbf{h}]$ is by definition just the Lie derivative $\mathscr{L}_{j^1X}\mathbf{h}$. Generally speaking, this vector-valued one-form takes values in $TJ^1\pi$. A straightforward calculation shows that this form has the following expression:

$$\mathscr{L}_{j^1X}\mathbf{h} = \left(\frac{\partial X^\nu}{\partial x^\mu} - \left(\frac{\partial}{\partial x^\mu}\right)^H (X^\nu)\right) \mathrm{d}x^\mu \otimes \frac{\partial}{\partial x^\nu} + \left(\Gamma^a_\nu \frac{\partial X^\nu}{\partial x^\mu} + j^1X(\Gamma^a_\mu) - \left(\frac{\partial}{\partial x^\mu}\right)^H (X^a)\right) \mathrm{d}x^\mu \otimes \frac{\partial}{\partial y^a} + (\ldots)\mathrm{d}x^\mu \otimes \frac{\partial}{\partial y^a_\nu}, \tag{A.1}$$

where we have used the following short-hand notation to denote the horizontal lift (with respect to the connection $\mathbf{h}$) of a vector field on X:

$$\left(\frac{\partial}{\partial x^\mu}\right)^H = \frac{\partial}{\partial x^\mu} + \Gamma^a_\mu \frac{\partial}{\partial y^a} + \Gamma^a_{\mu\nu} \frac{\partial}{\partial y^a_\nu}.$$

Expanding the terms between brackets now shows that the two first terms of (A.1) vanish, meaning that $[j^1X, \mathbf{h}]$ takes values in $V\pi_{1,0}$. □

References

1. E. Binz, M. de León, D. Martín de Diego and D. Socolescu, Nonholonomic Constraints in Classical Field Theories, *Rep. Math. Phys.* **49** (2002) 151–166.
2. A. Bloch, P. Krishnaprasad, J.E. Marsden and R. Murray, Nonholonomic mechanical systems with symmetry, *Arch. Rat. Mech. Anal.* **136** (1) (1996) 21–99.
3. F. Cantrijn, M. de León, J.C. Marrero and D. Martín de Diego, Reduction of constrained systems with symmetries, *J. Math. Phys.* **40** (2) (1999) 795–820.
4. I. Kolář, P. Michor and J. Slovák, *Natural operations in differential geometry* (Springer-Verlag, Berlin, 1993).
5. O. Krupková and P. Volný, Differential equations with constraints in jet bundles: Lagrangian and Hamiltonian systems, *Lobachevskii J. Math.* **23** (2006) 95–150.
6. O. Krupková, Partial differential equations with differential constraints, *J. Diff. Eq.* **220** (2) (2005) 354–395.
7. D.J. Saunders, *The Geometry of Jet Bundles* (London Mathematical Society Lecture Note Series, vol. 142, Cambridge University Press, 1989).
8. J. Vankerschaver, F. Cantrijn, M. de León and D. Martín de Diego, Geometric aspects of nonholonomic field theories, *Rep. Math. Phys.* **56** (3) (2005) 387–411.
9. J. Vankerschaver and D. Martín de Diego, Symmetry aspects of nonholonomic field theories, *J. Phys. A: Math. Theor.* **47** (4) (2008), 35401.
10. J. Vankerschaver, The momentum map for nonholonomic field theories with symmetry, *Int. J. Geom. Meth. Mod. Phys.* **2** (6) (2005) 1029–1041.
11. A. Vierkandt, über gleitende und rollende Bewegung, *Monatsh. f. Math.* **3** (1892) 31–54, 97–134.

LIST OF PARTICIPANTS

Toshiaki Adachi, Japan
Ilka Agricola, Germany
Alma Luisa Albujer, Spain
Teresa Arias-Marco, Spain
Andreas Arvanitoyeorgos, Greece
Akira Asada, Japan
Sándor Bácsó, Hungary
Vladimir Balan, Romania
Cornelia-Livia Bejan, Romania
Christina Beneki, USA
Ian Benn, Australia
Behroz Bidabad, Iran
Adara-Monica Blaga, Romania
Albert Borbely, Kuwait
Włodzimierz Borgiel, Poland
Alexander A. Borisenko, Ukraine
Ján Brajerčík, Slovakia
Miguel Brozos-Vázquez, Spain
Esteban Calviño-Louzao, Spain
José F. Cariñena, Spain
Marco Castrillón López, Spain
Marie Chodorová, Czech Republic
Ioannis Chrisikos, Greece
Hana Chudá, Czech Republic
Michael Crampin, Belgium
Balázs Csikós, Hungary
Lenka Czudková, Czech Republic
Pantelis A. Damianou, Cyprus
Manuel de León, Spain
Cemile Elvan Dinç, Turkey
Miroslav Doupovec, Czech Republic
Ugur Dursun, Turkey
Zdeněk Dušek, Czech Republic
Maria Vasilievna Dyachkova, Russia
Jürgen Eichhorn, Germany
Alberto Enciso, Spain
Jost-Hinrich Eschenburg, Germany
Younhei Euh, Korea
Vasyl Fedorchuk, Poland, Ukraine
Antonio Fernández Martínez, Spain
Agota Figula, Hungary
Thomas Friedrich, Germany
Eduardo García-Río, Spain
Peter Gilkey, USA
Simon Gindikin, USA
Vladislav Goldberg, USA
Hubert Gollek, Germany
Midori Goto, Japan
Xavier Gràcia, Spain
Galina Guzhvina, Germany
Izumi Hasegawa, Japan
Irena Hinterleitner, Czech Republic
Jaroslav Hrdina, Czech Republic
Cristina-Elena Hreţcanu, Romania
Dragos Hrimiuc, Canada
Sören Illman, Finland
Radu-Sorin Iordanescu, Romania
Pyotr Ivanshin, Russia
Josef Janyška, Czech Republic
Włodzimierz Jelonek, Poland
Igor Kanatchikov, Germany
Spiro Karigiannis, UK
Zuzana Kasarová, Czech Republic
Jerzy Kijowski, Poland
Hwajeong Kim, Germany
Kazuyoshi Kiyohara, Japan
Ivan Kolář, Czech Republic
Mayuko Kon, Japan
Anatoly Kopylov, Russia
Július Korbaš, Slovakia
Oldřich Kowalski, Czech Republic
László Kozma, Hungary

Boris Kruglikov, Norway
Demeter Krupka, Czech Republic
Olga Krupková, Czech Republic
Wojciech Krynski, Poland
Svatopluk Krysl, Czech Republic
Jan Kubarski, Poland
Jan Kurek, Poland
Miroslav Kureš, Czech Republic
Bavo Langerock, Belgium
Valentina B. Lazareva, Russia
Hong-Van Le, Czech Republic
Remi Léandre, France
Thomas Leuther, Belgium
Haizhong Li, China
Xingxiao Li, Germany
Andrea Loi, Italy
Sadahiro Maeda, Japan
Yoshinori Machida, Japan
Mikhail Malakhaltsev, Russia
Giovanni Manno, Italy
Michael Markellos, Greece
Hiroshi Matsuzoe, Japan
Roman Matsyuk, Ukraine
Vladimir S. Matveev, Germany
Sergei Merkulov, Sweden
Dimitrios Michalis, Greece
Josef Mikeš, Czech Republic
Włodzimierz M. Mikulski, Poland
Velichka Milousheva, Bulgaria
Oleg Mokhov, Russia
Giovanni Moreno, Italy
Piotr Mormul, Poland
Jana Musilová, Czech Republic
Péter T. Nagy, Hungary
Nobutada Nakanishi, Japan
Aleš Návrat, Austria
Stana Nikčević, Serbia
Takashi Okayasu, Japan
Peter Olver, USA
Bent Ørsted, Denmark
Radomír Paláček, Czech Republic
Marcella Palese, Italy
Martin Panák, Czech Republic
Jeong Hyeong Park, Korea
Aleš Paták, Czech Republic
Daniel Peralta-Salas, Spain
Monika Pietrzyk, Germany
Liviu Octavian Popescu, Romania
Andrey Popov, Russia
Serge Preston, USA
Geoff Prince, Australia
Fabrizio Pugliese, Italy
Christof Puhle, Germany
Morteza M. Rezaii, Iran
César Rodrigo, Portugal
Vladimir Rovenski, Israel
Jenny Santoso, Germany
Willy Sarlet, Belgium
David J. Saunders, Czech Republic
Esra Sengelen, Turkey
Artur Sergyeyev, Czech Republic
Alexander M. Shelekhov, Russia
Vadim V. Shurygin, Russia
Nils Schoemann, Germany
Jan Slovák, Czech Republic
Dana Smetanová, Czech Republic
Dalibor Šmíd, Czech Republic
Petr Somberg, Czech Republic
Vladimír Souček, Czech Republic
Karl Strambach, Germany
Martin Swaczyna, Czech Republic
Zoltán I. Szabó, USA
János Szenthe, Hungary
József Szilasi, Hungary
Masatomo Takahashi, Japan
Lajos Tamássy, Hungary
Dennis The, Canada
Radka Tománková, Czech Republic

Jiří Tomáš, Czech Republic
Jaroslav Trnka, Czech Republic
Zbyněk Urban, Czech Republic
Joris Vankerschaver, Belgium
Jiří Vanžura, Czech Republic
Alena Vanžurová, Czech Republic
Petr Vašík, Czech Republic
László Verhóczki, Hungary
Steven Verpoort, Belgium
Alexandre M. Vinogradov, Italy
Luca Vitagliano, Italy
Raffaele Vitolo, Italy
Jana Volná, Czech Republic
Petr Volný, Czech Republic
Jan Vondra, Czech Republic
Theodore Voronov, UK
Joanna Wełyczko, Poland
Robin Wilson, UK
Ekkehart Winterroth, Italy
Takahiro Yajima, Japan
Lenka Zalabová, Czech Republic
Igor Zelenko, Italy

PREVIOUS DGA CONFERENCE PROCEEDINGS

(1) O. Kowalski, ed., *Proceedings of the Conference on Differential Geometry and Its Applications*, Nové Město na Moravě, 1980 (Univerzita Karlova, Praha, 1982) 231 pp.
(2) O. Kowalski, ed., *Differential Geometry*, Proc. Conf. on Differential Geometry and Its Applications, Nové Město na Moravě, 1983, Part 1 (Charles University, Prague, 1984) 178 pp.
(3) D. Krupka, ed., *Geometrical Methods in Physics*, Proc. Conf. on Differential Geometry and Its Applications, Nové Město na Moravě, 1983, Part 2 (J.E. Purkyně University, Brno, 1984) 310 pp.
(4) D. Krupka and A. Švec, eds., *Differential Geometry and Its Applications*, Proc. Conf., Brno, 1986 (Kluwer, Dordrecht, 1987) 381 pp.
(5) D. Krupka and A. Švec, eds., *Differential Geometry and Its Applications, Communications*, Proc. Conf., Brno, 1986 (J.E. Purkyně University, Brno, 1987) 342 pp.
(6) J. Janyška and D. Krupka, eds., *Differential Geometry and Its Applications*, Proc. Internat. Conf., Brno, 1989 (World Scientific, Singapore, 1990) 465 pp.
(7) O. Kowalski and D. Krupka, eds., *Differential Geometry and Its Applications*, Proc. 5th Internat. Conf., Opava, 1992 (Mathematical Publications, Vol. 1, Silesian University at Opava, Opava, 1993) 540 pp.
(8) J. Janyška, I. Kolář and J. Slovák, eds., *Differential Geometry and Its Applications*, Proc. 6th Int. Conf., Brno, 1995 (Masaryk University, Brno, 1996) 658 pp.
(9) I. Kolář, O. Kowalski, D. Krupka and J. Slovák, eds., *Differential Geometry and Its Applications*, Proc. 7th Internat. Conf., Brno, Satellite Conf. of ICM in Berlin, 1998 (Masaryk University, Brno, 1999) 664 pp.
(10) O. Kowalski, D. Krupka and J. Slovák, eds., *Differential Geometry and Its Applications*, Proc. 8th Int. Conf., Opava, 2001 (Mathematical Publications, Vol. 3 Silesian University at Opava, Opava, 2001) 492 pp.
(11) *Diff. Geom. Appl.* **17** (2-3) (2002), special issue, O. Kowalski, D. Krupka, and J. Slovák, eds.[a]
(12) J. Bureš, O. Kowalski, D. Krupka and J. Slovák, eds., *Differential Geometry and its Applications*, Proc. 9th Int. Conf., Prague, 2004 (Matfyzpress, Charles University, Prague, 2005) 644 pp.

[a]Proc. 8th International Conference on Differential Geometry and Its Applications, Opava, 2001, Invited lectures.